T0211759

Lecture Notes in Computer Science 11944

Founding Editors

Gerhard Goos
Karlsruhe Institute of Technology, Karlsruhe, Germany
Juris Hartmanis
Cornell University, Ithaca, NY, USA

Editorial Board Members

Elisa Bertino
Purdue University, West Lafayette, IN, USA
Wen Gao
Peking University, Beijing, China
Bernhard Steffen
TU Dortmund University, Dortmund, Germany
Gerhard Woeginger
RWTH Aachen, Aachen, Germany
Moti Yung
Columbia University, New York, NY, USA

More information about this series at http://www.springer.com/series/7407

Sheng Wen · Albert Zomaya ·
Laurence T. Yang (Eds.)

Algorithms and Architectures for Parallel Processing

19th International Conference, ICA3PP 2019
Melbourne, VIC, Australia, December 9–11, 2019
Proceedings, Part I

 Springer

Editors
Sheng Wen
Department of Computer Science
and Software Engineering
Swinburne University of Technology
Hawthorn, Melbourne, VIC, Australia

Albert Zomaya
School of Computer Science
The University of Sydney
Camperdown, NSW, Australia

Laurence T. Yang
Department of Computer Science
St. Francis Xavier University
Antigonish, NS, Canada

ISSN 0302-9743   ISSN 1611-3349   (electronic)
Lecture Notes in Computer Science
ISBN 978-3-030-38990-1   ISBN 978-3-030-38991-8   (eBook)
https://doi.org/10.1007/978-3-030-38991-8

LNCS Sublibrary: SL1 – Theoretical Computer Science and General Issues

© Springer Nature Switzerland AG 2020
This work is subject to copyright. All rights are reserved by the Publisher, whether the whole or part of the material is concerned, specifically the rights of translation, reprinting, reuse of illustrations, recitation, broadcasting, reproduction on microfilms or in any other physical way, and transmission or information storage and retrieval, electronic adaptation, computer software, or by similar or dissimilar methodology now known or hereafter developed.
The use of general descriptive names, registered names, trademarks, service marks, etc. in this publication does not imply, even in the absence of a specific statement, that such names are exempt from the relevant protective laws and regulations and therefore free for general use.
The publisher, the authors and the editors are safe to assume that the advice and information in this book are believed to be true and accurate at the date of publication. Neither the publisher nor the authors or the editors give a warranty, expressed or implied, with respect to the material contained herein or for any errors or omissions that may have been made. The publisher remains neutral with regard to jurisdictional claims in published maps and institutional affiliations.

This Springer imprint is published by the registered company Springer Nature Switzerland AG
The registered company address is: Gewerbestrasse 11, 6330 Cham, Switzerland

Preface

Welcome to the proceedings of the 19th International Conference on Algorithms and Architectures for Parallel Processing (ICA3PP 2019). ICA3PP is with the series of conferences started in 1995 that are devoted to algorithms and architectures for parallel processing.

The conference of ICA3PP 2019 will be organized by Swinburne University of Technology, Australia, and was held in Melbourne, Australia. The objective of ICA3PP 2019 was to bring together researchers and practitioners from academia, industry, and governments to advance the theories and technologies in parallel and distributed computing. ICA3PP 2019 follows the traditions of the previous successful ICA3PP conferences held in Hangzhou, Brisbane, Singapore, Melbourne, Hong Kong, Beijing, Cyprus, Taipei, Busan, Melbourne, Fukuoka, Vietri sul Mare, Dalian, Japan, Zhangjiajie, Granada, Helsinki, and Guangzhou.

ICA3PP focuses on two broad areas of parallel and distributed computing: architectures, algorithms, and networks, and systems and applications. This conference is now recognized as the main regular event of the world that is covering the many dimensions of parallel algorithms and architectures, encompassing fundamental theoretical approaches, practical experimental projects, and commercial components and systems. As applications of computing systems have permeated in every aspect of daily life, the power of computing system has become increasingly critical. This conference provides a forum for academics and practitioners from countries around the world to exchange ideas for improving the efficiency, performance, reliability, security, and interoperability of computing systems and applications.

ICA3PP 2019 attracted 251 high-quality research papers highlighting the foundational work that strives to push beyond the limits of existing technologies, including experimental efforts, innovative systems, and investigations that identify weaknesses in existing parallel processing technology. Each submission was reviewed by at least two experts in the relevant areas, based on their significance, novelty, technical quality, presentation, and practical impact. According to the review results, 73 full papers were selected to be presented at the conference, giving an acceptance rate of 29%. We also accepted 29 short papers. In addition to the paper presentations, the program of the conference included three keynote speeches and two invited talks from esteemed scholars in the area, namely: (1) Y. Thomas Hou from Virginia Tech (USA), talking about "GPU-Based Parallel Computing for Real-Time Optimization," (2) Ying-Dar Lin from National Chiao Tung University (Taiwan), giving us a speech "5G Mobile Edge Computing: Research Roadmap of the H2020 5G-Coral Project," (3) Wanlei Zhou from University of Technology Sydney (Australia), giving us a talk "AI Security: A Case in Dealing with Malicious Agents," and (4) Hai Jin from Huazhong University of Science and Technology (China), giving us a talk "Evening Out the Stumbling Blocks for Today's Blockchain Systems." We were extremely honored to have had them as the conference keynote speakers and invited speakers.

ICA3PP 2019 was made possible by the behind-the-scene effort of selfless individuals and organizations who volunteered their time and energy to ensure the success of this conference. We thank all participants of the ICA3PP conference for their contribution. We hope that you will find the proceedings interesting and stimulating. It was a pleasure to organize and host the ICA3PP 2019 in Melbourne, Australia.

December 2019   Sheng Wen
                                                            Albert Zomaya
                                                         Laurence T. Yang

Organization

Honorary Chair

Yong Xiang   Deakin University, Australia

General Chairs

David Abramson   The University of Queensland, Australia
Yi Pan   Georgia State University, USA
Yang Xiang   Swinburne University of Technology, Australia

Program Chairs

Albert Zomaya   The University of Sydney, Australia
Laurence T. Yang   St. Francis Xavier University, Canada
Sheng Wen   Swinburne University of Technology, Australia

Publication Chair

Yu Wang   Guangzhou University, China

Publicity Chair

Jing He   Swinburne University of Technology, Australia

Steering Committee

Yang Xiang (Chair)   Swinburne University of Technology, Australia
Weijia Jia   Shanghai Jiaotong University, China
Yi Pan   Georgia State University, USA
Laurence T. Yang   St. Francis Xavier University, Canada
Wanlei Zhou   University of Technology Sydney, Australia

Program Committee

Marco Aldinucci   University of Turin, Italy
Pedro Alonso-Jordá   Universitat Politècnica de València, Spain
Daniel Andresen   Kansas State University, USA
Danilo Ardagna   Politecnico di Milano, Italy
Man Ho Au   The Hong Kong Polytechnic University, Hong Kong,
                         China
Guillaume Aupy   Inria, France

Joonsang Baek	University of Wollongong, Australia
Ladjel Bellatreche	LIAS/ENSMA, France
Siegfried Benkner	University of Vienna, Austria
Jorge Bernal Bernabe	University of Murcia, Spain
Thomas Boenisch	High performance Computing Center Stuttgart, Germany
George Bosilca	University of Tennessee, USA
Suren Byna	Lawrence Berkeley National Laboratory, USA
Massimo Cafaro	University of Salento, Italy
Philip Carns	Argonne National Laboratory, USA
Arcangelo Castiglione	University of Salerno, Italy
Tania Cerquitelli	Politecnico di Torino, Italy
Tzung-Shi Chen	National University of Tainan, Taiwan
Kim-Kwang Raymond Choo	The University of Texas at San Antonio, USA
Jose Alfredo Ferreira Costa	Federal University of Rio Grande do Norte, Brazil
Raphaël Couturier	University of Burgundy - Franche-Comté, France
Masoud Daneshtalab	Mälardalen University, KTH Royal Institute of Technology, Sweden
Gregoire Danoy	University of Luxembourg, Luxembourg
Saptarshi Debroy	City University of New York, USA
Casimer Decusatis	Marist College, USA
Eugen Dedu	FEMTO-ST Institute, University of Burgundy - Franche-Comté, CNRS, France
Frederic Desprez	Inria, France
Juan-Carlos Díaz-Martín	University of Extremadura, Spain
Christian Esposito	University of Napoli Federico II, Italy
Ugo Fiore	University of Napoli Federico II, Italy
Franco Frattolillo	University of Sannio, Italy
Marc Frincu	West University of Timisoara, Romania
Jorge G. Barbosa	University of Porto, Portugal
Jose Daniel Garcia	University Carlos III of Madrid, Spain
Luis Javier García Villalba	Universidad Complutense de Madrid, Spain
Harald Gjermundrod	University of Nicosia, Cyprus
Jing Gong	KTH Royal Institute of Technology, Sweden
Daniel Grosu	Wayne State University, USA
Houcine Hassan	Universitat Politècnica de València, Spain
Sun-Yuan Hsieh	National Cheng Kung University, Taiwan
Xinyi Huang	Fujian Normal University, China
Mauro Iacono	Università degli Studi della Campania Luigi Vanvitelli, Italy
Shadi Ibrahim	Inria Rennes Bretagne Atlantique Research Center, France
Yasuaki Ito	Hiroshima University, Japan
Edward Jung	Kennesaw State University, USA
Georgios Kambourakis	University of the Aegean, Greece

Helen Karatza	Aristotle University of Thessaloniki, Greece
Gabor Kecskemeti	Liverpool John Moores University, UK
Muhammad Khurram Khan	King Saud University, Saudi Arabia
Sokol Kosta	Aalborg University, Denmark
Dieter Kranzlmüller	Ludwig Maximilian University of Munich, Germany
Peter Kropf	University of Neuchâtel, Switzerland
Michael Kuhn	University of Hamburg, Germany
Julian Martin Kunkel	University of Reading, UK
Algirdas Lančinskas	Vilnius University, Italy
Che-Rung Lee	National Tsing Hua University, Taiwan
Laurent Lefevre	Inria, France
Kenli Li	Hunan University, China
Xiao Liu	Deakin University, Australia
Jay Lofstead	Sandia National Laboratories, USA
Paul Lu	University of Alberta, Canada
Tomas Margalef	Universitat Autònoma de Barcelona, Spain
Stefano Markidis	KTH Royal Institute of Technology, Sweden
Barbara Masucci	University of Salerno, Italy
Susumu Matsumae	Saga University, Japan
Raffaele Montella	University of Naples Parthenope, Italy
Francesco Moscato	University of Campania Luigi Vanvitelli, Italy
Bogdan Nicolae	Argonne National Laboratory, USA
Anne-Cécile Orgerie	CNRS, France
Francesco Palmieri	University of Salerno, Italy
Dana Petcu	West University of Timisoara, Romania
Salvador Petit	Universitat Politècnica de València, Spain
Riccardo Petrolo	Konica Minolta Laboratory Europe
Florin Pop	University Politehnica of Bucharest, National Institute for Research and Development in Informatics (ICI), Romania
Radu Prodan	University of Klagenfurt, Austria
Suzanne Rivoire	Sonoma State University, USA
Ivan Rodero	Rutgers University, USA
Romain Rouvoy	University of Lille, Inria, IUF, France
Antonio Ruiz-Martínez	University of Murcia, Spain
Francoise Sailhan	CNAM, France
Sherif Sakr	The University of New South Wales, Australia
Ali Shoker	HASLab, INESC TEC, University of Minho, Portugal
Giandomenico Spezzano	CNR, Italy
Patricia Stolf	IRIT, France
Peter Strazdins	The Australian National University, Australia
Hari Subramoni	The Ohio State University, USA
Frederic Suter	CC IN2P3, CNRS, France
Andrei Tchernykh	CICESE Research Center, Mexico
Massimo Torquati	University of Pisa, Italy

Tomoaki Tsumura   Nagoya Institute of Technology, Japan
Vladimir Voevodin   RCC MSU, Russia
Xianglin Wei   Nanjing Telecommunication Technology Research
                           Institute, China
Sheng Wen   Swinbourne University of Technology, Australia
Jigang Wu   Guangdong University of Technology, China
Roman Wyrzykowski   Czestochowa University of Technology, Poland
Ramin Yahyapour   GWDG, University of Göttingen, Germany
Laurence T. Yang   St. Francis Xavier University, Canada
Wun-She Yap   Universiti Tunku Abdul Rahman, Malaysia
Junlong Zhou   Nanjing University of Science and Technology, China
Albert Zomaya   The University of Sydney, Australia

Contents – Part I

Distributed and Parallel and Network-Based Computing

Big Data and Its Applications

Distributed and Parallel Algorithms

Applications of Distributed and Parallel Computing

Service Dependability and Security

IoT and CPS Computing

Performance Modelling and Evaluation

Contents – Part II

Big Data and Its Applications

Distributed and Parallel Algorithms

Applications of Distributed and Parallel Computing

Service Dependability and Security

IoT and CPS Computing

Performance Modelling and Evaluation

Parallel and Distributed Architectures

PPS: A Low-Latency and Low-Complexity Switching Architecture Based on Packet Prefetch and Arbitration Prediction

Yi Dai[⊠], Ke Wu, Mingche Lai, Qiong Li, and Dezun Dong

National University of Defense Technology, Changsha, China
{daiyi,wuke,laimingche,qiongli,dezundong}@nudt.edu.cn

Abstract. Interconnect networks increasingly bottleneck the performance of datacenters and HPC due to ever-increasing communication overhead. High-radix switches are widely deployed in interconnection networks to achieve higher throughput and lower latency. However, network latency could be greatly deteriorated due to traffic burst and microburst features. In this paper, we propose a Prefetch and prediction based Switch (PPS) which can effectively reduce the packet delay and eliminate the effect of traffic burst. By using dynamic allocation multiple queueing (DAMQ) buffer with data prefetch, PPS implements concurrent write and read with zero-delay, thus implementing full pipeline of the packet scheduling. We further propose a simple but efficient arbitration scheme, which completes a packet arbitration within one clock cycle meanwhile maintaining higher throughput. Moreover, by predicting the arbitration results and filtering the potential failed requests in the next round, our scheduling algorithm demonstrates indistinguishable performance from the iSLIP, but with nearly half of the iSLIP's area and 36.37% less logic units (LUTs). Attributing to the optimal schemes of DAMQ with control data prefetch and two-level scheduling with arbitration prediction, PPS achieves low-latency and high throughput. Also, PPS can easily extend the switching logic to a higher radix for the hardware complexity grows linearly with the number of ports.

Keywords: Switch architecture · DAMQ with prefetch · Speculative arbitration

1 Introduction

With the rapid development of artificial intelligence and deep learning, The volume of data communication between tens of thousands even a million nodes

This research was supported by 863 Program of China (2018YFB2202303, 2016YFB0200200), NSFC (61972412, 61832018), the national pre-research project (31511010202).

© Springer Nature Switzerland AG 2020
S. Wen et al. (Eds.): ICA3PP 2019, LNCS 11944, pp. 3–16, 2020.
https://doi.org/10.1007/978-3-030-38991-8_1

has increased dramatically in large-scale parallel computing and data processing systems, such as HPC and datacenters. With the extraordinary growth of computing and storage resources interconnected by an interconnect network, data movement between them largely determines HPC's performance. The routers or switches used for data routing and transmission for both off-chip or on-chip networks has an important impact on the network delay and bandwidth utilization. According to Dally's theory [1], packet's delay in an interconnect network can be divided into three parts:

$$T_0 = H_{min}T_r + \frac{D_{min}}{v} + \frac{L}{b} \tag{1}$$

where D_{min}/v is the link transmission latency, and L/b is the serialization latency, and H_{min} is the hop count. From this formula we can see the time T_r taken by the packet to traverse the switch plays a key role in the whole packet delay. T_r mainly includes the time for packet queueing in the buffer and that for packet passing through the crossbar. As a result, packet queueing and scheduling mechanisms with lower hardware complexity is very crucial to low-latency and high-throughput interconnect networks.

High-radix switches have been widely deployed in high performance interconnection networks to achieve lower latency and higher bandwidth [2,3]. It reduces the network diameter namely the number of hops H_{min} in formula (1), so that network latency can be considerably reduced. However, the complexity of the arbitration logic in of switches will increase quadratically with the number of ports, which restricts the switch to scale to a higher radix [4]. On the other hand, multi-VC design is the main way to provide Quality of Service (QoS) for mixed traffic [5] and solve deadlock problems caused by adaptive routing [6]. For example, IBA supports up to 16 VCs to provide 16 service levels, which permits a packet to operate at one of 16 service levels [7]. The increased number of VC could not only increase the memory overheads but the arbitration complexity. As the input queueing need provide dedicated buffer for each VC and the number of requests participating in arbitration is multiplied by VC number per port as well. With the ever increasing number of switch radix and Virtual Channels (VCs), it is very challenging to achieve low latency and 100% throughput with constrained memory overhead and fall-through latency.

In this paper, some simple but effective techniques are proposed for low latency and high throughput switch implementation, that makes PPS more scalable and energy-efficient. The main contribution of this paper is as follows:

(1) Propose an elaborate DAMQ (Dynamic Allocation Multiple Queueing) structure with data prefetch before the packet buffering in the queue, which reduces both the packet queueing and arbitration delay in the switch;
(2) Implement a result feedback mechanism for speculative. arbitration thus avoiding throughput loss resulted from failed arbitration;
(3) Propose an energy efficiency scheduling algorithm which significantly reduces area overhead meanwhile maintaining higher performance.

The remainder of this paper is organized as follows. In Sect. 2, we introduce the microarchitecture of PPS. The DAMQ multi-VC input buffer with data prefetch will be described in Sect. 3. In Sect. 4, we present the arbitration structure of PPS and then analyze PPS's scheduling algorithm in detail. Section 5 evaluates the performance and hardware overheads of PPS compared with conventional iSLIP [8] algorithm, finally the conclusion is available in Sect. 6.

2 Architecture

In this section, we present the microarchitecture of PPS as shown in Fig. 1. The PPS Switch consists of a crossbar, DAMQ buffers, VC arbiters, output port arbiters, and credit management modules for lossless packet scheduling. The function of each module is as follows:

DAMQ Input Buffers with Prefetch: Each arriving packet is stored in the corresponding VC buffer. We adopt a dynamic VC buffer allocation to improve the buffer utilization. All VC queues share a one-write-one-read SRAM and DAMQ divides the entire SRAM into private buffers one for each VC and public buffers shared by all VCs. DAMQ dynamically allocates buffer units according to the VC number of the packet. The prefetch top registers are used for timing compensation, thus pipelining data writing and reading with zero-delay. More Details of prefetch DAMQ are described in Sect. 3.

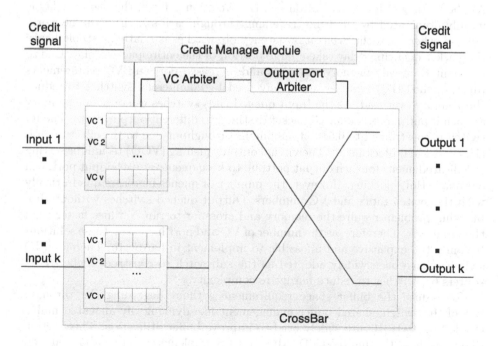

Fig. 1. PPS switching architecture for k input and output ports with v VCs per port.

VC Arbiters and Output Port Arbiters: The arbitration of the switch is divided into two stages. VC arbiters one for each DAMQ generate the output request from all head-of-line packets of each VC buffer to participate in the second stage port arbitration. The output arbiter one for each output port collects all arbitration request from k VC aribers and grants one request to output the corresponding packet. As the VC and output arbitration can be completed within one clock cycle the flit fall-through latency is just 2 clock cycles. The packet scheduling algorithm will be presented in Sect. 4 and evaluated in Sect. 5.

Credit Management Module: The credit module maintains the available size of each VC buffer of the downstream switch DAMQ and determines whether the packet can be sent, otherwise the packet can only be temporarily stored in the corresponding VC buffer.

The PPS, a typically multi-VC, DAMQ-based switching architecture, is similar to the majority of current switches. However, the elaborate packet queueing and scheduling schemes of PSS not only effectively reduce the hardware complexity but also achieve remarkable performance thus facilitating economical IC (Integrated Circuit) implementation for commercial switches. Meanwhile our enhancements in the burst abortion and speculative arbitration bring PPS low latency and high throughput, but with lower hardware overhead.

3   DAMQ Buffer with Data Prefetch

As the latency of link serialization and transmission is fixed, the packet delay is mainly dominated by the time traversing through each switch in the network. On the other hand the arbitration latency of switches is relatively stable hence the packet queueing delay take a large portion of the entire packet delay. Due to the limited on-chip memory resources and increased port and VC requirements Input queued (IQ) switches are widely used by commercial switches to reduce the memory overhead. As the Input queued (IQ) switches use only one memory at each input port to buffer all packet destined to different output ports. The IQ switch is constrained to 0.586 of maximum throughput due to the effect of HoL (Head-of-Line) blocking [9]. The virtual output queuing (VOQ) technique setting a dedicated queue for each output port, up to $k \times v$ queues at each input port, can eliminate HoL blocking. However, the number of queues increases quadratically with the switch radix and VC numbers. Output queued switches without HoL blocking problem require the memory and crossbar to run N times faster than the input link. Therefore, as the number of VC and port increases, these solutions becomes too expensive and infeasible to implement in hardware. Therefore, IQ switches have been widely adopted as the subswitch to construct higher-radix routers or switches to reduce hardware complexity.

To reduce the buffer space requirement without degrading the throughput of the switch, shared buffer management like dynamically allocated multi-queue (DAMQ) [10] are widely used to improve buffer utilization. This Section describes how the multi-VC DAMQ of PPS implements burst abortion and pipelines data writing and reading with zero-delay by data prefetching. Each

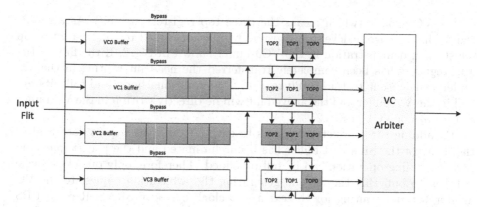

Fig. 2. Microarchitecture of PPS's input DAMQ buffer with data prefetch for 4 VCs, in which the buffer size of each VC is dynamically allocated on demand.

DAMQ associated with one input port consists of specified private buffers which can only be used by a particular VC and a common buffer that can be used by all VCs. By combining static and dynamic VC allocation schemes, we propose a fair credit management scheme that can efficiently assign buffers on demand among burst or uneven traffic [10]. Even some greedy VCs might occupy most of the shared buffers, the reserved private buffer ensures that VC starvation would not occur. Therefore, the micro-burst traffic on one single VC will be hidden by using shared buffer. Compared with allocating a fixed-size buffer to each VC, where some VC buffers may be always idle and others are insufficient at the same time, DAMQ can significantly improve the buffer utilization. The microstructure architecture diagram of the DAMQ buffer is shown in Fig. 2, three prefetch registers are set to eliminate the SRAM Read/Write latency. Generally, even using SARM with registered output (SRAM-R) to implement DAMQ buffer there could still be two clock latency for read operation, which means the head-of-line flit will be outputted on the third clock cycle although the read address has been asserted at the first cycle. For PPS, we use prefetch registers to eliminate the SRAM-R latency. As shown in Fig. 2 three TOP registers are set for flit prefetching and the width of the register is the length of single flit. When packets arriving at the input port, they will enter the corresponding VC buffer separately. There are three types of data might be registered in the TOP register (denoted as TOP0 in Fig. 2): the flit from the next level of the TOP register (shown as TOP1), the flit in the VC buffer and the bypass flit from the input port. When the VC buffer and top registers are empty as the VC3 buffer shown in Fig. 2, the flit from the input link will be directly transmitted to TOP0 by bypassing the VC buffer, and participate in VC arbitration. If this VC is temporarily blocked, the following flit will be stored in the three top registers in order. But when all those three top registers are filled, the subsequent flit will be written to the corresponding VC buffer, as the case of VC2 in Fig. 2. For the third circumstance, there are flits queueing in the VC buffer meanwhile the flit in the TOP0 register is granted by the VC arbiter (which means the three top registers are full). As a result, the flit in the current level top registers will be forwarded to the next level top register, as demonstration in the VC0 buffer shown in Fig. 2. If the flits in the Top registers has been completely transferred the head flit waiting in the VC buffer will be sent to the TOP0 register. The following body and tail flits are still in the VC buffer and the arrived flit will be directly written in the VC buffer as the case of VC1 in Fig. 2.

By adding three prefetch registers to each VC and the bypass data routes, the delay of the SRAM-R buffer has been eliminated and the packet queueing and scheduling operation can be full pipelined. Therefore, only one clock cycle is taken to buffering the flit and generating the arbitration request to the VC arbiter without waiting for the extra two clock cycles of SRAM-R read. PPS adopts DAMQ to improve the buffer utilization and prefetch registers to reduce the packet read delay that could also be used as a subswitch to construct parallel switch architectures to scale to higher radix.

4   Flow Control and Arbitration

4.1   Flow Control Scheme

Network flow control not only determines on-chip memory requirement but the way scheduling each packet which has great impact on switch throughput and latency. Flow control methods can be categorized into store-and-forward (SAF), wormhole (WH) and virtual-cut-through (VCT) [1].

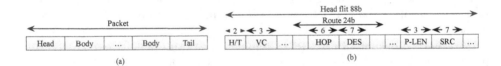

Fig. 3. (a) Packet is composed of multiple flits. (b) Head flit of a packet.

To improve the switching efficiency the variable-size packet is generally segmented into fixed-size flits including header flit, body flits and tail flit as shown in Fig. 3(a). In the simulation experiments carried out in Sect. 5 the length of a flit is 88 bits. The structure of the header flit including several switch-related fields are demonstrated in Fig. 3(b), such as Head/Tail indicator, VC number, hop counter and destination identification. For SAF [11] flow control, the whole packet must be stored in the buffer before proceeding to the next switch which results in higher packet latency for the packet queueing and scheduling are performed in a sequential way instead of a pipelined and parallel manner. For WH switching [12,13], the packet scheduling can be pipelined and the head flit is followed by other flits of the packet. As wormhole switching has no restrictions on the size of the buffer, it is widely adopted in the memory constrained environment such as NoC (Network on Chip). However, when the head flit is blocked due to the lack of free buffer, the whole packet would stop advancing and other packets can not be scheduled because of the blockage of the granted packet, which leads to throughput loss and higher packet latency. On the other hand, packet spanning across multiple routers results in new buffer dependency thus further complicating the implementation of the deadlock free adaptive routing [14]. As for VCT, as long as the next router has enough space to accommodate the packet, the head flit will be forwarded directly to the next router, thus the packet queueing is overlapped by the concurrent packet switching. Moreover, since there is sufficient space for the whole packet in the next router the process of packet scheduling would not be interrupted until the completion of the tail flit. Although VCT flow control needs each buffer accommodate the longest length packet it achieves higher throughput and lower latency compared with SAF and WH. In sum, by full pipelined and idle-free scheduling VCT can effectively reduce packet latency meanwhile avoiding throughput loss. Therefore, we optimize the packet queueing and scheduling mechanism of PPS based on VCT flow control.

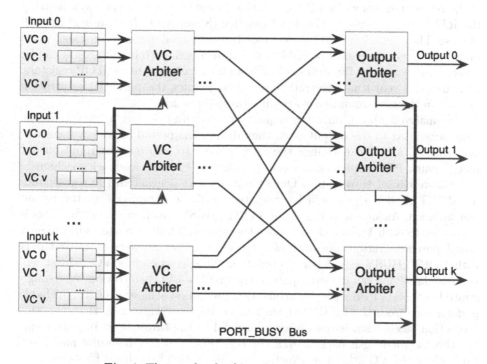

Fig. 4. The two-level arbitration structure of PPS.

4.2   Two-Level Arbitration Fabric with Failure Prediction

The PPS arbitration fabric is divided into two stages: the first-level VC arbitration performed at each input port and the second-level output arbitration executed at each output port as shown in Fig. 4. The VC arbiters select one port request from v VC buffers and the second level port arbiter grants one port request from k VC arbiters. Therefore all arbitrations work independently and asynchronously and the whole arbitration can be completed in one clock cycle. This distributed and parallel scheduling structure eliminates the complicated communication and global buffer-state maintenance performed in the centralized arbitration.

When the head flit enters the TOP registers of a VC buffer, the corresponding port request is generated based on the route field in the head flit. This request is sent to the VC arbier to participate in the VC arbitration. As we analyzed in Sect. 3, the zero DAMQ read latency ensures there is no null cycle between packet queueing and arbitration. The VC arbiter performs $v : 1$ arbitration based on the VC priority from v VC queues and send the granted request to its destination output arbiter. Consequently, each output arbiter receives up to k requests from all VC arbiters and eventually grants one to be scheduled. We evaluate the frequency of a PPS arbitration fabric ($v = 8, k = 8$) by synthesizing the RTL code in Synopsys Design Compiler (Synopsys DC) with a 28 nm cell library. These two-level arbitration can be completed within one clock cycle under the target frequency of 1GMHz which means the arbitration result can be determined exactly at the next cycle of head flit registering in the TOP registers. PPS use this real time arbitration result to predict the potential arbitration failures in the next round thus avoiding throughput loss.

As analyzed above, only the request granted by the port arbiter could be truly scheduled to the output port, then during the period of packet transmission the VC arbier can change the port request to improve the input-output match pairs. PPS implements a very simple but effective feedback mechanism to improve packet throughput. Once a output port is granted, the corresponding PORT_BUSY signal will be validated until the completion of the packet transmission. As shown in Fig. 4, the PORT_BUSY bus from k port arbiters is connected to each VC arbier to remove the potential failure requests in the next round port arbitration. For example, if a VC's packet requests an output port with PORT_BUSY signal being validated, the corresponding VC request will be invalidated and fail to participate in the first-level VC arbitration. As demonstrated in Sect. 5, PPS with this arbitration prediction achieves indistinguishable performance from the iSLIP, but with much less hardware cost. However, this prediction mechanism is not always beneficial to the throughput improvement. For the one-flit-length packet, even the real time arbitration results have been fed back to the VC arbiter at exactly the same cycle of the head flit being read out. This feedback signal might mislead to remove the VC requests which can be successfully scheduled in the next round, since the output port would be immediately released at the next cycle. The simulation results demonstrate that PPS with arbitration prediction works very well with long packets. However, a great portion of short packets especially single-flit-packets might lead to certain throughput loss.

4.3 Scheduling Strategy

Scheduling strategy determines the priority of all arbitration requests and has great impact on switch throughput. For VCT flow control, the basic scheduling unit is a whole packet and the priority will be updated after the scheduling of the tail flit. When the tail flit is in the TOP0 register the VC arbiter associated with an input port sends the tail signal to the requested output port, and next both the VC arbiter and the output arbiter simultaneously adjust scheduling priority based on recently granted lowest (RGL) policy. This strategy automatically skips unavailable VC requests that have no sufficient credit for next-level DAMQ or the requested port is busy. The recently granted VC is always adjusted to the lowest priority and other VC's priority is maintained in a cyclical manner which is different from Round Robin. An example of PPS's scheduling strategy is shown in Fig. 5.

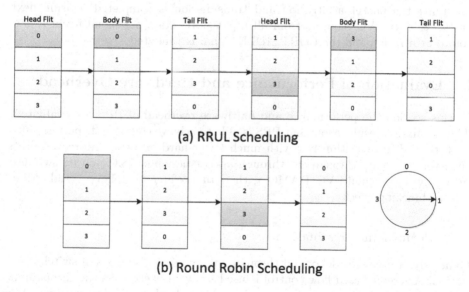

Fig. 5. Comparison between PPS's scheduling strategy and Round Robin policy for VC arbitration with v = 4.

The Round Robin scheduling is divided into three steps: request, grant, and accept. The iSLIP algorithm is a variant of round robin but with higher throughput. The difference between iSLIP and round robin is the update time of the grant pointer at the output port [8]. There are only two steps in the PPS scheduling strategy: request and grant. This is because the VC arbitration is performed before the port request, so that only one request is generated by the VC arbiter for each input port. Furthermore, the VC arbiter can be cascaded with the port arbiter by combinational logic which can be completed in a single clock cycle. As analyzed in Sect. 4.2 with arbitration prediction this simplified scheduling mechanism effectively reduces the hardware complexity of arbitration logic meanwhile maintaining high throughput and low latency.

As shown in Fig. 5(b), according to the round robin policy, the priority is adjusted to $VC1 > VC2 > VC3 > VC0$ after VC0 is selected for the first time. After the arbiter selecting VC3 for the second time, the priority sequence are adjusted to $VC0 > VC1 > VC2 > VC3$ according to the relative order of VCs. So that VC1 and VC2, which are not selected in the recent two rounds, still have no highest priority. According to PPS RGL policy, as shown in Fig. 5(a), the relative order between VCs is not maintained and the recently granted VC is set to the lowest priority. At the same time, the priorities of the remaining VCs are sequentially upgraded, which makes the VCs that have not been granted in recent rounds have higher priority thus ensuring fairness VC scheduling.

The priority adjustment policy of the output arbiter is the same as the VC arbiter. The input port priority is always updated at the same cycle as the transmission of the tail flit. At the end of this clock cycle, the entire VCT scheduling including the packet arbitration and transmission is completed. During next clock cycle, the VC requests from all input ports start participating in a new round arbitration and the PORT_IDLE signal is validated.

5   Evaluation of Performance and Hardware Overhead

In this section the performance and hardware overhead of PPS are evaluated. The simulation results show that PPS is able to achieve comparable performance with the iSLIP algorithm but with much lower hardware cost. Moreover, PPS surprisingly breaks through the throughput restriction of 58.6% on IQ switches by applying the multi-VC DAMQ with data prefetch to relieves head-of-line (HOL) blocking effect.

5.1   Performance Evaluation

Generally, average flit delay and throughput are used as metrics of switch evaluation. As credit-based flow control is used to avoid packet loss the throughput can be measured with input link usage or input loads in lossless network. We implement a PPS with $k = 8, v = 8$ in register transfer level (RTL) and conduct cycle-level simulation experiments to evaluate the delay and throughput.

Figure 6 shows the flit average delay of PPS and iSLIP under different traffic models. For the uniform traffic mode, both PPS and iSLIP achieve a maximum throughput to 85% as shown in Fig. 6(a). For the other three traffic modes as shown in Fig. 6(b), (c), (d), PPS can achieve a throughput about 70% which is apparently higher than the iSLIP with one iteration only achieving 50% throughput. The iSLIP algorithm with 2 or 3 iterations has comparable throughput with PPS.

When increasing the number of iterations per round iSLIP achieves lower latency and higher throughput as shown in Fig. 6. The iSLIP algorithm with 1 iteration has the poorest performance, PPS achieves delay and throughput performance comparable to iSLIP-2, and iSLIP-3 has the highest throughput and lowest delay. As shown in Fig. 6(a), (b), under the uniform and hotspot traffic mode, PPS surprisingly outperforms the iSLIP with two iteration when approaching the maximum load. In the traffic pattern of exponential and Poisson distributions, as shown in Fig. 6(c), (d), the latency between PPS and iSLIP with two iterations is almost the same. Besides, when the number of iterations is increased to 4, iSLIP performance remains almost the same. This is because iSLIP theoretically converges to the optimal performance after 3 iterations.

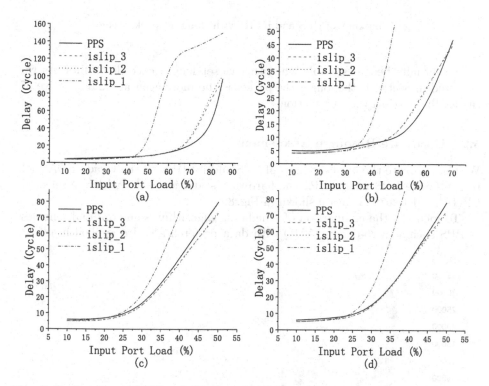

Fig. 6. Performance of PPS and iSLIP with 1/2/3 iterations in different traffic modes for k = 8 and v = 8. (a) Uniform (b) Hotspot (c) Exponential (d) Poisson

In the following, we will analysis the impact of packet length on latency performance. As shown in Fig. 7(b) the iSLIP algorithm with 2 or 3 iterations is less affected by the packet length in terms of packet delay which just slightly fluctuates with the increase of the packet length. The iSLIP latency reaches to the highest when the packet length is moderate. For example the iSLIP algorithm has a lower latency when the packet length is very short or long. In contrast, the latency of PPS gradually decreases with the increase of the packet length,

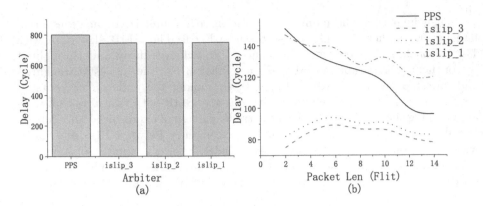

Fig. 7. Latency of PPS and iLSIP with different packet length.

and eventually becomes stable when the packet length exceeds 12 flits. As we analyzed in Sect. 4.1, the longer the packet is the more accurate and effective of the feedback arbitration prediction.

5.2 Hardware Resource Assessment

We also evaluate the hardware cost of PPS and iSLIP with the vivado EDA tool to synthesize PPS and iSLIP. The hardware statistic results with the selected FPGA model xc7vx415t are shown in Fig. 8.

To focus on the comparison of scheduling algorithm, some optimal schemes of PPS such as zero-delay DAMQ with data prefetch are also applied in iSLIP

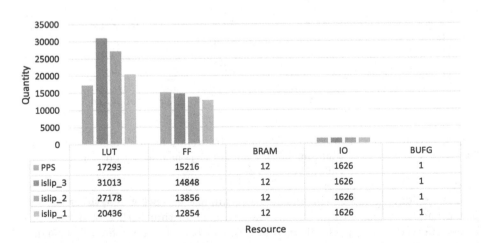

Fig. 8. Hardware cost comparison of PPS and iSLIP with 1–3 iterations

models. The only difference between them is the arbitration and scheduling strategy which mainly dominate the consumption of LUT resources. Figure 8 demonstrates that the amount of LUTs in PPS is almost half of the iSLIP with three iterations and it is also much less than that of iSLIP with one iteration. PPS has the same hardware overhead as iSLIP In terms of IO, BRAM and BUFG cost. This is because they adopt the same DAMQ buffers, flow control and routing algorithm only different arbitration fabric. In sum, PPS has lower hardware complexity and power consumption but provides comparable performance of iSLIP.

6 Conclusion

We propose a cost efficient switching architecture PPS for low power and high performance switch or Network on Chip (NoC) implementation. PPS provides an energy-efficient packet scheduling mechanism characterized by DAMQ prefetching and arbitration prediction, which makes PPS be easily applied to commercial IC and SoC design. These simple but effective optimal schemes of PPS such as full pipelined DAMQ buffer with zero read latency, two-stage speculative arbitration with RGL policy, can effectively reduce the area and power consumption meanwhile maintaining low latency and high throughput.

Ultra high-radix switches are generally constructed by interconnecting multiple subswitches via different topologies. For example, a 64-radix YARC switch is organized into 8×8 array of 8×8 subswitches [15]. SCOC uses 68 4×4 subswitches and a four-way TDM crossbar to implement a 136×136 switch [2]. PPS has been applied to non-blocking MBTR routers which achieves 100% throughput and a fall-through latency of just 30 ns [16].

References

1. Dally, W., Towles, B.: Principles and Practices of Interconnection Networks. Morgan Kaufmann Publishers Inc., San Francisco (2003)
2. Chrysos, N., Minkenberg, C., Rudquist, M., et al.: SCOC: high-radix switches made of bufferless clos networks. In: IEEE International Symposium on High Performance Computer Architecture, pp. 402–414. IEEE (2015)
3. Kim, J., Dally, W.J., Towles, B., et al.: Microarchitecture of a high-radix router. ACM SIGARCH Comput. Archit. News **33**(2), 420–431 (2005)
4. Ahn, J.H., Choo, S., Kim, J.: Network within a network approach to create a scalable high-radix router microarchitecture. In: IEEE International Symposium on High Performance Computer Architecture, pp. 1–12. IEEE (2012)
5. Vicente, A.M., Apostolopoulos, G., Alfaro, F.J., et al.: Efficient deadline-based QoS algorithms for high-performance networks. IEEE Trans. Comput. **57**(7), 928–939 (2008)
6. Ebrahimi, M., Daneshtalab, M.: EbDa: a new theory on design and verification of deadlock-free interconnection networks. In: International Symposium on Computer Architecture, pp. 703–715. ACM (2017)

7. InfiniBand Trade Association: InfiniBandTM Architecture Specification Volume 1, Release 1.0, October 2000. www.infinibandta.org
8. Mckeown, N.: The iSLIP scheduling algorithm for input-queued switches. IEEE/ACM Trans. Networking **7**(2), 188–201 (1999)
9. Karol, M., Hluchyj, M., Morgan, S.: Input versus output queueing on a space-division packet switch. IEEE Trans. Commun. **COM–35**(12), 1347–1356 (1987)
10. Zhang, H., Wang, K., Dai, Y., et al.: A multi-VC dynamically shared buffer with prefetch for network on chip. In: IEEE International Conference on Networking, Architecture and Storage, pp. 320–327. IEEE (2012)
11. Ni, L.M., Mckinley, P.K.: A survey of wormhole routing techniques in direct networks. Computer **26**(2), 62–76 (1993)
12. Duato, J.: A necessary and sufficient condition for deadlock-free routing in cut-through and store-and-forward networks. IEEE Trans. Parallel Distrib. Syst. **7**(8), 841–854 (1996)
13. Kermani, P., Kleinrock, L.: Virtual cut-through: a new computer communication switching technique. Comput. Netw. **66**(4), 4–17 (2014)
14. Chen, L., Pinkston, T.M.: Worm-bubble flow control. In: 2013 IEEE 19th International Symposium on High Performance Computer Architecture (HPCA 2013), pp. 366–377. IEEE (2013)
15. Abts, D., Abts, D., Kim, J., et al.: The BlackWidow high-radix clos network. In: International Symposium on Computer Architecture, pp. 16–28. IEEE (2006)
16. Dai, Y., Lu, K., Xiao, L., et al.: A cost-efficient router architecture for HPC interconnection networks: design and implementation. IEEE Trans. Parallel Distrib. Syst. **30**(4), 738–753 (2018)

SWR: Using Windowed Reordering to Achieve Fast and Balanced Heuristic for Streaming Vertex-Cut Graph Partitioning

Jie Wang and Dagang Li$^{(\boxtimes)}$

School of ECE, Shenzhen Graduate School, Peking University, Shenzhen, China
wang_jie@pku.edu.cn, dagang.li@ieee.org

Abstract. Graph partitioning plays a very fundamental and important role in a distributed graph computing (DGC) framework, because it determines the communication cost and workload balance among computing nodes. Existing solutions are mainly heuristic-based but unfortunately cannot achieve partitioning quality, load balance, and speed at the same time. In this paper, we propose Sliding-Window Reordering (SWR), a streaming vertex-cut graph partitioning algorithm, that introduces a pre-partitioning window to re-order incoming edges, making it much easier for a greedy strategy to maintain balance while optimizing edge assignment at a minimal computational cost. We analytically and experimentally evaluate SWR on several real-world and synthetic graphs and show that it achieves the best overall performance. Compared with HDRF, the state-of-the-art at present, the partitioning speed is increased by 3–20 times, and the partitioning quality is increased by 15% to 30% on average when achieving balanced load among all nodes.

Keywords: Graph partitioning · Streaming algorithms · Vertex-cut · Distributed graph-computing frameworks · Load balancing

1 Introduction

In recent years we have seen a fast growth in large-scale graph-structured data in various real-world applications. In order to process these large scale datasets, distributed graph-computing (DGC) frameworks become favorable which partition the graph across multiple machines that can compute in parallel using graph processing systems such as Pergel [16] or GraphLab [15]. Different graph partitioning (GP) strategies can greatly affect the overall performance of the

This work was supported by Shenzhen Peacock Innovation Program (KQJSCX20180323174744219), National Engineering Laboratory for Video Technology - Shenzhen Division and the Shenzhen Municipal Development and Reform Commission (Disciplinary Development Program for Data Science and Intelligent Computing).

© Springer Nature Switzerland AG 2020
S. Wen et al. (Eds.): ICA3PP 2019, LNCS 11944, pp. 17–32, 2020.
https://doi.org/10.1007/978-3-030-38991-8_2

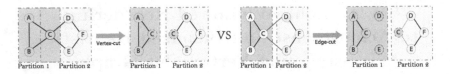

Fig. 1. Vertex-cut vs. edge-cut graph partitioning

DGC framework, because they determine the communication cost and workload balance among these machines.

Graph partitioning strategies can be categorized into either vertex-cut or edge-cut approaches. They both replicate some vertices into different partitions. Figure 1 shows how they work with a simple example. The number of edges in each partition determines the workload of each machine, and the total number of replicas determines the total communication cost to synchronize them.

Graph partitioning can be generally regarded as typical balanced k-way partitioning problem, while both vertex-cut and edge-cut variants are known to be NP-hard [2,7,13]. Although many algorithms have been proposed to solve the problem, most of them have high cost and cannot process large-scale real world graphs. Streaming graph partitioning has been proposed by Stanton and Kliot [19] to partition large scale graphs fast and efficiently. Since the quality of GP is crucial to the overall performance of DGC frameworks, adequate GP algorithms should be developed, taking three important aspects into consideration. The first is quality, mostly referring to the communication cost for synchronizing replicated items across different partitions during the computation phase. The second is balance. Large workload skews in a cluster of machines can greatly deteriorate the efficiency of the system. The third is speed. Although some existing GP algorithms can achieve quality partitions, if the GP phase itself takes too much time, the overall performance is still affected, for example off-line algorithms such as JA-BE-JA-VC [18], METIS [12], Ginger [3] and NE [21].

In this paper we focus on the edge distribution of the graph dataset since real world graphs are known to have skewed power-law degree distribution [8], and the order of incoming edges in the stream also show some common patterns. Several studies [1,14] have shown that edge-cut methods do not work well with power-law graphs, while vertex-cut is proven to be the better choice in theory [6] and in practice [8,9]. We will first analyze in depth two representative vertex-cut GP algorithms, greedy and HDRF, and then propose our *sliding-window reordering* (SWR) mechanism that introduces sliding-window based pre-reorder to the input edges as well as enhanced greedy rules, so the degree and neighboring information are better utilized to achieve better partitioning performance in terms of speed, quality and balance. The contributions are as follows:

1. A clustering model is provided to describe the characteristics of the input edges, and in-depth analysis on standard greedy and HDRF is provided to show how they are affected by the order of the edges.

2. A new streaming vertex-cut GP algorithm called SWR is proposed, which achieves better partition quality, speed and balance at the same time.
3. A new concept of dispersion factor is proposed and used to analyze theoretically the size of the window on the effectiveness of the SWR algorithm.

2 Problem Definition and Background

2.1 Balanced k-Way Vertex-Cut GP Problem

Consider a graph $G = (V, E)$ consisted of vertices $V = (v_1, v_2, \cdots, v_n)$ and edges $E \subseteq V \times V$. Vertex-cut GP will assign the edges into a series of pairwise disjoint sets $P = (p_1, p_2, \cdots, p_n)$, where for any $i \neq j$, $p_i, p_j \subseteq E$ and $p_i \cap p_j = \emptyset$.

Since in vertex-cut, cut vertices will have replicas in multiple partitions, we define $A(v)$ as the set of partitions where vertex v is replicated. During computation these replicas need to be synchronized which generates communication cost, therefore the main target of vertex-cut is to minimize replications while keeping partitions of each machine approximately the same size, so the balanced $|P|$-way vertex-cut GP problem can be formally defined as to solve:

$$minimize \frac{1}{|V|} \sum_{v \in V} |A(v)| \quad s.t. \quad \max_{p \in P} |p| < \delta \frac{|E|}{|P|} \tag{1}$$

where $\delta \geq 1$ is a small constant that defines the tolerance to partition imbalance. The object function in Eq. 1 is called **Vertex Replication Factor (VRF)** representing the average number of replicas per vertex.

2.2 Rule-Based and Score-Based Approaches

The core of a GP algorithm is the strategy to assign which edge to which partition. Most existing GP algorithms can be generally divided into two categories: rule-based and score-based. In a rule-based strategy, new edge is assigned to the partition that satisfies a number of simple rules, so the whole assignment process can be carried out with a series of if-else questions. For score-based strategy, matching scores need to be calculated between the new edge and all existing partitions, and the one with the highest score will be chosen, therefore much more calculation is involved in assigning each single edge. Rule-based approaches are generally faster than score-based ones but less optimal in partitioning quality. However they are not completely opposite to each other: rule-based strategy can be transformed to score-based, one example is the formal transformation of greedy in [17], but score-based algorithms are not always transformable to rule-based because score computation can be rather complicated.

2.3 Streaming Vertex-Cut Partitioning Algorithms

In streaming partitioning, a GP algorithm will process over the edge stream, which means that it needs to partition all the edges in a single pass. Streaming partitioning is fast and easy to be integrated to DGC frameworks because of its simplicity. However, good partitioning quality is more difficult to achieve because partitioning decisions are made on limited information, and minimizing vertex replication can easily end up in severe partition imbalance. If you want both quality and balance it will be too slow. To the best knowledge of the authors, there is not yet a solution that performs well in all the three important aspects.

Hashing Partitioning Algorithms: Hashing GP algorithms are rule-based. It can achieve fast and balanced partitioning but has no quality guarantee. The most known hashing GP are `hashing` in GraphX and DBH [20]. Because no partitioning history is considered in `hashing`, it results in very high VRF, especially on power-law graphs. DBH improves over pure hashing by taking vertex degree into consideration and cut higher degree vertices first, so the VRF is reduced.

Constrained Partitioning Algorithms: This category of GP algorithms are also rule-based. The core idea is to limit the candidate partitions a vertex can go within a *constrained set*, so as to restrain the VRF with a theoretical upper bound. The more constrained the rules are the better the quality could be but at higher risk of severe imbalance. Typical constrained GP algorithms are `grid` and `PDS` from GraphBuilder [11]. In `grid`, the partitions are represented in a matrix $P = M \times N$. Hash function $h(v)$ maps vertex v onto this matrix, and the constrained set $S(v)$ is the subset of partitions in P that are at the same row and column as v. The candidate partitions for the edge would be the two intersect partitions of $S(v_i)$ and $S(v_j)$ from its two vertices. `PDS` generates constrained sets using Perfect Difference Sets [10] which is more constrained.

Greedy Partitioning Algorithms: Greedy algorithms maintain some state information of the partitioning history to help later assignment. Simply put, vertex v is preferred to be assigned to a partition that it has been assigned before with earlier edges, and the smallest one is chosen when multiple partitions are qualified to improve balance. The `greedy` algorithm from PowerGraph [8] is rule-based and also the basis for many other more advanced greedy GP algorithms. HDRF is the state-of-the-art greedy algorithm which is score-based and can not be transformed to fast rule-based form. It follows the same greedy strategy but also take the vertex degree into consideration by calculating a matching score between the incoming edge and all existing partitions. A balance parameter λ is used to control the imbalance degree in the score calculation. ADWISE goes one step further by evaluating scores of *a window of* edges in the stream at the same time and assign the edge with the best score to the corresponding partition. Although score-based greedy algorithms may get improved partitioning quality, they are much slower, especially when existing partitions are big and plenty.

3   Edge Stream Model and Its Impacts on GP Algorithms

Natural graphs from the real-world are commonly found to be highly skewed with power-law degree distribution. Furthermore, the edge order very often resembles some specific patterns when loaded as edge stream, for example clustered in groups by the source vertex, or in the order of a BFS, etc., all tracing back to how the graph data were collected or stored in the first place. We found that these patterns in the edge order have strong impact on the partitioning results.

3.1   Clustering Model of Edge Stream

First we propose a clustering model on the ordering pattern of an edge stream to discuss its impacts on GP algorithms. A *dispersion factor* is defined to measure how strong consecutive edges are related to each other to resemble a pattern. The larger the dispersion factor the less clustering effect the edges have and the weaker the order pattern.

As a building block to the clustering model we define a **self-contained edge stream**, which can be represented by an ordered set of edges $E_{std} = \{e_1, e_2, \cdots, e_n\}$, where for any $j > 1$ exists $i < j$ satisfying $e_i \cap e_j \neq \emptyset$, meaning a later edge e_j always shares a common vertex with some earlier edge e_i. A self-constrained edge stream can be for example a sequence of edges collected by a single-thread breadth-first search from a connected graph. In this specific example of BFS, we further have: for any $e_j = (u_j, v_j), j > 1$ there exists $e_i = (u_i, v_i), i < j$ and $i \geq 1$, such that $v_i = u_j$. A self-contained edge stream shows very strong clustering effect since all the edges are related to some of the earlier edges in the stream, as if they are "dragged in" by their neighbors.

Considering that real-world graphs are most collected by multiple crawlers, combined from various sources or read from a distributed storage, although edges are more mixed, we still observe apparent clustering effect. To describe these more-general graphs, we can regard their edge stream as constructed from m intertwined self-constrained edge sub-streams. Essentially, edge stream from any graph dataset can be regarded this way, the only difference is how big m should be to describe it. Therefore, we choose m to define the **dispersion factor (DF)** of an edge stream. Formally put, any edge stream E can be defined as:

$$E = E_1 \cup E_2 \cup \cdots \cup E_m \tag{2}$$

where $E_i, i \in \{1, \cdots, m\}$ are self-contained edge streams.

Furthermore, if E and E_i, E_j are defined as

$$E_i = \{e_{i_1}, e_{i_2}, \cdots, e_{i_m}\}, E_j = \{e_{j_1}, e_{j_2}, \cdots, e_{j_k}\} \subset E = \{e_1, e_2, \cdots, e_n\} \tag{3}$$

then when $e_{i_m} \in E_i = e_t \in E$, there exists $i_k < i_m, r < t$ satisfying $e_{i_k} \in E_i = e_r \in E$, which means that the edges of E_i maintain their relative order in E. For edges from two different self-contained sub-streams E_i and E_j, we may have $e_i \in E_i \cap e_j \in E_j = \emptyset$ or not, since E_i and E_j may come from disconnected subgraphs, or different parts of a connected subgraph but only get to each other later on in the stream. They can not be merged according to the definition.

3.2   Impacts on Rule-Based Greedy

The rule-based greedy is fast and provides rather good partitioning quality, but for datasets with intrinsic small DF it may have very severe imbalance problem.

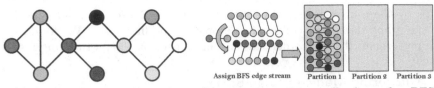

(a) A sample connected graph   (b) **greedy** partitions all edges of a BFS
                                          stream into a single partition

Fig. 2. A partitioning example with **greedy**. (Color figure online)

Let P_k be the k-th partition, $V(P_k)$ be its vertex set and $E(P_k)$ be its edge set. $e_i = (u_i, v_i)$ is the i-th edge of the edge stream.

We first analyze **greedy** on a self-contained edge stream E_{std} with DF of 1. For the first edge $e_1 \in E_{std}$, we can assign it to any of the empty partitions, let us say P_1, so we have $e_1 \to P_1$. For any later edge $e_j \in E_{std}$ and $j > 1$, according to the **greedy** rules, we need to see if any partition P_k already has u_j or $v_j \in P_k, k \in \{1, \cdots, m\}$. For the self-contained edge stream, there must exist $i < j$ satisfying $e_i \cap e_j \neq \emptyset$. Since the earlier edge e_i has already been assigned to a partition P_i, e_j will be assigned to that same partition, so we have $e_j \Rightarrow e_i \to P_i$. For the same reason, when e_i was assigned, there must have existed $h < i$, so that $e_i \Rightarrow e_h \to P_h$. In the end we can only have $e_j \Rightarrow e_i \Rightarrow e_h \Rightarrow \cdots \Rightarrow e_1 \to P_1$, and all edges of E_{std} are assigned to the same partition P_1. A demonstrative example is shown in Fig. 2(b), where the self-contained edge stream comes from a BFS starting from the red vertex in Fig. 2(a). The situation is less severe but similar for an edge stream with a larger DF of m. Since $E = E_1 \cup E_2 \cup \cdots \cup E_m$, if for any $i \neq j$, all $e_{i_r} \in E_i$ and $e_{j_t} \in E_j$ satisfy $e_{i_r} \cap e_{j_t} = \emptyset$, then every $E_k, k \in \{1, \cdots, m\}$ will have all its edges assigned to a separate partition of its own in the way described above, so the graph will be naturally partitioned into m partitions. If m is smaller than the target number of partitions n, we will end up with $(n - m)$ empty partitions; even when m is larger than n but not much, keeping the n partitions in balance is still difficult if possible.

Of course in reality, it is likely that some E_k's will have common vertices later on down the stream. Such situation will not change the DF m of the edge stream, but it may help to get more evenly distributed partitions, for some edges now have more choices during the assignment, **greedy** can use that to fill up smaller partitions when appropriate. That is why in practice **greedy** can still achieve acceptable balance on most natural graphs with moderate clustering effect.

3.3   Impacts on Score-Based Greedy

HDRF is the state-of-the-art GP algorithm that can achieve both low VRF and good balance. However, it handles the imbalance problem by introducing a balance parameter λ which enforces a brutal regulation to penalize edge assignment choices that may exceed the imbalance tolerance. Such methodology works but also brings new problems. First, a self-contained sub-stream may be brutally split

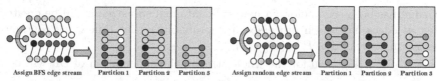

(a) HDRF has sub-optimal edge assignment.

(b) Re-ordering edges results in better partitioning.

Fig. 3. A partitioning example with HDRF and Re-ordering.

into multiple partitions so that its edges are sub-optimally assigned because of the balance enforcement. Second, the introduction of λ means that HDRF can only be score-based, and the involved score computation with all existing partitions to assign each new edge affects its speed a lot.

A simple example can demonstrate why sub-optimal edge assignment may happen. The same BFS edge stream from Fig. 2(a) is reused here, whose imbalance tolerance is set to 2. From Fig. 3(a) we can see much more balanced partitions, which is to be expected; however, the price paid is an increase in vertex replication, because now some edges with common vertices have to be assigned to different partitions to satisfy the balance requirement. As an indication, VRF also goes to 1.6 from 1 as in Fig. 2(b).

4 Design of SWR Algorithm

In this section we will introduce our SWR algorithm, which is designed for real-world big graph datasets that generally always possess a certain level of clustering effect. The discussion will be carried out from the aspects of load balance, quality and speed, respectively, and the pseudo code is shown in Algorithm 1.

The main idea is to introduce a moving-window mechanism on the edge stream, and by changing the order of the edges inside the window, we can attenuate the clustering effect and increase the DF, which makes a rule-based greedy heuristic to easily achieve partitioning balance while assigning edges optimally, taking the vertex degree and neighboring information into consideration.

4.1 Using Randomization to Keep Balance

Randomization can help to keep balance. To demonstrate the idea we refer back to the same graph in Fig. 2(a) but now pick the edges one by one at random to form a new edge stream, and use the same greedy algorithm to process them. The result is a set of balanced partitions as shown in Fig. 3(b) with a VRF of 1.2. Intuitively, a randomized edge stream has a smaller clustering effect and larger DF. We will analyze it theoretically.

To see that let us start with a completely random edge stream containing $|V|$ vertices and $|E|$ edges, and the average probability of vertex appearing in a new edge is $\frac{2}{|V|}$. During a streaming GP process, if none of the vertices of a new edge has been seen before, a new self-contained sub-stream will be created. When m edges have been processed, the probability that a new self-contained sub-stream is created is $P = (1 - \frac{2}{|V|})^{2m}$.

Let $c = |E|/|V|$, then for the whole stream, the expectation of DF will be

$$E' = \sum_{m=0}^{|E|-1} (1 - \frac{2}{|V|})^{2m} = \frac{|V|^2}{4|V| - 4}[1 - (1 - \frac{2}{|V|})^{2(|E|-1)}] \approx \frac{|V|}{4}(1 - e^{-4c}) \quad (4)$$

Algorithm 1. SWR sliding window randomization

1: Initalize $NumOfPartitions, AssignProportion, WinSize, WinMaxSize$ ...
2: **while** $|E| > 0$ **do**
3:     Fill W with edges from E.
4:     Sort edges in W by using the lower degree from two vertices in edge.
5:     Random.shuffe(W, 0, $|W|/2$) and Random.shuffe(W, $|W|/2$, $|W|$).
6:     **for** $i = 0 \rightarrow |W| * AssignProportion$ **do**   ▷ Start assigning edges
7:         $targetP \leftarrow SWRHeuristicRules(W[i])$
8:         $P_{targetP} \leftarrow P_{targetP} \cup \{W[i]\}$
9:     **end for**
10:    **if** $\dfrac{\max_{i \in n}\{|P_i|\} - \min_{i \in n}\{|P_i|\}}{\max_{i \in n}\{|P_i|\} + \min_{i \in n}\{|P_i|\}} > THRESHOLDUP$ **then**
11:        $WinSize \leftarrow min(MaxSize, WinSize * (1 + WindowSizeUpRate))$
12:    **end if**
13:    **if** $\dfrac{\max_{i \in n}\{|P_i|\} - \min_{i \in n}\{|P_i|\}}{\max_{i \in n}\{|P_i|\} + \min_{i \in n}\{|P_i|\}} < THRESHOLDDOWN$ **then**
14:        $WinSize \leftarrow WinSize/(1 + WindowSizeDownRate))$
15:    **end if**
16: **end while**

Generally $|E| > |V|$, therefore approximately we have $E' \approx \frac{|V|}{4}$. We have observed from various experiments that when the DF is order of magnitude higher than the target number of partitions, good balanced partitioning can be easily achieved with simple rule-based greedy algorithm. We don't single out the experimental result here due to space limit. And if a light-weight rule-based greedy algorithm can achieve good balance, then the heavier score calculating and additional imbalance adjustment will not be necessary any more. Our solution is a moving-window mechanism before the partitioning to reorder the incoming edges, so as to increase the observable DF. The basic idea is to keep a suitable window size allowed by the resource limit as long as the reordering (randomization) is sufficient to achieve good balance. For every round of assignment, the size of window will be adjusted according the balanced state of partitions.

Such mechanism can increase the DF effectively. Let us assume that the probability of a vertex with an earlier edge having same vertex is β. $\beta = \frac{2}{|V|}$ for a completely random edge stream. When $|E|$ is sufficiently large, the DF of the whole stream is $E' \approx \frac{1}{2\beta}$. When a window is used, assuming m edges have been processed, the probability that a vertex has appeared before is $\beta * \frac{m}{|E|}$, so at the n-th window, the probability that a vertex has appeared in front windows is $\beta * \frac{(n-1)*|W|}{|E|}$, and the expectation of the number of distinct vertices in the window is $|V|' = min(\frac{|W|}{c}, 2 + \frac{2}{\beta})$ so the expectation of the DF so far will be

$$E' = \sum_{n=1}^{\lceil \frac{|E|}{|W|} \rceil} [\sum_{m=0}^{m=|W|-1} (1 - \beta * \frac{(n-1)*|W|}{|E|} - \frac{2}{|V|'})^{2m}] \tag{5}$$

For the natural graph datasets that resemble certain ordering patterns, the value of β is generally large, so the DF of the whole edge stream is largely dependent on the first several windows, especially the first one. Therefore, we advise to use bigger windows to achieve higher DF. If the buffer size provided to the moving window is limited, we propose to use the technique of *partial dispatch*, which only assigns a fraction of edges in the window and leave the rest to mix with new edges, so as to effectively increase the range of randomization. Furthermore, changing the edge order to distribute the edges related to the same vertex more evenly over time, the partial degree of a vertex observed so far will be a much more reliable indicator when used for relative comparison among vertices during the partitioning, while in a stream of heavy clustering effect the relative partial degree may fluctuate a lot as edges of different vertices will come in waves.

4.2   Getting Better Quality

In this subsection we will discuss the reordering mechanism inside the window and the heuristic rules to further improve the partitioning quality.

It has been shown that for graphs with power-law degree distribution, preferring to cut high-degree vertices during a graph partitioning is a good strategy to lower VRF [4,5]. So the cost of cutting a low-degree vertex is more expensive than a high-degree one and avoid cut low-degree vertex will be helpful. And streaming greedy GP means later partitioning decisions are greatly affected by the existing edge assignment. Therefore in the window we want to move edges with low-degree vertices upfront so they are less likely to be cut into different partitions. We divide edges into two equal parts for getting most randomization.

We start from the rules of greedy and extend them with windowed randomization and degree considerations to design the heuristic rules. DBH and HDRF have applied "preferring to cut high-degree vertices" into hashing and score-based greedy GP, and integrating it into a rule-based greedy GP should also be beneficial. Another useful observation is regarding to the large scale graphs from the social network. The principle of triadic closure tells us that the more common neighbors two separate vertices have the more likely a new edge will form between them. When processing graphs in stream, the arrival of edges can be seen as analog to the formation of a social network, so if a vertex has more neighbors in one partition, it is more likely that a new edge connecting its neighbors in that partition will exist down the stream, so we'd better assign a new neighbor to the partition with most neighbors. Similar edge clustering effect were observed also in graph datasets of other origins, so the principle is also applicable.

From the discussions above we define four assignment rules in our heuristics as below. More specifically, with partition set P and $A(v)$ as defined before and the partial degree record δ, for a new edge $e = (u, v)$, the rules are:

1. If none of u and v has been assigned before, put e to the smallest partition in P.
2. If one of u and v has been assigned before, let it be t, among the smallest M partitions in $A(t)$, put e to the one with the most neighbors.
3. If both u and v have been assigned before, and $B = A(u) \cap A(v) \neq \emptyset$, among the smallest M partitions in B, put e to the one with the most neighbors of both u and v.
4. Otherwise, let $t = (\delta(u) < \delta(v) \,?\, u \,|\, v)$, among the smallest M partitions in $A(t)$, put e to the one with the most neighbors. A new vertex replica is created accordingly.

4.3   Complexity Discussion and Summary

The complexity of rule-based algorithms is generally $O(|E|)$. SWR is a rule-based approach and the added pay is the reorder cost R which is unrelated to the number of target partitions, so the complexity is $O(|E| + R)$. For the score-based HDRF, for each edge assignment $|P|$ scores need to be calculated so the complexity is $O(|E| * |P|)$. ADWISE also uses a sliding-window but only to extend the score calculation to multiple edges in order to increase the range of local optima in the edge assigning process. While the clustering effect behind the imbalance problem is actually not specifically handled, being a score-based algorithm the complexity of ADWISE is further increased to $O(|E| * |P| * |W|)$.

The algorithm of SWR can be easily extend from single-machine to work in parallel across multiple machines to accelerate graph partitioning. Just like the case of **greedy** and HDRF, only some shared status information will be needed. Since we have proven that windowed randomization can increase the DF and a large DF makes balanced partitioning easy to achieve for simple greedy approach, with revised assigning rules, our SWR algorithm can achieve stable partitioning results that are good in all three aspects of quality, balance and speed.

5   Performance Evaluation

In this section we will present the experimental results about the performance of SWR in comparison with other representative GP algorithms: **hashing**, DBH, **grid**, HDRF and ADWISE. Comparisons will be carried out on multiple real-world and synthetic graph datasets with regard to VRF, balance and speed, respectively. In the end, we will also analyze how window size affects the balance and VRF performance in SWR with respect to different DF datasets.

Table 1. Datasets used in the experiments

Dataset	$|V|$	$|E|$	Intrinsic DF
BA-BFS	500000	12663357	1
BA-DFS	500000	12663357	1
BA-RANDOM	500000	12663357	$\gg$10000
as-skitter	1696415	11095298	1
soc-pokec-relationships	1632803	30622564	2
soc-LiveJournal	4847571	68993773	1042
higgs-twitter	456626	14855842	1408
Tecent Weibo	1944589	50635143	$\gg$10000
Wiki-Topcats	1791489	28511807	$\gg$10000

5.1   Experimental Settings and Datasets

In our evaluations, three synthetic graph and six large real-world graph datasets were used to represent graphs with different DF corresponding to clustering effect from very strong, more general and rather weak as shown in Table 1 including the Tecent Weibo from KDD-Cup 2012. The others are from SNAP[1]. All the GP algorithms are implemented in Python except ADWISE, we use the source codes and parameters that are publicly available from the authors. For HDRF, we set the balancing factor $\lambda = 1.1$ as recommended by the authors and for ADWISE, the init window size is set to 100 and the same for SWR. The AssignProportion is set to 0.1 and the Neighbors is 3, meaning choosing from 3 smallest partitions with the most neighbors. Each experiment was repeated 10 times and the average was used as the result. The hardware platform is a workstation equipped with 12-core Intel Xeon CPUs and 64 GB physical memory.

5.2   Experimental Results and Discussions

Performance on Partitioning Quality, Balance and Speed

VRF: NumOfPartitions is set from 4 to 256, and the results are shown in Fig. 4(a)–(f). SWR is always the best performer. In particular, DBH always performs better than hashing with the only additional strategy on cutting high degree vertex first, showing the universal effectiveness of such strategy, which is especially strong on the Tencent Weibo dataset. However, ADWISE performs bad on this dataset although it is as good as HDRF on other datasets, which means that by considering more edges during the score calculation, the power-law degree distribution is not well handled. The performance of other algorithms is more consistent among different datasets. SWR makes better use on both the degree information as well as the neighboring information, so it achieves the overall best performance on all the tested datasets, which is 15% to 30% on

---

[1] http://snap.stanford.edu.

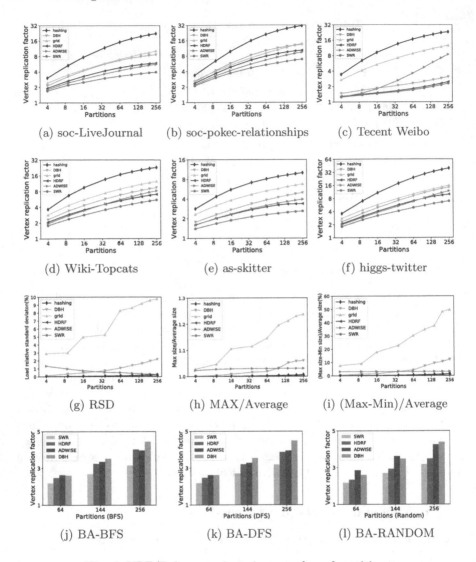

Fig. 4. VRF/Balance against given number of partitions.

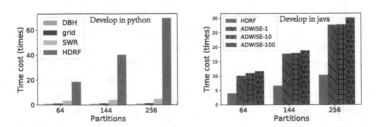

Fig. 5. Partitioning time against number of partitions. Time cost represents the times compared to `hashing`. For `ADWISE-X`, X represents the init window size.

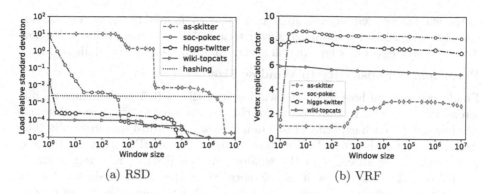

(a) RSD   (b) VRF

Fig. 6. Balance/VRF against different window sizes.

average better than HDRF or ADWISE. Figure 4(j)–(l) show the VRF of different algorithms on the same synthetic graph generated by BA model but collected by BFS, DFS and Random. HDRF performs the best on random graph, confirming the previous discussion of edge correlation influencing its performance. SWR outperforms HDRF even on random graph, because it not only solves the problem of edge correlation, but also takes better use of the vertex degree and neighboring information.

Balance: The experiments are carried out on Tecent Weibo dataset. Three measures were used to show how the algorithms perform from different angles. RSD represents the deviation of the partition size. Max, Min and Average represent the maximal, minimal and average size of all the partitions. The results are shown in Fig. 4(g)–(i). We can see that SWR and HDRF provide the best performance. The balance of hashing is a bit worse for purely depending on probability. DBH and ADWISE can only provide rather moderate balance. The balance performance of ADWISE is special: when the target number of partitions increases, RSD decreases but MAX/Average and (MAX-MIN)/Average keep constant. grid performs bad as the number of partitions grows, the imbalance increases fast until the partitioning result becomes unacceptable.

Speed: We compare the partitioning time of all the algorithms on Tecent Weibo datasets with 64, 144 and 256 partitions and the result is shown in Fig. 5. Since ADWISE was tested with the original source code implemented in Java while all the others were implemented in Python, we can not comparing them directly. Thankfully hashing and HDRF are implemented in both environment, so we use hashing as the reference to normalize the time to rule out the differences of environment. HDRF is used as the anchor when we compare ADWISE with the other algorithms we implemented. We can find SWR has the good feature as grid and DBH, whose time does not grow with the number of partitions, while HDRF and ADWISE need much more time which also grows linearly with the number of partitions. The differences of HDRF between Fig. 5(a) and (b) is due to the different implementation. ADWISE with different init window size behave similar as the window size will always adjust to 1000 as configured in the system. To our surprise, ADWISE is not too much slower than HDRF, for it uses a Lazy Window Traversal strategy, but they are still slower than all the rule-based algorithms.

Performance Sensitivity to Window Size

Finally we analyze how much the window size affects SWR performance on different DF datasets. In order to make the results reliable, we give a different window size for each experiment and don't allow it to adjust automatically. The AssignProportion is set to 1, Neighbors to 0, and NumOfPartitions to 100 and just randomizing all edges in the window. As the RSD of hashing is stable on different datasets we use it as reference. Since the DF of as-skitter dataset is 1, it's more difficult to achieve good balanced partitioning on it. As we can see in Fig. 6(a), when the window size is less than 1000, the windowed mechanism has no effect. This can be confirmed by the VRF of 1 in Fig. 6(b), which means that all the edges are assigned to the same partition. When the window size reaches 1000 and beyond, the windowed mechanism starts to help break the self-constrained edge stream to several sub-streams, there will be more but unbalanced partitions, so the VRF grows. When the window size reaches 10000, a sharp change happens and the RSD is in the level of 0.01 which is as good as hashing. All partitions get an acceptable balance and VRF become stable. This indicates that a window size of 10000 is generally sufficient for balanced partitioning and it is practically easy to release. As the window size keeps growing, both RSD and VRF declines, though smaller and smaller. This is because with larger window, randomization covers a larger range to break the clustering effect, and the degree and neighboring information also become more reliable. As the window size becomes very big, RSD and VRF converge to values that depend on the dataset itself. For soc-pokec-relationships dataset, a much small window size can make it balanced, it just needs a window of 100 to reach the balance level of hashing, although its DF is 2. A very small window can initialize a good balanced partitioning with SWR on it. The same phenomenon is observed on higgs-twitter dataset as well, it has a DF of 1408 and a window of 3 is enough for it to keep balance in many cases. As for wiki-topcats dataset, it has a very large DF, so it is balanced naturally with RSD below hashing. This indicates SWR can get better balance than hashing. Although maybe a small window can help to keep balance, we suggest a big init window size for lower VRF and RSD despite the window will change itself.

6   Conclusions

We found that the intrinsic edge order of natural graphs makes it difficult for existing graph partitioning algorithms to achieve balanced partitioning results. The reason is the clustering effect embedded in the intrinsic order. We propose a sliding window mechanism to reorder the edges, making it much easier for fast rule-based greedy approach to achieve balance. Considerations on vertex degree and neighboring information further improves the partitioning quality. Experimental results show that the proposed SWR algorithm achieves good performance in all aspects of quality, balance and speed, and outperforms all competing algorithms. Comparing to the state-of-the-art HDRF, partitioning speed is increased by 3–20 times, and quality is improved by 15% to 30% on average.

References

1. Abou-Rjeili, A., Karypis, G.: Multilevel algorithms for partitioning power-law graphs. In: Proceedings 20th IEEE International Parallel & Distributed Processing Symposium, 10 pp. IEEE (2006)
2. Andreev, K., Racke, H.: Balanced graph partitioning. Theory Comput. Syst. **39**(6), 929–939 (2006)
3. Chen, R., Shi, J., Chen, Y., Zang, B., Guan, H., Chen, H.: PowerLyra: differentiated graph computation and partitioning on skewed graphs. ACM Trans. Parallel Comput. (TOPC) **5**, 13 (2019)
4. Cohen, R., Erez, K., Ben-Avraham, D., Havlin, S.: Resilience of the internet to random breakdowns. Phys. Rev. Lett. **85**(21), 4626 (2000)
5. Cohen, R., Erez, K., Ben-Avraham, D., Havlin, S.: Breakdown of the internet under intentional attack. Phys. Rev. Lett. **86**(16), 3682 (2001)
6. Crucitti, P., Latora, V., Marchiori, M., Rapisarda, A.: Error and attack tolerance of complex networks. Phys. A **340**(1–3), 388–394 (2004)
7. Feige, U., Hajiaghayi, M., Lee, J.R.: Improved approximation algorithms for minimum weight vertex separators. SIAM J. Comput. **38**(2), 629–657 (2008)
8. Gonzalez, J.E., Low, Y., Gu, H., Bickson, D., Guestrin, C.: PowerGraph: distributed graph-parallel computation on natural graphs. In: Presented as Part of the 10th USENIX Symposium on Operating Systems Design and Implementation (OSDI 2012), pp. 17–30 (2012)
9. Gonzalez, J.E., Xin, R.S., Dave, A., Crankshaw, D., Franklin, M.J., Stoica, I.: GraphX: graph processing in a distributed dataflow framework. In: 11th USENIX Symposium on Operating Systems Design and Implementation (OSDI 2014), pp. 599–613 (2014)
10. Halberstam, H., Laxton, R.R.: Perfect difference sets. Glasgow Math. J. **6**(4), 177–184 (1964)
11. Jain, N., Liao, G., Willke, T.L.: GraphBuilder: scalable graph ETL framework. In: First International Workshop on Graph Data Management Experiences and Systems, pp. 1–6. ACM (2013)
12. Karypis, G., Kumar, V.: A fast and high quality multilevel scheme for partitioning irregular graphs. SIAM J. Sci. Comput. **20**(1), 359–392 (1998)
13. Kim, M., Candan, K.S.: SBV-Cut: vertex-cut based graph partitioning using structural balance vertices. Data Knowl. Eng. **72**, 285–303 (2012)
14. Leskovec, J., Lang, K.J., Dasgupta, A., Mahoney, M.W.: Community structure in large networks: natural cluster sizes and the absence of large well-defined clusters. Internet Math. **6**(1), 29–123 (2009)
15. Low, Y., Gonzalez, J., Kyrola, A., Bickson, D., Guestrin, C.: GraphLab: a distributed framework for machine learning in the cloud. arXiv preprint arXiv:1107.0922 (2011)
16. Malewicz, G., et al.: Pregel: a system for large-scale graph processing. In: Proceedings of the 2010 ACM SIGMOD International Conference on Management of Data, pp. 135–146. ACM (2010)

17. Petroni, F., Querzoni, L., Daudjee, K., Kamali, S., Iacoboni, G.: HDRF: stream-based partitioning for power-law graphs. In: Proceedings of the 24th ACM International on Conference on Information and Knowledge Management, pp. 243–252. ACM (2015)
18. Rahimian, F., Payberah, A.H., Girdzijauskas, S., Haridi, S.: Distributed vertex-cut partitioning. In: Magoutis, K., Pietzuch, P. (eds.) DAIS 2014. LNCS, vol. 8460, pp. 186–200. Springer, Heidelberg (2014). https://doi.org/10.1007/978-3-662-43352-2_15
19. Stanton, I., Kliot, G.: Streaming graph partitioning for large distributed graphs. In: Proceedings of the 18th ACM SIGKDD International Conference on Knowledge Discovery and Data Mining, pp. 1222–1230. ACM (2012)
20. Xie, C., Yan, L., Li, W.J., Zhang, Z.: Distributed power-law graph computing: theoretical and empirical analysis. In: Advances in Neural Information Processing Systems, pp. 1673–1681 (2014)
21. Zhang, C., Wei, F., Liu, Q., Tang, Z.G., Li, Z.: Graph edge partitioning via neighborhood heuristic. In: Proceedings of the 23rd ACM SIGKDD International Conference on Knowledge Discovery and Data Mining, pp. 605–614. ACM (2017)

Flexible Data Flow Architecture
for Embedded Hardware Accelerators

Jens Froemmer[1,2(✉)], Nico Bannow[1], Axel Aue[1], Christoph Grimm[2],
and Klaus Schneider[2]

[1] Robert Bosch GmbH, 70442 Stuttgart, Germany
{jens.froemmer,nico.bannow,axel.aue}@de.bosch.com
[2] University of Kaiserslautern, 67663 Kaiserslautern, Germany
{grimm,schneider}@cs.uni-kl.de

Abstract. In order to enable control units to run future algorithms, such as advanced control theory, advanced signal processing, data-based modeling, and physical modeling, the control units require a substantial step-up in computational power. In case of an automotive Engine Control Unit (ECU), safety requirements and cost constraints are just as important. Existing solutions to increase the performance of a microcontroller are either not suitable for a subset of the expected algorithms, or too expensive in terms of area. Hence, we introduce the novel Data Flow Architecture (DFA) for embedded hardware accelerators. The DFA is flexible from the concept level to the individual functional units to achieve a high performance per size ratio for a wide variety of data intensive algorithms. Compared to hardwired implementations, the area can be as little as 1.4 times higher at the same performance.

Keywords: Hardware accelerator · Data Flow Architecture ·
Coarse-Grained Reconfigurable Architecture · Embedded system ·
Automotive · Microcontroller

1 Introduction

A pivotal component of an Engine Control Unit (ECU) is its microcontroller, whose architecture comprises computational capabilities, memory size, and specialized peripherals. This paper proposes a digital hardware (HW) architecture called Data Flow Architecture (DFA) that targets the next ECU microcontroller generation. Other automotive as well as non-automotive control units and other applications that make use of the novel microcontroller and its supported algorithms will benefit from the DFA too.

Currently, most of the applied algorithms in an ECU are based on characteristic maps or perform simple signal processing, e.g. a knock detection. Increasing the computational capabilities in the next generation of ECUs enables the application of advanced algorithms, such as advanced control theory, advanced signal processing, data-based modeling, and physical modeling. These algorithms

© Springer Nature Switzerland AG 2020
S. Wen et al. (Eds.): ICA3PP 2019, LNCS 11944, pp. 33–47, 2020.
https://doi.org/10.1007/978-3-030-38991-8_3

often rely on large multidimensional data and parameter sets, are called at a high frequency, and require respective memory access patterns as well as elementary functions like the exponential function. The application in an ECU also implies safety criticality, real-time processing, and strong cost constraints. Existing architectural acceleration options to increase the performance of a microcontroller are either not automotive grade, not suitable for a subset of the expected algorithms, or too expensive in terms of area.

The DFA is inspired by data flow graphs, but not directly related to classical data flow architectures. Instead, the DFA is a novel combination of existing and new concepts regarding Coarse-Grained Reconfigurable Architectures (CGRAs) to deliver a high performance per size ratio that is close to a hardwired implementation. Due to features like its flexible (re)configurable functional units, diverse memory access patterns, loop level controlled behavior, and synchronization capabilities, a DFA HW accelerator can run a wide variety of future algorithms at a high utilization. The DFA supports the execution of algorithms not known at design time, considers non-functional automotive requirements, and offers a design space that enables scaling, especially with respect to the targeted algorithms, the performance, and the area.

1.1 Related Work

An ECU and its microcontroller face a high price pressure. Thus, the architectural design space of a HW accelerator is strictly constrained by costs, which can be said to be proportional to its area.

Hardwired Implementations. In case of real-time constraints coupled with high performance requirements, hardwired implementations complement the microcontroller already today. Hardwired implementations may either be integrated into the microcontroller or they may be designed as discrete Application-Specific Integrated Circuits (ASICs). One example is the Advanced Modeling Unit (AMU): integrated into the microcontroller, the AMU "support[s] the fast calculation of GPR [Gaussian Process Regression] models on the new ECU generation MDG" as stated by Diener et al. [4]. Naturally, hardwired implementations have the best performance per size ratio for a single algorithm, but are inflexible. Different ECUs implement different functionalities, but rely on the same microcontroller. Due to their inflexibility, hardwired implementations are not able to support a wide variety of future algorithms. In addition, hardwired implementations preclude further enhancements of algorithms during the lifetime of an ECU generation.

Architectural Options. Single Instruction, Multiple Data (SIMD) extensions, i.e. ARM NEON [9], and General-Purpose Computing on Graphics Processing Units (GPGPU) do not match the requirements of some of the advanced control theory and data-based modeling algorithms. These often include single elementary functions, such as the exponential function, that are not well parallelizable in terms of vector instructions.

Very Long Instruction Word (VLIW), Explicitly Parallel Instruction Computing (EPIC) [16], Tera-op, Reliable, Intelligently Adaptive Processing System (TRIPS) [15], WaveScalar [18] and Dynamically Specialized Datapaths for Energy Efficient Computing (DySER) [6] aim at the execution of arbitrary code or even at replacing a von Neumann processor. Hence, these approaches are not a perfect fit for the acceleration of selected automotive grade algorithms facing strong area constraints and high performance requirements.

Transport Triggered Architectures (TTAs) have been shown to be well suited for real-time applications [3], but imply a non-blocking network. Despite recent improvements [8], such a non-blocking network is expensive in terms of area compared to a sparse crossbar that is optimized for selected algorithms. The static scheduling and the non-configurable function units with a constant latency further limit the speedup or the area efficiency.

While offering the necessary performance and flexibility, Field-Programmable Gate Arrays (FPGAs) are too expensive in terms of the required area for a given functionality or require too much configuration time [11].

The preferred architectural acceleration option is a CGRA. Using coarse-grained function units decreases the flexibility or generality, but also the costs and the configuration time. Therefore, CGRAs are close to the performance per size ratio of a hardwired implementation. A CGRA is designed with respect to a certain set of algorithms or algorithm classes. Various coarse-grained architectures were already explored [7, 12, 19–21].

Related CGRAs. Accelerators tailored to medical applications, such as the Samsung Reconfigurable Processor (SRP) [10], share the increasing performance requirements because of more intelligent algorithms, but prioritize a low energy consumption. Similar to the Architecture for Dynamic Reconfigurable Embedded Systems (ADRES) [13], which was the basis for the SRP, and similar to the Expression-Grained Reconfigurable Array (EGRA) [1], the DFA provides a design space and means to explore it. Nevertheless, the actual goals differ: the DFA aims at the realization of a HW accelerator given certain cost and performance constraints as well as selected data intensive algorithms. EGRA opens up a much larger design space, since it targets the design space exploration itself at the levels of the functional units, the meshes, the memory interfaces and more. ADRES and SRP accelerate not only data flow intensive algorithms or algorithm parts, but also control flow intensive code by reusing the CGRA via VLIW.

The general approach of SoftBrain [14] is similar to the DFA concept: streaming the input data from the main memory into a CGRA with local memory as scratchpad memory. Additionally, the memory access unit(s) natively support different memory access patterns, but also allow for indirect memory address computation in order to realize a decoupled access/execute [17]. SoftBrain particularly targets the combination of an efficient parallel memory interface with an efficient parallel computation interface. In addition to the CGRA, SoftBrain comprises a control core, a scratchpad stream engine, a stream dispatcher, and three stream engines. The Extreme Processing Platform (XPP) [2] is closely related as well, but focuses on a sophisticated hierarchical configuration to minimize runtime reconfiguration costs and on a packet-oriented communication network. In contrast to SoftBrain and XPP, the DFA integrates the control logic and configuration options in its functional units, called base blocks, to meet the strong cost requirements of an ECU.

1.2   Outline

The second chapter *Requirements* introduces the premises and terms for the development of the DFA, especially the algorithmic requirements. The third chapter *Architecture* elaborates on the integration of the DFA into the microcontroller and the architectural elements of the DFA: base blocks, interconnect, and individual features. The fourth chapter *Experimental Results* provides a proof of concept. Finally, the fifth chapter *Conclusion and Discussion* discusses and summarizes the findings before looking out on future work.

2   Requirements

To meet the required performance and strict cost constraints, integrating the DFA into a microcontroller is preferred. The integration avoids a second ASIC and reduces the communication overhead between the HW accelerator, software (SW), and other resources of the microcontroller.

2.1   Programmability and Software Integration Requirements

The DFA is intended to accelerate selected computation-intensive algorithms, but is not required to support arbitrary algorithms. Thus, a library with highly optimized configurations for the selected algorithms is of much higher importance than a compiler supporting auto-vectorization. Still, the efficient mapping of new algorithms onto the DFA requires a modeling framework that initially abstracts from the targeted DFA HW accelerator.

2.2   Algorithmic Characteristics and Requirements

Algorithms targeted by the DFA:

- are data intensive,
- include little control flow,
- support Instruction Level Parallelism (ILP) or Data Level Parallelism (DLP),
- compute the exponential function and other elementary functions,
- access and reorganize data in memory access patterns,
- are safety-critical, and
- have strong real-time constraints.

Data reorganization and memory access patterns require respective load, store, and synchronization mechanisms. Furthermore, the computation and its duration must be well-defined and interruptibility must be ensured to meet safety and timing constraints.

2.3   Architecture Requirements

The following overview refers to the subsections of Sect. 3. In order to meet the requirements above, the DFA must comprise:

1. an efficient integration into the microcontroller,
2. an interconnection of the base blocks with minimal costs and minimal conflicts for the selected algorithms including allowance for future algorithms,
3. a base block structure and interface that allow for a simple extension and the integration of a SW core as well as other functionalities,
4-7. essential base blocks and features in line with above mentioned algorithmic characteristics and enabling the DFA to act intelligently and independently,
8. a design space, whose exploration allows for the optimization of the DFA with respect to selected algorithms, as well as
9. a programming interface and a workflow that lead to an optimal usage of the DFA resources and allow for the efficient implementation of new algorithms.

3   Architecture

As the DFA is an architecture for dedicated HW accelerators, it can in general be integrated into any microcontroller.

3.1   Microcontroller Integration

As depicted in Fig. 1, the DFA is mapped into the global address space and can be accessed by all masters. Additionally, the DFA has a master interface that allows for Direct Memory Access (DMA) with a high bandwidth. The software cores of the microcontroller may access the global configuration registers, the base block registers, and the local multi-bank SRAM of the DFA.

3.2   Interconnect

The base blocks are accessible through and connected by a sparse crossbar. As depicted in Fig. 2, the output ports of base blocks connect to routers. These routers are connected to arbiters, which connect to the input ports of the base blocks. The available connections, the number of buffers per connection, and the hierarchical structure of the interconnect are optimized with respect to the number of base blocks and targeted algorithms.

3.3   Base Block Structure and Properties

A base block consists of three stages: input stage, operation stage, and output stage. The input stage comprises input ports and registers, while the output stage comprises the output ports. The execution depends solely on the availability of the input arguments. Base blocks can write to their successors, but they cannot read from their predecessors. Hence, base blocks form a directed data flow

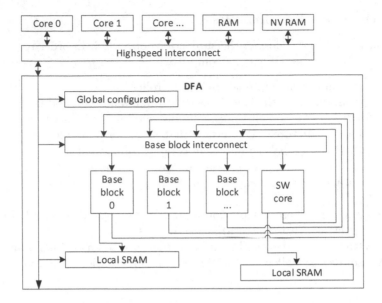

Fig. 1. The DFA is integrated into the microcontroller. Its global configuration, base blocks, and SRAM are accessible by all masters.

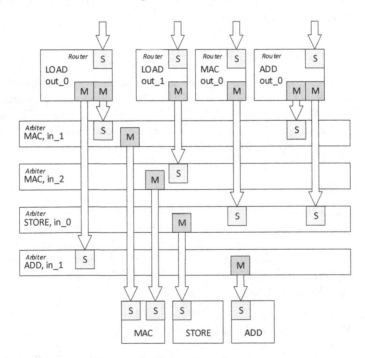

Fig. 2. A basic interconnect that connects input and output ports of a load, a MAC, an add, and a store base block. Data can flow from the load to the store base block through at least one of the mathematical base blocks: add and MAC base block.

that exploits temporal parallelism via pipelining and results in a decentralized intelligence. Thus, base blocks may have a varying latency and throughput, since their execution does not depend on time. The operation stage performs the actual computation, may include a state machine, and uses the configuration and the operation registers. The output stage supports writing to multiple targets, whose quantity is limited by the number of output ports and transfer cycles.

3.4 Base Block Types

In order to run independently, a DFA HW accelerator contains one or multiple Data Flow Control (DFC) units that handle configuring and reconfiguring the DFA. Load and store base blocks have access to the global memory and the local SRAM with high bandwidth. Furthermore, load and store base blocks are capable of address generation and synchronization to match the memory access patterns required by the algorithms. A number of base blocks for basic mathematical operations have been implemented, i.e. a MAC base block. The Coordinate Rotation Digital Computer (CORDIC) base block computes a variety of elementary functions and covers the complete value range. All base blocks can be scaled towards throughput, latency, accuracy, and area.

The well-defined base block interface and the internal structure simplify the development of new base blocks. Furthermore, they enable the integration of a SW core, a Sea-of-Gates, an FPGA, or other functionalities.

3.5 Extended Features

Base blocks may write to any register of their connected successors. Thus, base blocks can not only produce output data, but also compute configuration settings based on input data, e.g. to compute offset values used for an address generation.

Split/Combine Mode. A base block consists of one or multiple splits, where each split contains the full base block functionality. Multiple splits in one base block enable the optional combine mode, which is available only in selected base block types. The combine mode allows for using two splits to perform an operation with higher data width, e.g. combining two 32-bit addition base block splits leads to a combined 64-bit precision.

Multiple Register Sets. A base block split can have multiple register sets and dynamically switch between them. Hence, switching between different configurations during runtime is possible and highly performant, allowing for the efficient sharing of resources. Figure 3 shows the general concept. The address decoder forwards incoming data to the selected register of the respective set. Depending on the configuration and the available data, one of the sets is selected via the arbiter. While not shown in the simplified figure, the result of the operation stage is stored to an output register before being sent to one or multiple connected base blocks. The number of splits and the number of sets per split are configurable during design time.

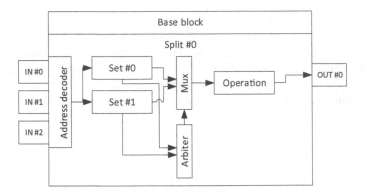

Fig. 3. Each base block may have multiple splits and sets per split as well as multiple input ports and output ports that are shared among the splits. An arbiter selects the active set based on the configuration and the available data.

Loop Level Information. Base blocks that compute offsets or addresses, i.e. a load base block, support nested loops. Based on the status of those loops, these base blocks can send a respective loop level. Every base block is able to evaluate, manipulate, and forward a loop level allowing for a loop level controlled behavior. A typical use case is a load base block sending data to a MAC base block: based on the attached loop level value, which is generated by the load base block, the MAC base block accumulates incoming values and then outputs the accumulated value, once a defined loop level has been received.

Offsets or Address Generation. Base blocks may compute addresses or indices by using multiple offsets in order to access memory, to induce loop level based behavior, or for computation. Hence, an arbitrary number of load and store operations can concurrently be performed, limited only by the number of corresponding base blocks and the memory bandwidth. The address generation can be scaled towards the area, the supported access patterns, and the number of loop levels.

Synchronization. Base blocks with address generation can send special data, so called *ready* data, dependent on the progress of their offsets. Combined with the address-based communication, this enables the base blocks to synchronize. E.g. once a store base block has written the last data, it might immediately start a preconfigured load base block.

3.6   Real-Time and Safety

In theory, the DFA could run multiple different algorithms in parallel. However, due to the real-time and safety requirements, running only one algorithm that utilizes as many resources as possible, is preferred. Eliminating the interference caused by running multiple algorithms in parallel ensures a constant execution time per algorithm, if only the local SRAM is used or a certain bandwidth to the global memory is guaranteed. State-of-the-art safety measures, such as Error-Correcting Code (ECC) and Built-In Self-Tests (BISTs), combined with analyses like the Failure Modes, Effects and Diagnostic Analysis (FMEDA), are applied to fulfill the highest automotive safety integrity level *ASIL D*.

3.7 Interruptibility

In order to enable the DFA to be used to capacity without loosing real-time capabilities, higher prioritized algorithms must be able to interrupt an ongoing computation. The DFC base block manages incoming requests to execute an algorithm based on a priority list of algorithms. This list contains all the required information about the algorithms, such as the interruptibility, and links the respective configurations. Interrupted cooperative algorithms do not lose their progress, but may be resumed. Because algorithms are executed as directed data flows, the state of an ongoing computation consists of the intermediate results in the local SRAM, the accumulated values within mathematical base blocks, and the progress in the memory access patterns of the load and store base blocks. The configuration of a cooperative algorithm includes respective sub-configurations to store and restore the progress of the computation. Base blocks with relevant internal values may send their status to the DFC base block. Additionally, the DFC base block is capable of tracking the progress independently. When the DFA is interrupted, the DFC base block stores the progress of the computation in the system memory.

3.8 Design Space Exploration

The following levels of the design space are to be explored when implementing a DFA HW accelerator for selected algorithms: the structure and the available connections of the interconnect, the types, the quantities, and the features of the base blocks, the SRAM size, the number of SRAM banks, and the DMA capabilities like bandwidth and buffering. The implementation of the DFA is modular and generic with respect to the design space exploration options: base block variants are described via a list of parameters selecting and enabling/disabling their features, the architecture via instantiations of the base block variants, the interconnect via a connection matrix, etc. An automated and tool supported design space exploration is future work.

3.9 Programming a DFA HW Accelerator

Programming a DFA HW accelerator equals configuring its base blocks, global registers, SRAM, and eventually debug features. Hence, using the DFA via SW comes down to writing a certain configuration.

Once started, the DFA can run independently and compute multiple subsequent algorithms. Additionally, the DFA supports partial configurations: inactive base blocks can be (re-)configured, while active base blocks protect their registers against write accesses during computations. Exploiting shared configuration values between subsequent algorithms reduces the configuration time. The DFA supports synchronous and asynchronous execution, but software preferably starts the DFA asynchronously with the exception of algorithms that require an as low as possible reaction time. The standard use case with respect to the usage in an ECU are calls to highly optimized library functions. Therefore, support for the automated implementation of new arbitrary algorithms is future work. Currently, the implementation of a new algorithm in a DFA HW accelerator consists of three steps [5]: modeling the data flow of the algorithm, mapping the data flow onto the DFA base blocks, and generating a configuration.

4   Experimental Results

Many algorithms expected for future automotive applications have been mapped onto the DFA and simulated successfully. Thus, the DFA provides the required functionality and flexibility. As an example, the implementation of a layer of a Multilayer Perceptron (MLP) is discussed below. Due to the scalability and the flexibility of the DFA, the performance and the supported algorithms of a DFA HW accelerator depend on its HW configuration. Therefore, Sect. 4.2 elaborates on the theoretical performance to enable runtime estimations for arbitrary HW configurations. Section 4.3 provides a first comparison based on actual silicon to evaluate the main goal: a performance per size ratio close to a hardwired implementation.

4.1   Implementation of a MLP-Layer

A feedforward artificial neural network may classify as a MLP, if it consists of three or more fully connected layers, whose nodes are neurons with nonlinear activation functions. In the example, the neurons of each layer use a bias of zero and a Rectified Linear Unit (ReLU) as the activation function. The output o of a neuron is based on the inputs x and the weights w, as described by the following equation:

$$o(\boldsymbol{x}) = max(0, \sum_{i=0}^{n}(x_i \times w_i)) \tag{1}$$

Figure 4 shows a configuration of the DFA that computes a single layer of a MLP consisting of m neurons computed as in Eq. 1. Each neuron expects n inputs. One load base block is configured to load the n inputs for all m neurons, while another or several other load base blocks load the n weights of all m neurons. The computation of a layer can be parallelized by multiple instances of the data path, which consists of the load base block loading the weights, the MAC base block computing the weighted sum, the activation function, and the store base block. The load base blocks send the inputs and weights with the attached loop level to a MAC base block. When all inputs or weights of one neuron have been loaded, the load base blocks increase the loop level from zero to one. The MAC base block blocks its output until the arrival of a loop level of one, while still multiplying and accumulating the incoming values. The ReLU is realized via a maximum base block that compares incoming values against a configured constant, which is set to zero. The connected store base block writes the received values to the local SRAM.

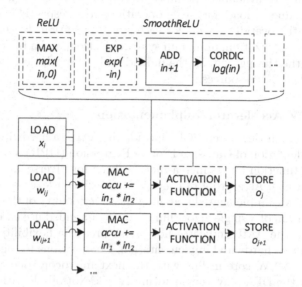

Fig. 4. Below the base block types, such as load, MAC, and store, the respective configurations are depicted. See also Eq. 1 for the referenced inputs x_i and w_{ij} as well as the referenced outputs o_j, where the index j refers to the j-th neuron.

Activation functions like the SmoothReLU require operations that are expensive in terms of area, e.g. the exponential and logarithmic functions. In case of parallelization, the base blocks configured to compute the activation function could have two ore more register sets sharing the same execution units in order to reduce the area of the respective DFA HW accelerator. Given a sufficient number of inputs n per neuron, the limited throughput of the activation function would not degrade the overall performance.

4.2   Theoretical Performance

Apart from the memory access conflicts, the performance of the DFA is well defined in case of (semi-) static configurations. Given a configuration of a DFA HW accelerator and the performance of its base blocks, the required cycles can be derived via a static analysis. The configuration time depends on the number of registers that are written and the number of registers that can be written simultaneously. The initial latency is determined by the longest path in the configured computation pipeline of the DFA: it is the sum of the latencies of the individual connected base blocks including the interconnect. Similarly, the throughput or number of cycles per result equals the minimum throughput of the individual base blocks in the critical path, which typically is the innermost loop. Sharing the operation stage of a base block between multiple register sets reduces the throughput of the base block in accordance to the frequency of the activation via newly available operands. If two or more data paths require a synchronization due to load and store operations, the respective latencies have to be added to the overall runtime for each synchronization. In general, spatial and temporal parallelization, minimum synchronization, and a balanced pipeline with regard to the throughputs of the individual base blocks lead to an optimal utilization of the DFA resources.

4.3   DFA HW Accelerator Implementation

Current mid- to high-tier Bosch ECUs include the AMU: a hardwired implementation of the calculation of Gaussian Process Regression (GPR) models that uses an exponential function and is increasingly utilized for diverse control/feedback control applications. As a proof of concept at the silicon level, a concrete DFA HW accelerator was implemented matching the functionality of the AMU. While the implementation of a single algorithm is not the intended use case of the DFA, comparing it to the AMU indicates how much area the flexibility of the DFA adds compared to a hardwired implementation. Comparisons to the performance and the area of a SW core in line with the next microcontroller (µC) generation, show how the DFA may benefit future ECUs. Not all algorithms share the requirement for safety or real-time capabilities. Hence, the performance and the area of a SW core of a microprocessor (µP), which is considered for high-tier ECUs, with and without a SIMD extension, are evaluated as well.

In order to achieve a realistic comparison, all assumed configurations are based on actual or planned products of Bosch and its business partners. The AMU as considered for the next ECU generation is the baseline. The DFA HW accelerator, the baseline AMU, and the microcontroller SW core were normalized to the same semiconductor technology and frequency. The AMU of the current ECU generation supports only half the frequency and uses a twice as large semiconductor scale. A microprocessor faces lower non-functional requirements, such as operating temperature and safety. Therefore, the frequency of the microprocessor SW core is assumed to be five times higher than the frequency of the baseline AMU, while the semiconductor scale is assumed to be half as large. In order to compare the area, only the logic without the SRAM was considered.

Comparison of Performance and Area. Because the usage in an ECU demands a high performance at low costs, the performance per size ratio is introduced as an important metric. All performance and size values are normalized to the baseline AMU as considered for the next ECU generation. As listed

Table 1. A performance and area comparison sorted by size and normalized to the performance and the area of the hardwired AMU implementation as considered for the next ECU generation.

Implementation	Performance	Size	Performance/Size
AMU cur. gen.	0.5	2	1 : 4
AMU	1	1	1 : 1
DFA	0.7	1	1 : 1.4
µC core	0.015	3.3	1 : 220
µP core	0.47	84	1 : 179
µP core + SIMD	2.3	84	1 : 37

in Table 1, the AMU of the current ECU generation is twice as large at half the performance. Hence the performance per size ratio of the current AMU is 1 : 4. The DFA implementation of the AMU delivers 70% of the performance at the same area. This results in a performance per size ratio of 1 : 1.4. A microcontroller core delivers only 1.5% of the performance of the AMU while requiring 3.3 times the area, which results in a performance per size ratio of 1 : 220. Five times the clock speed and half the semiconductor scale enable the microprocessor core to achieve 47% of the performance of the AMU, but at 84 times the size and at much higher costs that come along using a microprocessor in addition to the microcontroller. Thus, the performance per size ratio of 1 : 179 is still comparable to the one of the microcontroller core. Using the SIMD extension of the microprocessor core leads to a performance that is 2.3 times higher compared to the AMU and a performance per size ratio of 1 : 37 that falls in between the DFA and the microprocessor without the use of the SIMD extension.

These results confirm the initial hypothesis: The DFA as a novel CGRA achieves a similar performance compared to a hardwired solution, but adds flexibility without the overhead of an approach offering more generality, such as a SW core with a SIMD extension. This initial comparison bases solely on the AMU. Further studies regarding relevant algorithms and respective design space explorations are required and future work. However, the comparisons and estimations that are currently in progress coincide with the presented results.

5   Conclusion and Discussion

In order to enable advanced control theory, advanced signal processing, database modeling, physical modeling, and other future algorithms to run on the next ECU-generation, a flexible HW accelerator with a performance per size ratio close to a hardwired implementation is the preferred option. The DFA considers the safety and real-time requirements of an ECU and combines existing CGRA concepts in a novel way with the addition of features for the support of the required memory access patterns as well as elementary functions, such as the exponential function. The design space of a DFA HW accelerator especially comprises the types, quantities, and features of the base blocks, the interconnect, the number of SRAM banks, and the DMA capabilities.

Selected algorithms were successfully mapped onto the DFA and simulated, showing that the DFA provides the required functionality and flexibility. The DFA was validated at the silicon level by a successful implementation of the AMU algorithm as a DFA HW accelerator. The implementation of a single algorithm allows for a comparison regarding the area and the performance to determine the overhead caused by the flexibility of the DFA. The DFA implementation achieved 70% of the performance of the AMU at the same area leading to a performance per size ratio of 1 : 1.4. If required, the DFA can be scaled towards a higher performance without a meaningful reduction of the performance per size ratio. In comparison, the considered microprocessor core with lower non-functional requirements and a SIMD extension yielded a much lower performance per size ratio of 1 : 37. This indicates that the flexibility of the DFA compared to a hardwired solution comes at a reasonable cost in terms of area.

Further studies regarding the performance, the area, supported algorithms, and scalability are currently being conducted and show similar results. As the design space of the DFA may very well be too large to be explored manually when designing and optimizing a DFA HW accelerator for selected algorithms, an automated or interactive design optimization framework is future work.

References

1. Ansaloni, G., Bonzini, P., Pozzi, L.: EGRA: a coarse grained reconfigurable architectural template. IEEE Trans. Very Large Scale Integr. VLSI Syst. **19**(6), 1062–1074 (2011). https://doi.org/10.1109/TVLSI.2010.2044667
2. Baumgarte, V., Ehlers, G., May, F., Nückel, A., Vorbach, M., Weinhardt, M.: PACT XPP—a self-reconfigurable data processing architecture. J. Supercomput. **26**(2), 167–184 (2003). https://doi.org/10.1023/A:1024499601571
3. Bhagyanath, A., Schneider, K.: TTA as predictable architecture for real-time applications. In: 2014 International Conference on Science Engineering and Management Research (ICSEMR), November 2014. https://doi.org/10.1109/ICSEMR.2014.7043544
4. Diener, R., Hanselmann, M., Lang, T., Markert, H., Ulmer, H.: Data-based models on the ECU. In: Design of Experiments (DoE) in Powertrain Development, pp. 227–241. Expert-Verlag (2015)
5. Froemmer, J., Bannow, N., Aue, A., Grimm, C., Schneider, K.: Model-based configuration of a coarse-grained reconfigurable architecture. In: MBMV. VDE Verlag (2019)
6. Govindaraju, V., Ho, C.H., Sankaralingam, K.: Dynamically specialized datapaths for energy efficient computing. In: 2011 IEEE 17th International Symposium on High Performance Computer Architecture, February 2011. https://doi.org/10.1109/HPCA.2011.5749755
7. Hartenstein, R.: A decade of reconfigurable computing: a visionary retrospective. In: Proceedings of the Conference on Design, Automation and Test in Europe, DATE 2001. IEEE Press (2001)

8. Jain, T., Schneider, K., Jain, A.: An efficient self-routing and non-blocking inter-connection network on chip. In: Proceedings of the 10th International Workshop on Network on Chip Architectures, NoCArc 2017. ACM (2017). https://doi.org/10.1145/3139540.3139546

9. Jang, M., Kim, K., Kim, K.: The performance analysis of ARM NEON technology for mobile platforms. In: Proceedings of the 2011 ACM Symposium on Research in Applied Computation, RACS 2011. ACM (2011). https://doi.org/10.1145/2103380.2103401

10. Kim, C., Chung, M., Cho, Y., Konijnenburg, M., Ryu, S., Kim, J.: ULP-SRP: ultra low power Samsung Reconfigurable Processor for biomedical applications. In: 2012 International Conference on Field-Programmable Technology, December 2012. https://doi.org/10.1109/FPT.2012.6412157

11. Kuon, I., Rose, J.: Measuring the gap between FPGAs and ASICs. IEEE Trans. Comput. Aided Des. Integr. Circuits Syst. 26(2), 203–215 (2007). https://doi.org/10.1109/TCAD.2006.884574

12. Liang, C., Huang, X.: SmartCell: an energy efficient coarse-grained reconfigurable architecture for stream-based applications. EURASIP J. Embedded Syst. 2009, 1:1–1:15 (2009). https://doi.org/10.1155/2009/518659

13. Mei, B., Vernalde, S., Verkest, D., De Man, H., Lauwereins, R.: ADRES: an architecture with tightly coupled VLIW processor and coarse-grained reconfigurable matrix. In: Y. K. Cheung, P., Constantinides, G.A. (eds.) FPL 2003. LNCS, vol. 2778, pp. 61–70. Springer, Heidelberg (2003). https://doi.org/10.1007/978-3-540-45234-8_7

14. Nowatzki, T., Gangadhar, V., Ardalani, N., Sankaralingam, K.: Stream-dataflow acceleration. In: Proceedings of the 44th Annual International Symposium on Computer Architecture, ISCA 2017. ACM (2017). https://doi.org/10.1145/3079856.3080255

15. Sankaralingam, K., et al.: Exploiting ILP, TLP, and DLP with the polymorphous TRIPS architecture. In: Proceedings of the 30th Annual International Symposium on Computer Architecture, ISCA 2003. ACM (2003). https://doi.org/10.1145/859618.859667

16. Schlansker, M.S., Rau, B.R.: EPIC: explicitly parallel instruction computing. Computer 33(2), 37–45 (2000). https://doi.org/10.1109/2.820037

17. Smith, J.E.: Decoupled access/execute computer architectures. In: Proceedings of the 9th Annual Symposium on Computer Architecture, ISCA 1982. IEEE Computer Society Press (1982)

18. Swanson, S., Michelson, K., Schwerin, A., Oskin, M.: WaveScalar. In: Proceedings of the 36th Annual IEEE/ACM International Symposium on Microarchitecture, MICRO 36. IEEE Computer Society (2003)

19. Tanomoto, M., Takamaeda-Yamazaki, S., Yao, J., Nakashima, Y.: A CGRA-based approach for accelerating convolutional neural networks. In: 2015 IEEE 9th International Symposium on Embedded Multicore/Many-Core Systems-on-Chip, September 2015. https://doi.org/10.1109/MCSoC.2015.41

20. Tehre, V.: Survey on coarse grained reconfigurable architectures. Int. J. Comput. Appl. 48(16), 1–7 (2012)

21. Tessier, R., Pocek, K., DeHon, A.: Reconfigurable computing architectures. Proc. IEEE 103(3), 332–354 (2015). https://doi.org/10.1109/JPROC.2014.2386883

HBL-Sketch: A New Three-Tier Sketch for Accurate Network Measurement

Keyan Zhao[1], Junxiao Wang[1], Heng Qi[1(✉)] (iD), Xin Xie[1], Xiaobo Zhou[2], and Keqiu Li[2]

[1] Dalian University of Technology, Dalian, China
hengqi@dlut.edu.cn
[2] Tianjin University, Tianjin, China

Abstract. Network measurement is critical for many network functions such as detecting network anomalies, accounting, detecting elephant flow and congestion control. Recently, sketch based solutions are widely used for network measurement because of two benefits: high computation efficiency and acceptable error rate. However, there is usually a tradeoff between accuracy and memory cost. To make a reasonable tradeoff, we propose a novel sketch, namely the HBL (Heavy-Buffer-Light) sketch in this paper. The architecture of HBL sketch is three-tier consisting of heavy part, buffer layer and light part, which can be viewed as an improved version of Elastic sketch which is the state-of-the-art in network measurement. Compared to the Elastic sketch and other typical work, HBL sketch can reduce the average relative error rate by 55%–93% with the same memory capacity limitations.

Keywords: Network measurement · Sketch · Tradeoff · Average Relative Error

1 Introduction

Network measurement is often viewed as the basis of network anomaly detection, quality of service, capacity planning, accounting, congestion control, and so on. [5, 24, 30]. Therefore, network operators must use measurement tools to meet the growing demands on daily network operations and network management. Traditional network measurement methods are mostly based on packet sampling, such as sFlow [22] and NetFlow [2]. However, packet sampling will lose some information, which leads to the inaccuracy of network measurement. Several studies have shown that packet sampling is not sufficient for fine-grained network measurement [7,18]. Recently, sketching techniques have found a widespread use in network measurements such as flow sizes estimation [15,24,26], elephant flow detection [5,13,21,28], and flow number estimation [12,14,27]. Compared to the sampling methods, the sketch methods have higher accuracy [14,16].

In recent work, some typical sketch-based network measurement methods are proposed, such as CM sketches [9], CU sketches [11], counting sketches [8],

© Springer Nature Switzerland AG 2020
S. Wen et al. (Eds.): ICA3PP 2019, LNCS 11944, pp. 48–59, 2020.
https://doi.org/10.1007/978-3-030-38991-8_4

Bloom sketch [29], FlowRadar [15]. In existing work, state-of-the-art method named Elastic sketch [25] is proposed to provide with a more speedy processing and a less memory usage. The Elastic sketch consists of two parts: a heavy part and a light part. The heavy part mainly records elephant flows. Because the flow ID of the packet is recorded in heavy part, the number of packets in the flow can be accurately recorded. The light part is a simple CM sketch, which mainly records mouse flows, does not record the flow ID of the packet, and can only roughly estimate the number of packets in the flow.

In Elastic sketch, the incoming packet is first stored in heavy part by hashing. When a hash collision occurs in the heavy part, an elephant flow is also possibly removed from the heavy part. Then a part of the elephant flow leaves the precise heavy part and enters the rough light part, which we argue is negative for keeping up accuracy of measurement. On another hand, the part of the elephant flow which moved into the light part will destroy the correctness of mouse flows. By sharing position in the light part, all the mouses flows relating to the elephant flow are excessively recorded, which we argue is also harmful to accuracy of measurement.

The low accuracy of measurement will affect functions [3,17,19], which may cause network operators to make wrong decisions and affect the normal operation of the network. To improve the accuracy of Elastic sketch, additional memory is needed. However, network measurement is generally used in network devices such as switches and routers. Memory on these network devices is an extremely precious resource. Therefore, we should address a tradeoff issue between the accuracy and the memory cost.

In this paper, we propose a novel sketch based on the buffer layer and the Elastic sketch, namely the HBL (Heavy-Buffer-Light) sketch. The main idea of our HBL sketch is to leverage the buffer layer to avoid the part of the elephant flow muted in the light part, while avoiding the part of the elephant flow disturbing the correctness of the light part. With the HBL sketch, we can improve the accuracy of network measurement without additional memory cost.

In summary, the contributions of this paper are as follows:

(1) We design a novel sketch, namely HBL (Heavy-Buffer-Light) Sketch. With a buffer layer, not only the number of accesses to the light part is reduced, but also largely avoiding the heavy part wrongly passing the elephant flow records to the light part. The accuracy of network measurement can be improved by the HBL sketch.
(2) We implement a network measurement mechanism based on HBL Sketch, which includes insertion algorithm and query algorithm. Based on the proposed mechanism, we further implement two measurement modules which could be leveraged to estimate flow sizes and to detect elephant flows.
(3) We examine the HBL Sketch in terms of accuracy, memory usage and speed by plenty of experiments. The experimental results show that our HBL sketch significantly outperforms other sketches and decreases the error rate by 55%–93% on average with the same memory cost. Moreover, to obtain the same level of error rate, the memory usage of the HBL Sketch is about 1/5–2/3 of existing methods. In terms of processing speed, the HBL Sketch can match state-of-the-art method.

Code and Data Sharing. We implement HBL Sketch based on the code of Elastic Sketch[1]. We make the HBL Sketch data snapshot and code used for this paper available to the research community as well[2].

2   Related Work

Recently, sketch is widely used in network measurement, such as Elastic sketch [25], count-min (CM) sketch [9], conservative-update (CU) sketch [11], Count sketch [8], Bloom sketch [29], SF sketch [26] and FlowRadar [15].

Among them, CM Sketch, CU Sketch and Count sketch are three classic sketches with the same data structure. A CM sketch [9] consists of d arrays, each of which consists of w counters and corresponds with a hash function. When inserting a flow f, the CM sketch first computes the d hash functions and locates the d counters. Then it increments all the d hashed counters. When querying a flow f, the CM sketch reports the minimum value of the d hashed counters. The CU sketch [11] makes a slight but effective modification compared with the CM sketch. During insertions, the CU sketch only increases the smallest counter(s) among the d hashed counters while the query process keeps unchanged. The Count sketch [8] is also similar to the CM sketch except that each array uses an additional hash function to smooth the accidental errors. They all have the shortcomings of slow speed and poor accuracy when using small memory.

Bloom sketch [29] consists of sketch layers and Bloom filter [6] layers. SF sketch [26] is a two-stage sketch based on CM sketch. They all have faster processing speeds, but the accuracy is poor. FlowRadar [15] maps flows to counters through XOR operations, such that new flows can be reconstructed by repeatedly XOR-ing the counters with known flows. However, such extensions incur heavy computational overhead and its memory usage is still much higher than other sketches.

The state-of-the-art solution Elastic sketch [25] is proposed to provide with a more speedy processing and a less memory usage. And it can adapt to current traffic characteristics. However, Elastic sketch is composed of the precise heavy part and the rough light part. If there are too many flows, especially elephant flows, removed from heavy part and stored in light part, the error of network measurement will be greatly increased. Therefore, the design goal of this paper is to improve the accuracy of Elastic Sketch without increasing memory usage while keeping its advantages.

3   HBL Sketch

3.1   Data Structure

As shown in Fig. 1, the data structure consists of three parts: the heavy part, the buffer layer and the light part.

---

[1] https://github.com/BlockLiu/ElasticSketchCode.
[2] https://github.com/FlowAnalysis/HBLSketch.

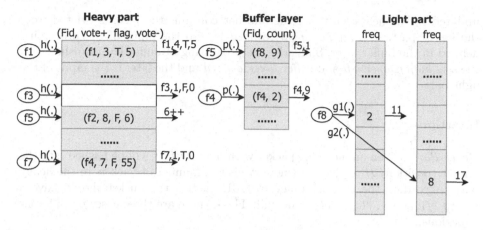

Fig. 1. The structure of our HBL sketch.

Heavy Part: The heavy part is a hash table associated with a hash function $h(.)$. Each position (bucket) of the heavy part records the information of a flow by: flow ID (Fid), positive votes (vote+), negative votes (vote−), and flag. Each packet arriving the system is calculated its Fid (according to five-tuples) and its position in the hash table. When the calculated Fid matches the one that belonging to the position, vote+ increases by 1, otherwise, vote− increases by 1. When the value of vote−/vote+ is more than a fixed threshold λ, the record originating on the position will be evicted from the heavy part. The flag indicates whether such an eviction has occurred on the position.

Buffer Layer: The buffer layer is a hash table associated with a hash function $p(.)$. Each position of the buffer layer records the information of a flow by: flow ID (Fid), packet counter (count). The record evicted from the heavy part is the input of the buffer layer. When the record arrives and the calculated Fid matches the one that belonging to the position, count increases by vote+, otherwise, the record originating on the position will be evicted from the buffer layer.

Light Part: The light part is a CM Sketch [9], consisting of d arrays. Each array is associated with one hash function, and is composed of w counters. Each counter records the information of a flow by: event frequency (freq). The record evicted from the buffer layer is the input of the light part and is eventually stored into the arrays.

3.2   HBL-Based Network Measurement Mechanism

Basically, Network measurement mechanism based on HBL sketch includes insertion and query. HBL sketch relies on a series of hash operations to update memory and to record the information of flows, just like working in a manner of pipeline processing. When the packet arrives the system, it's first processed by the hash operations defined in the heavy part. The processing finishes with two sets *heavy_part_left* and *heavy_part_evicted*. The first set contains the records updated in the heavy part and the another contains the records evicted from the heavy part. The set *heavy_part_evicted* is also the input of the processing defined in the buffer layer. Similarly, the processing of buffer layer finishes with the sets *buffer_layer_left* and *buffer_layer_evicted* and the later is the input of the light part.

Insertion

Heavy Part: Given an incoming packet with flow ID f, it will be hashed into the bucket (position) $H[h(f)\%N_1]$, where N_1 is the number of buckets in the heavy part. According the memory usage of HBL Sketch, the bucket should have a record $(f_1, vote+, flag_1, vote-)$ or null. Then, there are three cases by different situations,

Case 1, *the record is null*: Add $(f, 1, F, 0)$ into the set *heavy_part_left*, where $flags = F$ is a boolean value to indicate no eviction has happened in the bucket. *Case 2, f matches f_1*: Add $(f_1, sum(1, vote+), flag_1, vote-)$ into the set *heavy_part_left*. *Case 3, f mismatches f_1*: If $\frac{sum(1,vote-)}{vote+} < \lambda$, add $(f_1, vote+, flag_1, sum(1, vote-))$ into the set *heavy_part_left* and add $(f, 1)$ into the set *heavy_part_evicted*. Otherwise, add $(f, 1, T, 1)$ into the set *heavy_part_left* and add $(f_1, vote+)$ into the set *heavy_part_evicted*. The $flags = T$ indicates the bucket has happened eviction before flow f is elected.

Buffer Layer: For the flow ID f and the number of packets val evicted from the heavy part, it will be hashed into the bucket $B[p(f)\%N_2]$, where N_2 is the number of buckets in the buffer layer. The bucket is supposed to have a record $(f_1, count)$ or null. Then, there are three cases by different situations,

Case 1, *the record is null*: Add (f, val) into the set *buffer_layer_left*. *Case 2, f matches f_1*: Add $(f_1, sum(val, count))$ into the set *buffer_layer_left*. *Case 3, f mismatches f_1*: Add (f, val) into the set *buffer_layer_left*, and add $(f_1, count)$ into the set *buffer_layer_evicted*.

Light Part: For the flow ID f and the number of packets val evicted from the buffer layer, it will be d-hashed into d buckets (in d arrays) with records $[(freq)]_d$, and $[(sum(val, freq))]_d$ is added into the set *light_part_left*.

To further clarify the above mappings, let us get back to the example depicted in Fig. 1, where $\lambda = 8$. The packet of flow f_1, f_3, f_5 and f_7 arrive the heavy part, triggering the record eviction of flow f_4 and f_5. Then, they arrive the buffer layer and trigger the record eviction of flow f_8. At last, the evicted record arrives the light part and is updated by Count-Min Sketch [9]. Now, let's zoom in the processing conducted in the heavy part, buffer layer and light part. In heavy part, f_1 matches the Fid of the record $(f_1, 3, T, 5)$ in the hashed bucket, so the record in the bucket is updated to $(f_1, 4, T, 5)$. The record in the bucket hashed by f_3 is empty, so the record in the bucket is updated to $(f_3, 1, F, 0)$. f_5 does not match the Fid of the record $(f_2, 8, F, 6)$ in the hashed bucket and $\frac{6+1}{8} < \lambda$, so the record in the bucket is updated to $(f_2, 8, F, 7)$ and $(f_5, 1)$ is passed to the buffer layer. f_7 does not match the record $(f_4, 7, F, 55)$ in the hashed bucket and $\frac{55+1}{8} \geq \lambda$, so the record in the bucket is updated to $(f_7, 1, T, 0)$ and $(f_4, 7)$ is passed to the buffer layer. In buffer layer, f_5 does not match the record $(f_8, 9)$ in the hashed bucket, so the record in the bucket is updated to $(f_5, 1)$ and $(f_8, 9)$ is passed to the buffer layer. f_4 matches the Fid of the record $(f_4, 2)$ in the hashed bucket, so the record in the bucket is updated to $(f_4, 9)$. In light part, the value of all counters hashed by f_8 is increased by 9.

Query. For the flow recorded with HBL Sketch, since it might be stored in three parts, the system thus need to conduct (hash) query operations respectively in the heavy part, buffer layer and light part. The results returned from each part will be aggregated then to obtain the original record of the flow. Obviously, this can work well as a naive method. The problem is, how to optimise the query process with a less operations.

Optimized Query Process: The system begins the query process with the heavy part. For any flow f in the heavy part, first find the bucket by the hash mapping of flow f. Then, there are two cases by different situations,

Case 1, The flag of the bucket is false: The size of f is exactly equal to the value of vote+ (on the bucket), because no eviction has happened. All the packets of flow f shoot on the bucket, and no any other flows' packets touch the bucket.

Case 2, The flag of the bucket is true: The size of f is equal to the sum of the value of vote+, the value in the buffer layer and the value in the light part.

The value in the buffer layer is found by two steps. First, find the bucket by the hash mapping of flow f. Next, get the value of count (on the bucket). The value in the light part can be found by referring to the query of CM Sketch [9]. By d-hashed mapping, the $[(freq)]_d$ of d buckets (in d arrays) is found. Then, get the minimal freq (on the bucket).

3.3   Performance Analysis

Notice there should be no inaccuracy in the heavy part and the buffer layer if only no eviction happens both on them. But in practice it's often impossible. There are always some records evicted from the heavy part or the buffer layer and entering the light part. Since the light part doesn't record the flow ID, but rather the flow size, there will be some estimation error in the light part. Towards, we design the buffer layer to reduce the number of accesses to the light part and improve the accuracy of measurement.

Access to the Light Part. We use an example to demonstrate how the buffer layer reduces the number of accesses to the light part. Suppose now have 9 incoming packets evicted from the heavy part with flow IDs: f_1, f_2, f_3, f_1, f_3, f_1, f_2, f_1, f_3. If we just use memory with the heavy part and the light part (just like what has been done in Elastic Sketch [25]), the number of accesses to the light part is 9. In HBL Sketch, before the 9 packets enters the light part, they are first aggregated into 3 flows ($f_1 \times 4$, $f_2 \times 2$, $f_3 \times 3$) in the buffer layer and then with a fewer possibilities to be evicted to the light part. In this case, the light part can be accessed at most 3 times. This is very similar to the role of the multi-level caches in CPU, which has been widely proven its effectiveness [20]. By this way, we can significantly reduce the number of accesses to the light part.

Error Bound. To analyze the error bound of HBL Sketch, we refer to the following theorem, whose proof can be found in [9].

Theorem 1. *Given two parameters ε and δ, the reported size $\hat{f}_i$ by HBL sketch for flow i is bounded by*

$$\hat{f}_i \le f_i + \varepsilon \parallel f_{l_1} \parallel_1 < f_i + \varepsilon \parallel f_{l_2} \parallel_1 \tag{1}$$

with probability at least $1 - \delta$, where f_{l_1} and f_{l_2} denote the size vector of the substreams recorded by the light part in HBL sketch and Elastic sketch, respectively.

According to Theorem 1, the estimation error of HBL sketch is bounded by $\parallel f_{l_1} \parallel_1$, instead of $\parallel f_{l_2} \parallel_1$ which is provided by Elastic sketch. When the total memory and the memory occupied by the heavy part are fixed, the buffer layer can reduce the packets of a stream recorded in the light part. Therefore, $\parallel f_{l_1} \parallel_1$ is usually significantly smaller than $\parallel f_{l_2} \parallel_1$, and the HBL sketch has a much tighter error bound than Elastic sketch when the parameters are the same.

4　Experiments

4.1　Experimental Setup

Trace: We use the dataset used in the latest work Elastic sketch, which is the public traffic traces collected in Equinix-Chicago monitor from CAIDA [1]. There are 11 traces in the dataset, namely 0.dat, 1.dat, ...10.dat. Each trace is in dat format and contains about 2.5M packets and 110K flows. Each 13 bytes in a trace is 5-tuple in the format of (srcIP, srcPort, dstIP, dstPort, protocol).

Testbed: We performed all the experiments on a server with 6-core CPUs (12 threads, Intel®Core™ i7-8700K CPU @ 3.70 GHz) and 62.9 GiB total memory running Ubuntu 16.04.

Metrics: The evaluation metrics we used are as follows:

ARE (Average Relative Error): ARE denotes the difference between estimated value and real value, which is defined as

$$\frac{1}{n} \sum_{i=1}^{n} \frac{\mid f_i - \hat{f}_i \mid}{f_i} \tag{2}$$

where n is the number of values. For flow size estimation, f_i and $\hat{f_i}$ are the real flow sizes and estimated sizes. For elephant flow detection, f_i and $\hat{f_i}$ are the real amount of elephant flows and estimated amount. We use ARE to evaluate the accuracy of flow size estimation and elephant flow detection.

Throughput: We calculate the throughput with unit of million packets per second (Mpps). Throughput is used to evaluate the processing speed of the flow size estimation.

Average Latency: We calculate the average delay for each trace with unit of millisecond (ms). Average delay is used to evaluate the processing speed of the flow size estimation.

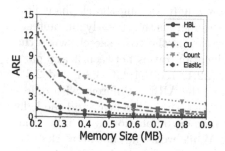

Fig. 2. ARE vs. memory sizes for flow size estimation.

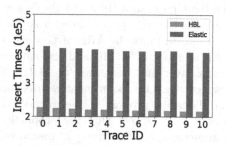

Fig. 3. Average insertion times for flow size estimation.

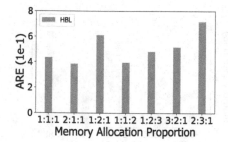

Fig. 4. ARE for flow size estimation under different memory allocations.

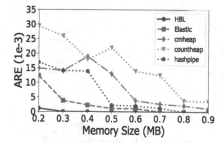

Fig. 5. ARE vs. memory sizes for elephant flow detection.

4.2 Experimental Results

Accuracy Comparison

Flow Size Estimation: We compare five methods: Elastic, CM, CU, Count, and our HBL sketch.

Figure 2 plots the AREs of flow size estimation for different methods on different memory sizes increasing from 0.2 MB to 0.9 MB with a step of 0.1 MB.

Our experimental results show that (1) As the total size of memory increases, the error rate decreases since the possibility of hash collision is getting smaller. (2) HBL Sketch outperforms other methods and reduces the error rate by 65%–93% on average. (3) Compared to state-of-the-art method Elastic Sketch, the error rate of HBL Sketch is 77% less than that in the extreme case.

Figure 3 plots the difference between HBL Sketch and Elastic Sketch with no buffer layer in terms of average insertion times (to the light part) when estimating flow sizes in different traces. It clearly shows that our solution with the buffer layer and the mechanism defined in Sect. 3 bring about 44% less average insertion times to the light part. This is in line with our motivation and also validates effectiveness of the buffer layer to significantly reduce the number of accesses to the light part, so as to avoid the heavy part wrongly passing the elephant flow record to the light part. The accuracy of measurement is thus greatly guaranteed.

Figure 4 plots the ARE of flow size estimation in the HBL sketch, when the total memory is 0.6 MB and the memory allocation ratios between heavy part, buffer layer and light part are 1:1:1, 2:1:1, 1:2:1, 1:1:2, 1:2:3, 3:2:1 and 2:3:1 respectively. Our experiment results show that the ARE is different with different allocation proportions between the heavy part, the buffer layer and the light part. It can be seen that the proportion of memory allocation is an influential parameter against accuracy of measurement. While we argue that network operators can employ a suitable proportion by some approaches of exploration and exploitation [10], nevertheless, the details of building such a parameter selection module are outside the scope of this paper. In the future, we thus aim to extend our study to this problem.

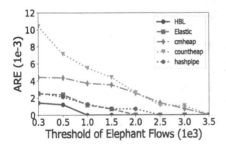
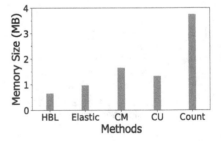

Fig. 6. ARE vs. threshold of elephant flows.

Fig. 7. Memory usage for flow size estimation along with fixed ARE = 0.2.

Elephant Flow Detection: We compare five methods: Elastic, cmheap, countheap, hashpipe, and our HBL sketch.

Figure 5 depicts the ARE of different methods for elephant flow detection, where the size of memory increasing from 0.2 MB to 0.9 MB with a step of 0.1 MB. It can be seen that HBL Sketch always achieves a more accurate detection against elephant flows. While some methods like cmheap, countheap and

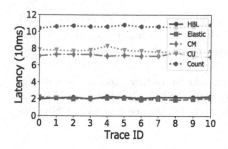

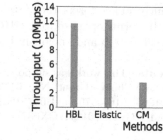

Fig. 8. Average delay for flow size estimation under different traces.

Fig. 9. Throughput of different methods for flow size estimation.

hashpipe are highly sensitive to the variations of memory size. In Fig. 5, we find that HBL Sketch reduces the error rate by 94%–99% on average.

Figure 6 depicts the ARE of different methods for elephant flow detection, where the thresholds of elephant flows are set to 300, 500, 1000, ..., 3500, respectively. Our experiment results show that HBL Sketch always achieves a more accurate detection against elephant flows. And it reduces the error rate by 61%–92% on average compared with other methods.

Memory Usage and Speed. We compare five methods: Elastic, CM, CU, Count, and our HBL sketch.

To show the memory usage among different methods at the same level of error rate, we plot Fig. 7 when the ARE for flow size estimation is 0.2. It clearly shows that HBL Sketch performs the best memory utilization. It has about 1/3 less memory usage compared with Elastic Sketch.

To quantify the impact of the buffer layer on the performance of processing speed, we compare HBL Sketch with other methods. Figures 8 and 9 plot the difference in terms of average latency and throughput. For latency, while HBL Sketch brings about 6% more overhead on average compared to state-of-the-art method Elastic Sketch, it overwhelms the rest of methods and decreases the latency by 70%–80% on average. This can also be observed for throughput. Our solution outperforms the methods except Elastic Sketch by 2.3–3.9×. This validates effectiveness of HBL Sketch to trade much accuracy improvement using very little loss in processing speed.

5 Conclusion

Sketch is one kind of useful data structure especially in the field of network measurement [4,23]. In this paper, we proposed a new sketch, namely the HBL sketch, which can improve the measurement accuracy without increasing memory. The key idea behind our proposed HBL sketch is to add a buffer layer and maintain the sketch composed of heavy part, buffer layer and light part to improve measurement accuracy. Our experimental results show that our HBL sketch significantly outperforms the-state-of-the-art in terms of accuracy and memory usage. In terms of speed, our HBL sketch is basically the same as the latest work Elastic sketch and obviously better than other methods.

Acknowledgments. This work was supported by the State Key Program of National Natural Science of China (Grant No. 61432002), NSFC Grant Nos. 61772112, U1836214, U1701263, 61672379, and 61751203, and the Science Innovation Foundation of Dalian under Grant 2019J12GX037.

References

1. The caida anonymized internet traces. http://www.caida.org/data/overview./
2. Cisco netflow. http://www.cisco.com
3. AlGhadhban, A., Shihada, B.: Flight: a fast and lightweight elephant-flow detection mechanism. In: 2018 IEEE 38th International Conference on Distributed Computing Systems (ICDCS), pp. 1537–1538. IEEE (2018)
4. Alipourfard, O., Moshref, M., Zhou, Y., Yang, T., Yu, M.: A comparison of performance and accuracy of measurement algorithms in software. In: Proceedings of the Symposium on SDN Research, p. 18. ACM (2018)
5. Ben Basat, R., Einziger, G., Friedman, R., Luizelli, M.C., Waisbard, E.: Constant time updates in hierarchical heavy hitters. In: Proceedings of the Conference of the ACM Special Interest Group on Data Communication, pp. 127–140. ACM (2017)
6. Bloom, B.H.: Space/time trade-offs in hash coding with allowable errors. Commun. ACM **13**(7), 422–426 (1970)
7. Brauckhoff, D., Tellenbach, B., Wagner, A., May, M., Lakhina, A.: Impact of packet sampling on anomaly detection metrics. In: Proceedings of the 6th ACM SIGCOMM Conference on Internet Measurement, pp. 159–164. ACM (2006)
8. Charikar, M., Chen, K., Farach-Colton, M.: Finding frequent items in data streams. In: Widmayer, P., Eidenbenz, S., Triguero, F., Morales, R., Conejo, R., Hennessy, M. (eds.) ICALP 2002. LNCS, vol. 2380, pp. 693–703. Springer, Heidelberg (2002). https://doi.org/10.1007/3-540-45465-9_59
9. Cormode, G., Muthukrishnan, S.: An improved data stream summary: the count-min sketch and its applications. J. Algorithms **55**(1), 58–75 (2005)
10. Črepinšek, M., Liu, S.H., Mernik, M.: Exploration and exploitation in evolutionary algorithms: a survey. ACM Comput. Surv. **45**(3), 35–68 (2013)
11. Estan, C., Varghese, G.: New directions in traffic measurement and accounting. ACM SIGCOMM Comput. Commun. Rev. **32**, 323–336 (2002)
12. Flajolet, P., Martin, G.N.: Probabilistic counting algorithms for data base applications. J. Comput. Syst. Sci. **31**(2), 182–209 (1985)
13. Gong, J., et al.: HeavyKeeper: an accurate algorithm for finding top-k elephant flows. In: 2018 USENIX Annual Technical Conference (USENIX ATC 2018), pp. 909–921 (2018)
14. Huang, Q., et al.: SketchVisor: robust network measurement for software packet processing. In: Proceedings of the Conference of the ACM Special Interest Group on Data Communication, pp. 113–126. ACM (2017)
15. Li, Y., Miao, R., Kim, C., Yu, M.: FlowRadar: a better NetFlow for data centers. In: 13th USENIX Symposium on Networked Systems Design and Implementation (NSDI 2016), pp. 311–324 (2016)

16. Liu, Z., Manousis, A., Vorsanger, G., Sekar, V., Braverman, V.: One sketch to rule them all: rethinking network flow monitoring with univmon. In: Proceedings of the 2016 ACM SIGCOMM Conference, pp. 101–114. ACM (2016)
17. Liu, Z., Gao, D., Liu, Y., Zhang, H., Foh, C.H.: An adaptive approach for elephant flow detection with the rapidly changing traffic in data center network. Int. J. Network Manage. **27**(6), e1987 (2017)
18. Mai, J., Chuah, C.N., Sridharan, A., Ye, T., Zang, H.: Is sampled data sufficient for anomaly detection? In: Proceedings of the 6th ACM SIGCOMM Conference on Internet Measurement, pp. 165–176. ACM (2006)
19. Poupart, P., et al.: Online flow size prediction for improved network routing. In: 2016 IEEE 24th International Conference on Network Protocols (ICNP), pp. 1–6. IEEE (2016)
20. Przybylski, S., Horowitz, M., Hennessy, J.: Characteristics of performance-optimal multi-level cache hierarchies. In: The 16th Annual International Symposium on Computer Architecture, pp. 114–121. IEEE (1989)
21. Sivaraman, V., Narayana, S., Rottenstreich, O., Muthukrishnan, S., Rexford, J.: Heavy-hitter detection entirely in the data plane. In: Proceedings of the Symposium on SDN Research, pp. 164–176. ACM (2017)
22. Wang, M., Li, B., Li, Z.: sFlow: towards resource-efficient and agile service federation in service overlay networks. In: Proceedings of the 24th International Conference on Distributed Computing Systems, pp. 628–635. IEEE (2004)
23. Wellem, T., Lai, Y.K., Chung, W.Y.: A software defined sketch system for traffic monitoring. In: Proceedings of the Eleventh ACM/IEEE Symposium on Architectures for Networking and Communications Systems, pp. 197–198. IEEE Computer Society (2015)
24. Wellem, T., Lai, Y.K., Huang, C.Y., Chung, W.Y.: A hardware-accelerated infrastructure for flexible sketch-based network traffic monitoring. In: 2016 IEEE 17th International Conference on High Performance Switching and Routing (HPSR), pp. 162–167. IEEE (2016)
25. Yang, T., et al.: Elastic sketch: adaptive and fast network-wide measurements. In: Proceedings of the 2018 Conference of the ACM Special Interest Group on Data Communication, pp. 561–575. ACM (2018)
26. Yang, T., et al.: Sf-sketch: a fast, accurate, and memory efficient data structure to store frequencies of data items. In: 2017 IEEE 33rd International Conference on Data Engineering (ICDE), pp. 103–106. IEEE (2017)
27. Yang, T., et al.: Empowering sketches with machine learning for network measurements. In: Proceedings of the 2018 Workshop on Network Meets AI & ML, pp. 15–20. ACM (2018)
28. Zhou, A., Zhu, H., Liu, L., Zhu, C.: Identification of heavy hitters for network data streams with probabilistic sketch. In: 2018 IEEE 3rd International Conference on Cloud Computing and Big Data Analysis (ICCCBDA), pp. 451–456. IEEE (2018)
29. Zhou, Y., Jin, H., Liu, P., Zhang, H., Yang, T., Li, X.: Accurate per-flow measurement with bloom sketch. In: IEEE INFOCOM 2018-IEEE Conference on Computer Communications Workshops (INFOCOM WKSHPS), pp. 1–2. IEEE (2018)
30. Zhou, Y., Liu, P., Jin, H., Yang, T., Dang, S., Li, X.: One memory access sketch: a more accurate and faster sketch for per-flow measurement. In: GLOBECOM 2017-2017 IEEE Global Communications Conference, pp. 1–6. IEEE (2017)

Accelerating Large Integer Multiplication Using Intel AVX-512IFMA

Takuya Edamatsu[1]([✉]) and Daisuke Takahashi[2][iD]

[1] Graduate School of Systems and Information Engineering, University of Tsukuba,
1-1-1 Tennodai, Tsukuba, Ibaraki 305-8573, Japan
edamatsu@hpcs.cs.tsukuba.ac.jp
[2] Center for Computational Sciences, University of Tsukuba, 1-1-1 Tennodai,
Tsukuba, Ibaraki 305-8573, Japan
daisuke@cs.tsukuba.ac.jp

Abstract. In this study, we implemented large integer multiplication with Single Instruction Multiple Data (SIMD) instructions. We evaluated the implementation on a processor with Cannon Lake microarchitecture, containing Intel AVX-512IFMA (Integer Fused Multiply-Add) instructions. AVX-512IFMA can compute multiple 52-bit integer multiplication and addition operations through one instruction and it has the potential to process large integer multiplications faster than its conventional AVX-512 counterpart. Furthermore, the AVX-512IFMA instructions take three 52-bit integers of 64-bit spaces as operands, and we can use the remaining 12 bits effectively to accumulate carries (*reduced-radix representation*). For multiplication in the context of larger integers, we applied the Karatsuba and Basecase methods to our program. The former is known to be a faster algorithm than the latter. For evaluation purposes, we compared execution times against extant alternatives and the GNU Multiple Precision Arithmetic Library (GMP). This comparison showed that we were able to achieve a substantive improvement in performance. Specifically, our proposed approach was up to approximately 3.07 times faster than AVX-512F (Foundation) and approximately 2.97 times faster than GMP.

Keywords: AVX-512 · IFMA · Large integer multiplication · Reduced-radix representation · Karatsuba method

1 Introduction

Multiplication is a fundamental arithmetic operation, along with addition, subtraction, and division. Most general-purpose processors cannot calculate numerical values more than 64 bits at a time. This kind of precision integer is called a multi-precision integer (large integer), and multiplication operations in these contexts are solved by software. These operations are important, for example, in the field of cryptography, and are used in public key systems such as Rivest-Shamir-Adleman (RSA) cryptography [21] and elliptic curve cryptography [19].

© Springer Nature Switzerland AG 2020
S. Wen et al. (Eds.): ICA3PP 2019, LNCS 11944, pp. 60–74, 2020.
https://doi.org/10.1007/978-3-030-38991-8_5

They are also applied in computer algebra systems, such as Maxima [6] and Wolfram Mathematica [5]. Since multiplication represents fundamental arithmetic, large integer multiplication is also a versatile operation. There are arbitrary precision arithmetic libraries such as the GNU Multiple Precision Arithmetic Library (GMP) [2], which support various multi-precision calculations including multiplication. Mathematica, for example, performs multi-precision operations using GMP [1].

In general, when performing multi-precision multiplication, a carry occurs in the intermediate process, and usually, we have to add extra processing for the carry separately from the original multiplication. However, the overhead attributable to this process is non-trivial, especially when the multiplier and the multiplicand are large, and this can thus significantly affect execution times. Therefore, it is desirable to perform multiplication without carry processing as much as possible. A method to avoid carry propagation is reported in [16]. This technique allows accumulation of carries by converting a large integer to a representation with fewer bits per word (*reduced-radix representation*) before proceeding to multiplication operations. Although the number of words for multipliers and multiplicands increases slightly, we can multiply large integers without carry processing during an intermediate calculation.

Modern processors can execute Single Instruction Multiple Data (SIMD) instructions, which process multiple data in parallel through one instruction. For example, instructions such as Intel SSE, SSE2, and SSE4.2 have 128-bit SIMD instructions, and Intel AVX and AVX2 have 256-bit SIMD instructions. In addition, processors that can run Intel AVX-512 (such as Xeon Phi Knights Landing and Skylake-X) have become available [3]. These instructions enable execution of 512-bit SIMD additions, multiplications, floating-point FMA (Fused Multiply-Add) operations, and so on. In recent years, processors with Cannon Lake microarchitecture have become available, enabling execution of AVX-512IFMA (Integer Fused Multiply-Add) instructions. This allows integer multiplication and addition to be processed through one instruction.

Herein, we evaluate the performance of large integer multiplication on a processor with Cannon Lake microarchitecture using the reduced-radix representation technique and AVX-512IFMA instructions. We believe that this study could serve as a basis for large integer multiplication using AVX-512IFMA instructions.

2   Related Works

Several studies have been conducted with the objective of speeding up large integer multiplication processing using AVX-512 instructions, such as in [18]. However, at the time that particular research was conducted, no processor was available that could execute AVX-512, and thus the Intel Software Development Emulator (SDE) was used to evaluate performance based on the number of instructions. Therein, implementation used a reduced-radix representation and when the number of words exceeded a certain threshold, carries were cleaned up (referred to as "cleanup" in [18]). Although clean-up processing was carried out, a 1.16 times improvement in performance was still recorded compared to GMP in the context of 2,048-bit multiplication. Therefore, the results in [18] suggest that AVX-512F instructions have the potential to speed up multiplication.

Gueron and Krasnov evaluated multiplication using AVX-512IFMA [14]. In that research, as well as in [18], the number of bits per word was reduced to prevent carry propagation. Since no processor was available that could execute AVX-512IFMA at the time that study was conducted, implementation was also evaluated using the Intel SDE in terms of instruction count. That research approached large integer multiplication using four sizes, namely multiplication of 1,024, 2,048, 3,072, and 4,096 bits. Performance evaluation focused on implementation using the conventional AVX-512F, GMP, and AVX-512IFMA. In spite of conversion to a reduced-radix representation and conversion back to normal representation, about one-eighth of the number of instructions was obtained in the context of 4,096-bit multiplication compared to GMP. Moreover, about one-quarter of the number of instructions was obtained compared with AVX-512F for the same size. Therefore, it was shown that AVX-512IFMA has the potential to process large integer multiplication operations faster than AVX-512F in terms of the number of instructions.

In [12], AVX-512F was used to evaluate the execution times of large integer multiplication operations. An Intel Xeon Phi Knights Landing processor was used for evaluation because it could execute AVX-512F instructions. A reduced-radix representation was applied and the study was designed flexibly to accommodate multiplication for up to 7,168 bits. Although there are overheads associated with flexibility, a performance improvement of up to approximately 2.5 times was still achieved in terms of execution time compared to GMP on the Xeon Phi processor. However, no significant difference was observed in small size cases. In summary, AVX-512F instructions have been shown to be advantageous through facilitating the processing of multi-precision multiplication at high speeds.

From another point of view, AVX-512IFMA was used to speed up modular squaring [11]. This study implemented squaring with Montgomery multiplication [20] (referred to as "Almost Montgomery Squaring" in [11]). Furthermore, the highest speed implementation among AVX-512IFMA instructions was also considered in terms of latency, throughput, and the number of instructions. The authors concluded that *Almost Montgomery Squaring* is faster than the usual calculation of modular exponentiation, and that the AVX-512IFMA instructions have the potential to speed up the calculation of modular exponentiation.

In summary, research has hitherto been conducted into speeding up multi-precision integer multiplication, combining AVX-512 instructions and a reduced-radix representation. However, the extent to which performance in terms of execution time can be further improved using AVX-512IFMA instructions, which is more powerful than AVX-512F, has not yet been explored. Since the Xeon and Xeon Phi processors that can execute AVX-512 instructions decrease operating frequency by several hundred MHz at high AVX frequencies [7,8], the time taken for one cycle is different compared to when these processors execute scalar instructions. It is unclear if this also applies in the case of Cannon Lake microarchitecture processors, although it is reasonable to assume a similar outcome. In addition, latency and throughput depend on instructions. Therefore, we cannot evaluate multiplication performance only in terms of the number of instructions. The Basecase method [9] has been used as the computation algorithm for multi-precision integer multiplication in previous studies. However, there are alternative algorithms (for example, the Karatsuba method [17], the Toom-Cook method [23], and methods based on fast Fourier transform (FFT) [10] such as Schönhage-trassen algorithm [22], Fürer's algorithm [13], and Harvey and van der Hoeven's algorithm [15]) that are asymptotically faster than the Basecase algorithm. Therefore, in this research, we combine AVX-512IFMA instructions and the reduced-radix representation technique, implement the Basecase and Karatsuba methods for multi-precision integer multiplication, and evaluate performance in terms of execution times.

3 Multiplication Algorithm

3.1 Reduced-Radix Representation

We describe a multiplier and a multiplicand as arrays of 64-bit unsigned integers. One 64-bit unsigned integer array represents one word. For example, a 1,024-bit integer can be represented by 16 (1,024/64) arrays. In general, an N-bit multi-precision integer A would be expressed as follows:

$$A = \sum_{i=0}^{\lceil N/64 \rceil - 1} a_i \beta^i, \tag{1}$$

where β is 2^{64} and a_i is one word.

Algorithm 1. BasecaseMultiply [9]

Input: $A = \sum\limits_{i=0}^{m-1} a_i \beta^i, B = \sum\limits_{j=0}^{n-1} b_j \beta^j$

Output: $C = AB := \sum\limits_{k=0}^{m+n-1} c_k \beta^k$

1: $C \leftarrow A \cdot b_0$
2: **for** $j \leftarrow 1$ to $n-1$ **do**
3:    $C \leftarrow C + \beta^j (A \cdot b_j)$
4: **end for**
5: return C.

Algorithm 1 expresses the Basecase multiplication method [9]. When we perform a multiplication, a carry may occur during the calculation. The calculation of digit carryover can be processed on the computer or manually. However, on a computer, if an operation result exceeds the bit width of the register, the carry disappears. In other words, the next digit would not be added a carry. For example, suppose that multiplication is performed with the 2^{64}-radix representation (i.e., 64 bits per word) as Eq. (1). During intermediate calculations, adding 64 bits and 64 bits results in up to 65 bits. However, since the bit width of general-purpose registers of most computers is 64 bits, they cannot hold 65 bits. Consequently, the overflowing bit will be lost, and thus no carry up of digits will occur. This would later lead to an erroneous calculation result. To prevent bit loss during calculations, we convert each operand into a representation with fewer than 64 bits prior to multiplication. Specifically, herein, we reduce from 64 bits to 52 bits. Intel AVX-512IFMA instructions take 2^{52}-radix numbers as operands. Thus, we convert a 2^{64}-radix number into a 2^{52}-radix number. Although the 2^{52}-radix representation creates redundancy in a 64-bit unsigned integer array, the remaining 12 bits of space can be used effectively for accumulating carries. With this representation, namely a *reduced-radix representation* [16], no carry propagation will occur through partial product addition.

The advantage of a reduced-radix representation is not only the prevention of loss of bits due to carry overflow; it also contributes to parallel processing. Whenever a carry-up occurs due to a carry during a calculation, we need to add the carry to the next word each time. In this case, we cannot vectorize using SIMD instructions, because of data dependence between words. However, in calculations with numbers in a reduced-radix representation, no carries can be generated, because each word accumulates carries in intermediate calculations. Therefore, we can process intermediate computations in parallel by SIMD instructions. Then, the accumulated carries are processed collectively. In summary, a reduced-radix representation has two advantages in that it prevents carry loss whilst also enabling parallelization by SIMD instructions. We use this representation for multiplication operations in this study because of these desirable properties.

We now discuss the feasibility of using a reduced-radix representation. We have 12-bit space to reduce from 64 bits to 52 bits per word, and we can execute $2^{12} - 1$ times 52 bits + 52 bits to accumulate carries here. Generally, when multiplying two M-word multi-precision integers using the Basecase method, $M - 1$ additions are performed. Therefore, with Basecase method, it is theoretically possible to perform multiplication without carry propagation, up to a maximum of 2^{12} words (i.e., 52 bits $\times$ 2^{12} = 212,992 bits).

3.2   Karatsuba Multiplication

In this paper, we use the Karatsuba multiplication algorithm [17]. Algorithm 2 describes the Karatsuba multiplication kernel.

The asymptotic computational complexity of the Karatsuba method is $O(n^{\log_2 3}) \approx O(n^{1.585})$, while that of the Basecase method is $O(n^2)$, where n is the number of operand words. Before we call **KaratsubaMultiply** for the first time, we adjust n so that it is always even until n satisfies $n \leq n_0$ and if the number of words in the multiplier and the multiplicand is less than or equal to n_0 in Algorithm 2, we call **BasecaseMultiply** and return. The essence of

---

Algorithm 2. KaratsubaMultiply

Input: $A = \sum_{i=0}^{n-1} a_i \beta^i, B = \sum_{j=0}^{n-1} b_j \beta^j$

Output: $C = AB := \sum_{k=0}^{2n-1} c_k \beta^k$

1: **if** $n \leq n_0$ **then**
2:     Borrow process
3:     return **BasecaseMultiply**(A, B)
4: **end if**
5: $k \leftarrow \lceil n/2 \rceil$
6: $(A_0, B_0) := (A, B) \bmod \beta^k$
7: $(A_1, B_1) := \lfloor (A, B) / \beta^k \rfloor$
8: $(A_2, B_2) := (A_0 - A_1, B_0 - B_1)$
9: $C_0 \leftarrow$ **KaratsubaMultiply**(A_0, B_0)
10: $C_1 \leftarrow$ **KaratsubaMultiply**(A_1, B_1)
11: $C_2 \leftarrow$ **KaratsubaMultiply**(A_2, B_2)
12: return $C := C_0 + (C_0 + C_1 - C_2)\beta^k + C_1 \beta^{2k}$.

Algorithm 2 is as follows: (i) splitting of both A and B into exactly half, as A_0, A_1 and B_0, B_1 respectively; (ii) separately multiplying the two lower parts, the two upper parts, and $(A_0 - A_1)$ and $(B_0 - B_1)$; (iii) adding each partial product at an appropriate position. This is based on the following equation.

$$(A_0 - A_1)(B_0 - B_1) = A_0 B_0 + A_1 B_1 - A_0 B_1 - A_1 B_0 \tag{2}$$

$$A_0 B_1 + A_1 B_0 = A_0 B_0 + A_1 B_1 - (A_0 - A_1)(B_0 - B_1) \tag{3}$$

Equation (2) shows the usual multiplication of AB when A and B are split into (A_0, A_1) and (B_0, B_1). In this expression, it is necessary to obtain four independent partial products of $A_0 B_0$, $A_1 B_1$, $A_0 B_1$, and $A_1 B_0$ to calculate AB. However, in Eq. (3), $A_0 B_1 + A_1 B_0$ can be obtained by calculating $(A_0 - A_1)(B_0 - B_1)$ after calculating $A_0 B_0 + A_1 B_1$. In other words, if the final additions are executed appropriately, it is possible to obtain AB through three multiplications.

We take 64×82 in decimal format as an example. Figure 1 shows the 64×82 process for the Basecase method (left) and the Karatsuba method (right). Using the Basecase method, we calculate the four multiplications of 4×2, 6×2, 4×8, and 6×8 to obtain 5,248. On the other hand, when using the Karatsuba method, 5,248 is obtained by the three multiplications of 4×2, 6×8, and $(48 + 8) - (6 - 4)(8 - 2)$. The Karatsuba method yields correct results even if either or both of $A_0 - A_1$ and $B_0 - B_1$ are negative. Therefore, using the Karatsuba method, we can obtain the same result with fewer multiplications than when using the Basecase method. The Karatsuba method appears somewhat more complex for such a simple example. However, in computational terms, multiply instructions are more expensive than addition and subtraction. Furthermore, in the case of multi-precision integer multiplication, the number of multiply instructions increases. Thus, reducing this expensive operation is important for the purposes of decreasing execution time.

$$
\begin{array}{r}
6\,4 \\
\times \quad 8\,2 \\
\hline
8 \\
1\,2 \\
3\,2 \\
4\,8 \\
\hline
5\,2\,4\,8
\end{array}
\qquad
\begin{array}{r}
6\,4 \\
\times \quad 8\,2 \\
\hline
8 \\
4\,8 \\
4\,4 \\
\hline
5\,2\,4\,8
\end{array}
$$

Fig. 1. Multiplication using the Basecase method (left) and the Karatsuba method (right)

Since we can vectorize lines 3, 8, and 12 in Algorithm 2, we use AVX-512 instructions for these processes. A and B are in the 2^{52}-radix redundant representation, and the arguments of **BasecaseMultiply** must be within 52 bits. Therefore, before calling **BasecaseMultiply**, we have to process borrows generated in $A_0 - A_1$ and $B_0 - B_1$. We need not handle borrows in the case of non-subtracted arguments, because no borrow occurs for these arguments, and therefore, we skip the second line of Algorithm 2 to avoid unnecessary processing in these cases. In our implementation, by designing to pre-adjust the length of operands A and B so that the highest word of A_1 and B_1 is zero, we avoid the case of $(A_0 - A_1)$ and $(B_0 - B_1)$ becoming negative as much as possible. If we still expect one of two subtractions to result in a negative value, we switch the order of the subtraction to avoid a negative value and then change $(C_0 + C_1 - C_2)$ to $(C_0 + C_1 + C_2)$ as appropriate.

4   Implementation

4.1   Intel AVX-512IFMA

Intel AVX-512IFMA is a set of integer fused multiply-add instructions using 512-bit SIMD registers. AVX-512IFMA consists of vpmadd52luq and vpmadd52huq instructions. vpmadd52luq returns the lower 52 bits of the 104-bit integer obtained by performing a multiply-add operation of three 52-bit integers, and vpmadd52huq returns its upper 52 bits.

We used vpmuludq to perform large integer multiplication until we could execute the AVX-512IFMA instructions. However, this instruction takes a 32-bit multiplier and a 32-bit multiplicand to make 64-bit partial products in parallel. Later, AVX-512DQ (Doubleword and Quadword) instructions became available, and vpmullq was introduced, which takes two 64-bit integer operands. However, vpmullq returns only the lower 64-bit part of a 128-bit integer obtained by multiplication, and we do not have any way to obtain the upper 64-bit part. Therefore, AVX-512IFMA instructions offer the following three advantages over vpmuludq and vpmullq: (i) takes operands greater than 32 bits; (ii) calculates addition and multiplication through one instruction; (iii) returns both low and high partial products. By virtue of the above advantages, these instructions are useful for multi-precision integer multiplication.

AVX-512IFMA exploits fractions of floating-point arithmetic for FMA operation, and thus, operands of these instructions must be 52 bits. This is the reason why each instruction name contains the string "52". To use these instructions, we need to convert integers as operands into 52 bits per word in advance. Importantly, AVX-512IFMA has high affinity to this representation for large integer multiplication because we can also leverage it simultaneously as a reduced-radix representation.

To use the AVX-512IFMA instructions, we coded with Intel Intrinsics [4]. These are built-in functions to manipulate SIMD instructions instead of using assembly language directly. Because the same effect as writing low-level code can be obtained by calling C intrinsic functions, readability and availability of source code are higher. Intrinsic functions support AVX-512 including AVX-512IFMA as well as AVX, SSE, and MMX. The two intrinsic functions of the AVX-512IFMA instructions used in this research are as follows [4].

_mm512_madd52lo_epu64(_m512i a, _m512i b, _m512i c)

Calculate multiply-add in 512-bit SIMD registers and this function is converted into vpmadd52luq. Multiply packed unsigned 52-bit integers b and c and get 104-bit intermediate products. Then, add 52-bit integer a and the 104-bit intermediate products and generate results. This function (instruction) returns the lower 52 bits of the result.

_mm512_madd52hi_epu64(_m512i a, _m512i b, _m512i c)

Same as _mm512_madd52lo_epu64 except that it returns the upper 52 bits of the results. This function makes vpmadd52huq instruction.

4.2   Implementation of Multiplication Kernel

We implement multi-precision integer multiplication using the Karatsuba method as stated in Subsect. 3.2. The sixth and seventh lines of Algorithm 2 require modular arithmetic and division, respectively. However, herein, since a large integer is expressed with 64-bit unsigned integer arrays, (A_0, B_0) is passed the start address of each operand, and we give (A_1, B_1) the midpoint address of arrays. Therefore, we can perform modular arithmetic and division by manipulating the address references.

Since subtractions are performed on the eighth line of Algorithm 2 and vectorization is possible, we vectorize these processes. We can accumulate borrows as well as carries because of the reduced-radix representation, and thus we can subtract without concern for data dependencies. Unfortunately, however, the AVX-512IFMA instructions ignore the 53rd bit to the 64th bit of the operand, which are storing the borrows. Hence, we cannot proceed to **BasecaseMultiply** with the borrows and we clean up in advance. We skip over the second line in the case of non-subtracted arguments and proceed to **BasecaseMultiply** as stated in Subsect. 3.2. This circumvents extra processing.

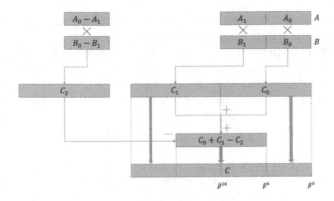

Fig. 2. Calculating AB large integer multiplication using the Karatsuba method

At line 12 of Algorithm 2, we calculate $(C_0 + C_1 - C_2)$ and add each partial product to the appropriate place as shown in Fig. 2. Finally, we get the result of C. In those processes, we also vectorize with SIMD instructions, using AVX-512 instructions here as well. As noted in Subsect. 3.2, we can perform fewer multiplications with the Karatsuba method compared to with the Basecase method.

5   Evaluation

For evaluation purposes, we used a host machine with an Intel Core i3-8121U (2.2 GHz, 2 Cores, 4 Threads, Cannon Lake microarchitecture), 4 GB DDR4 memory, and the Ubuntu Server 18.04.2 LTS operating system. All our experimental programs were implemented in the C language and compiled through the Intel C Compiler `icc` version 19.0.4.243 with `-O3 -Wall -xCANNONLAKE -mtune=cannonlake -std=c99` options. The comparison targets were programs from [14] and [12], and the GNU Multiple Precision Arithmetic Library (GMP) version 6.1.2. These were also built with `icc`. During the evaluation, we executed all programs with a single core and a single thread. We performed the multiplications 3,000 times and averaged their execution times. To generate the multiplier and multiplicand, we use the `rand` function with a pseudo-random seed generated by the system time.

5.1   Comparison with Basecase Multiplication

As discussed, in this research we implemented a large integer multiplication kernel using the Karatsuba method. Although the amount of computation is small for the Karatsuba method, the Basecase method offers shorter execution times if the operand size is small because the Karatsuba method uses Basecase multiplication several times. Therefore, we compare execution times of each algorithm. Through this comparison, we can determine the optimal size for switching the algorithms. Indeed, GMP also switches algorithms according to operand size.

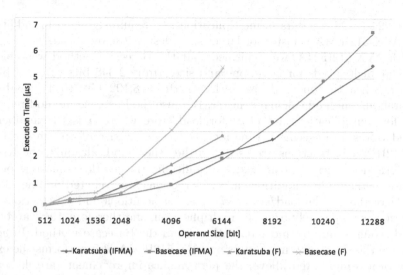

Fig. 3. Execution times associated with the Karatsuba and Basecase methods for 512-bit to 12,288-bit operands

In the experimental environment of this study, the GMP uses the Karatsuba method when operand size equals or exceeds 1,792 bits.

Figure 3 compares execution times for each operand size between the Karatsuba and Basecase methods with AVX-512F (interrupted due to the limitation of the 2^{28}-radix representation) and AVX-512IFMA. The vertical axis represents execution time (microseconds) and the horizontal axis represents multiplier and multiplicand size (bits). Since each method uses reduced-radix representations, we focused on comparing execution times of the multiplication kernel.

Using the Basecase method, AVX-512IFMA is faster than AVX-512F at all sizes. On the other hand, focusing on the Karatsuba method, the execution time is similar for small sizes. Although AVX-512F is superior to AVX-512IFMA at 2,048 bits, AVX-512IFMA is faster at larger sizes. The Karatsuba method with AVX-512IFMA requires borrow processing according to the second line in Algorithm 2. In contrast, this method with AVX-512F does not need the borrow processing, because vpmuludq takes 32-bit operands at full while vpmadd521(h)uq takes only 52 bits in 64-bit space for operands. We cannot ignore the AVX-512IFMA-specific overhead for small operand sizes, especially at operations below one-microsecond. Furthermore, due to the overhead, the kernel with AVX-512IFMA divides one operand into two parts, while the kernel with AVX-512F splits it into up to four parts. This is because the former increases the number of the extra borrow processing as the number of partitions increases. Therefore, there is also a difference between the two in terms of actual complexity for the same operand. However, AVX-512IFMA deals with 416 ($= 52 \times 8$) bits in multiplication and addition through one instruction. Meanwhile, AVX-512F handles 224 ($= 28 \times 8$) bits in multiplication and requires vpaddq for addition. Thus, AVX-512IFMA is faster for larger sizes despite the overhead.

With AVX-512IFMA, we confirmed that the Basecase method was able to process at high speeds for small operand sizes (from 2,048 bits to 6,144 bits). However, when the operand size is larger (from 8,192 bits to 12,288 bits), the Karatsuba method calculates faster. Because the Basecase method requires $O(n^2)$ for multiplication and $O(n)$ for load/store where n is the number of operand words, the execution time increases with the size of operands even if AVX-512IFMA instructions are used. On the other hand, although the Karatsuba method reduces the complexity of multiplication to three-quarters of the Basecase method, $O(n)$ additions and subtractions are executed on lines 8 and 12 of Algorithm 2. The load/store processes for additions and subtractions are also performed separately from the multiplication, and therefore, the Karatsuba method requires more memory operations than the Basecase method. In addition, since the Karatsuba method using AVX-512IFMA also performs the extra borrow processing stated above, the performance improvement rate decreases relative to the Basecase method using the same instruction.

In summary, with AVX-512IFMA, the Basecase method is more advantageous when the size of the multiplier and multiplicand is small because the Karatsuba method cannot ignore the overheads due to the extra borrow processing, function calls, and load/store. For larger operands, however, the Karatsuba method is faster than the Basecase method. Based on the results of this experiment, we configured a switch from the Basecase method to the Karatsuba method when operand size equals or exceeds 7,168 bits.

5.2   Comparison with Related Works

We compared the proposed implementation with programs from [14] and [12]. See Fig. 4 for a visualization of the differential execution times. The execution times for [12] and our method consist of Kernel, Split, and Combine. "Split" refers to the conversion from a 2^{64}-radix number to a 2^{52}-radix number in our approach and from a 2^{32}-radix number to a 2^{28}-radix number for [12]. "Combine" refers to the opposite of Split.

First, compared to [12] which is implemented with AVX-512F, our approach was faster at all operand sizes, representing up to approximately 3.07 times improvement in performance. From this result, it can be determined that AVX-512IFMA is superior to AVX-512F for large integer multiplication in terms of execution time. Furthermore, there are also differences in the execution times associated with Split and Combine between our implementation and [12]. Our program for conversion was implemented with AVX-512BW (Byte and Word) and AVX-512VBMI (Vector Byte Manipulation Instructions). Thus, we can observe that AVX-512BW and AVX-512VBMI are useful for accelerating the program.

Second, the program of [14] which is also implemented with AVX-512IFMA is superior to the proposed program, although the execution times are converging. The AVX-512IFMA programs of [14] are specialized for 1,024-bit, 2,048-bit,

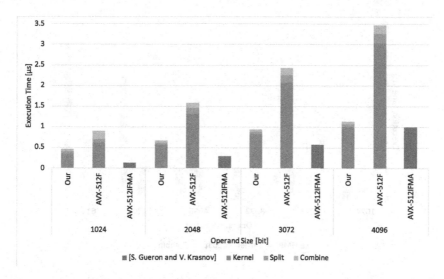

Fig. 4. Execution times associated with the Basecase method for 1,024-bit to 4,096-bit operands compared to related works

3,072-bit, and 4,096-bit operands, and the codes for Split, Combine, carry processing, and Kernel are put together in one module, which is similar to a monolithic kernel design. Therefore, programs of [14] are implemented using registers as much as possible to circumvent memory accesses. In contrast, we implemented programs for Split, Combine, carry processing, and Kernel independently and called these modules to support various operand sizes, which is similar to a microkernel design. Due to the modularization, costs are incurred associated with size-dependent branches, memory access overheads for calling modules, and load/store data. Especially for processing less than one microsecond, we cannot ignore the execution time of those overheads. However, at the same time, it is clear that the monolithic design of large integer multiplication with AVX-512IFMA is very fast for certain applications. In short, AVX-512IFMA has the potential to carry out multiplication operations faster.

5.3   Comparison with GNU MP

We compared the execution time of the GMP and our implementation in 1,024 bits to 12,288 bits (Figs. 5 and 6). Based on the results in Subsect. 5.1, our program is compared with GMP using the Basecase method in Fig. 5 and the Karatsuba method in Fig. 6. The execution times at 1,024 bits are almost equal, and our proposed program is superior to the GMP for larger multiplication. Although Kernel processing is faster than the GMP at 1,024 bits, with the addition of Split and Combine processing, the GMP calculates slightly faster. However, for other sizes, our implementation is faster even with Split and Combine processing, and we obtained up to approximately 2.97 times improvement in performance.

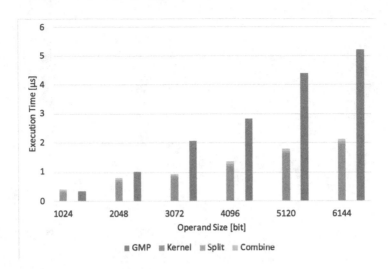

Fig. 5. Execution times associated with the multiplication function using AVX-512IFMA for 1,024-bit to 6,144-bit operands compared to GMP

The GMP does not use SIMD instructions for multiplication, and thus, this result suggests that vector instructions are faster than scalar instructions for large integer multiplication.

In the environment of this study, the GMP switches from the Karatsuba method to the Toom-Cook method (more specifically, Toom 3-way multiplication) at the boundary of 5,184 bits (81 words). Nevertheless, Karatsuba multiplication with AVX-512IFMA instructions is faster. Therefore, we concluded from this result that the AVX-512IFMA instructions are very effective features for speeding up multi-precision integer multiplication.

6   Conclusions and Future Work

In this study, we aimed to speed up multi-precision integer multiplication. To achieve this, we utilized a reduced-radix representation and AVX-512IFMA. Furthermore, to speed up multiplication in large size contexts, we implemented a multi-precision integer multiplication program using algorithms with the Karatsuba method as well as with the conventional Basecase method. Since this study focuses on general-purpose multiplication, we designed it to be able to handle various sizes with an emphasis on flexibility. We evaluated the implemented program on a processor with Cannon Lake microarchitecture and compared its performance with [12,14], and the GMP in terms of execution times.

Compared to [12] using AVX-512F, our program was faster at all operand sizes: we obtained up to approximately 3.07 times improvement in performance. Furthermore, from these results, we found that AVX-512BW and AVX-512VBMI, as well as AVX-512IFMA, are effective for speeding up. In comparison with [14] using AVX-512IFMA, unfortunately, our program was slower

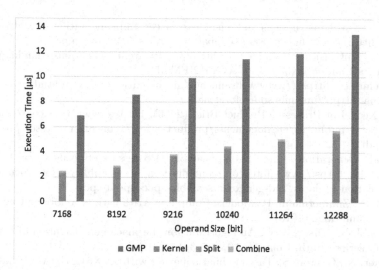

Fig. 6. Execution times associated with the multiplication function using AVX-512IFMA for 7,168-bit to 12,288-bit operands compared to GMP

at all sizes. We conclude that the main cause of this result is the difference between how a multiplication kernel is implemented (monolithic design versus microkernel-like design). In other words, we found through this experiment that the AVX-512IFMA instructions can process multiplication extremely fast when the application is specialized. Finally, in comparison with GMP, although execution times were similar when the operand size was as small as 1,024 bits, our program is faster for multiplications with larger sizes equaling or exceeding 2,048 bits. As a result, we obtained up to approximately 2.97 times improvement in performance. In summary, we were able to successfully speed up large integer multiplications and found that the AVX-512IFMA is useful for this purpose.

Although our implementation used the Karatsuba method as well as the Basecase method, the Toom-Cook method [23] is known to be one of the faster algorithms. The Toom-Cook method reduces the burden of multiplication processing by dividing one operand into three or more, while one operand is divided into two in the Karatsuba method. GMP performs multiplication by increasing the number of divisions to 3-way and 4-way when the size of the multiplier and multiplicand exceeds some threshold. Therefore, in future research, it will be desirable to implement multi-precision integer multiplication with the Toom-Cook method using the AVX-512IFMA instructions, and investigate which operand sizes the Toom-Cook method can process faster than the Karatsuba method.

References

1. GMP source code used in Mathematica. http://www.wolfram.com/LGPL/GMP
2. The GNU MP. https://gmplib.org/

3. Intel AVX-512 Instructions and Their Use in the Implementation of Math Functions. http://www.fit.vutbr.cz/~iklubal/IPA/AVX-512_Cornea.pdf
4. Intel Intrinsics Guide. https://software.intel.com/sites/landingpage/IntrinsicsGuide/#avx512techs=AVX512IFMA52
5. Mathematica. https://www.wolfram.com/mathematica/index.en.html
6. Maxima. http://maxima.sourceforge.net
7. Intel Xeon Phi Processor Product Brief (2016). https://www.intel.com/content/dam/www/public/us/en/documents/product-briefs/xeon-phi-processor-product-brief.pdf
8. Second Generation Intel Xeon Scalable Processors Specification Update, May 2019. https://www.intel.com/content/dam/www/public/us/en/documents/specification-updates/2nd-gen-xeon-scalable-spec-update.pdf
9. Brent, R., Zimmermann, P.: Modern Computer Arithmetic. Cambridge University Press, Cambridge (2010)
10. Cooley, J.W., Tukey, J.W.: An algorithm for the machine calculation of complex Fourier series. Math. Comput. **19**, 297–301 (1965)
11. Drucker, N., Gueron, S.: Fast modular squaring with AVX512IFMA. In: Latifi, S. (ed.) 16th International Conference on Information Technology-New Generations (ITNG 2019). AISC, vol. 800, pp. 3–8. Springer, Cham (2019). https://doi.org/10.1007/978-3-030-14070-0_1
12. Edamatsu, T., Takahashi, D.: Acceleration of large integer multiplication with Intel AVX-512 instructions. In: 2018 IEEE 20th International Conference on High Performance Computing and Communications (HPCC), pp. 211–218 (2018)
13. Fürer, M.: Faster integer multiplication. SIAM J. Comput. **39**, 979–1005 (2009)
14. Gueron, S., Krasnov, V.: Accelerating big integer arithmetic using Intel IFMA extensions. In: Proceedings of the 2016 IEEE 23rd Symposium on Computer Arithmetic (ARITH), pp. 32–38 (2016)
15. Harvey, D., van der Hoeven, J.: Integer multiplication in time $O(n \log n)$ (2019). https://hal.archives-ouvertes.fr/hal-02070778
16. Intel Corporation: Using Streaming SIMD Extensions (SSE2) to Perform Big Multiplications, version 2.0. Technical report AP-941 (248606–001) (2000). https://software.intel.com/sites/default/files/14/4f/24960
17. Karatsuba, A., Ofman, Y.: Multiplication of multidigit numbers on automata. Sov. Phys. Dokl. **7**, 595 (1963)
18. Keliris, A., Maniatakos, M.: Investigating large integer arithmetic on Intel Xeon Phi SIMD extensions. In: Proceedings of the Design & Technology of Integrated Systems in Nanoscale Era (DTIS), pp. 1–6 (2014)
19. Miller, V.S.: Use of elliptic curves in cryptography. In: Williams, H.C. (ed.) CRYPTO 1985. LNCS, vol. 218, pp. 417–426. Springer, Heidelberg (1986). https://doi.org/10.1007/3-540-39799-X_31
20. Montgomery, P.L.: Modular multiplication without trial division. Math. Comput. **44**, 519–521 (1985)
21. Rivest, R.L., Shamir, A., Adleman, L.: A method for obtaining digital signatures and public-key cryptosystems. Commun. ACM **21**, 120–126 (1978)
22. Schönhage, A., Strassen, V.: Schnelle Multiplikation großer Zahlen. Computing **7**, 281–292 (1971)
23. Toom, A.L.: The complexity of a scheme of functional elements realizing the multiplication of integers. Sov. Math. Doklady **3**, 714–716 (1963)

A Communication-Avoiding Algorithm for Molecular Dynamics Simulation

Bei Wang[1(✉)], Yifeng Chen[1], and Chaofeng Hou[2]

[1] EECS, Peking University, Beijing 100871, China
{wangbei.sei,cyf}@pku.edu.cn
[2] Institute of Process Engineering, Chinese Academy of Sciences, Beijing, China
cfhou@ipe.ac.cn

Abstract. Molecular dynamics and many similar time-dependent computing tasks are defined as simple state updates over multiple time steps. In recent years, modern supercomputing clusters have enjoyed fast-growing compute capability and moderate-growing memory bandwidth, but their improvement of network bandwidth/latency is limited. In this paper, we propose a new communication-avoiding algorithmic model based on asynchronous communications which, unlike BSP, records and handles multiple iterative states together. The basic idea is to let computation run in small regular time steps while communications over longer dynamic time steps. Computation keeps checking inaccuracies so that the intervals between communications are small in volatile scenarios but longer when dynamics is smooth. This helps reduce the number of data exchanges via network communication and hence improve the overall performance when communication is the bottleneck. We test MD simulation of condensed covalent materials on the Sunway TaihuLight. For best time-to-solution, the general-purpose supercomputer Sunway TaihuLight performs 11.8 K steps/s for a system with 2.1 million silicon atoms and 5.1 K steps/s for 50.4 million silicon atoms. This time-to-solution performance is close to those of state-of-art hardware solution. A software solution using general-purpose supercomputers makes the technology more accessible to the general scientific users.

Keywords: BSP · Parallel-in-time · Molecular dynamics

1 Introduction

Molecular dynamics (or MD) is defined as simple state updates over multiple time steps. A typical MD simulation requires simulating a large number of particles interacting over millions of steps. Today's general-purpose supercomputers have succeeded on compute-intensive or memory-bandwidth-intensive applications, but the improvement of network bandwidth/latency is fairly limited.

Supported by National Key R&D Program of China under Grants No. 2017YFB0202000.

© Springer Nature Switzerland AG 2020
S. Wen et al. (Eds.): ICA3PP 2019, LNCS 11944, pp. 75–88, 2020.
https://doi.org/10.1007/978-3-030-38991-8_6

Traditional parallel MD algorithms partition particles into groups that are distributed over processes. To compute a new state of some particle, a process must receive the states of all particles that interact with it. Thus two interacting particles on different processes must exchange states in every time step via network.

MD in real applications are often highly non-linear: mostly smooth movements mixed with occasional bursts of sudden changes. Standard MD simulation methods use a constant length for all time steps in the entire simulation. The most compute-intensive operation for inter-particle forces is only performed at each step's time point. The simulated system is assumed to follow certain state-updating model of integration (*e.g.* leap-frog etc) between the time points of adjacent two steps. Shorter time steps are more accurate but require performing more time steps and lead to poor time-to-solution. The precision of a simulation can be measured separately by computing many extremely short steps (much shorter than a particle's oscillation cycles) between two adjacent time points. That maximum value difference between the one-step coarse simulation and the multi-step one is a good indication about the inaccuracy of the coarse simulation as "error per time step".

The uniformly constant time step must be small enough to allow accurate simulation of those sudden movements. Inaccurate simulation may lead to later divergent movements that violate physical laws. That also means the chosen length of each step is often unnecessarily small for smooth parts of the simulation. A quick thought would be to use changing time steps: shorter steps for highly non-linear parts and longer ones for near-linear parts. The difficulty is, of course, to determine when it is "safe" to increase the time length. Unless there is a way to ensure precision, it will be unsafe to increase it. On the other hand, if there does exist a mechanism to keep inaccuracies in check, it will allow us to try many heuristic methods to improve performance/precision.

In this paper, we propose a new communication-avoiding algorithmic model for MD computation and similar computing tasks. The basic idea is to compute in small regular time steps but communicate in longer changeable time steps. Each *communication step* consists of multiple *computation steps*, and only the state of the last is broadcast to other processes. The result of each computation step is checked against a certain state-updating model (usually the same model with only larger time steps). The communication step only occurs after several computation steps if new states deviate from the model of the longer communication steps. Generally, computation steps keep checking the inaccuracies and communication occurs when the difference exceeds certain threshold.

The first contribution of the communication-avoiding algorithmic model is that it makes sure that the MD simulation can adaptively slow down and communicate more frequently if some atoms occasionally collide in highly nonlinear trajectories. The second contribution is that it enables the simulation to correct what has been computed after receiving more accurate data from communications and ensures the precision of MD simulation. The third contribution is that it helps significantly improve reduce the number of data exchanges in network communication and hence improve the overall performance when communication is the bottleneck.

These contributions shows that, by adopting the communication avoiding algorithm method, these supercomputers can also achieve competitive time-to-solution for communication-intensive MD simulations. On the Sunway Tai-huLight general-purpose supercomputer, the "time-to-solution" optimization reaches a speed of 11.8 K steps/s for a system with 2.1 million silicon atoms and 5.1 K steps/s for 50.4 million silicon atoms. The speeds match that of the special ASIC system Anton II which, in comparison, performs 16.7 K steps/s or 3.6 μs per day with each step at $2.5fs$ simulating 2.2 million atoms of a protein molecule. The new computational algorithm model reduces communication to about 1/5.1 of the standard MD method. Compared to the special hardware solution, a software solution on general-purpose supercomputers is more accessible to the general scientific users.

2 Related Work

The current state-of-art solutions for MD simulation depend on the application requirement and the computational methods. Given some acceptable level of targeted precision, hardware and software solutions are comparable on scale (number of atoms) and speed (steps/s).

High-performance MD computation tends to have two extreme setups. For short-period phenomena, each processor computes as many atoms as possible so that a large atomic system close to realistic physical sizes can be simulated. The main performance optimization most likely lies with memory bandwidth, while the speed of simulation rarely exceeds 1000 steps/s (see Table 1). For long-period phenomena, however, the main objective of performance optimization is to maximize the speed of simulation(in steps/s).

Table 1. Performance comparisons of hardware and software solutions for MD

Processor	#atoms	#nodes	#atoms/node	steps/s	time/day	fs/step
Anton II 2014 [1]	2.2 M	512	4 K	16.7 K	3.6 μs	2.5
Anton I 2007 [2]	23.6 K	512	46	67.3 K	14.5 μs	2.5
KNL 2016 [3]	512 K	1	512 K		8 ns	
CPU+GPU 2015 [4]	140 K	128	1.2 K	1.12 K	250 ns	2.5
2CPU+GPU 2014 [5]	12 M	1024	12 K		54 ns	
Cray XK7 2014 [6]	224 M	16384	14 K	0.13 K	23 ns	2.0

Hardware method such as special ASIC system Anton [1] uses the standard computational method and is currently the only system that can perform more than 10 K steps per second when incorporating network communication (connecting 512 processors). A number of special hardware features are designed to shorten the data paths between processors. For example, a short-distance, fast and low-latency network is used on Anton. L2 cache and other unnecessary overhead-incurring interfaces are removed. States are stored and updated within L1 cache to minimize latencies. The network's links are connected directly into the processors. These breakthrough designs are particularly suitable for molecular dynamics. Compared to Anton, a GPU-accelerated modern supercomputer would better support throughput computing but have much longer data paths from CPU cores, to memory controller, GPU device memory, PCI, main memory, network card and optical-fabric routers. The speed of data exchanges between GPU cores on different servers is one or two magnitudes slower than Anton. However, a special hardware solution is costly to build and may encounter scalability issues with longer-distance network communications if the number of processors further increases.

State-of-art software methods, instead, focus on reducing communications, sometimes by allowing additional precision errors that are kept small relative to the inherent precision errors caused by temporal discretization.

NAMD based on Charm++ [7] tries to maintain the globally synchronous parallel computation of molecular dynamics while taking advantage of the sparseness of force interactions and allowing communications to overlap with computation as much as possible in a locally synchronous executional model.

Bulk Synchronous Parallel (BSP [8]) algorithm consists of some supersteps separated by globally synchronized barriers. The algorithmic complexity can be characterized by flops of computation, bytes of communication and number of synchronization. Many systems implement BSP algorithms using a (non-BSP) data-flow executional model to overlap communication and synchronization (*e.g.* for graphs). Nevertheless, all varieties of BSP techniques observe a common restriction: at anytime local computation only records and processes the state of one superstep, and each process always receives the latest results of the last superstep. Such semantic restriction inevitably affects its scalability on modern architectures.

Parallel-in-Time (PiT [9]) algorithms like Parareal [10] promise better scalability for time-dependent computation by speculatively performing **fixed-length** long time steps and then using many separate small-step simulations in parallel from the generated coarse states to adjust and improve precision. When the simulated system is small with only a few atoms, Parareal works fine and tends to converge in a few rounds of coarse-fine iteration. For a realistic system with a few thousand or more atoms, the precision results are poor. The trajectories after some time steps become chaotic. For a targeted level of precision (*e.g.* 10^{-7} relative error per atom per step), the performance of PiT is often significantly worse than that of the standard synchronous computation.

Asynchronous Iterative Algorithms (AIA [11]) are becoming popular in HPC applications like deep learning if consecutive time steps are only weakly dependent. For example, in deep learning, out-of-order adjustment of weights can be tolerated to some extent. The idea of AIA is simply to compute a **fixed number** of steps before any communication in the hope that using the states of other processors received a few steps before are still accurate enough. AIAs have been successful in many practical applications, though the main concern remains on the algorithm's uncontrollable precision loss. This is particularly concerning for highly nonlinear systems that generally exhibit smooth state changes most of the time but may have occasional sudden state changes (*e.g.* collision of atoms in MD). As the number of non-communicated local computation of time steps is constant, the precision loss may be tiny compared to the inherent computational errors most of the time but too large occasionally.

3   Solution: Communication-Avoiding Algorithm for MD

3.1   The Standard Parallel Algorithm

The standard parallel computation (Algorithm 1) of MD is defined as state updates to acceleration, velocity and positions of an atom on each processor. Here we consider exactly one atom per processor. The case with more atoms per processor is a straightforward generalization. Communications occur between processes with neighboring atoms. The ratio of computation vs communication is 1(step):1(step). For a typical 3D neighborhood communication pattern (with each processor communicating with 6 neighboring processors), if the network can at most perform 3200 rounds of data exchanges, then the maximum possible speed is 3200 steps/s, no matter how fast the CPUs are.

Process p communicating with other processes $p_1, \cdots, p_k$

$n \leftarrow 0$
while do

  $\mathbf{x}^p[n+1] \leftarrow f(\mathbf{x}^p[n], \mathbf{x}^{p_1}[n], \cdots, \mathbf{x}^{p_k}[n])$
  Send $\mathbf{x}^p[n+1]$ to neighboring processes
  $n \leftarrow n+1$
end

Algorithm 1: Standard parallel algorithm for MD where $\mathbf{x}$ is a combined vector of acceleration, velocity and positions, and f is the state-updating function that computes acceleration and use it to update velocity and positions in every step.

A common method called "leap frog" is commonly used for state updates. Let $\mathbf{x}^p[n]$ be the combined vector of acceleration $\mathbf{a}$, velocity $\mathbf{v}$ and positions $\mathbf{s}$ of the atom on a process p at step n (though more conveniently shifted by half time step in representation). The update of velocity and position after computing $\mathbf{a}$ is defined as

$$\mathbf{v}[m+l] = \mathbf{v}[m] + \mathbf{a}[m]\tau$$
$$\mathbf{s}[m+l] = \mathbf{s}[m] + \mathbf{v}[m]\tau + \mathbf{a}[m]\tau^2/2$$

where $\tau = l\,dt$. The definition also "predicts" a future state at step $m+l$ from the previous state at step m. The mathematical model for such "prediction" is the same as the leap-frog interpolation of the state update of the standard method, only with longer time steps. The standard Algorithm 1 always uses $m = n$ and $l = 1$ for both computation and communication.

3.2 The Communication-Avoiding Algorithm

The precision of a state update of an atom depends much more on its last precise state than the positions of the neighboring atoms. For example, instead of using a uniform time step everywhere, experiment shows that an atom's position can be less frequently communicated to other atoms than the local update of its state. A new communication-avoiding algorithm model is proposed to take advantage this and only perform a data exchange after several computational steps. Algorithm 2 computes in rounds. Communications only occur after each round of communication-free steps.

Process p communicating with $p_1, \cdots, p_k$:

$r \leftarrow 0$ and $m^p[0] \leftarrow 0$

while do
  $\quad m \leftarrow \min(m^p[r], m^{p_1}[r], \cdots, m^{p_k}[r])$
  $\quad$ **for** l **from** 0 **to** l_{max} **do**
    $\quad\quad n \leftarrow m + l$
    $\quad\quad \mathbf{x}^p[n+1] \leftarrow f(\mathbf{x}^p[n], \overline{\mathbf{x}}_r^{p_1}[n], \cdots, \overline{\mathbf{x}}_r^{p_k}[n])$
    $\quad\quad$ **if** $diff(\mathbf{x}^p[n+1], \overline{\mathbf{x}}_r^p[n+1]) \geq \varepsilon$ **then**
      $\quad\quad\quad$ break
    $\quad\quad$ **end**
  $\quad$ **end**
  $\quad m^p[r+1] \leftarrow n+1$
  $\quad$ Send $\mathbf{x}^p[n+1]$, $m^p[r+1]$ to other processes
  $\quad r \leftarrow r+1$
end

Algorithm 2: Communication-avoiding algorithm for MD where l_{max} is the maximum number of steps to be computed during each round, $m^{p_i}[r]$ is the final step of atom p_i at the end of the last round, the state $\overline{\mathbf{x}}_r^{p_i}[n]$ of atom p_i at step n is "predicted" from its state at step $m^{p_i}[r]$, and ε is a parameter for precision control.

The first idea of this new algorithm is to use different time steps instead of a uniform time step for computation and communication. Communication only occurs after a few computational steps. This has been tested in other HPC applications like deep learning but not in MD. The key second idea is to allow computation to go back to an earlier step and re-compute some of the previous steps. This mechanism enables a simulation to correct what has been computed after receiving more accurate data from communications. The third idea is to determine when to stop to communicate and return the computation to an earlier step. This condition makes sure that the simulation can adaptively slow down and communicate more frequently if some atoms occasionally collide in highly nonlinear trajectories. In summary, the communication-avoiding algorithm model reduces communications (*i.e.* the number of messages) in MD by allowing some small additional precision errors (compared to that of the standard method).

The key of this new algorithm is the condition for precision control:

$$diff\left(\mathbf{x}^p[n+1],\ \overline{\mathbf{x}}_r^p[n+1]\right) \geq \varepsilon,$$

which is used to cut short the communication-less computation of each round if the locally computed new states deviate too much from what can be "predicted" from the last state sent out via communication. As a special case, if $\varepsilon = 0$, the condition is always satisfied after computing a step, then Algorithm 2 behaves exactly like the standard Algorithm 1; or in the other extreme case, if $\varepsilon = \infty$, the condition is never satisfied, and the algorithm always completes a fixed number l_{max} of steps, then the Algorithm 2 behaves exactly like AIA (see Sect. 2). For $0 < \varepsilon < \infty$, the parameter controls the precision of each computation round. This mechanism is particularly important in slowing down a simulation when necessary and preventing divergences.

3.3 Precision

The precision of a simulation here is measured by relative errors after $1\,\mathrm{K}$ steps compared to the calibration run of $100\,\mathrm{K}$ steps with each time step adjusted to $1/100$ of the original time step. The relation between performance and precision is illustrated in Fig. 1.

Algorithm 1 always performs one communication after every step of computation. The ratio between computation and communication is 1:1. A larger time step leads to less running time but larger precision errors. On the other hand, experiments show that Algorithm 2 will compute more steps but conduct fewer communications. The ratio is around 5:1 in average for the silicon material. That means, on average, only one round of communication is needed for every 5 converged steps. Experiments show that this important ratio depends only on local atomic interactions and is not affected by the size of the whole system. The ratio can be improved by more accurate mathematical model of interpolation. For instance, the above ratio 5:1 is achieved by adding three additional higher orders of differentials calculated from previous values of (second-order) acceleration of the Taylor series.

On Sunway TaihuLight, the average time of computing one step for 512 silicon atoms in one CPU is $1.92 \times 10^{-5}\,\mathrm{s}$, while the time to conduct one round of 3D communication over 21952 CPUs is $3.8 \times 10^{-4}\,\mathrm{s}$. Thus reducing communication is more important than reducing computation. For the same target precision, the combined running time of computation and communication of the two algorithms can be compared. Figure 1 clearly shows that Algorithm 2 outperforms Algorithm 1. Note that when time step is 1.27fs, the relative precision error is 0.1×10^{-3} when the communication-avoiding algorithm outperforms the standard algorithm by 6.7x.

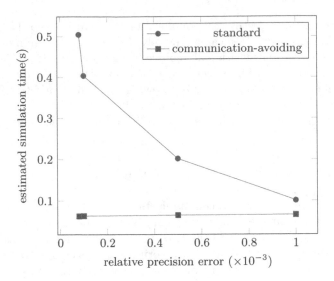

Fig. 1. The running time of the communication-avoiding algorithm compared to that of the standard synchronous algorithm for silicon simulation, given a target level of precision.

4   Profiling: Application and Environment

4.1   Application

Since MD simulation is capable of depicting microscopic details within the atomic vibration period, it has been a powerful tool for studying the thermophylic properties of materials. However, the computational costs at full atomic scales are extremely high both spatially and temporally and require efficient and scalable algorithms and implementation, especially for some semiconductor materials with complex interatomic many-body potential. Silicon is the most important material for information and clean energy industries and has diverse potential applications in field-effect transistors, photovoltaic and thermoelectric devices, biological and chemical sensors, etc. [12,13]. However, their performance and stability are sensitive to the thermal properties and heat dissipation process right from the atomistic scales up to micron scales [14,15]. Thanks to the computationally intensive BOP model of Tersoff [16,17], their thermal conductivity can be measured in simulations now, but mostly in the nanometer range far below the scale-independent limit [18–20]. Furthermore, solidification of silicon is a vital fabrication technology for highly pure single crystalline silicon, where the nucleation, growth and coarsening processes of nucleus in the matrix of liquid silicon need a much longer time to simulate [21,22]. Compared to the atomic model of protein folding, a silicon atom has fewer interacting neighbors but requires more computation of special functions [23].

In MD simulations, the system size can be readily enlarged by large-scale parallelization of processors, However, the temporal scale is much harder to promote. The typical time steps in MD simulations are in femto-seconds while some molecular processes, such as prediction of transport properties and statistics of some physical quantities including free energy of solid nanomaterials [18,24,25] may last from microsecond to milliseconds. And, condensed nanomaterials are more challenging due to the strong nonlinear scale-dependence of their properties. Some thermal properties of pure bulk materials do not stabilize under micron scales. Correspondingly, the time scale required for such simulations increases along with spatial scale, typically beyond μs for spatial scales at the $\mu m \sim mm$ range. That means, the time steps required are beyond 10^{12} which is too long for current technologies.

4.2   System and Environment

The algorithms are implemented on a system at National Supercomputing Center of China in Wuxi with up to 40960 many-core processors SW26010 [26]. Figure 2 illustrates the architecture. Each CPU consists of 256 slave cores and 4 master cores. The clock of master and slave cores run at 1.45 GHz. The floating-point performance of a code can be measured by a profiling tool called SWPerf. Each CPU has an infiniband network port connected to the optical-fabric network. The peak bandwidth of peer-to-peer communication is 56 Gb/s. The network is configured in a fat-tree structure at two levels. The upper level has 1/4 the combined bandwidth of the lower level. As the four core groups share the same Infiniband port, to minimize the number of messages in communication optimization, the code must be implemented for all 256 slave cores. The data path from each slave core goes through a crossbar that connects the core groups (see Fig. 2). As optimal time-to-solution is pursued, it is important to benchmark the communication network. Atoms are arranged in an $N \times N \times N$ 3D-circular cubic mesh with each processor holding typically 512 (or up to 2048) atoms. The communication pattern first exchanges data between processors in two opposite

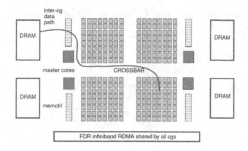

Fig. 2. Architecture of the SW26010 manycore processors

directions of dimension X, then communicate along dimension Y and finally along dimension Z. The average message length is 60 KB. Some atoms in the corner of a $8 \times 8 \times 8$ block are received from one dimension, copied to and re-sent to another dimension. Table 2 shows the results of the benchmarking. Without communication avoidance, a simulation cannot exceed the speed of 1.7 K steps/s for a cluster of 32768 servers, no matter how fast the computation is.

Table 2. Benchmarking the fat-tree communication network in a 6-way 3D communication pattern

size	#nodes	rounds/s
$2 \times 2 \times 2$	8	7.4 K
$8 \times 8 \times 8$	512	5.9 K
$16 \times 16 \times 16$	4096	3.2 K
$28 \times 28 \times 28$	21952	2.6 K
$32 \times 32 \times 32$	32768	1.7 K

5   Performance Results

5.1   Single CPU Performance

A single SW26010 CPU's performance depends on the number of atoms assigned to its 256 slave cores. Table 3 shows the single CPU performance for different numbers of atoms per core. Performance is tested from a scale of 4 atoms per core, which achieves the maximum simulation speed of 120 K steps/s, up to 384 atoms per core, which allows a maximum number 98 K of atoms per CPU and achieves the maximum floating-point utilization 15%.

Table 3. Single CPU performance (CPU peak at 3Tflops/s)

#atoms	steps/s	Gflops/s	FP utilization
64	120K	152	5%
512	52K	264	9%
2K	15K	305	10%
98K	2.2K	450	15%

5.2 Overall Performance

Table 4 shows the speed of simulation in steps/s for different numbers of atoms on Sunway TaihuLight compared with Anton II. The algorithm reaches a speed of 11.8 K steps/s for a system with 2.1 million silicon atoms and 5.1 K steps/s for 50.4 million silicon atoms. The speeds match that of the special ASIC system Anton II which, in comparison, performs 16.7 K steps/s or 3.6µs per day with each step at 2.5fs simulating 2.2 million atoms of a protein molecule. Moreover, it can scale to simulate more atoms. Using the communication-avoiding algorithm, the number of communications has been reduced to 1/5 of the total steps. Figure 3 illustrates the weak scaling of performance from 512 CPUs to 32768 CPUs (or 80% of the full Sunway TaihuLight system).

Table 4. Comparison of software solution on Sunway TaihuLight and the hardware solution on Anton II

Processor	#atoms	#nodes	comm:comp	steps/s
Anton II	2.2M	512	1:1	16.7K
Sunway TaihuLight	0.26M	512	5:1	17.8K
Sunway TaihuLight	0.9M	1728	5:1	15.8K
Sunway TaihuLight	2.1M	4096	5:1	11.8K
Sunway TaihuLight	4.1M	8000	5:1	10.9K
Sunway TaihuLight	7.1M	13824	5:1	10.8K
Sunway TaihuLight	14.2M	13824	5:1	8.2K
Sunway TaihuLight	22.4M	21952	5:1	7.7K
Sunway TaihuLight	33.6M	21952	5:1	6.6K
Sunway TaihuLight	50.4M	32768	5:1	5.1K

5.3 Optimizing Latencies and Overheads

Communication latency and various overheads are the main target for performance optimization besides usual optimization techniques such as vectorization. To minimize a processor's response time to newly arrived communications, a few optimization techniques are used. Instead of storing states of atoms in a regular array of states, some of the states are directly allocated in send and receive RDMA buffers so as to reduce state copying.

5.4 Computational Results

Preliminary computational results of heat transfer of bulk silicon and solidification of molten silicon are illustrated in Fig. 4. The scale of the simulation domain is 34.74 nm^3 of 2.1 million atoms.

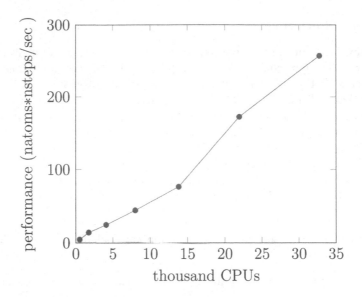

Fig. 3. Weak scaling of performance

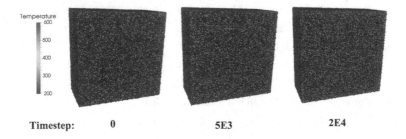

Fig. 4. Simulation of heat transfer of bulk silicon

6   Conclusion

The highly efficient and scalable molecular dynamics (MD) simulation framework with a novel communication-avoiding algorithm is proposed. Today's general-purpose supercomputers have succeeded on compute-intensive or memory-bandwidth-intensive applications. This contribution shows that, by adopting communication-avoiding algorithm model, these supercomputers can also achieve competitive time-to-solution for communication-intensive MD simulations. On the Sunway TaihuLight general-purpose supercomputer, this time-to-solution performance reaches a speed of 11.8 K steps/s for a system with 2.1 million silicon atoms and 5.1 K steps/s for 50.4 million silicon atoms. The speed matches those of state-of-art hardware solution (*i.e.* the special ASIC system Anton II) and can scale to simulate more atoms. The algorithm model reduces communication to about 1/5.1 of the standard MD method and shows the best time to solution so far for general-purpose systems. Generally, compared to the special hardware solution, a software solution on general-purpose supercomputers is more accessible to the general scientific users. This proposed algorithm framework is expected to be widely applicable to other long time-dependent simulations such as the simulation of turbulent flows.

The setup of this contribution in MD simulation is overwhelmingly communication-intensive. The competitive results on performance and scalability are achieved on a fat-tree network. Torus networks [6] are known to scale better than fat trees. The performances are expected to improve on those supercomputing platforms.

References

1. Shaw, D.E., Grossman, J.P., Bank, J.A., et al.: Anton 2: raising the bar for performance and programmability in a special-purpose molecular dynamics supercomputer. In: International Conference for High Performance Computing, Networking, Storage and Analysis, SC 2014, New Orleans, LA, USA, 16–21 November, 2014, pp. 41–53 (2014)
2. Shaw, D.E., Deneroff, M.M., Dror, R.O., et al.: Anton, a special-purpose machine for molecular dynamics simulation. In: 34th International Symposium on Computer Architecture (ISCA 2007), San Diego, California, USA, 9–13 June, 2007, pp. 1–12 (2007)
3. Höhnerbach, M., Ismail, A.E., Bientinesi, P.: The vectorization of the tersoff multibody potential: an exercise in performance portability. In: Proceedings of the International Conference for High Performance Computing, Networking, Storage and Analysis, SC 2016, Salt Lake City, UT, USA, 13–18 November, 2016, pp. 69–81 (2016)
4. Abraham, M.J., Murtola, T., Schulz, R., et al.: GROMACS: high performance molecular simulations through multi-level parallelism from laptops to supercomputers. Softwarex 1–2(C), 19–25 (2015)
5. Páll, S., Abraham, M.J., Kutzner, C., et al.: Tackling exascale software challenges in molecular dynamics simulations with GROMACS, CoRR, vol. 70, pp. 3–27 (2014)
6. Phillips, J.C., Sun, Y., Jain, N., et al.: Mapping to irregular torus topologies and other techniques for petascale biomolecular simulation. In: International Conference for High Performance Computing, p. 81 (2014)
7. Phillips, J.C., et al.: Scalable molecular dynamics with namd. J. Comput. Chem. 26(16), 1781–1802 (2005)
8. Valiant, L.: A bridging model for parallel computation. Commun. ACM 33(8), 103–111 (1990)
9. Baffico, L., Bernard, S., Maday, Y., et al.: Parallel-in-time molecular-dynamics simulations. Phys. Rev. E 66(2), 057701 (2002)
10. Lions, J.-L., et al.: Resolution EDP par unschema en temps parareal. C. R. Acad. Sci. Numer. Anal. 332(7), 661–668 (2001)
11. Bahi, J.M., Contassot-Vivier, S., Couturier, R.: Evaluation of the asynchronous iterative algorithms in the context of distant heterogeneous clusters. Parallel Comput. 31(5), 439–461 (2005)

12. Boukai, A.I., Bunimovich, Y., Tahir-Kheli, J., et al.: Silicon nanowires as efficient thermoelectric materials. Nature **451**(7175), 168–171 (2008)
13. Tian, B., Kempa, T.J., Lieber, C.M.: Single nanowire photovoltaics. Chem. Soc. Rev. **38**(1), 16–24 (2009)
14. Yang, N., Zhang, G., Li, B.: Violation of fourier's law and anomalous heat diffusion in silicon nanowires. Nano Today **5**(2), 85–90 (2010)
15. Schelling, P.K., Phillpot, S.R., Keblinski, P.: Comparison of atomic-level simulation methods for computing thermal conductivity. Phys. Rev. B **65**(14), 144306–144317 (2002)
16. Tersoff, J.: New empirical approach for the structure and energy of covalent systems. Phys. Rev. B **37**(14), 6991–7000 (1988)
17. Tersoff, J.: Empirical interatomic potential for silicon with improved elastic properties. Phys. Rev. B **38**(14), 9902–9905 (1988)
18. He, Y., Savic, I., Donadio, D., Galli, G.: Lattice thermal conductivity of semiconducting bulk materials: atomistic simulations. Phys. Chem. Chem. Phys. **14**(47), 16209–16222 (2012)
19. Cruz, C., Termentzidis, K., Chantrenne, P., Kleber, X.: Molecular dynamics simulation for the prediction of thermal conductivity of bulk silicon and silicon nanowires: influence of interatomic potentials and boundary conditions. J. Appl. Phys **110**(3), 34309–34316 (2011)
20. Park, M., Lee, I., Kim, Y.: Lattice thermal conductivity of crystalline and amorphous silicon with and without isotopic effects from the ballistic to diffusive thermal transport regime. J. Appl. Phys. **116**(4), 43514–43522 (2014)
21. Krzeminski, C., Brulin, Q., Cuny, V., et al.: Molecular dynamics simulation of the recrystallization of amorphous Si layers: comprehensive study of the dependence of the recrystallization velocity on the interatomic potential. J. Appl. Phys. **101**(12), 6336-4 (2011)
22. Lee, B.M., Baik, H.K., Seong, B.S., et al.: Molecular-dynamics analysis of the nucleation and crystallization process of Si. Phys. B Condens. Matter **392**(1–2), 266–271 (2007)
23. Hou, C.F., Xu, J., Wang, P., et al.: Petascale molecular dynamics simulation of crystalline silicon on Tianhe-1A. Int. J. High Perform. C. **184**(5), 1364–1371 (2013)
24. Perez, D., Huang, R., Voter, A.F.: Long-time molecular dynamics simulations on massively parallel platforms: a comparison of parallel replica dynamics and parallel trajectory splicing. J. Mat. Res. **33**(7), 813–822 (2018)
25. Elber, R.: Perspective: computer simulations of long time dynamics. J. Chem. Phys. **144**(6), 98–103 (2016)
26. Fu, H., Liao, J., Yang, J., et al.: The sunway TaihuLight supercomputer: system and applications. Sci. China Inf. Sci. **59**(7), 072001 (2016)

Out-of-Core GPU-Accelerated Causal Structure Learning

Christopher Schmidt[(⊠)], Johannes Huegle, Siegfried Horschig,
and Matthias Uflacker

Enterprise Platform and Integration Concepts, Hasso Plattner Institute,
University of Potsdam, Potsdam, Germany
{christopher.schmidt,johannes.huegle,matthias.uflacker}@hpi.de,
siegfried.horschig@student.hpi.de

Abstract. Learning the causal structures in high-dimensional datasets enables deriving advanced insights from observational data. For example, the construction of gene regulatory networks inferred from gene expression data supports solving biological and biomedical problems, such as, in drug design or diagnostics. With the adoption of Graphics Processing Units (GPUs) the runtime of constraint-based causal structure learning algorithms on multivariate normal distributed data is significantly reduced. For extremely high-dimensional datasets, e.g., provided by The Cancer Genome Atlas (TCGA), state-of-the-art GPU-accelerated algorithms hit the device memory limit of single GPUs and consequently, execution fails. In order to overcome this limitation, we propose an out-of-core algorithm for GPU-accelerated constraint-based causal structure learning on multivariate normal distributed data. We experimentally validate the scalability of our algorithm, beyond GPU device memory capacities and compare our implementation to a baseline using Unified Memory (UM). In recent GPU generations, UM overcomes the device memory limit, by utilizing the GPU page migration engine. On a real-world gene expression dataset from the TCGA, our approach outperforms the baseline by a factor of 95 and is faster than a parallel Central Processing Unit (CPU)-based version by a factor of 236.

Keywords: GPU-acceleration · Out-of-core · Causal structure learning · PC algorithm

1 Introduction

Learning causal structures from observational data is an active field of research in statistics and data mining. Discovering the causal relationships between observed variables in complex systems fosters new insights and is of particular interest in the context of high-dimensional settings such as in personalized medicine. For example, in genetic research, the construction of gene regulatory networks inferred from gene expression data supports drug design or diagnostics [19].

© Springer Nature Switzerland AG 2020
S. Wen et al. (Eds.): ICA3PP 2019, LNCS 11944, pp. 89–104, 2020.
https://doi.org/10.1007/978-3-030-38991-8_7

Causal graphical modeling is a well-known concept for the formalization of causal structures [7,17,26], where directed edges between nodes represent causal relationships between the observed variables. Algorithms for learning the structure of causal graphical models from observational data build upon two main approaches: score-based and constraint-based methods. Score-based approaches treat structure learning as an optimization problem over a score function on all possible graphical models which raises an NP-hard problem [3]. Constraint-based approaches apply a series of statistical tests to identify the conditional independence constraints of the causal structure as a first step and an orientation step that incorporates deterministic rules as a second step. While this approach may be exponential to the number of observed variables in the worst case, constraint-based algorithms such as the PC algorithm introduced by Spirtes et al. [26] run in polynomial time in sparse settings. Nevertheless, the algorithm's long runtime hinders its application in practice, in particular for high-dimensional data [13].

With recent advances in hardware technology, i.e., multi-core, many-core, and Graphics Processing Unit (GPU)-accelerated systems parallel versions of the algorithm have been developed [13,15,23,24,28]. GPU-accelerated causal discovery implementations report a significant speedup of factors between 700 to 1,300 compared to Central Processing Unit (CPU)-based implementations under the assumption of multivariate normal distributed data [23,28]. These approaches support the application in practice reporting fast learning of causal structures for datasets with up to 5,361 observed variables. While both implementations focus on the PC-algorithm, the GPU-accelerated adjacency search is also applicable to other constraint-based causal structure learning algorithms, such as FCI, RFCI, and CCD [5,20,26]. Yet, both implementations are limited by the GPU device memory and execution fails once data structures exceed available memory. Given, that memory requirement is quadratic to the number of observed variables in the multivariate normal distributed case, the limitation of current GPU-accelerated implementations is reached for extremely high-dimensional data. In particular, in genetic research, high-dimensional datasets are being collected and made available for research, e.g., see The Cancer Genome Atlas (TCGA) [1]. Resulting gene expression datasets from TCGA contain information on more than 55,000 observed genes [18], exceeding the device memory of recent GPU generations.

In order to benefit from GPU-acceleration beyond device memory limitations, we propose an out-of-core causal structure learning algorithm, which enables a GPU-accelerated adjacency search for extremely high-dimensional datasets, i.e., as available in TCGA. In our approach, we split data into smaller-sized blocks that fit into device memory and yield enough statistical tests to occupy the compute cores of the GPU. In the worst case, the adjacency search of the PC Algorithm requires to conduct statistical tests between all observed variables, thereby requiring data from multiple blocks. Therefore, we incorporate knowledge of data dependencies across blocks for higher-order statistical tests and manually manage data transfer. By overlapping computation and data transfer using different CUDA Streams[1] we hide any transfer overhead. We compare our block-based implementation with manual memory management to a version based on automatic memory management utilizing the default page migration engine available in current NVIDIA V100 GPUs. Thereby, we show the scalability of our block-based implementation beyond device memory limitations for higher-order statistical tests within the PC algorithm and a speedup of the runtime on a gene expression dataset from TCGA by a factor of 95.

The remainder of this paper is structured as follows. Section 2 provides the necessary background on constraint-based causal structure learning algorithms, in particular, the PC algorithm. In Sect. 3, we discuss related work on parallel constraint-based causal structure learning and general out-of-core GPU-accelerated algorithms. Afterward, in Sect. 4, we present our GPU-accelerated block-based implementation of the adjacency search, which allows execution on datasets exceeding device memory limits. The evaluation of the implementation is provided in Sect. 5 before we conclude our work in Sect. 6.

2 Background

In this section, we provide necessary background on constraint-based causal structure learning, in particular, the PC-stable algorithm. Further, we introduce an existing GPU-accelerated version of the adjacency search in the context of the PC-stable algorithm's skeleton estimation.

2.1 Constraint-Based Causal Structure Learning

In the framework of causality according to Pearl [17] and Spirtes et al. [25], causal relationships between N observed variables V_i, $i = 1, \ldots, N$, can be represented in a Directed Acyclic Graph (DAG) $\mathcal{G}$. In this DAG $\mathcal{G} = (\mathbf{V}, \mathbf{E})$, the vertices $\mathbf{V} = \{V_1, \ldots, V_N\}$ represent the observed variables and the directed edges $\mathbf{E} \in \mathbf{V} \times \mathbf{V}$ between the vertices denote direct causal relationships, i.e., $V_i \to V_j$ for $i, j = 1, \ldots, N$. Constraint-based methods for causal structure learning exploit the factorization properties of the joint distribution P of $\mathbf{V}$ and apply conditional independence (CI) tests to determine the Markov equivalence class of the DAG $\mathcal{G}$ that is uniquely described by a Complete Partially Directed Acyclic Graph (CPDAG) [2]. In particular, the undirected skeleton $\mathcal{C}$ of $\mathcal{G}$ together with all unshielded colliders $V_i \to V_j \leftarrow V_k$ for non-adjacent V_i and V_j with $i, j, k = 1, \ldots, N$ in $\mathcal{G}$ entail the CI information of the joint distribution P of $\mathbf{V}$. Accordingly, based on the d-separation criterion [17], two variables V_i and V_j are conditionally independent given a set of variables $\mathbf{S}$ if and only if the vertices V_i and V_j are d-separated by the set $\mathbf{S} \subset \mathbf{V} \setminus \{V_i, V_j\}$ for $i, j = 1 \ldots, N$. Hence, under the assumption of causal sufficiency and causal faithfulness of the joint distribution P of $\mathbf{V}$ the application of consistent CI tests enables to derive the Markov equivalence class of the true underlying DAG $\mathcal{G}$, e.g., see [4,9,25].

---

[1] http://docs.nvidia.com/cuda/cuda-c-programming-guide/index.html#streams-cdp.

Constraint-based algorithms, such as the well-known PC algorithm introduced by Spirtes et al. [26], follow this theoretical foundation. The algorithm first determines the undirected skeleton $\mathcal{C}$ of $\mathcal{G}$ in an adjacency search through the repeated application of CI tests between variables given increasing separation sets $\mathbf{S}$ chosen from the set of adjacent vertices in $\mathbf{V}$. In a second step, the repeated application of deterministic orientation rules on the skeleton $\mathcal{C}$ orients edges, by orienting the detected unshielded colliders in $\mathcal{C}$ based on the examined separation set $\mathbf{S}$. Moreover, edges can be oriented through the application of further orientation rules such that neither additional unshielded colliders nor cycles are present in the resulting graph which results in the CPDAG.

For multivariate normal distributed variables $V_1, \ldots, V_N$, two variables V_i and V_j are conditionally independent given a set of variables $\mathbf{S} \subseteq \{V_1, \ldots, V_n\} \setminus \{V_i, V_j\}$ if and only if the corresponding partial correlation coefficient is equal to zero [12]. Hence, a consistent conditional independence test can be derived upon the corresponding sample partial correlation coefficient $\hat{\rho}(V_i, V_j | \mathbf{S})$ following standard statistical decision theory [14]. Moreover, in high-dimensional multivariate normal distributed settings, the PC algorithm is established as a computationally feasible and provable correct estimation procedure of the equivalence class of a sparse DAGs $\mathcal{G}$ [9].

While the original version of the PC algorithm depends on the order of the variables set $V_1, \ldots, V_N$ the PC-stable algorithm [4] is an order-independent

---

Algorithm 1. Adjacency search of PC-stable algorithm [4]
Input: Vertex set V, tuning parameter α
Output: Estimated skeleton $\mathcal{C}$, separation sets **Sepset**

1: Start with fully connected skeleton $\mathcal{C}$ and $l = -1$
2: **repeat**
3:     $l = l + 1$
4:     **for all** Vertices V_i in $\mathcal{C}$ **do**
5:         Let $a(V_i) = adj(\mathcal{C}, V_i)$;
6:     **end for**
7:     **repeat**
8:         Select a pair of variables V_i and V_j adjacent in $\mathcal{C}$ with $|a(V_i) \setminus \{V_j\}| \geq l$
9:         **repeat**
10:             Choose separation set $\mathbf{S} \subseteq a(V_i) \setminus \{V_j\}$ with $|\mathbf{S}| = l$.
11:             **if** $p(V_i, V_j | \mathbf{S}) \geq \alpha$ **then**
12:                 Delete edge $V_i - V_j$ from $\mathcal{C}$;
13:                 Save $\mathbf{S}$ in **Sepset**;
14:             **end if**
15:         **until** edge $V_i - V_j$ is deleted in $\mathcal{C}$
16:         or all $\mathbf{S} \subseteq a(V_i) \setminus \{V_j\}$ with $|\mathbf{S}| = l$ have been chosen
17:     **until** all adjacent vertices V_i, and V_j in $\mathcal{C}$ such that
18:     $|a(V_i) \setminus \{V_j\}| \geq l$ have been considered
19: **until** each adjacent pair V_i, V_j in $\mathcal{C}$ satisfy $|a(V_i) \setminus \{V_j\}| \leq l$
20: **return** $\mathcal{C}$, **Sepset**

extension that is the basis for efficient parallel adaptions [13,24], also in the context of GPU-accelerated implementations [23,28]. This order-independent version of the adjacency search of the PC-stable algorithm is outlined in Algorithm 1.

Following the previously introduced concepts the adjacency search of the PC-stable algorithm depicted in Algorithm 1 starts with a complete undirected skeleton $\mathcal{C}$ and uses CI tests with an increasing size of separation set $\mathbf{S}$ of adjacent vertices in the corresponding level l to subsequently remove edges between vertices that are determined as being independent (see lines 8–15).

For every level l, the adjacency sets $a(V_i) = adj(\mathcal{C}, V_i)$ of vertices V_i within the current skeleton $\mathcal{C}$ are computed and stored (see lines 4–6). Thus, at each level l, the algorithm marks edges for removal and deletes the edges only when entering the subsequent level $l + 1$ which guarantees the order-independence.

For $l = 0$, all edges between adjacent vertices $V_i, V_j \in \mathbf{V}$ are deleted if the corresponding independence test, i.e., given an empty set $\mathbf{S} = \emptyset$, rejects the null-hypothesis of dependence against independence. In this sense, independence between V_i and V_j can be concluded if the p-value $p(V_i, V_j | \emptyset)$ is greater than the significance level α and the empty set $\emptyset$ can be stored as separation set in **Sepset** (see lines 11–13 in Algorithm 1). Note, that the significance level α can be treated as a tuning parameter influencing the sparsity of the estimated skeleton. After testing all pairs of vertices the algorithm proceeds to the next level $l = 1$.

For $l = 1$, the algorithm applies the CI tests for variables V_i and V_j given a separation set of size 1 if they remained adjacent in the skeleton $\mathcal{C}$ after $l = 0$. Therefore, it is now examined whether it holds for the corresponding p-value that $p(V_i, V_j | \mathbf{S}) \geq \alpha$ with subset $\mathbf{S} \subset adj(\mathcal{C}, V_i) \setminus V_j$ of size 1 until either all other subsets in $adj(\mathcal{C}, V_i) \setminus V_j$ have been considered or the variables V_i and V_j are found to be conditionally independent. If the variables V_i and V_j are found to be conditionally independent, the corresponding edge $V_i - V_j$ is removed, and the corresponding separation set $\mathbf{S}$ is stored in **Sepset**. Once all pairs of variables V_i and V_j that are adjacent in the current skeleton $\mathcal{C}$ are tested the algorithm proceeds to the next level $l + 1$. The same procedure is repeated until l reaches the maximum size of the adjacency sets of the vertices $max_{V_i \in \mathbf{V}} |adj(\mathcal{G}, V_i) \setminus \{V_j\}|$ in the underlying DAG.

2.2 GPU-Accelerated Adjacency Search

For the case of multivariate normal distributed variables $V_1, \ldots, V_N$, approaches exist that allow the processing of the corresponding consistent CI tests on the basis of partial correlations in parallel on the GPU [23,28]. In order to benefit from the parallel processing capabilities of GPUs, the CI tests are distributed to threads following the Single Instruction Multiple Threads (SIMT) model.

For level $l = 0$, a mapping of CI tests to threads is straightforward as, given an empty separation set $\mathbf{S} = \emptyset$ for each pair of vertices V_i and V_j, each thread is processing a single CI test which launches N^2 threads [23]. For subsequent level $l \geq 1$, a CI test given a subset $\mathbf{S}$ of size l is considered. Hence, testing a pair of vertices V_i and V_j for conditional independence within a single level l may require multiple CI tests given appropriate adjacent subsets S in the current skeleton C. Therefore, current implementations join a fixed number of threads to blocks for each pair of vertices V_i and V_j, e.g., launching a single thread block with a number of threads. In this case, each thread within a thread block is responsible for processing several CI tests for the corresponding pair of vertices V_i and V_j given the appropriate subsets S. This distribution of CI tests for the same pair of vertices V_i and V_j to threads within the same thread block allows sharing data structures using shared memory and improves performance [28]. Furthermore, early termination is possible through synchronization within the thread block such that unnecessary CI tests are avoided. This ensures that the separation sets S examined are consistent for GPU-based and CPU-based implementations [23].

3   Related Work

Constraint-based causal structure learning is an active field of research, in particular in the context of the PC algorithm [26]. Given the algorithm's long execution time due to its computational complexity several parallel adaptions and implementations have been proposed [13,15,22–24,28]. Whereas the majority of work addresses CPUs as execution units, recently also two GPU-accelerated versions for multivariate normal distributed datasets have been proposed [23,28]. Both GPU-accelerated versions state a significant speed-up of three to four orders of magnitude compared to existing CPU versions.

In the work of Schmidt et al. [23] a GPU-accelerated adjacency search for levels 0 and 1 is proposed. As most CI tests on publicly available gene expression datasets are executed on levels $l = 0, 1$ they introduce an implementation of GPU-accelerated kernels for these two cases only. Another limitation of their implementation is the applicability to datasets that fit in the device memory.

The work of Zare et al. introduces a GPU-accelerated implementation, called cuPC [28], which is capable of processing any level l on the GPU. An extension cuPC-S enables the sharing of intermediate results during the computation of CI tests, in particular, parts of matrix inversions for the calculation of the sample partial correlation coefficients are shared. They show that the sharing of intermediate results can further reduce the execution time in higher levels $l \geq 1$. Moreover, they add a compact step after each level, for better assignment of threads to edges which leads to a reduction of the required memory. Hence, it helps to lift the device memory limitation for extremely high-dimensional datasets, yet, in the worst case, i.e., a dense graphical model, it also fails to process.

Approaches to overcome the device memory limitation of a GPU are subject to research [6,8,16,21,27,29]. Application-specific out-of-core algorithms avoid hitting device memory limits by splitting the specific task or data and processing the parts independent from each other [6,8]. These approaches either require a dividable-task [27] or a redesign of the algorithm [8]. Furthermore, the out-of-core algorithms require manual data management, i.e., data transfers, and orchestration, i.e., kernel launches. Generic frameworks to overcome the memory limitation provide the developer with API functions to allocate memory that is managed by the framework [16, 21]. This enables automatic data transfers, according to the memory regions requested by the GPU during kernel execution, using the GPU driver and the memory management unit of the GPU [16]. Thus, the available memory capacity is effectively extended to the capacity of DRAM in the system [21] or going even beyond to nonvolatile memory (NVM) [16]. In the context of UM, this capacity increase is achieved through the addition of 49-bit virtual addressing and on-demand page migration within recent GPU generations [21]. Yet, previous performance evaluations of UM on Kepler GPUs indicate that the utilization of UM introduces performance overhead with marginal improvement in code complexity [11].

4   An Out-Of-Core GPU-Accelerated Parallel Adjacency Search for the PC Algorithm

In the following section, we introduce two implementation strategies for an out-of-core GPU-accelerated adjacency search. The first implementation which can be treated as a baseline integrates the concept of UM into the GPU-accelerated skeleton discovery of Schmidt et al. [23]. Thus, it relies on a generic framework to overcome the device memory limitation. For the second implementation, we propose block-based algorithm of the adjacency search. In contrast to the first approach, the second requires manual data management and orchestration, i.e., data transfer and kernel launches.

4.1   Unified Memory Based Adjacency Search

With the introduction of a page migration engine on recent NVIDIA GPU generations [21], it is possible to use CUDA API calls to address larger memory than available on-chip. Specifically, a call to `cudaMallocManaged()` allocates memory, which is accessible from both the CPU and the GPU. In order to incorporate page migration into the kernels for levels $l = 0, 1$, for detail see [23], we adapt our implementation in the following way. Data structures required during kernel execution are allocated using `cudaMallocManaged()`, making them accessible from both CPU and GPU. Once, the GPU kernels are launched data is managed by the GPU driver and is transferred to the device transparently via the page migration engine. Thus, API calls for the explicit allocation of memory on the GPU and API calls to initiate data transfer to and from the device are removed. The source code of the kernels launched on the GPU remains unchanged.

4.2   Block-Based Adjacency Search

The block-based adjacency search follows the idea of splitting the data processed during the adjacency search into smaller blocks, that are independent of each other, to overcome the device memory limitation. The implementation

Algorithm 2. Block-based Adjacency Search of PC-stable algorithm

Input: Vertex set V, correlation matrix Cor, tuning parameter α, block size bs

Output: Estimated skeleton C, separation sets **Sepset**

1: Start with fully connected skeleton C and $l = -1$
2: **repeat**
3:     $l = l + 1$
4:     **for all** Vertices V_i in C **do**
5:         Let $a(V_i) = adj(C, V_i)$;
6:     **end for**
7:     $blocks = Split(V, \mathrm{Cor}, C, \mathbf{Sepset}, bs)$
8:     **for all** b in $blocks$ **do**
9:         Transfer b to GPU
10:        **if** $l == 0$ **then**
11:            $BlockCITest(b, \alpha)$
12:        **else**
13:            $sepsetblocks = SepSetCombination(b, l, blocks)$
14:            **for all** s in $sepsetblocks$ **do**
15:                Transfer s to GPU
16:                $BlockCITest(b, s, \alpha)$
17:            **end for**
18:        **end if**
19:         Transfer b from GPU
20:     **end for**
21:     $Merge(blocks)$
22: **until** each adjacent pair V_i, V_j in C satisfy $|adj_{out}(V_i) \setminus \{V_j\}| \le l$
23: **return** C, **Sepset**

aims to avoid any overhead introduced by UM [11] and aims to avoid dependency on novel GPU features. Following previous GPU-accelerated implementations [23,28], our block-based adjacency search assumes multivariate normal distributed data that yields to consistent CI tests on the basis of the corresponding sample partial correlation coefficients $\hat{\rho}(V_i, V_j | \mathbf{S})$. As the sample partial correlation coefficients can be derived from the sample correlation coefficients between the variables V_i, V_j and $\mathbf{S}$, our implementation operates on the correlation matrix Cor that contains the precomputed sample correlation coefficients $\hat{\rho}(V_i, V_j)$ for all $i, j = 1, \ldots, N$, for more information we refer to [23].

Algorithm 2 sketches the block-based adjacency search executed on a GPU, which extends the PC-stable algorithm (see Algorithm 1). Within each level l, the relevant data structures are split into a set of blocks b in $blocks$ (see line 7 in Algorithm 2). Note, the input parameter bs determines the sizes of data structures within a block b in $blocks$. In this sense, the operation Split() returns an iterable list $blocks$ of blocks b, for which each block b contains disjunct $bs \times bs$-submatrices of the following data structures: the correlation matrix Cor, the skeleton C. Moreover, it returns corresponding subsets with cardinality bs of the current the adjacency set $a(V_i)$, the separation sets **Sepset**, and auxiliary data structure containing the calculated p-values $p(V_i, V_j | \mathbf{S})$ as well as a mapping

$m(b)$ of the positions of the $bs \times bs$ submatrices of Cor and C in b to their original positions. Zero-padding is applied in the case that the dimension of the dataset is not a multiple of bs. Next, the list $blocks$ of blocks b is iterated and each block b is processed. First, all data structures in b are transferred to the GPU (see line 8 and 9). If the algorithm operates on level $l = 0$, the kernel can be launched directly as depicted in lines 10 and 11 of Algorithm 2. In this case, the algorithm operates on the fully connected skeleton C, such that all CI tests between V_i and all V_j from the subset with cardinality b of the adjacency set $a(Vi)$ can be executed on the basis of the $bs \times bs$-submatrix Cor that incorporates the corresponding sample correlation coefficients. Processing of the edges in the corresponding undirected bs-dimensional subgraph of C in BlockCITest() follows the standard procedure within the lines 8 and 18 of Algorithm 1. Afterwards, the corresponding result of block b, e.g., the separation sets and p-values as well as the updated part of C, is transferred back to the CPU where it is merged together with all other block results within $blocks$ (see line 19–21).

For all remaining levels $l \geq 1$, a CI test between a pair of variables V_i and V_j adjacent in the b-dimensional subgraph of C requires a separation set $\mathbf{S}$ of size l drawn from the subset with cardinality b from $a(V_i)$. Note that, in contrast to level $l = 0$, the derivation of the p-values $p(V_i, V_j | \mathbf{S})$ corresponding to CI test may not be restricted to the sample partial correlation coefficient that are available in $bs \times bs$-submatrix of Cor within block b. Thus, additional $bs \times bs$-submatrices of Cor, which we call separation set blocks are required. The number of separation set blocks increases with l. For a given block b, the list of sets of required separation set blocks, $sepsetblocks$, is determined within the function SepSetCombination() (see line 13). The list contains all $bs \times bs$ disjunct submatrices of Cor that contain at least one sample correlation coefficient that is needed to derive the p-value $p(V_i, V_j | \mathbf{S})$ of the corresponding CI-test between the variables V_i and V_j given a subset $\mathbf{S}$ with $|\mathbf{S}| = l$ of the subset of $a(V_i) \setminus \{V_j\}$ with cardinality bs. As depicted in the lines 14 till 17 in Algorithm 2, each element s in the list $sepsetblocks$ is transferred to the GPU such that a GPU kernel can be launched, which conducts all necessary CI tests between the variables V_i and V_j given the corresponding set $\mathbf{S}$. Afterward, the block b contains all required conditional independence results, e.g., separation sets and p-values as well es the corresponding bs-dimensional subgraph of $Skel$, and is transferred from the GPU into the host memory (see line 19). Once, all blocks b in $blocks$ have been processed the results are merged in line 21. This process is repeated for an increasing level l until the same termination criterion in line 19 of the PC-stable algorithm sketched in Algorithm 1 is satisfied (compare line 22 of Algorithm 2).

Following previous work [23], we provide an implementation of the block-based algorithm for levels $l = 0, 1$. This enables to investigate the performance improvements of the introduced concepts for CI tests with ($l = 1$) and without ($l = 0$) a separation set. Note, that according to cupc [28] the behaviour for levels $l \geq 2$ is similar to level $l = 1$. In the implementation of Algorithm 2, we incorporate the following optimizations. First, in level $l = 0$, we do not transfer the adjacencies $a(V_i)$ and the separation sets **Sepset** reducing the memory

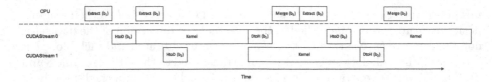

Fig. 1. Overlapping execution on CPU with data transfer and execution on GPU within the adjacency search of the block-based PC algorithm.

footprint during kernel execution. In the level $l = 0$, the deletion of an edge can be carried out on the skeleton $\mathcal{C}$ directly, without influencing any other CI test. Thus, the adjacencies $a(V_i)$ are not necessary. Furthermore, an empty separation set is used in level $l = 0$, which we assume as the initial value in the data structure **Sepset** for each pair of variables V_i, V_j. Hence, setting it within the kernel for $l = 0$ is not required. Note, that the same optimization applies to the kernel for level $l = 0$ in the UM-based adjacency search. Second, we overlap data extraction and merging on the CPU, as well as, data transfer to the GPU with kernel execution, as shown in Fig. 1, in order to reduce the overall runtime of the block-based algorithm. We realize the overlap by using two separate CUDA streams and adapt the proposed Algorithm 2, accordingly. In the implementation we first obtain the mapping for the blocks b, thus enabling to extract the appropriate data structures within the for loop in line 8 of Algorithm 2 for each block b independently. The same applies to the separation set blocks. Hence, all required data structures to launch a kernel on block b are extracted on the CPU and transferred to the GPU on one CUDA stream, while at the same time a GPU kernel is executed in a second, separate, CUDA stream, processing a second block from *blocks*. Once the kernel execution is finished, data is transferred back to the host and merged on the CPU, within the for loop.

5   Evaluation

For the evaluation of the proposed block-based adjacency search executed on a GPU, we conduct the following two experiments. First, we compare the performance of our block-based approach to the baseline implementation that applies the concept of UM with a page migration engine to overcome the device memory limit. Here, we focus on an examination of the approaches with regards to their scalability with respect to an increasing number of multivariate normal distributed variables, assuming a fully connected underlying causal graphical model. Since the assumption of a fully connected causal graphical model is not realistic demonstrating the worst case, we conduct a second experiment. Therefore, we examine a real-world gene expression dataset from the TCGA project that has been used in the context of integrative gene selection approaches [18]. Thereby, we investigate the performance benefits of the block-based approach compared to a CPU-based implementation and its applicability to real-world high-dimensional datasets.

The experiments are executed on an NVIDIA V100 card, with 32 GB of high bandwidth memory, inside an enterprise-grade server with 2 Intel® Xeon® Gold 6148 CPU with 20 cores each. The GPU card is connected via PCI-E version 4. Furthermore, the server is equipped with 1.5 TB of RAM, allowing to keep all data in memory during the execution of the experiments. The operating system is an Ubuntu 18.04 and the NVIDIA driver version 410.79 is installed with CUDA version 9.1. For measurements of the runtime we utilize the system_clock from the C++ standard library chrono. This allows to include all CPU-based operations, such as splitting and merging of blocks, kernel launches and data transfer in our measurements. In particular, we measure the runtime for each level $l = 0, 1$ separately. If not stated differently, we report the median value of 10 measured executions for dataset sizes until 40, 000 variables, for higher dimensional datasets we report the median value of 3 measured executions, due to long runtimes. For the block-based approach, we choose a block size bs of 2, 560, which maps to the hardware specifications of the NVIDIA V100. For brevity we omit measurement results, which confirmed the chosen block size.

5.1 Experiments on Scalability Properties

In the first experiment, we examine the scalability with regards to the number of variables in the dataset. We consider datasets with 5, 000 to 55, 000 variables, increasing with a step size of 5, 000. Hence, we scale beyond the dimensions of previously used datasets for evaluation of GPU-accelerated causal structure learning algorithms [23, 28], which did not exceed the available device memory. Based on available gene expression datasets from TCGA [18], we choose 55, 000 variables as our upper limit for the experiment. Under the assumption of double-precision values, all data structures required during kernel execution result in a maximum memory footprint of 0.47 GB for 5, 000 variables to 56.35 GB for 55, 000 variables in level $l = 0$. Respectively, the maximum memory footprint in level $l = 1$ for 5, 000 variables is 0.65 GB and for 55, 000 variables is 78.88 GB. Note, the difference is due to two memory optimizations in level $l = 0$ applicable to approaches, as explained in Sect. 4. Furthermore, in order to eliminate any influence of the underlying graph structure on our measurements, we assume that all possible CI tests have to be conducted within each level l. This coincides with a fully connected graph and sketches a worst case for the number of CI tests to be conducted. Hence, in level $l = 0$, over 12 million CI tests are conducted for 5, 000 variables and over a billion CI tests are conducted for 55,000 variables. For level $l = 1$ this results in over 60 billion CI tests for 5, 000 variables and around 83 trillion CI tests for 55, 000 variables.

The measurements presented in Fig. 2 display the median execution times for the adjacency search in level $l = 0$ on the left and the adjacency search for level $l = 1$ on the right. The vertical blue line marks the device memory limit of the GPU used in the experiments. Note, previous implementations would fail on datasets whose number of variables is located to the right of the vertical line. Furthermore, the measurements of the block-based adjacency search is drawn

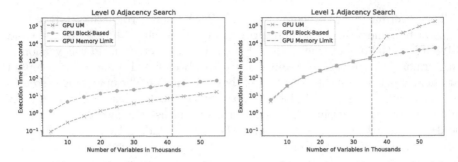

Fig. 2. Median execution times of the GPU UM-based (green) and the GPU Block-based (yellow) implementation for level 0 (left) and level 1 (right) of the skeleton discovery with an increasing number of variables; with all possible CI tests conducted (Color figure online)

in a yellow line denoted with GPU Block-Based and the UM-based adjacency search drawn in a green line denoted with GPU UM.

Considering the adjacency search for level $l = 0$ our measurements show that the block-based approach has a continuously higher execution time compared to the UM-based approach. In fact, the UM-based approach outperforms the block-based approach by a factor of up to 15 for datasets with few variables to a factor of 4.4 for datasets with 55,000 variables. In-depth analysis with the `nvprof` profiler has shown that the total data transfer time exceeds the kernel execution time. Additional time has to be added for the block extraction and merge on the CPU, which is not efficiently overlapped. Both approaches scale well, beyond the device memory limit. As no separation sets are required in level $l = 0$ and coalesced memory accesses is used such that the number of page faults scales linearly with the dataset size. Thus, the UM-based approach shows good performance beyond device memory limitations. Note, that the performance gap between both approaches decreases with an increasing number of variables.

For the adjacency search in level $l = 1$ the runtime of both approaches is similar up to a dataset with 35,000 variables. For datasets with a number of variables that scales beyond this number, the device memory limit is exceeded. The block-based approach shows similar scalability on these higher-dimensional datasets compared to the lower-dimensional datasets. In contrast, the UM-based approach's performance drops significantly beyond the GPU memory limit. In fact, the block-based approach outperforms the UM-based approach by factors from 12.4, for 40,000 variables up to 34, for 55,000 variables. Profiling with `nvprof` revealed that for the UM-based approach the number of page faults increases drastically once the device memory limit is hit. For 35,000 variables the number of page faults is around 140,000 and increases to almost 12 million for 40,000 variables. The performance difference is accounted for by memory stalls, for page accesses that result in page faults. When conducting a CI test with a separation set of size $l = 1$, the values for the separation set can be arbitrarily scattered across pages within the required data structures.

Table 1. Execution times of the CPU- and the two GPU-based implementations over both levels $l = 0, 1$ on TCGA dataset consisting of 55,572 variables with α of 0.01.

CPU pcalg (on 32 cores)	GPU UM	GPU block-based (2560 block size)
107.6 h	43.4 h	0.46 h

Hence, the page migration engine reloads previously evicted pages, as caching is not trivial.

Comparing the measured execution times for both levels, it is evident that the runtime in level $l = 1$ is significantly higher, due to the larger amount of CI tests that are conducted. Yet, for the smallest dataset in level $l = 1$ approximately 40 times more CI tests are conducted compared to the largest dataset in level $l = 0$, in less time. We account that for overhead in data transfer in level $l = 0$, which is also shown in [23] and results in a poorer ratio of computation per memory access.

5.2 Experiments on Real-World Gene Expression Data

In our second experiment, we examine the performance of the approaches on a real-world gene expression dataset from TCGA. The dataset contains 55,572 variables with 3,189 observations [18]. We set the tuning parameter α to 0.01, which is a common option found in literature [4]. In this setting, one-third of all edges are removed within level $l = 0$ leading to fewer tests in level $l = 1$. In addition to both GPU-accelerated implementations of the PC-stable algorithm's adjacency search, we executed a CPU-based OpenMP-enabled version chosen from the well-known R-package pcalg [10] (version 2.6). We conducted the CPU-based measurements on a separate system equipped with Intel® Xeon® E7-4850 v4 CPUs with 16 cores and 2 TB of DRAM. For the execution, we allowed OpenMP to scale to 32 cores and limited the execution to levels $l = 0, 1$, by setting the parameter of the $m.max = 1$. Using the skeleton method stable.fast an underlying C++ extension is used in the adjacency search, in which we integrate the time measurements. In Table 1, we state the measured execution times for the three approaches. The GPU block-based approach executes the adjacency search for level $l = 0, 1$ in 27.35 min. The GPU UM-based approach requires 43.4 h to finish, suffering from a large number of page faults. In comparison, the CPU-based version runs for over 4 days. Thus, the GPU block-based approach outperforms, the GPU baseline using UM by a factor of 95 and the CPU-based implementation by a factor of 236.

6 Discussion and Conclusion

In this paper, we proposed an out-of-core GPU-accelerated adjacency search to overcome the GPU device memory limitation. The adjacency search is a substantial part of constraint-based causal structure learning algorithms, such as the

PC-algorithm. It is used for the estimation of the Markov equivalence class of the underlying causal graphical model $\mathcal{G}$ from observational data. Our proposed algorithm splits the correlation matrix of multivariate normal distributed data and other relevant data structures, e.g., the skeleton, into small blocks such that datasets that usually exceed device memory can be processed efficiently on the GPU. We compare the approach to a baseline implementation using the concept of UM to overcome the memory limit. The baseline implementation relies on the page migration engine available in recent NVIDIA GPU-generations, which automatically transfers data between host and device memory. This baseline is easy to implement and shows good performance for level $l = 0$ of the adjacency search. In this case, it outperforms the block-based approach, which suffers from the overhead of block extraction and merging. In contrast, we show that the baseline suffers severely from page faults in case of arbitrarily scattered memory accesses, which occur in the adjacency search for level $l \geq 1$. For these cases, the performance gain of the block-based approach is significant, with factors up to 34. Furthermore, on real-world gene expression data, we show that the block-based approach outperforms the naive baseline by a factor of 95 and a parallel CPU-based version by a factor of 236. While our current implementation is limited to levels $l = 0, 1$, we assume similar behavior for higher levels $l \geq 2$. In future work, we aim to extend our implementation to these higher levels to evaluate limitations with regards to the number of required separation set blocks. Furthermore, adding a compact procedure, similar to the one described in cuPC [28], could be of interest since implementations may benefit from the condensed memory layout. Summarizing, we conclude that the block-based approach is well suited to extend GPU-accelerated causal structure learning algorithms to extremely high-dimensional datasets, which exceed the available device memory and could not benefit from the parallel processing capabilities of the GPU before.

References

1. Cancer Genome Atlas Research Network, Weinstein, J.N., et al.: The cancer genome atlas pan-cancer analysis project. Nat. Genet. **45**(10), 1113–1120 (2013)
2. Chickering, D.M.: Learning equivalence classes of Bayesian-network structures. J. Mach. Learn. Res. **2**, 445–498 (2002)
3. Chickering, D.M., Heckerman, D., Meek, C.: Large-sample learning of Bayesian networks is NP-hard. J. Mach. Learn. Res. **5**, 1287–1330 (2004)
4. Colombo, D., Maathuis, M.H.: Order-independent constraint-based causal structure learning. J. Mach. Learn. Res. **15**(1), 3741–3782 (2014)
5. Colombo, D., Maathuis, M.H., Kalisch, M., Richardson, T.S.: Learning high-dimensional DAGs with latent and selection variables. In: Proceedings of the Twenty-Seventh Conference on Uncertainty in Artificial Intelligence, UAI 2011, p. 850. AUAI Press, Arlington (2011)
6. Endo, T.: Realizing out-of-core stencil computations using multi-tier memory hierarchy on GPGPU clusters. In: 2016 IEEE International Conference on Cluster Computing (CLUSTER), pp. 21–29, September 2016
7. Heckerman, D., Geiger, D., Chickering, D.M.: Learning Bayesian networks: the combination of knowledge and statistical data. Mach. Learn. **20**(3), 197–243 (1995)

8. Kabir, K., Haidar, A., Tomov, S., Bouteiller, A., Dongarra, J.: A framework for out of memory SVD algorithms. In: Kunkel, J.M., Yokota, R., Balaji, P., Keyes, D. (eds.) ISC 2017. LNCS, vol. 10266, pp. 158–178. Springer, Cham (2017). https://doi.org/10.1007/978-3-319-58667-0_9

9. Kalisch, M., Bühlmann, P.: Estimating high-dimensional directed acyclic graphs with the PC-algorithm. J. Mach. Learn. Res. **8**, 613–636 (2007)

10. Kalisch, M., Mächler, M., Colombo, D., Maathuis, M.H., Bühlmann, P.: Causal inference using graphical models with the R package pcalg. J. Stat. Softw. **47**(11), 1–26 (2012)

11. Landaverde, R., Zhang, T., Coskun, A.K., Herbordt, M.: An investigation of unified memory access performance in CUDA. In: 2014 IEEE High Performance Extreme Computing Conference (HPEC), pp. 1–6, September 2014

12. Lauritzen, S.L.: Graphical Models, vol. 17. Clarendon Press (1996)

13. Le, T., Hoang, T., Li, J., Liu, L., Liu, H., Hu, S.: A fast PC algorithm for high dimensional causal discovery with multi-core PCs. IEEE/ACM Trans. Comput. Biol. Bioinform. (2015)

14. Lehmann, E.L., Romano, J.P.: Testing Statistical Hypotheses. Springer, Heidelberg (2006). https://doi.org/10.1007/0-387-27605-X

15. Madsen, A.L., Jensen, F., Salmerón, A., Langseth, H., Nielsen, T.D.: A parallel algorithm for Bayesian network structure learning from large data sets. Know.-Based Syst. **117**(C), 46–55 (2017)

16. Markthub, P., Belviranli, M.E., Lee, S., Vetter, J.S., Matsuoka, S.: DRAGON: breaking GPU memory capacity limits with direct NVM access. In: Proceedings of the International Conference for High Performance Computing, Networking, Storage, and Analysis, pp. 32:1–32:13. IEEE Press, Piscataway (2018)

17. Pearl, J.: Causality: Models, Reasoning and Inference, 2nd edn. Cambridge University Press, New York (2009)

18. Perscheid, C., Grasnick, B., Uflacker, M.: Integrative gene selection on gene expression data: providing biological context to traditional approaches. J. Integr. Bioinform. (2018)

19. Rau, A., Jaffrézic, F., Nuel, G.: Joint estimation of causal effects from observational and intervention gene expression data. BMC Syst. Biol. **7**(1), 111 (2013)

20. Richardson, T.: A discovery algorithm for directed cyclic graphs. In: Proceedings of the Twelfth International Conference on Uncertainty in Artificial Intelligence, pp. 454–461. Morgan Kaufmann Publishers Inc., San Francisco (1996)

21. Sakharnykh, N.: Beyond GPU Memory Limits with Unified Memory on Pascal, December 2016. https://devblogs.nvidia.com/beyond-gpu-memory-limits-unified-memory-pascal/

22. Schmidt, C., Huegle, J., Bode, P., Uflacker, M.: Load-balanced parallel constraint-based causal structure learning on multi-core systems for high-dimensional data. In: Proceedings of Machine Learning Research, vol. 104, pp. 59–77. PMLR, Anchorage, 05 August 2019

23. Schmidt, C., Huegle, J., Uflacker, M.: Order-independent constraint-based causal structure learning for gaussian distribution models using GPUs. In: Proceedings of the 30th International Conference on Scientific and Statistical Database Management, SSDBM 2018, pp. 19:1–19:10. ACM, New York (2018)

24. Scutari, M.: Bayesian network constraint-based structure learning algorithms: parallel and optimized implementations in the bnlearn R package. J. Stat. Softw. **77**(2), 1–20 (2017)

25. Spirtes, P.: Introduction to causal inference. J. Mach. Learn. Res. **11**, 1643–1662 (2010)

26. Spirtes, P., Glymour, C.N., Scheines, R.: Causation, Prediction, and Search. MIT Press, Cambridge (2000)
27. Wu, J., JáJá, J.: Achieving native GPU performance for out-of-card large dense matrix multiplication. Parallel Process. Lett. **26**(2), 1–17 (2016)
28. Zare, B., Jafarinejad, F., Hashemi, M., Salehkaleybar, S.: cuPC: CUDA-based parallel PC algorithm for causal structure learning on GPU. CoRR (2018)
29. Zheng, T., Nellans, D., Zulfiqar, A., Stephenson, M., Keckler, S.W.: Towards high performance paged memory for GPUs. In: 2016 IEEE International Symposium on High Performance Computer Architecture (HPCA), pp. 345–357, March 2016

Software Systems and Programming Models

Accelerating Lattice Boltzmann Method by Fully Exposing Vectorizable Loops

Bin Qu, Song Liu$^{(\boxtimes)}$, Hailong Huang, Jiajun Yuan, Qian Wang, and Weiguo Wu$^{(\boxtimes)}$

School of Computer Science and Technology, Xi'an Jiaotong University, Xi'an, Shaanxi 710049, China
{liusong,wgwu}@mail.xjtu.edu.cn
{qbqbqb,hhl15015970612,plusss,Rebeccamango}@stu.xjtu.edu.cn

Abstract. Lattice Boltzmann Method (LBM) plays an important role in CFD applications. Accelerating LBM computation indicates the decrease of simulation costs for many industries. However, the loop-carried dependencies in LBM kernels prevent the vectorization of loops and general compilers therefore have missed many opportunities of vectorization. This paper proposes a SIMD-aware loop transformation algorithm to fully expose vectorizable loops for LBM kernels. The proposed algorithm identifies most potential vectorizable loops according to a defined dependence table. Then, it performs appropriate loop transformations and array copying techniques to legalize loop-carried dependencies and makes the identified loops automatically vectorized by compiler. Experiments carried on an Intel Xeon Gold 6140 server show that the proposed algorithm significantly raises the ratio of number of vectorized loops to number of all loops in LBM kernels. And our algorithm also achieves a better performance than an Intel C++ compiler and a polyhedral optimizer, accelerating LBM computation by 147% and 120% on average lattice update speed, respectively.

Keywords: Lattice Boltzmann Method · Auto vectorization · Performance · SIMD · Loop transformation algorithm

1 Introduction

Lattice Boltzmann Method (LBM) [8] is a numerical simulation of physical phenomena. It is one of the most important CFD methods and is widely used for single-phase/multiphase flow, kinetics of surface and kinetics of crystallization and so on. LBM simulation plays an important role in aviation, water conservancy and thermal engineering and so on. Hence, accelerating the computation of LBM leads to the decrease of computing simulation costs for many industries. Previous studies on LBM computing are mainly about simulation algorithm. Qian et al. proposed a Lattice Bhatnagar-Gross-Krook (LBGK) model [20] to reduce computation by replacing collision matrix with single relaxation time

© Springer Nature Switzerland AG 2020
S. Wen et al. (Eds.): ICA3PP 2019, LNCS 11944, pp. 107–121, 2020.
https://doi.org/10.1007/978-3-030-38991-8_8

coefficients. So far, LBGK is one of the fastest LBM models based on uniform grid scheme. As the complexity of simulation grows, the number of grids increases rapidly, which results in massive storage and temporal overhead. Therefore, later studies present many large-eddy simulation methods based on nonuniform grid scheme [16] to reduce spacial and temporal complexity.

Since the first generation of commercial vector processor—Pentium II was released in 1996 [21], vector computing becomes a necessary function of modern processors, especially at the age when AI applications are rapidly growing. Meanwhile, hardware vendors have also designed many SIMD instruction sets (also known as intrinsic instructions) for programmers to implement vector computing in their codes, such as MMX, SSE and AVX. More conveniently, major modern C/C++ and FORTRAN compilers provide auto vectorization function that automatically translates appropriate loops to SIMD instructions. Because of the natural data-level parallelizability, it is valuable to accelerate LBM by using vector computing. In terms of software running on general purpose processors, previous studies on optimization of vector computing are mainly about data organization and data layout schemes. Struct-of-Array (SoA) scheme [1, 22] and data alignment [14] are the most common techniques that are applied for SIMD optimization. However, the loop-carried dependency (a statement in one iteration of a loop depends on a statement in a different iteration of the same loop) [9, 11] roosted in LBM kernels is against the rule of parallel and vector computing, and thus it prevents auto-vectorization for the innermost loops of LBM kernels. Unfortunately, the data organization and data layout optimizations can not solve this problem.

The LBM kernels, as the hot spots of program, generally consist of multiple nested loops (space dimensional loops) within a time dimensional loop. Loop transformations are considered as effective techniques to optimize LBM codes. In recent decade, the polyhedron theories [5, 7, 12, 13, 17, 24] have been rapidly developed to boost the rise of many optimizer frameworks that guide performing efficient loop transformations. These transformations enable the parallel execution of loop codes with dependency preservation. In previous studies, the objectives of loop transformations are mainly about exploiting parallelism for outer-dimensional loops, minimizing synchronization and enhancing data locality, but vectorization for innermost loops (vectorizable loop in a nested loop) is rarely considered. Due to the phase-ordering problem, some affine schedule algorithms in optimizer frameworks even transform the innermost loop to a non-vectorizable loop, which seriously wastes the computing power of vector units and is adverse to computing performance. A few studies focus on SIMD optimization based on polyhedron model. Kong et al. proposed a polyhedral compiling framework [15] which aims for integrated data locality, multi-core parallelism and SIMD execution. In this framework, loop codes are blocked and plain codes in loop blocks are translated to SIMD codelets (code blocks). Experiments demonstrate that this framework achieves times of speedup than previous loop transformation methods. However, they do not present any clear algorithm to guide automatic SIMD codelet generation.

In this paper, we propose a novel SIMD-aware loop transformation (SLT) algorithm which identifies the plain loops that have potentiality to be vectorized and guides these loops to be transformed into vectorizable loops for LBM kernels. The identification is based on dependency analysis on a defined dependency table. This algorithm is easier to understand than the polyhedron-based loop transformation method since it does not require much knowledge of convex analysis and linear programming. The SLT algorithm has been validated for LBM kernels on an Intel Xeon Gold 6140 server which has processors with the Skylake microarchitecture and supports AVX-512 SIMD instruction set. Our algorithm can detect much more potentially vectorizable loops than Intel C++ compiler and a state-of-the-art polyhedral optimizer - PLuTo [4], and thus it can significantly accelerate the computation of LBM codes. Besides, since LBM is a kind of typical stencil computation, the proposed algorithm is also suitable for any other LBM-like numerical computations, such as FDTD, Gauss-Seidel iteration and convolutional neural network. In summary, this paper makes the following contributions.

- We propose a novel SIMD-aware algorithm which can identify all possible potential vectorizable loops that may be missed by compilers and polyhedral optimizers due to the loop-carried dependencies within LBM kernels. The identification is based on dependency analysis on a defined dependency table.
- The proposed algorithm can guides the identified loops to automatically perform loop transformations and array copying techniques to generate new LBM kernels with compiler-identifiable vectorizable loops. The algorithm has a polynomial-time solution.
- Experimental results demonstrate that the proposed algorithm can significantly raise the ratio of number of vectorized loops to number of all loops in the LBM kernels and achieves a better performance than a polyhedral optimizer PLuTo.

The rest of the paper is organized as follows. Section 2 presents related work about LBGK model and auto vectorization. Section 3 illustrates the details of our SIMD-aware loop transformation algorithm. Section 4 presents the experimental results and analysis. Finally, Sect. 5 concludes this paper and points out the future work.

2   Related Work

2.1   LBGK Model

LBGK model is one of the most popular LBM models. Previous work [18] has revisited the basic principle of LBGK. LBGK model includes $DnQb$ models where a particle is collided with b surrounding particles (including itself) in an n-dimensional space, such as D2Q9 model and D3Q19 model. Generally, in the LBGK model, the procedure of a single time step of LBM is divided into two phases:

– Collision

$$f_i^*(\boldsymbol{x}, t) = f_i(\boldsymbol{x}, t) - \frac{1}{\tau}[f_i(\boldsymbol{x}, t) - f_i^{(eq)}(\boldsymbol{x}, t)], \tag{1}$$

– Streaming (or propagation)

$$f_i(\boldsymbol{x} + \boldsymbol{e}_i \Delta t, t + \Delta t) = f_i^*(\boldsymbol{x}, t), \tag{2}$$

where i is the direction of particle velocity; $\boldsymbol{x}$ is the location of particle; t is current time step; Δt is a time slot; $\boldsymbol{e}_i$ is the particle velocity in direction i; τ is relaxation time; $f_i(\boldsymbol{x}, t)$ is the distribution function of particle in direction i; and $f_i^*(\boldsymbol{x}, t)$ is the equilibrium distribution function after collision. In D3Q19 model, $i \in [0, 18]$, the directions of particle velocity $\boldsymbol{e}_i$ are from $\boldsymbol{e_0}$ to $\boldsymbol{e_{18}}$ (as shown in Fig. 1). In Eq. (1), the equilibrium distribution function $f_i^{(eq)}(\boldsymbol{x}, t)$ is defined as

$$f_i^{(eq)}(\boldsymbol{x}, t) = w_i \rho [1 + \frac{3e_i u}{c^2} + \frac{9(e_i u)^2}{wc^4} - \frac{3u^2}{2c^2}], \tag{3}$$

where ρ denotes fluid density, u denotes fluid velocity, and c denotes lattice speed. And in D3Q19 model, the weighing factor w_i is defined as

$$w_i = \begin{cases} \frac{1}{3}, \ i = 0 \\ \frac{1}{18}, \ i = 2, 4, 6, 8, 9, 14 \\ \frac{1}{36}, \ i = 1, 3, 5, 7, 10, 11, 12, 13, 15, 16, 17, 18. \end{cases} \tag{4}$$

Obviously, the computation of LBGK model is suitable for parallel execution. However, in LBGK model, a lattice references its neighboring lattices, which indicates loop-carried dependencies that prevent auto vectorization for loop codes. In this paper, we use the D3Q19 LBGK model to illustrate and validate our algorithm for LBM codes.

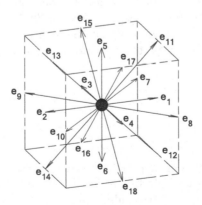

Fig. 1. Particle velocity of D3Q19 model.

2.2 Auto Vectorization

Auto vectorization is a commonly-used technique embedded in most modern compilers. When this technique is activated in compilation, it scans each loop in LBM kernels and tries to translate vectorizable loops into SIMD instructions. Since not all kinds of loops are vectorizable, a compiler that supports auto vectorization must identify vectorizable loops based on certain features. If a loop can be automatically vectorized, it must satisfy the following conditions [2, 15, 25]

1. **Countable**: the loop trip count must remain constant in the duration of the loop.
2. **Single entry & exit**: the loop must be a perfect loop with only one entry and one exit.
3. **Containing straight-line code**: branches are not permitted, but the $if\{...\}$ statement is permitted as long as it can be implemented as masked assignments.
4. **Innermost loop of a nest**: the loop must be an innermost loop of a nest or a one-dimensional loop.
5. **Without function calls**: no function calls in loop body.

In addition, if a loop contains any loop-carried dependencies, parallel execution of multiple iterations may lead to error results. Hence, vectorizable loop does not allow the existence of loop-carried dependency. Most previous loop transformation methods are effective to both preserve dependencies and implement parallel execution by exploiting pipeline/wavefront parallelism. However, the transformed innermost loops are usually uncountable and are with multiple entries & exits and branches. Therefore, they can not be vectorized by compilers.

The expected performance of vectorization should also be considered, and it helps to decide whether vectorization is profitable. Inappropriate data organization, such as Array-of-Struct (AoS) and unaligned data, may lead to unsatisfied performance of vectorization. Trifunovic et al. have proposed a vectorization cost model to estimate speedup [23]. The cost model computes the expected speedup by comparing the total execution time of vectorized loops with the execution time of scalar loops. The cost model is also applied in modern compilers. If the expected speedup of a vectorizable loop is not greater than one, the loop will not be vectorized by compiler.

3 SIMD-aware Loop Transformation (SLT) Algorithm

For simplicity, the loop referred is related to the innermost loop of a nested loop and the dependency referred is related to the loop-carried dependency of innermost loop without specification. All pseudo codes are written in C-language-like format.

3.1   Representation of Loop Domain

We use a triple (l, i, s) to represent the domain of each loop of innermost loops, where l denotes the sequence number of one loop of innermost loops, i denotes the iteration variable of current loop and s denotes the sequence number of a statement in innermost loops. For instance, Fig. 2 shows an example of innermost loops. There are 2 innermost loops and 3 statements in the example. Given that $i = 2$, then triple $(1, 2, 1)$ denotes statement $S1$ in innermost loop $L1$ and triple $(2, 2, 3)$ denotes statement $S3$ in innermost loop $L2$.

```
for ...
    for ...
    {
        L1: for(i=1;i<=4;i++)
        {
            S1: B[i]=A[i];
            S2: A[i]=A[i+1];
        }
        L2: for(i=1;i<=4;i++)
            S3: C[i]=A[i];
    }
```

Fig. 2. Example of innermost loops.

Let $\boldsymbol{x}_t(l_t, i_t, s_t)$ and $\boldsymbol{x}_s(l_s, i_s, s_s)$ respectively denote the target and source iteration instances of a loop-carried dependency among innermost loops, the dependence vector $\boldsymbol{d}$ of the loop-carried dependency is defined as

$$\boldsymbol{d} \equiv \boldsymbol{x}_t - \boldsymbol{x}_s = (\Delta l, \Delta i, \Delta s). \tag{5}$$

The loop-carried dependencies are detected based on Bernstein Conditions [6].

Theorem 1 (Bernstein Conditions). *Given two references, there exists a dependency between them if the three following conditions hold*

1. *they reference the same array (cell);*
2. *one of this access is a write;*
3. *the two associated statements are executed.*

According to Bernstein Conditions, we can find out all dependence vectors of loop-carried dependencies and construct a dependence table with these vectors. The dependence table is used for identification of vectorizable loops and guidance of loop transformations. For instance, Table 1 shows the dependence table D of the example of Fig. 2.

3.2   Identification of Vectorizable Loops

L.N Pouchet has pointed out that no dependence between points of a hyperplane indicates parallelism on the hyperplane [19]. In Fig. 2, loop $L1$ can not be vectorized, while loop $L2$ is vectorizable. Iterations of loop $L1$ can not be parallel

Table 1. Dependence table D of example loops.

Dependency	Type	$x_s(l_s, i_s, s_s)$	$x_t(l_t, i_t, s_t)$	$d(\Delta l, \Delta i, \Delta s)$
d_1: S1(RHS)→S2(LHS)	anti	$(1, i, 1)$	$(1, i+1, 2)$	$(0, 1, 1)$
d_2: S2(RHS)→S2(LHS)	anti	$(1, i, 2)$	$(1, i+1, 2)$	$(0, 1, 0)$
d_3: S2(LHS)→S3(RHS)	flow	$(1, i, 2)$	$(2, i, 3)$	$(1, 0, 1)$

executed due to the dependencies d_1 and d_2 (as shown in Table 1) between points of hyperplane $[1, 0, 0] \cdot [l, i, s]^T = 1$. Based on Pouchet's work, the constrains of vectorizable loop can be summed up as Lemma 1.

Lemma 1 (Constrains of Vectorizable Loop). *The lth innermost loop is vectorizable if $\forall d(0, \Delta i, \Delta s) \in D(l_s = l, l_t = l)$:*

$$\Delta i = 0.$$

A similar goal is to find dependency that prevents vectorization (illegal dependency) or not (legal dependency). The features of these two kinds of dependencies can be summed up as Lemmas 2 and 3.

Lemma 2 (Illegal Dependency). *A dependency $\forall d(\Delta l, \Delta i, \Delta s) \in D$ prevents vectorization if*

$$\Delta l = 0 \quad and \quad \Delta i \neq 0.$$

Lemma 3 (Legal Dependency). *A dependency $\forall d(\Delta l, \Delta i, \Delta s) \in D$ does not prevent vectorization if*

$$\Delta l \neq 0 \quad or \quad \Delta i = 0.$$

Based on Lemma 1, we can identify vectorizable loops. For potentially vectorizable loops, we can find out the illegal dependencies that prevent vectorization. Then we try to transform illegal dependencies to legal dependencies by modifying Δl.

3.3   Loop Transformations for Vectorization

Statement Rearrangement. Statement rearrangement is a considerable way to transform a potentially vectorizable loop to a vectorizable one. It is able to modify Δl of illegal dependency and does not incur uncountable loop trip, multiple entries & exits and branches which dose not satisfy the requirements of auto vectorization. For instance, to legalize dependency d_1 in Table 1, we can perform statement rearrangement on loop $L1$ and move statement $S2$ to $L2$. Hence, d_1 is legalized to $(1, 1, 1)$ and loop $L1$ becomes vectorizable. The example codes after the rearrangement are shown in Fig. 3. If statements in the last loop need to be rearranged, we create a new loop and move statements to the new loop.

```
for ...
    for ...
    {
        L1: for(i=1;i<=4;i++)
            S1: B[i]=A[i];
        L2: for(i=1;i<=4;i++)
        {
            S2: A[i]=A[i+1];
            S3: C[i]=A[i];
        }
    }
```

Fig. 3. Example loops after rearrangement.

Array Copying. Statement rearrangement is unable to legalize the dependency whose source statement and target statement are the same ($\Delta s = 0$). To solve this problem, we need to separate write operation and read operations on the same array into different statements. Therefore, we should perform array copying [10] before performing statement rearrangement to ensure more dependencies can be legalized.

We use a buffer array to replace the left side of a statement and create a following statement that reads data from the buffer array and writes it to original array. For instance, the example loops after performing array copying on statement $S2$ are shown in Fig. 4. The original statement $S2$ is divided into statements $S2$ and $S3$. Dependency d_2 becomes $(0, 1, 1)$ and it can be legalized by further performing statement rearrangement.

```
for ...
    for ...
    {
        L1: for(i=1;i<=4;i++)
        {
            S1: B[i]=A[i];
            S2: buf[i]=A[i+1];
            S3: A[i]=buf[i];
        }
        L2: for(i=1;i<=4;i++)
            S4: C[i]=A[i];
    }
```

Fig. 4. Example loops after array copying.

However, to preserve dependency, array copying should not be performed on the statement whose read operation depends on its written value in previous iteration. The dependence vector of the corresponding dependency is like

$$\mathbf{d}(0, \Delta i, 0), \Delta i < 0.$$

The expected speedup of array copying should also be considered, since array copying doubles the complexity of computation to gain profit from vectorization.

The expected speedup of array copying is roughly calculated by

$$speedup_{array\_copying} = \frac{speedup_{vectorization}}{2} \tag{6}$$

where the $speedup_{vectorization}$ is estimated by compiler. If $speedup_{array\_copying} \leq$ 1, array copying will be not launched.

Algorithm Guiding Loop Transformations. The procedure of performing loop transformations can be divided into two phases—array copying and statement rearrangement. The detail of the procedure is shown in Algorithm 1.

Algorithm 1: SIMD-aware Loop Transformation

Input:
Plain LBM kernel codes
Output:
Transformed codes

1 Scan input codes and generate dependence table D
2 **while** $\exists d(0, \Delta i, 0) \in D$ **and** $\Delta i > 0$ **do**
3   Try to find the first dependency like $d(0, \Delta i, 0) \in D, \Delta i > 0$ and its source/target statement S
4   **if** *such S is found* **then**
5   **if** $speedup_{array\_copying}(S) > 1$ **then**
6   /* Perform array copying */
7   Use a buffer array to replace the written array of S and create a new statement where the buffer array is read next to S
8   Update D
9   **end**
10   **end**
11 **end**
12 **while** $\exists d(0, \Delta i, \Delta s) \in D$ **and** $\Delta i \neq 0$ **and** $\Delta s \neq 0$ **do**
13   Try to find the first dependency like $d(0, \Delta i, \Delta s) \in D, \Delta i \neq 0, \Delta s \neq 0$ and its target statement S_t
14   **if** *such S_t is found* **then**
15   **if** $speedup_{vectorization}(S_t) > 1$ **then**
16   /* Perform statement rearrangement */
17   Move S_t to the next loop (if the next loop does not exist, create a new one)
18   Update D
19   **end**
20   **end**
21 **end**

The input of the algorithm is plain LBM kernel codes and the output is transformed codes. The algorithm is able to legalize most dependencies except dependencies with $d(0, \Delta i, 0)$ where $\Delta i < 0$. Since many loops with illegal dependencies will not be analyzed and further transformed for vectorization by most general compilers, a variety of vectorization opportunities have been missed. Whereas, our algorithm can effectively avoid this case and generate more vectorizable loops. Therefore, we achieve the acceleration of LBM computing by taking full use of vector units on processors. In addition, Algorithm 1 is not only suitable for optimizing LBM, but also suitable for other LBM-like computation.

Algorithm 1 is a polynomia-time solution. Given that there are n statements in a plain code, the temporal complexity of generating a dependence table D, scanning each d in D, updating D, performing array copying and performing statement rearrangement are $T_1 = O(n)$, $T_2 = O(n)$, $T_3 = O(1)$, $T_4 = O(n)$ and $T_5 = O(n)$, respectively. Hence, the total temporal complexity of Algorithm 1 is

$$T = max(T_1, T_2, T_3 \times T_4, T_3 \times T_5) = O(n)$$

which indicates that the algorithm is a polynomial-time solution for SIMD-aware loop transformation problem.

4   Experiments

4.1   Experimental Setup

We carried out experiments on an Intel(R) Xeon(R) Gold 6140 server. The test server has 36 cores with Skylake microarchitecture and it supports the most advanced SIMD instruction set AVX-512. The LBM benchmark is openLBM-flow [3] based on LBGK D3Q19 model. We divided the benchmark into 4 different versions: baseline-no-vec (baseline benchmark without auto-vectorization compiler option), baseline-vec (baseline benchmark with auto-vectorization compiler option), Pluto (codes generated by PLuTo and with auto-vectorization compiler option) and SLT (codes transformed by our SLT algorithm with auto-vectorization compiler option). The compiler is Intel C++ Compiler version 19.1.

4.2   Comparison of Ratio of Vectorized Loops

The vectorization reports generated by compiler provide detailed information about the number of loops and the number of auto-vectorized loops by compiler for four tested versions of LBM kernel codes. Based on the information, we calculated the ratio of number of vectorized loops to number of all loops (RVL). Table 2 shows the vectorization information of each version of benchmark.

One-quarter of innermost loops in baseline version of codes is vectorized. That is, even with no optimization, the RVL of baseline-vec reaches 25%. The PLuTo optimizer performs pipeline/wavefront parallelism to legalize loop-carried dependencies, and the RVL of Pluto reaches 50%. Since the polyhedron-based transformations incur uncountable, multiple entries & exits and branched loops, half innermost loops of Pluto can not be vectorized by compiler. As our SLT algorithm created new loops to legalize loop-carried dependencies, the number

Table 2. Vectorization information

Benchmark	# of loops	# of vectorized loops	RVL
baseline-no-vec	4	0	0%
baseline-vec	4	1	25%
Pluto	4	2	50%
SLT	53	52	98%

of loops has been increased. However, the majority of transformed loops are vectorized due to the advantages of our algorithm which maintains the new loops satisfying the auto-vectorization conditions. After all, 98% of loops in the version of SLT codes are automatically vectorized by compiler.

Theoretically, the update speed of LBM kernels is positively correlated with the value of RVL. Therefore, the descending order of four versions of benchmarks sorted by update speed should be: SLT > Pluto > baseline-vec > baseline-no-vec.

4.3   Performance Comparison

We tested the update speed of million lattice updates per second (MLUPS) and the speedup of four versions of LBM codes to evaluate the performance of the proposed SLT algorithm. Different grid sizes and numbers of threads are used in our tests. The grid sizes are $64 \times 64 \times 64$, $128 \times 128 \times 128$, $192 \times 192 \times 192$, $256 \times 256 \times 256$ and $320 \times 320 \times 320$, respectively. And the numbers of threads are 8, 16, 24 and 32, respectively. The vectorization of loops was automatically realized by compiler with AVX-512.

Figure 5 shows the results of update speed comparison. It is clear that the SLT achieves the fastest update speed for all kinds of test codes in Fig. 5. The maximum update speed is about 56 MLUPS with grid size of $64 \times 64 \times 64$ and 16 threads in our tests. And the average update speed of SLT is about 36 MLUPS for all test codes, which is 174%, 147% and 120% faster than the average update speed of baseline-no-vec, baseline-vec and Pluto, respectively. According to the RVL values in Sect. 4.2, the results also indicate that the update speed of LBM codes is positively correlated with the RVL values. We can also observe that the update speed decreases along with the growth of grid sizes in Fig. 5. This is because we do not perform any other optimizations on tested codes except automatic vectorization of compiler. And the limitation of memory bandwidth, i.e. the "memory wall" issue, has arisen when grid sizes are growing. However, the solution to the memory optimization is beyond the scope of this paper.

The speedups are calculated by respectively comparing the execution time of baseline-vec, Pluto and SLT with the execution time of baseline-no-vec. Figure 6 shows the results of speedup comparison. The results are similar to the update speed.

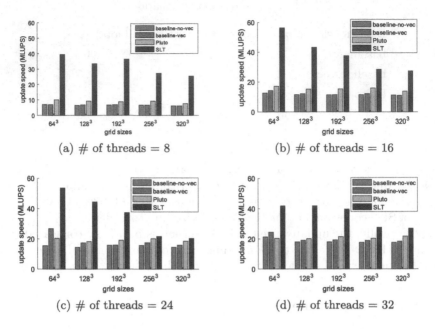

Fig. 5. Update speeds comparison.

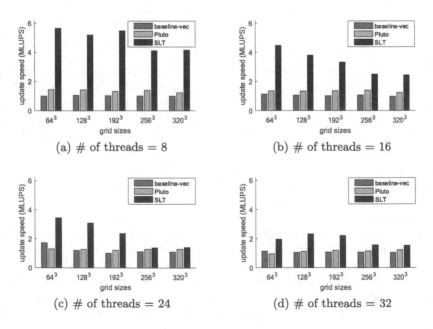

Fig. 6. Speedups comparison.

The speedups of SLT are much higher than the speedups of baseline-vec and Pluto for all tested codes. SLT achieves an average speedup of 3.1 in our tests, which is 186% and 145% higher than the average speedups of baseline-vec and Pluto, respectively. The downtrend of speedups is also observed in Fig. 6 when grid sizes and the number of threads increase. As explained before, the memory bandwidth becomes the major performance bottleneck and the performance profit from vectorization decreases. Besides, when the number of threads grows, the contention of memory bandwidth exacerbates the problem and leads to a further decline of speedup. However, despite of these factors, the experimental results can demonstrate that our SLT algorithm is effective to accelerate LBM computation by fully exposing vectorizable loops.

5   Conclusion

In this paper, we propose a SIMD-aware loop transformation algorithm to accelerate the computation of LBM codes by making much of vectorization. The proposed SLT algorithm is able to identify most potential vectorizable loops that are ignored by general compilers based on a defined dependence table. And it also provides a solution to transform these loops into automatically identifiable vectorizable loops by compilers, which performs array copying and statement rearrangement for LBM kernels. Compared with polyhedron-based loop transformation techniques, SLT algorithm maintains the conditions of vectorization for loops. Experimental results show that SLT algorithm can significantly raise the ratio of number of vectorized loops to number of all loops for LBM kernels. And our algorithm gets better performance than a polyhedral optimizer and the Intel C++ compiler in vectorization. It also indicates that the proposed algorithm should be effective in the acceleration of other LBM-like computations.

In future work, we will combine loop tiling techniques and SLT algorithm to further enhance data locality and reduce the adverse impact of memory bandwidth. We will also apply SLT algorithm to other LBM-like applications to mine the potential of vectorization power on modern processors.

Acknowledgement. This work was supported in part by the National Key Research and Development Program of China under Grant No. 2016YFB0201800, the National Natural Science Foundation of China under Grant No. 91630206 and 61672423.

References

1. AOS and soa. https://en.wikipedia.org/wiki/AOS_and_SOA. Accessed 1 Apr 2019
2. Intel® c++ compiler 19.0 developer guide and reference. https://software.intel.com/en-us/cpp-compiler-developer-guide-and-reference-vectorization-and-loops. Accessed 6 June 2019
3. openlbmflow. https://sourceforge.net/projects/lbmflow. Accessed 15 June 2019
4. Pluto - an automatic parallelizer and locality optimizer for affine loop nests. http://pluto-compiler.sourceforge.net. Accessed 7 June 2019

5. Acharya, A., Bondhugula, U.: PLUTO+: near-complete modeling of affine transformations for parallelism and locality. In: Proceedings of the 20th ACM SIGPLAN Symposium on Principles and Practice of Parallel Programming, PPoPP 2015, San Francisco, CA, USA, 7–11 February, 2015, pp. 54–64 (2015)
6. Bernstein, A.J.: Analysis of programs for parallel processing. IEEE Trans. Electron. Comput. **5**, 757–763 (1966)
7. Bondhugula, U., Hartono, A., Ramanujam, J., Sadayappan, P.: A practical automatic polyhedral program optimization system. In: ACM SIGPLAN Conference on Programming Language Design and Implementation (PLDI), June 2008
8. Chen, S., Doolen, G.D.: Lattice boltzmann method for fluid flows. Ann. Rev. Fluid Mechan. **30**(1), 329–364 (1998)
9. Devan, P.S., Kamat, R.: A review-loop dependence analysis for parallelizing compiler. Int. J. Comput. Sci. Inf. Technol. **5**(3), 4038–4046 (2014)
10. Di, P., Ye, D., Su, Y., Sui, Y., Xue, J.: Automatic parallelization of tiled loop nests with enhanced fine-grained parallelism on gpus. In: 2012 41st International Conference on Parallel Processing, pp. 350–359. IEEE (2012)
11. Du, X., et al.: Comparative study of distributed deep learning tools on supercomputers. In: Vaidya, J., Li, J. (eds.) ICA3PP 2018. LNCS, vol. 11334, pp. 122–137. Springer, Cham (2018). https://doi.org/10.1007/978-3-030-05051-1_9
12. Feautrier, P.: Some efficient solutions to the affine scheduling problem. i. one-dimensional time. Int. J. Parallel Program. **21**(5), 313–347 (1992)
13. Feautrier, P.: Some efficient solutions to the affine scheduling problem. part ii. multidimensional time. Int. J. Parallel Program. **21**(6), 389–420 (1992)
14. Feng, Y., Tang, J., Wang, C., Xie, J.: CuAPSS: a hybrid CUDA solution for all pairs similarity search. In: Vaidya, J., Li, J. (eds.) ICA3PP 2018. LNCS, vol. 11334, pp. 421–436. Springer, Cham (2018). https://doi.org/10.1007/978-3-030-05051-1_29
15. Kong, M., Veras, R., Stock, K., Franchetti, F., Pouchet, L., Sadayappan, P.: When polyhedral transformations meet SIMD code generation. In: ACM SIGPLAN Conference on Programming Language Design and Implementation, PLDI 2013, Seattle, WA, USA, 16–19 June, 2013, pp. 127–138 (2013)
16. Krafczyk, M., Tölke, J., Luo, L.S.: Large-eddy simulations with a multiple-relaxation-time lbe model. Int. J. Modern Phys. B **17**(01n02), 33–39 (2003)
17. Lim, A.W., Lam, M.S.: Maximizing parallelism and minimizing synchronization with affine transforms. In: Proceedings of the 24th ACM SIGPLAN-SIGACT Symposium on Principles of Programming Languages, pp. 201–214. ACM (1997)
18. Liu, S., Zou, N., Cui, Y., Wu, W.: Accelerating the parallelization of lattice boltzmann method by exploiting the temporal locality. In: 2017 IEEE International Symposium on Parallel and Distributed Processing with Applications and 2017 IEEE International Conference on Ubiquitous Computing and Communications (ISPA/IUCC), pp. 1186–1193. IEEE (2017)
19. Pouchet, L.N.: Interative optimization in the polyhedral model. Ph.D. thesis, University of Paris-Sud 11, Orsay, France, January 2010
20. Qian, Y., d'Humières, D., Lallemand, P.: Lattice BGK models for navier-stokes equation. EPL (Europhys. Lett.) **17**(6), 479 (1992)
21. Shanley, T.: Pentium Pro and Pentium II System Architecture. Addison-Wesley Professional, Boston (1998)
22. Tran, N.P., Lee, M., Choi, D.H.: Memory-efficient parallelization of 3D lattice boltzmann flow solver on a gpu. In: 2015 IEEE 22nd International Conference on High Performance Computing (HiPC), pp. 315–324. IEEE (2015)

23. Trifunovic, K., Nuzman, D., Cohen, A., Zaks, A., Rosen, I.: Polyhedral-model guided loop-nest auto-vectorization. In: 2009 18th International Conference on Parallel Architectures and Compilation Techniques, pp. 327–337. IEEE (2009)
24. Xue, J.: Loop Tiling for Parallelism, vol. 575. Springer Science & Business Media, New York (2012). https://doi.org/10.1007/978-1-4615-4337-4
25. Zhang, W., Zhang, L., Chen, Y.: Asynchronous parallel Dijkstra's algorithm on intel xeon phi processor. In: Vaidya, J., Li, J. (eds.) ICA3PP 2018. LNCS, vol. 11334, pp. 337–357. Springer, Cham (2018). https://doi.org/10.1007/978-3-030-05051-1_24

A Solution for High Availability Memory Access

Chunjing Gan[1], Bin Wang[2,3], Zhi-Jie Wang[1(✉)], Huazhong Liu[4],
Dingyu Yang[5,7], Jian Yin[1], Shiyou Qian[2], and Song Guo[6]

[1] School of Data and Computer Science, Sun Yat-Sen University, Guangzhou, China
`ganchj3@mail2.sysu.edu.cn`, {`wangzhij5,issjyin`}`@mail.sysu.edu.cn`
[2] Department of Computer Science and Engineering, Shanghai Jiao Tong University,
Shanghai, China
`binqbu2002@gmail.com`, `qshiyou@sjtu.edu.cn`
[3] Department of Computer Science, Ulster University, Coleraine, UK
[4] Huazhong University of Science and Technology, Wuhan, China
`sharpshark_ding@163.com`
[5] Alibaba Group, Hangzhou, China
`dingyu.ydy@alibaba-inc.com`
[6] Department of Computing, Hong Kong Polytechnic University, Hong Kong, China
`song.guo@polyu.edu.hk`
[7] Guangdong Key Laboratory of Big Data Analysis and Processing,
Guangzhou, China

Abstract. Nowadays, in-memory computing has plenty of applications like artificial intelligence, databases, machine learning, etc. These applications usually involve with the frequent access to memory. On the other hand, memory components typically become error-prone over time due to the increase of density and capacity. It is urgently important to develop solutions for high-availability memory access. Yet, existing solutions are either lack of flexibility, or consistently more expensive than native memory. To the end, this paper presents a solution called SC2M. It is a software-controlled, high-availability memory mirroring solution. Our solution can flexibly set the granularity of the memory areas for various levels. Furthermore, it can perform duplication of the user-defined data structures in a high-availability version. The systematic instruction-level granularity for memory duplication reduces the overheads for backup, and lowers the probability of data loss. Experiment results demonstrate the feasibility and superiorities of our solution.

Keywords: High availability · Hardware virtualization · System architecutre

1 Introduction

In modern artificial intelligence (AI) system, it is known that vicious attacks to memory modules are common [27,33]. An error-prone AI system may cause serious issues in terms of service quality and computation efficiency [15]. Besides the AI system, many other applications are also significantly depended on the

© Springer Nature Switzerland AG 2020
S. Wen et al. (Eds.): ICA3PP 2019, LNCS 11944, pp. 122–137, 2020.
https://doi.org/10.1007/978-3-030-38991-8_9

correctness of memory. For example, in recent years in-memory computing has been more and more applications such as *Redis* [23], *Memcached* [1], *Spark* [34].

The above applications and many others generate large mission-critical workloads, which need to frequently access memory or cache. Particularly, these applications have strict requirements on the correctness and stability of memory. Yet, memory components typically become error-prone over time, due to the increase of density and capacity [13]. Eventually, they may no longer guarantee *high availability* (HA). Moreover, although some errors from the *dynamic random access memory* (DRAM) can be detected by the hardware, they usually cannot be corrected instantly. These *uncorrected errors* often incur system crash [5, 18, 29]. All of the above facts have led to a sharp increase in the demand for high-availability memory access.

To provide high-availability memory access, different hardware vendors have implemented their own product-related solutions. For example, the bit-retrieval approach [14] has been widely used in industry, including the HP corporation [13], IBM corporation [7]. In these industries, a handful of motherboards have integrated the *error correcting code* (ECC) memory into their servers [14]. This approach, however, is often ineffective for the block failure. Moreover, another well-known hardware-based approach is *mirror memory* [14]. This approach uses the dual chip (i.e., double chips) to backup data on-the-fly. Although this dedicated approach is useful for many applications, it often generates too much overhead, and is consistently more expensive than native memory [30].

As for software-based solutions, a common method is by simulating hardware checking. For example, software ECC [12] works quite similarly as hardware ECC. Another useful software-based approach is to duo-backup at the application level [5, 18]. This can be witnessed in the Google and Amazon services [27]. Although duo-backup solutions show low latency and fewer interruptions, they usually only consider the application's tolerance, which cannot provide transparent high-availability for operating systems. On the other hand, some *virtual machine* (VM)-based solutions do effectively deal with this shortcoming [30, 31]. For example, a system named Remus [5] uses virtualization checkpointing technology to backup an entire VM. However, checkpointing technology does not backup the system extemporaneously, so data between two checkpoints may be lost during failure; in addition, the overhead of such solutions like Remus is more than 100%, compared to the native memory [5].

Instead of directly repairing above approaches, this paper presents a solution that is a software-controlled memory mirroring (known as SC2M), based on the principles of static binary translation and hardware virtualization. Here static binary translation technology enables our solution to provide a software-controlled high-availability, while virtualization technology uses software/firmware that divides physical hardware equipments into multiple independent virtual instances, it enables our solution to support multiple VMs on the same physical machine. In brief, SC2M explores a redundant memory space (called the *mirror space*) in the physical host machine; it injects instructions into the mirror space to backup the multi-level memory writes, where the *static binary translation* is used. Particularly, to backup data, it implements mirror instructions not only for user mode codes, but also for kernel mode codes. Therefore, when errors happen in the original memory space, the compromised data can be recovered from the mirror space, where only low-cost memory is required to recover data. Our solution allows the application to specify the data structure to be duplicated, and it is flexible since it can support memory mirroring at different levels (ranging from data level, application level, to system level), and the mirror memory can be easily set to support N-modular redundancy for some specialized, business-critical applications. In addition, it is a lightweight and real-time mirroring solution, compared against traditional methods that use memory-mapped files [27]. Particularly, our solution is implemented by integrating the Intel *shadow page table* (SPT) [35]; this provides even more flexibility in distributed environments or other similar environments, where the *operating system* (OS) level access cannot be perceived by users. To summarize, the main contributions of this paper are: (i) We develop a software-controlled memory mirroring solution, called SC2M, that can achieve high availability, and is cost-effective. (ii) We conduct extensive experiments to evaluate its performance. The experimental results demonstrate the feasibility and effectiveness of our solution.

In next section, we review the related work. Section 3 presents the system architecture of SC2M. The workflow and implementation of our solution are covered in Sect. 4. We evaluate the performance of our solution in Sect. 5. Finally, Sect. 6 concludes this paper.

2   Related Work

Existing solutions for high availability (HA) memory can be generally classified into two categories: hardware-based and software-based solutions. We next review prior works related to these two categories.

2.1   Hardware-Based HA Memory

Initially, hardware providers adopt extra bits to check and correct memory errors, e.g., the well-known techniques are like *parity checking* [3] and *error correcting codes* (ECC) [16]. Some hardware vendors also promote ECC to support their motherboard services. For example, HP Advanced ECC [13], Google ECC [6], and IBM Chipkill [7]. Besides above approache, there are also various solutions that further improve ECC technique. For example, Odd-ECC [19] is used for conventional 2D DRAMs, DIMMs, or even to 3D-stacked DRAMs. Another notable ECC-based scheme is also proposed, and it introduces the In-DRAM ECC architecture [2]. On the other hand, bit-checking method for large area failures is also investigated [28]. Overall, these methods are effective to deal with limited bit errors, yet they are not suitable for handling massive block failures. To retrieve massive block failures, some works use the mirrored hardware. For example, HP's mirrored channel [13] provides full protection against single-bit and multi-bit errors. The subsystem writes identical data to two channels simultaneously. In case of errors, the system is able to automatically recovery the data from the mirrored memory. In [14], two optimization techniques are developed: *lazy-migration* and *adaptive-activation*. The lazy-migration technique increases the utilization of the rental memory via the volatile page allocations, while the adaptive-activation technique saves the active pages in the rental memory during the reallocation. To some extent, these hardware-based solutions are usually not cost-effective, and may lack the flexibility.

2.2   Software-Based HA Memory

In the literature, there are also many software-based solutions for HA memory. For example, SWIFT [24], a software-only fault-detection technique, duplicates a program's instructions by inserting explicit validation codes, and compares the results of original instructions and their corresponding duplicates. Later, CRAFT [25] adopts the extra hardware structures to improve the SWIFT technique. Compared to these methods, our solution mainly duplicates the memory write instructions, avoiding the extra overhead. Another typical software-based solution is the dual-machine VM replication [18]. As for this method, a backup server is synchronized to the host machine. Besides the methods mentioned above, there are also a lot of works that use VM migration and/or replication [5,10]. For example, Remus [5], in which the state of the primary VM is frequently recorded and transmitted to the backup server during execution. Remus achieves the high availability memory, yet the compile time of the Linux kernel is doubled. In order to improve the performance and scalability of the Remus, the ReNIC system provides an architectural extension to the Single Root I/O Virtualization (SR-IOV) system that achieves efficient I/O replications [10]. This method requires some new hardware-assisted I/O virtualization (such as SR-IOV). Moreover, a system called Memvisor [9,22] is proposed, which uses the direct page table (DPT) technique and is tailored for HA memory in cloud environments. Later, a system called kMemvisor [30] is also developed for HA memory in cloud environments and uses also DPT technique. However, this architecture requires the large modifications on the guest OS, while our solution does not need to perform those complicated modifications. In addition, our solution employs the shadow page table, which provides much more flexibility, and so it could be used for more environments besides cloud environments.

2.3   Others

Besides the above works, we also note that there are some works (e.g., *memory error prediction* and *algorithm-based recovery*) that could be complementary to our solution. For example, the authors in [29] propose a scalable and fault tolerant HPL, called SKT-HPL, and validate their method on two large-scale systems. Moreover, a system called Jenga [20] is proposed for protecting 3D DRAM, specifically high bandwidth memory (HBM), from failures in bits, rows, and blocks in the memory. On the other hand, memory error prediction is also benefit to improve the high availability of memory. For example, the work [17] introduces the cache persistence analysis into memory backup for self-powered non-volatile processors. One can integrate this prediction technique to improve the reliability of memory.

3   System Architecture

Figure 1 shows the architecture of sc2m. At the bottom of our architecture, two kinds of memories are deployed: (i) a native memory, and (ii) a mirror memory. The former is mainly used to achieve the high available VMs that can run mission-critical applications; here the VMs can be configured at the data-level, application-level or even to system-level, according to the mirror requirements. Correspondingly, the latter is mainly used to backup the native VM. Generally, sc2m is implemented with a *copy-on-write* manner. That is, whenever a write operation occurs in the native memory, the same write operation should be done in the mirror memory. At the top of our architecture, there are two major modules, which are *memory management module* and *code translation management module*. These two modules shall interact with the components at the middle level of our architecture. Next, we discuss more details about these two modules.

▶ *Memory Management Module.* It monitors the operations related to page table, and maintains both native and mirror *page table entries* (PTEs). Same to the memory management of modern OS (i.e., operating system), here page table is used to map virtual addresses into physical addresses, so that address space can be extended in memory. A PTE can be created using "syscalls" such as `mmap()`, `malloc()` methods in the kernel space. The memory allocator [4] in the hypervisor maintains the relationship between the *guest physical address* and the *guest virtual address* for each VM, and intercepts all PTE operations. In our solution, we implement the virtual memory allocator via Intel SPT (i.e., shadow page table) technique [35], which is a memory virtualization technology used in full virtualization where host OS can run multiple guest page tables, and the guest OS do not need to be modified [4,35]; this provides even more flexibility in distributed environments or other similar environments, where the *operating system* (OS) level access cannot be perceived by users.

Denote by Add_{mva} and Add_{nva} the *mirror virtual address* and the *native virtual address*, respectively. The relationship between them is established by the equation: $Add_{mva} = Add_{nva} + offset$, where the *offset* is a constant value. In order to provide data and application-level mirror memory, sc2m defines interfaces by wrapping up the memory area with `getter()` and `setter()` methods. Actually, setting the offset is the most direct way to build the mirror address, since there is no need to modify the memory layout. Notice that, an appropriate value should be carefully selected for the offset, in order to avoid the address conflict and also to ensure the correct creation of the mirror address.

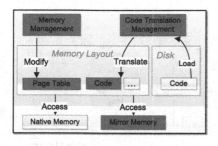

Fig. 1. Architecture of SC2M.

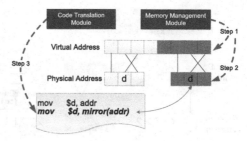

Fig. 2. Workflow of mirror initialization.

▶ *Code Translation Management Module.* It inserts mirror instructions, identifies all memory write instructions, and replicates them. As we know, in the traditional X86 architecture, an instruction "write destination" is usually translated into a virtual address. Yet, injecting mirror instructions is complicated when some instructions do not have explicit write destinations. For example, as for the atomic and privilege operations, it is difficult to mirror them, due to the reasons above. SC2M alleviates these problems by using two ideas together: (i) translating the explicit write destinations into all write destinations using *static binary translation*; and (ii) generating mirrored instructions at the instruction compile stage. An extra benefit of the above strategy is that it also reduces the runtime overhead. Specifically, the mirrored instructions are inserted when OS loads the program from the disk (see Fig. 1 again).

4   Workflow and Implementation of SC2M

There are several important issues needing to be clarified in the implementation. They are: (i) the creation of mirror page table; (ii) the mapping from physical addresses to virtual addresses; (iii) data synchronization; and (iv) mirroring data for high-availability applications. Before we discuss these issue in detail, we first introduce the workflows of mirror initialization and data recovery, which could be helpful to understand the rest of the paper.

4.1   Mirror Initialization and Data Recovery

As for the mirror initialization, SC2M shall do the following steps (see Fig. 2).

Step 1 — Reserve physical memory. When a VM (i.e., virtual machine) starts up, SC2M checks its configuration. When the VM is configured as HA-type, both native and mirror memory are to be created for this VM, and the sizes of the two memories are the same. In other words, in this step a block of physical memory will be reserved when the VM starts up.

Step 2 — Create mirror page table. SC2M intercepts the operations related to page table from the VM, and simultaneously creates mirror page tables. That is, when native PTEs are updated, related mirror PTEs are created as well. In our

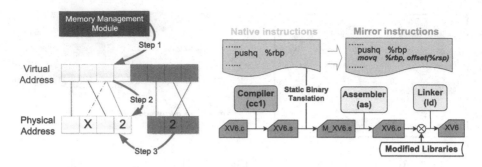

Fig. 3. Workflow of data recovery.   **Fig. 4.** Data synchronization of SC2M

solution, Intel SPT technique is used, which provides us the *application program interfaces* (APIs) to implement this step.

Step 3 — Write redundant data. The mirror write instruction is replicated through the *static binary translation*, and the redundant data is written using the mirror instruction. In other words, native instructions and mirror instructions will write the same data to different addresses.

As for data recovery, it is mainly for recovering the corrupted data based on the data in the mirror space. Specifically, when a memory failure occurs, the hardware detection module (e.g., parity checking or ECC) notifies SC2M by invoking a *machine check exception*. Unlike the normal response of restarting the OS (i.e., operating system), SC2M quickly and effectively retrieves the corrupted data through the following steps (see Fig. 3).

Step 1 — Allocate a new page. In the native memory, the memory module allocates a new page from the free zone, blanks the options of the new page, and marks it with "writable"; here the new page will be allocated to the VM.

Step 2 — Remap the new page. The corrupted PTE is rewritten and mapped to the new page (allocated in Step 1). Here the virtual address will be mapped to the new page.

Step 3 — Recovery corrupted data. Data is copied from the *mirror virtual address* to the *native virtual address*. After that, the program continues to execute.

4.2   Creation of Mirror Page Table

As mentioned earlier, we employ the SPT (i.e., shadow page table) technique when creating *mirror page table*. In general, SPT is a technique that maintains the real mapping in the hardware when executing the guest instructions. Here the SPT module is invoked when the guest OS modifies the CR3 register, which is one of control registers. It contains the base address of the page table. In our implementation, SC2M translates the native page table to a new table, and resets the CR3 such that it still points to this new page table. Meanwhile, during the translation, SC2M creates the mirror PTEs (i.e., page table entries) in the new table. The above is the general implementation of creating the mirror page table.

We may need to mention that, in the full virtualization mode of our SC2M, the CPU shall conduct a permission check when an instruction is executed. Specifically, if the check reports an illegal operation, the hypervisor executes a predefined procedure to handle the problem. On the other hand, if a sensitive instruction is not privileged, then a guest OS may obtain (or even modify) resources belonging to the host (or other VMs), yet it does not inform the hypervisor. An immediate example is the mov() instruction, which is not privileged in X86-64, and so the hypervisor cannot trap any guest PTE operation from mov() instruction. Compared to the instructions like move(), updating the CR3 is a privileged instruction that can be used to track the operation of the guest PTEs. In our implementation, the CR3 operations are intercepted by the SC2M. Naturally, our SC2M can easily read the PTE of the guest OS. This way, a new page can be translated to a new space in the SPT.

Additionally, we observe that OS may not always use the new information changed from the translation. This implies that, during the SPT implementation, there may not need to do a complete translation for every change in the CR3. To improve the performance of SPT, one can set the P (i.e., present) bit to 0 in SPT entries. This way, when a guest OS accesses this address, it causes a page fault due to the nature of the x86 architecture. This page fault can be also captured by SC2M. In this case, SC2M shall find the original value in the CR3, according to the SPT. Then, it determines whether the page fault is caused during the translation. If so, SC2M translates the PTE (i.e., page table entry) to finish the exception handling. Otherwise, it indicates that the page fault is produced by the guest OS, and so SC2M forwards it to the guest OS that shall handle it.

4.3 P2V Mapping

In the virtualization mode, it is necessary to build a mapping between physical addresses and virtual addresses. In our implementation, we rebuild the "vm_struct", which is a special data structure that stores the mapping between the virtual addresses and the physical addresses [29,31]. Our modification is mainly on the layout of guest memory. In our implementation, each virtual memory area has its own mirrored area. The native and the mirrored virtual areas are mapped to different physical memories to ensure that data is replicated physically. This modification allows us to free address space more efficiently. For example, when a process is killed, the virtual address and its related physical address can be released simultaneously. On the other hand, as for the layout of host memory, we exploit the mirror area in both the kernel and user spaces, and so each part of the memory shall reside in different locations of the physical memory. This way, it allows the system to perform physical replication of the data easily. Note that, in our implementation we use the simple memory layout for both guest and host ends. Nevertheless, one can also use more complicated layouts, which are depending on the specific application requirements.

4.4 Data Synchronization

In order to guarantee the data synchronization, the *GNU compiler collection* (GCC) procedure is modified to perform the analysis of the assembly source files and the *static binary translation*, before the assembler handles them. In our implementation, SC2M calculates the mirror memory addresses to determine where to replicate the data. Then, mirror instructions are injected into the source files, which are processed by the assembler as executable files. Note that, since assembling files (created by the GCC) have target addresses represented as labels, these instructions do not cause any problems for indirect **branches** or indirect **jumps**. In addition, once the executable files are generated, they shall have the ability to write data into both native and mirror memories. This way, when the programs run, the data is replicated instantly.

Nevertheless, we may need to note that, there are two types of mode codes, which need to be considered carefully. We next address them respectively.

▶ *User Mode Code.* In the user mode, some special instructions can change the execution sequence. So, in some cases they can be incorrectly mirrored even if the native instructions are correct; this incurs the inconsistent data. For example, the method **call()** saves procedure linking information on the stack, and branches to the called procedure; in this case, if the mirror instructions are inserted right after the native ones, they will not be executed until the called procedure returns. This creates a long-time data inconsistency between the native and mirror memories. To remedy this, we employ a **copy-on-call** method to replicate the linking information, by inserting mirror instructions at the very start of the called procedure. Since a procedure may be called several times but needs to be implemented only once, the **copy-on-call** method reduces the overall number of mirror instruction calls.

▶ *Kernel Mode Code.* Codes in the kernel mode are much more complicated than codes in the user-mode programs, although some instructions can be mirrored according to the user mode codes. The complexity is mainly because we need to take into account the atomic operations and privilege changes in the kernel mode. Generally, when an interrupt occurs, the kernel stops the normal processor loop, and starts to process the execution of a new sequence, called an *interrupt handler*. When an interrupt occurs, the processor first saves several registers such as *eip* and *esp*. This is mainly for restoring, when it returns from the interrupt.

4.5 Mirroring Data for High-Availability Applications

The SC2M can achieve fine-grained replication, e.g. data-level memory mirroring. This means that, SC2M can choose a certain address area for the application rather than using the entire VM memory. This can protect the important memory areas from more overhead. In our implementation, SC2M performs data-level replication by wrapping the memory area with **getter** and **setter** methods. Generally, SC2M provides four interfaces for data-level memory mirroring:

- *create.* The *create* operation generates a memory area for mirroring.
- *get.* The *get* operation fetches the content of the mirror memory.
- *set.* The *set* operation configures the mirror memory according to the requirements of the application.
- *destroy.* The *destroy* operation deletes all the mirror PTEs and releases the mirror space.

The steps of these operations can be briefly described as follows. First, developers use the *create* command to create a memory area and then use the *get* and the *set* to access this memory area. Later, they use the *destroy* command to invalidate the memory area. To reduce the complexity of the program, SC2M also provides an approach to mirror the user-defined format of the memory area, based on the user-defined data structure. In this case, users need to declare the structure, implement the *create, get, set* and *destroy* methods based on specific requirements. SC2M wraps the four methods with binary translation so that every memory-write operation in the memory area is duplicated to the mirrored area. More importantly, SC2M can also mirror some structures and classes in certain libraries by translating libraries to a redundant version. Therefore, this technique can be used by developers to create high-availability STL *map* and *list* in their applications. In that case, the map and list are used to build parts of the infrastructure library.

5 Performance Evaluation

In this part, we first present the experimental settings (Sect. 5.1), and then present our main experimental results (Sect. 5.2), and finally discuss the limitations and summarize our findings (Sect. 5.3).

5.1 Experiment Setup

As mentioned earlier, the SC2M is developed for high availability. In our experiments, we use the gzip and gunzip software sets to conduct evaluation on the data-level high-availability. We test the *gzip* and the *gunzip* programs with eight different file sizes. They range from 100 MB to 800 MB. Following prior works [5,11], we use the execution time to measure the performance. The execution time can easily represent the overhead. The execution time for these two applications refer to the compressing/decompressing time. Similar to prior works [5,11,14], we compare our proposed solution with the original (native) implementation. As for the former, it is a double write, while the latter a single write. In our experiments we did not extensively compare with existing solutions, since most of them are hardware-based solutions, which are hard to implement in our experimental platform. As for the software-based solutions, most of them are developed for the application level, or rely on disk. Our solution relies on memory and can provide high availability for data-, application-, and system-levels. Its superiorities are obvious, since (i) it is clearly faster than them (due to the difference between

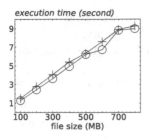
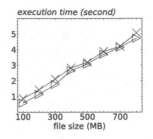
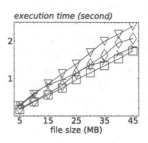

Fig. 5. Gzip application   **Fig. 6.** Gunzip application   **Fig. 7.** Sequential read

memory and disk), and (ii) it has more service domains, ranging from data-, application-, to system-level. Nevertheless, in Sect. 5.3 we shall give a discussion between our solution and the one most closest to ours.

We use XV6 [32] as the guest OS, and test the syscall `mmap()` function in the kernel space using or without using the SC2M, for evaluation on the system-level high-availability. To compare, we choose three write back frequencies for `mmap()`, they are 2048, 4096, and 8192; and we create an array of characters and a memory-mapped area to test the sequential read performance of our solution and the native system call command `mmap()` with different frequencies.

Our solution SC2M generates mirror instructions at the compile stage. It is necessary and also interesting to investigate the number of inserting mirror instructions, which can reflect the application-level high-availability, to some extent. To evaluate this performance, we use several applications like Memcached [1], Parsec Ferret [21] and SPECjbb [26] as testing benchmarks. Additionally, since our SC2M can support an extra interface that allows us to analyse the mirror instructions in detail, we also present the results related to this feature.

Our goal in this work does not optimize the performance inside the native memory, we shall mainly test how much overhead shall be used for our solution. Ideally, if our overhead is less than two times of the overhead used by native, then our solution should be effective. Our tests are run on a DELL PowerEdgeT30 with a 16-core 3.7 GHz Intel Xeon E3-1225v5 CPU with 8 GB DDR3 RAMs with ECC and a 1 TB SATA Disk. Our SC2M is implemented on an Xen-4.6. We use Linux (kernel version 4.3.31) as the host OS, and deploy a lightweight system, Busybox-1.19.2 [8]. Each VM is allocated two virtual CPUs and memories.

5.2   Experimental Results

Data-Level. As we know, the gzip program does not occupy too many CPU resources while it is memory-intensive. We first test gzip application to see whether our solution is favourable.

Figure 5 shows the results when the gzip application is used. In this figure, the "our-gzip" means that the hypervisor has already been patched with SC2M, while the "native-gzip" means that the benchmark was run without any modification inside the hypervisor. That is, it does not use the memory mirror and thus no write backup operations are involved. From this figure, we can see that these two curves are almost coincident, although our solution employs the memory mirror that needs the write backup operations and the extra space for mirroring. This result essentially implies that our solution is slightly affected by the mirror space and also the write backup operations. In other words, for the gzip application, adding the mirror leads to only a minor performance degradation. This should be pretty positive and optimistic. The reader could be curious why the overhead of SC2M is too low, instead of two times of the native overhead. The main reasons are (i) SC2M uses redundant memory to backup native memory, avoiding additional I/O reads and writes; and (ii) most of mirror instructions generated by the static binary translation are explicit instructions, SC2M can read the destination address directly, which significantly speed up the process of mirror writes and reads.

Correspondingly, Fig. 6 shows the results when the gunzip application is used. In this figure, the "our-gunzip" and "native-gunzip" have the similar meanings with those mentioned in the previous paragraph. We can also see that the overhead of our solution is also close to that of the benchmark. This further demonstrates the feasibility of our solution. Moreover, one can see that when the file size is equal to 800 MB, our solution presents the worst performance, compared against the benchmark. Nevertheless, the gap is still very small. Specifically, the extra overhead incurred by our solution is less than 10% overhead in the *gunzip* benchmark. This minor extra overhead is fully acceptable and reasonable for most of real applications, since our solution (supporting the memory mirror) provides high-availability memory access.

System-Level. As for evaluation on the system-level high-availability, we test the sequential read performance. The comparative results of our tests are plotted in Fig. 7. In this figure, the "our-read" means our solution, while "mmap-2048", "mmap-4096" and "mmap-8192" refer to the native system call method *mmap()* with different frequencies. The results show that the execution time of our solution lies between mmap-2048 and mmap-8192. The performance is reasonable since our solution uses the memory mirror that can achieve high-availability. In contrast, for the benchmark, although we execute system calls periodically (every 2048, 4096 and 8192 bytes), the memory-mapped file could still not guarantee that all memory states are retained before the next write sequence. In this case, once a memory error occurs, the new data, which has not been written in the last call, could be lost. This reflects from another perspective that our solution is much more reliable and reasonable than the *mmap* alternative.

Application-Level. This experiment is used to study the application-level high availability. Figure 8 shows the number of inserting instructions at the compile stage for the *Memcached* benchmark. In this figure, the "native" denotes the compile test using the original GCC, while "our" denotes the compile test using our solution, SC2M. It can be seen that, on average, our solution at compile stage only leads to about 20% increase, in terms of the number of compile instructions (notice: the famous Remus system leads to 70% increase, as stated in [5]). This essentially reflects that our solution should be favourable. As mentioned earlier, our SC2M can support an extra interface that allows us to analyse the mirror

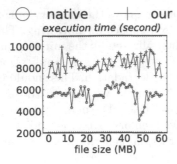

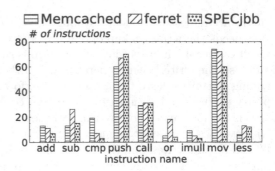

Fig. 8. Compile test　　　　　**Fig. 9.** Mirror instructions detail

instructions in detail. Note that, since mirror instructions correspond to the native ones, they essentially also provide detailed information for the native programs, to some extent. Figure 9 shows the detailed information of mirror instructions in Memcached, Ferret and SPECJbb, respectively. We can see that, the numbers of `mov` and `push` instructions alone are about 80% of the total number of instructions. In addition, it can be seen that, among all the memory write instructions, the majority of instructions are comprised of explicit write instructions (e.g., `mov`). The implicit write instructions are fewer (e.g., `imull`, `sub, or, les`) but usually are more complicated, which easily incur the extra overhead.

5.3　Discussion and Summary

Although all the experimental results show the feasibility and benefits of our solution, we would like to point out that, our solution may not be absolutely better than some existing (hardware-based or software-based) solutions. This is mainly because different solutions rely on different configurations, they may have their own advantages. A good example could be Remus [5]. It backs up the whole virtual machine in disk, has the full checkpoints and recovery mechanisms. In this regard, our solution does not always perform better than Remus. Yet, Remus generates 100% (double write) overhead using duo-backup, which is significantly larger than our overhead. Our solution uses directly the memory to backup memory data, it avoids additional disc I/O read and write. In this regard, our solution has its largest superiority than almost all the previous memory fault-tolerant solutions. Besides, our solution is also equipped with the following advantages: (i) it can retrieve the data quickly because it does not need any I/O operation from external devices; (ii) a hypervisor in the physical host does not need to launch new VMs for the backup of native VMs; and (iii) redundancy is narrowed down to one general machine or single server, so it does not involve networking resilience or migration maintenance; that is to say, large-scale deployment of our proposed system affects bandwidth utilization only minimally. Therefore, on the whole our solution is still competitive and attractive.

Summary. We find that (i) although our solution employs the memory mirror that needs the write backup operations for mirroring, it only leads to a minor performance degradation (e.g., executing gzip and gunzip applications), instead of two times of the native overhead. Specifically, in most of cases it curs about only 1.5% performance degradation. Such a performance is very positive, since our solution uses the memory mirror that can achieve high-availability for data level. (ii) Our solution can efficiently perform system-level operations (e.g., sequential read), and its performance is optimistic. Specifically, its performance is close to native system call method, yet it is much more reliable than native method. (iii) As for application-level (e.g., *Memcached*), our solution incurs a little more (about 20%) inserting instructions at the compile stage, and most of instructions are explicit write instructions (e.g., mov), which usually take less overhead, compared implicit instructions (e.g., imull).

6 Conclusion

In this paper, we proposed sc2M, which provides software-controlled memory mirroring based on hardware virtualization and static binary translation. sc2M duplicates memory by implementing mirror instructions in both user and kernel modes. It is able to simultaneously support high-availability and native virtual machines. For each VM one can choose to use data-, application-, or system-level high availability. sc2M can also increase the memory copies on demand, and so is more flexibility than hardware-based approaches. We conducted experiments to validate the feasibility and superiorities of our solution.

Acknowledgement. This work was partially supported by the National Key R&D Program of China (No. 2018YFB1004400), the NSFC (No. U1811264, U1501252, U1611264, U1711261, U1711262, U1711263, 61972425, 61702320, 61867002, 61772334, 61872310), and the Key R&D Program of Guangdong Province (No. 2018B010107005, 2019B010120001), and the Opening Project of Guangdong Province Key Laboratory of Big Data Analysis and Processing (No. 201807).

References

1. ACME Laboratories. memcached-ahigh-performance, distributed memory object caching system (2018). http://www.memcached.org/about
2. Cha, S., et al.: Defect analysis and cost-effective resilience architecture for future dram devices. In: HPCA, pp. 61–72 (2017)
3. Chen, C., Hsiao, M.Y.B.: Error-correcting codes for semiconductor memory applications: a state-of-the-art review. IBM J. Res. Dev. **28**(2), 124–134 (1984)
4. Chisnall, D.: The Definitive Guide to the Xen Hypervisor. Pearson Education, London (2008)
5. Cully, B., Lefebvre, G., Meyer, D., Feeley, M., Hutchinson, N., Warfield, A.: Remus: high availability via asynchronous virtual machine replication. In: NSDI, pp. 161–174 (2008)
6. Deegan, J., Gower, K.: High reliability memory subsystem using data error correcting code symbol sliced command repowering. US Patent App. 10/723,055 (2005)

7. Dell, T.J.: Ecc-on-simm test challenges. In: ITC, pp. 511–515 (1994)
8. Vlasenko, D.: BusyBox: The Swiss Army Knife of Embedded Linux (2013). http://www.busybox.net/
9. Dong, H., et al.: Memvisor: application level memory mirroring via binary translation. In: CLUSTER, pp. 562–565 (2012)
10. Dong, Y., Chen, Y., Pan, Z., Dai, J., Jiang, Y.: Renic: architectural extension to sr-iov i/o virtualization for efficient replication. ACM Trans. Archit. Code Optim. **8**(4), 40:1–40:22 (2012)
11. Ferraro-Petrillo, U., Grandoni, F., Italiano, G.F.: Data structures resilient to memory faults: an experimental study of dictionaries. ACM J. Exp. Algorithmics **18**, 1–6 (2013)
12. Fiala, D., Ferreira, K.B., Mueller, F., Engelmann, C.: A tunable, software-based DRAM error detection and correction library for HPC. In: Euro-Par Workshops, pp. 251–261 (2011)
13. HP Corporation: HP advanced memory protection technologies (2013). http://h18000.www1.hp.com/products/servers/technology/memoryprotection.html
14. Jeong, J., Kim, H., Hwang, J., Lee, J., Maeng, S.: Rigorous rental memory management for embedded systems. ACM Trans. Embed. Comput. Syst. **12**(1s), 43:1–43:21 (2013)
15. Khan, S., Paul, D., Momtahan, P., Aloqaily, M.: Artificial intelligence framework for smart city microgrids: state of the art, challenges, and opportunities. In: FMEC, pp. 283–288 (2018)
16. Levine, L., Myers, W.: Special feature: semiconductor memory reliability with error detecting and correcting codes. IEEE Comput. **9**(10), 43–50 (1976)
17. Li, J., Zhao, M., Ju, L., Xue, C.J., Jia, Z.: Maximizing forward progress with cache-aware backup for self-powered non-volatile processors. In: DAC, pp. 1–6 (2017)
18. Liu, H., Xu, C.-Z., Jin, H., Gong, J., Liao, X.: Performance and energy modeling for live migration of virtual machines. In: HPDC, pp. 171–182 (2011)
19. Malek, A., Vasilakis, E., Papaefstathiou, V., Trancoso, P., Sourdis, I.: Odd-ECC: On-demand dram error correcting codes. In: MEMSYS, pp. 96–111 (2017)
20. Mappouras, G., Vahid, A., Calderbank, R., Hower, D.R., Sorin, D.J.: Jenga: efficient fault tolerance for stacked dram. In: ICCD, pp. 361–368 (2017)
21. Parsec: Parsec - a unit of measure (2019). http://parsec.cs.princeton.edu/
22. Qi, Z., Dong, H., Sun, W., Dong, Y., Guan, H.: Multi-granularity memory mirroring via binary translation in cloud environments. IEEE Trans. Netw. Serv. Manage. **11**(1), 36–45 (2014)
23. Redis: redis - an open source, BSD licensed, advanced key-value cache and store (2014). http://www.redis.io/
24. Reis, G.A., Chang, J., Vachharajani, N., Rangan, R., August, D.I.: Swift: software implemented fault tolerance. In: CGO, pp. 243–254 (2005)
25. Reis, G.A., Chang, J., Vachharajani, N., Rangan, R., August, D.I., Mukherjee, S.S.: Design and evaluation of hybrid fault-detection systems. SIGARCH Comput. Archit. News **33**(2), 148–159 (2005)
26. SPEC: SPEC benchmark (2018). https://www.spec.org/benchmarks.html
27. Sridharan, V., Liberty, D.: A study of dram failures in the field. In: SC, pp. 1–11 (2012)
28. Stuart, S., Loh, G.H., Karin, S., Doug, B.: Use ECP, not ECC, for hard failures in resistive memories. SIGARCH Comput. Archit. News **38**(3), 1–12 (2010)
29. Tang, X., Zhai, J., Yu, B., Chen, W., Zheng, W., Li, K.: An efficient in-memory checkpoint method and its practice on fault-tolerant hpl. IEEE Trans. Parallel Distrib. Syst. **29**(4), 758–771 (2018)

30. Wang, B., Qi, Z., Guan, H., Dong, H., Sun, W., Dong, Y.: kMemvisor: flexible system wide memory mirroring in virtual environments. In: HPDC, pp. 251–262 (2013)
31. Wang, B., Qi, Z., Ma, R., Guan, H., Vasilakos, A.V.: A survey on data center networking for cloud computing. Comput. Netw. **91**, 528–547 (2015)
32. XV6: XV6 Doc (2011). http://pdos.csail.mit.edu/6.828/2011/xv6.html
33. Ye, K., Liu, Y., Xu, G., Xu, C.-Z.: Fault injection and detection for artificial intelligence applications in container-based clouds. In: CludCom, pp. 112–127 (2018)
34. Zaharia, M., Das, T., Li, H., Hunter, T., Shenker, S., Stoica, I.: Discretized streams: fault-tolerant streaming computation at scale. In: SOSP, pp. 423–438 (2013)
35. Zheng, H., Zhu, Z., Dong, X., Chen, B., Liu, C.: Studying shadow page cache to improve isolated drivers' performance. Concurr. Comput. Pract. Exp. **29**(10), e4081 (2017)

Verification of Microservices Using Metamorphic Testing

Gang Luo[1], Xi Zheng[1(✉)], Huai Liu[3], Rongbin Xu[4], Dinesh Nagumothu[2], Ranjith Janapareddi[2], Er Zhuang[1,5], and Xiao Liu[2]

[1] Department of Computing, Macquarie University, Sydney, NSW 2109, Australia
`james.zheng@mq.edu.au`
[2] School of Information Technology, Deakin University, Geelong, VIC 3220, Australia
[3] Department of Computer Science and Software Engineering,
Swinburne University of Technology, Melbourne, Australia
[4] School of Computer Science and Technology, Anhui University, Hefei, China
[5] School of Information and Electronic Engineering,
Zhejiang University of Science and Technology, Hangzhou, China

Abstract. Microservices architecture is drawing more and more attention recently. By dividing the monolithic application into different services, microservices-based applications are more flexible, scalable and portable than traditional applications. However, the unique characteristics of Microservices architecture have also brought significant challenges for software verification. One major challenge is the oracle problem: in the testing of microservices, it is often very difficult to verify the test result given a test input, due to the features of wide distribution, heterogeneity, frequent changes, and numerous runtime behaviors. To tackle such a challenge, in this paper, we investigate how to apply metamorphic testing into the verification of microservices-based applications, which is a simple yet effective approach to oracle problem. Empirical studies are conducted to evaluate the performance of metamorphic testing based on real-world microservice applications, against the baseline random testing technique with a complete oracle. The results show that in the absence of oracles, metamorphic testing can deliver relatively high failure-detection effectiveness. Our work demonstrates that metamorphic testing is both applicable and effective in addressing the oracle problem for the verification of microservices, similar to many other application domains.

Keywords: Automatic test case generation · Microservice · Metamorphic Testing · Mutation testing · Software verification

1 Introduction

In recent years, Microservices architecture emerged as a new paradigm in software architecture patterns. Many large companies like Amazon [23], Netflix [28], Gilt [18], LinkedIn [22] and SoundCloud [4] have adopted Microservices to build their services. By dividing large and complex applications into a set of services,

© Springer Nature Switzerland AG 2020
S. Wen et al. (Eds.): ICA3PP 2019, LNCS 11944, pp. 138–152, 2020.
https://doi.org/10.1007/978-3-030-38991-8_10

microservices achieve significant improvement in scalability, deployment, and portability than the traditional monolithic applications.

While Microservices architecture is gaining more and more popular today, its unique characteristics have brought some challenges in software verification. The heterogeneity of microservices requires testing tools to be specific to each service's programming language and runtime environment. Failure-recovery logic rather than business logic becomes the main focus due to the rapidly evolving code [20]. No unified model of system validation can support continuous integration of microservices [36]. Among all these issues, in this paper, we focus on the following oracle problem in the testing of microservices.

Moving from monolithic architecture to Microservices architecture usually requires to decompose the applications, then to distribute and process the data into hundreds of microservices [18]. Keeping each test specification up-to-date with such a large number of service interfaces is very difficult or even practically infeasible [5]. The lack of latest and comprehensive test specifications makes it almost impossible to precisely verify the test result given any possible test input, that is, the oracle problem exists. The interactions among services are more complicated compared to the traditional Service-Oriented Architectures (SOAs). The goal of traditional SOAs is to integrate different software assets to realize business processes, while microservices are aiming at improving the characteristics of software. Thus, it is a challenge to track the test dependencies within each microservice [19]. With the size and complexity growing in deployments, the behavior of each underlying microservice and how microservices are configured to interact also change. All these challenges make finding oracles (e.g., invariants) even more untenable for microservices-based applications [31]. In other words, the oracle problem becomes even more serious in Microservices architecture, as compared with SOA, not to mention other traditional application domains. To ensure the correctness of such application, an effective way of testing microservices without oracle is urgently needed.

Aiming to address the challenges in the verification of microservices-based applications, this paper mainly investigates into the state-of-the-art testing tools and models. Throughout the investigation, we find that all the existing microservice testing tools and models assume (explicitly or implicitly) the presence of oracle. If the oracle problem exists, the effectiveness of these techniques is significantly hindered. Moreover, these techniques may even become useless if the test results cannot be verified.

In the context of software testing, metamorphic testing [8,37] has emerged as a mainstream approach to the oracle problem. It simply utilizes some necessary properties of the software under test, termed as metamorphic relations, among inputs and outputs of multiple executions of the software under test for verifying the rest results. It has been justified that by simply using a small number of diverse metamorphic relations, metamorphic testing is able to detect a similar number of bugs as a complete orale [27]. Metamorphic testing has been successfully applied to detect various faults in a variety of application domains, including bioinformatics [9], compilers [25] and web services [38,39]. It is thus natural to consider the usage of metamorphic testing in addressing the predominant oracle problem for the verification of microservices-based applications. In this paper, we investigate the applicability and effectiveness of metamorphic testing via a series of experiments on real-world microservices-based applications.

There are three contributions as follows.

(1) For the first time, metamorphic testing is presented to detect and identify the failures in microservices, which has been evaluated as an effective approach in various application domains.
(2) A series of metamorphic relations are built to validate the correctness of each application, which has been verified as an effective variable from software specification.
(3) Compared with random test results. Prove that using metamorphosis tests in microservices is an effective method.

The remainder of this paper is organized as follows: Sect. 2 provides an overview of Microservices architecture and compares it with the traditional monolithic architectures to identify the difficulties in software verification brought by the characteristics of Microservices architecture. Section 3 presents some state-of-the-art microservice testing tools and models. Section 4 explains the metamorphic testing approach and justifies why we choose it for the verification of microservices. Section 5 evaluates metamorphic testing against a few real-world microservices-based applications and presents various convincing results. Section 6 concludes the paper and points the future work.

2   Microservice Preliminary

Microservice is aimed to decompose large, complex applications into a set of services to achieve a better performance in the development and deployment of various services [26]. Each service implements some distinct features and functionalities of the application. However, microservice works as an individual mini-application, which means each service can have its architecture, enabling the technical possibilities to be implemented more flexibly. Some microservices would expose an API consumed by other microservices or by the client of applications. Other microservices might implement a web user interface (UI). When a request is received, the relevant microservices work together as a whole to give back a response.

Microservices have many benefits. By dividing complex monolithic applications into a set of more manageable services, it is much faster to develop and much easier to understand individual services [35]. Also, each service can be developed independently by a team focusing on that service. As long as the functionality of this service is correctly implemented, any technology could be used. One service may be implemented by JAVA, and another service may be programmed by C++. The communication between services will be achieved by a predetermined protocol. At present, the entities provided by each service are identified by a universal Uniform Resource Identifier (URI) [3], and Hypertext

Transfer Protocol [15] is used as the communication protocol. In [16], the representational state transfer (REST) has been presented as the general conceptual framework for API designing, which becomes a popular practice for microservice APIs. Moreover, each microservice can be deployed and scaled independently, enabling the application to achieve better performance with heterogeneous resources. For example, for a frequently used service, more resources will be allocated to it than a less popular service. In [41], a case study is conducted where an enterprise application is developed and deployed in the cloud with a monolithic approach and a microservice approach. These results show that the microservices approach has a lower infrastructure cost but a higher response time.

However, microservice has its drawbacks. The Microservices architecture will require more effort from developers to build applications due to the complexity of the distributed system [26]. Besides, how to partition the existing system into microservices requires much more consideration [29]. Multiple services call for stronger coordination in the development teams. And the complicated inter-service communication brings big challenges for software verification. Therefore, microservice also has oracle problems because of the system structure of microservice, such oracle problems in microservice programs are more serious than those in traditional single-chip microcomputer structure.

3   Related Work

Though the microservice is a rising concept, it shares many similarities with traditional web services. Many testing tools and models used in web services are also applicable to microservice testing. In [2], an automatic test case generation method is presented with the Web Service Description Language (WSDL) in webservices. In [30], an improved test case generation method is proposed for web services with the pairwise testing technique. In [6], a method is proposed to test RESTful services using a formal specification of the web services to generate the test cases. In [33], a model-based approach is presented to test RESTful webservices with the UML protocol state machines. In [14], a model-driven testing approach of RESTful APIs is presented. In [24], property-based testing with the RESTful web service is proposed to achieve promising results. In [45,46], a native language based specification is provided to verify the monitored applications at runtime with bounded resource usage. However, none of them can address the verification challenge (i.e., testing without oracle) we are raising in this paper. Next, we will investigate some most promising testing tools and models for microservices.

The Directed Automated Random Testing (DART) is a testing tool proposed in [17]. With DART, unit testing can be performed completely and automatically on any program without any test driver or harness code. The static source-code analysis is applied to automatically extract the interface of a program with its external environment. A test driver for this interface can be automatically generated to perform random testing to simulate the most general environment that the program can operate in. Moreover, how the program behaves under random testing is analyzed dynamically to automatically generate new test inputs to systematically direct the execution along alternative program paths.

The symbolic execution plays an important role in the DART. Starting with random input, a DART-instrumented program calculates an input vector during each execution for the next execution. This vector contains values that are the solutions of symbolic constraints gathered from predicates in-branch statements during the previous execution. The new input vector attempts to force the execution of the program through a new path. By repeating this process, a directed search attempts to force the program to sweep through all its feasible execution paths [17]. DART makes automatic test case generation possible, which solves the first challenge. However, DART can only be applied at the unit testing level. Also, the computational expense of running tests with DART is huge due to the symbolic execution.

In [1], EvoMaster is a prototype microservice testing tool that is proposed aiming to generate test case for RESTful API automatically in the white-box test. In this tool, the APIs are directly retrieved from the JSON [10] definition generated by Swagger, which is a popular REST documentation generation tool available for more than 25 different programming languages. Then Genetic Algorithm (GA) is applied to generate the test cases for a given test problem. The Whole Test Suite approach is adapted with the extra usage of a test archive to evolve the test suites. A set of randomly initialized test cases with variable size and length are treated as GA individuals. A test suite's fitness will be the total fitness of all its test cases. When new offspring is generated, the crossover operator will mix test cases from two-parents sets.

In EvoMaster, each test case will be modified by the mutation operator, such as increasing or decreasing a numeric variable by 1. This supports all valid types in JSON. When a test is executed, the tool will check all the targets it covers. If a new target is covered, then the test will be duplicated from the test suite and put into an archive to save the target during the search process. At the end of the search, all the tests in the archive with the removed tests are checked to find out the minimized suite to be written as a test class file. Regarding the verification aspect, HTTP status codes are used to check the behavior of the system. For example, an HTTP invocation with the returned status of 403 (unauthorized) can imply that the authorization check is wrongly relaxed. EvoMaster can automatically generate test cases for higher-level testing. However, services are assumed in isolation, and it is a white-box testing which requires access to the source code of the services. Besides, code coverage results are relatively low compared to the existing, manually written results. The tool is restricted to JAVA, while many other programming languages such as C++ are used to build the microservices-based application.

In [42], a REST API testing model is proposed with an expressive description language based on JSON to solve the challenges like multidimensional API validation requirements, call sequencing, organization of test cases and data dependency between test cases. The test model description language has a good expressing capability and makes it possible to automatically generate test cases through an interface. Following the thought of traditional programming languages, the whole test plan is considered as a program and the test case is considered as a function, which are the basic units of the whole test plan to be executed.

The test suite is also considered as a function, which is the call of a function group. A function has some parameters and a return value. Parameters indicate what function depends on from the external environment; the return value indicates what influences the function will make on the external environment. Similarly, if one test case depends on the data generated by a previously executed test case, the data are input as parameters of the current test case. If the subsequent test case depends on the data generated by the current test case, the data generated by the current test case is, as an input, transmitted to the subsequent test case. The idea of this testing model is straightforward and effective. Implementing this model will solve the challenge of how to organize the test sequence of test cases and the test data dependency between test cases.

After the investigation of the existing testing tools and models for microservices, we can observe that much attention has been paid to the automatic test case generation under the assumption of the presence of oracle. However, none of the existing approaches has addressed the oracle problem, which is a common problem in microservices testing and verification. In this study, we aim to apply a novel approach, namely metamorphic testing, to address the oracle problem.

4   Metamorphic Testing

Most previous studies in test case generation assumed the availability of a test oracle, which can verify whether the result is correct or not for any given test cases [21]. However, in many practical situations, it is very difficult, or even impossible, to verify the test result, given a test input. Such a problem, termed as the oracle problem, is a fundamental problem in software testing.

More specifically, in the scenarios of microservices, as explained before, a test oracle either does not exist or is impractical to be used. This oracle problem thus becomes a fundamental challenge for microservice verification, because it significantly restricts the applicability and effectiveness of most test strategies [27]. Similar to many other domains, the applicability and effectiveness of many existing test case generation strategies are greatly hindered due to the oracle problem. In particular, the outcomes of many microservices can only be vaguely described without precise fixed values.

Metamorphic Testing (MT) [7] has been empirically justified as a cost-effective solution to address the oracle problem in many areas. The core part of MT is a set of Metamorphic Relations (MRs), which are identified based on the necessary properties of the system under test. Each MR actually represents the verifiable relationship among multiple inputs and their corresponding expected outputs. Once an MR is identified, it can be used to transform some existing test cases (called source test cases) to generate test cases (called follow-up test cases). Different from the traditional way of verifying test results against the existing oracles, MT verifies the source and follow-up test cases against the corresponding MR. For instance, consider the mathematical function $\min(a, b)$ that calculates the minimum value of two integers a and b. The order of the inputs should not influence the output, which can be expressed as the following metamorphic relation: $\min(a, b) = \min(b, a)$. In this MR, (a, b) is called the source test case, and (b, a) is called the follow-up test case. Let P_{min} be a program implementing the minimum function. P_{min} can be tested against the MR by running some metamorphic tests where specific input values are used. For instance, we can first run $P_{min}(2, 3)$ and then run $P_{min}(3, 2)$ and then check whether these two outputs are equal or not. If the outputs of a source test case and its followup test case(s) violate the MR, we can conclude that the program under test contains a bug.

In this paper we use MT to alleviate the oracle problem in microservice verification under a similar intuition: even if we cannot determine the correctness of each test case created for microservices, it might still be possible to use MRs to identify the failures in Microservices [8].

Following the general procedure of MT [27], the basic steps for implementing MT in microservices can be summarized as follows:

- Identify necessary properties of the microservices under test and use MRs to represent these properties.
- Generate the source and follow-up test cases based on each MR, and execute these test cases.
- The test outputs are evaluated against the corresponding MRs to check whether MRs are satisfied or violated. Any violation of MRs indicates a program failure in the observed microservices.

In the following section, we will evaluate the effectiveness of MT in the verification of microservices.

5   Evaluation

For evaluating the effectiveness of the metamorphic testing, some real-world microservices-based applications are selected and tested against random testing with oracle for comparison. The research questions that need to be addressed are as following:

- Can metamorphic testing alleviate the oracle problem in microservice verification?
- How effective is metamorphic testing when compared with random testing with oracle?

5.1   Objects

As one of the advocates to develop and update the applications rapidly using microservices, Eventuate is one platform that develops the sample microservices which make the business logic easy to be implemented [13]. Numerous microservices that are developed by Eventuate and the selected applications are three applications across three different areas. As shown in Table 1, one microservice application is Transactional Service [40] in the Event-Sourcing + CQRS example application [12], which involves mathematical operations for business transactions. Another microservice application is Restaurant Management Application [34], which comprises search and return functions. The third application is the Customers and Orders Management Application [11] that is made of both mathematical and search functions.

Table 1. Basic statistics of the sample microservice

Selected application in eventuate	Application area	Function
Business transactions	Event-Sourcing + CQRS	Mathematical operations for business transactions
Restaurant management application	API gateways, orchestrated APIs	Search and return functions
Customers and orders management application	Message buses/brokers, direct calls	Mathematical and search functions

5.2 Implanting Mutants to Source Microservice Application

We use mutation to plant errors for these three microservices-based applications. For testing purposes, we create four mutant versions of each application and the number of mutants implanted in each mutant version is shown in Table 2. We implant mutants for the above three microservice programs from three aspects, which are internal implementation of microservice (Version 1), microservice communication (Byzantine failure and Omission failure, Version 2), microservice data (Version 3) (thus each microservices-based application has three different mutated versions), and a version combination with all these errors combined (Version 4). For internal mutations in microservices, we use the automated mutation generation tool LittleDawina [32]. Call mutations between microservices

Table 2. The number of mutants implanted in each mutant version

Application object	Version 1	Version 2	Version 3	Version 4
Business transactions	5	6	5	12
Restaurant management application	4	5	3	10
Customers and orders management application	3	5	3	6

are achieved by changing the relationship between publication and subscription of events (publisher and subscriber modes are the communication mechanism among microservices [43,44]). Data mutations are achieved by changing the order, by which the communication data are stored in each microservice.

- MR1.1 - The sum balance of debtors and creditors before the transfer should be equal to the values after the transfer. $\sum M[i]$ (before) $= \sum M[i]$ (after), where M[i] is the account balance in customer i's account and $i = 1, 2, 3, \ldots$, X, where X is the total number of the debtors and creditors.
- MR1.2 - The sum balance of creditors before the transfer divided by the sum balance of the creditors after the transfer will always be less than one. Mb[i] (before)/Mb[i] (after) < 1, where Mb[i] is the account balance in creditor i's account and $i = 1, 2, 3, \ldots$, Y, where Y is the total number of the creditors.
- MR1.3 - The sum balance of debtors before the transfer divided by the sum balance of the debtors after the transfer will always be greater than one. Md[i] (before)/Md[i] (after) > 1, where Md[i] is the account balance in debtor i's account and $i = 1, 2, 3, \ldots$, Z, where Z is the total number of the debtors.

For the Restaurant Management Application, the restaurants are created along with the corresponding details like Zip Code, Opening Time, Closing Time and the Day of the week. The microservice GET function can fetch the results of the restaurants when any of those details are entered as inputs. The metamorphic relations we created for this microservice are as follows:

- MR2.1 - The restaurant details we get when only "zip code" is entered must be a superset of the details we get when both "Zip Code" and "day of the week" are entered.
- MR2.2 - The restaurant details we get when only "zip code" is entered must be a superset of the details we get when both "Zip Code" and "hour(time)" are entered.
- MR2.3 - The restaurant details we get when only "day of the week" is entered must be a superset of the details we get when both "day of the week" and "hour(time)" are entered.

For the customers and orders application, the developed metamorphic relation is that the status of one order is not equal to the status of another order where the credit limit and order amount are inversed.

- MR3.1 - If f(x,c) defines the order creation service, where x is the order amount and c is the credit limit, then the status of f(x,c) $\neq$ f(c,x).

5.3   Validating the Relations

After identifying a Metamorphic relation (MR), we validate the correctness of MR by reading the software specification, analyzing the source program of the relevant part of the MRs, and performing unit testing on this part of the program. To further validate our MR from the software specification, we also conduct source code analysis and unit testing of the microservices. We have assigned different people for MR creation and verification to avoid the validity issues.

5.4   Variables

Independent Variables. As we are going to measure the effectiveness of the metamorphic testing, we have considered the effectiveness as one independent variable. There is a comparison for measuring the effectiveness, and random testing with oracle has been chosen as a baseline technique. If both these variables show similar results or if the metamorphic testing outperforms the random testing, we can deduce the metamorphic testing as a reliable technique.

Dependent Variables. The metric of evaluation is the dependent variable in this case, where we consider a number of failures identified. For a given number of test cases in all mutated versions of three microservices we have tested, we compare the number of detected failures by random and metamorphic testing. This metric is used in measuring the effectiveness of metamorphic testing.

5.5   Generation of Test Cases

Random Test Case Generation. Random testing has been performed on the transactional microservice by creating random users and adding several self-accounts and third-party accounts for each user. The transactions are carried out by transferring some random amount of money that is generated by a random function $randi(imax)$, which can return an equally distributed random number from 1 to $imax$. For the restaurant management application, the restaurants are created with some random attributes. Then, queries are created using "zipCode", "hour" and "dayOfweek" by randomizing each value. The customers and orders application comprises creating customers and placing orders. The customers have the attributes like credit limit and orders have bill amount. Both of them are generated using a random function. The orders are queried using the order id randomly for the verification of response. By using the random function, we generate random tests for each version of each object. Here, the number of random test cases created for each object is different and it is related to the function of the application. Random test cases need to be written by considering functional coverage. So there will be fewer test cases with fewer features.

Metamorphic Test Case Generation. For executing the metamorphic tests, we need to use the metamorphic relation to generate follow-up test cases based on the original test cases (source test cases). Along with execution of both source and follow-up test cases, the test results are compared with the metamorphic relation for finding the fault. If the relation is not satisfied, then it can be considered that the program contains bugs and needs to be modified. Iterative testing consists of repeated application of the relation to generate various follow-up test cases based on source test cases [13]. For the transactional service application, we randomly generated source test cases and use the metamorphic relation to generate follow-up test cases to verify the application. We have generated random numbers as the balances for debtors and creditors accounts and other random numbers which are the transfer amounts. For the restaurant management service application, we have created some restaurants randomly using the random input values for various attributes in each restaurant. For generating the source test cases, we have created random Zip Code for a list of restaurants. To generate the follow-up test cases for the first metamorphic relation, we have selected Zip Code and Day of the week to verify the list of those restaurants which are open. We have created the source and follow-up test cases randomly for the other relations in a similar fashion, by selecting the Zip Code, Hour, and Day of week respectively to verify the list of those restaurants which are open. For the order application, we also use random values for credit limit and order amount. The source test cases are generated using random testing and the follow-up test cases are generated by reversing the credit limit and order amount.

5.6   Evaluation Environment

To run these microservices for our evaluation, we have chosen to use the Ubuntu Linux operating system and have installed Docker-Compose in it. We use the personal computer with Intel $Core^{TM}$ i7-7500U CPU @ 2.70 GHz 2.90 GHz along with 12 GB DDR4 2133 MHz SODIMM RAM and Intel HD Graphics 620, NVIDIA GeForce 920MX. The microservices are compiled using the Gradle provided in the applications and these compiled applications are uploaded to the Docker, which runs in the Oracle Virtual Box. The applications that are uploaded to the Docker can be accessed from any browser, and we have used the built-in Firefox for the evaluation in this case.

5.7   Results

As illustrated in Fig. 1a, we have run 30 test cases for all the mutated versions (four versions) of the transaction service application, and it is found that 24 failed test cases have been identified by using the random testing for the first mutated version with a fully developed oracle. On the other hand, the metamorphic testing has achieved 26 total failed test cases out of 30 without a comprehensive oracle. The situations for the other 3 mutated versions get 30, 24 and 23 failures respectively while the metamorphic testing gets 20, 8 and 20 failed cases without a completely developed oracle. It can demonstrate that the metamorphic testing outperforms the random testing for the first version where as it performs on par for the 2nd version and 4th version. This justifies the effectiveness and validity of metamorphic testing.

As illustrated in Fig. 1b, we have run 48 test cases for all the mutated versions (four versions) of the restaurant management service application, and it is found that 45 failed test cases have been identified by using the random testing for the first mutated version with a fully developed oracle. On the other hand, the metamorphic testing has achieved 46 total failed test cases out of 48 without a comprehensive oracle. The situations for the other 3 mutated versions get 48, 44 and 42 failures respectively while the metamorphic testing gets 44, 35 and 43

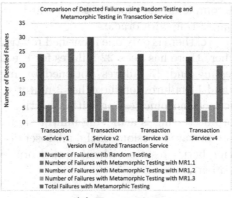

(a) Transaction

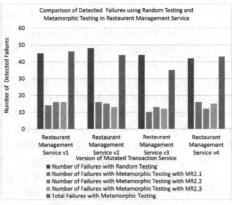

(b) Restaurant Management

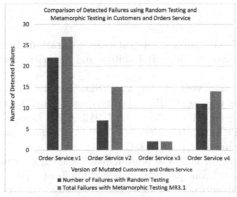

(c) Customer and Order Service

Fig. 1. Comparison of failure detection using RT and MT

failed cases without a completely developed oracle, which implies a comparable performance to random testing with oracle.

As illustrated in Fig. 1c, there is only one identified relation for the Customers and Orders service, where there has been 22 failures for the first version of order service by random testing and 27 failures obtained by metamorphic testing. There have been 7, 2 and 11 failures for the other 3 mutated versions, and 15, 2 and 14 failures for the metamorphic testing with a partial oracle. These results indicate that metamorphic testing is achieving almost similar results to that of random testing in many mutated versions of the microservice based applications and sometimes achieving better results without proper oracle where the random testing is tested with a perfect oracle.

6   Conclusion and Future Work

In this paper, a major verification challenge of testing microservices without oracle has been identified and some state-of-the-art microservice testing tools and models have been studied to highlight the contributions we made by adopting metamorphic testing approach. Then, after walking through the Microservices architecture and metamorphic testing methodology, we presented our empirical study and results by applying metamorphic testing on three real-world microservice based applications. The empirical results have been clearly shown that metamorphic testing is able to alleviate the oracle problem in microservice verification and has a relatively high effectiveness in identifying problem failures compared with a state-of-the-art testing strategy with perfect oracle. Our next step is to evaluate the metamorphic testing against much larger real-world microservice based applications to identify further research direction in applying metamorphic testing to aid the verification of microservices-based applications.

References

1. Arcuri, A.: Restful API automated test case generation. In: IEEE International Conference on Software Quality, Reliability and Security (QRS), pp. 9–20. IEEE (2017)
2. Bai, X., Dong, W., Tsai, W.T., Chen, Y.: WSDL-based automatic test case generation for web services testing. In: IEEE International Workshop on Service-Oriented System Engineering (SOSE 2005), pp. 207–212. IEEE (2005)
3. Berners-Lee, T., Fielding, R., Masinter, L.: Uniform resource identifier (URI): generic syntax. Technical report (2004)
4. Calçado, P.: Building products at soundcloud—Part I: dealing with the monolith (2014). https://tinyurl.com/jxy8yl9. Accessed 23 June 2019
5. de Camargo, A., Salvadori, I., Mello, R.D.S., Siqueira, F.: An architecture to automate performance tests on microservices. In: Proceedings of the 18th International Conference on Information Integration and Web-Based Applications and Services, pp. 422–429. ACM (2016)
6. Chakrabarti, S.K., Rodriquez, R.: Connectedness testing of restful web-services. In: Proceedings of the India Software Engineering Conference, pp. 143–152. ACM (2010)

7. Chen, T.Y., Cheung, S.C., Yiu, S.M.: Metamorphic testing: a new approach for generating next test cases. Technical report, HKUST-CS98-01, Department of Computer Science, Hong Kong (1998)

8. Chen, T.Y., et al.: Metamorphic testing: a review of challenges and opportunities. ACM Comput. Surv. (CSUR) **51**(1), 4 (2018)

9. Chen, T., Ho, J.W., Liu, H., Xie, X.: An innovative approach for testing bioinformatics programs using metamorphic testing. BMC Bioinform. **10**(1), 24 (2009)

10. Crockford, D.: The application/json Media Type for JavaScript Object Notation (JSON). Technical report (2006)

11. Customers and orders management application. https://github.com/eventuate-tram/eventuate-tram-sagas-examples-customers-and-orders. Accessed 23 June 2019

12. Event-sourcing + CQRS example application. https://tinyurl.com/y3fqzz8y. Accessed 23 June 2019

13. Eventuate.io microservice provider. https://eventuate.io/. Accessed 23 June 2019

14. Fertig, T., Braun, P.: Model-driven testing of restful APIs. In: Proceedings of the 24th International Conference on World Wide Web, pp. 1497–1502. ACM (2015)

15. Fielding, R., et al.: Hypertext Transfer Protocol - HTTP/1.1. Technical report (1999)

16. Fielding, R.T., Taylor, R.N.: Architectural styles and the design of network-based software architectures, vol. 7. Irvine Doctoral dissertation, University of California (2000)

17. Godefroid, P., Klarlund, N., Sen, K.: DART: directed automated random testing. ACM SIGPLAN Not. **40**, 213–223 (2005)

18. Goldberg, Y.: Scaling gilt: from monolithic ruby application to distributed Scala micro-services architecture (2014)

19. Heinrich, R., et al.: Performance engineering for microservices: research challenges and directions (2017)

20. Heorhiadi, V., Rajagopalan, S., Jamjoom, H., Reiter, M.K., Sekar, V.: Gremlin: systematic resilience testing of microservices. In: International Conference on Distributed Computing Systems (ICDCS), pp. 57–66. IEEE (2016)

21. Hierons, R.M.: Oracles for distributed testing. IEEE Trans. Softw. Eng. **38**(3), 629–641 (2011)

22. Ihde, S., Parikh, K.: From a monolith to microservices + REST: the evolution of LinkedIn's service architecture (2015)

23. Kramer, S.: The biggest thing Amazon got right: the platform. Gigaom, 12 October 2011

24. Lamela Seijas, P., Li, H., Thompson, S.: Towards property-based testing of RESTful web services (2013)

25. Le, V., Afshari, M., Su, Z.: Compiler validation via equivalence modulo inputs. ACM SIGPLAN Not. **49**, 216–226 (2014)

26. Lewis, J., Fowler, M.: Microservices (2014). martinfowler.com

27. Liu, H., Kuo, F., Towey, D., Chen, T.Y.: How effectively does metamorphic testing alleviate the oracle problem? IEEE Trans. Softw. Eng. **40**(1), 4–22 (2013)

28. Mauro, T.: Adopting microservices at netflix: lessons for architectural design (2015). Acesso em 8 2016

29. Namiot, D., Sneps-Sneppe, M.: On micro-services architecture. Int. J. Open Inf. Technol. **2**(9), 24–27 (2014)

30. Noikajana, S., Suwannasart, T.: An improved test case generation method for web service testing from WSDL-S and OCL with pair-wise testing technique. In: International Computer Software and Applications Conference, vol. 1, pp. 115–123. IEEE (2009)
31. Panda, A., Sagiv, M., Shenker, S.: Verification in the age of microservices. In: Proceedings of the 16th Workshop on Hot Topics in Operating Systems, pp. 30–36. ACM (2017)
32. Parsai, A., Murgia, A., Demeyer, S.: LittleDarwin: a feature-rich and extensible mutation testing framework for large and complex Java systems. In: Dastani, M., Sirjani, M. (eds.) FSEN 2017. LNCS, vol. 10522, pp. 148–163. Springer, Cham (2017). https://doi.org/10.1007/978-3-319-68972-2_10
33. Pinheiro, P.V.P., Endo, A.T., Simao, A.: Model-based testing of restful web services using UML protocol state machines. In: Brazilian Workshop on Systematic and Automated Software Testing, pp. 1–10 (2013)
34. Restaurant management application. https://github.com/eventuate-examples/eventuate-examples-restaurant-management. Accessed 23 June 2019
35. Richardson, C.: Microservices: decomposing applications for deployability and scalability. InfoQ **25**, 15–16 (2014)
36. Savchenko, D., Radchenko, G.: Microservices validation: methodology and implementation. In: 1st Ural Workshop on Parallel, Distributed, and Cloud Computing for Young Scientists, pp. 21–28 (2015)
37. Segura, S., Fraser, G., Sanchez, A.B., Ruiz-Cortés, A.: A survey on metamorphic testing. IEEE Trans. Softw. Eng. **42**(9), 805–824 (2016)
38. Segura, S., Parejo, J.A., Troya, J., Ruiz-Cortés, A.: Metamorphic testing of restful web APIs. IEEE Trans. Softw. Eng. **44**(11), 1083–1099 (2017)
39. Sun, C.A., Wang, G., Mu, B., Liu, H., Wang, Z., Chen, T.Y.: Metamorphic testing for web services: framework and a case study. In: 2011 IEEE International Conference on Web Services, pp. 283–290. IEEE (2011)
40. Transactional service application. https://github.com/cer/event-sourcing-examples. Accessed 23 June 2019
41. Villamizar, M., et al.: Evaluating the monolithic and the microservice architecture pattern to deploy web applications in the cloud. In: 2015 10th Computing Colombian Conference (10CCC), pp. 583–590. IEEE (2015)
42. Hu, W., Yu, H., Liu, X., Chen, X.: Study on REST API test model supporting web service integration. In: 2017 IEEE 3rd International Conference on Big Data Security on Cloud (BigDataSecurity), IEEE International Conference on High Performance and Smart Computing (HPSC), and IEEE International Conference on Intelligent Data and Security (IDS), pp. 133–138. IEEE (2017)
43. Yu, D., Jin, Y., Zhang, Y., Zheng, X.: A survey on security issues in services communication of microservices-enabled fog applications. Concurr. Comput. Pract. Exp. **31**, e4436 (2018)
44. Zheng, T., et al.: SmartVM: a SLA-aware microservice deployment framework. World Wide Web **22**(1), 275–293 (2019). https://doi.org/10.1007/s11280-018-0562-5
45. Zheng, X., Julien, C., Podorozhny, R., Cassez, F.: BraceAssertion: runtime verification of cyber-physical systems. In: 2015 IEEE 12th International Conference on Mobile Ad Hoc and Sensor Systems, pp. 298–306. IEEE (2015)
46. Zheng, X., Julien, C., Podorozhny, R., Cassez, F., Rakotoarivelo, T.: Efficient and scalable runtime monitoring for cyber-physical system. IEEE Syst. J. **12**(2), 1667–1678 (2016)

A New Robust and Reversible Watermarking Technique Based on Erasure Code

Heyan Chai[1]([✉]), Shuqiang Yang[1,2], Zoe L. Jiang[1,2], Xuan Wang[1], Yiqun Chen[2], and Hengyu Luo[1]

[1] School of Computer Science and Technology, Harbin Institute of Technology, Shenzhen, Shenzhen 518055, China
chaiheyan@stu.hit.edu.cn, sqyang9999@126.com
[2] Cyberspace Security Research Center, Peng Cheng Laboratory, Shenzhen, China

Abstract. With the growth of Information and Communication Technology, data sharing plays a pivotal role in all parts of the Internet. A major issue that needs to be tackled urgently is to protect the ownership of data during the sharing process. Digital watermarking technology is a major solution to ownership protection. Many reversible watermarking approaches are proposed to achieve the function of ownership protection, but most methods can only resist malicious attacks to a certain extent and can not regenerate the complete watermarks in case part of watermarks are destroyed. In this paper, a robust and reversible watermarking technique (RRWEC) applied in relational database has been proposed, which utilizes the Erasure Codes Technique and watermarking in groups. The watermarks embedded into data come from meaningful characters of identification information of a database owner. The grouping method based on clustering, without dependent primary keys, is utilized to group the data, which can resist attacks on the data structure. The erasure code technique is devised to regenerate the complete watermarking information when some sub-watermarks embedded into data are maliciously modified or removed. The results of experiments verify the effectiveness and robustness of RRWEC against malicious attacks, such as data structure attack, subset addition attack, subset alteration and deletion attack, and show that when half of the watermarks are destroyed, the watermark regeneration algorithm can still recover the complete watermarking information.

Keywords: Reversible watermarking · Robustness · Ownership protection · Erasure code · Watermark regeneration

Supported by The National Key Research and Development Program of China (No. 2017YFB0803002), Key Research and Development Program of Guangdong Province (No. 2019B010136003), Peng Cheng Laboratory Project of Guangdong Province (PCL2018KP004).

© Springer Nature Switzerland AG 2020
S. Wen et al. (Eds.): ICA3PP 2019, LNCS 11944, pp. 153–168, 2020.
https://doi.org/10.1007/978-3-030-38991-8_11

1   Introduction

With the rapid development of the Internet industry, the transmission and communication of digital information has become much easier than before. However, the process when the data is generated, transferred or shared incurs a series of digital information security issues, such as data theft, data illegal duplication and copyright violation. Indeed, outsourced data is easy to be tampered and distributed without permission due to the powerful copyability and convenience to be modified of digital information. Therefore, the issue of ownership protection of digital information has become a hot issue that needs to be solved urgently.

To solve this issue, the technique of digital watermarking applied in digital information, especially relational database watermarking, has been proposed, which can achieve ownership protection. Relational data, unlike pixels in an image, are highly independent and discontinuous, so that even a slight fluctuation of the internal values in the tuples will change the whole database greatly [9]. However, these watermarking approaches modified the data to a very large extent and can not recover the original data completely, which result in more distortion of data. Moreover, with this approach, data will not resist malicious attacks so that embedded watermarks are liable to get broken then. Reversible watermarking techniques are designed for recovering original data. Meanwhile, it also recovers the complete watermarks from the extracted partial sub-watermarks to achieve ownership protection.

In this paper, a new robust and reversible watermarking technique based on erasure codes (RRWEC) has been proposed that the identification information of a data owner will be treated as watermarks and embedded into the data after some preprocessing. A watermark regeneration algorithm is devised to recover complete watermarks in case of the extracted watermarks incomplete, which greatly improves the robustness of RRWEC. The major contributions of this paper are as follows:

- The mechanism of multi-bit watermark embedding was provided for embedding the meaningful copyright information into data through multi-bit mapping technique.
- In case that the partial sub-watermarks are lost or modified, the algorithm of watermark regeneration can ensure that embedded sub-watermarks are completely restored.
- A new robust and reversible watermarking technique that is resilient against data structure attack, subset addition attack, subset alteration and deletion attack.

The subsequent sections of the paper are organized as follows: Sect. 2 elucidates the work related to this paper. Section 3 elaborates the basic ideas and processes of the proposed watermarking scheme. The experiments and robustness analysis of proposed scheme are described in detail through simulating various attack experiments in Sect. 4. Finally, conclude Sect. 5 summarizes the paper and discusses the future work.

2   Related Work

Literature is rife with significant research in the field of digital watermarking for numerical data, which is to protect ownership of data owner. Previously, Khanna et al. [14] gave a concept to protect the map information of a database through using digital watermarking technique in 2000. Agrawal and Kiernan [1], in 2002, firstly implemented digital watermarking methods in the relational database. Since then many watermarking methods applied in database have been proposed one after another.

According to the modifications introduced in the original data, the reversible watermarking technique can be divided into two categories: distortion-based digital watermarking methods [21,22,27] and distortion-free digital watermarking methods [3,11,16]. In the following description, we will focus on reversible watermarking technique with data distortion. The Reference [5,6] proposed a Difference Expansion Watermarking (DEW), which can recover original data. Jawad [10] used the genetic algorithm (GA) to select the position to generate watermarks, and then embedded the watermarking into the data by using DEW [6]. In 2013, Farfoura et al. [4] proposed a new watermarking method, Prediction-Error Expansion Watermarking (PEEW), which can protect the quality of data to a small extent against malicious watermarking attacks. There are some drawbacks in their methods: no effectiveness for embedding multi-bit watermarking and no robustness against data structure attacks. The watermarking capacity is too small to embed more meaningful information representing data owner ownership. When the structure of data was destroyed by attackers, watermarks embedded in the data cannot be completely extracted. Therefore, these methods cannot protect the ownership of data owner.

In some recent researches, many new methods have been proposed to improve existing problems. The Reference [2,12,13] proposed a multipurpose watermarking scheme that can protect the ownership of data owner and verify data integrity. Iftikhar et al. [9] proposed a Robust and Reversible Watermarking (RRW), which can generate the watermarks by using Genetic Algorithm (GA) and select the candidate attributes through calculating mutual information (MI) of all the features. In 2018, Donghui et al. [8] used GA to encode the watermarks and then embedded encoded watermarking into data by using Histogram Shifting Technique (GAHSW). The method can minimize distortion and effectively resist some malicious attacks like subset addition, subset deletion and alteration. However, the GAHSW cannot extract all the watermarks information under data structure attacks and the quality of data reduces drastically. In 2019, Li et al. [15] proposed a reversible watermarking method with low distortion using histogram gap (HGW), which improves the algorithm of GAHSW in finding position for watermark embedding and shifting direction of histogram. However, all of these methods proposed above have the following serious shortcomings: (1) the multi-bit watermarks can not be embedded into the data; (2) these methods are not robust and reversible against data structure attack; (3) the error of extracted watermarks cannot be detected and corrected, that is to say that watermarks extracted from the watermarked data are different from that originally embedded in the data. To our knowledge, there is no watermarking technique that can detect and correct the sub-watermarks modified by attackers. When watermarks were maliciously destroyed, the probability of extracting the correct watermarks reduces drastically, and even the complete watermarks cannot be extracted correctly.

To address these issues above, we proposed a robust and reversible watermarking technique using erasure codes, which can detect the wrong sub-watermarking and regenerate original watermarks based on partial correct sub-watermarks. In our approach, a grouping method based on clustering was devised to address the issue of primary key dependent, and then our scheme can resist data structure attack. In RRWEC, data owner identification information is treated as watermark and then embedded into the data after series of processes of watermark preprocessing. In the following section, the details of proposed scheme will be elucidated.

3   The Proposed RRWEC Technique

This section elaborates the basic ideas of RRWEC for relational numerical database. RRWEC proposes a new algorithm to regenerate the broken watermarks and improves watermark robustness against malicious attacks. Figure 1 presents the main architecture of RRWEC. It consists of three parts shown in the following: (1) watermark preprocessing; (2) watermark embedding; (3) watermark extraction and data recovery.

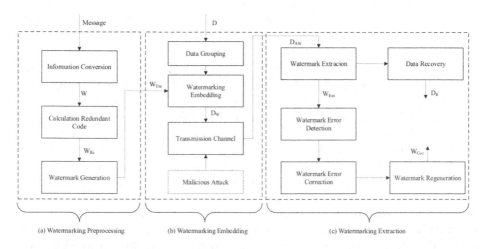

Fig. 1. Main architecture of RRWEC, where *message* is the copyright information, D is the database, W_{Enc} is the encoded watermark bit, D_{AW} is the watermarked database after attacks, W_{Ext} is the extracted watermarks, W_{Cor} is the corrected watermarks, ans D_R is the recovered data.

In the process of watermark preprocessing, the message representing identity of data owner should be converted into the watermark bits. And erasure code

Table 1. Notations used in the paper

Symbol	Description	Symbol	Description
D	Original data	$message$	Copyright information of data owner
W	Watermarking bits	W_{Rc}	Redundant code of watermark
W_{Enc}	Encoded watermark bit	D_W	Watermarked database
D_{AW}	The watermarked database after attacks	W_{Ext}	Extracted watermarks
D_R	The recovered database	W_{Cor}	Regenerated watermarks
$2T$	Length of the redundant check code	X	Single tuple of database
A_i	The i-th attribute of database	β	Number of candidate attributes
$Attrs_{imp}$	List of important attributes	γ	Number of least significant bit (LSB) of fractional part
k	Length of original watermarks	n	Length of watermark codes

method is utilized to calculate redundant code of watermark bits. Finally, we use watermark generation algorithm to watermark bits which are embedded into the data in the phase of watermark embedding.

The key point of watermark embedding phase is to embed encoded watermark bit to protect the quality of data, which it is imperceptible for user to detect the embedded watermarks.

After watermarking, the watermarked data can be delivered to intended user through open communication circumstance. The data may undergo data structure attacks [10], subset addition, subset alteration and deletion attacks [26], resulting in watermarks embedded into data modified or removed. The effectiveness of RRWEC is presented by robustness analysis for above attacks. The watermark extraction phase mainly extracts embedded watermarks and then recovers the complete watermarks through watermark regeneration algorithm. Data recovery phase successfully recovers the original data from watermarked data. For a quick reference, Table 1 gives the notations used in this paper.

3.1 Watermark Preprocessing

In the watermark preprocessing stage, two important tasks should be accomplished: (1) convert the copyright information into watermark bits; (2) calculate redundant code of watermark and encode the watermark bits. The procedure is presented in Fig. 1(a). Assume the scheme of relational database is $D(A_0, A_1, ..., A_{\beta-1})$ where $A_0, A_1, ..., A_{\beta-1}$ are the β attributes. For minimizing the distortion, the encoded watermarks will be embedded into the least significant bit (LSB) of fraction part of candidate attributes that are the most important attributes including key information.

Information Conversion. A meaningful string treated as ownership information is converted into a stream of decimal digits, which are the watermarks. These watermarks will be embedded into data after encoding, which has no effect on data quality. The table of ASCII was used to convert the single alphabet to value of ASCII. The information conversion is given by

$$W(w_1, w_2, \ldots, w_k) = ASCII(message) \tag{1}$$

where k is the length of *message*, it means the number of characters.

Watermark Encoding. In order to recover complete watermarks from broken sub-watermarks, the erasure code [18] is devised for watermark to detect and correct the wrong sub-watermarks. The encoder, called Reed-Solomon Codes, divides the original watermarks information into watermark blocks of $K = N - 2T$ sub-watermarks [19]. Each watermark block is equivalent to a watermarking polynomial of degree $K - 1$, called as $m(x)$. Then, the encoded sub-watermark is formed by simply appending $2T$ redundant encoded blocks to the end of the K-sub-watermarks block, as shown in Fig. 2. The redundant sub-watermarks are a parity-check code.

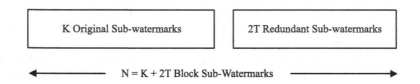

$$N = K + 2T \text{ Block Sub-Watermarks}$$

Fig. 2. Encoded watermarking structure

The redundant watermarking code are calculated by redundant polynomial $p(x)$, which is the remainder calculated by dividing $x^{2T}m(x)$ by generator polynomial $g(x)$:

$$p(x) = \left(x^{2T}m(x)\right) \bmod g(x) \tag{2}$$

where the most frequently used generator polynomial is chosen:

$$g(x) = (x + \alpha^{p_1})(x + \alpha^{p})(x + \alpha^{p_3})\ldots(x + \alpha^{p_{2T}})$$
$$g(x) = (x + \alpha)(x + \alpha^2)(x + \alpha^3)\ldots(x + \alpha^{2T}) \tag{3}$$

where α is a primitive defined under $GF(2^m)$. The code-watermark polynomial $c(x)$ is defined as follows:

$$c(x) = x^{2T}m(x) + p(x) \tag{4}$$

Since all operation is under Galois Field $(GF(2^m))$ algebra, the code watermark is equal to the polynomial $x^{2T}m(x)$ minus its remainder under division by $g(x)$. Therefore, $c(x)$ is a multiple of $g(x)$. After systematic encoding, we can obtained the encoded watermarks (W_{Enc}) with length of $N = K + 2T$, which includes $2T$ redundant check codes.

The encoding algorithm can be regarded as a polynomial operation, in which all operations are performed under the $GF(2^w)$ to ensure reversibility. Algorithm 1 describes the process of watermark encoding. $gfPolyMul$ and $gfMul$ means multiply two polynomials or real number, inside Galois Field.

Algorithm 1. The Watermark Encoding Algorithm

Input: The list of all sub-watermarks $W(w_1, w_2, \cdots, w_k)$; Size of redundant code $2T$;
Output: W_{Enc};
1: $gen = [1]$
2: **for** $i = 1 \rightarrow n$ **do**
3:     // Generate an irreducible generator polynomial
4:     $gen = gfPolyMul(gen, [1, gfExp[i]])$;
5: **end for**
6: $W_{Enc} = [0] * (len(W) + 2T - 1)$;
7: $W_{Enc}[: len(W)] = W$;
8: **for** $i = 0 \rightarrow len(W)$ **do**
9:     $coef = W_{Enc}[i]$;
10:     **for** $j = 1 \rightarrow m$ **do** //calculate the redundant check code.
11:         $W_{Enc}[i + j] = W_{Enc}[i + j] + gfMul(gen[j], coef)$;
12:     **end for**
13: **end for**
14: $W_{Enc}[: len(W)] = W$;
15: **return** W_{Enc};

3.2 Watermark Embedding

The process of watermarking embedding is presented in Fig. 1(b), which mainly comprises two parts of data grouping and watermarking embedding. Some important attributes are selected to be candidate attributes to embedded watermark. These attributes cannot be removed by attackers because of key information of data included in it. If attackers deleted these important attributes, the ownership protection of data will become meaningless.

Data Grouping. For the purpose of making watermark embedding or extraction independent of the way that the database is stored, we perform a tuple grouping method based on clustering prior to watermark embedding. The data grouping operation generates a set which includes $n = k + 2T$ non-overlapping groups $\{D_i\}_{i=1,...,n}$ by using the ideas of clustering. The improved Euclidean distance of every tuple is treated as feature to group data. The feature of every tuple can be calculated by given Eq. (5). Then, a sort algorithm is used to sort all tuples according to feature calculated by Eq. (5). After that, all tuples will group into n groups that every group has the same size.

$$Dis(X) = \sqrt{\sum_{A_i \in Attrs_{imp}} (X.A_i.int)^2} \tag{5}$$

where $X.A_i.int$ represents the integer part of A_i attributes of tuple X. $Attrs_{imp}$ is the list of important attributes.

The one-way hash cryptographic function [20] is used to selected attributes and the position of watermark embedding from candidate attributes. A multi-bit mapping method is utilized to embed selected sub-watermark into every tuple of one data group. For considering security factors, the requirements of parameters of one-way hash function is selected from a large enough key space. Therefore, the attackers cannot guess the key of hash function. The formula of hash function is $H = Hash(key||label)$, where key is selected secret key from secret key space, $||$ denotes concatenation, $label$ is group id of every tuple that is calculated by $Group()$ function proposed in the phase data grouping. $LSB(key||label||i)$ is utilized to select position in the fractional part of candidate attribute to embed sub-watermark, where i is id of candidate attributes. The detailed watermark embedding algorithm is presented in Algorithm 2.

Algorithm 2. The Watermarking Embedding Algorithm

Input: The sub-watermarks list W_{Enc}, The secret key key, D, n;
Output: The watermarked data D_W;
 1: group data by using Eq. (3)
 2: **for** each $tuple$ in D **do**
 3:     $label = Group(tuple)$;
 4:     $H = Hash(key||label)$;
 5:     $i = H\%\beta$; // select the attributes
 6:     $floatP = getFrac(tuple[i])$;
 7:     $intP = getInt(tuple[i])$;
 8:     $x = H\%LSB(key||label||i)$; // select the position
 9:     $index = (label + len(key))\%n$; //select watermark
10:     $floatWm = str(floatP[0 : x]) + W_{Enc}[index] + str(floatP[x : -1])$;
11:     $newTup = int(str(intP) + floatWm)$;
12:     $tuple = newTup$;
13: **end for**
14: **return** D_W;

3.3 Watermark Extraction Phase

In this phase, there are three important tasks to be accomplished: (1) extracting watermarks; (2) recovering the original data; (3) correcting/regenerating watermarks from extracted watermarks. The procedure is presented in Fig. 1(c).

Watermark Extraction and Data Recovery. Watermark extraction and data recovery stage mainly comprises correct extraction of embedded watermark and successful recovery of the original data. In the process of watermark extraction and data recovery, some information in previous section should be used, such as grouping information $Group()$, the length of sub-watermark γ. The function, $Group()$, described in the embedding phase is utilized to calculate the group-id of tuple. The group-id of tuple is the same as in the watermark embedding method due to identical distribution of one-way hash function when using the same secret key. Therefore, the attributes and position of watermark embedded are also the same as in the process of watermark embedding. We can extract embedded watermarks by using Algorithm 3 and then recover the original data after removing all the embedded sub-watermarks. However, the sub-watermarks extracted from the same data group may be different due the dynamic update of database or watermark attacks. To solve above issue, a majority voting mechanism is used to determine the sub-watermark which is embedded into this group in the watermark embedding phase. The detailed steps are shown in Algorithm 3.

Algorithm 3. Watermarking Extraction and Data Recovery

Input: n; γ; Watermarked dataset D_W; The secret key key;
Output: Extracted watermarks W_{Ext}; Recovery dataset D_R
  1: $WmExtDict = dict()$ //stored the extracted watermarks
  2: $D_R = D_W.copy()$
  3: **for** each tuple in D_W **do**
  4:    $label = Group(tuple)$;
  5:    $H = Hash(key||label)$;
  6:    $i = H\%\beta$; // select the attributes
  7:    $floatP = getFrac(tuple[i])$;
  8:    $intP = getInt(tuple[i])$;
  9:    $x = H\%LSB(key||label||i)$; // select the position
10:    $index = (label + len(key))\%n$; //select watermark
11:    $floatOri = str(floatP[0:x]) + str(floatP[x+\gamma:-1])$;
12:    $WmExtDict[index] \leftarrow str(floatP[x:x+\gamma])$;
13:    $newTup = float(str(intP) +'.' + floatOri)$;
14:    $D_R.tuple[i] = newTup$; // update data
15: **end for**
16: $W_{Ext} \leftarrow MajorVote(WmExtDict)$ //using the majority voting to find final watermarking information.
17: **return** D_R, W_{Ext};

Watermark Regeneration. To our knowledge, the extracted sub-watermarks are often not exactly the same as the watermark we originally embedded in the data due to insecure transmission channel. There are some errors in the extracted watermarks (W_{Ext}). Therefore, we should find the errors which is called *error detection*, and correct them by using watermarks decoding that is called *error correction*, and then the original watermarks will be regenerated successfully. Conventionally, the Petersen-Gorenstein-Zierler (PGZ) algorithm [23] is utilized to decode the encoded watermarks, which comprises of three parts: (1) syndromes calculation; (2) error location calculation; (3) error correction [19]. These three parts are briefly described as follows.

Syndromes Calculation. According to sub-watermarks extracted from water-marked data, the extracted watermarking polynomial is reconstructed, denoted as $r(x)$. The extracted watermarking polynomial consists of correct code $c(x)$ and error polynomial $e(x)$:

$$r(x) = c(x) + e(x) \tag{6}$$

The general form of the error polynomial is given by:

$$e(x) = e_{i_0} x^{i_0} + e_{i_1} x^{i_1} + e_{i_2} x^{i_2} \cdots, e_{i_{(v-1)}} x^{i_{(v-1)}} \tag{7}$$

where $i_{0,1,\cdots,(v-1)}$ denotes the indices of error location, v is the actual number of errors. And the $2T$ syndromes are calculated through evaluating the polynomial $r(x)$ at the $2T$ field points: $\alpha, \alpha^2, \alpha^3 \ldots, \alpha^{2T}$. We define the form of syndromes : $S_1, S_2, \ldots, S_{2T}$. Since $c(x)$ is a multiple of $g(x)$ according to the process of watermark encoding, the form of $c(x)$ is given by:

$$c(x) = q(x)g(x) \tag{8}$$

where $q(x)$ is from extracted watermark polynomial. $\alpha, \alpha^2, \alpha^3 \ldots, \alpha^{2T}$ are the roots of $g(x)$. Hence $c(x)$ vanishes at the $2T$ points. And the syndromes $S_1, S_2, \ldots, S_{2T}$ can calculated by:

$$S_i = e(\alpha^i), \quad i \in \{1, 2, 3, \ldots, 2T\} \tag{9}$$

If the syndrome is zero, that is to say that all syndromes vanish, that indicates that no errors have occurred or errors that can not be detected. Then we define the $2T$ syndromes again according to Eqs. (7) and (9), which is $S_i = Y_1(X_1)^i + Y_2(X_2)^i + \ldots + Y_v(X_v)^i$. The $S_1 - S_{2T}$ can be expressed by:

$$\begin{aligned} S_1 &= Y_1 X_1 + Y_2 X_2 + Y_3 X_3 \ldots Y_v X_v \\ S_2 &= Y_1 (X_1)^2 + Y_2 (X_2)^2 + Y_3 (X_3)^2 \ldots Y_v (X_v)^2 \\ S_3 &= Y_1 (X_1)^3 + Y_2 (X_2)^3 + Y_3 (X_3)^3 \ldots Y_v (X_v)^3 \\ &\cdots \\ S_{2T} &= Y_1 (X_1)^{2T} + Y_2 (X_2)^{2T} + Y_3 (X_3)^{2T} \ldots Y_v (X_v)^{2T} \end{aligned} \tag{10}$$

where v is the actual number of errors, x^{i_k} as X_k and e^{i_k} as Y_k is the error values.

Error Location Calculation. The Berlekamp-Massey algorithm [25] is used to calculated the error-location polynomial $\Lambda(x)$, defined as:

$$\Lambda(x) = (1 + xX_1)(1 + xX_2) \ldots (1 + xX_v) \equiv 1 + \lambda_1 x + \lambda_2 x^2 + \ldots + \lambda_v x^v \tag{11}$$

$\Lambda(x)$ has at most v different roots. i_k is the error location index. And the syndromes:

$$S_{j+v} + \lambda_1 S_{j+v-1} + \lambda_2 S_{j+v-2} \ldots \lambda_v S_j = 0 \qquad (12)$$

The Berlekamp-Massey algorithm can find a minimum-degree polynomial that satisfies the Newton identities for any j. In this phase, the decoder can detect up to T errors.

Error Correction. The Forney algorithm [7] is utilized to calculate the error values. When we obtained the error location X_k in the previous step, we can calculate the error values Y_k through v syndromes equation:

$$\begin{bmatrix} X_1 & X_2 & \ldots & X_v \\ \ldots & \ldots & \ldots & \ldots \\ X_1^v & X_2^v & \ldots & X_v^v \end{bmatrix} \times \begin{bmatrix} Y_1 \\ \ldots \\ Y_v \end{bmatrix} = \begin{bmatrix} S_1 \\ \ldots \\ S_v \end{bmatrix} \qquad (13)$$

Since all operation are under $GF(2^m)$, the Forney algorithm can be used to invert matrix and calculate the error value of $Y_1, Y_2, \ldots, Y_k$, which is the original sub-watermarks denoted $W\{w_1, w_2, \ldots, w_k\}$ in the preprocessing phase. And then the original watermarks are completely regenerated. Algorithm 4 describes the steps of detecting and correcting error of extracted watermark, then regenerating the original watermarks. Algorithm 4 implementation refers to [24].

Algorithm 4. Watermarking Regeneration Algorithm

Input: k; $2T$; n; Extracted watermarks W_{Ext}; Generator 2;
Output: Regenerated watermarks W_{Cor};
  1: $synd = rs\_calc\_syndromes(msg, 2T, 2)$; //calculate the syndromes
  2: $pos = rs\_find\_error(synd, n)$; //find the error location by using BM algorithm
  3: $W_{Cor} = rs\_correct\_msg(W_{Ext}, synd, pos)$; // error erasure and regenerate the original watermarks.
  4: **return** W_{Cor};

---

4   Experimental Results and Analysis

In this phase, the experimental results of this study are presented in following subsection. The main purpose of experiments is to analysis robustness of RRWEC against heavy watermark attacks and Imperceptibility of watermark embedding for user. The experiments are validated on a desktop PC with Intel Core i7 with 2.2 GHz cpu and RAM of 16 G, which is running MAC operation system. The all algorithms are implemented by using python 3.7.1. The experimental dataset is the Forest Cover Type given by University of California. The dataset contains 581,012 tuples and 54 attributes. And 1,000 tuples and 10 numerical attributes are selected to validate the experiments, which includes 5 important attributes that cannot be removed due to containing key information of dataset. The selected subset do not contain the primary key.

4.1   Watermark Regeneration Analysis

In the previous section, the Reed-Solomon (RS) code is used to encoded watermarks and then the encoded watermarks are embedded into data. Therefore, the generated watermark code has the same properties as Reed-Solomon code, which includes that RS codes match the Singleton bound [7]. The watermark code generated in this paper also has the property that the number of watermarks it can recover match singleton bound. According to singleton bound, we can calculated the bound of our watermark code. If the errors number of extracted watermark codes exceed t, the watermark code cannot recover original watermark information [17]. t can calculated by:

$$t = (n - k)/2 \tag{14}$$

where t is half the number of the redundant code, also denoted T. Hence, we can completely regenerate the original watermark when the number of broken watermark is less than half the number of redundant codes. When all the watermarks were destroyed, our approach still regenerate the original watermarks when the number of redundant matches $t > 2k$.

4.2   Robustness Analysis

In this section, the robustness of RRWEC against malicious watermark attacks is reported. The attack experiments are demonstrated with the robustness. We mainly performs these four types of attack experiments: (1) data structure attack; (2) tuples deletion; (3) tuples addition; (4) subset alteration. For a better analysis of robustness of RRWEC, we compare RRWEC with the state-of-the-art watermarking technique like DEW, GADEW, PEEW, RRW, HGW. The robustness of RRWEC is evaluated by extracted watermarking accuracy (EWA). The ratio of extracted sub-watermarks will be used to evaluate watermarking accuracy. The EWA is calculated by Eq. (15). Where N_{get} is the watermarks after watermark regeneration operation, k is the length of the original watermarks.

$$EWA = \frac{N_{get}}{k} \times 100\% \tag{15}$$

Data Structure Attack. The first experiment to evaluate robustness of RRWEC is data structure attack of watermarked data. This type of attack deletes or modifies several attributes of the database to destroy the watermarks. When attackers get a database through an illegal route, they can modify several attributes in the database is to remove the watermarks that is embedded by data owner. After that, the attacker can take the database for himself. We delete the attributes for $10\%, 20\%, \ldots 100\%$, which are not the important attributes. If the important attributes are deleted, the watermarking technique for ownership protect will be meaningless. The experimental result show that all watermarks can be extracted correctly and completely after removing 100% attributes, including the primary key of the database. Therefore RRWEC is robust against data structure attack.

Subset Deletion Attack. In this type of attack, tuples are deleted randomly or sequentially to destroy watermark. In the first deletion scenario, the tuples of different ratio are deleted randomly. Figure 3(a) shows the EWAs of the extracted watermarks of RRWEC, DEW, PEEW, GADEW, GAHSW, HGW methods against subset deletion attack. And the watermark recovery accuracy of RRWEC is 100% whereas the DEW, GADEW, PEEW, GAHSW and HGW methods present less accuracy when a larger number of tuples are randomly deleted as shown in Fig. 3(a).

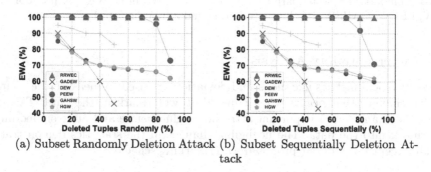

(a) Subset Randomly Deletion Attack (b) Subset Sequentially Deletion Attack

Fig. 3. Comparison of watermark recovery accuracy EWA of RRWEC with GADEW, DEW, PEEW, GAHSW, HGW after deletion attacks.

In the second deletion scenario, we continuously delete 10, 20, 30, up-to 90% of the tuples. Figure 3(b) shows the results of this attack in this scenario. Watermark recovery accuracy is 100% in RRWEC when up to 90% tuples are removed. And DEW, GADEW, PEEW, GAHSW and HGW present less accuracy in original watermark recovery from watermarked database.

Subset Addition Attack. In this attack, a large number of tuples are created randomly and inserted randomly into data to damage embedded watermarking information. We compared RRWEC with state-of-the-art watermarking approaches for accuracy of recovering the original watermark against this attack. The Fig. 4(a) illustrates the EWA of RRWEC ups to 100% when a larger number of tuples are removed. Therefore, The RRWEC approach is robust to watermark addition attack.

Subset Alteration Attack. This types of attack modifies value of attributes of the data randomly in order to damage the embedded watermarks. Experimental results show that the watermark is recovered with 100% accuracy when up to 90% tuples of the data are removed. And other well known watermarking methods give less accuracy as is shown in Fig. 4(b).

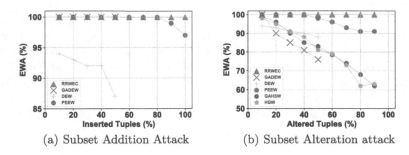

(a) Subset Addition Attack   (b) Subset Alteration attack

Fig. 4. Comparison of watermark recovery accuracy EWA of RRWEC with GADEW, DEW, PEEW, GAHSW, HGW after addition and alteration attacks.

4.3 Imperceptibility Analysis

Some statistical metrics like mean and deviation are used to measure the quantitative changes that occurs when embedding watermark in the data. Table 2 shows the changes of 5 important attributes value before and after embedding watermark into the data. The negligible changes in the statistics demonstrate the imperceptibility of our proposed watermarking scheme.

Table 2. Effect of watermarking on different attributes

Attribute name	Mean	Variance	Mean changes	Variance changes
A1	2.87e+03	3.02e+04	−6.70e−03	−1.19e−01
A2	1.42e+02	1.18e+04	−3.64e−02	−5.41e−01
A3	1.13e+01	3.63e+01	2.72e−04	2.28e−04
A4	2.37e+02	3.61e+04	−5.68e−14	−1.46e−11
A5	3.10e+01	1.39e+03	−5.21e−10	−2.27e−13

5   Conclusions

In this paper, we proposed a new robust and reversible watermarking technique based on erasure code. A watermarking encoding method based on erasure code was devised to recover the original watermarks from damaged watermarks. The proposed approach can correctly recover all watermarks even if all watermarks embedded in the data were destroyed. In our proposed scheme, the meaningful copyright information was viewed as watermarks and then embedded into the data. Meanwhile, a grouping method was devised to embed watermark which makes the watermarking formation embedded in the data redundant enough to be more robust against malicious attacks. The results of experiments above show the robustness of our proposed approach after various malicious attacks and successful regeneration the original watermarks from damaged watermarks. In our future work, we will devise a more convenient watermark embedding algorithm, and then apply it to the actual scenario.

References

1. Agrawal, R., Kiernan, J.: Watermarking relational databases. In: VLDB 2002: Proceedings of the 28th International Conference on Very Large Databases, pp. 155–166. Elsevier (2002)
2. Chen, J., Zheng, P., Guo, J., Zhang, W., Huang, J.: A privacy-preserving multipurpose watermarking scheme for audio authentication and protection. In: 2018 17th IEEE International Conference on Trust, Security and Privacy in Computing and Communications/12th IEEE International Conference on Big Data Science and Engineering (TrustCom/BigDataSE), pp. 86–91. IEEE (2018)
3. Dwivedi, A.K., Sharma, B., Vyas, A.: Watermarking techniques for ownership protection of relational databases. Int. J. Emerg. Technol. Adv. Eng. **4**(1), 368–375 (2014)
4. Farfoura, M.E., Horng, S.J., Wang, X.: A novel blind reversible method for watermarking relational databases. J. Chin. Inst. Eng. **36**(1), 87–97 (2013)
5. Gupta, G., Pieprzyk, J.: Reversible and blind database watermarking using difference expansion. In: Proceedings of the 1st International Conference on Forensic Applications and Techniques in Telecommunications, Information, and Multimedia and Workshop, p. 24. ICST (Institute for Computer Sciences, Social-Informatics and Telecommunications Engineering) (2008)
6. Gupta, G., Pieprzyk, J.: Database relation watermarking resilient against secondary watermarking attacks. In: Prakash, A., Sen Gupta, I. (eds.) ICISS 2009. LNCS, vol. 5905, pp. 222–236. Springer, Heidelberg (2009). https://doi.org/10.1007/978-3-642-10772-6_17
7. Guruswami, V., Rudra, A., Sudan, M.: Essential coding theory (2012). http://www.cse.buffalo.edu/atri/courses/coding-theory/book
8. Hu, D., Zhao, D., Zheng, S.: A new robust approach for reversible database watermarking with distortion control. IEEE Trans. Knowl. Data Eng. **31**(6), 1024–1037 (2018)
9. Iftikhar, S., Kamran, M., Anwar, Z.: RRW – a robust and reversible watermarking technique for relational data. IEEE Trans. Knowl. Data Eng. **27**(4), 1132–1145 (2015)
10. Jawad, K., Khan, A.: Genetic algorithm and difference expansion based reversible watermarking for relational databases. J. Syst. Softw. **86**(11), 2742–2753 (2013)
11. Khan, A., Husain, S.A.: A fragile zero watermarking scheme to detect and characterize malicious modifications in database relations. Sci. World J. **2013**, 1–16 (2013)
12. Khanduja, V., Chakraverty, S., Verma, O.P.: Enabling information recovery with ownership using robust multiple watermarks. J. Inf. Secur. Appl. **29**, 80–92 (2016)
13. Khanduja, V., Chakraverty, S., Verma, O.P., Tandon, R., Goel, S.: A robust multiple watermarking technique for information recovery. In: 2014 IEEE International Advance Computing Conference (IACC), pp. 250–255. IEEE (2014)
14. Khanna, S., Zane, F.: Watermarking maps: hiding information in structured data. In: Proceedings of the Eleventh Annual ACM-SIAM Symposium on Discrete Algorithms, pp. 596–605. Society for Industrial and Applied Mathematics (2000)

15. Li, Y., Wang, J., Ge, S., Luo, X., Wang, B.: A reversible database watermarking method with low distortion. Math. Biosci. Eng. **16**, 4053–4068 (2019)

16. Mohanpurkar, A., Joshi, M.: Applying watermarking for copyright protection, traitor identification and joint ownership: a review. In: 2011 World Congress on Information and Communication Technologies, pp. 1014–1019. IEEE (2011)

17. Oz, J., Naor, A.: Reed Solomon encoder/decoder on the StarCore^{TM} SC140/SC1400 cores, with extended examples (2003)

18. Plank, J.S.: T1: erasure codes for storage applications. In: Proceedings of the 4th USENIX Conference on File and Storage Technologies, pp. 1–74 (2005)

19. Reed, I.S., Solomon, G.: Polynomial codes over certain finite fields. J. Soc. Ind. Appl. Math. **8**(2), 300–304 (1960)

20. Schneier, B.: Applied Cryptography: Protocols, Algorithms, and Source Code in C. Wiley, Hoboken (2007)

21. Shehab, M., Bertino, E., Ghafoor, A.: Watermarking relational databases using optimization-based techniques. IEEE Trans. Knowl. Data Eng. **20**(1), 116–129 (2008)

22. Sion, R., Atallah, M., Prabhakar, S.: Rights protection for relational data. IEEE Trans. Knowl. Data Eng. **16**(12), 1509–1525 (2004)

23. Srinivasan, M., Sarwate, D.V.: Malfunction in the Peterson-Gorenstein-Zierler decoder. IEEE Trans. Inf. Theory **40**(5), 1649–1653 (1994)

24. tomerfiliba: A pure-python Reed Solomon encoder/decoder (2015). https://github.com/tomerfiliba/reedsolomon

25. Wikipedia: Berlekamp–Massey algorithm – Wikipedia, the free encyclopedia (2019). https://en.wikipedia.org/w/index.php?title=Berlekamp

26. Xie, M.R., Wu, C.C., Shen, J.J., Hwang, M.S.: A survey of data distortion watermarking relational databases. Int. J. Netw. Secur. **18**(6), 1022–1033 (2016)

27. Zhang, Y., Yang, B., Niu, X.M.: Reversible watermarking for relational database authentication. J. Comput. **17**(2), 59–66 (2006)

Exit-Less Hypercall: Asynchronous System Calls in Virtualized Processes

Guoxi Li$^{(\boxtimes)}$, Wenhai Lin, and Wenzhi Chen

College of Computer Science, Zhejiang University, Hangzhou 310027, China
{guoxili,linwh,chenwz}@zju.edu.cn

Abstract. Many projects of virtualized processes are emerging for less overhead than traditional virtual machines and more isolation than containers. A virtualized process uses hardware virtualization to provide a process abstraction. The virtualized processes are deemed as inefficient compared against native processes using system calls since hypercalls they use cause high-overhead context switches.

However, current performance of system calls is severely damaged by Kernel Page Table Isolation (KPTI) while hypercalls are unaffected. Unexpectedly, that gives hopes for virtualized processes to reach competitive performance against native processes.

In this paper, we propose and implement Exit-Less Hypercall, a new style of execution framework in virtualized processes by introducing asynchronity, new thread models and adaptive migration.

We evaluate the prototype and make a detailed analysis on the impacts of context switches from the native and virtualized processes with KPTI. Moreover, the experiments also show that Exit-Less Hypercall achieves a good performance improvement of up to 121% on virtualized processes using legacy hypercalls and even outperforms native processes using legacy system calls with KPTI by 81%.

Keywords: Asynchronous system calls · Virtualization · Operating system

1 Introduction

Cloud computing takes an important role in current large scale companies making transactions and business that use resources delivered through the Internet from remote servers. Virtualization technology provides platforms on which companies or individuals could rent multiple virtual machines. Each VM runs as a full-featured computer with a standard operating system kernel on which unmodified applications are deployed with increased storage, flexibility and cost reduction.

However, rooted in their design, all aforementioned traditional OS kernels are intended to be as general-purpose as possible for meeting various demands from various tenants. Unfortunately, that would cause some sub-optimization

© Springer Nature Switzerland AG 2020
S. Wen et al. (Eds.): ICA3PP 2019, LNCS 11944, pp. 169–182, 2020.
https://doi.org/10.1007/978-3-030-38991-8_12

in hypervisors since traditional OSes came into being long before concepts of virtualization.

Therefore, more lightweight virtualization technologies such as Docker and LXC are catching more attention. However, less isolation due to sharing one single kernel earns containers rebukes of security from both industry and academia. As a result, recent researches are focusing on a kind of virtualization technology that can make a better trade-off between isolation and performance.

In order to solve these problems, researchers revisit an operating system design idea—library OS which can date back to 1994 [1–3]. A lot of works, such as Unikernel [8,16,17] and Dune [7], divide traditional OS into many functional components and link them as libraries to provide specific applications exactly with what they need and to eliminate the unused parts. These new works are implemented in many different way, but what they have in common is a dedicated secure process with a minimal kernel linked as library. In this paper, they are called virtualized processes since they achieve isolation with the help of hardware virtualization and compatibility with the support of traditional host OSes.

These processes have lower performance than native. Although they achieve great hardware compatibility through hypervisors' stable APIs, the context switches of the new processes are much heavier in the old name of VM exits or Hypercalls. A VM exit often involves more context states to be saved and CPU pipeline to be flushed. After VM exits, the IPC of the application drops significantly. They are the key source of performance degradation in a virtualized process. However, KPTI patches into OS kernel bring higher overhead of system calls for native processes but impact slightly on virtualized equivalents. We believe that it shows hopes for virtualized processes to achieve competitive performance even against the native ones.

We propose Exit-Less Hypercall, a new framework of thread and hypercall model in virtualized processes. To reduce frequencies of VM exits in many cases, Exit-Less Hypercall uses:

1. A custom version of C library to intercept and dispatch hypercalls.
2. A new thread-to-fiber mapping model with adaptive migration.
3. A runtime for batching dispatched hypercalls and scheduling fibers.

We implement this design through a prototype to proof our idea. We apply it to a typical virtualized process, Dune [7] which comes with its own hardware virtualization support of Linux kernel module, and modify a version of glibc [14] for binary compatibility.

We first show how to reduce the overhead of VM exits by asynchronize "hypercalls" of virtualized processes. We then show how the prototype system uses new thread models, asynchronous hypercalls and adaptive migration to improve throughput and compatibility.

Our contributions are as follows: (1) a detailed analysis on the impact of context switches of the native and virtualized processes from KPTI; (2) an innovative way to divide invocation and execution of VM exits in virtualized processes (3) a transparent scheduler of M:N thread model with adaptive migration for intercept and dispatch hypercalls in virtualized processes to improve performance and throughput. (4) a less intrusive way of applying the design to current applications without recompiling.

Challenges and limitations will be presented in further sections.

2   Background

2.1   Kernel Page Table Isolation or KPTI

In 2018, Meltdown [4] and Spectre [5] take industries by storm. Meltdown and Spectre are critical vulnerabilities in modern processors. They are lurking right there in hardware and are ignored by us for long. Therefore, the revelation of them subvert the assumption of security from people almost completely. Meltdown destroys the most fundamental isolation between user space application and kernel space code. Spectre breaks the isolation between different applications. Up to now, more and more variations of them have come out.

Only for Meltdown, there exists a software patch called Kernel Page Table Isolation (KPTI or KAISER). It enforces a strict kernel and user space isolation. Until the revelation of Meltdown, kernel space is mapped into user space and isolation between them is thought to be guaranteed by hardware. This mechanism makes applications run faster since the permanently-mapped kernel eliminates trouble to flush the processor's TLB when making context switches between user and kernel space.

However, the patch splits the page tables of one process into two sets. The one for kernel has mappings of the full address space for the process before while the other one has only the addresses of user space and some trusted entry/exit and interrupt related codes of kernel space. Therefore, the KPTI damages performance greatly for workloads having heavy system calls like IO and networking.

2.2   Context Switch and Cache Contention

Overheads of context switch are caused in two ways. Switch codes of saving and restoring registers, TLB flush and CPU pipeline drain contribute to a fixed amount of overheads. We call this direct cost. There is also another variable cost. Context switch and memory access patterns of all processes sharing the same cache lead to the cache contention. However, the degree varies as combination of different patterns vary. We call this indirect cost.

Moreover, the cache contention can be divided into two cases. Code and data of kernel and user space interfere with each other by system calls or interrupts, and address spaces of different processes contend for the limited cache. Many researches [2, 18] suggest that costs by user-user interference are higher than the user-kernel one since user and kernel spaces are mapped in one set of page table. However, is this conclusion holds true after KPTI which destroys the premise it is based on?

Before we answer the question, we consider another situation. When virtualization technologies are applied, context switches are becoming even more complicated. We have to take VM entry and VM exit into consideration. VM entry and VM exit are both context switches between guest OS and VMM but in the mutually reverse directions. VM exits happen when guest OS executes sensitive instructions or external interruptions are triggered. On the other hand, VM entries happen when VMM launches VMs or emulates sensitive instructions after VM exits. These context switches are much heavier than user-kernel context switches of processes. A single VM exit or entry needs to save and restore more registers and many other non-register states, resulting hundreds of thousands of cycles.

2.3   Virtualized Process

On the cloud, infrastructure providers guarantee isolation between VMs or containers that applications from mutually-mistrusting users cannot get informations from others. Easy deployment and styles of microservices [6] for these applications lead to a trend of single application VM or container.

A traditional virtual machine contains a general purpose operating system as guest OS. When applications within perform I/O operations, the data and control paths go along complicated layers through Guest OS and Host OS.

On the other hand, containers like Docker seem to address the problem of heavyweight virtual machines as they get rid of multiple layers of abstractions and make system call directly to the host like normal processes. However, the shared kernel between containers exposes non-negligible threats although mechanism behind the technique like namespace and cgroup provide some limited isolation.

Recently researchers focus on a middle ground between VMs and OSes in order to achieve good isolation in a lighter-weight way. A lot of pioneering works like **Dune** [7], **Unikernel** [8,16,17], **Kata Containers** [9] and **gVisor** [10] emerge. They minimize the guest OS and use virtualization hardware for good performance and stronger isolation. They don't claim themselves as processes. However, most of them do not require hardware abstraction, and they are more like processes than VMs. In this paper, we call them virtualized processes.

2.4   KPTI Impacts on Legacy Hypercalls of Virtualized Processes

Things get worse when processes are virtualized using hardware virtualization like Dune [7] and Unikernel [8,16,17]. These virtualized processes request kernel services through hypercall. One hypercall triggers a VM exit. A VM exit have more overheads as it involves more context state save and restore. A null exit in which the VMM does nothing but resumes execution immediately will take around 2000 cycles (Our experiment uses benchmarks that invoke null hypercall through QEMU and KVM. The results has a minimum of 2005 cycles and an average of 2142). Therefore, virtualized processes are inferior to normal processes when running workloads with frequent system calls or IO requests, which hinders more widespread appliance of the technology.

Unexpectedly, the advent of Meltdown and Spectre becomes a perplexing turn of events for virtualized processes. As KPTI is merged into Linux kernel, context switches even for kernel and user will take longer time of 1521 cycles instead of 925 cycles before (a detailed analysis is in Sect. 6.1). The overhead gap between normal processes and virtualized processes in kernel-user context switches narrows enormously. We see hope for filling the gap completely, and trends towards running virtualized processes with proper optimization.

3   Design

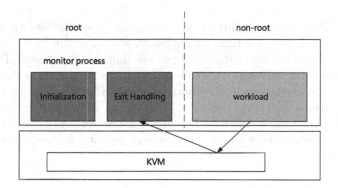

Fig. 1. Typical architecture of a virtualized process.

3.1   Typical Architecture of Virtualized Process

Figure 1 shows the typical architecture of virtualized processes, using hardware virtualization for isolation. In this paper, we limit the discussion to the Intel x86-64 VT-x without losing generality.

There is a monitor process. It runs in a less-restricted mode called root mode in VT-x. The function of it is similar to QEMU without hardware emulation, but combined with minimal kernel. It has two tasks: **Initialization** and **Exit Handling**. Firstly, the minimal kernel performs memory setup (allocating page table, GDT, IDT, etc.) and virtual CPU setup for transition to a restricted execution mode called non-root mode in VT-x.

During execution in non-root mode, the virtualized process will exit to the monitor via hypercall, like performing network or IO operations. For example, the virtualized process invokes a system call and switches to the minimal kernel. The minimal kernel handles the request if possible or replaces it with a hypercall which cause a VM exit and transfer the control to the hypervisor.

The isolation comes from the minimal kernel that sits between the application (non-root mode) and the host (root mode). When virtualized processes make system calls, the minimal kernel catches and checks the sanity of the request and handles it at its best. When more privileges are required, the kernel makes hypercalls with great care and without violation of hypervisor security. However, as mentioned before this is also where the performance penalty comes from.

3.2 Exit-Less Hypercall

To overcome the performance impact of the legacy synchronous hypercalls inherited from normal processes, we propose a new mechanism related to virtualized processes called **Exit-Less Hypercall**. Exit-Less Hypercall is a mechanism for requesting hypervisor services asynchronously that does not require one call per VM exit and reduce frequencies of VM exits greatly, thus improving performance.

Hypercall Dispatch and Batching. To make hypercalls asynchronous, we need to decouple invocation from execution. A shared region exists in memory which is visible to both the virtualized process and the hypervisor. To be exitless, where the hypercall happens is to be replaced with operations that just put the requests with arguments in the shared region without triggering VM exits.

Another case is batching. Batching is wrapping multiple hypercalls into a big hypercall. Applying of batching hypercalls allows for fewer context switches which means fewer register saves, restores and pipeline flushes, thus reducing the direct cost.

User Mode OS Services. Traditional hypervisor services like I/O and network can be implemented as dedicated services in different userland threads. These service threads use their CPU cores exclusively out of the reach from operating system scheduler, which means fewer cache contentions and fewer scheduling costs

4   Challenge

We first present the case study of a typical virtualized process Dune and present challenges new to our situation in order to better step towards the prototype of Exit-Less Hypercall.

4.1   Dune

Dune [7] is developed by Stanford University to provides applications with direct but safe access to hardware features while preserving the existing OS interface for processes. It uses virtualization hardware to provide a process instead of a machine abstraction.

Dune contains a library and kernel module. It does not have a monitor process to load the actual workload code into its memory. Unlike typical virtualized processes, it implements its own kernel module that leverages the VT-x features without resorting to the KVM module.

Therefore, a Dune program has to depend on the system loader to load its code into memory, and then it enters into the restricted execution mode through explicitly invoking a function provided by the library. In the restricted execution mode, when it invokes system calls, the control will transfer to the system call entry function that is appointed by a special register called MSR (Model Specific Register). The entry function provided by the library will invoke a hypercall which triggers a VM exit directly to the VMM. The VMM handles the hypercall in the way as the kernel handles system calls from normal processes. The overhead of one system call into one hypercall can be over 5000 cycles.

Thread model is the key to asynchronous hypercalls. One challenge is that the worker threads for executing hypercall cannot be implemented using kernel threads as previous works [11,12] do. Kernel threads in term of Linux are referred as threads that are completely running in the kernel mode but can control and access the user address and files of its parent process. However, this feature has been disabled by Linux kernel [15] already. Currently, only a kernel thread can spawn a kernel thread.

One reason for not using them is that kernel threads handing system calls on different cores than system calls are issued, which inevitably cause cache line ping pong. Moreover, the issuing side has to wait when kernel threads handle system calls if on the same core. Therefore, we believe an general asynchronous API cannot avoid exception completely. At least, one blocking system call should be provided and let user space decide the context to do whatever users want to do.

Another good reason is that kernel threads on the side of hypervisor cannot be run in a less privileged mode. A virtualized process in less privileged mode spawn threads in more privileged mode shows security issues obviously and is infeasible.

Therefore, implementation of Exit-Less Hypercall needs a new thread model and hypercall dispatching mechanism to address the challenges.

5 Implementation

5.1 Architecture

Exit-Less Hypercall's goal is to reduce VM-exit-causing system calls in virtualized processes. The architecture is presented in Fig. 2, which is divided into five components: (1) a custom version of C library to intercept and dispatch hypercalls; (2) a new thread-to-fiber mapping model; (3) a runtime for batching dispatched hypercalls and scheduling fibers; (4) a hypervisor part for executing batched hypercalls; (5) shared regions for storing batched hypercalls.

Our implementation uses Linux kernel version of 4.9, glibc version of 2.24 and the newest Dune at https://github.com/project-dune/dune.

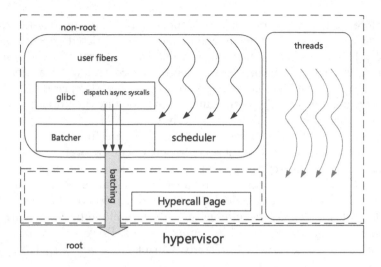

Fig. 2. Architecture of Exit-Less Hypercall.

5.2 Dispatcher and Shared Region

Intercepting and Dispatching System Calls. Dune issues system calls through hypercalls, and glibc provides C wrappers for these system calls. In order to intercept them, we make a custom glibc based on the version of 2.24. Every spot where glibc issues system calls is replaced with a hook. It is exported by glibc and users can install their own custom system call dispatching mechanism. The hook has a default implementation. If users do not install their own hook, glibc just issues system calls as before.

Shared Region. Firstly, Exit-Less Hypercall makes shared regions for each CPU core. Secondly, it installs a hook that captures all available arguments used for issuing system calls and dispatches them to the shared region on the target core. What the shared regions store are the system calls ready to issue.

5.3 New Thread Model and Scheduler

As mentioned in Sect. 4.1, kernel threads for handling hypercalls cannot be used any more. Userland threads will be used instead. For convenience, let us define both kernel threads and userland threads as kernel level threads or simply as threads which are supported and managed directly by the kernel. Userland threads execute their workloads in user space while kernel threads run their jobs in kernel space. Let us also refer user level threads as fibers which are supported and managed by runtime from user level library.

Unlike previous works [11,12], Exit-Less Hypercall applies a combination of M:N threading model (M fibers executing on N kernel level threads where N is no more than maximum cores) along with adaptive migration to avoid starving of other fibers and two modes for serving system calls:

1. No more than N fibers are scheduled to run on the underlying kernel level threads simultaneously. Fibers are not preempted and have to non-passively yield CPU when calling system call wrappers provided by glibc. After yielding, the context is saved and the fiber is put into a waiting queue. The scheduler will get a fiber from the ready queue to resume where it stops last time.
2. After all fibers scheduled to run on one kernel level thread are in the waiting queue, the scheduler issues a hypercall using the aforementioned shared region containing ready-to-issue system calls. These batched system calls are issued once and thus only cause one VM exit, greatly reducing overheads of frequent context switches. Since no fibers can progress, we make the underlying kernel level thread wait for results of batched system calls.
3. The first mode is to use the same kernel level thread in root mode for handling these batched system calls. When finished, fibers are put into the ready queue and the non-root part of the corresponding kernel level thread is resumed. During the whole process only one VM exit happens.
4. The second mode is to use other worker threads to handling them and resume the waiting thread when finished. System calls involving OS services like network or file system can be implemented in user space. Since the performance is much dependent on the specific implementation of these OS services which are out of scope of this paper, we leave an evaluation of complicated applications using these user space OS services for future work.

In the first mode, there is one case requiring special attention. In one batch of system calls, if the system call issued by one fiber needs sleeping to wait for the result the whole kernel level thread freezes and other fibers running on it starve. Therefore, we introduce one mechanism called adaptive migration.

We need one extra pool of kernel level threads besides N kernel level threads in our M:N thread model. We call them IO threads. When issuing batched system calls, the scheduler checks for system calls expected to freeze the underlying thread and migrates the corresponding fiber to one of idle IO threads. The rest is like mentioned before except that the fiber will be migrated back when sleeping is over. The checks for freezing or blocking system calls are empirical. For example, in network communications of sequential read system calls, very possibly the first one blocks and others return immediately. Currently, we use simple algorithms that recognize the obvious blocking system calls like `nanosleep` or `poll`. We leave more sophisticated algorithms for future works.

5.4   Compatibility

Our user level thread implement has the same interface as NPTL [13], the default thread implementation of Linux. Moreover, the initialization can be invoked before **main** function through C library runtime. For dynamically linking program, we can transparently apply the new thread models and asynchronize system calls, which achieves binary compatibility.

6   Evaluation

We evaluate Exit-Less Hypercall using a variety of user applications and benchmarks, comparing against the performance of Dune [7] which is state-of-the-art. Experiments are performed on one machine with 4 Intel Core i5 3.30 GHz CPUs and 8 GB memory. We use Linux 4.9 and CentOS 9 for the host environment.

First, we use microbenchmarks that show overheads of basic Exit-Less Hypercalls of our system and the new thread model with adaptive migration. Finally, we show performance of a typical echo server using our user level thread implementation.

In figures below, we represent baseline configurations using synchronous interfaces in environment of native and Dune as **"native"** and **"dune"**. Our implementations can be used with or without Dune which are shown as **"exitless"** and **"asyscall"**.

6.1   Overheads

First, we want to show how Exit-Less Hypercall reduces the overhead of VM exits in virtualized processes that involve heavy hypercalls. To measure this, we create microbenchmarks using virtualized processes that successively issue system calls to be replaced with hypercalls. The system call we choose is getppid, since this system call is simple and not optimized by glibc, and the time measured is the direct cost of VM exits which are the aim of our experiment.

Figure 3 shows the results when KPTI is on and off. The baselines are horizontal lines which show the exact cost of system calls or hypercalls with KPTI on or off. We can see that KPTI makes a single context switch 64% slower in native while it merely affects within 9% on virtualized processes.

In **"exitless"** and **"asyscall"**, we increase the number of batched system calls or hypercalls to show that less overheads can be accomplished in asynchronous ways. We can see that from beginning, both **"exitless"** and **"asyscall"** are slower than their corresponding baselines. However, when no less than 10 batched system calls or hypercalls, overheads are gone in both cases.

When batching number is more than 50 in native cases, we observe up to 580% improvement with KPTI on and up to 300% with KPTI off. Since KPTI makes context switches worse, asynchronous ways can invoke more potentials of performance.

Whether KPTI is on or off, 1700% faster executions in virtualized processes are shown. In virtualized processes, Exit-Less Hypercall can improve performance more significantly than in native processes as VM exits matter more than KPTI does.

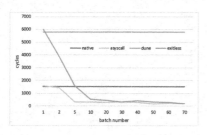 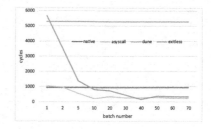

(a) Exit-less Hypercall with KPTI on.   (b) Exit-less Hypercall with KPTI off.

Fig. 3. Overheads of Exit-Less Hypercall

6.2   Thread Model

In Sect. 5.3, we present our thread models of Exit-Less Hypercall in two modes. We want to measure the performance when the thread model using worker threads is applied. For the same reason, we choose getppid. The second mode uses other user level threads to handle system calls or hypercalls and resumes the waiting thread when finished. The time measured is the direct cost of user-kernel context switch plus indirect cost of user-user cache contention.

Figure 4 shows the results when KPTI is on and off. The baseline is the same as Sect. 6.1. When batch number is 1, both **"exitless"** and **"asyscall"** are slower than the baselines, but we can get the cost of a single user-user interference. We see that user-user interference has no more than 4% difference while user-kernel interface has more cost with KPTI on as expected. As batching number is increasing, there is no overhead for both asynchronous cases. Exit-Less Hypercall gets the result of 750% faster than Dune in both case of KPTI on and off, showing less impact of KPTI in virtualized processes.

On the other hand, **"asyscall"** runs up to 220% faster than **"native"** with KPTI on. However, only 90% faster execution is achieved when KPTI is off. KPTI makes user-kernel interference worse while user-user remains almost the same. That results in the dominant impact of user-kernel interference against user-user interference, which explains the reason of better performance when reducing frequency of kernel-user context switch preferably to user-user cache contention. Finally, we can go back to Sect. 2.2 and answer the question: after KPTI, the user-kernel interference brings higher overhead. The cost of it in native processes even get paralleled with those of virtualized processes. It is more advisable to optimize hypercall frequency to get competitive performance against native processes.

6.3   Adaptive Migration and Network Applications

We use a typical echo server to evaluate the thread model in the first mode with adaptive migration on 1 Gbps cables. Our thread model is binary compatible with NPTL. With the help of dynamic linker, we can apply our M:N model

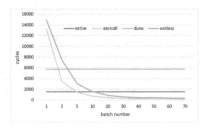   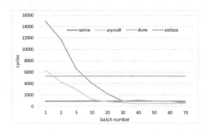

(a) The second mode with KPTI on.   (b) The second mode with KPTI off.

Fig. 4. Overheads of Exit-Less Hypercall in the second mode

transparently using the same binary. For convenience, we only test them with KPTI on.

Figure 5 shows the results of the echo server when running on 1, 2 and 4 cores. For "**exitless**", there is one extra thread for adaptive migration. The results show that, except for one concurrent request, Exit-Less Hypercall outperforms Dune by a wide margin. On 1, 2 and 4 cores, throughput improvements of up to 80%, 100% and 121% are observed.

With adaptive migration disabled, 1 core case is the worst as freezing the only underlying kernel level starves all other working fibers, reducing the throughput by 91%. Adaptive migration shows great effectiveness.

Moreover, with concurrent requests increasing, not only "**exitless**" has no overhead than "**native**" but also reaches the best cases with 23%, 42% and 81% faster on 1, 2 and 4 cores. The result shows that reducing the frequency of VM exits using exit-less ways of hypercalls can get the notoriously inefficient virtualized processes to compete with or even outperform native processes with KPTI.

7   Related Work

Virtualized processes lie between VMs and OSes in order to achieve good isolation in a lighter-weight way. Dune [7] makes most of virtualization technology to access hardware feature directly and safely, but provides a process rather than a machine abstraction. We use it to prototyping Exit-Less Hypercall.

Unikernel [8,16,17] has its standalone kernel combined with the single-purpose application during compile time prevented from modification after deployment. gVisor [10] is a new open-source project for sandboxing application. It uses **ptrace** and **KVM** to intercept system calls.

FlexSC [11] uses exception-less system calls to eliminate the direct and indirect cost in native processes, which inspires us in terms of the asynchrony of VM exits in virtualized processes. However, in the case of virtualized processes there are different challenges needed to address. We use a new thread model and enhance it with adaptive migration preventing fibers from starving.

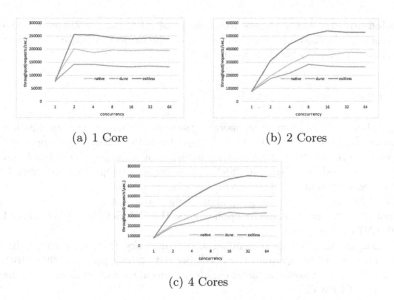

(a) 1 Core   (b) 2 Cores

(c) 4 Cores

Fig. 5. Throughput of the echo server on 1, 2 and 4 cores

VirtuOS [12] uses virtualization to slice the exiting OS kernels vertically into separate service domains, which inspires us to design a new thread model in the second mode where worker threads are applied and system calls are handled in different threads.

8   Conclusions

Virtualized processes lie in a middle ground between VMs and OSes in term of performance and isolation. However, virtualized processes communicate with hypervisors through synchronous interface of hypercalls which cause VM exits. VM exits bring much more overhead compared to those through interface of system calls.

The advent of KPTI narrows the performance gap between native and virtualized processes, especially for context switches like system calls for native and hypercalls for virtualized.

Through experiments, we make a detailed analysis on the impact of context switches brought by KPTI and come to conclude that kernel-kernel interference becomes dominant in heavy-system-call workloads. We also propose a design of Exit-Less Hypercall to address these problem by using asynchronous ways of issuing hypercalls. Leveraging batching and new thread models with binary compatibility and adaptive migration, the prototype of our design achieves higher performance of up to 121% than typical virtualized processes using synchronous interface and even outperforms native processes with KPTI by up to 81%.

Acknowledgments. Many thanks to members of ARC Lab of Zhejiang University for their constructive comments and helps during the project. We would like to thank the anonymous reviewers for their feedback.

References

1. Engler, D., et al.: Exokernel: an operating system architecture for application-level resource management. ACM SIGOPS Oper. Syst. Rev. **29**(5), 251–266 (1995)
2. Cheriton, D.R., Duda, K.J.: A caching model of operating system kernel functionality. In: Proceedings of the 1st USENIX Conference on Operating Systems Design and Implementation, p. 14. USENIX Association (1994)
3. Leslie, I.M., et al.: The design and implementation of an operating system to support distributed multimedia applications. IEEE J. Sel. Areas Commun. **14**(7), 1280–1297 (1996)
4. Lipp, M., et al.: Meltdown: reading kernel memory from user space. In: 27th USENIX Security Symposium (USENIX Security 2018), pp. 973–990 (2018)
5. Kocher, P., et al.: Spectre attacks: exploiting speculative execution. arXiv preprint arXiv:1801.01203 (2018)
6. Dragoni, N., et al.: Microservices: yesterday, today, and tomorrow. In: Mazzara, M., Meyer, B. (eds.) Present and Ulterior Software Engineering, pp. 195–216. Springer, Cham (2017). https://doi.org/10.1007/978-3-319-67425-4_12
7. Belay, A., et al.: Dune: safe user-level access to privileged CPU features. In: Presented as Part of the 10th USENIX Symposium on Operating Systems Design and Implementation (OSDI 2012), pp. 335–348 (2012)
8. Madhavapeddy, A., et al.: Unikernels: library operating systems for the cloud. In: Proceedings of the ACM International Conference on Architectural Support for Programming Languages and Operating Systems (ASPLOS), pp. 461–472 (2013)
9. Kata Containers - The speed of containers, the security of VMs. https://katacontainers.io
10. gvisor - Container Runtime Sandbox. https://github.com/google/gvisor
11. Soares, L., Stumm, M.: FlexSC: flexible system call scheduling with exception-less system calls. In: OSDI, vol. 10, pp. 1–8 (2010)
12. Nikolaev, R., Back, G.: VirtuOS: an operating system with kernel virtualization. In: Proceedings of the Twenty-Fourth ACM Symposium on Operating Systems Principles, pp. 116–132. ACM (2013)
13. Drepper, U., Molnar, I.: The native POSIX thread library for Linux (2005). http://www.akkadia.org/drepper/nptl-design.pdf
14. glibc - The GNU C library. https://www.gnu.org/software/libc/
15. Al Viro: Al Viro's new execve/kernel_thread design. https://lwn.net/Articles/520227/
16. Madhavapeddy, A., et al.: Unikernels: the rise of the virtual library operating system. Commun. ACM **57**(1), 61–69 (2014)
17. Williams, D., et al.: Unikernels as processes. In: Proceedings of the ACM Symposium on Cloud Computing, pp. 199–211. ACM (2018)
18. Fromm, R., Treuhaft, N.: Revisiting the cache interference costs of context switching. Computer Science Division, University of California-Berkeley (1996)

Automatic Optimization of Python Skeletal Parallel Programs

Frédéric Loulergue[1]([⊠])[iD] and Jolan Philippe[1,2][iD]

[1] School of Informatics Computing and Cyber Systems,
Northern Arizona University, Flagstaff, AZ, USA
frederic.loulergue@nau.edu
[2] IMT Atlantique, Inria, LS2N, UBL, Nantes, France
jolan.philippe@imt-atlantique.fr

Abstract. Skeletal parallelism is a model of parallelism where parallel constructs are provided to the programmer as usual patterns of parallel algorithms. High-level skeleton libraries often offer a global view of programs instead of the common Single Program Multiple Data view in parallel programming. A program is written as a sequential program but operates on parallel data structures. Most of the time, skeletons on a parallel data structure have counterparts on a sequential data structure. For example, the map function that applies a given function to all the elements of a sequential collection (e.g., a list) has a map skeleton counterpart that applies a sequential function to all the elements of a distributed collection.

Two of the challenges a programmer faces when using a skeleton library that provides a wide variety of skeletons are: which are the skeletons to use, and how to compose them? These design decisions may have a large impact on the performance of the parallel programs. However, skeletons, especially when they do not mutate the data structure they operate on, but are rather implemented as pure functions, possess algebraic properties that allow to transform compositions of skeletons into more efficient compositions of skeletons. In this paper, we present such an automatic transformation framework for the Python skeleton library PySke and evaluate it on several example applications.

Keywords: Algorithmic skeletons · Program transformation · Python

1 Introduction

Context and Motivation. Most computing devices are now parallel architectures, and more and more data is produced and analyzed. However, writing programs for parallel architectures remains difficult compared to writing sequential programs. Parallel programming may increase programs performances but also hinders programming productivity.

The skeletal approach proposed by Cole [13] consists in defining a program using computational patterns that have parallel implementations. In other words, programmers do not have to think about parallel aspects of their program, but only about how to write their program using already implemented patterns.

© Springer Nature Switzerland AG 2020
S. Wen et al. (Eds.): ICA3PP 2019, LNCS 11944, pp. 183–197, 2020.
https://doi.org/10.1007/978-3-030-38991-8_13

Skeletons are often implemented as higher-order functions, providing both flexibility and expressivity for the developers. Generally, a skeleton on a distributed data-structure is associated with a sequential function for a corresponding sequential data-structure. Such a high-level approach makes parallelism more affordable, even for non-expert parallel developers. However, most of the libraries handling parallel data-structures have been developed in order to be used by advanced users: the use of low-level languages or low-level primitives for parallelism, *etc.* For example, SkeTo [15], SkePu [16], OSL [18], or Muesli [11] are skeleton libraries written in C++. Their main advantage is performance. However, C++ itself is a complex language compared to Python and its dynamic typing and high-level features.

PySke [20], whose name is composed by *Python* and *Skeleton*, aims at providing computational patterns for lists and trees in both sequential and parallel, making the parallel aspects of programs as abstract as possible. PySke targets users without a deep background in computer science. There are two main challenges for such developers when using a skeleton library that provides a wide variety of skeletons: the choice of skeletons to compose the application, and the way these skeletons are composed. These application design choices may have a big impact on the performance of the application. Nonetheless, compositions of skeletons, especially on immutable data-structures, obey some algebraic laws that can be used for optimizing skeleton compositions.

Contributions. In [21] we evaluated such optimizations. However, the transformations were not automated but hand-written. In this paper, we propose a fully automatic optimizer of PySke programs based on program transformation, and we evaluate it on some example applications.

Outline. The paper is organized as follows. In Sect. 2, we discuss work related to our contribution. We give an overview of PySke in Sect. 3. Section 4 is devoted to the design of the new program transformation feature of PySke. We evaluate the performance of optimized example programs in Sect. 5. We conclude and discuss future work in Sect. 6.

2   Related Work

The transformation rules we use to optimize programs come from the Bird-Merteens Formalism (BMF) tradition [6,7]. For example, the map function/skeleton – that applies the same function to all the elements of a list – satisfies the following property: The composition of two map can be transformed into a single use of map. Considering two functions f and g, $(map\ f) \circ (map\ g)$ can be transformed into $map\ (f \circ g)$, where $\circ$ is function composition. This transformation removes the allocation and transversal of an intermediate list, thus optimizes the initial program.

To our knowledge, PySke is the only algorithmic skeleton library for Python. But there are many skeleton libraries or eDSL for other programming languages as well as programming languages specifically designed around the concept of algorithmic skeletons. In this section, we discuss only libraries and languages where program transformation is used as a means to optimize parallel programs.

For skeletons libraries, there are two main categories:

- Libraries where the optimization is done at compile time using meta-programming techniques either provided directly by the host language, or extensions, or abusing some other language features;
- Libraries where the optimization is done at execution time, either by encoding the skeleton expressions in a data structure similar to an abstract syntax tree and perform the optimization on this structure, or by using the introspection/reflection features of the host language.

In the first category, the Delite [22] framework for Scala can be considered as a skeletal parallelism approach. It features a set of data structures and mostly classical skeletons on them: map and variants, reduce and variants, filter, sort; and one less usual skeleton: group-by, as Delite has dictionaries as one of its supported data structures. Delite provides compile-time optimization through staged programming. Delite targets heterogeneous architectures CPU/GPU but only shared memory architectures. Several C++ libraries [15,17,18] are also in this category: they abuse the template feature of C++ to provide compile-time optimization. Note that in these libraries, the transformations are quite ad-hoc, and the main goal is to remove intermediate data-structures and loop traversals. The map transformation rule given above is actually such a transformation. But more complex or high-level transformations are usually not handled by these libraries that often only optimize the sequential parts of the programs but stop their optimizations as soon as there are communications. To our knowledge, there is no compile-time staging framework for Python thus such an approach cannot be used for PySke without writing a specific compiler (possibly a source-to-source compiler).

In the second category, Accelerate [10, 12] – a Haskell framework for programming GPUs – features classical data-parallel skeletons (map and variants, reduce, scan and permutation skeletons) on multi-dimensional arrays and streams. Optimization of the GPUs programs are done at runtime. PySke also belongs to this category: we optimize compositions of skeletons at runtime, and we rely on a data-structure representing these compositions to perform the transformations. In this category, another approach is to rely also on the reflection features of the host language. Lithium [1] is a structured parallelism framework that leverages Java reflection to optimize programs. Introspection and reflection also exist in Python, and we do rely on such features in the extension of PySke considered in this paper.

Finally, there are skeleton languages. The main advantage of such an approach is that the way optimizations are performed is not limited to what the host language allows. The main drawbacks are that specific compilers should be designed and implemented, and new languages do not have as large libraries as mainstream languages do, which hinders their adoption. An intermediate approach is to have a specific language for writing the composition of skeletons and use a usual sequential language for writing the arguments to these skeletons. This is less flexible than having a completely new language but mitigates the drawbacks mentioned. P3L [19] associated with the FAN transformation framework [2,4] is a representative of this approach. While being very efficient, this kind of approach is much more complex to deploy than a Python library and requires a significant effort from the programmer.

3   An Overview of PySke

PySke is a library for Python currently implemented on top of MPI and *mpi4py* [14]. Note that the programming model of PySke is independent of the underlying communication library.

PySke offers a *global view* of programs. A PySke program is written (and read) as a sequential program but it operates on parallel data structures. This aspect is very different from the SPMD paradigm of MPI where most of the time a program is actually parametrized by the process identifier (returned by the method Get_rank in mpi4py), and the global parallel program should be understood as the parallel composition of instantiations – for all possible process identifiers – of this parametrized sequential program. This "par of seq" structure is more complicated to deal with than the "seq of par" structure that global view offers [8].

For readers familiar with MPI, PySke can be thought as a library of collectives. There is a major difference: arguments to MPI collectives have regular C types but the collection on all processors of the values of these sequential types may be thought as a parallel data structure whereas in PySke there are classes dedicated to parallel data structures. In PySke, the type of a value indicated whereas this value is sequential or parallel. In MPI there is no parallel type, and it may be difficult (and it is an undecidable problem in general) to know if a value is sequential or should be thought as being part of a distributed data structure.

```
n   = data.length()
avg = data.reduce(add) / n
def f(x): return (x-avg) ** 2
var = data.map(f).reduce(add) / n
```

Fig. 1. Variance in PySke

When possible, PySke offers the same methods for both a sequential data structure and the corresponding parallel data structure. The code in Fig. 1 computes the variance of a discrete random variable data. This code is valid when data is either an instance of SList, i.e. a sequential list, or an instance of PList, i.e. a parallel list distributed on all the processors of the parallel machine running the program.

This example shows two classical skeletons: map that applies the same function (here f) to all the elements of a data-structure, and reduce that uses a binary *associative* operation (here add as defined in the Python module operator) to "sum" all the elements of a list using the binary operation. For their implementations in PList, map does not require any communication to be executed and reduce does require some communication: first, the partial sums are computed on each processor, then these partial sums are sent to all processors (total exchanged), and finally, the final sum is computed.

The user of PySke can see a PList as a usual list (SList extends the Python lists with additional methods). The implementation of PList is however an MPI one and therefore follows the SPMD approach. On each processor, a PList is actually composed of several fields: the content (a sequential list specific to each processor), the global size (same value on all processors), the distribution (the same list on each processor, this list contains the lengths of all local contents), and the index of the first element of the local list in the global list (for example at processor 2, this value would be 9 if processor 0 has a local list of 5 elements, and processor 1 a local list of 4 elements). All the methods provided by `PList` ensure that the content of these fields stay what we have just described: This may require communications for some skeletons which intuitively do not need them to perform the intended computations on the content of the parallel list.

For example, the `filter` skeleton – that takes as argument a predicate p and returns a parallel list where only the values satisfying the predicate p are kept – does not require any communication to compute the content of the returned parallel list. It is enough to perform a sequential `filter` on each of the local contents. However, if the skeleton only performed this local filtering, the global size, the local index, and the distribution would no longer represent the actual global size, local index, and distribution.

The current set of list skeletons contains [20]: `reduce`, `map` and variants (`mapi`, `map2`, `zip`), `scan` and variants, `get_partition`, `flatten`, and `balance`.

For example, `SList([1,2,3]).scan(add)` is `[0, 1, 3, 6]`.

`get_partition` makes the way the data-structure is distributed (or partitioned) visible in the value itself. For example if the global view of a parallel list `pl` is `[1, 2, 3, 4, 5, 6]` and the distribution on 4 processors is `[2, 2, 1, 1]`, then the global view of `pl.get_partition()` is `[[1, 2], [3, 4], [5], [6]]`. `flatten` is the inverse operation (and requires communications).

These two skeletons can be used to implement a `filter` skeleton:

```
def filter(self, p):
  return self.get_partition().
          map(lambda l: l.filter(p)).
          flatten()
```

After a call to `filter` the resulting parallel list may be unbalanced, i.e., some processors may contain much more elements than some others. The `balance` skeleton redistributes values such that the parallel list is evenly distributed, i.e., any processor has at most one more element than each other processor.

PySke also features a set of skeletons on trees. There are three tree data-structures: binary trees, linearized trees, parallel trees. The two first structures are sequential structures, while the last one is a parallel data structure. All of them represent binary trees.

Figure 2 is the code necessary to count the number of elements of a parallel tree that satisfy a predicate p. The code differs for a value of class `BTree` of sequential binary trees (variable `bt`) and a value of class `PTree` of parallel trees (variable `pt`). This is due to the fact that in order to be executed in parallel the reduction of a tree using an operator $\oplus$, this operator should satisfy a property called the *closure* property that allows expressing the operator using 4 auxiliary functions. This property is in a way what corresponds to associativity for lists. `map` on trees needs two arguments: a function to apply to the leaf values and a function to apply to the values at the nodes. There are two functions on trees that correspond to `scan` on lists: downwards accumulation (`dacc`) and upwards accumulation (`uacc`). More details are provided in [20].

```
def id(x): return x
def sum(x, y, z): return x + y + y
def f(x): return 1 if p(x) else 0
count_bt = bt.map(f,f).reduce(sum)
count_pt = pt.map(f,f).reduce(sum, id, sum, sum, sum)
```

Fig. 2. Count on trees in PySke

```
1   from pyske.core.list.plist import PList as PL
2   from pyske.core.opt.list import PList
3   # ...
4
5   pl1 = PL.init(rand, size)
6   pl2 = PL.init(rand, size)
7
8   def dot_product1(pl1, pl2):
9       dot = pl2.zip(pl1).map(uncurry(mul)).reduce(add, 0)
10      return dot
11
12  def dot_product2(pl1: PL, pl2: PL):
13      pl1 = PList.wrap(pl1)
14      pl2 = PList.wrap(pl2)
15      return dot_product1(pl1, pl2).run()
```

Fig. 3. PySke with automatic optimization

4 PySke with Automatic Optimization

In order to support automatic optimization of PySke applications, the user programming interface needs to slightly change with respect to what we presented in the previous section. We present these changes in Sect. 4.1. The remaining sub-sections are devoted to the design of the automatic optimization mechanism of PySke which is rule-based. It therefore relies on concepts from term rewriting systems [3]: we give a short overview in Sect. 4.2, and explain how we implemented support for such systems in a generic way in Python in Sect. 4.3. This generic framework is then instantiated to deal with PySke skeletons (Sect. 4.4).

4.1 User Programming Interface

We illustrate the differences between the previous API and the new one through the example of Fig. 3: the computation of the dot product of two vectors represented as two parallel lists.

dot_product1 is the version written using only the features presented in Sect. 3. It uses the data structure PList (imported as PL in this example) of the module pyske.core.list.plist and its skeletons.

dot_product2 is the version written using the new API: the PList class of the module pyske.core.opt.list. First note that dot_product2 can still take as input usual parallel lists: we just need to use a wrapper that transforms a PL into a PList (Lines 13–14). If dot_product were to take as argument parallel lists of type PList then no wrapping would be necessary. The only other difference with the previous API is that it is necessary to call the method run to launch the optimization of the skeleton composition and the execution of the optimized version. Note that to implement dot_product2 we can reuse dot_product1: this makes the transition to the new version easier.

uncurry is a higher-order function that transforms a function taking as input two formal parameters into a function taking only one formal parameter, this parameter being a pair. curry does the inverse transformation. Compositions of such functions are optimized too.

The performances of this example are discussed in Sect. 5.

4.2 Term Rewriting

Basically, PySke programs are skeleton expressions. Therefore, they can be represented as terms, and we can use rules to transform these terms. Our optimization mechanism is thus a term rewriting system [3].

Terms A *signature* Σ is a set $\mathcal{F}$ of function symbols with a function $ar : \mathcal{F} \to \mathbb{N}$ that for each function symbol f gives its *arity*. We call symbol functions of arity 0 *constants*.

We assume a countable set $\mathcal{X}$ of variables such that $\mathcal{X} \cap \mathcal{F} = \emptyset$.

The set $\mathcal{T}$ of terms over $\mathcal{X}$ and Σ is the smallest set such that:

- for all variable $x \in \mathcal{X}$, $x \in \mathcal{T}$,
- for all function symbol $f \in \mathcal{F}$, number n such that $n = ar(f)$, and for all terms $t_1, \ldots, t_n$ then $f(t_1, \ldots, t_n) \in \mathcal{T}$.

The set of variables of a term t can be computed by the following function $\mathcal{V}$:

$$\begin{cases} \mathcal{V}(x) & = \{x\} \\ \mathcal{V}(f(t_1, \ldots, t_n)) = \cup_{k=1}^{n} \mathcal{V}(t_k) \end{cases}$$

A *closed* term is a term t that does not contain any variable, i.e. $\mathcal{V}(t) = \emptyset$. A term is said *linear* if each $x \in \mathcal{V}(t)$ appears at most once in t.

A *substitution* σ is a partial function from $\mathcal{X}$ to $\mathcal{T}$. It can be extended to $\mathcal{T}$ as follows: for any $t \in \mathcal{T}$, either $t \in \mathcal{X}$ and we can directly use σ, or there exist a function symbol f of arity n, and n terms $t_1, \ldots, t_n$ such that $t = f(t_1, \ldots, t_n)$. We can then define: $\sigma(t) = f(\sigma(t_1), \ldots, \sigma(t_n))$.

$\sigma_\perp$ denotes the substitution undefined everywhere, and for a substitution σ, a variable x and a term t, the update $\sigma[x \mapsto t]$ is the substitution defined by:

$$\sigma[x \mapsto t](y) = \begin{cases} \sigma(y) & \text{if } y \neq x \\ t & \text{if } y = x \end{cases}$$

Rules. A *rule* is a pair of terms, denoted by $l \rightarrow r$, such that $\mathcal{V}(r) \subseteq \mathcal{V}(l)$. A rule is *linear* if both l and r are linear.

Assuming t is a closed term and p is a linear term, we define the function $match(t, p)$ as follows:

$$\begin{cases} match(t, x) & = \sigma_\perp[x \mapsto t] \\ match(f(t_1, \ldots, t_n), g(t'_1, \ldots, t'_m)) = \\ \quad merge(match(t_1, t'_1), \ldots, match(t_n, t'_n)) & \text{if } f = g \text{ and } n = m \\ match(f(t_1, \ldots, t_n), g(t'_1, \ldots, t'_m)) = \perp & \text{otherwise} \end{cases}$$

The function $merge$ returns $\perp$ if one of it arguments is $\perp$, otherwise it merges the substitutions in arguments. As p is supposed to be linear, each substitution maps variables that are different from the variables mapped by each other substitution.

The application of a rule $l \rightarrow r$ at the root of a closed term t proceeds as follows:

1. we check if the left-hand side of the rule matches t, i.e. we compute $\sigma = match(t, l)$,
2. if $\sigma = \perp$ then it means l does not match t, so the rule cannot be applied,
3. if σ is a valid substitution, the rule can be applied and then we obtain the term $\sigma(r)$.

A rule can be applied to sub-terms of a term t. In this case, if the matching succeeds, it is the matched sub-term that is replaced by $\sigma(r)$.

Example. Let us consider the following signature: $\mathcal{F} = \{\texttt{map}, \texttt{reduce}, \texttt{pl}, \texttt{incr}, +, \circ\}$ with $ar(\texttt{map}) = 2$, $ar(\texttt{reduce}) = 2$, $ar(\texttt{pl}) = 0$, $ar(\texttt{incr}) = 1$, $ar(+) = 2$, $ar(\circ) = 2$, and we use $+$ and $\circ$ as infix operations. They represent respectively the addition and function composition.

Let us consider the following term:

$$t = \texttt{reduce}(\texttt{map}(\texttt{map}(\texttt{pl}, \texttt{incr}), \texttt{incr}), +)$$

that can be thought as a representation of the following PySke program:

```
pl.map(incr).map(incr).reduce(add)
```

and the following rule:

$$\texttt{map}(\texttt{map}(l, f), g) \rightarrow \texttt{map}(l, g \circ f)$$

where l, f, and g are variables.

The rule cannot be applied to the root of t, but the left-hand side of this rule matches the sub-term $t' = \texttt{map}(\texttt{map}(\texttt{pl}, \texttt{incr}), \texttt{incr})$ of t:

$$match(\texttt{map}(\texttt{map}(\texttt{pl}, \texttt{incr}), \texttt{incr}), \quad \texttt{map}(\texttt{map}(l, f), g))$$
$$= [l \mapsto \texttt{pl}, f \mapsto \texttt{incr}, g \mapsto \texttt{incr}]$$

Applying the obtained substitution to the right-hand side of the rule gives:

$$\texttt{map}(\texttt{pl}, \texttt{incr} \circ \texttt{incr})$$

and replacing t' in t by this new term, we finally obtain:

$$\texttt{reduce}(\texttt{map}(\texttt{pl}, \texttt{incr} \circ \texttt{incr}), +)$$

This skeleton expression is likely to be more efficient because it avoids to create an intermediate list in-between the two calls to $\texttt{map}$. We use such rules in PySke to optimize skeleton compositions.

4.3 Term Rewriting in Python

Our framework first provides features that are close to the theoretical term rewriting system framework presented in the previous sub-section. These are provided in one Python module: `terms`.

The `terms` module offers two main classes: `Var` that just extends `str` and that corresponds to variables in the theoretical framework, and `Term` than implements non-variable terms. This latter class has two main attributes: `function` and `arguments` that basically correspond to the function symbol and arguments in a theoretical term.

We however take advantage of the flexibility of Python type system. The `function` attribute can be either a string, or a Python function. The arguments are represented as a list. Its elements can be either values of type `Term`, strings (in this case representing a class), or Python values of arbitrary types. The two latter cases can be thought as constants in the theoretical framework.

Substitutions in the theoretical framework are implemented as Python dictionaries. Applying a substitution to a term t (in Python a value either of type `Var` or of type `Term`) is implemented as a Python function `subst(t, s)` where s is a dictionary.

One important feature of the theoretical framework is the *match* function. It is implemented in Python as a method of the class `Term` that takes as argument a value either of type `Var` or of type `Term`. This function either returns a dictionary in case the matching succeeds, or `None` if the matching fails (`None` thus corresponds to $\perp$ of the theoretical framework).

Finally rules are implemented as a Python `namedtuple`, so essentially records. These records contain of course two terms, `left` and `right`, but also a name for the rule, and a `type`. This `type` is actually a sub-class of `Term` as we will explain in Sect. 4.4.

The module `terms` contains also a function to apply a rule at the root of a term, and a function `inner_most_strategy(t)` that repeatedly tries to apply all the available rules to every sub-terms of t until it is no longer possible to apply any rule. The strategy to apply the rules is discussed in the next section.

As an example, Fig. 4 presents the encoding of the rule presented in the previous section. This example illustrates the fact that the `function` attribute of a term can be either a string (most of the cases) or a Python function. This is the case for `compose` (Line 4) which is a function defined as:

```
def compose(f, g):
    return lambda x: f(g(x))
```

Similarly the elements of `arguments`, the second parameter to the constructor of `Term` can be either variables, terms, or arbitrary Python values (this latter case is not present in the example).

4.4 Rule-Based Skeleton Compositions Optimization

We have a framework to represent generic terms, and rules to transform such terms. There remain two main questions to use such a framework to automatically optimize skeleton compositions:

```
1  Rule(left=Term('map',
2                  [Term('map', [Var('PL'), Var('f')]), Var('g')]),
3        right=Term('map',
4                  [Var('PL'), Term(compose,
5                                    [Var('f'), Var('g')])]),
6        name="map_map",
7        type=_List)
```

Fig. 4. A rewriting rule in Python

1. How to represent PySke skeleton compositions as instances of `Term` in a way that is mostly transparent to the user with respect to the former PySke API?
2. How to evaluate/execute such terms?

The design principles for a solution to (1) are:

– inherit from `Term`, and
– use `Python` introspection capabilities to obtain a concise implementation.

Thus essentially, for each class that provided a data structure and associated skeletons in PySke we define a "wrapper" class that features the same methods than the initial class, but these methods build terms instead of executing parallel code.

A class such as `Plist` has static methods (and we identify the class constructor to a static method in our framework) and methods. For both kind of methods, the representation is such that:

– the method name is represented as the function symbol of the term (in class `Term`, it is attribute `function`),
– the class or object that is the target of the method call is represented as the *first* argument of the term (first element of the list `arguments`),
– the remaining of the method arguments are represented as additional elements in the list `arguments`.

Python offers many introspection features. In particular it is possible to capture any call to non-static methods that are not defined in a class. Therefore to define a `PList` wrapper class that inherits from `Term` we have basically to define static methods that correspond to the initial class static methods, and a generic way to catch calls to non-static methods to build the corresponding terms.

The initial `PList` class features the `init(f, size)` static method. The wrapper `PList(Term)` class thus features the following static method:

```
@staticmethod
def init(f, size):
    return PList('init', ['PList', f, size])
```

As indicated before, the value for the attribute `function` is a string that is the name of the corresponding static method in the initial PySke `PList` class, and the `arguments` list, in addition to the arguments to `init` that are `f` and `size`, contains as first element the name of the class (here `PList`) that contains the `init` static method.

As explained before, for non-static methods, we do not re-implement each method but rather rely on a feature of Python that allows to capture any call to non-defined methods in a class (technically we redefine the __getattr__ attribute). We refer to the source code[1] for more technical details. An instance of this generic code for method map follows:

```
def map(self, f):
    return PList('map', [self, f])
```

Using this wrapper class PList is therefore exactly the same than using the initial PList but nothing is executed: instead a term representing the skeleton expression is built. The only difference are: the raw method that allows to embedded a value of the initial class into the wrapper class as illustrated in Sect. 4.1, and launching the optimization and execution of such a term.

Indeed, *in fine* it is necessary to execute the skeleton composition. This is done using a method called run(). It proceeds in two steps:

1. first the term is transformed using rules such as the rule presented in Fig. 4,
2. then the optimized term is executed.

The optimization and execution features are not specifically implemented for each wrapper class. They are actually offered by two methods of the class Term:

- opt() transforms a term using all the available rules by calling the transformation function inner_most_strategy that applies all the available rules as many times as possible on all the sub-terms of the current term;
- eval() executes a term. We refer to the code for details, but basically Python allows to retrieve attributes of a class (including methods) by their names represented by a string, and to apply any function or method to a list of values in such a way that this list is considered as the effective parameters of the function or method. We use both this features to retrieve the target class and methods in the initial PySke classes from objects of the wrapper classes.

Strategy. The transformation rules are applied using an innermost strategy: the rules are first applied (when possible) to the leaves of the tree representing the composition of skeletons, and then in an upwards fashion. When two rules may be applied to the same sub-expression of the skeleton expression, priorities associated to rules are used. As future work we plan to extend this default behavior as discussed in Sect. 6.

5   Applications and Experiments

Currently, our framework contains a dozen of optimization rules:

- optimization of a composition of maps,
- optimization of compositions of map and reduce (using and internal map_reduce skeleton),

---
[1] Available at https://github.com/pyske/PySke.

- optimization of compositions of map and zip to map2,
- optimization of compositions of map and reduce using boolean operators as their arguments (a generalized De Morgan rule),
- optimization of compositions of higher-order functions such as uncurry and curry (for example, the composition of these two functions is the identity function, and the identity function is a neutral element for function composition).

We evaluated our framework on three examples:

1. the dot product presented in Fig. 3,
2. the conjunction of the negation of a list l of Boolean values:

```
l.map(operator.not_).reduce(operator.and_, True)
```

3. the computation of the normalized average of vectors: a parallel list of vectors (with 10 components) is first normalized (the norm of each vector is computed and the vector is multiplied by a scalar that is the inverse of its norm), then the average of these lists is computed:

```
smul(1 / l.length(), l.map(normalize).reduce(vadd, vzero))
```

The experiments were conducted on a shared memory machine (256 Gb), with two Intel Xeon E5-2683 v4 processors each having 16 cores at 2.10 GHz. Each example has been run 30 times. We used the following software: CentOS 7, Python 3.6.3, mpi4py 3.0.2, OpenMPI 2.6.4.

For the dot product example, we run four versions, on a parallel list of 5×10^7 elements:

- direct corresponds to dot_product1,
- wrapper builds the term representation of the skeleton composition, then executes it without transformation. This version allows to check the overhead introduced by the building of the intermediate representation and the overhead introduced by using such a representation for execution,
- optimized is the same than wrapper but the skeleton composition is optimized using all available rules,
- finally hand_written is the version using directly the PList but the composition is optimized manually.

The results are presented Fig. 5. First of all, there is no noticeable overhead due to the wrapper classes with respect to the use of the direct implementation. That is true both for the non-optimized code (direct and wrapper) and the optimized code (optimized and hand-optimized). Secondly, the automatic optimization of PySke provides, in this case, a significant speedup of about 2.5.

For the list of Boolean values example, the average speed-up of the automatically optimized version with respect to the direct implementation is 85%. For the vector average after normalization, the average speed-up is more modest at 16%.

In all cases, the speed-up is there for only a slight change in the program with respect to the former API.

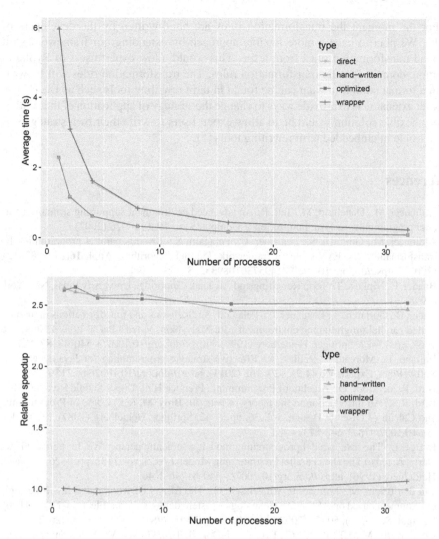

Fig. 5. Dot product example

6   Conclusion and Future Work

Skeletal parallel programming is a productive way to write parallel programs. With PySke we propose a library simple to use for non-expert programmers. However even with algorithmic skeletons, two difficulties remain: the choice of skeletons when the library is large, and the way these skeletons are composed to form an application. These choices have a large impact on the performance of the parallel program. To further eases the development of efficient parallel programs with PySke, we propose automatic program transformation based on high-level transformation rules. Evaluation of this new mechanism in PySke shows it can provide significant improvements to skeletal programs.

For the moment the transformation rules are hand-written by the developers of PySke. We plan to have a more flexible approach by extending our framework so it can read transformation rules from a file. This would allow expert users of PySke to add more domain specific transformation rules. The transformation rules will be written in a format similar to what can be found in term rewriting tools such as ELAN [9]. Another extension is to provide ways to change the strategy of application of these rules. A very flexible solution would be to allow expert users to write their own strategies as it is possible in embedded term rewriting tools [5].

References

1. Aldinucci, M., Danelutto, M., Teti, P.: An advanced environment supporting structured parallel programming in Java. Future Gener. Comput. Syst. **19**, 611–626 (2002)
2. Aldinucci, M., Gorlatch, S., Lengauer, C., Pelagatti, S.: Towards parallel programming by transformation: the FAN skeleton framework. Parallel Algorithms Appl. **16**(2–3), 87–121 (2001). https://doi.org/10.1080/01495730108935268
3. Baader, F., Nipkow, T.: Term Rewriting and All That. Cambridge University Press, New York (1998)
4. Bacci, B., Gorlatch, S., Lengauer, C., Pelagatti, S.: Skeletons and transformations in an integrated parallel programming environment*. In: Malyshkin, V. (ed.) PaCT 1999. LNCS, vol. 1662, pp. 13–27. Springer, Heidelberg (1999). https://doi.org/10.1007/3-540-48387-X_2
5. Balland, E., Moreau, P., Reilles, A.: Effective strategic programming for Java developers. Softw. Pract. Experience **44**(2), 129–162 (2014). https://doi.org/10.1002/spe.2159
6. Bird, R., de Moor, O.: Algebra of Programming. Prentice Hall, Upper Saddle River (1996)
7. Bird, R.S.: An introduction to the theory of lists. In: Broy, M. (ed.) Logic of Programming and Calculi of Discrete Design, vol. 36, pp. 5–42. Springer, Heidelberg (1987). https://doi.org/10.1007/978-3-642-87374-4_1
8. Bougé, L.: The data parallel programming model: a semantic perspective. In: Perrin, G.-R., Darte, A. (eds.) The Data Parallel Programming Model. LNCS, vol. 1132, pp. 4–26. Springer, Heidelberg (1996). https://doi.org/10.1007/3-540-61736-1_40
9. van den Brand, M., Moreau, P., Ringeissen, C.: The ELAN environment: a rewriting logic environment based on ASF+SDF technology - system demonstration. Electron. Notes Theor. Comput. Sci. **65**(3), 50–56 (2002). https://doi.org/10.1016/S1571-0661(04)80426-2
10. Chakravarty, M.M.T., Keller, G., Lee, S., McDonell, T.L., Grover, V.: Accelerating Haskell array codes with multicore GPUs. In: Workshop on Declarative Aspects of Multicore Programming (DAMP), pp. 3–14 (2011). https://doi.org/10.1145/1926354.1926358
11. Ciechanowicz, P., Poldner, M., Kuchen, H.: The Münster Skeleton Library Muesli - A Comprenhensive Overview. Technical report Working Paper No. 7, European Research Center for Information Systems, University of Münster, Germany (2009)
12. Clifton-Everest, R., McDonell, T.L., Chakravarty, M.M., Keller, G.: Streaming irregular arrays. In: Proceedings of the 10th ACM SIGPLAN International Symposium on Haskell, pp. 174–185. ACM (2017). https://doi.org/10.1145/3122955.3122971
13. Cole, M.: Algorithmic Skeletons: Structured Management of Parallel Computation. MIT Press, Cambridge (1989)
14. Dalcin, L.D., Paz, R.R., Kler, P.A., Cosimo, A.: Parallel distributed computing using Python. Adv. Water Resour. **34**(9), 1124–1139 (2011). https://doi.org/10.1016/j.advwatres.2011.04.013. New Computational Methods and Software Tools

15. Emoto, K., Matsuzaki, K.: An automatic fusion mechanism for variable-length list skeletons in SkeTo. Int. J. Parallel Prog. **42**, 546–563 (2013). https://doi.org/10.1007/s10766-013-0263-8

16. Enmyren, J., Kessler, C.: SkePU: a multi-backend skeleton programming library for multi-GPU systems. In: 4th Workshop on High-Level Parallel Programming and Applications (HLPP). ACM (2010)

17. Falcou, J., Sérot, J., Chateau, T., Lapresté, J.T.: Quaff: efficient C++ design for parallel skeletons. Parallel Comput. **32**, 604–615 (2006)

18. Légaux, J., Loulergue, F., Jubertie, S.: Managing arbitrary distributions of arrays in Orléans Skeleton Library. In: International Conference on High Performance Computing and Simulation (HPCS), Helsinki, Finland, pp. 437–444. IEEE (2013). https://doi.org/10.1109/HPCSim.2013.6641451

19. Pelagatti, S.: Structured Development of Parallel Programs. Taylor & Francis, Milton Park (1998)

20. Philippe, J., Loulergue, F.: PySke: algorithmic skeletons for Python. In: International Conference on High Performance Computing and Simulation (HPCS), Dublin, Ireland, pp. 40–47. IEEE (2019)

21. Philippe, J., Loulergue, F.: Towards automatically optimizing PySke programs (poster). In: International Conference on High Performance Computing and Simulation (HPCS), Dublin, Ireland, pp. 1045–1046. IEEE (2019)

22. Sujeeth, A.K., et al.: Delite: a compiler architecture for performance-oriented embedded domain-specific languages. ACM Trans. Embed. Comput. Syst. **13**, 134:1–134:25 (2014). https://doi.org/10.1145/2584665

Distributed and Parallel and Network-Based Computing

Impromptu Rendezvous Based Multi-threaded Algorithm for Shortest Lagrangian Path Problem on Road Networks

Kartik Vishwakarma and Venkata M. V. Gunturi$^{(\boxtimes)}$

IIT Ropar, Punjab, India
{2017csm1001,gunturi}@iitrpr.ac.in

Abstract. Input to the shortest lagrangian path (SLP) problem consists of the following: (a) road network dataset (modeled as a time-varying graph to capture its temporal variation in traffic), (b) a source-destination pair and, (c) a departure-time (t_{dep}). Given the input, the goal of the SLP problem is to determine a fastest path between the source and destination for the departure-time t_{dep} (at the source). The SLP problem has value addition potential in the domain of urban navigation. SLP problem has been studied extensively in the research literature. However, almost all of the proposed algorithms are essentially serial in nature. Thus, they fail to take full advantage of the increasingly available multi-core (and multi-processor) systems. However, developing parallel algorithms for the SLP problem is non-trivial. This is because SLP problem requires us to follow *Lagrangian reference frame* while evaluating the cost of a candidate path. In other words, we need to relax an edge (whose cost varies with time) only for the time at which the candidate path (from source) arrives at the head node of the edge. Otherwise, we would generate meaningless labels for nodes. This constraint precludes use of any label correcting based approaches (e.g., parallel version of Delta-Stepping at its variants) as they do not relax edges along candidate paths. Lagrangian reference frame can be implemented in label setting based techniques, however, they are hard to parallelize. In this paper, we propose a novel multi-threaded label setting algorithm called IMRESS which follows *Lagrangian reference frame*. We evaluate IMRESS both analytically and experimentally. We also experimentally compare IMRESS against related work to show its superior performance.

1 Introduction

The problem of path finding in road networks has been of great importance. Its value addition potential gets boosted even further when the path-finding algorithms start to consider the traffic congestion patterns. A recent report from McKinsey [12] estimates that we can save billions of dollars (annually) by helping vehicles avoid traffic congestion. Preliminary evidence of potential of a

© Springer Nature Switzerland AG 2020
S. Wen et al. (Eds.): ICA3PP 2019, LNCS 11944, pp. 201–222, 2020.
https://doi.org/10.1007/978-3-030-38991-8_14

traffic-aware path-finding algorithm has been already highlighted in the research literature (e.g., [18]). These results [18] indicate that a traffic-aware path-finding algorithm can suggest routes which have (on average 30%) less travel-time.

Given the overall significance of the problem area, several researchers (e.g., [3,5,6,10,15]) have been exploring it from different aspects. Among these, the most fundamental being: computing the fastest path between a source and destination for a given departure-time (at source). In this problem, the underlying road network (and its time-varying traffic congestion) is conceptualized as a time varying graph [8] (more details later). Given a time-varying graph representation of the road network, a source s, a destination d and, a departure-time (t_{dep}); the goal is to determine a journey $\mathcal{L}$ between s and d which departs from s at time t_{dep}. The key property of $\mathcal{L}$ being that there does not exist any other journey $\mathcal{L}'$ between s and d which departs from s at time t_{dep} but arrives at destination earlier than $\mathcal{L}$. This optimal journey $\mathcal{L}$ has been referred to as a *shortest lagrangian path* in some literature [10,11] (more details later). Note that though the other researchers (e.g., [3,5,15]) did not use the term "Lagrangian" in their papers, but they essentially use the same concept.

Over the years, several works proposed solutions for the shortest lagrangian path problem (e.g., [3,5,7,15]). However, all of the proposed techniques are essentially serial in nature and thus, they fail to take full advantage of the increasingly available multi-core and multi-processor machines. Moreover, the serial nature of current solutions also leads to slower response times for queries where the source and destination are far apart, unless a significant amount of pre-computation is done (e.g., [4]). Excessive pre-computation is not desirable any real-life navigation system as any real road network is bound have edge updates due to events like road blocks, construction, etc. Solutions which rely heavily on pre-computed data-structures may force us to make the navigational system offline in order to update the required pre-computed data structures in case of changes to the underlying road network.

Challenges: Given that we have a time-varying graph, the cost of any particular edge would vary with time. Now, if an edge needs to be relaxed, we need to *specify the time-coordinate* at which this particular edge needs to be relaxed. And SLP problem requires us to relax an edge *only for the time when a candidate path* (from source) arrives at the head node of the edge. In other words, if we want to relax an edge $e = (u, v)$, then we should do it only for the time at which the candidate path (under consideration) arrives at node u. Failure to do so would lead to meaningless (and multiple, if the edge is relaxed for all times) labels for the tail node v. As mentioned earlier, this temporal constraint on edge relaxation has been followed by almost all the well cited works in the area (e.g., [3,5,7,15]). Some researchers [10] choose to give this constraint a unique name called as the *Lagrangian reference frame.*

Limitations of the Related Work: Though there have been several works in the area of parallel algorithms for the shortest paths problem (e.g., [1,2, 14,16,17]), they cannot be extended for our SLP problem because they cannot be modified to follow the *Lagrangian reference frame.* Parallel algorithms

[1,2] based on label correcting based approaches are in principle *not suitable* for implementing Lagrangian reference frame. This is because label correcting based approaches do not follow any notion of candidate paths. Edges are repeatedly relaxed until a criteria is met. Therefore, we cannot adapt a label correcting based approach to relax an edge for a particular time-coordinate as required by the Lagrangian reference frame. Label setting based approaches (e.g., Dijkstra's and its variants) are better suited for implementing Lagrangian reference frame, however, they are hard to parallelize.

Sub-graph centric works (e.g., [14,16,17]) parallelize shortest path computation by *distributing the work across a set of independent sub-graphs*. Each sub-graph is assigned to a different processor. Then, a label setting algorithm is applied on each sub-graph. The final shortest path is computed by stitching the partial paths computed on sub-graphs via synchronization. These techniques (e.g., [14,16]) also cannot consider the Lagrangian reference frame as we cannot pre-determine the time-coordinates required for relaxing the edges in different sub-graphs. Edges (in different sub-graphs) could be somewhere in the middle of the optimal path and thus, there is no way to pre-determine when the candidates paths from the source node would arrive at these edges. Algorithms based on contraction hierarchies [4] are also not suitable for the SLP problem as they heavily rely on a intricately pre-computed data-structure (short-cuts for different times).

Our Contributions: This paper makes the following contributions: (a) propose a novel multi-threaded label setting algorithm (referred to as the *IMRESS* algorithm) for the shortest Lagrangian path (SLP) problem; (b) evaluate the performance of IMRESS analytically and; (c) conduct extensive experimental analysis and compare with the alternative approaches.

Novelty of the proposed approach: IMRESS internally uses multiple threads for computing a solution for the given SLP problem instance. IMRESS is adaptive in the sense that given p threads, it can automatically distribute the load across p threads uniformly. It would use one thread for the forward search, one thread of termination check, and remaining $p - 2$ for the backward search (referred to as trace search in this paper) from the destination node. Also, IMRESS can be trivially modified to solve the latest departure path problem [11]. It is important to note that though there have been works (e.g., [5,15]) which use the notion of a bi-directional search in road network datasets, our technique is different from them for the following two reasons:

1. The backward search in the IMRESS algorithm is executed on the reverse of the input time-varying graph itself. This is unlike the current works (e.g., [5,15]) where the backward search is done on the underlying static graph.
2. IMRESS uses multiple trace search instances to the cover the *temporal search space* of potential "departure-times" at the destination. This search space is implicit in any bi-directional search strategy on time-varying graphs. Existence of this *temporal search space* is due to the fact that, unlike the forward search, we dont know the correct "departure-time" at which the backward search should started from the destination.

Outline: The rest of the paper is organized as follows. Section 2 presents the basic concepts and the formal problem definition. Details of our proposed IMRESS algorithm are given in Sect. 3. In Sect. 4, we present an analytical evaluation of the IMRESS algorithm. Section 5 presents the experimental evaluation of IMRESS and the alternative approaches.

2   Basic Concepts

Spatio-Temporal (ST) Graph: This paper models the road network as a time-varying graph. The specific model used in this paper is formally called as a *spatio-temporal graph (ST graph)* [8]. We use the same term in rest of the paper. In a ST graph, road intersections are modeled as nodes and the road segments are modeled as directed edges. Each edge is associated with a time-series, called as its *arrival time-series*, which stores the arrival time at the tail node of the edge for different departure-times at the head node of the edge.

Consider Fig. 1 which illustrates a sample road network modeled as a spatio-temporal graph. Here, nodes S,A,B,C and D are the road intersections. Time series shown next to the edges denote their respective *arrival time-series*. For instance, edge S-A is associated with the times series [1 2 3 5]. This means that if one

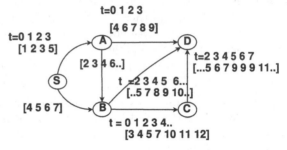

Fig. 1. A sample road network.

were to depart from node S at times $0, 1, 2$ and 3, then they would arrive at node A at times $1, 2, 3$ and 5 respectively. Temporal granularity of arrival time-series is one minute.

Lagrangian Reference Frame: While considering an edge $e = (u, v)$ in any candidate path P in ST graphs, we need to take the cost of e for the time when P arrives at the head node (in this case u) of the edge. Therefore, while determining the total cost of a path P (for a given departure time t_{dep}), we need to consider the total cost of a *journey* on P which departs from its start node at t_{dep}. This requirement on choosing appropriate time-coordinate while computing the cost of a candidate path (or an edge) is called as *Lagrangian reference frame*.

For instance, consider a journey on the path S-A-D in Fig. 1 for a departure-time of $t = 1$ at S. This journey starts at node S at time $t = 1$, reaches at node A at time $t = 2$ (via edge (S,A)), then departs from node A at time $t = 2$ and finally reaches node D at time $t = 7$ (via edge (A,D)). We considered the cost of the edge (A,D) for time $t = 2$, i.e., the time when we arrived at head node of the edge (A,D). Also, time at which we depart from any intermediate node (e.g., A) is greater than (or equal to) the time at which we arrive at that particular intermediate node. The total cost (computed via Lagrangian reference frame) of the path S-A-D is 7 time (when we depart from S at $t = 1$).

Lagrangian path is a valid specification of a *journey on a physical path (e.g., S-A-D) for a specific departure-time* at the start node of the path. The total cost of a Lagrangian path is computed via Lagrangian reference frame. When the context is clear, we may drop the term "Lagrangian" from "Lagrangian path".

Shortest Lagrangian Path: Consider a Lagrangian path $\mathcal{L}$ between two nodes u and v which departs from node u at time $t = \alpha$ and arrives at node v at time $t = \beta$. Now, if there *does not exist* any other Lagrangian path between u and v which also departs at $t = \alpha$ but reaches v before $t = \beta$; then $\mathcal{L}$ would be termed as the *shortest Lagrangian path* between u and v for the departure-time $t = \alpha$.

2.1   Problem Definition: Shortest Lagrangian Path (SLP) Problem

Input of the problem consists of the following:

- A road network represented as a ST graph $STG = \{V, E\}$.
- Each edge $e = (u, v) \in E$ ($u \in V$ and $v \in V$) is associated with an arrival time series $\Gamma_{(u,v)}()$. $\Gamma_{(u,v)}(t_d)$ denotes the time of arrival at node v, for a departure at time $t = t_d$ from u.
- A source $s \in V$, a destination $d \in V$, and departure-time at source t_{dep}.

Output

- A shortest Lagrangian path between s and d which departs from s at t_{dep}.

Scope: We assume that all the arrival time series $(\Gamma_{(u,v)}())$ in the input ST graph are FIFO in nature. This implies that an earlier departure (on any path) leads to an earlier arrival at the tail node of the path. In other words, given any particular path P, waiting at any node of P does not lead to an *earlier arrival at the tail node of P*. And this would be true for all departure-times at the start node of P. ST graphs which contain only FIFO arrival time series are called FIFO graphs. FIFO graphs is a common assumption in almost all the well respected works in this area (e.g., [3,5,6]).

3   Proposed Approach

The proposed approach consists of following three core ideas: (a) Forward Search; (b) Trace Search; and (c) Impromptu Rendezvous condition. Forward search starts exploring candidate paths form the source node. Whereas, the trace search starts exploring candidate paths from the destination. Trace search is done on the reverse ST graph which contains reverse of all the edges present in the original ST graph. Unlike the forward search, an appropriate "departure-time" is not well defined for the trace search starting at the destination. Note that the phrase "departure-time" in context on trace search actually implies the target arrival-time at the destination (detailed later in the section). Thus, to disambiguate, we would use the term *departure-time* in the context of forward search, and *trace-time* in the context of trace search. And lastly, we use *impromptu rendezvous* condition to stitch the paths generated by forward and trace searches.

Remainder of this Section is organized as follows. We first discuss the basic ideas of forward search, trace search and impromptu rendezvous condition in Sect. 3.1. Following this, we present details of our proposed algorithm (in Sect. 3.2) which puts together all the core ideas along with a termination condition.

3.1   Basic Concepts

Forward Search is similar to running an instance of Dijsktra's algorithm from the given source node on the input ST graph. As Dijkstra's was traditionally designed to execute on static graphs, it would have to be slightly modified to execute on a ST graph. The primary modification being the following: while relaxing an edge (u, v), we would pick the edge cost for the time when the optimal path reaches the head node u of the edge (u, v). This modification was proposed in [7] where the modified Dijkstra's algorithm (called SP-TAG) was able to determine a shortest Lagrangian path between a source and a destination for a given departure-time t_{dep}. We use this modified Dijkstra's for our forward search. Forward search has the following loop invariant.

Loop Invariant of Forward Search: Consider any intermediate stage during the execution of the forward search. Assume that set Ψ contains all the nodes which were closed[1] so far by the search. The following loop invariant holds: we have a shortest Lagrangian path from the source node to all the nodes in set Ψ, and each of these Lagrangian paths start from source at t_{dep}. Note that t_{dep} is the departure-time at source given as input. Readers may refer to [7] for details on correctness proof and pseudo-code. These were omitted due to lack of space.

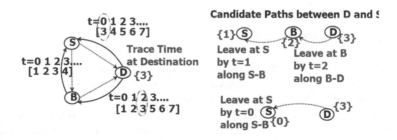

Fig. 2. Sample Trace search on a ST graph.

---

[1] A node v is added to the closed set Ψ when v is the result of extract-min from the priority queue used in the Dijkstra's implementation.

Trace Search is similar to forward search in terms of its path exploration strategy. This is in the sense that trace search also uses Dijsktra's style of enumeration for computing optimal paths. However, trace search differs from the forward search in the following three ways. First, search starts from the destination node. Second, search is performed on the reverse of the original ST graph. And lastly, during the search, the algorithm travels backwards in time starting from destination at a specific time called *trace-time*. In trace search, the label of any node v maintained during the search denotes the *latest departure-time* that one can take (at v) in order to arrive at the destination by time $t = trace - time$ (along the reverse of the current best path from destination to node v). Consequently, when a node v (with a label α) is closed by the trace search, it implies that we have a Lagrangian path from v to the destination which has the following two properties: (a) It starts as late as possible (at $t = \alpha$) from the v and arrives at destination by $t = trace - time$; (b) There does not exist any other path between v and destination on which we can start later than $t = \alpha$ but arrive at destination on or before $t = trace - time$.

Figure 2 illustrates trace search on a sample ST graph. Here, edges drawn using dotted lines represent the original edges in the ST graph and edges drawn with bold lines represent their corresponding reverses. Consider the original edge (S,D) and its reverse (D,S). The original edge had the arrival time series [3 4 5 6 7] for departure-times $t = 0, 1, 2, 3$ and 4 (at S). In trace search, we would read inverse of this arrival time series from the perspective of the node D. In other words, given a trace-time, e.g., 3 at D, we ask the following question: when is the latest one can leave from S so as to arrive at D on (or before) time $t = 3$?

Consider again the reserve ST graph shown in Fig. 2. Now, given a trace-time of $t = 3$ at D, the goal of the trace search is to determine *latest departure* Lagrangian paths from each of the other nodes (i.e., nodes S and B) such that they arrive at D by time $t = 3$. In other words, from each of the other nodes (nodes S and B), we determine a shortest Lagrangian path such that it arrives at D by time $t = 3$ and with an added constraint that we would like to depart at late as possible. Now, in the example illustrated in Fig. 2, we have two paths between D and S in the reverse ST graph. One being D-B-S, and other

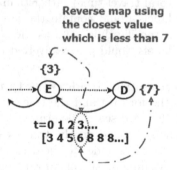

Reverse map using the closest value which is less than 7

$\{3\}$

E .... D $\{7\}$

t=0 1 2 3...
[3 4 5 6 8 8 8...]

Fig. 3. Illustrating case where we cannot reverse-map a trace-time.

one D-S (direct connection). As the figure shows, if we would like to travel via D-S (i.e., S-D in the original ST graph), then we have to depart from S at time $t = 0$. In contrast, if we travel via D-B-S (i.e., S-B-D in the original ST graph), then we can depart from S at time $t = 1$. Therefore, the trace-search would output D-B-S as the *latest departure* Lagrangian path for node S (for arriving at D by $t = 3$).

**Map to the latest
departure-time at E**

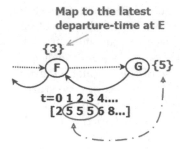

Fig. 4. Illustrating case of ambiguity in reverse-mapping a trace-time.

It is important to note that we may not always be able to reverse-map the target arrival times on nodes. Figures 3 and 4 illustrate two such cases. In Fig. 3, we need to reverse-map the target arrival-time of 7 on node D along the edge D-E in the reverse ST graph. However, the arrival time series of original edge is [3 4 5 6 8 8 8...] (for $t = 0, 1, 2, 3, 4, \ldots$). In such a case, we would reverse-map 7 to a departure-time of 3 (at E) which is the latest departure-time at E to reach D by $t = 7$. Now consider Fig. 4. Here, we need to reverse-map an arrival-time of 5 on node G along the edge G-F in the reverse ST graph. However, the arrival time series of the original edge is [2 5 5 5 6 8...] (for $t = 0, 1, 2, 3, 4, \ldots$). Similar to previous case, we would reverse-map 5 to a departure-time of 3 (at F) which is the latest departure-time at F to reach G by $t = 5$. With the intention of keeping the discussion concise, we are not providing a detailed pesudo-code of the trace search. This algorithm would would be very similar to the forward search with just following three changes:

1. Use reverse of the original ST graph.
2. Relax the edges using the reverse-mapping techniques discussed previously.
3. Priority queue is ordered on the absolute value of path lengths (in terms of total travel-time) rather than the node labels. This is because, now the node labels would contain the latest departure-time to reach the destination by a specific time. And as the algorithm proceeds, the absolute values of these labels would progressively decrease (as path-lengths increase).

Impromptu Rendezvous Condition is used to stitch the paths determined by the forward and the trace searches. As explained earlier, in each iteration, the forward search determines a shortest Lagrangian path to a node (from the source). And similarly, trace search determines a latest departure Lagrangian path from a node (to the destination).

Assume a pair of forward and trace searches running in parallel. Forward search is processing paths for the departure-time t_{dep} (at source) and, trace search is processing for trace-time t_{arr} (at destination). Consider a node v which is closed by the forward search with a label α. In a similar fashion, consider a case when the trace search also closes the same node v with a label α. This event of *closing of a common node v with same label* by both forward and trace searches is called as an *impromptu rendezvous event*.

Now, let P_{for}^v be the shortest Lagrangian path to v (from source at $t = t_{dep}$) as determined by the forward search. And similarly, let P_{trace}^v be the path (with edges reversed) determined by trace search between v and destination such that it arrives at destination by $t = t_{arr}$. Consider the path Q obtained by the joining P_{for}^v and P_{trace}^v at node v. This is a valid Lagrangian path which departs from source at time $t = t_{dep}$ and arrives at destination at time $t = t_{arr}$. In other words, each occurrence of an impromptu rendezvous event gives us a source-destination Lagrangian path which leaves source at the designated departure-time (t_{dep}) and arrives at the destination by corresponding trace-time (t_{arr}) of the trace search. Note that the notion of impromptu rendezvous was first proposed in [10] in a different context and for a different problem definition. We are generalizing it for the shortest Lagrangian path problem.

Figure 5 illustrates an instance of impromptu rendezvous condition. The figure contains a sample ST graph on the left and illustrates fully constructed forward and trace search trees on the right. The forward search tree was constructed with S as the source node and $t = 0$ as the departure-time. On the other hand, the trace search tree was constructed with D as the destination and $t = 7$ as the trace-time. Note that though the figure shows fully constructed forward and trace search trees, in our algorithm these would be constructed in parallel while checking for impromptu rendezvous condition via a shared memory. Impromptu rendezvous condition is checked after the extract-min operation (i.e., closing a node). In the example shown in Fig. 5, the optimal arrival-time of the forward search at the destination node is 7 (via the path S-A-B-D), which also happens to be our trace-time for the trace search starting from D. On comparing both the search trees, one may notice that both searches close the intermediate nodes A and B with same labels. In other words, there was a possibility of impromptu rendezvous (had the two searches were executing in parallel) on the nodes which were actually on the shortest Lagrangian path between S and D (for $t_{dep} = 0$).

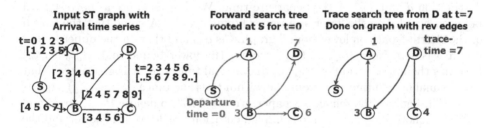

Fig. 5. Sample Forward and Trace search on a ST graph. In the search trees, numbers written next to the nodes represent their optimal labels as determined by the respective algorithms.

Previous example illustrates the fact that if the trace-time at the destination is indeed the optimal arrival-time of the forward search at the destination, then there is a chance of impromptu rendezvous along the optimal path. However, if the chosen trace-time is not the optimal arrival then we can have two cases: (1) trace-time > optimal arrival-time and, (2) trace-time < optimal arrival-time. In the first case, the forward and the trace searches can have an impromptu rendezvous along a path which is longer than the optimal path (for the given problem instance). On the other hand, we cannot have any impromptu rendezvous in the second case. Note that an impromptu rendezvous in the second case would imply that we have a source-destination Lagrangian path which starts from the source at the designated departure-time but *arrives at the destination before* our "optimal arrival-time" (a contradiction).

3.2   Details of Algorithm

This section presents details of our proposed *IMpromptu REndezvous based Shortest lagrangian path Solver* (IMRESS algorithm). The IMRESS algorithm puts together the previously proposed ideas of forward search, trace search and the impromptu rendezvous condition. Following two challenges need to be addressed in this regard:

1. **Challenge A: Trace-time at the destination is not known apriori.**
2. **Challenge B: Termination condition.** Termination condition should put together the forward and the trace searches while ensuring efficiency.

Addressing Challenge (A): We address this challenge by determining a lower bound and a upper bound on the optimal arrival time at the destination from the source node. We determine the bounds in the following way. We first derive a *lower bound graph* from our input ST graph by replacing the arrival time series of all edges with their corresponding *lowest* possible travel-time (on that particular edge). Similarly, we derive a *upper bound graph* by replacing the arrival time series with their corresponding *highest* possible travel-time (on that edge). Now consider an instance of the shortest lagrangian path problem with source s, destination d and, t_{dep} as the departure-time. We first compute a shortest path between s and d on the lower and upper bound graphs. Assume that the length of the shortest path on lower bound graph was α_{sd} and that on the upper bound graph was ω_{sd}. Now, $(t_{dep} + \alpha_{sd})$ becomes the lower bound and, $(t_{dep} + \omega_{sd})$ becomes the upper bound on the optimal arrival time at d. It is guaranteed that the optimal arrival time at d would lie within the time range $[(t_{dep}+\alpha_{sd})$ $(t_{dep}+\omega_{sd})]$. Throughout the remaining paper, we use LB_{tr} to denote $(t_{dep} + \alpha_{sd})$ and UB_{tr} to denote $(t_{dep}+\omega_{sd})$. In our implementation, we derived the lower and the upper bound graphs and, pre-computed the shortest distance between all pairs of nodes in each of the graphs. These results were stored for further use during processing of the queries. Comments on the suitability of this pre-computation in a real life navigational system are given later in the section.

Addressing Challenge (B): Termination condition of the IMRESS algorithm has the following facets: (a) termination of the Forward search instance; (b) termination of the Trace search instance(s) and; (c) termination of the overall algorithm. We now detail these facets.

Forward Search Termination: Consider an instance of the problem with source s, destination d and, t_{dep} as the departure-time (at source). We would first compute lower $(t_{dep}+\alpha_{sd})$ and upper bound $(t_{dep}+\omega_{sd})$ on the optimal arrival times at the destination using the previously described procedure. Following this, we would spawn instances of forward and trace search. Here, we would be having only one instance of the Forward search which starts exploring paths for $t = t_{dep}$. Definition 1 presents the termination condition of forward search.

Definition 1. *Given an instance of problem with source s, destination d and, t_{dep} as the departure-time. It is sufficient to terminate the forward search instance of the IMRESS algorithm when it closes d, i.e., the result of the extract-min is node d.*

Trace Search Termination: Trace search has two very unique aspects to consider with regards to termination. First, we would require several instances of Trace search executing in parallel, each with a different trace-time. And second, termination of any one particular instance of the trace search sometimes depends on the status of other instances. We now detail these aspects.

Consider again our previous problem instance with source s, destination d and, t_{dep} as the departure-time. Now ideally, we would need $((t_{dep} + \omega_{sd}) - (t_{dep} + \alpha_{sd}) + 1)$ number of Trace search instances to be executed in parallel, each with a unique trace-time from the interval $[(t_{dep} + \alpha_{sd})\ (t_{dep} + \omega_{sd})]$. However, such an approach would require a lot of CPU cores in the system. Moreover, thread scheduling becomes a major challenge when the number threads exceed the number of available CPU cores. This would in turn effect the termination of the algorithm. To this end, we set the maximum number of Trace search instances to be run in parallel (parameter Λ) according to the number of CPU cores available in the system.

Following this the entire horizon of interest $[(t_{dep} + \alpha_{sd})\ (t_{dep} + \omega_{sd})]$ is divided among Λ threads. For instance, thread 1 would instantiate trace search for trace-times $(t_{dep} + \alpha_{sd}), (t_{dep} + \alpha_{sd} + \Lambda), (t_{dep} + \alpha_{sd} + 2\Lambda), \ldots$, etc. Similarly thread 2 would instantiate trace search for trace-times $(t_{dep} + \alpha_{sd} + 1), (t_{dep} + \alpha_{sd} + \Lambda + 1), (t_{dep} + \alpha_{sd} + 2\Lambda + 1), \ldots$, etc. Note that each thread would process its respective trace-times sequentially in the increasing order (of trace-times). But, at any particular instant we would be having only Λ instances of trace search running in parallel. Definition 2 formally presents the termination condition of any particular instance of trace search.

Definition 2. *Consider an instance of problem with source s, destination d and, t_{dep} as the departure-time at source. Now, given Λ instances of trace search executing in parallel, any particular trace search instance (with trace-time t_α) is terminated when any one of the following four conditions is satisfied.*

1. *Source node s is closed by that instance.*
2. *Label of the node in the top of heap (μ) in the trace search instance is less than t_{dep} (i.e., $t_{dep} > \mu$). In other words, this trace search instance can never get a Lagrangian path which starts from s at time t_{dep}.*
3. *It encounters an impromptu rendezvous event.*
4. *Another instance of trace search with smaller trace-time t_β (i.e., $t_\beta < t_\alpha$) encounters impromptu rendezvous event.*
5. *The forward search closes d.*

Note that conditions 4 and 5 defined in Definition 2 are used to prune later instances of trace searches as well. Reasoning behind condition 5 is trivial. Regarding condition 4, if a trace search instance terminates via this condition, then (as discussed earlier) it implies that we already have a Lagrangian path which starts from s at t_{dep} and arrives at d by t_β. Thus, there is no need to explore the existence of longer Lagrangian paths (by considering a trace search with trace-time $> t_\beta$).

Overall Termination of IMRESS Algorithm: IMRESS algorithm has one instance of forward search whose termination condition is governed by Definition 1. And it has $((t_{dep} + \omega_{sd}) - (t_{dep} + \alpha_{sd}) + 1)$ number of trace search instances (executing Λ instances at a time) whose termination is governed by Definition 2. Under these conditions, the termination condition stated in Proposition 1 is guaranteed to provide a correct answer. Note that in case IMRESS terminates by condition (2) of Proposition 1, then the path resulting from the impromptu rendezvous of the youngest trace-search (i.e., lowest trace-time) would be returned as the answer. The only exception to this being: presence of a trace search instance which closed s with a label t_{dep}, in which case that would be the answer.

Proposition 1. *Given an instance of the problem with source s, destination d and departure-time t_{dep}. Consider an instance of IMRESS algorithm where forward search termination is defined by Definition 1 and trace search termination is defined by Definition 2. It is sufficient to terminate the algorithm when one of the following two things happen:*

1. *Forward search terminates according to condition stated in Definition 1.*
2. *All $((t_{dep} + \omega_{sd}) - (t_{dep} + \alpha_{sd}) + 1)$ trace search instances terminate according to Definition 2.*

Data Structures Required in Implementation: For implementing Proposition 1, we need some global data-structures as detailed next. Firstly, the closed lists of the forward and the trace search(s) are stored in a globally shared memory. Secondly, we need to maintain a globally shared location which stores the details of any impromptu rendezvous event happened so far. This location stores the *trace-time* of the *youngest trace search* which encountered an impromptu rendezvous event. We refer to this trace-time as the *imr-trace-time*. Recall that occurrence of an impromptu rendezvous event does not guarantee optimality. It just guarantees the presence of a path which starts from source at time t_{dep} and arrives at destination by time *imr-trace-time*. Thus, as the algorithm proceeds, we would lower the value of *imr-trace-time* via subsequent impromptu rendezvous events. As IMRESS algorithm employs multiple threads, several threads may attempt to write this memory location concurrently. To this end, we maintain a critical section around it which would allow only one thread to update it at a time. While updating this memory location, the updating thread would be allowed to make the change iff the current value of *imr-trace-time* is greater than the new value. In other words, once an younger trace search (i.e., lower trace-time) encounters a impromptu rendezvous, all the older trace searches (i.e., those having larger trace-time) are effectively invalidated (item 4 of Definition 2).

In our implementation, only the trace search instances keep checking for an impromptu rendezvous event. We made this design decision to avoid impromptu rendezvous checks (by forward search) against Λ closed lists of trace searches. The check is performed immediately after every extract-min operation in the trace search. In case an impromptu rendezvous event happens, then it checks if the current value of $imr\text{-}trace\text{-}time$ can updated.

Practical Aspects in Implementing Proposition 1: It is important to note that though Proposition 1 guarantees theoretical correctness, it may not always guarantee better performance over the serial algorithm [7]. This can primarily happen because of the following two reasons: (i) the given source and destination nodes in the problem instance are very close to each other and, (ii) lower number of cores available. We now detail these aspects. In case (i), the extra time which is needed for creating the Λ threads starts to outweigh any benefits produced by the bi-directional search. As a result the algorithm is likely to terminate when the forward search closes the destination (condition (1) of Proposition 1). Note that, though this will give correct answer, we would not get any speed-up over the serial algorithm [7] which (as we mentioned earlier) is essentially the forward search algorithm. Thread creation overhead is dependent on the thread library used and is thus independent of the details of the proposed algorithm.

Now consider the case when the given value of Λ happens to be very small (case (ii)). In this case, the trace search will now take a lot of time to go through its $((t_{dep}+\omega_{sd})-(t_{dep}+\alpha_{sd})+1)$ instances. And once again, the algorithm would terminate when the forward search closes the destination, thereby no speed-up over the serial algorithm. We address case (ii) though the following design decision. The algorithm would wait only a bounded amount of time (parameter ϕ) after the first impromptu rendezvous before declaring the result. The rationale behind the parameter ϕ is the following. Recall that an impromptu rendezvous event just gives us a $s\text{-}d$ path which departs from s at time t_{dep} and arrives at d by the designated trace-time of the trace search instance involved in the impromptu rendezvous event. And, this path $s\text{-}d$ path may not be optimal. Now, if the algorithm waits for ϕ time units after the first impromptu rendezvous event, then more impromptu rendezvous events can happen and potentially improve the quality of the solution by lowering the value of $imr\text{-}trace\text{-}time$ as much as possible. Note that $imr\text{-}trace\text{-}time$ cannot go below a optimal value which is indeed the optimal arrival time of the forward search at d. In case there was no impromptu rendezvous, then the algorithm would terminate via the following conditions: (a) forward search closing d or, (b) a trace search closing s with a label t_{dep} (the departure-time given in the problem instance). Definition 3 formally presents an adapted version of Proposition 1 after including the previously discussed parameter ϕ.

Definition 3. *Given an instance of the problem with source s, destination d and departure-time t_{dep}. Consider an instance of IMRESS algorithm where forward search termination is defined by Definition 1 and trace search termination* *is defined by Definition 2. IMRESS terminates when one of the following three things happen:*

1. *Forward search terminates according to condition stated in Definition 1.*
2. *A trace search instance closes d with a label t_{dep}.*
3. *ϕ time units have elapsed since the first impromptu rendezvous event.*

IMRESS implements Definition 3 for termination. Note that if the IMRESS terminates as per condition (1) or (2) in Definition 3, then we are guaranteed to a get the correct answer. However, we may not get any speed-up over the serial algorithm. In case of condition (3), we may get incorrect answer if the thread corresponding to the optimal trace-time does not get an impromptu rendezvous (due to CPU scheduling or lower value of Λ) within ϕ time units of the first impromptu rendezvous event. However, we are extremely likely to get a speed-up over the serial algorithm in this case. As one can imagine, a larger the value of ϕ would push us closed to the optimal answer. However, a larger value of ϕ also means lower speed-up over the serial algorithm. This trade-off between correctness and efficiency is explored in detail in the experiments. Our experiments reveal that by just having $\phi = 10$ millisecs, **IMRESS was able to bring down the total deviation (from the optimal answer) to under 4 min (in all our datasets), while still maintaining a speed-up over the serial algorithm**.

Algorithm 1 details a pseudo-code of the IMRESS algorithm. In our implementation, we pre-compute the all-pair shortest distances on the lower and upper bound graphs. After initialization of data structures, the algorithm starts a parallel section at line # 10. In this parallel section, we create $\Lambda + 2$ threads. Out of these, Λ threads are for trace search instances, one for forward and, one for checking the termination condition. The thread which is responsible for checking termination implements Definition 3.

Regarding Pre-computation Done in IMRESS: Events such as road construction and traffic blocks (due to accidents) may increase the lower and upper bound travel-times. Increased lower bound is not an issue as the older value of lower bound would still theoretically continue to be a lower bound. However, if the current optimal travel-time becomes larger that our stored upper bound, then we would not get any impromptu rendezvous, hence no speed-up over the serial algorithm. In that case we need to re-compute the upper bound shortest distances. One can even start with slightly higher upper bounds to be on the safer side. Overall, we argue that, in any system based on IMRESS, changes to the underlying road network *would not require us to make the system offline* for applying the updates. We can just re-compute the new upper and lower bound distances separately and then update the required data-structures by putting a temporary lock on the required fields.

Algorithm 1. IMRESS algorithm for computing shortest Lagrangian paths

Input: Source node s, Destination node d, Departure-time t_{dep}, Max # Trace search threads (Λ), Max time to wait after first Impromptu Rendezvous (ϕ)
Output: Shortest Lagrangian path between s and d for time t_{dep}

1: Compute α_{sd} and ω_{sd} on the lower and upper bound graphs.
2: $LB_{tr} \leftarrow t_{dep} + \alpha_{sd}$
3: $UB_{tr} \leftarrow t_{dep} + \omega_{sd}$
4: Initialize closed list for Forward search
5: Initialize Λ #closed lists corresponding to Λ Trace search instances
6: **for all** threads $t_i \in [1 \ \Lambda]$ **do**   /*Thread 0 is reserved for forward search*/
7:     $\lambda_{t_i} \leftarrow$ ordered list of trace-times for thread t_i
8:     Set λ_{t_i} to $[(LB_{tr} + t_i - 1), (LB_{tr} + t_i - 1 + \Lambda), (LB_{tr} + t_i - 1 + 2\Lambda)\ldots]$
9: **end for**
10: Start parallel section by creating $\Lambda + 2$ threads /*Thread# $t_i \in [0, 1, \ldots, \Lambda + 2]$*/
11: **if** $t_i == 0$ **then**
12:     Assign forward search from s (for time t_{dep}) to thread 0
13: **else if** $t_i \geq 1$ and $t_i \leq \Lambda$ **then**
14:     **for all** trace-times tr_{arr} in increasing order from λ_{t_i} **do**
15:         Thread t_i starts trace-search from d with trace-time tr_{arr}
16:         **if** next tr_{arr} in $\lambda_{t_i} <$ trace-time which had impromptu rendezvous **then**
17:             Break from loop
18:         **end if**
19:     **end for**
20: **else**   /*$t_i == \Lambda + 2$*/
21:     **while** Termination condition in Definition 3 not met **do**
22:         Thread t_i checks for conditions in Definition 3 and waits for ϕ time units after the first impromptu rendezvous.
23:     **end while**
24:     Output the source-destination path.
25: **end if**
26: Kill all $\Lambda + 2$ threads
27: End parallel section started on Line 10

4   Analytical Evaluation

Lemma 1. *Pre-processing cost: Consider an ST graph $\{V, E\}$ with n nodes and m edges, where each edge $e = (u, v) \in E$ is associated with a arrival-time series $(\Gamma_{(u,v)})$. Additionally, assume that each of the arrival-time series is of length $|\Gamma|$. Total cost of pre-processing in IMRESS is at most be $O(n^3 + m|\Gamma| + k|\Gamma|^2)$, where k is number of edges (if any) which exhibit non-FIFO behavior.*

Query cost: Consider an instance of our problem on this ST graph with source s, destination d, departure-time t_{dep} (at source). Assume that LB_{tr} and UB_{tr} are the lower and upper bounds on trace-times. An instance of the IMRESS algorithm could in worst case have $O((UB_{tr} - LB_{tr})(m + n) \log n)$ amount of work over its execution (including the multi-threaded part).

Proof. **Pre-processing Cost:** Computing the lower and upper bound graphs of the input ST graph would take $O(m|\Gamma|)$ time. Note that we would have to convert the arrival-time series into travel-time series for computing the lower and upper bound graphs. This is done by subtracting the departure-time from the arrival-time which would take another $O(|\Gamma|)$ per edge. After computing the lower and upper bound graphs, we compute the shortest distance between all pair of vertices in these graphs. This entails a cost of $O(n^3)$. Results of these pre-computations are stored. And lastly, if any $\Gamma_{(u,v)}$ is non-FIFO in nature (checking takes $O(|\Gamma|)$ time per edge), then it would be converted into FIFO via the earliest arrival time transformation [11] which takes $O(|\Gamma|^2)$ time. If k of the m edges show non-FIFO behavior, then the total pre-processing cost in worst case would be $O(n^3 + 3m|\Gamma| + k|\Gamma|^2)$, which is $O(n^3 + m|\Gamma| + k|\Gamma|^2)$.

Query Cost: Now, consider an instance of the SLP problem. We are given $(\Lambda + 2)$ threads for executing the IMRESS algorithm. Steps 1–3 of the IMRESS algorithm (Algorithm 1) determine the lower and upper bound of trace-times at d. Let LB_{tr} be the lower bound and UB_{tr} upper bound on trace-times. Following this, Steps 4 and 5 initialize the closed lists which take a total cost of $O(n\Lambda)$. In Steps 6–9, IMRESS creates lists of trace-times to be handled by each of the Λ trace search instances. Note that even though Algorithm 1 includes steps (7 and 8) on initializing trace-time lists for all threads, we can get a faster implementation by just storing the first trace-time of each thread and its increment value (Λ). This would total to $O(\Lambda)$ cost. Algorithm 1 spawns the prescribed $(\Lambda + 2)$ number of threads in step 9. Following this, tasks of forward search, $(UB_{tr} - LB_{tr} + 1)$ trace search instances, and checking the termination, are distributed among the $(\Lambda + 2)$ threads. We now detail the overall complexity of each of these tasks.

Consider the cost of forward search first. As mentioned earlier, forward search in IMRESS is very similar to running a temporal generalization of Dijsktra's (SP-TAG [7]) on the input ST graph. Thus, the total work done by the thread assigned to the forward search would be $O((m + n) \log n)$ (if priority queue is implemented as a min-heap).

Now consider the cost of a single trace search instance. Algorithmically, trace search is also very similar to a Dijkstra's algorithm. The only exception to this statement being the five termination conditions mentioned in Definition 2, each of which has to be checked after every extract-min operation. However, each of these conditions would require only $O(1)$ for checking. Thus, the overall cost of a single trace search instance remains $O((m + n) \log n)$. IMRESS may have to execute $(UB_{tr} - LB_{tr} + 1)$ trace search instances in the worst case. Thus, the total amount of work across the Λ threads (designated for trace searches) can in worst case be $O((UB_{tr} - LB_{tr})(m + n) \log n)$ as $\Lambda \leq (UB_{tr} - LB_{tr} + 1)$. Finally, consider the work done by the thread responsible for checking the overall termination given in Definition 3. Checking the conditions given in Definition 3 would take only $O(1)$ time.

Thus, the worst case time complexity of IMRESS is $O((UB_{tr} - LB_{tr})(m + n) \log n + n\Lambda)$, which is $O((UB_{tr} - LB_{tr})(m + n) \log n)$. Note that this absolute worst case is possible only when the thread corresponding to the forward search gets descheduled for a long time.

Proposition 2. *The total space requirement of the IMRESS consists of the following: (a) Storing the input ST graph which requires $O(m|\Gamma| + n)$ space; (b) storing the all-pair shortest distances in lower and upper bound graphs which requires $O(n^3)$ space and; (c) Priority queue, closed lists and predecessor information in $(\Lambda + 1)$ threads which totals to $O(\Lambda n)$ space. Thus, the total space complexity of IMRESS is $O(n^3 + \Lambda n + m|\Gamma|)$. $|\Gamma|$ is the length of the arrival time-series.*

5 Experimental Analysis

5.1 Datasets

	%edges whose travel time changes in rush hour	Max % increase in travel-time during rush-hour
Dataset I	35	40
Dataset II	35	50
Dataset III	45	40

Fig. 6. Details of different datasets used in experiments.

We used the San Francisco road network [13] which contained **50,000** nodes and **99,999 edges**. We created three different datasets by simulating the traffic (in different ways) on this underlying road network. In each of these datasets, we basically control the following two parameters: (a) percentage of edges whose travel-time changes during the rush hours (morning and evening) and; (b) maximum increase in the travel-time that the edges would experience during the rush hours. Figure 6 details the variation of these parameters in the datasets used in our experiments. Morning rush hours were taken to be between 7:30am – 9:30am, and evening rush hours were taken to be between 6:00pm – 8:00pm. This trend was assumed to repeat Monday through Friday. Each of the datasets mentioned in Fig. 6 were created in the following way. Consider Dataset I for instance. Here, we first select 35% of edges (of the dataset) in a uniformly random fashion. Following this, for each of the selected edges (in previous step), we gradually increase its travel-time (from the baseline) during the rush hours (morning and evening) and then bring it down to the baseline travel-time by the end of the rush hour. The baseline travel-time of an edge is computed as $\frac{edge-length}{Speed-limit}$. Travel-time was measured in minutes and was represented as integers. Temporal granularity of the arrival time-series in ST graph is also one minute. Note that we did vary the size of the underlying road network as well. However, those experiments did not yield any different results and, thus were omitted from the paper. This behavior was expected given that IMRESS is a label setting algorithm. In any label setting algorithm, the total run-time is usually proportional to the number of nodes closed (and edges relaxed) until the destination is reached.

5.2   Candidate Algorithms

IMRESS was compared against the following three alternative approaches:

1. **SP-TAG Algorithm** [7]: SP-TAG is a single threaded algorithm for computing shortest Lagrangian paths on ST graphs. This algorithm is essentially a temporal generalization of Dijkstra's algorithm for ST graphs.
2. **Parallel SP-TAG (Para-SP-TAG):** In this algorithm, we use multiple threads in the inner loop of the SP-TAG algorithm. One may recall that, after the extract-min operation, a Dijkstra's styled algorithm (and thus SP-TAG) relaxes the out-going edges from the result of extract-min in its inner loop. In parallel SP-TAG, we use multiple threads for this loop.
3. **SP-TAG with Distributed Heap (SP-TAG-Dist-Heap):** This technique is based on the idea of distributing priority queue [9]. Here, we divide the vector of current labels into p threads. Each thread maintains a local priority queue (implemented as a min-heap) of $\frac{n}{p}$ labels, where n is the number of nodes in the input ST graph. Each of the p threads perform extract-min locally, then the global minimum is determined through synchronization. Global minimum is shared with all threads. Following this edge relaxations are done locally.

Table 1. Comparing SP-TAG, Parallel SP-TAG and SP-TAG with Distributed Heap. Running times are in milli-seconds. Dataset I, rush-hour departure-times.

Path length (mins)	SP-TAG	Para-SP-TAG	SP-TAG-Dist-Heap 3 threads	SP-TAG-Dist-Heap 10 threads
120	100.2	395.8	1350.3	1750.5
140	96.09	561.9	1288.6	1690.9
160	175.8	887.1	1382.1	1792.8
180	228.9	953.9	1319.8	1702.8
200	308.7	1076.3	1316.0	1683.6

5.3   Parameters Varied

Departure-Time: We chose following two types of departure-times in our experiments: (a) Non-Rush hours and (b) Rush hours. Non-rush hours were taken during late night (after 9:00pm) and mid afternoon.

Path-Length: Running time of algorithms is greatly effected by the length (total travel-time) of the shortest Lagrangian path (SLP) between the given source and destination. However, as one can imagine, the total travel-time of the SLP between the same source-destination pair changes with time. In our experiments, we set the value of this parameter according to the length of the optimal path at mid-night. In other others, given any source-destination pair,

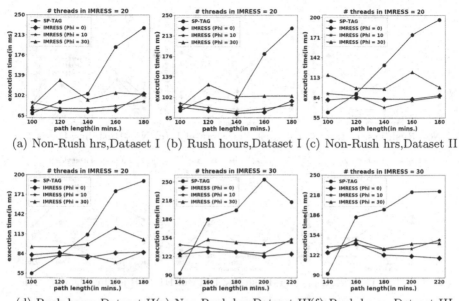

(a) Non-Rush hrs,Dataset I (b) Rush hours,Dataset I (c) Non-Rush hrs,Dataset II

(d) Rush hours,Dataset II (e) Non-Rush hrs,Dataset III(f) Rush hours,Dataset III

Fig. 7. Running time of IMRESS and SP-TAG. Y-axis Value for each path-length is an average of run-times obtained for 10 different source-destination pairs having the particular path-length. Running times are in milli-seconds.

we categorize it according to the length (total travel-time) of the SLP for a midnight departure at source. The choice of the midnight departure-time (for measuring path-length) is solely based on the intention to have interpretability in results.

Number of Threads (Λ) and ϕ: In our experiments, we varied the number of threads in SP-TAG-Dist-Heap. And in case of IMRESS algorithm, we varied both its ϕ and Λ ($\Lambda = \{20, 30\}$) parameters.

5.4   Metrics Measured and Test Environment

Following metrics were measured: (a) query execution time and, (b) Deviation (in travel-time) from the optimal result. Item (b) is measured only for the IMRESS algorithm. Query execution time does not include the time required for pre-processing tasks. Algorithms were implemented in C++. OpenMP was used for handling threads. Experiments were conducted on a Ubuntu machine which had 64 GB RAM and two 18-core processors (Intel Xeon E7-8870 v3 2.40 GHz).

5.5   Experimental Results

Performance of Para-SP-TAG and SP-TAG-Dist-Heap: Results of this experiment are shown in Table 1. We used Dataset I and rush hours departure

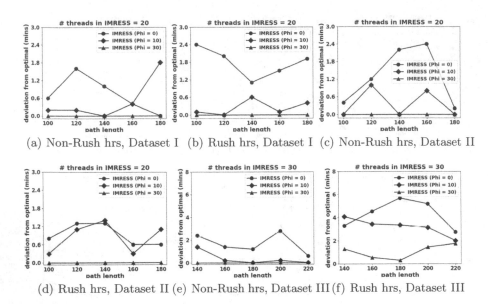

(a) Non-Rush hrs, Dataset I (b) Rush hrs, Dataset I (c) Non-Rush hrs, Dataset II

(d) Rush hrs, Dataset II (e) Non-Rush hrs, Dataset III (f) Rush hrs, Dataset III

Fig. 8. Solution quality of IMRESS. Y-axis value denotes the average of total deviation (in terms of total travel-time) from the optimal answer observed for 10 different source-destination pairs of any particular path-length (X axis).

times (e.g., 7:30 am). Each of the running time values in Table 1 is an average of 10 different source-destination pairs selected at random. As the results show both para-SP-TAG and SP-TAG-Dist-Heap take significantly more time than SP-TAG. Similar trends were seen for other values of varied-parameters as well. Therefore, we removed both of these algorithms from further experiments.

Comparing SP-TAG and IMRESS Algorithms: Results of these experiments are shown in Figs. 7 and 8. Figure 7 illustrates the running time of SP-TAG and IMRESS for different path-lengths. Solution quality of IMRESS is illustrated in Fig. 8. Here, we plot the total deviation (in terms of total travel-time) from the optimal answer on the Y-axis. Following conclusions can be drawn from these results: (a) IMRESS performs better than SP-TAG only when the path-length is long (120 min in case of Dataset I & II and 140 min in case of Dataset III). The likely reason for this being that if the source-destination pairs are very close to each other, then the thread creation overhead outweighs any benefit of bi-directional search. (b) Execution time of IMRESS grows much more slowly than SP-TAG because the total search space explored per thread in IMRESS is much less than SP-TAG. (c) As ϕ increases the solution of IMRESS is usually much closer to the optimal answer. However, in a extremely rare situation if the trace-search thread (whose trace-time was closer to optimal) gets descheduled for a very long time, then we may have poor solution even if we have longer ϕ. (d) Datasets having more rush hour traffic (e.g., Dataset III) require a larger value of Λ for getting speed-up over SP-TAG.

6 Conclusion

The SLP problem is of importance in the domain of urban navigation. While several researchers explored it from different aspects, all of the proposed solutions are essentially serial in nature. Parallel algorithms developed for the shortest path problem cannot be generalized for the SLP problem. We propose a novel multi-threaded algorithm called IMRESS. IMRESS was able to get better performance over the serial algorithm (while maintaining quality of solution) specially when the source destination nodes are far apart. In future, we would like to explore the A* based generalizations of the IMRESS algorithm.

Acknowledgement. Authors would like to thank DST SERB (grant# ECR/2016 /001053) and IIT Ropar for their support in this work.

References

1. Chakaravarthy, V.T., et al.: Scalable single source shortest path algorithms for massively parallel systems. IEEE Trans. PDS **28**(7), 2031–2045 (2017)
2. Davidson, A., et al.: Work-efficient parallel GPU methods for single-source shortest paths. In: Proceedings of the IPDPS, pp. 349–359 (2014)
3. Delling, D.: Time-dependent SHARC-routing. In: Halperin, D., Mehlhorn, K. (eds.) ESA 2008. LNCS, vol. 5193, pp. 332–343. Springer, Heidelberg (2008). https://doi.org/10.1007/978-3-540-87744-8_28
4. Delling, D., et al.: Phast: hardware-accelerated shortest path trees. J. Parallel Distrib. Comput. **73**(7), 940–952 (2013)
5. Demiryurek, U., Banaei-Kashani, F., Shahabi, C., Ranganathan, A.: Online computation of fastest path in time-dependent spatial networks. In: Pfoser, D., et al. (eds.) SSTD 2011. LNCS, vol. 6849, pp. 92–111. Springer, Heidelberg (2011). https://doi.org/10.1007/978-3-642-22922-0_7
6. Ding, B., et al.: Finding time-dependent shortest paths over large graphs. In: Proceedings of the 11th International Conference on Extending Database Technology, pp. 205–216. ACM (2008)
7. George, B., Kim, S., Shekhar, S.: Spatio-temporal network databases and routing algorithms: a summary of results. In: Papadias, D., Zhang, D., Kollios, G. (eds.) SSTD 2007. LNCS, vol. 4605, pp. 460–477. Springer, Heidelberg (2007). https://doi.org/10.1007/978-3-540-73540-3_26
8. George, B., Shekhar, S.: Time-aggregated graphs for modeling spatio-temporal networks. In: Spaccapietra, S., et al. (eds.) Journal on Data Semantics XI. LNCS, vol. 5383, pp. 191–212. Springer, Heidelberg (2008). https://doi.org/10.1007/978-3-540-92148-6_7
9. Grama, A., et al.: Introduction to Parallel Computing. Pearson (2003)
10. Gunturi, V., et al.: A critical-time-point approach to all-departure-time lagrangian shortest paths. IEEE trans. KDE **27**(10), 2591–2603 (2015)
11. Gunturi, V.M.V., Shekhar, S.: Spatio-Temporal Graph Data Analytics. Springer, Cham (2017). https://doi.org/10.1007/978-3-319-67771-2. 978-3-319-67770-5
12. Henke, N., et al.: The age of analytics: competing in a data-driven world, Mckinsey Global Institute, December 2016. https://tinyurl.com/yb7vytkg

13. Karduni, A., et al.: A protocol to convert spatial polyline data to network formats and applications to world urban road networks. Sci. Data **3**, Article no. 160046 (2016)
14. Maleki, S., et al.: DSMR: a shared and distributed memory algorithm for single-source shortest path problem. In: Proceedings of the 21st PPoPP, pp. 39:1–39:2 (2016)
15. Nannicini, G., et al.: Bidirectional a* search on time-dependent road networks. Networks **59**(2), 240–251 (2012)
16. Simmhan, Y., et al.: Distributed programming over time-series graphs. In: 2015 IEEE International Parallel and Distributed Processing Symposium, pp. 809–818 (2015)
17. Simmhan, Y., et al.: *GoFFish*: a sub-graph centric framework for large-scale graph analytics. In: Silva, F., Dutra, I., Santos Costa, V. (eds.) Euro-Par 2014. LNCS, vol. 8632, pp. 451–462. Springer, Cham (2014). https://doi.org/10.1007/978-3-319-09873-9_38
18. Yuan, J., et al.: T-drive: driving directions based on taxi trajectories. In: Proceedings of the SIGSPATIAL International Conference on Advances in GIS, GIS 2010, pp. 99–108 (2010)

FANG: Fast and Efficient Successor-State Generation for Heuristic Optimization on GPUs

Marcel Köster[✉], Julian Groß, and Antonio Krüger

Saarland Informatics Campus, Campus D3.2, 66123 Saabrücken, Germany
{marcel.koester,julian.gross,antonio.krueger}@dfki.de

Abstract. Many optimization problems (especially nonsmooth ones) are typically solved by genetic, evolutionary, or metaheuristic-based algorithms. However, these genetic approaches and other related papers typically assume the existence of a neighborhood or successor-state function $N(x)$, where x is a candidate state. The implementation of such a function can become arbitrarily complex in the field of combinatorial optimization. Many $N(x)$ functions for a huge variety of different domain-specific problems have been developed in the past to solve this general problem. However, it has always been a great challenge to port or realize these functions on a massively-parallel architecture like a Graphics Processing Unit (GPU). We present a GPU-based method called *FANG* that implements a generic and reusable $N(x)$ for arbitrary domains in the field of combinatorial optimization. It can be customized to satisfy domain-specific requirements and leverages the underlying hardware in a fast and efficient way by construction. Moreover, our method has a high scalability with respect to the number of input states and the complexity of a single state. Measurements show significant performance improvements compared to traditional exploration approaches leveraging the CPU on our evaluation scenarios.

Keywords: Heuristic search · Combinatorial optimization · Successor-state generation · Neighborhood exploration · Massively-parallel processing · Graphics processing units · GPUs

1 Introduction

There are many different optimization algorithms for a huge variety of problems. Every problem can be assigned to a category and different methods are used to solve a problem–even in the scope of a single problem category. Convex optimization problems have the advantage that a local optimum is equal to the global optimum. However, so called *nonsmooth optimization problems (NSPs)* are typically assumed to be non-convex. Furthermore, it is not possible in general to determine the direction into which an optimizer has to continue from a certain

This work was funded by the Germany Federal Ministry for Economic Affairs and Energy (BMWi): Project BloGPV (grant number 01MD18001B).

© Springer Nature Switzerland AG 2020
S. Wen et al. (Eds.): ICA3PP 2019, LNCS 11944, pp. 223–241, 2020.
https://doi.org/10.1007/978-3-030-38991-8_15

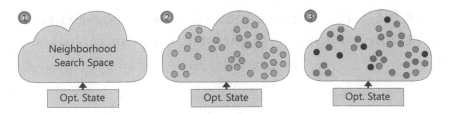

Fig. 1. The first image shows the abstract and large neighborhood search space around a single optimization state (1). We have to identify possible (with respect to constraints) and beneficial (with respect to the cost function) neighboring successor states within this very large search space (2). Since it is typically not possible to explore all potential successors, the available possibilities are rated according to heuristics (3). Once the different ratings are available, a selection strategy chooses possibly beneficial successor states out of this set.

point in order to find a better solution. Such problems are often optimized by different genetic, evolutionary or in general heuristic-based algorithms.

In order to apply existing optimization algorithm for heuristic search, a local neighborhood or successor-state function is required to enumerate all states nearby (Fig. 1). Without such a function, possible candidate states of a single source state cannot be determined. This functionality is typically referred to by $N(x)$, where x is a candidate state in the scope of the optimizer [4]. However, $N(x)$ can become arbitrarily complex and difficult to implement, even when focusing on the pure algorithms and logics. If we consider an implementation on a high-performance massively-parallel processor like a *Graphics Processing Unit (GPU)*, a default CPU-based implementation of $N(x)$ has to be manually adapted and tuned to the target hardware. Since many heuristic optimizers track multiple candidate states at the same time, a simple way to parallelize a successor-state generation would be to simply invoke a sequential existing $N(x)$ function for every state in parallel. However, this typically sacrifices a large amount of performance since it often leads to non-optimal memory-access patterns and does not pay attention to other hardware-specific peculiarities like *single-instruction multiple threads (SIMT)* units[1]. Hence, developers have to adjust their implementation to these hardware characteristics, which is error-prone and time consuming [6]. Consequently, an arbitrary $N(x)$ function on a GPU, for instance, becomes even more sophisticated to implement.

In this paper we present a new method to implement domain-independent $N(x)$ functions for combinatorial heuristic optimization on massively-parallel architectures using SIMT units. It is designed to achieve a high utilization of the available processing power and scales well with the number of variables, possibilities per variable and the number of states. This allows for an application to large optimization problems that significantly benefit from parallel processing. Furthermore, it enables the design and the implementation of GPU-based

---

[1] We will refer to a single SIMT unit as warp in the scope of this paper.

optimizers that can perform nearly all required steps in parallel without high communication overhead to the CPU (the transfer of state information during the optimization process, for example). In order to guide $N(x)$ to enumerate possibly interesting states, we offer so called *local heuristics* that guide the generation of successors. Since our method only determines the direct neighboring or potential successor states for a given set of states, it can be easily integrated into any existing optimization algorithm like Tabu Search. We demonstrate several use cases in the evaluation section and show the significant improvements in terms of performance and scalability.

In the remainder of this paper, we focus on related work from the field of parallel neighborhood exploration in the context of heuristic optimization. We introduce our method in Sect. 3 and present a detailed explanation how the general algorithm works (Subsects. 3.1 and 3.2) and can be implemented on GPUs (Subsect. 3.4). The evaluation (Sect. 4) shows different optimization problems that were optimized using our $N(x)$ implementation.

2   Related Work

There has been a lot of work on using GPUs for solving optimization problems in general. We focus on a selection of papers that involve $N(x)$ realizations in favor of purely parallelized optimization algorithms.

Campeotto et al. [3] focus on parallelizing constraint solving using GPUs. They use a hybrid design in which they switch between CPU and GPU in every solver step: The actual constraint propagation and consistency checks are executed on the GPU, whereas the main solver runs on the CPU. This can be seen as a parallel evaluation of several potential states in the scope of an abstract $N(x)$ function, which is primarily evaluated on the CPU side. For general work on local search and constraint programming we refer the interested reader to [4].

A follow-up paper by Campeotto et al. [2] goes into more detail about the neighborhood processing functionality. The CPU selects subsets of variables to explore and copies the required information to the GPU. Afterwards, the GPU can process the different sets and explore the resulting states in parallel. They use different strategies (called *local search strategies*) to select these potentially interesting subsets. This is very related to our approach; however, we perform the whole $N(x)$ evaluation in parallel on the GPU without the need for additional CPU communication.

Munawar et al. [14] investigated solving of MAX-SAT problems on GPUs using efficient genetic algorithms and local search. The neighborhood exploration is based on a 4D virtual grid which yields four neighbor possibilities (2D) for every individual and four possibilities for every population (2D) in the scope of the genetic algorithm. This can be directly mapped to GPUs using multiple thread groups that process all possibilities in parallel. This is very similar to our approach: the whole neighborhood exploration is performed on the GPU. In contrast to their $N(x)$ implementation, our method can work with arbitrary local-search criteria and is not tied to a particular mathematical optimization model (MAX-SAT in this case).

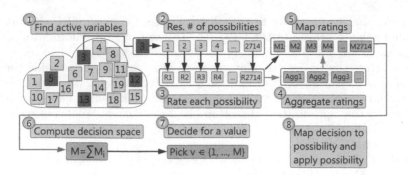

Fig. 2. High-level view of our method when it is applied to a single optimization state.

A combination of the previously presented approaches is the one by Abdelkafi et al. [1]. They leverage OpenCL to evaluate the neighborhood in parallel by assigning different threads to different neighbors. Afterwards, they evaluate each neighbor sequentially in the scope of their optimization system. They investigated the knapsack problem and the travelling salesman problem (TSP) using their method by leveraging customized data structures during neighbor generation. In comparison to our approach, they are limited to manually adjusting their method based on the domain and do not support any kind of variable-assignment ratings. The latter one is particularly important when exploring extremely large neighborhoods that cannot be expanded in memory (see Fig. 1). Similar to these general approaches is the one by Lam et al. [12]. They investigated a parallelization of TSP on CPUs and GPUs using simulated annealing. As before, the proposed neighborhood generation is wired to and specialized for the underlying mathematical optimization model which makes it difficult to reuse it in a different context.

Luong et al. [8,9] use the GPU to explore the neighborhood of a single state and to evaluate the consequences of several decisions. In this context, they focus on binary problems using different Hamming Distances. Again, this approach lacks generality: a generic problem of an unknown domain requires a much more sophisticated concept to explore the neighborhood. They also applied their own successor-generation approach in a parallelized implementation of Tabu Search [7]. Ghorpade et al. [5] used the method by Luong et al. to perform a parallel evaluation of all neighboring states in the domain of TSPs using the 2-opt algorithm. They adapted the method in such a way that they have encoded their own structures and strategies to generate successor states according to their use case. In a follow-up work from Melab et al. [11], they extended Luong et al.'s method to perform a whole framework-based approach. They use GPU acceleration for neighborhood exploration based on local-search metaheuristics as before. However, the basic exploration approach in the paper stays the same as before.

Novoa et al. [15] created a parallel search implementation for the quadratic assignment problems on GPUs. They use permutations to generate successor states in the scope of their GPU kernels. Their algorithm is based on a binary decision structure that allows variables to be 0 or 1. This significantly simplifies the permutation process in this domain compared to our generic approach.

The probably most related work is the one by Rashid et al. [13]. They discuss challenges of designing neighborhood generation on GPUs in general and give several proposals. Furthermore, they provide high-level algorithms to realize a GPU optimization system based on S- and P-metaheuristics [19]. However, they do not give a detailed explanation regarding neighborhood generation. Their high-level approach already relies on a notation of $N(x)$. Once they have this abstract concept at hand, they can build upon this and apply their algorithms.

3   FANG

As previously mentioned, a heuristic optimization system often tracks multiple candidate states at the same point in time. Thereby, a single optimization state consists of state-dependent information and several variables λ_k that are part of the optimization problem, where $k \in \{1, \ldots, |\lambda|\}$ and $|\lambda|$ refers to the number of variables. A common task is to find an assignment of all variables λ_k to values V_l according to the individual constraints of every variable in order to find the best state according to a given cost function, where $l \in \{1, \ldots, |V|\}$ and $|V|$ refers to the number of possibilities. Our target are large-scale optimization problems that require the evaluation of hundreds of possibilities per variable and a large number of candidate states. As previously mentioned, the number of neighbor states might be very large, and thus, cannot be simply generated or returned by $N(x)$. Instead of creating and iterating over all neighbor states, we have to limit the range and want to enumerate only potentially "interesting" successors of a given state. In order to explore the neighborhood of every variable, we build upon the basic ideas for random-based local-exploration by Munawar et al. [14] and Campeotto et al. [2]. However, instead of randomly choosing and assigning variables, we investigate assignment possibilities of all variables and determine probabilities for every possible assignment (see below). We chose a single possibility based on a random value that is determined using a uniform random distribution.

We propose the concept of *local heuristics* that guide the successor generation in such a way that they rate variable assignments (see Fig. 1). In other words, they answer the question: Is an assignment of variable λ_k to value V_l a good choice? They are called *local*, since the question is answered *locally* per variable and assignment: the rating of the assignment $\lambda_k \rightarrow V_l$ takes the current state x into account but does not pay attention to other possible assignments of this variable. This leads to a major benefit: all ratings can be considered in parallel since every potential assignment is treated on its own. However, assignments of other variables have to be considered during the actual successor construction. There might be a constraint that hinders the assignments of two variables λ_k and λ_l to the same value V_p, for instance. For this reason, we assign different variables sequentially in general. This avoids race conditions during variable assignments and simplifies the rating process: Every rating can be computed by accessing all other already assigned variables since the optimization state is read only at this point in time. This is not a strict limitation and can be relaxed if multiple variables can be assigned without any interference. We further distinguish between different *variable types* (referred to as λ^T). This is useful for many domain-specific problems that leverage different heuristics for distinct variables in order to simplify the modeling process.

3.1   High-Level Algorithm

From a high-level point of view, we distinguish between *active variables* that have to be assigned to a *value* and *inactive ones* that do not require a new assignment. Whether a variable is active or not during successor creation is typically determined by the surrounding optimization system. We assume that the decision step already happened and our task is to find all active variables that require an assignment in this step (Fig. 2-1).

First, we have to pick an active variable and have to resolve the number of possibilities to assign the current one (Fig. 2-2)[2]. Second, we can rate each possibility using a *local heuristic* (Fig. 2-3). Thereby, a rating value R has a user-defined value type that is opaque to our algorithm. Since the initial rating itself is designed as a local process that handles every possibility independently, we added the opportunity to define a custom rating aggregator that accumulates global information about all ratings (Fig. 2-4). An aggregator can carry multiple aggregation values that are combined with the help of custom aggregation functions. This step happens during and immediately after all possibilities have been rated.

Afterwards, we can perform a mapping operation to convert each user-defined rating R using the globally aggregated information into a mapped rating $M \in \mathbb{N}$ (Fig. 2-5). From a theoretical point of view, a mapped rating can be seen as a probability of the associated assignment possibility. Note that the value 0 of a mapped rating M corresponds to a probability of 0. This allows to easily forbid possibilities that cannot/should not be selected for some reason. In practice, however, it is much easier to convert the initial rating into another value $\in \mathbb{N}$ that can be converted to an assignment possibility (see Fig. 3): A larger mapped rating indicates a higher probability with respect to the sum of all mapped ratings M (Fig. 2-6). Next, we pick a uniformly-distributed random value v between 1 and M (Fig. 2-7). We have to remap the chosen value v to an associated mapped rating M, and thus, to its original possibility l it was computed from (Fig. 2-8). Finally, we can apply the selected possibility l and its associated (domain-specific) value V_l by assigning the variable λ_k. The whole process will be repeated until no active variable can be found any more.

3.2   Low-Level View

The whole algorithm can be realized with the help of a single GPU kernel. Every thread group (consisting of N threads) on the GPU handles a single optimization state and assigns all active variables in a cooperative way. The

---

[2] The number of possibilities can be seen as a subset of all possible successor states from Fig. 1-2.

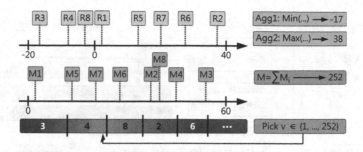

Fig. 3. Sample rating mapping that realizes a non-trivial rating process (see Subsect. 3.1 for more details). Eight different possibilities are individually rated according to a user-defined rating function (top). In this sample, positive ratings directly correspond to their final mapped value (middle). Negative values that are computed by the rating function should be more important than the largest positive value. For this reason, the custom aggregator stores the maximum value of all ratings which can be directly used to remap negative values. All mapped ratings are accumulated in order to derive their individual probability (bottom). Larger intervals correspond to higher probabilities. We can then choose our decision value v out of the computed set of values. This value is then mapped to a target interval, and thus, to a target value to assign.

variables of a single state are managed with the help of Boolean sets (Fig. 4-1, see Subsect. 3.8). Starting with these sets, we have to iterate over all active variables (Fig. 4-2). The number of possibilities per variable is resolved by all threads of the group in parallel. This avoids expensive group-synchronization operations by preferring operations on registers. Alternatively, the number of possibilities could be resolved by the first thread of the group and can be made accessible by all other threads via shared memory. It reduces the number of active warps and leaves more opportunities to the warp dispatcher. However, we have not seen any computationally or memory-expensive implementations that made it necessary to rely on shared memory and group synchronization in practice.

Every possibility will be processed by a single thread using a group-stride loop (Fig. 4-3). This ensures coalesced memory accesses in the scope of the heuristic h and the *rating storage*. The latter one stores all computed intermediate user-defined rating values R from the heuristic in global memory. Storing computed ratings avoids expensive re-computations during the mapping process. If a re-computation of all possibilities is much cheaper than storing the values in global memory, the rating storage can be omitted. However, storing the values in global memory does not impose a significant overhead when processing a large number of states since the memory latency could be hidden by the GPU.

The rating aggregators are locally maintained in register space in the scope of every thread. All computed ratings are directly accumulated using the local aggregators which avoids synchronization with other threads. After computing and storing all R values, the individual intermediate aggregators are combined into a globally available aggregator (Fig. 4-4). The global aggregator can be

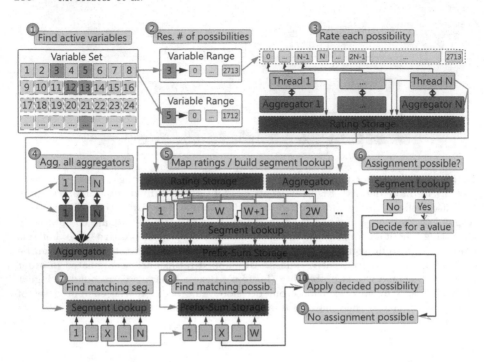

Fig. 4. Low-level view of our method that is closely related to Algorithm 1. Black arrows indicate memory accesses. Red/organge arrows indicate global memory accesses to allocated regions our approach requires. Blue/green arrows indicate logical associations. Red arrows mark decisions. Dashed boxes are intermediate information that is available to all threads. (Color figure online)

stored in shared memory, registers or local memory depending on the individual requirements and the user-defined heuristic implementation (since it is readonly after the aggregation step).

Next, we map all rating values R to their mapped counterparts M using the previously computed aggregator information (Fig. 4-5). We conceptionally split all mapped ratings into *segments* of the size of a single warp. This enables us to build an acceleration structure that significantly speeds up the possibility resolving step in the end. The acceleration structure is completely stored in shared memory and consumes a single 64bit unsigned integer per segment. Leveraging shared memory ensures high-performance random-access lookups and avoids consumption of additional global memory. The disadvantage of this concept is the limited number of possibilities per variable that can be stored in shared memory: Assuming 24 kb of shared memory per thread group[3] and a warp size of 32 [17], yields a total number of $\frac{24 \cdot 1024}{8} \cdot 32 = 98304$ possibilities.

All individually computed mapped ratings M will be processed using a group-stride loop. They are on-the-fly accumulated in the form of a prefix sum that is computed for the whole group after every thread has mapped its associated

---

[3] Without limiting the parallel execution of multiple groups per multiprocessor.

rating value R. Using a prefix sum allows us to represent the different probabilities that are implied by all M values (see Fig. 3). The resulting prefix-sum values will be written into a *prefix-sum storage* that resides in global memory. Again, this avoids expensive prefix-sum re-computations (see above). In the same run, the first thread of every warp stores the right boundary value of all threads in the warp by writing its value into the shared-memory segment lookup.

In the 6th step (Fig. 4-6) all threads determine whether an assignment is possible or not. This breaks down to a simple check whether the value of the last segment in the used lookup is zero or not. If no assignment is possible, we will have to deactivate the current variable λ_k (Fig. 4-9) and continue with the next active one. If an assignment is possible, every thread in the group picks the same decision value v that is less than the maximum prefix sum value of the last segment.

The last two phases resolve the possibility that belongs to the chosen decision value v. First, we have to identify the target segment that narrows the search space to the number of threads in a warp (Fig. 4-7). We do not leverage any advanced search algorithm to find the target segment (like a binary-search, for instance). Instead, we simply check each segment using a single thread and group-stride loop. Once the target segment has been found, the first warp of every group investigates all potential matches in the target segment in parallel (Fig. 4-8). Only one thread can win during the comparison of its neighboring prefix-sum values to the chosen value v, Finally, this thread applies the decided possibility V_l to the current variable λ_k and deactivates it. Afterwards, we continue with the next active variable by repeating the whole assignment process.

3.3 Successor Generation

The actual successor generation process happens in three steps (see Fig. 5). We leverage a double-buffer approach: a source and a target buffer containing all state information. First, the source states from the readonly source buffer will be cloned into the target buffer. A user defined parameter specifies the number of successors per input state. A common scenario in this context is a single input state that will be expanded several times in the target buffer. Second, the successor indices will be generated using an external random number generator

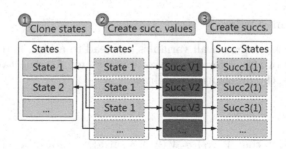

Fig. 5. The visualized successor generation process

and assigned to every new cloned state. The successor index is used as random seed for all subsequent assignment decisions in the scope of the associated state. Third, the already presented variable-assignment algorithm will be applied to every target state resulting in different successor states. Finally, the source and target buffers are swapped to delete all source states and to "free" memory for the next successor-generation step.

3.4   Algorithm

The main algorithm can be found in Algorithm 1 which represents a single GPU kernel and the actual implementation of our $N(x)$ function. It is designed in a way that it can be directly converted into code with minor adjustments (see Subsect. 3.8). The kernel is launched with a group size that is a multiple of the warp size in order to achieve high occupancy and to ensure that the segment lookup works as intended. The term *group index* refers to the index of the i-th thread inside the group.

The algorithm takes a state x and a set of heuristics h^T for the different variable types λ^T that can occur in the scope of the problem instance. We allocate the required amount of shared memory, initialize the random seed based on the successor index and iterate over all active variables sequentially. Lines 6–8 and Algorithm 2 correspond to the FANG steps 1–4 in Fig. 4. Lines 9–18 and Algorithm 3 correspond to the FANG steps 5,6 and 9. Note that we also call the user-defined heuristic in line 14 that can perform state-dependent adjustments in the case of no assignment before deactivating the variable. Lines 20–31 reflect the remaining FANG steps 7–10. Again, we execute user-defined code to enable customized assignment logic. Finally, we store the updated random-number-generator seed in the optimization state. Upcoming assignment steps will then use the updated seed to generate "new" random numbers.

Note that there cannot be any race-conditions between different variable assignments. Only one thread at a time executes the actual (no-)assignment, activation or deactivation procedures. The thread barriers in line 17 and 33 ensure that the changes of these threads will be visible to all other threads in the group after an assignment loop. This also guarantees that the influence of a variable assignment can be taken into account by all other variables that are assigned afterwards.

3.5   Assignment Order

The sequential assignment of all variables can lead to a bias, and thus, to unintended behavior of the general optimization system. Assume two variables λ_1 and λ_2 that have the following possibilities:

- $\lambda_1 \Rightarrow \{1\}$ and
- $\lambda_2 \Rightarrow \{1\}$,

with the constraint that $V(\lambda_1) \neq V(\lambda_2)$. Consider a state in which both variables are active at the same time. Let us also assume, we assign both variables sequentially in a pre-defined order, e.g. λ_1 before λ_2. Then, variable λ_2 cannot

---

Algorithm 1: The FANG Algorithm

Input: state x, heuristics h^T for the variables types λ^T

1 sharedSegments := **shared memory rating type**[#ratings per group];
2 sharedPrefixSum := **shared memory int64[group size]**;
3 seed := SuccessorValue(x);
4 **foreach** $\lambda \in$ *active variables* **do**
5   h := DetermineHeuristic(λ based on its type λ^T);
6     $|R|$:= GetNumberOfPossibilities(h, x, λ);
     /* Algorithm 2   */
7   agg := RateAllPossibilities(x, h, sharedPrefixSum);
     /* Algorithm 3   */
8   MapAllRatings(x, h, agg, sharedSegments, sharedPrefixSum);
     /* Check for a possible assignment   */
9   globalRating := PrefixSumValue(last segment);
10   **if** *globalRating < 1* **then**
11   **if** *is first lane in warp* **then**
          /* No assignments possible   */
12   CouldNotAssign(h, x, λ);
13   DeactivateVariable(λ);
14   **end**
        /* Wait for all threads...   */
15   **group barrier**;
16   **continue**;
17   **end**
     /* Find decision value v */
18   v, seed := DrawRandomNumber() $\in [0, \ldots,$ globalRating $- 1]$;
     /* Algorithm 4   */
19   segment := FindMatchingSegment(x, v,
20   sharedSegments, sharedPrefixSum);
     /* Select possibility based on v */
21   **if** *is first warp* **then**
22   p := segment * **warp size** + **group index**;
23   leftSum := prefixSumStorage(x)[p - 1];
24   rightSum := prefixSumStorage(x)[p];
25   **if** *leftSum $\leq v \wedge$ rightSum $< v$* **then**
26   Assign(h, x, λ);
27   DeactivateVariable(λ);
28   **end**
29   **end**
     /* Wait for all threads...   */
30   **group barrier**;
     /* Store updated successor value   */
31   **if** *is first thread in group* **then**
32   SuccessorValue(s) := seed;
33   **end**
34 **end**

Algorithm 2: RateAllPossibilities

Input: state x, heuristic h, shared memory
1 agg := CreateAggregator(h);
   /* Iterate over all possibilities and aggregate information   */
2 **for** i := **group index**; $i < |R|$; i += **group size do**
3   R_i := ComputeRating(h, x, λ, i);
4   ratingStorage(x)[i] := R_i;
5   agg := Aggregate(agg, R_i);
6 **end**
   /* Wait for all threads...   */
7 **group barrier**;
   /* Aggregate all thread-local aggregators   */
8 agg := **reduce** (sharedPrefixSum, agg);
9 **return** agg;

Algorithm 3: MapAllRatings

Input: state x, heuristic h, aggregator, shared memory
1 lowerBound := 0;
   /* Iterate over ratings in the rating storage   */
2 **for** i := **group index**; $i < |R|$; i += **group size do**
3   initialRating := ratingStorage(x)[i];
4   M_i := MapRating(h, initialRating, agg);
5   prefix := lowerBound + **prefix sum** (sharedPrefixSum, initialRating);
       /* Store prefix sum, update lower bound, and update segment
           lookup   */
6   prefixSumStorage(x)[i] := prefix;
7   lowerBound := PrefixSumValue(last thread in group);
8   **if** *is first lane in warp* **then**
9   sharedSegments$\left[\dfrac{i}{\textbf{warp size}}\right]$:= lowerBound;
10   **end**
11 **end**
   /* Wait for all threads...   */
12 **group barrier**;

---

be assigned in any case to the value $V(\lambda_2) = 1$. Instead the variable assignment will always yield $V(\lambda_1) = 1$ and $V(\lambda_2) = \bot$ (could not be assigned). This can be seen as the intended behavior that was desired by the user. If not, we will also have to change the order in which we assign variables from time to time. This can be directly achieved by randomly permuting the order in which we iterate over all active variables.

3.6 Duplicate States

A disadvantage of our parallel-processing method is the (potential) generation of duplicate successor states from a single state. This is caused by our random selection process that chooses from different assignment possibilities. The probability

---

Algorithm 4: FindMatchingSegment

Input: state x, value v, shared memory
1 ResetSharedMemory();
2 targetSegmentIdx := -1;
   /* Iterate over segment lookup in shared memory   */
3 **for** $i :=$ **group index**; $i < \lceil \frac{|R|}{\text{warp size}} \rceil$; i += **group size do**
4 | leftBoundary := sharedMemory[i - 1];
5 | rightBoundary = sharedMemory[i];
6 | match := leftBoundary $\leq v \wedge$ rightBoundary $< v$;
7 | targetSegmentIdx := **reduce** (sharedPrefixSum, match ? i : -1);
8 | **if** *target segment match* **then**
9 | | **break**;
10 | **end**
11 **end**

---

that this actually happens depends on the optimization domain, e.g. the number of variables, individual constraints, the rating functionality and the number of possibilities per assignment. This can be safely neglected in general, since the probability that two identical successors will end up in the same state after n successor generation steps is p^n, where $p \in [0, \ldots, 1]$ refers to the probability that a duplicate state emerges.

3.7   Memory Consumption

Our algorithm requires a temporary array to store all custom rating values and all prefix-sum information. The size of a single array entry in bytes is $|entry| = \text{sizeof(rating type)} + 8$, since we store all accumulated mapped ratings as unsigned 64bit integers to avoid overflows. Segment-lookup information is stored in shared memory on the multiprocessor and does not require additional global memory. However, this array is required per state. Consequently, the global memory consumption is

$$\max \left(|R(\lambda_0)|, \ldots, |R(\lambda_{n-1})| \right) \cdot |X| \cdot |entry|, \tag{1}$$

where $|R(\lambda_i)|$ is the maximum number of possibilities of the variable λ_i and $|X|$ is the number of states.

If we process multiple variables in parallel, we require several instances of the array in memory to store all intermediate values. Hence, Eq. 1 has to be adapted in order to reflect the additional memory consumption:

$$\left(\sum_i |R(\lambda_i)| \right) \cdot |X| \cdot |entry|, \tag{2}$$

if we assume that all variables can be assigned in parallel. Since we disable parallel variable assignment in our real-world applications, the memory consumption is the one from Eq. 1.

Table 1. Influence of the number of states, the number of possibilities and the number of variables on the overall run time. GPUs: GeForce GTX 1080 Ti and GeForce GTX Titan X. CPUs: Intel Core i9 7940X and AMD Ryzen 7 2700X.

Load	$|X|$	$|\lambda|$	$|R|$	1080 Ti	σ	Titan X	σ	i9 7940X	σ	R. 2700X	σ
1	1024	8	1224	2.20	0.09	3.04	0.12	3.26	0.72	4.61	0.20
*	*	*	4488	3.34	0.14	4.73	0.10	10.42	1.16	14.95	1.21
*	*	*	9800	5.05	0.33	7.66	0.10	22.28	1.43	31.06	1.12
*	*	32	1224	6.97	0.63	10.85	0.44	11.06	1.40	16.03	0.91
*	*	*	4488	9.85	0.89	17.13	0.13	40.02	1.29	56.18	1.82
*	*	*	9800	16.90	1.22	28.49	0.22	87.31	1.81	121.75	2.64
*	4096	8	1224	6.48	0.57	11.54	0.11	12.08	1.44	16.73	0.49
*	*	*	4488	9.84	0.71	17.75	0.12	40.18	1.54	55.50	1.40
*	*	*	9800	16.94	0.98	30.13	0.10	85.09	1.79	118.61	2.25
*	*	32	1224	23.27	1.47	44.33	0.44	42.21	1.79	60.32	1.31
*	*	*	4488	37.78	2.19	70.08	0.69	154.70	1.96	217.85	3.86
*	*	*	9800	66.30	1.75	118.56	0.53	336.196	3.96	471.87	6.80

3.8   Implementation Details

We have implemented our algorithm in C++ using Cuda for all GPU kernels. Variables are managed with the help of bit-sets in form of unsigned integers. They also act as acceleration structures to skip over larger regions of inactive variables. Active variables are identified with the help of hardware bit-manipulation instructions. For performance reasons, we typically leverage an *XorShift** or an *XorShift1024** random-number generator [10]. They yield excellent results on our common optimization domains and are not too expensive to compute in every assignment step.

We further leverage template specialization to instantiate different assignment kernels for each heuristic. Based on our experience, the majority of heuristics use uniform control flow that does not diverge into too many distinct sections. This results in specialized GPU kernels for each variable kind that benefit from the common uniform-control-flow pattern of each heuristic, which significantly reduces thread divergences. Moreover, we assign variables of different types typically sequentially to ensure that decisions from previous categories are visible to the heuristic of the next variable kind.

All memory buffers are allocated before the optimization process starts, since the required memory size is already known during initialization. This avoids unnecessary dynamic memory allocations during runtime. We use warp shuffles to improve performance of all prefix-sum and reduction computations [16]. Note that the loops in Algorithms 3 and 4 have thread divergences the way they are described in the paper. In order to leverage group barriers inside reduce and prefix-sum computations, these loop bounds are padded to avoid any divergences in our implementation. Furthermore, accesses to shared memory in Algorithm 4 and to the prefix-sum storage in Algorithm 1 are modified to avoid out-of-bounds accesses.

Table 2. Influence of the compute load on the overall runtime. Please note that the table is not meant to compare the theoretical computational power of GPUs against CPUs. GPU: GeForce GTX 1080 Ti. CPU: Intel Core i9 7940X.

Load	$|X|$	$|\lambda|$	$|R|$	1080 Ti	σ	i9 7940X	σ
1	1024	8	1224	**2.20**	0.09	**3.26**	0.72
8	*	*	*	**1.88**	0.05	**7.81**	0.11
16	*	*	*	**1.82**	0.02	**14.05**	0.15
1	*	*	4488	**3.34**	0.14	**10.42**	1.16
8	*	*	*	**3.40**	0.13	**27.39**	0.47
16	*	*	*	**2.75**	0.02	**50.11**	0.55
1	4096	*	1224	**6.48**	0.57	**12.08**	1.44
8	*	*	*	**6.26**	0.03	**29.42**	0.28
16	*	*	*	**6.26**	0.02	**53.70**	0.44
1	*	*	4488	**9.84**	0.71	**40.18**	1.54
8	*	*	*	**10.00**	0.89	**104.61**	0.86
16	*	*	*	**9.91**	0.70	**193.39**	0.98

4   Evaluation

The evaluation section does not cover benchmarks of our $N(x)$ algorithm in the scope of different optimization systems using hard-to-reproduce and difficult to understand optimization benchmarks. Instead, we evaluated the pure successor-state generation process using a well known heuristic from the field of shortest-path optimization. We consider a discrete 2D grid using the Manhattan distance, which is given by

$$d(p, q) = \sum_{i=1}^{n} |p_i - q_i|. \tag{3}$$

We iteratively move a point p to its neighboring cell according to the evaluation result of the Manhattan distance. Hence, we consider neighboring grid cells in 2D as potential target points for the next step during heuristic evaluation. The number of variables $|\lambda|$ indicates the number of points that we want to compute a shortest path for. Consequently, every variable λ_i corresponds to a single current point p and a corresponding goal point q that we want to reach. We choose $|\lambda|$ to be $\in \{8, 32\}$ to demonstrate the effect of a small number of variables on the overall run time. The number of possibilities $|R|$ for every variable is derived from the number of neighboring cells that we can move to in a single step. This number can be computed using

$$|R| = j \cdot j - 1, \tag{4}$$

where j refers to the number of neighboring cells in one dimension. Since we are interested in large-scale optimization problems, we chose $j \geq 35$ to have some reasonable number of possibilities per variable ($j \in \{35, 67, 99\}$). Choosing

j to be smaller does not make any sense for a GPU-based optimizer since the number of possibilities is too small. The number of states $|X|$ is chosen to be $\in \{1024, 4096\}$ to create some workload. However, in reality $|X| \ll |R|$, which also avoids duplicate states. For the sake of completeness, we also included performance measurements of such cases in which $|X| > |R|$. In order to simulate more- and less-expensive rating computations based on the Manhattan distance, we introduce the *load* factor. It indicates the number of $d(p, q)$ computations that are performed using different goal positions q inside a loop. As baseline we chose *load* to be 1, which is the worst-case for our algorithm since the workload of every single possibility evaluation is extremely small.

Our CPU implementation is derived from an object-oriented design, that instantiates state objects encapsulating bit-fields of active variables. During successor generation, these states are cloned and are assigned in parallel to leverage all cores. Schulz et al. [18] reported on many benchmarks in the field of discrete optimization that GPU and CPU comparisons of algorithms often lack comparability of the results. For this reason, our CPU version uses the same method as shown in Fig. 2 to ensure that the solutions from all scenarios are identical on the CPU and the GPU.

We used two GPUs from NVIDIA (a GeForce GTX 1080 Ti and a GeForce GTX Titan X) and two CPUs: one from Intel (a Core i9 7940X) and one from AMD (a Ryzen 7 2700X). Every performance measurement is the median execution time of 100 algorithm executions. Moreover, all variables λ_i were considered to be active at the same time to increase the number of assignment steps per state. The CPU code was compiled with *msvc v19.16.27030* with all compiler optimizations and the AVX2 instruction set enabled. The Cuda code was compiled with *nvcc v10.1.105* with all compiler optimizations enabled.

Table 1 shows the main evaluation table demonstrating the impact of the number of states, possibilities and variables on the overall run time. As previously mentioned, the load factor was set to 1 in order to measure the worst case of our method. On the GPU side, the 1080 Ti is roughly between $1.5\times$ and $2\times$ as fast on all benchmarks compared to its older generation counterpart. Both GPUs scale well with the overall complexity of the assignment problem. However, they are heavily influenced by the number of active variables $|\lambda|$, which are assigned sequentially. This holds also true for the CPUs: doubling $|\lambda|$ roughly results in a doubled execution time, which corresponds to the expected behavior. Fixing $|\lambda|$ and changing the number of possibilities $|R|$ per variable results in a very good scalability of our method on the GPUs: increasing the number of possibilities by a factor of 8 (from 1224 to 9800) results in an increase of the run time of roughly a factor of 4. This is caused by the parallel evaluation of many possibilities in the scope of a single thread group, which is very work efficient. The CPUs have a dramatic slowdown that is tightly coupled to $|R|$ since they are processing the assignment possibilities one by one. Fixing $|\lambda|$ and $|R|$ while changing $|X|$ yields comparable results: The GPUs scale well when increasing the number of states, whereas the CPUs register a bad scalability. This is due to the fact that the GPU scheduler can choose between more thread groups to hide memory latency and improve the overall occupancy.

Table 2 shows the impact of the compute load on the run time of the assignment process. The measurements clearly show the great scalability of our algorithm on the GPUs when it comes to more complex rating functions. A larger computational load shifts the focus from a memory-dependent execution to a computation-dependent execution. This affects the behavior of the GPU scheduler, such that the scheduling overhead can be significantly reduced to spent more time on the actual computations. Hence, an increased load does cause any slowdowns on the GPUs. The CPU versions suffer dramatically from the additional load, as their maximum occupancy was already reached.

The general speedup that can be achieved using our algorithm on the GPU over traditional CPU versions yields speedups between 7.7× on problems with small computation load and up to 27× using more workload depending on the actual hardware. In general, our method provides great scalability that significantly outperforms the CPU versions.

5   Conclusion

We present a new method to implement a generic and reusable $N(x)$ function for heuristic optimization systems. The neighborhood is explored using *local heuristics* that rate all assignment possibilities of every variable. These ratings are converted to probabilities which form the basis to find a decision value. The value itself is resolved using a random-number generator that has a unique seed for every optimization state. The decision is mapped to a variable-assignment possibility using specially designed lookup tables that can be stored in fast on-chip-memory.

Our approach scales very well with the complexity/cost of the rating functions. It also scales excellently with the number of states and assignment possibilities per variable. For instance, an assignment of 32 variables that are active at the same time in 1024 states with 4488 possibilities each require only ≈10 ms on current hardware to complete. Comparing the performance to traditional $N(x)$ implementations on the CPU yields significant speedups of up to 27× on our evaluation scenarios. Moreover, it allows the design and implementation of fully GPU-based heuristic-driven optimization systems without the need to perform neighbor search or successor-state generation on the CPU. This makes it a perfect extension for every modern heuristic-based optimizer.

Probably the main downside of our approach is the high memory consumption. We require a single array entry of at least 12 bytes in memory for every assignment possibility in every state that should be processed in parallel. However, we do not believe that this is a major limitation in practice since large optimization problems require huge amounts of memory anyway.

In the future we would like to extend the concept to support parallel assignment of variables. This would require specific compiler extensions to automatically determine whether or not some variables can be assigned in parallel. In addition, we want to experiment with locally cached ratings in shared memory, since the recent trend has shown increasing sizes of on-chip-memory.

Acknowledgments. The authors would like to thank Wladimir Panfilenko and Thomas Schmeyer for their suggestions and feedback regarding our method. Furthermore, we would like to thank Gian-Luca Kiefer for additional feedback on the paper. Special thanks to Wladimir at this point for adding the concept of integer-based bit sets for active variables in a single state. This reduces global-memory consumption and improves performance of searching for active variables.

References

1. Abdelkafi, O., Chebil, K., Khemakhem, M.: Parallel local search on GPU and CPU with OpenCL language. In: Proceedings of the First International Conference on Reasoning and Optimization in Information Systems, September 2013
2. Campeotto, F., Dovier, A., Fioretto, F., Pontelli, E.: A GPU implementation of large neighborhood search for solving constraint optimization problems. In: Proceedings of the Twenty-First European Conference on Artificial Intelligence (2014)
3. Campeotto, F., Dal Palù, A., Dovier, A., Fioretto, F., Pontelli, E.: Exploring the use of GPUs in constraint solving. In: Flatt, M., Guo, H.-F. (eds.) PADL 2014. LNCS, vol. 8324, pp. 152–167. Springer, Cham (2014). https://doi.org/10.1007/978-3-319-04132-2_11
4. Focacci, F., Laburthe, F., Lodi, A.: Local search and constraint programming. In: Milano, M. (ed.) Constraint and Integer Programming. Operations Research/Computer Science Interfaces Series, vol. 27, pp. 293–329. Springer, Boston (2004). https://doi.org/10.1007/978-1-4419-8917-8_9
5. Ghorpade, S., Kamalapur, S.: Solution level parallelization of local search meta-heuristic algorithm on GPU. Int. J. Comput. Sci. Mob. Comput. (2014)
6. Köster, M., Leißa, R., Hack, S., Membarth, R., Slusallek, P.: Code refinement of stencil codes. Parallel Process. Lett. (PPL) **24**, 1–16 (2014)
7. Luong, T.V., Loukil, L., Melab, N., Talbi, E.: A GPU-based iterated tabu search for solving the quadratic 3-dimensional assignment problem. In: ACS/IEEE International Conference on Computer Systems and Applications (AICCSA) (2010)
8. Luong, T.V., Melab, N., Talbi, E.G.: Large neighborhood local search optimization on graphics processing units. In: Workshop on Large-Scale Parallel Processing (LSPP) in Conjunction with the International Parallel & Distributed Processing Symposium (IPDPS) (2010)
9. Luong, T.V., Melab, N., Talbi, E.G.: Neighborhood structures for GPU-based local search algorithms. Parallel Process. Lett. **20**, 307–324 (2010)
10. Marsaglia, G.: Xorshift RNGs. J. Stat. Softw. **8**, 1–6 (2003)
11. Melab, N., Luong, T.V., Boufaras, K., Talbi, E.G.: ParadisEO-MO-GPU: a framework for parallel GPU-based local search metaheuristics. In: 11th International Work-Conference on Artificial Neural Networks (2011)
12. Ming Lam, Y., Hung Tsoi, K., Luk, W.: Parallel neighbourhood search on many-core platforms. Int. J. Comput. Sci. Eng. **8**, 281–293 (2013)
13. Mohammad Harun Rashid, L.T.: Parallel combinatorial optimization heuristics with GPUs. Adv. Sci. Technol. Eng. Syst. J. **3** (2018)
14. Munawar, A., Wahib, M., Munetomo, M., Akama, K.: Hybrid of genetic algorithm and local search to solve MAX-SAT problem using nVidia CUDA framework. Genet. Program. Evolvable Mach. **10**, 391 (2009)
15. Novoa, C., Qasem, A., Chaparala, A.: A SIMD tabu search implementation for solving the quadratic assignment problem with GPU acceleration. In: Proceedings of the 2015 XSEDE Conference: Scientific Advancements Enabled by Enhanced Cyberinfrastructure (2015)

16. NVIDIA: Faster Parallel Reductions on Kepler (2014)
17. NVIDIA: CUDA C Programming Guide v10 (2019)
18. Schulz, C., Hasle, G., Brodtkorb, A.R., Hagen, T.R.: GPU computing in discrete optimization. Part II: survey focused on routing problems. EURO J. Transp. Logistics **2**, 159–186 (2013)
19. Talbi, E.G.: Metaheuristics: From Design to Implementation. Wiley, Hoboken (2009)

DETER: Streaming Graph Partitioning via Combined Degree and Cluster Information

Cong Hu[1], Jiang Zhong[1,2(✉)], Qi Li[1], and Qing Li[1]

[1] Chongqing University, Chongqing 400044, People's Republic of China
`zhongjiang@cqu.edu.cn`
[2] Key laboratory of Dependable Service Computing in Cyber Physical Society, Chongqing University, Chongqing 400044, People's Republic of China

Abstract. Efficient graph partitioning plays an important role in distributed graph processing systems with the rapid growth of the scale of graph data. The quality of partitioning affects the performance of systems greatly. However, most existing vertex-cut graph partitioning algorithms only focused on degree information and ignored the cluster information of a coming edge when assigning edges. It is beneficial to assign an edge to a partition with more neighbors because keeping a dense subgraph in one partition would reduce the communication cost. In this paper, we propose DETER, an efficient vertex-cut streaming graph partitioning algorithm that takes both degree and cluster information into account when assigning an edge to one partition. Our evaluations suggest that DETER algorithm owns the ability to efficiently partition large graphs and reduce communication cost significantly compared to state-of-the-art graph partitioning algorithms.

Keywords: Graph partitioning · Vertex-cut · Streaming · Distributed graph computing

1 Introduction

In the last few years, a large amount of information has been produced on the Internet such as large graph data. For example, Facebook owns 2.34 billion users and 346 billion social connections since 2018 [3]. Besides social networks, road networks, biological networks are also big graph data in the real world. Consequently, most real-word graph should be distributed stored and computed

Supported in part by National Key Research and Development Program of China under Grant 2017YFB1402400, in part by graduate research and innovation foundation of Chongqing, China under Grant CYB18058, in part by the Fundamental Research Funds for the Central Universities under Grant 2018CDYJSY0055, in part by the Frontier and Application Foundation Research Program, PR China of CQ CSTC under Grant cstc2017jcyjAX0340.

© Springer Nature Switzerland AG 2020
S. Wen et al. (Eds.): ICA3PP 2019, LNCS 11944, pp. 242–255, 2020.
https://doi.org/10.1007/978-3-030-38991-8_16

due to the capacity of a single machine. To achieve the distributed storage and computation of these large graphs. We need to partition these large graphs across machines for faster localized processing and less communication. Meanwhile, the balance of each partition should also be kept. This problem is called balanced k-way graph partitioning problem.

Graph partitioning is a pre-work for diverse real-world applications and fields, like knowledge graph, recommendation system and so on. It aims to improve the computation speed and reduce the communication cost between different machines. Traditional graph partitioning algorithms are mainly dynamic method such as METIS [4] and Mizan [8] etc. For these algorithms, the whole graph must be loaded into memory when partitioning and this requires huge memory capacities when the graph is very large. Besides, it's scalability is bad because it can not process new added vertexes and edges. These drawbacks affect the efficiency of graph computation tasks such as PageRank [16], SSSP (Single-Source Shortest Path) [21] and so on.

Due to these drawbacks, Stanton [20] proposed Streaming graph partitioning, aiming to provide a fast and efficient method with less memory size by assigning the arriving edge or vertex into one of the partitions instead of operating on the entire graph. It puts all edges or vertexes into a graph stream randomly or by DFS or by BFS method. For each edge or vertex in the graph stream, there exists a heuristic method to compute a score according to a score function. The partitioner will assign the edge or vertex to the corresponding partition according to the score. Streaming graph partitioning method doesn't require loading whole graph into memory and can significantly reduce the execution time of the graph partitioning. Moreover, in real graphs such as social graphs like twitter, new users are created or deleted every time. Therefore it is urgent to have a graph partitioning method for dynamic graphs. Since the scale of graph data is rapidly growing. It became very popular and many researchers focused on it.

There are two kinds of approaches for graph partitioning: Vertex-cut partitioning and Edge-cut partitioning. Vertex-cut partitioning algorithm assigns each edge to a single machine and it would cause replica vertex. Edge-cut partitioning algorithm assigns each vertex to a single machine and causes cut-edges across machines. Several studies have been focused on the two approaches for streaming graph partitioning [5,12,15,17,23,24]. There also came out several distributed graph computing frameworks, such as GraphLab [10], GraphX [25], Pergel [11], PowerGraph [5] and so on. GraphX and PowerGraph adopt vertex-cut partitioning strategy, GraphLab and Pergel adopt edge-cut partitioning strategy. However, most vertex-cut partitioning algorithms like HDRF [17] and DBH [24] only use the degree information and ignore the cluster information, both of them assign edges to a partition according to the degree of vertexes.

In this paper, we propose DETER (Degree and Cluster), a streaming graph partitioning algorithm to reduce the replica factor while keeping the balance of each machine. The core of DETER is combining both degree information and cluster information when allocating edges because most vertex-cut algorithms only focus on degree information of vertex and ignore the cluster information which has significant affection on the performance. The contributions of this paper are as follows:

- We propose DETER, a new vertex-cut streaming graph partitioning algorithm that outperforms than other algorithms. DETER takes both degree information and cluster information into consideration when assigning edges.
- We employ a method to compute the cluster information from a different perspective to speed up the execution of graph partitioning. This method uses an extra table that records the neighbors' partition information to avoid time-consuming computation. It could reduce the graph partitioning time by up to 90% than without using this acceleration method.
- We do comprehensive experiments on various large-scale graph datasets. The results show the efficiency of DETER. The replica factor reduced by up to 56% than other algorithms.

This paper is organized by the following structure: In Sect. 2, we discuss the related work. In Sect. 3, we give the definition of streaming graph partitioning and introduce two streaming graph partitioning algorithms that stimulate DETER. In Sect. 4, we explain our proposed streaming graph partitioning algorithm and the speeding up method in detail. Comprehensive experiments and analysis are stated in Sect. 5. Finally, we discuss the conclusion of this work.

2   Related Work

Balanced graph partitioning is an NP-hard problem [1]. Streaming graph partitioning has attracted many researchers' studies since it was proposed by Stanton [20] due to it's better performance and fast execution time compared with offline graph partitioning method like METIS [4]. Besides, streaming graph partitioning doesn't require loading the whole large-scale graph into memory and doesn't need large memory capacity. Streaming k-way balanced graph partitioning is the most common method, before the distributed graph computing framework PowerGraph [5] was proposed, most studies were focused on edge-cut partitioning methods which assign vertexes to the corresponding partition, it aims to reduce the cut edges across machines while keeping that each partition holds the approximate vertex numbers.

Stanton [20] proposed several heuristic streaming graph partitioning strategies, such as Weighted Randomized Greedy, Linear Deterministic Greedy (LDG), etc. Among these heuristics, the LDG algorithm performs better than others. LDG is greedy and assigns a vertex to the partition which has most neighbors of this vertex while respecting a capacity constraints to keep balance. Fennel [23] is another heuristic partitioning algorithm with balancing goals, it interpolates between two heuristics, one is maximizing the neighbor vertexes [18] and the other is minimizing the non-neighbor vertexes [2]. Fennel keeps the load balance by setting a threshold of the maximum capability of partitions. Planted Partition Model [22] using higher length walks for streaming graph partitioning and achieves better quality. Martella proposed Spinner [12], a scalable graph partitioning algorithm based on label propagation. Zheng [28] proposed ARGO based on the findings that in modern multicore clusters, resource contention on memory and homogeneous network costs would affect the speed of graph partitioning. Nishimura [15] proposed a re-streaming algorithm. It performs well on the graph which not change frequently.

For vertex-cut partitioning, Degree-based Hashing (DBH) finds that if high-degree vertexes become the replica vertexes, the replica rate will be reduced. According to these findings, DBH allocates edges by hashing the lower-degree vertex which is one of the end-points of an edge. Greedy [5] assigns edges according to a series of rules and uses the history of the assignments to make the current decision. Grid [7] is another vertex-cut partitioning algorithm that an edge only could be assigned to a constrained subset of partitions. The drawback of Grid is that the partition number is limited to a logarithmic degree. Petroni [17] proposed High Degree Replica First (HDRF). Its main idea is that most real-world graphs have few high-degree vertexes and many low-degree vertexes, so it's suitable for power-law graphs. HDRF has been integrated into GraphLab now. Hu [6] presented a method to estimate graph vertex count, graph workload, and graph processing time in vertex-cut systems. Mofrad [14] proposed Revolver, which uses reinforcement learning and label propagation to carry out vertex-cut graph partitioning and achieve better partitioning performance. Mayer [13] proposed GrapH, a graph processing system considering both vertex traffic and homogeneous network costs. Most of these vertex-cut partitioning algorithms, however, do not take the cluster information into account.

3   Streaming Algorithms

3.1   Problem Definition

We consider a graph $G = (V, E)$, with a set of vertices $V = \{v_1, ..., v_n\}$ and a set of edges $E = \{e_1, ..., e_m\}$, $|V| = n$, $|E| = m$. $e <u, v>$ means edge e connecting vertex u and v. We also define a set of partitions $P = p_1, p_2, ..., p_k$, each partition contains a subset of edges and there doesn't exist same edges on different partitions, so that $p_1 \cup p_2 \cup ... \cup p_k = E$. Let $A(v) \in P$ be the set of partitions that vertex v is replicated. It means that if v has been replicated on p_1 and p_2, that is $A(v) = \{p_1, p_2\}$.

When processing a graph on a distributed system, the replica vertex v needs to communicate with its other replicas on different partitions. So the goal of vertex-cut graph partitioning is to minimize the number of replication vertexes while keeping the balance of each machine. This can be formulated as follows:

$$min \frac{1}{|V|} \sum_{v \in V} |A(v)| \tag{1}$$

$$s.t. \ |p_i| < (1 + \alpha)\frac{|E|}{k}, \forall i \in \{1, ...k\}$$

where α is a parameter to keep the balance and it is a non-negative real number. The larger the α is, the more imbalanced the partitions are. This objective function in Eq. (1) is called *replication factor (RF)*, which will be used as a metric in our evaluations in Sect. 5.

3.2   Streaming Graph Partitioning

Streaming graph partitioning algorithms could be divided into two main approaches. One is vertex-cut graph partitioning; the other is edge-cut graph partitioning. The partitioner of vertex-cut graph partitioning mainly uses greedy heuristic information such as assigning to the partition containing most of the vertex's neighbors while keeping partition balance. The partitioner of edge-cut graph partitioning mainly uses degree information to assign edges like DBH [24] and HDRF [17].

Linear Deterministic Greed (LDG). There exist several heuristic greedy methods for edge-cut graph partitioning. LDG achieves the best performance among these methods. It tries to assign neighbor vertexes into the same partition to reduce the cut-edge. In other words, the partitioner of LDG assigns vertex v to a partition holding the most neighbors of v. Meanwhile, it sets a penalty to keep the balance of partitions. LDG's heuristic could be presented as follows:

$$f(v) = argmax_{i \in [i,k]} |P_i \cap N(v)| w(i) \tag{2}$$

$$w(i) = 1 - \frac{|P_i|}{C} \tag{3}$$

where $C = \frac{n}{k}$ is the capacity of partitions, $N(v)$ is the neighbors set of v. LDG assigns vertex to the partition that maximizes $|P_i \cap N(v)|$ while keeping partition balance by the capacity constraint C. For a coming vertex from the stream, LDG will compute $f(v)$ and select the partition with the highest score as the target partition. $w(i)$ could control the balance of each partition, avoiding the extreme case that most vertexes are assigned to the same partition.

High Degree Replica First (HDRF). Most real-word graphs are power-law graphs, owning a few of high-degree vertexes and a lot of low-degree vertexes. HDRF [17] is particular for power-law graphs. HDRF cuts high-degree vertexes and replicates them on many different partitions, aiming to allocate a strongly connected component with low-degree vertexes into a single partition. High-degree vertexes are few so the replication factor would be reduced by this method.

HDRF maintains a partial degree table for the vertexes with the processing of the graph. For a coming edge $e \in E$ connecting v_i and v_j, HDRF retrieves the degree table and gets the partial degree information of vertexes. After assigning the edge e, the degree table would be updated. The partial degree of v_i could be notated as $\delta(v_i)$ and the partial degree of v_j could be notated as $\delta(v_j)$. To avoid an extreme case that the degree is very large, the degree would be normalized:

$$\theta(v_1) = \frac{\delta(v_1)}{\delta(v_1) + \delta(v_2)} = 1 - \theta(v_2) \tag{4}$$

The score function of HDRF is defined as the following Equation:

$$C^{HDRF}(v_i, v_j, p) = C_{REP}^{HDRF}(v_i, v_j, p) + C_{BAL}^{HDRF}(p) \tag{5}$$

$$C_{REP}^{HDRF}(v_i, v_j, p) = g(v_i, p) + g(v_j, p) \tag{6}$$

$$g(v, p) = \begin{cases} 1 + (1 - \theta(v)), & \text{if } p \in A(v) \\ 0, & \text{otherwise} \end{cases} \tag{7}$$

$$C_{REP}^{HDRF}(p) = \lambda \cdot \frac{maxsize - |p|}{\epsilon + maxsize - minsize} \tag{8}$$

where $A(v)$ is a set of partitions that contain vertex v, $maxsize$ is the size of the maximum-load partition. On the contrary, $minsize$ is the size of the minimum-load partition. ϵ is a constant, and the parameter λ is used to control the load imbalance. For the coming edge e, HDRF computes a score $C^{HDRF}(v_i, v_j, p)$ (Eq. (5)) for all partitions $p \in P$, and assign e to the partition p with the max score C^{HDRF}. If there exist two or more same maximum score values, HRDF will randomly choose one as the target partition.

4   The Deter Algorithm

In this section, we present DETER, a greedy algorithm that not only considers the degree but also considers the neighbors of vertexes for vertex-cut graph partitioning. Besides, we propose a method to accelerate the computation by storing the history vertexes assigning information.

4.1   DETER

Most real-world graphs are power-law graphs, which have few high-degree vertexes and many low-degree vertexes. If the high-degree vertexes are replicated, the replica factor will reduce significantly. Several algorithms are based on this characteristic. DBH [24] employs hashing strategy and assigns an edge to the partition according to the hash of the vertex with lower degree. Greedy [5] assigns the edge to the partition according to a series of simple rules. These rules are based on the history of the edge assignments. HDRF is quite similar to Greedy but HDRF performs better than Greedy. All of these vertex-cut approaches above only consider the degree information and ignore the cluster information of an edge's endpoint vertexes. That means previous work doesn't consider the cluster information of the vertexes. Graph clustering algorithms can identify the

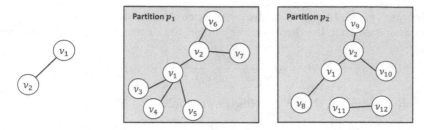

Fig. 1. An example of cluster score: Assigning edge e (left) to Partition p_1 (middle) or Partition p_2 (right).

partial part with high density in G and if edge $e <v_i, v_j>$ is assigned to the partition which contains more neighbors of v_i and v_j. The vertexes in a partition will cluster more densely and the partition quality will be better.

If partition p holds more neighbors of vertex v_i or v_j, then the edge $e <v_i, v_j>$ is more likely to be assigned to p. Inspired by LDG's score function, the cluster score of v_i could be notated as $|p \cap N(v_i)|$ and the cluster score of v_j could be notated as $|p \cap N(v_j)|$. Consequently, the cluster score of $e <v_i, v_j>$ is $|p \cap N(v_i)| + |p \cap N(v_j)|$. To avoid some extreme cases, we normalize the cluster score as:

$$C_{CLU}^{DETER}(v_i, v_j, p) = \frac{|p \cap N(v_i)| + |p \cap N(v_j)|}{|N(v_i)| + |N(v_j)|} \tag{9}$$

DETER's score function is based on HDRF. Taking cluster score into consideration, we could get the final score function of DETER as follows:

$$C^{DETER}(v_i, v_j, p) = C_{REP}^{DETER}(v_i, v_j, p) + C_{CLU}^{DETER}(p) + C_{BAL}^{DETER}(p) \tag{10}$$

$$C_{REP}^{DETER}(v_i, v_j, p) = g(v_i, p) + g(v_j, p) \tag{11}$$

$$g(v, p) = \begin{cases} 1 + (1 - \theta(v)), & \text{if } p \in A(v) \\ 0, & \text{otherwise} \end{cases} \tag{12}$$

$$C_{CLU}^{DETER}(v_i, v_j, p) = \frac{|p \cap N(v_i)| + |p \cap N(v_j)|}{|N(v_i)| + |N(v_j)|} \tag{13}$$

$$C_{BAL}^{DETER}(p) = \lambda \cdot \frac{maxsize - |p|}{\epsilon + maxsize - minsize} \tag{14}$$

In Eq. (10), the score of DETER consists of three parts: the replication score, the cluster score, and the balance score. The λ parameter in Eq. (14) is used to control the load imbalance of partitions. When $\lambda \leq 1$, some extremely unbalanced cases maybe occur and the algorithm is similar to the greedy heuristic strategy if the input stream order is in depth-first search or breadth-first search. Therefore, this will incur imbalanced partitions. When $\lambda > 1$, the importance of balance is proportional to this parameter. Especially, if $\lambda \to \infty$, the partitioner assigns the edges randomly and it becomes a random hashing partitioning strategy.

We illustrate how the cluster score works in Fig. 1. For edge $e <v_1, v_2>$ and partitions p_1, p_2, the vertexes v_1 and v_j have been replicated in p_1 and p_2 so the replication score of both partitions are the same and it doesn't work. Besides, both partitions hold same number of edges so the balance score doesn't affect the assignment. The edge e has 5 neighbors (v_3, v_4, v_5, v_6, v_7) in p_1 and has 3 neighbors (v_8, v_9, v_{10}) in p_2. So the cluster score of p_1 is higher than p_2 and the edge e should be assigned to p_1.

4.2   Acceleration of Computation

DETER algorithm considers the cluster information. When processing the coming edge $e <v_i, v_j>$, DETER requires to know how many neighbor vertexes of v_i and v_j have been replicated on each partition. If we directly compute $|p_i \cap N(v_i)|$, it is time-consuming and the time complexity of this operation is $O(|p_i| * |N(v_i)|)$. When the size of $|p_i|$ is very large, it will cost a lot of time on this operation.

Based on what we analyzed above, DETER uses a method to get the result of $|p_i \cap N(v_i)|$ from another perspective instead of computing directly. This method can reduce the time of computing significantly. We design an extra table to record the neighbors' partition IDs of v_i and v_j. This table consists of two columns: the first column records the vertexes, and the other column records the IDs of partitions that the vertex's neighbors have been assigned to. The partition IDs could be repeated because different vertexes could be stored on the same partition.

Table 1. Vertex's neighbors partitioning information

Vertex	Neighbors' partition ID
v_1	p_1, p_2, p_3
v_2	p_2, p_2, p_4
v_3	p_5
v_4	p_3, p_5

As shown in Table 1, the first row could be explained that the neighbors of v_1 have been assigned to p_1, p_2 and p_3 up to now. When edge $e <v_i, v_j>$ arriving from the stream, we first retrieval the table to find that whether v_i and v_j are in the table. If v_i is not in table, it indicates that the neighbors of v_i had never been assigned to any partitions before. In other words, the intersection of $N(v_i)$ and p_i is empty for all partitions, it means $\sum_i^k |p_i \cap N(v_i)| = 0$; If v_i is in the table, it indicates that the neighbors of v_i had been assigned to the partitions recorded in the second column. So the value of $|p_i \cap N(v_i)|$ is the number of p_i in the second column. In this way, we avoid calculating $|p_i \cap N(v_i)|$ directly and reduce the computing time. This method could also be used in other partitioning algorithms which require the cluster information such as LDG. We verify the efficiency of the acceleration method on DETER through experiments in Sect. 5.

5   Evaluation

In this section, we do comprehensive experiments to discover the performance of our proposed streaming graph partitioning method. We verify whether DETER could reduce the replica factor and improve the performance when processing large-scale graphs.

5.1   Experiment Setup

In our evaluation, we used four real-world large-scale graphs which are various types and have different scales. As shown in Table 2, these graphs are Cite-patents [9], Wiki-topcats [27], Orkut [26], Human Brain [19]. These graphs are all available on the Internet.

Table 2. Graph datasets for evaluations

Name	$|V|$	$|E|$
Cite-patents	3.8M	16.5M
Wiki-topcats	1.8M	28.5M
Orkut	3.1M	117.2M
Human Brain	861.6K	169.4M

All of our graph partitioning experiments ran on a single machine with 12-core *Intel Xeon* CPUs and 96 GB of memory. To test the execution time of real graph applications after graph partitioning, we used GraphX to run PageRank task on a cluster consisting of 6 nodes with 12 core *Intel Xeon* CPUs and 32 GB of memory for each node. We compared our method against three different streaming graph partitioning methods: *DBH* [24], *Greedy* [5], *HDRF* [17]. These methods are some better-performing methods so far. In addition. We also compared the execution time between adopting the acceleration method and not adopting the acceleration method. The number of partition were 8, 16, 32, 64, 128 and 256. There are three data input orders: BFS, DFS and random. We used random input order in our experiments.

We evaluate our methods by two metrics for the partitioning result: replica factor and the execution time. The replica factor has been mentioned in Sect. 3. The execution time includes graph partitioning time T_p, graph computing time T_c and total execution time T_{total}. T_{total} is the sum of T_p and T_c.

$$T_{total} = T_p + T_c \tag{15}$$

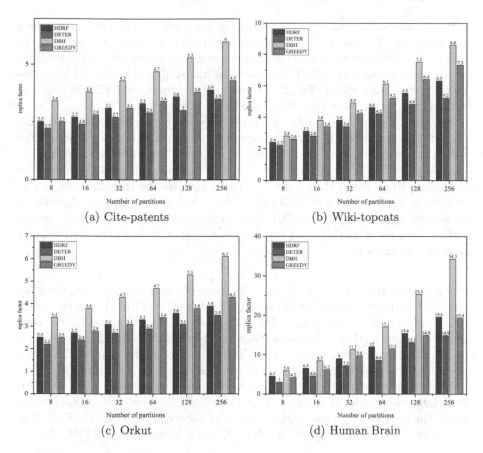

(a) Cite-patents

(b) Wiki-topcats

(c) Orkut

(d) Human Brain

Fig. 2. Replica factor for different numbers of partition on various graph datasets.

Only a graph paritioning algorithm which could reduce both replica factor and T_{total} has practical significance for the distributed graph system.

5.2   Performance Evaluation

In this section, we first compare the result of the experiments that shows the good performance of our algorithm on the replica factor metric. Next, we compare the partitioning time between DETER with acceleration and DETER without acceleration. The result shows that the partitioning time has been reduced significantly. Last, we compare the total execution time T_{total} and graph computing time T_c. It shows that the total time could be reduced if using DETER.

Replica Factor Comparison. Figure 1 shows the comparison of the replica factor between HDRF, DBH, Greedy and the DETER algorithms on different various real-world graph datasets. We set different numbers of partition, ranging from 8 to 256 and the number of each partition is a power of 2.

It is obvious that the DETER algorithm outperforms than the other three algorithms for all considered graphs and the replica factor increased with the growth of partition number. DBH performs worst than the other three algorithms and the performance of HDRF and Greedy are similar. Greedy achieves less replica factor than HDRF for the Human Brain graph and HDRF gets less communication cost than Greedy for the other three graphs.

For the Cite-patents graph, the replica factor of our algorithm DETER reduced 12% than HDRF when the number of partition is 64. For the Wiki-topcats graph, the replica factor reduced by almost 17% than HDRF. The replica factor is almost 14% smaller for the Orkut graph and almost 33% smaller for the Human Brain graph than HDRF. In particular, for the Human Brain graph, the average replica factor of DETER is 22% smaller than Greedy algorithms, and almost 56% smaller than DBH. In the experiment of replica factor, we use random data input method and set the parameter λ as 1, so the load balance could be guaranteed.

Acceleration Result. To verify the efficiency of our acceleration method, we compared the partitioning time between DETER and *DETER–Acc* on the Orkut graph and Human Brain graph. *DETER–Acc* means DETER without using the acceleration method. The comparison result is illustrated in Fig. 2. DETER reduces the partitioning time significantly for the two graphs than *DETER–Acc* due to that it avoids computing the intersection of p_i and $N(v)$ directly.

When the number of partition is 32, the partitioning time T_p of DETER for Orkut graph reduced at most 85% than *DETER–Acc*. For Human Brain graph, the partitioning time reduced at most 90% compared to *DETER–Acc*. The average partitioning speed of DETER is 8× faster than *DETER–Acc* for

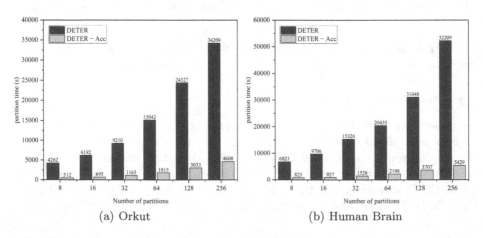

(a) Orkut   (b) Human Brain

Fig. 3. Partitioning time T_p of DETER and DETER–Acc on Orkut and Human Brain graph datasets.

Orkut graph. That means this acceleration method can reduce partitioning time of DETER dramatically.

Graph Computing Time. Based on the partitioning result, we executed the PageRank on GraphX. Graph partitioning aims to reduce the communication cost between machines and reduce the graph computing time. A good partitioning algorithm could lead to less graph computing time. So we compared the time to execute PageRank based on the partitioning results provided by HDRF, DETER, and DBH. The PageRank algorithm was executed 100 iterations and the number of partitions was 32. The result is plotted in Fig. 3.

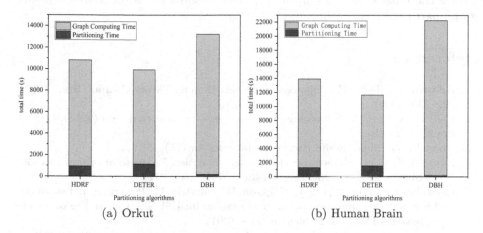

(a) Orkut   (b) Human Brain

Fig. 4. The partitioning time T_p, the graph computing time T_c for PageRank algorithm and the total execution time T_{total} on the Orkut and Human Brain graph datasets, with 32 partitions.

DETER's score function has an additional cluster score. This leads to the partitioning time of DETER is a little longer than HDRF. But the DETER performs better in replica factor. DETER reduces communication cost and speeds up the graph computing time. As shown in Fig. 4, even though DETER consumes a little more time in graph partitioning than DBH and HDRF, the execution of PageRank consumes less time. The total time T_{total} of DETER is lower than the other two partitioning algorithms. For PageRank execution time, DETER consumes 12% lower time than HDRF on the Orkut graph and 20% lower time than HDRF on the Human Brain graph. In total time, DETER reduces 32% time for the Orkut graph and 54% time for the Human Brain graph compared to DBH. This result indicates that DETER achieves better performance in real distributed graph applications.

6   Conclusion

Fast and scalable graph partitioning is the base of distributed graph computing systems. Good graph partitioning leads to fast graph computing and low response. This paper proposed DETER, a new vertex-cut streaming graph partitioning algorithm. DETER is based on HDRF and obeys the rules that high-degree vertexes should be replicated in priority. Besides, DETER also takes the cluster information into consideration because putting the high-density subgraph into the same partition could also reduce the communication cost. In addition, DETER adopts an acceleration method to speed up the partitioning. Experiment results show the effectiveness of the DETER algorithm. DETER could reduce the replica factor by up to 12%–56%, compared to the state-of-the-art graph partitioning algorithms.

References

1. Andreev, K., Racke, H.: Balanced graph partitioning. Theory Comput. Syst. **39**(6), 929–939 (2006)
2. Bali, P., Kalavri, V.: Streaming graph analytics framework design (2015). http:// urn.kb.se/resolve
3. Donnelly, G.: Super-useful Facebook statistics for (75) (2018)
4. Fineschi, S., et al.: Metis: a novel coronagraph design for the solar orbiter mission. Proc. SPIE - Int. Soc. Opt. Eng. **8443**(8), 457–469 (2012)
5. Gonzalez, J.E., Low, Y., Gu, H., Bickson, D., Guestrin, C.: Powergraph: distributed graph-parallel computation on natural graphs. In: USENIX Conference on Operating Systems Design & Implementation (2012)
6. Hu, K., Zeng, G., Jiang, H., Wang, W.: Partitioning big graph with respect to arbitrary proportions in a streaming manner. Future Gener. Comput. Syst. **80**, 1–11 (2018)
7. Jain, N., Liao, G., Willke, T.L.: Graphbuilder: scalable graph ETL framework. In: International Workshop on Graph Data Management Experiences & Systems (2013)
8. Kalnis, P., Awara, K., Jamjoom, H., Khayyat, Z.: Mizan: optimizing graph mining in large parallel systems. Technical report, King Abdullah University of Science and Technology (2012)
9. Leskovec, J., Kleinberg, J., Faloutsos, C.: Graphs over time: densification laws, shrinking diameters and possible explanations. In: Proceedings of the Eleventh ACM SIGKDD International Conference on Knowledge Discovery in Data Mining, pp. 177–187. ACM (2005)
10. Low, Y., Bickson, D., Gonzalez, J., Guestrin, C., Kyrola, A., Hellerstein, J.M.: Distributed graphlab: a framework for machine learning and data mining in the cloud. Proc. VLDB Endow. **5**(8), 716–727 (2012)
11. Malewicz, G., et al.: Pregel: a system for large-scale graph processing. In: Proceedings of the 2010 ACM SIGMOD International Conference on Management of Data, pp. 135–146. ACM (2010)
12. Martella, C., Logothetis, D., Loukas, A., Siganos, G.: Spinner: scalable graph partitioning in the cloud. In: IEEE International Conference on Data Engineering (2017)

13. Mayer, C., Tariq, M.A., Mayer, R., Rothermel, K.: Graph: traffic-aware graph processing. IEEE Trans. Parallel Distrib. Syst. **29**(6), 1289–1302 (2018)
14. Mofrad, M.H., Melhem, R., Hammoud, M.: Revolver: vertex-centric graph partitioning using reinforcement learning. In: 2018 IEEE 11th International Conference on Cloud Computing (CLOUD), pp. 818–821. IEEE (2018)
15. Nishimura, J., Ugander, J.: Restreaming graph partitioning: simple versatile algorithms for advanced balancing. In: ACM SIGKDD International Conference on Knowledge Discovery & Data Mining (2013)
16. Page, L., Brin, S., Motwani, R., Winograd, T.: The pagerank citation ranking: bringing order to the web. Technical report, Stanford InfoLab (1999)
17. Petroni, F., Querzoni, L., Daudjee, K., Kamali, S., Iacoboni, G.: HDRF: stream-based partitioning for power-law graphs. In: Proceedings of the 24th ACM International on Conference on Information and Knowledge Management, pp. 243–252. ACM (2015)
18. Prabhakaran, V., Wu, M., Weng, X., McSherry, F., Zhou, L., Haradasan, M.: Managing large graphs on multi-cores with graph awareness. In: Presented as Part of the 2012 USENIX Annual Technical Conference (USENIX ATC 2012), pp. 41–52 (2012)
19. Rossi, R.A., Ahmed, N.K.: The network data repository with interactive graph analytics and visualization. In: Proceedings of the Twenty-Ninth AAAI Conference on Artificial Intelligence (2015). http://networkrepository.com
20. Stanton, I., Kliot, G.: Streaming graph partitioning for large distributed graphs. In: Proceedings of the 18th ACM SIGKDD International Conference on Knowledge Discovery and Data Mining, pp. 1222–1230. ACM (2012)
21. Thorup, M.: Undirected single-source shortest paths with positive integer weights in linear time. J. ACM (JACM) **46**(3), 362–394 (1999)
22. Tsourakakis, C.: Streaming graph partitioning in the planted partition model. In: Proceedings of the 2015 ACM on Conference on Online Social Networks, pp. 27–35. ACM (2015)
23. Tsourakakis, C., Gkantsidis, C., Radunovic, B., Vojnovic, M.: Fennel: streaming graph partitioning for massive scale graphs. In: Proceedings of the 7th ACM International Conference on Web Search and Data Mining, pp. 333–342. ACM (2014)
24. Xie, C., Yan, L., Li, W.J., Zhang, Z.: Distributed power-law graph computing: theoretical and empirical analysis. In: International Conference on Neural Information Processing Systems (2014)
25. Xin, R.S., Gonzalez, J.E., Franklin, M.J., Stoica, I.: GraphX: a resilient distributed graph system on spark. In: First International Workshop on Graph Data Management Experiences and Systems, p. 2. ACM (2013)
26. Yang, J., Leskovec, J.: Defining and evaluating network communities based on ground-truth. In: IEEE International Conference on Data Mining (2012)
27. Yin, H., Benson, A.R., Leskovec, J., Gleich, D.F.: Local higher-order graph clustering. In: ACM SIGKDD International Conference on Knowledge Discovery & Data Mining (2017)
28. Zheng, A., Labrinidis, A., Chrysanthis, P.K., Lange, J.: Argo: architecture-aware graph partitioning. In: IEEE International Conference on Big Data (2017)

Which Node Properties Identify
the Propagation Source in Networks?

Zhong Li[1,2], Chunhe Xia[1,2,4], Tianbo Wang[1,3(✉)], and Xiaochen Liu[1,2]

[1] Beijing Key Laboratory of Network Technology, Beijing, China
{liz0827,xch,wangtb,xcliu}@buaa.edu.cn
[2] School of Computer Science and Engineering, Beihang University, Beijing, China
[3] School of Cyber Science and Technology, Beihang University, Beijing, China
[4] School of Computer Science and Information Technology,
Guangxi Normal University, Guilin, China

Abstract. Malignant propagation events in networks, such as large-scale diffusion of computer viruses, rumors and failures, have caused massive damage to our society. Thus, it is critical to study how to identify the propagation source. However, existing source identification algorithms only quantify the impact mechanisms of part of the factors that affect the Maximum Likelihood Estimator (MLE) of propagation source, which result in reduced source identification accuracy. In this paper, through constructing a mathematical model for propagation process, we derive two node properties, called Average Eccentricity and Infection Force, which quantify the impact mechanisms of all the factors that affect the MLE of propagation source. And then, we design an AEIF source identification algorithm based on the above two node properties, which make AEIF algorithm has improved accuracy and lower time complexity than existing algorithm. Finally, in the experimental part, extensive simulations on various synthetic networks and real-world networks demonstrate the outperformance of AEIF algorithm than existing algorithms, and based on the experimental results, some assignment suggestions of parameters in AEIF algorithm are given.

Keywords: Propagation source identification · Complex network · Average Eccentricity · Infection Force · AEIF algorithm

1   Introduction

In the modern world, rapid development of network technologies has brought great convenience to our lives and work, and the earth is still getting smaller. However, the networking of the world does not always bring benefits, it also exposes us to various network risks, and these network risks have a common characteristic that isolated risks are amplified because they can spread quickly

This work was supported by the National Natural Science Foundation of China [U1636208, No. 61862008, Grant No. 61902013].

© Springer Nature Switzerland AG 2020
S. Wen et al. (Eds.): ICA3PP 2019, LNCS 11944, pp. 256–270, 2020.
https://doi.org/10.1007/978-3-030-38991-8_17

and on a large scale through underlying networks. For example, computer viruses can infect millions of computers over the Internet [15]; rumors can spread quickly to a large number of users over social networks at an incredible speed [3]; in smart grids, isolated failures can lead to large-scale power outages through cascading effects [16]. More seriously, the development of 5G and IoT technologies will further expand the impact of these risks. Therefore, it is of great significance to study how to quickly and accurately identify the source of propagation events in various networks for ensuring social stability in the network era and reducing economic losses when large-scale malicious propagation events occurred.

In the literature published in 2010, Shah et al. systematically analyzed the source identification problem of network propagation events firstly [14], and proposed the idea of finding true source by solving the Maximum Likelihood Estimator (MLE) of propagation source, and given an equivalent estimator, the so-called Rumor Center, in regular tree networks where propagation follows homogeneous Susceptible-Infected (SI) model. Following their ideas, many kinds of literature have studied source identification problems in more complex scenarios. For instance, [4,5,12] proposed some algorithms for identifying multiple sources that simultaneously initiate network propagation events, [8,9,11] researched the source identification problem when propagation follow other propagation models such as Susceptible-Infected-Recovered (SIR) model and Susceptible-Infected-Susceptible (SIS) model, [1,13] designed source identification schemes for different network observation methods such as snapshot and sensor observation, [7] proposed a method for finding propagation source in dynamic networks. However, since solving the MLE of propagation source is a complex computational problem, existing source identification algorithms obtain time efficiency's improvement at the expense of identification accuracy, and the network structure or propagation process is simplified during the derivation of equivalent source estimator. Thus, only the impact mechanisms of part of the factors that affect the MLE of propagation source are quantified in existing algorithms. In this paper, our motivation is to propose an improved solution, which makes the source identification algorithm quantify the impact mechanisms of all the factors related to the MLE of propagation source, and at the same time has the efficiency that is not significantly lower than existing algorithms. The work and contributions of this paper is in following:

1. Based on the mathematical formalization of propagation process given in this paper, we get two node properties (called Average Eccentricity and Infection Force) that directly and uniquely affect the probabilities of different node as propagation source, by summarizing and deriving all the factors that affect the MLE of propagation source. To the best of our knowledge, this paper is the first work to thoroughly analyze the relationship between the MLE of propagation source and various factors, and provides a basis for other researchers to design better source identification estimators and algorithms.

2. A new source identification algorithm, called AEIF, is designed based on the two proposed node properties. Due to the comprehensive quantification of the impact mechanisms of all the factors that affect the MLE of

propagation source, AEIF algorithm has a theoretically higher source identification accuracy. Furthermore, by reasonably simplifying the calculation of the two proposed node properties, AEIF algorithm has the same or even lower time complexity than existing algorithms.

3. Through extensive simulations on synthetic networks and real-world networks, we not only verify that AEIF is significantly better than existing source identification algorithms, but also find the assignment patterns of AEIF algorithm's parameters, which provides guidance for the application of AEIF algorithm.

The rest of this paper is organized as follows: Sect. 2 gives a formulation of the problem we studied; Sect. 3 detailly introduces the method we proposed; Sect. 4 presents the experiment results; Sect. 5 summarizes.

2   Problem Formulation

2.1   SI Model for Network Propagation Events

As with much previous research work, this paper solves the source identification problem of propagation events that follow Susceptible-Infected (SI) model. Consider an undirected graph $G = (V, E)$, where V is the set of nodes and E is the set of edges. Each node $v \in V$ has two possible states: Susceptible (Sus) state indicating that a computer virus, rumor, or failure has not yet been spread to the node, and Infected (Inf) state indicating that the node has been infected by a computer virus, has accepted a rumor, or has failed (For the sake of convenience, in the following, we refer to computer viruses, rumors, and faults collectively as information spreading in underlying networks). Then, a time-slot system is assumed. At the initial moment, only the propagation source is in Inf state, and other nodes are in Sus state. Thereafter, in each time slot, if a node $u \in V$ is in Inf state, it has a probability of $\eta_{(u,v)}$ propagating information to its neighbor node v through edge (u, v), and the information propagations occurring on each edge is independent of each other. Figure 1 illustrates the state of one SI model-compliant information propagation process in a network at initial five time-slots.

2.2   Propagation Source Identification Problem

When a (bad) information has spread to a certain scale in the underlying network, the regulators (perhaps government, or network administrators) will find it and take appropriate measures. While, finding the most likely propagation source based on the state of all nodes obtained from monitoring is an important step in the measures taken by regulators. If using $\mathcal{O}$ to represent the state of all nodes in underlying network obtained from monitoring (referred to as monitoring status), the propagation source identification problem can be formulated as the following mathematical problem:

$$v^* = arg \max_{v \in V_I} P(s = v | o = \mathcal{O})$$

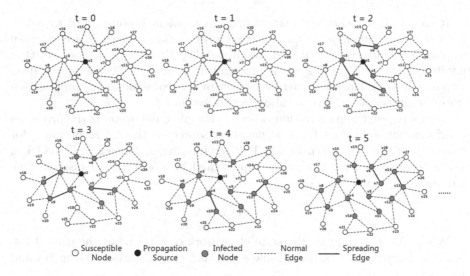

Fig. 1. One SI model-compliant information propagation process in a network

While, V_I represents the set of nodes in Inf state at monitoring time-slot, s and o represent the propagation source random variable and monitoring status random variable, respectively.

As in previous research work, by assuming that each node $v \in V$ has the same prior probability $\cdot P(s = v)$ to be the propagation source, this paper converts the original problem into a Maximum Likelihood Estimator (MLE) problem as follows:

$$v^+ = arg \max_{v \in V_I} P(o = \mathcal{O}|s = v) \tag{1}$$

Next, for designing a more accurate source identification algorithm, we will start by analyzing the above MLE problem, and then derive the node properties that affect the Maximum Likelihood Estimate $P(o = \mathcal{O}|s = v)$ of each node $v \in V$ as propagation source.

3   Methodology

3.1   Mathematical Formalization of Propagation Processes

For the purpose of solving the key term in Eq. (1), that is, the Maximum Likelihood Estimate $P(o = \mathcal{O}|s = v)$ of each node $v \in V$ as propagation source, we introduce the concept of propagation process to decompose this term as follows:

$$P(o = \mathcal{O}|s = v) = \sum_{\mathcal{P} \in S_v^{Pcs}} [P(o = \mathcal{O}|p = \mathcal{P}, s = v)P(p = \mathcal{P}|s = v)] \tag{2}$$

Where p represents the propagation process random variable, $\mathcal{P}$ represents specific propagation process, $S_{v,\mathcal{O}}^{Pcs}$ represents the set of propagation processes starting from node v.

It can be found that the core of original problem becomes how to calculate the probability $P(p = \mathcal{P}|s = v)$ that propagation process $\mathcal{P}$ occurs when propagation source is node v, and the probability $P(o = \mathcal{O}|p = \mathcal{P}, s = v)$ that monitoring status $\mathcal{O}$ occurs when propagation process is $\mathcal{P}$ and propagation source is node v. Thus, we start derivation by the construction of propagation process's mathematical formalization in this section.

In this paper, the propagation process's complete definition is the processes of information spreading from propagation source to the entire network, for example the process shown in Fig. 1. Therefore, when following SI model, a specific propagation process can be represented by a sequence of discrete-time parameters $T \in \{0, 1, 2, 3, \cdots\} = \mathcal{N}_0$, i.e.:

$$\mathcal{P} = \{X_G(t), t \in \mathcal{N}_0\}$$

Where $X_G(t)$ represents the state of network $G = (V, E)$ at the time-slot t, and it is formalized by a multi-tuple consisting of the states of all nodes and states of all edges, i.e.:

$$X_G(t) = <X_{v_1}(t), X_{v_2}(t), \ldots, X_{v_{|V|}}(t), X_{e_1}(t), X_{e_2}(t), \ldots, X_{e_{|E|}}(t)>$$

Where the state $X_v(t)$ of each node $v \in V$ at each time-slot t is divided into Susceptible (Sus) and Infected (Inf) according to whether it has been infected before t, and the state $X_e(t)$ of each edge $e = (u, v) \in E$ at each time-slot t is divided into two different states based on whether or not information spreading occurs between nodes u and v within $(t-1, t]$ period: Spreading (Srd) state means u spreads information to v or v spreads information to u during $(t-1, t]$ period, and No Action ($NAct$) state means that no information spreading action occurs between u and v during $(t-1, t]$ period.

3.2 Occurrence Probability of Propagation Processes

With the above mathematical definition, it is easy to deduce that the occurrence probability of a specific propagation process $\mathcal{P}$ can be calculated as follows:

$$P(p = \mathcal{P}) = P[X_G(0)] \prod_{t \in \mathcal{N}_0} P[X_G(t+1)|X_G(t)] \tag{3}$$

Furthermore, since the propagation source is the only infected node at time-slot 0, so when node v is the source of propagation process $\mathcal{P}$, Eq. (3) can be decomposed into the following two terms:

$$P[X_G(0)] = P(s = v)$$

$$\prod_{t \in \mathcal{N}_0} P[X_G(t+1)|X_G(t)] = P(p = \mathcal{P}|s = v) \tag{4}$$

While, the term we care about is Eq. (4), it can be converted into the following form:

$$P(p = \mathcal{P}|s = v) = \prod_{t \in \mathcal{N}_0} P[X_V(t+1)|X_E(t+1), X_V(t)]P[X_E(t+1)|X_V(t)] \quad (5)$$

For the multiplication term in Eq. (5), it is easy to know that $P[X_V(t+1)|X_E(t+1), X_V(t)]$ will always be 1, and according to SI model, $P[X_E(t+1)|X_V(t)]$ can be decomposed into the product of the occurrence probabilities of that each edge is in corresponding state at time-slot $t+1$, i.e.:

$$P[X_E(t+1)|X_V(t)] = \prod_{e=(u,w)\in E} P[X_e(t+1)|X_u(t), X_w(t)]$$

$$= \prod_{e \in E_1^{\mathcal{P}}(t+1)} \eta_e \cdot \prod_{e \in E_2^{\mathcal{P}}(t+1)} (1 - \eta_e)$$

Where $E_1^{\mathcal{P}}(t+1)$ and $E_2^{\mathcal{P}}(t+1)$ are respectively the following collections:

$$E_1^{\mathcal{P}}(t+1) = \{e = (u, w)|X_e(t+1) = Srd, X_u(t) = Inf, X_w(t) = Sus\}$$
$$E_2^{\mathcal{P}}(t+1) = \{e = (u, w)|X_e(t+1) = NAct, X_u(t) = Inf, X_w(t) = Sus\}$$

Therefore, a computable form of Eq. (5) is as follow:

$$P(p = \mathcal{P}|s = v) = \prod_{t \in \mathcal{N}_0} [\prod_{e \in E_1^{\mathcal{P}}(t+1)} \eta_e \cdot \prod_{e \in E_2^{\mathcal{P}}(t+1)} (1 - \eta_e)] \quad (6)$$

3.3 Occurrence Probability of Monitoring Status

In this section, we analyze and derive the second core problem in Eq. (2). To make the probability $P(o = \mathcal{O}|p = \mathcal{P}, s = v)$ of that monitoring status $\mathcal{O}$ occurs when propagation process is $\mathcal{P}$ non-zero, firstly, propagating process $\mathcal{P}$ must make the state of all nodes in the underlying network the same as monitoring state $\mathcal{O}$ at a certain period $[T_{\mathcal{P},\mathcal{O}}^{min}, T_{\mathcal{P},\mathcal{O}}^{max}]$; secondly, the time-slot $T_{\mathcal{O}}$ regulators monitoring the underlying network must be in time period $[T_{\mathcal{P},\mathcal{O}}^{min}, T_{\mathcal{P},\mathcal{O}}^{max}]$. Thus, if the probability of that regulators' monitoring network action occurs at T is $P(t_{\mathcal{O}} = T)$, then:

$$P(o = \mathcal{O}|p = \mathcal{P}, s = v) = \begin{cases} \sum_{T=T_{\mathcal{P},\mathcal{O}}^{min}}^{T_{\mathcal{P},\mathcal{O}}^{max}} P(t_{\mathcal{O}} = T), \exists X_G(t) \in \mathcal{P}, X_V(t) = \mathcal{O} \\ 0, \quad \forall X_G(t) \in \mathcal{P}, X_V(t) \neq \mathcal{O} \end{cases} \quad (7)$$

So far, we have given a complete systematic calculation method of propagation source's MLE based on the proposed mathematical formulation of the propagation process. However, due to the huge number of propagation processes, it is still complicated to calculate the MLE of propagation source directly using Eqs. (2), (6), and (7). Therefore, next we will derive a more efficient source identification solution based on the above equations.

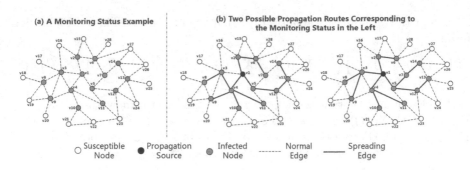

Fig. 2. Example of monitoring status and propagation routes

3.4 Average Eccentricity of Nodes

In this section, to alleviate the problems caused by a large number of propagation processes, we first introduce the concept of propagation route. We define a propagation route as a connected subgraph formed by the edges, on which information spread from Inf nodes to Sus nodes within the time period from initial time-slot to the monitoring time-slot $\mathcal{T}_{\mathcal{O}}$. For example, as shown in Fig. 2b, there are two possible propagation routes corresponding to the monitoring status shown in Fig. 2a.

Obviously, one propagation route corresponds to multiple possible propagation processes, and the propagation processes involved in different propagation routes do not intersect each other. Thus, Eq. (2) can be converted into the following form:

$$P(o = \mathcal{O}|s = v) = \sum_{\mathcal{R} \in S_{\mathcal{O}}^{Rot}} \sum_{\mathcal{P} \in S_{v,\mathcal{R}}^{Pcs}} [P(o = \mathcal{O}|p = \mathcal{P}, s = v)P(p = \mathcal{P}|s = v)] \tag{8}$$

Where $S_{\mathcal{O}}^{Rot}$ represents the propagation routes' set when monitoring status is $\mathcal{O}$ (noted that propagation routes are connected subgraph of underlying network. Therefore, the set of propagation routes is the same when different nodes are propagation source), $S_{v,\mathcal{R}}^{Pcs}$ represents the propagation processes' set when propagation route is $\mathcal{R}$ and propagation source is node v.

Further, for the simplification of Eq. (8), we classify all propagation processes with same sequence value before they reach the monitoring status into the same class, and define propagation process bundle as follows:

$$\overline{\mathcal{P}} = \{\mathcal{P}|\mathcal{T}_{\mathcal{P},\mathcal{O}}^{min} = \mathcal{T}_{\overline{\mathcal{P}},\mathcal{O}}, \mathcal{P}(0, \mathcal{T}_{\overline{\mathcal{P}},\mathcal{O}}) = \{X_G(t), t = 0, 1, \cdots, \mathcal{T}_{\overline{\mathcal{P}},\mathcal{O}}\}\}$$

Where $\mathcal{T}_{\overline{\mathcal{P}},\mathcal{O}}$ represents the time-slot when propagation process bundle $\overline{\mathcal{P}}$ reaches monitoring status $\mathcal{O}$, $\mathcal{P}(0, \mathcal{T}_{\overline{\mathcal{P}},\mathcal{O}})$ represents the subsequence of propagation process $\mathcal{P}$ in time period $[0, \mathcal{T}_{\overline{\mathcal{P}},\mathcal{O}}]$.

Then, substituting Eqs. (6) and (7) to the accumulated term in Eq. (8) yields:

$$
\sum_{\mathcal{P} \in S_{v,\mathcal{R}}^{Pcs}} [P(o = \mathcal{O}|p = \mathcal{P}, s = v)P(p = \mathcal{P}|s = v)]
$$

$$
= \sum_{\mathcal{P} \in S_{v,\mathcal{R}}^{Pcs}} \{ \sum_{T=T_{\mathcal{P},\mathcal{O}}^{min}}^{T_{\mathcal{P},\mathcal{O}}^{max}} P(t_{\mathcal{O}} = T) \cdot \prod_{t \in \mathcal{N}_0} [\prod_{e \in E_1^{\mathcal{P}}(t+1)} \eta_e \cdot \prod_{e \in E_2^{\mathcal{P}}(t+1)} (1 - \eta_e)]\}
$$

$$
= \sum_{\overline{\mathcal{P}} \in S_{v,\mathcal{R}}^{PcsBud}} \{ P(t_{\mathcal{O}} = T_{\overline{\mathcal{P}},\mathcal{O}}) \cdot \prod_{t=0}^{T_{\overline{\mathcal{P}},\mathcal{O}}} [\prod_{e \in E_1^{\overline{\mathcal{P}}}(t)} \eta_e \cdot \prod_{e \in E_2^{\overline{\mathcal{P}}}(t)} (1 - \eta_e)]
$$

$$
- \sum_{i=1}^{+\infty} [(1 - P(t_{\mathcal{O}} = T_{\overline{\mathcal{P}},\mathcal{O}} + i)) \cdot \prod_{e \in E_{\mathcal{O}}^{Bnd}} (1 - \eta_e)^i]\}
$$

(9)

Where $S_{v,\mathcal{R}}^{PcsBud}$ represents the propagation process bundles' set when propagation route is $\mathcal{R}$ and propagation source is v, $E_{\mathcal{O}}^{Bnd}$ is defined as follow ($V_{\mathcal{O}}^{Inf}$ represents the Inf nodes' set under monitoring status $\mathcal{O}$):

$$
E_{\mathcal{O}}^{Bnd} = \{e = (u,v)|u \in V_{\mathcal{O}}^{Inf}, v \notin V_{\mathcal{O}}^{Inf}\}
$$

Since the faster the information spreads, the more likely it is to have a large impact, and the easier it is to be discovered by regulators, namely, the larger the value of $P(t_{\mathcal{O}} = T_{\overline{\mathcal{P}},\mathcal{O}})$ and $P(t_{\mathcal{O}} = T_{\overline{\mathcal{P}},\mathcal{O}} + i)$. Thus, it is can be found that the smaller the time information spreading from a node v to monitoring status, the larger the value of Eq. (9), and based on this knowledge, we design Average Eccentricity of each node $v \in V$ as follows:

$$
F^{AE}(v) = \frac{1}{|S_{\mathcal{O}}^{Rot}|} \sum_{\mathcal{R} \in S_{\mathcal{O}}^{Rot}} D_{v,\mathcal{R}}^{Ecc}
$$

Where $D_{v,\mathcal{R}}^{Ecc}$ represents the eccentricity value [17] of node v in propagation route $\mathcal{R}$.

By the above design, the smaller Average Eccentricity of a node v, the smaller the distance between v and other nodes over each propagation route, and the smaller the time for the propagation processes starting from v to reach monitoring status, and the Maximum Likelihood Estimate of node v as propagation source is larger.

3.5   Infection Force of Nodes

Average Eccentricity characterizes the effect of the time information spreading from each node to monitoring status, next we will analyze the rest parts of

Eq. (9). Firstly, we remove the time-related terms in Eq. (9) and simplify it as follows:

$$
\sum_{\overline{\mathcal{P}} \in S_{v,\mathcal{R}}^{PcsBud}} \{ \prod_{t=0}^{T_{\overline{\mathcal{P}},\mathcal{O}}} [\prod_{e \in E_1^{\overline{\mathcal{P}}}(t)} \eta_e \cdot \prod_{e \in E_2^{\overline{\mathcal{P}}}(t)} (1-\eta_e)] \}
$$

$$
= \prod_{e \in E_{\mathcal{R}}} \eta_e \times \sum_{\overline{\mathcal{P}} \in S_{v,\mathcal{R}}^{PcsBud}} \prod_{t=0}^{T_{\overline{\mathcal{P}},\mathcal{O}}} \prod_{e \in E_2^{\overline{\mathcal{P}}}(t)} (1-\eta_e)
$$

(10)

Where $E_{\mathcal{R}}$ represents the set of edges in propagation route $\mathcal{R}$.

It can be found that the values of term $\prod_{e \in E_{\mathcal{R}}} \eta_e$ in Eq. (10) are the same when different nodes are propagation source under each propagation route. Thus, we only need to deal with the rest parts of Eq. (10), and do transformation as follow (assume $e = (u, w)$):

$$
\sum_{\overline{\mathcal{P}} \in S_{v,\mathcal{R}}^{PcsBud}} \prod_{t=0}^{T_{\overline{\mathcal{P}},\mathcal{O}}} \prod_{e \in E_2^{\overline{\mathcal{P}}}(t)} (1-\eta_e)
$$

$$
= \sum_{\overline{\mathcal{P}} \in S_{v,\mathcal{R}}^{PcsBud}} \{ \prod_{e \in E_1^{\mathcal{R},\mathcal{O}}} (1-\eta_e)^{|T_{u,\overline{\mathcal{P}}}^{Inf} - T_{w,\overline{\mathcal{P}}}^{Inf}| - 1} \cdot \prod_{e \in E_2^{\mathcal{R},\mathcal{O}}} (1-\eta_e)^{|T_{u,\overline{\mathcal{P}}}^{Inf} - T_{w,\overline{\mathcal{P}}}^{Inf}|}
$$

$$
\cdot \prod_{e \in E_3^{\mathcal{R},\mathcal{O}}} (1-\eta_e)^{|T_{\overline{\mathcal{P}},\mathcal{O}} - max\{T_{u,\overline{\mathcal{P}}}^{Inf} - T_{w,\overline{\mathcal{P}}}^{Inf}\}|} \}
$$

(11)

Where $T_{v,\overline{\mathcal{P}}}^{Inf}$ represents the time-slot node v infected when propagation process bundle is $\overline{\mathcal{P}}$, $E_1^{\mathcal{R},\mathcal{O}}$, $E_2^{\mathcal{R},\mathcal{O}}$ and $E_3^{\mathcal{R},\mathcal{O}}$ are respectively defined as follow:

$$
E_1^{\mathcal{R},\mathcal{O}} = E_{\mathcal{R}}
$$

$$
E_2^{\mathcal{R},\mathcal{O}} = \{e = (u, w) | u, w \in V_{\mathcal{O}}^{Inf}, e \notin E_{\mathcal{R}}\}
$$

$$
E_3^{\mathcal{R},\mathcal{O}} = \{e = (u, w) | u \in V_{\mathcal{O}}^{Inf}, w \notin V_{Inf}\}
$$

Then, laws begin to appear in Eq. (11), namely, the contribution of the adjacent edges of the nodes closer to propagation source in each propagation route is greater to Eq. (11), since that they are infected earlier, the index of the corresponding term in Eq. (11) is larger. Based on this knowledge, we design Infection Force of each node $v \in V$ as follows:

$$
F^{IF}(v) = [\prod_{\mathcal{R} \in S_{\mathcal{O}}^{Rot}} \prod_{d=1}^{H_{v,\mathcal{R}}} \prod_{u \in V_{v,\mathcal{R},d}} \prod_{e \in E_u^{Ngb}} (C_{v,u,e,\mathcal{R}}^{\lambda^{-d}})]^{\frac{1}{|S_{\mathcal{O}}^{Rot}|}} = [\prod_{\mathcal{R} \in S_{\mathcal{O}}^{Rot}} F_{\mathcal{R}}^{IF}(v)]^{\frac{1}{|S_{\mathcal{O}}^{Rot}|}}
$$

Where $\lambda \in (0,1)$ is a constant, used to control the attenuation rate of the contribution of different nodes' adjacent edges to v's Maximum Likelihood Estimate, as the distance between the nodes and v increases. The assignments of λ can be adjusted as needed, and will be discussed in the experimental section detailedly. If using $D_{v,u,\mathcal{R}}$ represents the distance between node v and u in propagation route $\mathcal{R}$, then $C_{v,u,e,\mathcal{R}}$ is defined as follow:

$$C_{v,u,e,\mathcal{R}} = \begin{cases} (1 - \eta_e), & e = (u,w), D_{v,u,\mathcal{R}} < D_{v,w,\mathcal{R}} \\ 1, & e = (u,w), D_{v,u,\mathcal{R}} > D_{v,w,\mathcal{R}} \end{cases}$$

By the above design, the larger a node v's Infection Force value, the closer the distance between v and the nodes with large values of $C_{v,u,e,\mathcal{R}}$ on adjacent edges in each propagation route, since the index λ^{-d} corresponding to these edges is larger. Therefore, the Maximum Likelihood Estimate of v as propagation source is larger, conforming to the law in Eq. (11).

3.6   AEIF Source Identification Algorithm

In order to reduce computational complexity, we first do some simplification for Average Eccentricity and Infection Force before designing the AEIF algorithm. Firstly, learning from the idea in [2], the probability multiplication operations in Infection Force of specific propagation routes is converted into additive operations by logarithmically as follows:

$$-logF_{\mathcal{R}}^{IF}(v) = \sum_{d=1}^{H_{v,\mathcal{R}}} \sum_{u \in V_{v,\mathcal{R},d}} \sum_{e \in E_u^{Ngb}} [\lambda^{-d} \times (-logC_{v,u,e,\mathcal{R}})]$$

Secondly, due to finding all the propagation routes in underlying networks is a highly computationally complex problem, and BFS tree and DFS tree are respectively the propagation routes that minimize and maximize the eccentricity value of corresponding root node. Thus, we use the following two new node properties to approximate Average Eccentricity and Infection Force:

$$\widetilde{F}^{AE}(v) = [D_{\mathcal{R}_v^{BFS}}^{Ecc}(v) + D_{\mathcal{R}_v^{DFS}}^{Ecc}(v)]/2$$

$$\widetilde{F}^{IF}(v) = \{[-logF_{\mathcal{R}_v^{BFS}}^{IF}(v)] + [-logF_{\mathcal{R}_v^{DFS}}^{IF}(v)]\}/2$$

Where $\mathcal{R}_v^{BFS}$ and $\mathcal{R}_v^{DFS}$ respectively represent the propagation route corresponding to the BFS tree and DFS tree when v is the root node.

Based on the above two node properties, we give the indicator $F^{AEIF}(v)$, which is inversely proportional to the probability of v as the propagation source, used for the source identification algorithm designed in this paper as follows:

$$F^{AEIF}(v) = \alpha \widetilde{F}^{AE}(v) + \beta \widetilde{F}^{IF}(v)$$

Where α and β are constants and satisfy $\alpha + \beta = 1$. The assignments of these two parameters can be adjusted as needed, and will be discussed in the experimental section detailedly.

Finally, Algorithm 1 gives the pseudo-code of AEIF source identification algorithm. It is easy to observe that the running time of AEIF algorithm is mainly consumed in finding BFS trees and DFS trees in underlying networks. Thus, its overall time complexity is $O(|V|^2 + |V||E|)$, and is better than the most existing source identification algorithms' time complexity $O(|V|^3)$ or $O(|V|^2 log|V|)$ [6].

Algorithm 1. AEIF Source Identification Algorithm

Require: G, $\mathcal{O}$

Ensure: v^+ (the estimator of propagation source)

  Set v^+ with nothing;

  Set $F_{min}^{AEIF} = MAX\_VALUE$;

  Set subgraph $G_{\mathcal{O}}^{Inf} = <V_{\mathcal{O}}^{Inf}, E_{\mathcal{O}}^{Inf}>$ to be a subgraph of G induced by $\mathcal{O}$;

  for each $v \in V_{\mathcal{O}}^{Inf}$ **do**

    Finding the BFS tree T_v^{BFS} and DFS tree T_v^{DFS} with v as root node in $G_{\mathcal{O}}^{Inf}$;

    Set $D_{v,\mathcal{R}_v}^{Ecc}{}_{BFS} = GetHeight(T_v^{BFS})$;

    Set $D_{v,\mathcal{R}_v}^{Ecc}{}_{DFS} = GetHeight(T_v^{DFS})$;

    Set $F_{v,\mathcal{R}_v}^{IF}{}_{BFS} = GetSpreadingForce(T_v^{BFS})$;

    Set $F_{v,\mathcal{R}_v}^{IF}{}_{DFS} = GetSpreadingForce(T_v^{DFS})$;

    Set $\widetilde{F}^{AE}(v) = (D_{v,\mathcal{R}_v}^{Ecc}{}_{BFS} + D_{v,\mathcal{R}_v}^{Ecc}{}_{DFS})/2$;

    Set $\widetilde{F}^{IF}(v) = (F_{v,\mathcal{R}_v}^{IF}{}_{BFS} + F_{v,\mathcal{R}_v}^{IF}{}_{DFS})/2$;

    Set $F^{AEIF}(v) = \alpha \widetilde{F}_{AE}(v) + \beta \widetilde{F}_{IF}(v)$;

    if $F^{AEIF}(v) < F_{min}^{AEIF}$ **then**

      Set $v^+ = v$;

      Set $F_{min}^{AEIF} = F_v^{AEIF}$;

    end if

  end for

  return v^+;

---

4 Evaluations and Discussions

In this section, we compare AEIF algorithm with six typical source identification algorithms: Rumor Centrality [14], Jordan Centrality [17], Concentrality [2], Closeness Centrality [5], Dynamic Message Passing (DMP) [11] and SFT [9].

4.1 Evaluation Metrics

There are two main evaluation metrics for source identification algorithms' evaluation in existing work: distance D_{v^+,v^*} between propagation source estimator v^+ and actual source v^*, and ranking R_{v^*} of actual propagation source's indicator value (note that in addition to providing a source estimator, source identification algorithms can also rank the nodes in infection graph according to corresponding indicator values, such as Rumor Centrality, Jordan Centrality and the proposed indicator $F^{AEIF}(v)$). In this paper, to avoid the impact of network size on the evaluation results, we apply the normalized values of the above two metrics, called Distance Ratio $\gamma_{Distance}$ and Ranking Ratio $\gamma_{Ranking}$, for algorithm evaluation, and define them respectively as follows:

$$\gamma_{Distance} = \frac{D_{v^+,v^*}}{D_{G_{\mathcal{O}}^{Inf}}}$$

$$\gamma_{Ranking} = \frac{R_{v^*}}{|V_{\mathcal{O}}^{Inf}|}$$

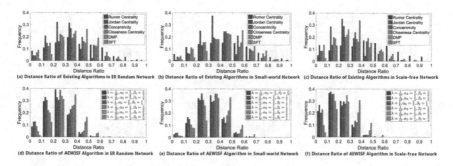

Fig. 3. Frequency histogram of the Distance Ratio of existing algorithms and AEIF algorithm under synthetic networks

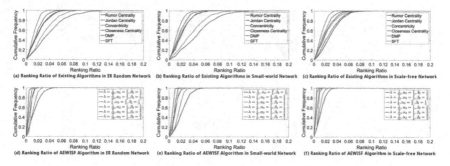

Fig. 4. Cumulative frequency curve of the Distance Ratio of existing algorithms and AEIF algorithm under synthetic networks

Where $D_{G_{\mathcal{O}}^{Inf}}$ represents the diameter of subgraph consisting by Inf state nodes in monitoring status $\mathcal{O}$.

4.2 Evaluation on Synthetic Networks

In this section, we perform algorithm evaluations in three popular synthetic networks: ER random networks, Small-World networks and Scale-Free networks, and their synthesis method in each experiment are respectively as follows:

- ER random networks, randomly generating 1000 nodes and 2500 edges.
- Small-World networks, constructing by WS model with 1000 nodes, 20 initial edges per node and edge reconnection probability is 0.5.
- Scale-Free networks, constructing by BA model with 1000 nodes and power index is 2.5.

As in other literature, we conduct 400 experimental samplings for each type of network, and then count the occurrence frequency of different values of each algorithm's Distance Ratio and Ranking Ratio under a different type of networks. Concretely, in each experiment, the information spreading probabilities of each

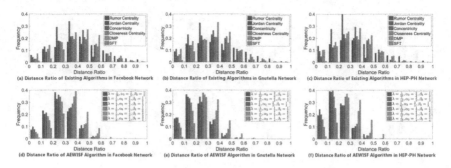

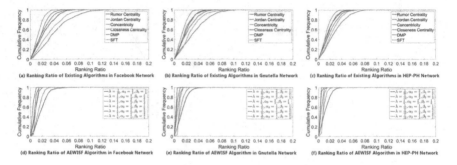

Fig. 5. Frequency histogram of the Distance Ratio of existing algorithms and AEIF algorithm under three real-world networks

Fig. 6. Cumulative frequency curve of the Distance Ratio of existing algorithms and AEIF algorithm under three real-world networks

edge are generated by distribution $N(0.5, 0.2^2)$, then one random node is select as propagation source to perform information propagation simulation, and finally each algorithm is run to identify the propagation source to obtain Distance Ratio and Ranking Ratio in a single experiment.

The experimental results are shown in Figs. 3 and 4. Noted that since all algorithms have small Ranking Ratio evaluation metric values and relatively large Distance Ratio values. Thus, Fig. 3 shows the frequency of Distance Ratio in each interval using frequency histograms, and Fig. 4 uses cumulative frequency curves to show the frequency when Ranking Ratio is less than different values. In addition, due to the parameters of the AEIF algorithm, λ, α, and β, have a wide range of values, we only show the corresponding experimental results of six assignment that can reflect the trend.

It can be observed from the experimental results that the accuracy of AEIF algorithm is obviously higher than existing algorithms, especially when assigning λ, α, and β suitably. Furthermore, the impact of λ, α, and β's assignment for the performance of AEIF algorithm presents four distinct patterns, which can guide the parameters' assignment in algorithm application, namely: (1) AEIF's Distance Ratio is minimally affected by λ; (2) the greater the assignment of α, the better AEIF's Distance Ratio; (3) when the assignment of λ is close to the diameter of underlying networks, AEIF's Ranking Ratio performs better (statistical results indicate that average diameters of the above ER random networks, Small-World networks, and Scale-Free networks are 10, 6, and 12, respectively); (4) when the assignments of α and β are respectively near 1/3 and 2/3, AEIF's Ranking Ratio can achieve an optimal value.

4.3   Evaluation on Real-World Networks

In this section, we perform algorithm evaluations in three real-world networks [10] shown in Table 1.

Table 1. Parameters of tree real-world networks sample

Name	Nodes	Edges	Diameter
Facebook network	4039	88234	8
Gnutella peer-to-peer network	62586	147892	11
High-energy physics citation network	34546	421578	12

Using the same experimental method in Sect. 4.2, experimental results shown in Figs. 5 and 6 are obtained. It can be seen that the accuracy of AEIF algorithm under real-world networks also better than existing algorithms, and the patterns for the assignment of λ, α, and β are consistent with those found in synthetic networks. Therefore, when applying AEIF algorithm, it is recommended to set λ as underlying network's diameter, set α larger when need a high Distance Ratio and β larger when need a high Ranking Ratio.

5   Conclusions

In this paper, through constructing a mathematical model for propagation process, we firstly derive two node properties, called Average Eccentricity and Infection Force, which can quantify the impact mechanisms of all the factors that affect the MLE of propagation source under SI model; and then, we present an efficient AEIF source identification algorithm based on the proposed two node properties; finally, extensive simulations on synthetic networks and real-world networks in experimental part demonstrate the outperformance of AEIF algorithm than the existing source identification algorithms, and give several suggestions for assignment of AEIF algorithm's parameters.

References

1. Antulovfantulin, N., Lancic, A., Stefancic, H., Sikic, M., Smuc, T.: Statistical inference framework for source detection of contagion processes on arbitrary network structures (2013)

2. Dirk, B., Dirk, H.: The hidden geometry of complex, network-driven contagion phenomena. Science **342**(6164), 1337–1342 (2013)
3. Doerr, B., Fouz, M., Friedrich, T.: Why Rumors Spread So Quickly in Social Networks (2012)
4. Fabrizio, A., Alfredo, B., Luca, D., Alejandro, L.C., Riccardo, Z.: Bayesian inference of epidemics on networks via belief propagation. Phys. Rev. Lett. **112**(11), 118701 (2013)
5. Jiang, J., Sheng, W., Shui, Y., Yang, X., Zhou, W.: K-center: an approach on the multi-source identification of information diffusion. IEEE Trans. Inf. Forensics Secur. **10**(12), 2616–2626 (2015)
6. Jiang, J., Sheng, W., Shui, Y., Yang, X., Zhou, W.: Identifying propagation sources in networks: state-of-the-art and comparative studies. IEEE Commun. Surv. Tutor. **19**(1), 465–481 (2017)
7. Jiang, J., Sheng, W., Shui, Y., Yang, X., Zhou, W.: Rumor source identification in social networks with time-varying topology. IEEE Trans. Dependable Secure Comput. **PP**(99), 166–179 (2018)
8. Kai, Z., Lei, Y.: A robust information source estimator with sparse observations. Comput. Soc. Netw. **1**(1), 1–21 (2014)
9. Kai, Z., Lei, Y.: Information source detection in networks: possibility and impossibility results. In: IEEE Infocom - The IEEE International Conference on Computer Communications (2016)
10. Leskovec, J.: Stanford large network dataset collection. http://snap.stanford.edu/data/index.html. Accessed 4 July 2019
11. Lokhov, A.Y., Mézard, M., Ohta, H., Zdeborová, L.: Inferring the origin of an epidemic with a dynamic message-passing algorithm. Phys. Rev. E Stat. Nonlinear Soft Matter Phys. **90**(1), 012801 (2014)
12. Luo, W., Tay, W.P., Leng, M.: Identifying infection sources and regions in large networks. IEEE Trans. Signal Process. **61**(11), 2850–2865 (2013)
13. Luo, W., Tay, W.P., Leng, M.: How to identify an infection source with limited observations. IEEE J. Sel. Top. Signal Process. **8**(4), 586–597 (2014)
14. Shah, D., Zaman, T.: Detecting sources of computer viruses in networks. ACM Sigmetrics Perform. Eval. Rev. **38**(1), 203–214 (2010)
15. Wang, Y., Wen, S., Xiang, Y., Zhou, W.: Modeling the propagation of worms in networks: a survey. IEEE Commun. Surv. Tutor. **16**(2), 942–960 (2014)
16. Yan, Y.Q.Y., Qian, Y., Sharif, H., Tipper, D.: A survey on smart grid communication infrastructures: motivations, requirements and challenges. IEEE Commun. Surv. Tutor. **15**(1), 5–20 (2013)
17. Zhu, K., Ying, L.: Information source detection in the sir model: a sample path based approach. IEEE/ACM Trans. Netw. **24**(1), 408–421 (2016)

t/t-Diagnosability of BCube Network

Yuhao Chen[1,2], Haiping Huang[1,2(✉)], Xiping Liu[1], Hua Dai[1], and Zhijie Han[3]

[1] Nanjing University of Posts and Telecommunications, Nanjing 210023, China
hhp@njupt.edu.cn
[2] Jiangsu High Technology Research Key Laboratory for Wireless Sensor Networks,
Nanjing 210023, China
[3] Henan University, Kaifeng 475001, China

Abstract. BCube network is one of the most classical structures in the server-centered data center networks. The diagnosability of BCube is very important for the network reliability. However, there are few complete theoretical studies in this area. In this paper, the t/t-diagnosis strategy is adopted for the BCube network system under the Preparata, Metze, and Chien's (PMC) model. And meanwhile, the corresponding diagnosis algorithm is designed to be applied in the scenarios with more wrong nodes. Finally, we prove that the t/t-diagnosability for $BCube_{k,n}$ is $(2k + 1)(n - 1) - 2$ for $n \geq 2$, $k \geq 3$ based on theoretical analysis and simulation experiments.

Keywords: t/t-Diagnosability · Data center network · BCube network · PMC diagnostic model

1 Introduction

Data center has a crucial impact on the performance of cloud computing. Various businesses from Google, Amazon, Microsoft and other famous companies have heavily depended on data center network. According to statistics, the number of global data centers has reached 8 million, and the number is still growing. Therefore, many researchers devote to designing new data center network structures to improve the efficiency of network system. Among them, the server-centered data center network structure has become the hot research topic in recent years where BCube network [1] is one of the most classical structures.

BCube network is a scalable, fault-tolerant network structure with high bandwidth proposed by Guo et al. [1]. BCube consists of millions of servers and multiple port switches with recursive structure. It has the advantages of large network capacity, small diameter and strong fault tolerance. Compared with traditional network structure, BCube only needs ordinary switches and does not need expensive core ones, used for supporting fault-tolerant routing algorithm [2]. Based on the BCube network structure, RCube [3], BCCC [4], GBC3 [5] and other BCube-like [2] network structures are proposed, which inherit superior features of BCube network. However, most of the current research on BCube focuses on routing and tree embedding, while less on fault-tolerant reliability.

© Springer Nature Switzerland AG 2020
S. Wen et al. (Eds.): ICA3PP 2019, LNCS 11944, pp. 271–282, 2020.
https://doi.org/10.1007/978-3-030-38991-8_18

As the number of servers in a data center network increases, system designers must consider the network reliability, where automatic fault tolerance has become the most important way to maintain the reliability. Fault tolerance consists of two main steps: the first is fault identification, in which the faulty server is diagnosed; the second is called system configuration, in which the faulty server is replaced or the other servers in the system are redistributed to execute tasks that are running on the faulty one. In automatic fault diagnosis, the system-level diagnosis model is usually employed. Therefore, in order to better evaluate the reliability of BCube network, we use the most classical diagnosis system model PMC (Preparata, Metze, and Chien's) [6] which can accurately obtain the degree of network diagnosis. Under PMC model, a complete and correct diagnosis is undoubtedly the ideal diagnosis of network system, where complete diagnosis means that all fault nodes can be identified and correct diagnosis means that no fault-free node can be diagnosed as a fault one. According to the test results obtained by [6], the basic t-diagnosable system automatically realizes a complete and correct diagnosis. However, for most t-diagnosable network systems with n nodes, the value of the diagnosability t, which is the maximal number of faulty nodes that a system can guarantee to diagnose, is much smaller than n. In other words, if the number of wrong nodes exceeds the network's diagnosability, then the network realized by t-diagnosable may not be effective. In order to solve this problem, many methods have been put forward in the past decades. The most classic method is to increase the diagnosability of a network system by allowing some nodes to be incorrectly identified, such as the t/t-diagnosability [7].

As we known, there is still no systematic proof for the t/t-diagnosability of BCube network under PMC model. Therefore, this paper will focus on t/t-diagnosis of BCube in order to improve the diagnostic capability and verify the reliability of network system.

The rest of this paper can be organized as follows. Section 2 introduces the latest development of diagnostic research and the basic knowledge of diagnostic theory. Section 3 describes the relevant definitions and lemmas of BCube. Section 4 proves the t/t-diagnosability of BCube. Section 5 further verifies the correctness of this scheme through simulation experiments. Finally, we conclude in Sect. 6.

2 Related Work

Research on the characteristics of BCube network is also helpful for the study of BCube-like network structures. To date, a lot of research on the routing and embedding of BCube network have been done. For example, Du et al. proved the Hamiltonian of BCube [8], and Pan et al. proved the existence of CISTs structure in BCube [9]. However, research on reliability and diagnostic of BCube network are rare.

Under PMC model, t-diagnosis is the most basic diagnosis of network structure. In previous studies, t-diagnosis of many network structures has been proved, such as BC graph [10], Mobius cubes [11], and BCube [12] etc., and t-diagnosis can be used as the basis for other diagnostic studies involved the t/t-diagnosis.

Kavianpour et al. [13] proposed a new theory of fault diagnosis at the expense of some accuracy of diagnosis, which prompted the creation of t/t-diagnosis. A network system is said to be t/t-diagnosable if it can locate a t-node set containing all faulty nodes provided that the number of faulty nodes is no more than t [14]. Chwa et al.

[15] extended the above conclusion and proved that at most one fault-free node can be identified as a faulty one. As can be derived from the t/t-diagnosis of several main interconnection networks [10, 11, 15–18], the results of t/t-diagnosis is about twice as large as the classical t-diagnosis. The conclusions of some diagnostic studies are shown in Table 2.

Table 1. The precise diagnosabilities of several main interconnection networks

Interconnection networks	t-diagnosis (PMC)	t/t-diagnosis (PMC)
Q_n	$n(n \geq 5)$ [15]	$2n - 2$ [18]
TQ_n	$n(n \geq 5)$ [16]	$2n - 2$ [11]
FH_n	$n + 1(n \geq 5)$ [16]	$2n - 2$ [11]
MQ_n	$n(n \geq 3)$ [17]	$2n - 2$ [10]

In this article, we used the classic PMC diagnostic model to study the diagnosability of BCube network. Our contribution is to choose and test appropriate properties of BCube network and then proves that the t/t-diagnosability for $BCube_{k,n}$ under the PMC model is $(2k + 1)(n - 1) - 1$ for $n \geq 2$, $k \geq 3$.

3   Preliminaries

A data center network system is modeled as an undirected graph $G = (V(G), E(G))$ whose nodes represent servers and edges represent communication links. $G = (V(G), E(G))$ is a pair comprised of a node set $V(G)$ and an edge set $E(G)$, where $V(G)$ is a finite set and $E(G)$ is a subset of $\{(u, v) \mid (u, v)$ is an unordered pair of distinct elements in $V(G)\}$ [19].

For a set of nodes S, the notation $G - S$ denotes the graph obtained by deleting all the nodes belonging to S from G. The components of a graph G refer to its maximal connected subgraphs. A component is trivial if it has no edges; otherwise, it is nontrivial. A connected graph is one with a single component. Throughout this article, we only consider simple and connected graphs.

The k-dimensional BCube can be defined recursively as follows:

Definition 3.1 [1]: $BCube_{0,n}$ is composed of n servers connected to an n-port switch (a.b. $B_{0,n}$). The k-dimensional BCube network consists of $n(k - 1)$-dimensional BCube networks $(k \geq 1)$ and n^k n-port switch (a.b. $B_{k,n}$). There are n^{k+1} servers and $(k + 1)n^k$ switches in $B_{k,n}$.

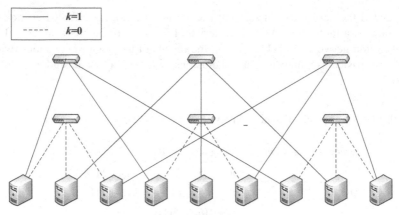

Fig. 1. Network structure of $B_{1,3}$

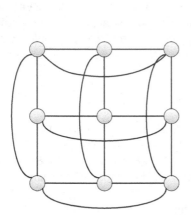

Fig. 2. Network structure of $B_{1,3}$

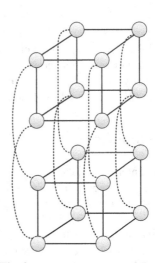

Fig. 3. Network structure of $B_{3,2}$

Since BCube [1] is a server-centered data center network structure, the server has routing function while the switch can be regarded as a transparent link. According to this principle, $B_{1,3}$ in Fig. 1 can be transformed into Fig. 2, where all nodes are servers. Similarly, Fig. 3 shows the network architecture after transformation of $B_{3,2}$. In the following proofs, the transformed network structure will be adopted.

Definition 3.2 [1]: *Each server in BCube has an address represented by a number array* $u_k, \ldots, u_0 \left(u_p \in [0, n-1], p \in [0, k] \right)$.

Given that two servers $u = (u_k, \ldots, u_0)(a_p \in [0, n-1], p \in [0, k])$ and $v = (v_k, \ldots, v_0)(v_p \in [0, n-1], p \in [0, k])$, when u and v are adjacent if $\sum_{p=0}^{k} u_p \oplus v_p = 1$ ($\oplus$ means exclusive or); otherwise, u and v are not adjacent. Obviously, we can derive the following properties from Definition 3.2.

Property 3.1: *u and v are n-hop adjacent if* $\sum_{p=0}^{k} u_p \oplus v_p = n$.

Lemma 3.1: *When only servers are considered as nodes, $B_{k,n}$ is a $(k+1)(n-1)$-regular graph* [10].

Proof: For $\forall u = (u_k, u_{k-1}, \ldots, u_0)(u_p \in [0, n-1], p \in [0, k])$ in $B_{k,n}$, all adjacent nodes of u form a set $N_u.N_u$ can be partitioned into $k+1$ subsets, where each subset represents $n-1$ adjacent nodes of u at the same level, i.e. $N_u = \{(u_k, u_{k-1}, \ldots, u_0') \mid u_0' \in [0, n-1] \wedge (u_0' \neq u_0)\} \cup \{(u_k, u_{k-1}, \ldots u_1', u_0) \mid u_1' \in [0, n-1] \wedge (u_1' \neq u_1)\} \ldots \cup \{(u_k', u_{k-1}, \ldots u_1, u_0) \mid u_k' \in [0, n-1] \wedge (u_k' \neq u_k)\}$. Therefore, $|N_u| = (k+1)(n-1)$.

The t/t-diagnosability of the diagnostic capability of a graph is defined as follows:

Definition 3.3 [7]: *A graph G is t/t-diagnosable if all the faulty nodes can be isolated to a set of t nodes having at most one fault-free node, provided that the number of faulty nodes at any given time is at most t. The t/t-diagnosability of G is the maximum number t such that G is t/t-diagnosable.*

The neighborhood of a vertex u in a subgraph $S \subseteq G$, is denoted by $N_S(u)$. Similarly, the set of neighbor nodes of any subset S in graph G is denoted by $N_G(S)$.

Lemma 3.2 [7]: *Let G be a network system with n nodes. And then, G is t/t-diagnosable if and only if for each $S \subset V(G)$ with $|S| = 2i$ and $i \in \{1, 2, \cdots, t\}$, $|N_G(S)| > t-i+1$.*

4   t/t-Diagnosability of BCube Network

Before proving the t/t-diagnosability of BCube Network, we have listed some conclusions that are useful to our proof.

Lemma 4.1 [20, 21]: *Let $G = (V(G), E(G))$ be a l-regular graph, $S \subseteq V, S \neq \phi$.*

(i)   *If $|V - S| \leq l - 1$, then $N_G(S) = V - S$.*
(ii) *If $|V - S| \geq l$. then $N_G(S) \geq l$.*

Lemma 4.2: *For two nodes u, v in $B_{k,n} |N_{B_{k,n}}(\{u, v\})| \geq (2k+1)(n-1) - 1$.*

Proof: Two cases will be discussed based on Definition 3.1 and Lemma 3.1.

Case 1: u and v are adjacent (seen from Fig. 4), there is one and only one different digit between two addresses. We can get their address number arrays $u = (u_k, \ldots, u_0)(u_p \in [0, n-1], p \in [0, k])$ and $v = (v_k, \ldots, v_0)$ $(v_q \in [0, n-1], q \in [0, k])$. Therefore, u and v in $B_{k,n}$ have common neighbors in $B_{k,n}$ and $|N_{B_{k,n}}(u) \cap N_{B_{k,n}}(v)| = n - 2$. Let $S = \{u, v\}$, we obtain $|N_{B_{k,n}}(\{u, v\})| = |N_{B_{k,n}}(u) \cup N_{B_{k,n}}(v) - S| = |(N_{B_{k,n}}(u) - S) \cup (N_{B_{k,n}}(v) - S)| = |N_{B_{k,n}}(u) - S| +$

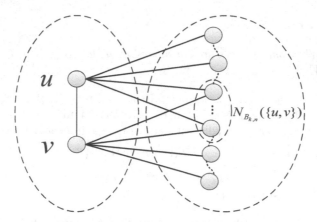

Fig. 4. Illustration of Lemma 4.2 in Case 1

$\left|N_{B_{k,n}}(v) - S\right| - \left|\left(N_{B_{k,n}}(u) - S\right) \cap \left(N_{B_{k,n}}(v) - S\right)\right| = (k+1)(n-1) - 1 + (k+1)(n-1) - 1 - (n-2) = (2k+1)(n-1) - 1$.

Case 2: u and v are not adjacent. Let $S = \{u, v\}$, we obtain $\left|N_{B_{k,n}}(\{u, v\})\right| = \left|N_{B_{k,n}}(u) \cup N_{B_{k,n}}(v) - S\right| = \left|\left(N_{B_{k,n}}(u) - S\right) \cup \left(N_{B_{k,n}}(v) - S\right)\right| = \left|N_{B_{k,n}}(u) - S\right| + \left|N_{B_{k,n}}(v) - S\right| - \left|\left(N_{B_{k,n}}(u) - S\right) \cap \left(N_{B_{k,n}}(v) - S\right)\right| \geq (k+1)(n-1) - 1 + (k+1)(n-1) - 1 - 2 == 2(k+2)(n-1) - 2$ (Fig. 5).

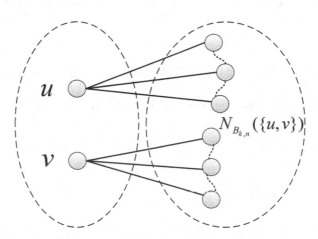

Fig. 5. Illustration of Lemma 4.2 in Case 2

The proof is completed based on the preceding several cases and the conclusion is $\left|N_{B_{k,n}}(\{u, v\})\right| \geq (2k+1)(n-1) - 1$.

Theorem 4.1: *For* $k \geq 3, n \geq 2$, $B_{k,n}$ *is* $((2k+1)(n-1) - 2)/((2k+1)(n-1) - 2)$-*diagnosable.*

Proof: By Lemma 3.2, we only need to prove the following conclusions: for each $i \in \{1, 2, \cdots, (2k + 1)(n - 1)\}$. It's not hard to figure out each $S \subset V(B_{k,n})$ with $|S| = 2i, |N(S)| > (2k + 1)(n - 1) - 2 - i + 1 = (2k + 1)(n - 1) - 1 - i$.

From Definition 3.1, we can get $|V(B_{k,n})| = n^{k+1}|$ from Lemma 3.2 we can conclude that $l = (k + 1)(n - 1)$. According to these conditions, we can discuss the t/t-diagnosis s of BCube network according to the following situations:

Case 1: $i = 1$.

In this case, $|S| = 2$, by Lemma 3.1, $|N_G(2)| \geq (2k + 1)(n - 1) - 1$, and according to Lemma 3.2, the result is true.

Case 2: $2 \leq i \leq (2k + 1)(n - 1) - 1$.

Since $|V(B_{k,n}) - S| = n^{k+1} - 2((2k + 1)(n - 1) - 2)$, for $k \geq 3, n \geq 2$, we can conclude that $|V(B_{k,n}) - S| \geq l = (k + 1)(n - 1)$, by Lemma 4.1, we know that $|V(B_{k,n}) - S| \geq l$ is always holds.

In case 2, as S increases, $N_G(S)$ also increases, while $t-i + 1$ monotonically decreases; so when $i = 2, S = 4$, according to $|N_G(S)| > t - i + 1$ in Lemma 3.2, the left side of this inequality is the minimum while the right side of that is the maximum. And in this case, $N_G(4) > N_G(2) \geq t - i + 1$. The conclusion is true.

The proof is completed based on the preceding several cases.

The t/t-diagnosability of $B_{k,n}$ is the maximum value satisfying t/t-diagnosable degree [22], therefore, we can draw theorem 4.2.

Theorem 4.2: *For $k \geq 3, n \geq 2$, the t/t-diagnosability of $B_{k,n}$ is $(2k + 1)(n - 1) - 1$.*

Proof: We just need to prove that $B_{k,n}$ is not $((2k + 1)(n - 1) - 1)/((2k+1)(n-1)-1)$-diagnosable. Consider a pair complement nodes u and v, if $\sum_{p=0}^{k} u_p \oplus v_p = 1$, then nodes u and v are adjacent. According to Lemma 4.2, we have that $|N_{B_{k,n}}(\{u, v\})| = (2k + 1)(n - 1) - 1 \leq ((2k + 1)(n - 1) - 1 + 1)$. Based on Lemma 3.2, $B_{k,n}$ is not $((2k + 1)(n - 1) - 1)/((2k + 1)(n - 1) - 1)$-diagnosable.

Based on these topological properties and diagnostic results, we propose a fast fault diagnosis algorithm for BCube network. The main idea of this algorithm is traversing the entire BCube network with breadth-first search strategy, and then all nodes are classified into different sets: put fault nodes into the faulty set F and non-fault nodes into the fault-free set T. According to the conclusions in [20, 21] and Table 1, we can judge whether a node set includes faulty nodes. Algorithm 4.1 is designed to determine whether a node is faulty or fault-free.

Algorithm 4.1. t/t-diagnosis algorithm for BCube network under PMC model.

Algorithm 4.1

Fault-free node set $T = \{t_i \in V \mid t_i$ is a fault-free node$\}$;
Faulty node set $F = \{f_i \in V \mid f_i$ is a faulty node$\}$;

The set of all neighbour nodes of node u is $N(u)$; $N(u) = \{v \in V \mid (u, v) \in E(B_{k,n})\}$;

Select any node x as a misdiagnosis node, x does not carry on the diagnosis work.

$\sigma(u, v)$ is the test result of node u to test v;
Set the parameter *count*=0, and the counter will be incremented by 1 for special cases.

Input: Enter a BCube network $B_{k,n}$, for $n \geq 2, k \geq 3$, set fault number which cannot exceed $(2k + 1)(n - 1) - 2$.

Output: Whether t/t-diagnosis is feasible and the faulty node set F.

1: Initialize the input information of BCube network, create the network topology and output the BCube diagram, select a random node u in the BCube network as the start node, set $l = 1$, $i = 1$, *count* = 0 and $S_i(0) = u$.

2: According to the property 3.1, all neighbour nodes of node u are found and saved in the neighbour node set $N(u)$ of node u.

3: Node u tries to find its adjacent nodes from $N(u)$ which are still not diagnosed according to the sequence of breadth first search of BCube network; and then, these nodes will be tested in line with their respective patterns under PMC model. However, if $|N(u)|=0$ or $N(u)=\emptyset$, it means that all nodes have been diagnosed.

    (1) If node u has been detected as a fault-free node, determine whether the *count* is exactly the node degree of node u; if it is established go to Step 8, otherwise go to Step 5.

    (2) If node u has been detected as a faulty node, then go to Step 5.

4: The node u and its neighbor undiagnosed node v are tested to diagnose each other. The test results can be classified into four cases:

    (1) If $\sigma(u,v) = 0$ && $\sigma(v,u) = 0$, u and v belong to the same set, $S_i(1) = v, u = v, l = l + 1$, *count*=0, then back to Step 3.

    (2) If $\sigma(u,v) = 0$ && $\sigma(v,u) = 1$, the diagnostic node u is definitely faulty, which means that all elements in S_i are faulty. Put the elements in S_i into the fault set F, proceed to test the next adjacent node with node v, that is $S_i(0) = v, u = v$, *count*=0, then go to Step 3.

    (3) If $\sigma(u,v) = 1$ && $\sigma(v,u) = 0$, the diagnostic node v is definitely faulty. Put v into the fault set F, proceed to test the next adjacent node with node u, *count*=0, then go to Step 3.

    (4) If $\sigma(u,v) = 1$ && $\sigma(v,u) = 1$, it is unable to determine the properties of v, u, proceed to test the next adjacent node with node u, then go to Step 3, *count* plus 1.

5: Search for a node w that is still not diagnosed. if w exists, let $u = w$, go to Step 3, otherwise go to Step 6.

6: A series of node sets $S_1, S_2, \ldots S_l$ generated by the above steps can be judged as fault sets or fault-free sets according to the properties of fault sets, and then elements in S_l can be put into the fault set F or the fault-free set T correspondingly.

7: Output fault set F.

8: If unable to continue diagnosis, output the existing faulty set F.

Next, we will analyze the time complexity of Algorithm 4.1 under the PMC model.

For $B_{k,n}(n \geq 2, k \geq 3)$, on Step 3 and Step 4, traverse the entire BCube network according to the breadth first search strategy. We use the data structure of adjacency list for storage, and in the worst case, the time complexity is $O(n^{k+1})$. Furthermore, the time complexity of Step 5 is $O(1)$, and the time complexity of Step 6 is also $O(1)$. In conclusion, the time complexity of this algorithm is $O(n^{k+1})$.

5 Simulation Experiments

Since t/t-diagnosability of BCube network have not been proved before, we can only verify the above proof process according to the original nature.

The experiments verify the topological property and t/t-diagnosis property of $B_{k,n}$, respectively; the simulation tests of topology property are composed of node coding and topology property data statistics. First, all nodes of $B_{k,n}$ are encoded. According to Definitions 3.1 and 3.2, each node has a $(k-1)$-bit n-ary number, and there are n^{k+1} nodes in total. Meanwhile, the node degree is calculated according to Lemma 3.1. The experimental data of the topological property of $B_{k,n}$ is shown in Table 2. From the statistical data in Table 2, the total number of nodes and the number of node degrees are respectively consistent with the results of theoretical derivations.

Table 2. BCube network topology data

k	n	Nodes	Degrees
1	3	9	4
3	2	16	6
3	4	256	12
4	5	3125	20

According to the simulation results of BCube network, *num* faulty nodes are randomly added into the generated topology to determine whether t-diagnosis is valid or not. As long as a failure occurs, it means that the t-diagnosis of BCube network fails.

Aiming at different n, k values, the corresponding BCube network topologies are generated, and *num* fault nodes are randomly added in these topologies. In this scenario, we will compare the success number of t-diagnosis tests and t/t-diagnosis tests.

Taking the BCube network with $n = 3$ and $k = 4$ as an example, 6 fault nodes are randomly added, and 100 times t-diagnosis tests are performed and the number of success will be recorded. And then, another one fault node (7 fault nodes totally) is randomly added, and 100 times tests are performed for the statistics of success number. Repeat the above process until the count of added fault nodes reaches 17. The test results are displayed in Table 3.

Similarly, we achieve t/t-diagnosis tests according to the same test scenario as t-diagnosis. The test results are shown in Table 4.

From the results in Tables 3 and 4, we can find that t-diagnosis starts to fail when the number of fault nodes is just 9, while t/t-diagnosis starts to fail when $num = 17$. It can be seen that t/t-diagnosis is more satisfactory than t-diagnosis especially for the scenario with more fault nodes.

Table 3. t-diagnosis tests of $B_{4,3}$

num	6	7	8	9	10	11
Number of success	100	100	100	99	100	99
num	12	13	14	15	16	17
Number of success	98	95	94	90	88	76

Table 4. t/t-diagnosis tests of $B_{4,3}$

num	6	7	8	9	10	11
Number of success	100	100	100	100	100	100
num	12	13	14	15	16	17
Number of success	100	100	100	100	100	99

However, only a set of k, n values is not convincing for the above conclusion. Therefore, we select different combination of values of k and n, and observe the threshold values of fault nodes when t/t-diagnosis and t-diagnosis begin to fail. The statistical results are shown in Table 5.

Table 5. The threshold values of t/t-diagnosability test

$k(k \geq 3)$	$n(n \geq 2)$	t-Diagnosability	t/t-Diagnosability
3	2	3	6
3	3	7	13
4	3	9	17
5	4	17	32

According to the experimental results in Table 5, the t/t-diagnosability of BCube network obtained by our experiments is consistent with the conclusion we have proved before, and meanwhile it demonstrated that the t/t-diagnosis of BCube network can detect more faulty nodes, and the number is about twice as many as the t-diagnosis degree.

6 Conclusions

The diagnosability of a network system based on diagnosis strategies refers to the maximum number of faulty nodes identified correctly by the system. Under the PMC model, t/t-diagnosis strategy was adopted in BCube network system in order to improve the diagnosability especially for the scenario with more faulty nodes. We prove the conclusion that under the t/t-diagnosis strategy, the diagnosability of $B_{k,n}$ is $(2k + 1)(n - 1) - 2$ for $k \geq 3, n \geq 2$. Moreover, Algorithm 4.1 is designed to test the diagnosis degree for t/t-diagnosis. Simulation experiments further demonstrate the effectiveness of t/t-diagnosis for BCube network.

In the future, a long-term study is to prove the t/s-diagnosability and t/k-diagnosability of BCube network. In addition, research on the diagnostic properties of BCube-like network structures such as BCCC and GBC3 will become one of the important future work.

Acknowledgement. This work was supported by the National Natural Science Foundation of P. R. China (No. 61672297), the Key Research and Development Program of Jiangsu Province (Social Development Program, No. BE2017742).

References

1. Guo, C., Lu, G., Li, D., et al.: BCube: a high performance, server-centric network architecture for modular data centers. ACM SIGCOMM Comput. Commun. Rev. **39**(4), 63–74 (2009)
2. Guo, D.: Aggregating uncertain incast transfers in BCube-like data centers. IEEE Trans. Parallel Distrib. Syst. **28**(4), 934–946 (2017)
3. Li, Z., Yang, Y.: RCube: a power efficient and highly available network for data centers. In: 2017 IEEE International Parallel and Distributed Processing Symposium (IPDPS), pp. 718–727. IEEE (2017)
4. Li, Z., Guo, Z., Yang, Y.: BCCC: an expandable network for data centers. IEEE/ACM Trans. Netw. **24**(6), 3740–3755 (2016)
5. Li, Z., Yang, Y.: GBC3: a versatile cube-based server-centric network for data centers. IEEE Trans. Parallel Distrib. Syst. **27**(10), 2895–2910 (2016)
6. Pretarata, F.P., Metze, G., Chien, R.T.: On the connection assignment problem of diagnosis systems. IEEE Trans. Comput. **16**(12), 848–854 (1967)
7. Yang, C.L.: On fault isolation and identification in t1/t1-diagnosable systems. IEEE Trans. Comput. **100**(7), 639–643 (1986)
8. Du, X., Huangfu, Y.: The hamiltonian property analysis and proof of BCube topology. In: 2018 IEEE 4th International Conference on Big Data Security on Cloud (BigDataSecurity), IEEE International Conference on High Performance and Smart Computing (HPSC) and IEEE International Conference on Intelligent Data and Security (IDS), pp. 151–154. IEEE (2018)
9. Pan, T., Cheng, B., Fan, J., et al.: Toward the completely independent spanning trees problem on BCube. In: 2017 IEEE 9th International Conference on Communication Software and Networks (ICCSN), pp. 1103–1106. IEEE (2017)
10. Fan, J., Lin, X.: The t/k-diagnosability of the BC graphs. IEEE Trans. Comput. **54**(2), 176–184 (2005)
11. Fan, J.: Diagnosability of the Mobius cubes. IEEE Trans. Parallel Distrib. Syst. **9**(9), 923–928 (1998)
12. Huang, H., Chen, Y., Liu, X., et al.: t-Diagnosability and conditional diagnosability of BCube networks. In: 2019 IEEE 21st International Conference on High Performance Computing and Communications (HPCC). IEEE Computer Society (2019, in press)
13. Kavianpour, A., Kim, K.H.: Diagnosabilities of hypercubes under the pessimistic one-step diagnosis strategy. IEEE Trans. Comput. **40**(2), 232–237 (1991)

14. Chwa, K.Y.: On fault identification in diagnosable systems. IEEE Trans. Comput. **100**(6), 414–422 (1981)
15. Karunanithi, S., Friedman, A.D.: Analysis of digital systems using a new measure of system diagnosis. IEEE Trans. Comput. **2**, 121–133 (1979)
16. Chang, G.Y., Chang, G.J., Chen, G.H.: Diagnosabilities of regular networks. IEEE Trans. Parallel Distrib. Syst. **16**(4), 314–323 (2005)
17. Armstrong, J.R., Gray, F.G.: Fault diagnosis in a Boolean n cube array of microprocessors. IEEE Trans. Comput. **8**, 587–590 (1981)
18. Lee, C.W., Hsieh, S.Y.: Determining the diagnosability of (1, 2)-matching composition networks and its applications. IEEE Trans. Dependable Secure Comput. **8**(3), 353–362 (2011)
19. Liang, J., Zhang, Q.: The t/s-diagnosability of hypercube networks under the PMC and comparison models. IEEE Access **5**, 5340–5346 (2017)
20. Gu, M.M., Hao, R.X., Xu, J.M., et al.: Equal relation between the extra connectivity and pessimistic diagnosability for some regular graphs. Theoret. Comput. Sci. **690**, 59–72 (2017)
21. Chang, N.W., Hsieh, S.Y.: Conditional diagnosability of (n, k)-star graphs under the PMC model. IEEE Trans. Dependable Secure Comput. **15**(2), 207–216 (2018)
22. Liang, J., Chen, F., Zhang, Q., et al.: t/t-Diagnosability and t/k-Diagnosability for augmented cube networks. IEEE Access **6**, 35029–35041 (2018)

Big Data and Its Applications

Strark-H: A Strategy for Spatial Data Storage to Improve Query Efficiency Based on Spark

Weitao Zou, Weipeng Jing[✉], Guangsheng Chen, and Yang Lu

College of Information and Computer Engineering, Northeast Forestry University,
Harbin 150040, China
weipeng.jing@outlook.com

Abstract. In this paper, we propose Strark-H, a storage and query strategy for large-scale spatial data based on Spark, to improve the response speed of spatial query by considering the spatial location and category keywords of spatial objects. Firstly, we define a custom InputFormat class to make spark natively understand the content of Shapefile, which is a common file format to store spatial data. Then, we put forward a partition and indexing method for spatial storage, based on which spatial data is partitioned unevenly according to the spatial position, which ensures the size of each partition does not exceed the block in HDFS and preserve the spatial proximity of spatial objects in the cluster. Moreover, a secondary index is generated, including global index based on spatial position for all partitions as well as local index based on category of spatial objects. Finally, we design a new data loading and query scheme based on Strark-H for spatial queries including range query, K-NN query and spatial join query. Extensive experiments on OSM show that Strark-H can be applied to Spark to natively support spatial query and storage with efficiency and scalability.

Keywords: Distributed computing · Spark · HDFS · Spatial data · Spatial query

1 Introduction

Spatial data is a collection of multiple data, including raster data (tiff, jpg) and vector data (ShapeFile, KML/KMZ). The vector data containing location information of spatial objects and their feature attributes such as the length and width, is commonly used for spatial statistics and analysis. Among the formats of the vector data file, ShapeFile is the most important in practical application, which consists of a main file (*.shp*) including the geometry data, an index file (*.shx*) consisting of the offset and length of the records in main file, and a dBase table (*.dbf* [1]) storing the attributes of each spatial object.

During the past several decades, spatial data has been widely used in the area of navigation [2], smart forestry [3]. And with the advancement of technology such as remote sensing and mobile Internet, the ability of spatial data acquisition is significantly enhanced. Traditional DBMS and file systems based on single physical machine cannot effectively handle large scale spatial data with complex structure. Therefore, it has become an urgent and tough issue in the field of spatial information to improve the storage scalability and query efficiency at low-costs [4]. Recently, big data technologies [5]

© Springer Nature Switzerland AG 2020
S. Wen et al. (Eds.): ICA3PP 2019, LNCS 11944, pp. 285–299, 2020.
https://doi.org/10.1007/978-3-030-38991-8_19

including distributed storage and distributed computing are introduced for the storage and retrieval of spatial data, which improve the efficiency of managing large scale spatial data.

Among the current mainstream distributed computing frameworks, Hadoop [6] and Spark [7] have been jointly improved and developed by contributors from all over the world as the top open source projects of the Apache community, which are widely used in practical production. They are both frameworks with master-slave structure based on the MapReduce [8] paradigm, and provide friendly programming interfaces, so the underlying network communication, resource allocation, and resource scheduling are transparent for developers. As one of the most important components of the Hadoop framework, HDFS [9] is a distributed file system for big data storage, which supports a variety of distributed computing frameworks. Similar to the MapReduce, HDFS is with master-slave architecture. HDFS cluster has two types of nodes, which are Namenode and Datanode, and responsible for storage of metadata and data, respectively.

There are other distributed file systems and NoSQL such as Ceph [10] and HBase [11], which is commonly applied to real-time calculation. However, HDFS is more suitable for batch analytics with high performance. To facilitate transmission and storage in a distributed cluster, when storing spatial data into HDFS, the files need to be split into chunks called data blocks, which are the smallest unit in the file system with fix block. The data blocks are then assigned to different data nodes by the scheduling module. Unfortunately, it is not suitable for the retrieval of spatial data, because the data format is not directly supported by Hadoop. Moreover, Spark is an in-memory computing framework, so it is typically applied into iterative jobs. During spatial query based on Spark, jobs are divided into tasks and assigned to nodes where spatial data is stored to reduce the network communication burden caused by frequent data transmission among nodes. However, Spark does not support the format of spatial data, either, and it is not designed for analysis of data with multi dimension.

This paper describes Strark-H that is a storage strategy for big spatial data in a distributed cluster. We design a scheme to allow ShapeFile to be directly loaded into Spark. For spatial storage, spatial data is unevenly partitioned and indexed according to the spatial information, and spatial objects in each partition are clustered and indexed according to their category keywords, which ensures that spatial objects in the same partition close to each other and can be retrieved without traversing all spatial objects in the partition. Based on Strark-H, a method for spatial query is designed by filtering data according to the index in the stage of data loading, and data is further calculated on the basis of its storage structure.

The remainder of the paper is arranged as follows. In Sect. 2, a brief introduction to the background and related research work involved in this article. In Sect. 3, we introduce our proposed data storage and query strategy, as well as data support solutions. In Sect. 4, we evaluate our work through public datasets. Section 5 is the summary of the work in this paper.

2   Related Work

It is not reasonable to directly store ShapeFile into HDFS, because it is indivisible and the structure will be destroyed in distributed file System without preprocessing.

Besides, the frameworks such as Hadoop and Spark do not natively support the format of ShapeFile. In Spark-GIS [12], spatial data are converted into text format for further process and analysis based on Spark. GS-Hadoop [13] ensures that Hadoop can load the entire ShapeFile data by re-extending the ShapeFile format. However, there is not an effective method for Hadoop to directly understand the content of the files. GeoSpark [14] is designed to support the directly loading ShapeFile. However, the data loaded into memory is only applicable to the self-defined RDDs in GeoSpark, which makes it is not suitable for handling spatial objects of multiple categories within one RDD of spark, and it cannot guarantee that spatial objects closed to each other's locate in the same node.

Moreover, the current distributed computing frameworks are not designed to support the calculation of multidimensional data. It is generally considered to reorganize the storage structure of spatial data in order to improve the efficiency of spatial query and analysis. Ordinarily, spatial data storage consists of two stages, which are partitioning and indexing. In the stage of partitioning, the number of partitions should be determined at first based on the size of data block. The entire dataset is then partitioned according to the shape and location of the spatial objects contained in the data. For example, Eldawy et al. [15] proposed SpatialHadoop, which partitioned the input files into n partitions and spatially nearby objects are assigned to the same partition. However, the spatial objects are spatially unevenly distributed, the amount of the data within a partition is most likely to exceed the size of a data block. In response to this problem, a strategy for recursive partitioning of spatial data is generally considered [16]. Furthermore, Yao et al. [17] proposed a partitioning algorithm for spatial data, the idea of which is that spatial data is unevenly partitioned according to density of spatial objects. However, it is not easy to encode the partitions based on their algorithm.

Spatial index is a storage structure to improve the efficiency of the spatial query, which is generally consistent with the master-slave model of the distributed computing framework. It is also called secondary index [15, 18], that the global index is built for all partitions of the whole dataset, as well as the local index is built for the spatial objects in each partition. There are many index structures for spatial data, such as R-tree [19], PR-tree [20], and Geohash [21], based on which the time cost of spatial query can be reduced. And for data in memory, building index consumes a certain amount of computing resources, so it is a tradeoff between the time cost of building spatial index and its ability to improve query efficiency. For example, a mechanism for building index on demand is provided in SparkGIS [12]. However, in the storage structure of the current frameworks only the spatial properties of data are taken into account, while the category of data is also an important factor in spatial query.

Spatial query consists of two stages including data loading and calculation. In frameworks such as GeoSpark [22] and LocationSpark [23], the entire dataset is loaded into memory during the system initialization. Although this can reduce the frequency of the disk IO, it is not suitable for analysis of large-scale data and spatial query of multiuser because of the requirements of the large amount of memory. The data can also be filtered according to the query conditions and index when loading into memory and converting into spatial objects, which can decrease memory space and the burden of disk IO. For different spatial queries such as range query [24], K-NN query [25] and spatial join query [26], the lopping operation based on index in memory is critical to further refine the spatial object. Therefore, the query tasks are scheduled in cluster and spatial objects in each partition are judged if the query condition meets.

Although the current frameworks based on Hadoop and Spark can support the storage and query of spatial data, there is still shortcoming in query efficiency, because the data structure are designed without considering the category keywords of spatial objects. Additionally, it is necessary to provide a method to load and transform the Shapefile directly for Spark.

3   Methodology

As shown in Fig. 1, it is a review of Strark-H. To directly load spatial data into memory with Spark, we propose a scheme for Spark to support the format of ShapeFile, which can load spatial objects with multiple categories into an RDD. The data loaded into Spark is converted, partitioned and indexed, then stored into HDFS. Data is stored based on spatial location and category keywords of spatial objects. Based on Strark-H, a new method for data loading and querying is proposed to improve the efficiency of spatial query.

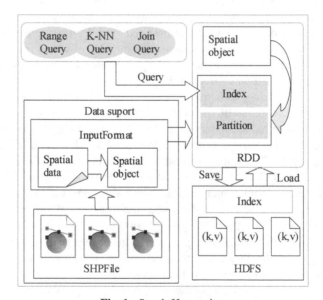

Fig. 1. Strark-H overview

3.1   Support the Format of ShapeFile

When data is loaded into RDD from a local disk or a distributed file system, the *Input-Format* class describes the input-specification for a Spark job. There are two important methods in *InputFormat*: (1) *getSplits()* method is responsible for splitting up the input files into logical splits, each of which is then assigned to an individual task; (2) *createRecordReader()* method returns an instance object of *RecordReader* which is used to glean records from logical splits for further processing. To make Spark support the format of Shapefile, we define a custom *InputFormat* inheriting the default *InputFormat* in Spark. In *NewInputFormat*, the *getSplits()* calls the same method of its parent class, and the *createRecordReader()* method is overridden and returns *SHPReader*, a self-defined class inheriting the default *RecordReader* in Spark which assumes the responsibility of converting byte-oriented ShapeFile into Spatial objects. Then, three methods of *RecordReader* including *initialize()*, *nextKeyValue()* and *close()* are overridden. Figure 2 shows the correspondence between the two custom classes and their parent classes.

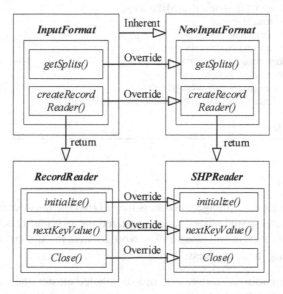

Fig. 2. The relationship between the custom classes and their parent classes

The *initialize()* method of *RecordReader* is only called once at initialization and within this method, the components for data loading are initialized. The *nextKeyValue()* method is called every time a record of the target files is read and the record will be converted into a tuple and sent to RDD. Before the stage of data loading is finished, the *close()* is called to recycle computation resources such as memory and CPU. Algorithm 1 shows the modification for these three methods.

The *SpatialObject* is a custom class with two properties, including *geometry*, an instance object of *Geometry* from the JTS Topology Suite [27] which stores the shape of the spatial objects, as well as a *Map* object for the feature attributes of the spatial objects. And the *SpatialIterator* is an iterator used to traverse spatial objects in the target dataset. In the method of *initialize()*, an instance object of *ReadSHP* is created to convert to the main file of ShapeFile into *Geometry* objects based on the index file of ShapeFile, because it stores the offset of the main file's records, and records of individual spatial object saved in the dBase is converted into a *Map* object, both the two objects are used to construct a *SpatialObject*. In the *nextKeyValue()* method, one spatial object is gotten at a time from the *iterator*. Moreover, the object is transformed to key/value pair: the key is a string for category of the object; and the value is a Json string for the *SpatialObject*. After all files loading into memory, the *close()* method is called to close the iterator. As a result, the overridden *RecordReader* and *NewInputFormat* ensures that spatial objects with different categories can be loaded into the same RDD without modifying the class.

Algorithm 1. SHPReader

```
01:   Initiation. Define the following global variables:
02:   │  SpatialObject spatialobject
03:   │  SpatialIterator iterator
04:   function initialize
05:   │  ReadSHP readSHP ←new ReadSHP(path)
06:   │  iterator ← readSHP.read
07:   function nextKeyValue
08:   │  if iterator has next element do
09:   │  │  Text key, value← new Text
10:   │  │  spatialobject ← iterator.next
11:   │  │  String category← getCategory(path)
12:   │  │  key.set(category) value.set(spatialobject →jsonString)
13:   function close
14:   │  iterator.close
```

3.2 Data Storage

In the stage of spatial data storage, the data is loaded into memory based on the method mentioned above and spatial objects are reorganized through partitioning and indexing through the spatial location and category key words. Then, the spatial objects are stored in HDFS supporting random access.

Data Partition Encoding. To store data in a distributed cluster, the data is partitioned and encoded with Z-order curve. If the order of the curve is N, the space covered by the dataset is divided into $2^N * 2^N$ grid cells with the same size. In space division, a sampling is carried out by calling the method of *RDD.sample()*, where the number of elements is denoted as $count_{sample}$, and the total size of spatial objects is denoted as $size_{sample}$. The average size of the spatial objects is $size_{obj}$ and $size_{data}$ denotes an estimate for the total size of spatial objects in the original RDD.

$$size_{obj} = \frac{size_{sample}}{count_{sample}} \tag{1}$$

$$size_{data} = size_{obj} * count_{RDD} \tag{2}$$

Where the $count_{RDD}$ is the number of elements in the original RDD. As a result, the initial value of the coding order is:

$$N = \left\lceil \log_2 \frac{size_{data}}{size_{block}} \right\rceil \tag{3}$$

The $size_{block}$ is the size of the data block of HDFS which is generally 128 or 64 MB in default. Formula 3 ensures that each logical spatial grid contains as many spatial objects as possible but the total size of them does not exceed the size of a block. After that, $global\_Envelope$ which is the minimum bounding box (MBR) of all spatial objects in the RDD, is calculated by traversing the elements in RDD, the length and width of which is represented by $length_{Spatial\_Grid_i}$, and $width_{Spatial\_Grid_i}$. According to the result of (3), the space is evenly divided into spatial grids and named as $Spatial\_Grid$. And for each gird cell, $Spatial\_Grid_i$, the length and width are:

$$length_{Spatial\_Grid_i} = \frac{length_{global\_Envelope}}{2^N} \tag{4}$$

$$width_{Spatial\_Grid_i} = \frac{width_{Spatial\_Grid_i}}{2^N} \tag{5}$$

Spatial objects in the RDD are partitioned according to the $Spatial\_Grid$, and each grid cell is encoded as $Prefix\_i$ with Z-order curve. However, the spatial objects are unevenly distributed in space, it is normal that the amount of data in a partition exceeds the size of the HDFS block. W_i is the threshold for repartitioning:

$$W_i = \frac{size_{block}}{size_{obj}} - count_{partition_i} \tag{6}$$

Where the $count_{partition_i}$ represents the number of spatial objects of $partition_i$. If the size of the object in the partition exceeds the size of HDFS block, the partition is divided into N_{split} sub-partitions and encoded with Z-order curve.

$$splitorder = \left\lceil \log_2 \left[\sqrt{\left\lceil \frac{count_{partition_i} * size_{obj}}{size_{block}} \right\rceil} \right] \right\rceil \tag{7}$$

$$N_{split} = 2^{splitorder} * 2^{splitorder} \tag{8}$$

As shown in Fig. 3, the partition is divided according to the algorithm, and each sub partition is encoded as $suffix\_j$. The ultimate code of a sub partition consists of the $suffix\_j$ and the code of their farther partition i.e. $Prefix\_i$, as a result, it is finally encoded as $Prefix\_i\_ suffix\_j$.

After completing the logical partitioning, Spatial data in RDD is physically partitioned according to Algorithm 2. In the stage of initiation, the whole dataset is loaded into RDD based on the method in Sect. 3.1. And the partition boundaries are stored in a Map object with key/value pair, where the key is the code of each partition and the value is the MBR of grid cell represented by each partition. Then, the spatial objects in the RDD are physically partitioned by replicating and assigning them to each intersecting partition as shown in Algorithm 2. The ID of the partition is the same as the code of the logical partition.

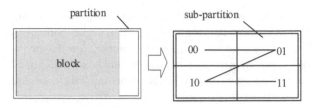

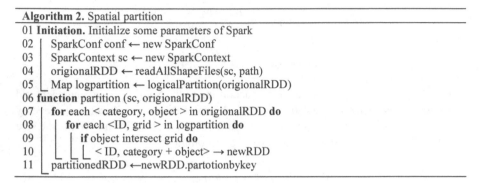

Fig. 3. The partition is divided according to the repartition algorithm

---

Algorithm 2. Spatial partition

01 **Initiation.** Initialize some parameters of Spark
02   SparkConf conf ← new SparkConf
03   SparkContext sc ← new SparkContext
04   origionalRDD ← readAllShapeFiles(sc, path)
05   Map logpartition ← logicalPartition(origionalRDD)
06 **function** partition (sc, origionalRDD)
07   **for** each < category, object > in origionalRDD **do**
08   **for** each <ID, grid > in logpartition **do**
09   **if** object intersect grid **do**
10   < ID, category + object> → newRDD
11   partitionedRDD ←newRDD.partotionbykey

Data Index and Storage. Based on the physical partitioned spatial objects, we build a secondary index and store the data and index in HDFS with SequenceFile with the advantage of supporting random access. Data in each partition is stored in a separate file and the file is named after the partition ID. However, the data in each partition is disordered, which increases the burden of disk IO in spatial query. As a result, a local index based on the category keywords of the spatial object is built.

The spatial objects are sorted by the category within each partition, so as to achieve the purpose of clustering. The data in each partition is store into a separate file and local index file for the partition is generated. The index file stores the offset of the first element for each category. The data files and index files for each partition are stored in a directory named with the partition ID. A global index is built for all these files based on the partition boundaries and the partition IDs.

3.3 Spatial Query

After the spatial data is saved in HDFS based on Srark-H, the query efficiency can be improved by designing a new query strategy for the storage structure. When the client sends a spatial query request to the spark cluster, the data is loaded into memory based on index, and further filtered and refined based on Spark. In this section, we only focus on range query, K-NN query and spatial join query, which are representative in the field of GIS.

Algorithm 3. Range query

```
1  function rangeQuery(range, category)
2      rangeRDD ← loadData(range, category)
3      for each partition in rangeRDD do
4          for each spatialObject in partition do
5              if spatialObject.MBR intersect range do
6                  if spatialObject intersect range do
7                      spatialObject → resultRDD
```

Range Query. There are two parameters need by range query: (1) spatial query area; (2) category of query objects. Algorithm 3 shows the process of implementing range query in Spark. When loading data, access the global index and find all blocks in HDFS overlapping the query area, then the data of the target category is selected based on the local index for each partition. The data in an RDD maintains the same logical partition as that of blocks in HDFS. For spatial objects in each partition of the RDD, the query is processed on MBR of the objects. If an object's MBR intersect with the query area, it is further judged using its real shape.

Algorithm 4. K-NN query

```
01  function knnQuery(query_point, category)
02      Stage 1:
03          range ← getRangeofPartition(query_point)
04          initialRDD← loadData(range, category )
05          initialKnn ←getKnn(initialRDD)
06          maxdist ← getmaxdistDist(initialKnn)
07      Stage 2:
08          wholeRDD← loadData(condtructCircle(query_point, maxdist ), category)
09          if partitionNumber of wholeRDD ==1 do
10              return initialKnn
11          for each partition in wholeRDD do
12              for each spatialObject in partition do
13                  (dist, spatialObject) → knnPartition
14              getKnn(knnPartition) → knnRDD
15          result ← getTopK(knnRDD)
```

K-NN Query. The task of K-NN query is to find the k spatial objects with a category closest to the spatial point *query_point* from the dataset. To minimize disk IO overheads associated with loading data into memory, the query strategy composed of two stages. As shown in Algorithm 4, we describe the K-NN query based on Spark. In stage 1, the k nearest spatial objects to *query_point* is selected within the same data block. And distance from the furthest neighbor to query point is *maxdist*. In stage 2, a circle centered at *query_point* with a radius of *maxdist* is constructed. If the circle only intersects with the boundary of one partition, the result calculated in stage 1 is the final result. Otherwise, all spatial objects within the circle are loaded into RDD and the k nearest neighbors of the query point are selected as final result.

Spatial Join Query. A spatial join query takes two sets of spatial objects A and B with different categories and a join predicate θ (such as intersections) as input, and returns the set of all pairs $<a, b>$ where $a \in A$, $b \in B$, and $(a \; \theta \; b)$ is *true*. The spatial join query processing logic is listed in Algorithm 5. The categories of spatial objects are *category_1, category_2*, and predicate is *join_pred*.

Algorithm 5. Join query

```
1   function joinQuery(category_1, category_2, join_pred)
2       joinRDD ←loadData(category_1, category_2)
3       for each partition in joinRDD do
4           for each spatialObject in partition do
5               if category of spatialobject is category_1 do
6                   spatialobject → cary_1_partition
7               else do
8                   spatialobject → cary_2_partition
9           for each so_1 in cary_1_partition and so_1 in cary_2_partition do
10              if so_1 join_pred so_2 do
11                  (so_1, so_2) → resultRDD
```

The spatial objects of *category_1, category_2* are loaded into RDD at first, which preserves the same logical partition as block in HDFS, so that spatial objects are in the same partition if they are close to each other. Then the RDD is divided into two RDD according to the category, and partitions of the two new RDDs are still the same. As a result, the spatial join query is applied over each two partition with the same partition ID.

4 Experiments

4.1 Experimental Environment

Our evaluations are conducted on a cluster of 5 servers. Four equipped with Intel Xeon E5-2407 processor and 8 GB of memory are virtualized by XenServer6.2 into 16 hosts as compute nodes. The other one is equipped with 16G memory and Intel Xeon E7-4807 as the master node. Red Hat Enterprise Linux Server release 6.2 with the kernel version 2.6.32 is installed on each host. Moreover, each host is equipped with Hadoop-2.7.2 and Spark-2.4.1.

4.2 Datasets

The experiment uses a public dataset provided by OpenStreetMap (OSM) [28]. OSM is a large-scale map project through extensive collaborative contribution from a large number of community users. It contains spatial representation of geometric features such as lakes, forests, buildings and roads. Spatial objects are represented by specific types such as points, lines and polygons. Two sub datasets are picked from OSM, one covers the area of Taiwan with the data size of 0.3 GB, the other contains the spatial data for the area of France, the data size of which is up to 19.9 GB. It is worth noting that the second dataset consists of several smaller sub datasets, each of which covers a region of France. Moreover, for each sub datasets, data with five categories including roads, buildings, land usage, POIs and waterways is selected for spatial storage and query.

4.3 Query Efficiency Analysis

In this section, the query efficiency of range query, K-NN query and join query is compared amongst Strark-H and GeoSpark [22]. And the spatial data for these queries based on Strark-H has been already stored into HDFS. To evaluate the impact of different data size, these queries are executed on dataset with the size of 2 GB and 6 GB in a cluster of 16 nodes. And to compare the impact of cluster size, we execute these queries for 2 GB data in the spark cluster with 4 nodes and 16 nodes, respectively. It is remarkable that in our experiments the spatial object can be considered to belong to a range if it overlaps with the geometry of the range, while its position is still determined by its centroid when calculating the distance from spatial objects to a point.

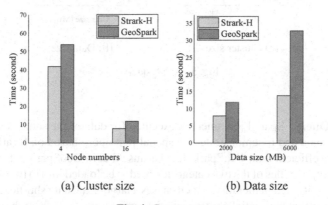

(a) Cluster size   (b) Data size

Fig. 4. Range query

Range Query. In range query, the coincidence is ensured by selecting the same size of query area. As shown in Fig. 4a, the query time remarkably decreases when the number of nodes increased, because of more computing resources can be used for query tasks. In Fig. 4b, both Strark-H and GeoSpark exhibit good performance, and Strark-H is more efficiency with the size of data increased. This is because during the stage of range query based on Strark-H, data is not and no need to be fully loaded into memory and traversed which highly utilizes the secondary index in HDFS. When the size of dataset increases, only the data block overlapping the query area with the target category will be loaded into memory for further processing.

K-NN Query. In the control experiment, the value of k in K-NN query remains the same for each time and the query point is randomly picked within the dataset boundaries. In Fig. 5a, it shows that Strark-H out performs GeoSpark in the cluster with different nodes. And the result in Fig. 5b shows that the execution time of Strark-H does not significantly increase when the size of data changed from 2 GB to 6 GB, but the same parameter of GeoSpark increases obviously. This is because of GeoSpark has to load the whole dataset into memory and traverse all spatial objects to select the result, which is memory exhausting and time-consuming. However, the K-NN query based on Strark-H only need to load data in one or several blocks from HDFS into memory based on the 2-stage query strategy proposed in Sect. 3.3, which significantly reduce the disk IO burden, memory space and calculation time.

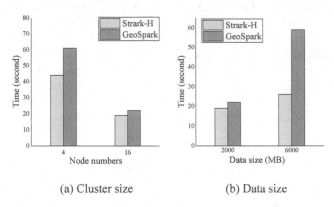

(a) Cluster size   (b) Data size

Fig. 5. K-NN query

Spatial Join Query. Spatial join query is executed for data of the same two categories in OSM. Figure 6a and b demonstrate that spatial join query based on Strark-H for this dataset is more efficient than GeoSpark. It is because that in GeoSpark, when executing spatial join query, the files of the two categories need to be loaded into to two SRDDs, and then they are partitioned in memory, which causes a large-scale data shuffle. Moreover, it is not guaranteed that the spatial objects with the same partition ID from the two SRDDs reside on the same node. Therefore, there are many inter-node communications which will lead to the problem of bottleneck in spatial join query. However, in Strark-H, the data with different categories is stored in the same data block if they are close to each other. As a result, the data loaded into RDD preserves data locality. When spatial join query is executed, there will be much fewer inter-node communications to accelerate the query efficiency.

In a summary, spatial query based on Strark-H performs better in compute-intensive analytical tasks in terms of time efficiency. When the size of computing nodes increases, the time spent on query is significantly reduced. Therefore, it is more suitable for query task with a large amount of data.

4.4 Storage Scalability Analysis

Figure 7 demonstrates the scalability of the storage strategy proposed in this paper. We first analyze the time delay to store data into HDFS as the number of the cluster nodes increasing. We use the data of 2 GB and 6 GB, and the result is shown in Fig. 7a. The experiments show that the speed for processing the data increases almost linearly when using more nodes. For example, the running time for the same size of data reduced nearly to half with the number of nodes changing from 2 to 4, which demonstrates that approximate inverse proportions between query time and size of data nodes exists. We also study the running time with different size of data in a 2-nodes cluster. We select

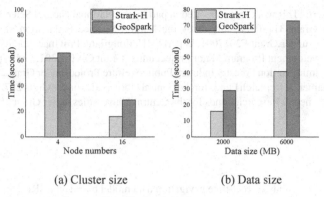

(a) Cluster size   (b) Data size

Fig. 6. Spatial join query

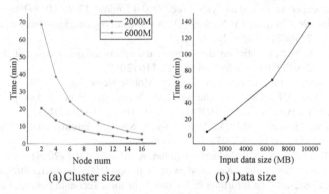

(a) Cluster size   (b) Data size

Fig. 7. As the nodes number and the data size vary, the trend of data storage time

several sub datasets from the OSM, the size of which are 0.3 GB, 2 GB, 6 GB, and 10 GB. As shown in Fig. 7b, the running time changes roughly linearly with the data size. It also indicates that the proposed strategy for data storage enables scalability for input data of different size. As a result, the Strark-H can deal with data of different size in a cluster of different nodes.

5 Conclusion

In this paper, we study the storage and query strategies of spatial data in distributed clusters and propose Strak-H, which is a strategy for spatial data storage and query for the distributed cluster. Because of that spark does not natively support the file format of spatial data, we propose a method to allow spatial data directly loaded into RDD. Then, we design a new storage structure for spatial data based on the spatial location and category keywords. Moreover, we optimize the range query, K-NN query and join query based on Strark-H. Experiments show that the strategy is suitable for the storage of large-scale spatial data, and can effectively improve the query efficiency.

Acknowledgment. This work was supported in part by the National Natural Science Foundation of China under Grant 31770768, in part by the Natural Science Foundation of Heilongjiang Province of China under Grant F2017001, in part by Heilongjiang Province Applied Technology Research and Development Program Major Project under Grant GA18B301, and in part by China State Forestry Administration Forestry Industry Public Welfare Project under Grant 201504307, in part by the Fundamental Research Funds for the Central Universities under Grant 2572018AB33, in part by the Fundamental Research Funds for the Central Universities under Grant 2572019BH01.

References

1. Xia, K., Wei, C.: Study on real-time navigation data model based on ESRI shapefile. In: 2008 International Conference on Embedded Software and Systems Symposia, pp. 174–178 (2008)
2. Tong, Y., She, J., Ding, B., Chen, L., Wo, T., Xu, K.: Online minimum matching in real-time spatial data: experiments and analysis. Proc. VLDB Endow. **12**(9), 1053–1064 (2016)
3. Zou, W., Jing, W., Chen, G., Lu, Y., Song, H.: A survey of big data analytics for smart forestry. IEEE Access **7**, 46621–46636 (2019)
4. Jiang, H., et al.: Vector spatial big data storage and optimized query based on the multi-level hilbert grid index in HBase. Information **5**(9), 116 (2018)
5. Chen, M., Mao, S., Liu, Y.: Big data: a survey. Mobile Netw. Appl. **2**(19), 171–209 (2014)
6. Lee, K.H., Lee, Y.J., Choi, H., Chung, Y.D., Moon, B.: Parallel data processing with MapReduce: a survey. ACM SIGMOD Rec. Arch. **40**(4), 11–20 (2012)
7. Veith, A.D.S., Assunção, M.D.D.: Apache Spark. Springer International Publishing, Cham (2018). https://doi.org/10.1007/978-3-319-63962-8
8. Dean, J., Ghemawat, S.: MapReduce. Commun. ACM **1**(51), 107 (2008)
9. Kala Karun, A., Chitharanjan, K.: A review on Hadoop - HDFS infrastructure extensions. In: 2013 IEEE Conference on Information & Communication Technologies, pp. 132–137. IEEE (2013)
10. Weil, S., Brandt, S., Miller, E., Long, D., Maltzahn, C.: Ceph: a scalable, high-performance distributed file system. In: Proceedings of the 7th Symposium on Operating Systems Design and Implementation, pp. 307–320. USENIX Association (2006)
11. Chang, F., et al.: Bigtable. ACM Trans. Comput. Syst. **2**(26), 1–26 (2008)
12. Baig, F., Vo, H., Kurc, T., Saltz, J., Wang, F.: SparkGIS: resource aware efficient in-memory spatial query processing. In: Proceedings of the 25th ACM SIGSPATIAL International Conference on Advances in Geographic Information Systems, pp. 1–10. ACM (2017)
13. Abdul, J., Alkathiri, M., Potdar, M.B.: Geospatial Hadoop (GS-Hadoop) an efficient MapReduce based engine for distributed processing of shapefiles. In: 2016 2nd International Conference on Advances in Computing, Communication, & Automation (ICACCA), pp. 1–7 (2016)
14. Yu, J., Zhang, Z., Sarwat, M.: Spatial data management in apache spark: the GeoSpark perspective and beyond. GeoInformatica **1**(23), 37–78 (2019)
15. Eldawy, A.: SpatialHadoop: towards flexible and scalable spatial processing using MapReduce. In: Proceedings of the 2014 SIGMOD PhD Symposium, pp. 46–50. ACM (2014)
16. Aji, A., et al.: Hadoop GIS. Proc. VLDB Endow. **11**(6), 1009–1020 (2013)
17. Yao, X., et al.: Spatial coding-based approach for partitioning big spatial data in Hadoop. Comput. Geosci. **106**, 60–67 (2017)
18. Xiao, F.: A big spatial data processing framework applying to national geographic conditions monitoring. In: ISPRS - International Archives of the Photogrammetry, Remote Sensing and Spatial Information Sciences, (XLII-3), pp. 1945–1950 (2018)

19. Beckmann, N., Kriegel, H., Schneider, R., Seeger, B.: The R*-tree: an efficient and robust access method for points and rectangles. In: ACM SIGMOD International Conference on Management of Data, Atlantic City, New Jersey, USA. ACM (1990)
20. Arge, L., Berg, M.D., Haverkort, H., Yi, K.: The priority R-tree. ACM Trans. Algorithms 1(4), 1–30 (2008)
21. Jiajun, L., Haoran, L., Yong, G., Hao, Y., Dan, J.: A geohash-based index for spatial data management in distributed memory. In: 2014 22nd International Conference on Geoinformatics, pp. 1–4 (2014)
22. Yu, J., Wu, J., Sarwat, M.: A demonstration of GeoSpark: a cluster computing framework for processing big spatial data. In: 2016 IEEE 32nd International Conference on Data Engineering, pp. 1410–1413. IEEE (2016)
23. Tang, M., Yu, Y., Malluhi, Q.M., Ouzzani, M., Aref, W.G.: LocationSpark. Proc. VLDB Endow. 13(9), 1565–1568 (2016)
24. Pagel, B., Six, H., Toben, H., Widmayer, P.: Towards an analysis of range query performance in spatial data structures, pp. 214–221. ACM (1993)
25. Iwerks, G., Samet, H., Smith, K.: Maintenance of K-nn and spatial join queries on continuously moving points. ACM Trans. Database Syst. (TODS) 2(31), 485–536 (2006)
26. You, S., Zhang, J., Le, G.: Large-scale spatial join query processing in Cloud. In: 2015 31st IEEE International Conference on Data Engineering Workshops, Seoul, South Korea, pp. 34–41. IEEE (2015)
27. Davis, M.: JTS Topology Suite (2018)
28. OSM. https://www.openstreetmap.org. Accessed 2019

Multitask Assignment Algorithm Based on Decision Tree in Spatial Crowdsourcing Environment

Dunhui Yu[1,2] , Xiaoxiao Zhang[1](✉) , Xingsheng Zhang[1] , and Lingli Zhang[1]

[1] College of Computer and Information Engineering, Hubei University,
Wuhan 430062, China
yumhy@hubu.edu.cn, zxxhubu@163.com, Zhangzero@163.com,
zll02100210@163.com
[2] Education Informationization Engineering and Technology Center, Hubei,
Wuhan 430062, China

Abstract. To improve the resource utilization rate of a platform and increase worker profit, addressing the problem of a limited suitable range in a single-task assignment in a spatial crowdsourcing environment, this paper provides a single-worker multitask assignment strategy. A candidate worker-selection algorithm based on location entropy minimum priority is proposed. Candidate tasks are selected by calculating their location entropy within a selected area. A candidate worker is obtained based on the Manhattan distance between the candidate task and the worker, completing the single-task assignment to the single worker. Then a multitask assignment algorithm based on a decision tree is designed, which builds a multitask screening decision tree and calculates the candidate tasks' time difference, travel cost ratio, coincidence rate of route, and income growth rate of workers. We filter out the most appropriate task and assign it to a worker to complete the multitasking assignment. Experimental results show that the proposed algorithm can effectively reduce the average travel cost, reduce the idle rate of workers, and improve their income, which has better effectiveness and feasibility.

Keywords: Spatial crowdsourcing · Multitask assignment · Location entropy · Decision tree

1 Introduction

The sharing economy brings new forms of value innovation by centralizing, reusing, and distributing the vast amount of idle resources scattered throughout society, greatly improving the efficiency and quality of life [1–3]. Due to the characteristics of time and space, spatial crowdsourcing is the best technology to realize the sharing economy. With the help of spatial crowdsourcing, many sharing applications are emerging, e.g., Didi Travel, Uber, Hungry, and Meituan Takeaway [4]. Their goal is to connect as many vehicles as possible through the internet and to continuously introduce innovative travel services to build the world's leading artificial intelligence traffic engine and brain.

© Springer Nature Switzerland AG 2020
S. Wen et al. (Eds.): ICA3PP 2019, LNCS 11944, pp. 300–314, 2020.
https://doi.org/10.1007/978-3-030-38991-8_20

Spatial crowdsourcing promotes healthy development of transportation toward sharing, intelligence, and new energy. The use of spatial crowdsourcing in taxi applications is steering cities into a green "carpooling era," which puts a high requirement on algorithms in applications [5–7]. We must ensure that every seat in the car is properly and efficiently shared and utilized. When a user places a task on a taxi, the platform quickly analyzes information about it and matches it to the most suitable driver [8]. This provides users with the optimal distribution results to meet their needs. In summary, spatial crowdsourcing task assignment has become an urgent problem to be solved.

Many scholars have researched spatial crowdsourcing task assignment. Taxonomy was introduced to spatial crowdsourcing to solve the problem of tasks that must be completed at specific locations [9]. However, the lack of dynamic analysis of tasks is not conducive to their real-time assignment, and the assignment efficiency is low. Algorithms were proposed to determine the optimal path of workers while maximizing the number of tasks [10]. For complex types of spatial crowdsourcing tasks, it is not enough to rely solely on a single crowdsourcing worker, in which case a team is necessary. In response to such problems, the recommendation problem of the top-k best crowdsourcing teams was studied, and an approximation algorithm and pruning strategy based on the task-specific skill requirements to recommend suitable workers were proposed [11]. However, no optimization method has been proposed for the assignment strategy between single workers and multitasking. How to plan tasks for a worker on a crowdsourcing platform given the budget costs of workers was studied [12], thereby maximizing the number of tasks completed or the utility of completing a task. But this took only the perspective of the workers, and there was no method to ensure the common profits of many parties. A spatial crowdsourcing task dynamic assignment method combining dual-objective optimization with multi-arm gambling machine aimed to maximize task reliability while minimizing travel cost was proposed [13]. However, the active range of workers and the scopes of tasks were not considered. The online task assignment problem was solved by considering the task, worker, and location of the crowd in space and time, and an efficient greedy algorithm, random threshold algorithm, and adaptive stochastic threshold algorithm were proposed [14]. The results obtained by the adaptive random threshold algorithm were proved optimal. The situation of individual workers in multitasking situations was ignored.

There is little research on multitask assignment in spatial crowdsourcing. To reduce the idle rate of platform resources, improve workers' income, and better solve the problem of multitask allocation, we present a candidate worker-selection algorithm based on location entropy minimum priority (LEMP) and a multitask assignment algorithm based on decision tree (MTADT) for multitask assignment. We first select the task with the lowest location entropy as the first task. We select the worker nearest to the task to execute it. So, this worker can be assigned multitasks. A multitask screening decision tree by building calculates the candidate tasks' time difference, travel cost ratio, coincidence rate of route, and income growth rate of workers. Then we filter out the most appropriate tasks for assignment to workers to complete the multitask assignment. We conduct experiments on three simulated and real datasets.

2 Preliminaries

2.1 Definition

Definition 1. Spatial crowdsourcing tasks.

The spatial crowdsourcing task is defined as $t = <i_t, e_t, r_t, p_t, d_t, f_t>$, where i_t is the starting position of task t, e_t is the end position of task t, r_t is the radius of the range of task t, p_t is the task publishing time, d_t is the task publishing deadline, and f_t is the compensation for task t.

Definition 2. Spatial crowdsourcing workers.

Spatial crowdsourcing workers are those who specifically perform crowdsourcing tasks, defined as $w = <l_w, a_w, d_w, r_w, s_w>$, where l_w is the current position of worker w, which indicates the coordinates on the map, a_w is the time to log in to the platform, d_w is the time to leave the platform, an r_w is the worker w who can accept the range radius of the task, and s_w is the historical task success rate of w.

2.2 Description of the Problem

The problem is to design a multitask assignment method to maximize the benefits of workers and assign tasks as soon as possible.

Assume that when the platform is performing task assignment, the set of tasks satisfies the condition T and a group of workers satisfies the condition W. The goal is to find a multitasking assignment $A = (w, t_1, \ldots \ldots, t_i)$ to maximize the profit of a single worker w when completing multiple tasks task $t_i (t_i \in T)$. $I(t_i, w)$ indicates the income of the worker w to complete the task t_i. First, to meet the multitask assignment of spatial crowdsourcing requirements, tasks and workers must simultaneously be online. It must satisfy

$$a_w \leq p_t \leq d_t \leq d_w. \tag{1}$$

At the same time, the distance between task t_i and worker w in task assignment A should satisfy

$$|l_t - l_w| < r_w. \tag{2}$$

The objective is to maximize the total benefits of workers in multitask assignment, which can be formally expressed as the following optimization problems:

$$\text{MaxSum}(A) = \sum\nolimits_{t \in T, w \in W} I(t_i, w) \tag{3}$$

First is the time constraint; tasks can be allocated when tasks and workers are online at the same time. The platform should produce the task assignment results before crowdsourcing task deadline d_t. The second constraint is d_t. Second is the spatial constraint. Task t_i in task assignment A must be within the region with center l_w and radius r_w. The constraints of invariance must also be satisfied. Once crowdsourcing tasks are allocated, the assignment results cannot be changed.

3 Candidate Workers Selection Algorithm Based on LEMP

We present a candidate workers selection algorithm based on LEMP, using the spatial characteristics of the current environment. Each task in a spatial environment corresponds to a certain spatial location. The more workers are where a task is located, the easier the task is to complete. We refer to the sum of the crowdsourcing workers in the task area and their relative proportions passing through the area for a period in the future as the location entropy. If more workers pass through the same area, then it has higher location entropy [15]. However, tasks with low location entropy may be completed with low opportunity.

We use location entropy to quantify the possibility that a task is completed; this represents the probability that the task will be allocated within its region. To improve the assignment of all tasks in a region, we give a higher priority to tasks with low location entropy, so the tasks will quickly be matched in the region. To achieve the optimal global assignment as often as possible, we prioritize crowdsourcing tasks with low location entropy for the next iteration of the algorithm.

R represents the range of the task t, U_R is the set of all workers who have visited area R, W_R is the set of worker who access area R, $|U_R|$ is the total number of times area R is accessed, and $|U_{w,R}|$ is the number of times worker w accesses area R. The probability that worker w will appear in area R is

$$P_{R(w)} = \frac{|U_{w,R}|}{|U_R|}. \tag{4}$$

Location entropy measures the total number of crowdsourcing workers in the task area and their relative proportions through the area and is calculated by

$$E(r) = -C \sum_{w \in W_R} P_{R(w)} \times \log_2 P_{R(w)} \tag{5}$$

where r represents all possible workers in the area where task t is located, and C is any constant associated with unit selection. C is an arbitrary constant related to unit selection.

The Manhattan distance is more suitable for calculating driving directions. We rank tasks in ascending order according to location entropy. Then we assign tasks to the workers who are the smallest in Manhattan [16].

4 Multitask Assignment Algorithm Based on Decision Tree

In the previous section, we assigned tasks to the workers. We now study the problem of assigning multiple tasks to workers who are already on a task, and propose a multitask assignment algorithm based on a decision tree. Although workers are assigned to a single task, many unallocated tasks are emerging around them. Multiple tasks are expected to be assigned to workers while considering time differences, travel cost ratios, coincidence rates of routes, and income growth rates of workers by a multitask screening decision tree. Workers are assumed to be traveling at a constant speed, and the current position of the task publisher is the same as the starting position of the issued task. At this time, the influence of road conditions, traffic lights, and other factors on the distance and speed of workers is ignored.

4.1 Build the Multitask Screening Decision Tree

We build a multitask screening decision tree to calculate and set time differences, travel cost ratios, coincidence rates of routes, income growth rates of workers, and other quantities [17–19].

(1) The root node is generated as the input node, and its child node is constructed as the time-difference feature node.

To enable workers to be assigned to multitasks, tasks must be assigned before the start time [20]. Specifically, when new tasks appear, we must ensure that workers can arrive at the starting position of a new task before the task publishing deadline. This can be expressed as

$$D(t, t_{now}) = (d_{ti} - CT) - \max t_d \geq 0 \tag{6}$$

where $D(t, t_{now})$ is the time difference between a task being executed and a new task, $(d_{ti} - CT)$, is the difference between the current time and the task publishing deadline, and $\max t_d$ is the greatest travel time for a worker to move from the current position to the start position of a new task.

(2) Travel cost ratio feature node

Since the spatial positions of workers constantly change over time, the distances between the tasks to be assigned and the workers also change. Detours will inevitably occur during multitask assignment, and this increases the cost of the crowdsourcing task [21]. Hence, we calculate the ratio of the distance between the worker and the task to be assigned to the distance traveled by the worker at the time of the task (we assume that the worker is traveling at a constant speed). This ratio can be recorded as the wear-and-tear ratio of the task to determine whether the task can be added to the set of candidate tasks. The ratio is

$$L(t, t_{now}) = \frac{d(l_w - i_t)}{d(e_t - i_t)} \tag{7}$$

Where $d(l_w - i_t)$ is the distance difference between the current position of the worker and the starting position of the task to be assigned, and $d(e_t - i_t)$ is the distance difference between the end position of the assigned task and the starting position, i.e., the travel distance of the task.

Then the judgment condition dynamic threshold λ is set, and the left child is constructed as the coincidence rate of the route feature node, and the right child is constructed as a leaf node to tasks that do not meet the condition of the dynamic threshold.

(3) Coincidence rate of route feature node

The modified Hausdorff distance (MHD) algorithm can be used to calculate the matching degree between point sets, but worker trajectories are composed of small line segments that represent the Manhattan distance [22–24]. Therefore, to more accurately obtain the coincidence rate of the route, we will improve the MHD algorithm and use it to calculate the shortest distance between a crowdsourcing worker location point set and the line segment in the set of crowdsourcing task locations to be allocated.

Suppose that the location point collection of the crowdsourcing worker is $U = \{u_1, u_2, \cdots, u_m\}$, the line segment set of the crowdsourcing task to be selected is $V = \{v_1, v_2, \cdots, v_n\}$, and the distance from a crowd point worker's location point set U to the line segment set V of the task to be assigned can be expressed as

$$h(U, V) = \frac{1}{f} \sum_{u_i \in U} h(u_i, V) \tag{8}$$

where f is the starting position, the ending position and the number of waypoints of the entire journey of the crowd-carrying workers in the set U (here, for the convenience of calculation, we set the position of the traffic light in the crowd's driving route to the waypoint position), $h(u_i, V)$ represents the shortest distance from point u_i in set U to all line segments in line set V.

In the task assignment process, we must also consider the direction between the task to be assigned and the workers, so we designed a weight function to reflect the angle between the direction of the workers and that of the route to be assigned for the trend of the coincidence rate of routes, which is expressed as

$$\text{weight} = \alpha \cos \beta \tag{9}$$

where β represents the angle between the direction of travel of a worker and the direction of the route of the task to be assigned, where $\alpha \in [0, 1]$, $\beta \in [0, \pi]$, weight $\in [0, 1]$. This formula can reflect the influence of the angle on the weight.

The weight function is set on the one hand to ensure the crowdsourcing task publisher's experience of the worker's current task, so as to sacrifice less time, and on the other hand to minimize the travel cost caused by the crowdsourcing worker.

After considering the above factors, the coincidence rate of routes can be expressed by

$$R_c(t, t_{now}) = (1 - \frac{h(U, V)}{d(e_t - i_t)}) \cdot \text{weight} \tag{10}$$

Since the coincidence rate of routes is inversely related to the time spent on the crowd-sourcing task, we sort the results of the above formula in descending order and select the y tasks with the highest coincidence rates of routes $R_c(t, t_{now})$ as the candidate tasks. The number of candidate tasks y can be dynamically adjusted according to the difference between the currently calculated coincidence rate of routes, which can be expressed as

$$y = \left\lceil \frac{C}{V(T, t_{now})} \right\rceil \tag{11}$$

where C is a constant, and $V(T, t_{now})$ is the variance of the coincidence rate of the route.

Therefore, the node inputs a judgment condition, and selects the top y tasks with the lowest coincidence rates, the left child is constructed as the income growth rate of the worker's feature node, and the right child is constructed as a leaf node to store the tasks that do not meet the condition [25, 26].

(4) Income growth rate of worker's feature node

For multitasking assignments of workers, the ideal goals are to give workers more than one task and to achieve a lower cost than if the tasks were allocated separately. We assume that workers are currently assigned tasks with a t_1 travel distance of d_1. The distances of the y tasks with higher smoothness in the candidate tasks selected by the above steps are d_i, the detour distance is d_a, and the total distance is d, which can be obtained as

$$d = \sum d_a + \min\{d_1, d_i\} \tag{12}$$

where d_a is the distance between the worker and the candidate task plus the distance between the two task ends.

While workers are multitasking, task initiators will sacrifice a certain amount of time. A discount rate is set to compensate for this expense. The discount rate is inversely proportional to the detour distance and can be expressed as

$$b = 1 - \frac{d_a}{\delta \cdot d(e_t - i_t)} \tag{13}$$

Crowdsourcing workers must meet the following conditions when a task is assigned:

$$I = b \cdot per \cdot (d1 + di) - per \cdot cw \cdot \sum da - per \times d1 > 0 \tag{14}$$

where I is the income of workers, and per is the unit price of the task.

The above formulas can be used to express the growth rate G_I of the workers as

$$G_I(t, t_{now}) = \frac{b \cdot per \cdot (d_1 + d_i) - per \cdot c_w \cdot \sum d_a - per \cdot d_1}{per \cdot d_1} \tag{15}$$

In the income growth rate of a worker node, the input judgment condition takes the maximum value. The left child is constructed as the output node if the condition is met, otherwise the right child is constructed as the leaf node to store a task that does not meet the condition.

Finally, to quickly and accurately assign multitasks, in each timestamp, the data of new tasks, candidate workers, and their assigned tasks are read when new tasks appear. The most appropriate tasks that are filtered out by the multitask screening decision tree are assigned to the candidate workers who are executing the first task; otherwise it waits for the new task that appears in the next timestamp and filters again [27–29].

4.2 Algorithm Execution

The process is shown in Algorithm 1.

Algorithm 1: Multitask assignment algorithm based on decision tree

---

Input: candidate tasks are new task set T, worker w, task currently executed t_{now}, cost ratio threshold λ
Output: final assigned task t
1: input data to decision tree
2: $enableT \leftarrow \varnothing$
3: **for** t in T
4: **if** $D(t,t_{now}) \geq 0$ and $L(t,t_{now}) \leq \lambda$ **then**
5:     $enableT \leftarrow enableT \cup \{t\}$, t goes to the left child node
6: **end if**
7: end for
8: sort elements in $enableT$ in descending order of $R_c(t,t_{now})$
9: $y \leftarrow \left\lceil \dfrac{C}{D(T,t_{now})} \right\rceil$
10: $i \leftarrow 0$
11: $tempG \leftarrow 0$
12: $r \leftarrow \varnothing$
13: **for** t in $enableT$
14: **if** $i < y$ and $G_l(t,t_{now}) > tempG$ **then**
15:     $tempG \leftarrow G_l(t,t_{now})$, t goes to the left child node
16:     $r \leftarrow t$
17:     $i \leftarrow i+1$
18: **end if**
19: **end for**
20: return r

4.3 Examples

The coordinates of task T are shown in in Fig. 1. Since all distances in this chapter are Manhattan distances, a grid in the figure can represent a unit distance. (We previously assumed that workers travel at a constant speed, so in this case, a grid can also be viewed as a unit of time, and we take ten minutes as the time).

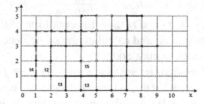

Fig. 1. Schematic diagram of tasks and workers

The information about T and W is shown in Table 1. Assume that the dynamic threshold of the task cost ratio is $\lambda = 0.2$. For ease of calculation, we set the number of path points for each task to be the same, i.e., $f = 3$.

Table 1. Task and worker information

Object	Starting point coordinates	Endpoint coordinates	Start time	End time	Total distance
w1			9:00		
t1	(3, 0)	(8, 5)	9:00	9:10	10
t2	(2, 1)	(5, 5)	9:20	9:50	7
t3	(4, 0)	(9, 3)	9:20	9:50	8
t4	(1, 1)	(5, 4)	9:15	9:35	7
t5	(4, 2)	(6, 5)	9:10	9:25	5

Let $w_1, t_1, t_2, t_3, t_4, t_5$ be the current area and (w_1, t_1) the single-task assignment result completed according to the algorithm in Sect. 3. Worker w_1 has started the route according to t_1. When t_2, t_3, t_4, t_5 appear in succession, we can calculate time difference, travel cost ratio, coincidence rate of routes, and income growth rate of workers according to the MTADT.

(1) Time difference feature node

Crowdsourcing worker w_1 starts at t_1 9:00, and when w_1 travels to (3, 1), t_5 appears at time $D_5(t, t_{now}) = (d_{t5} - CT) - \max t_{d(5,1)} = -5 < 0$. Since t_5 does not meet the condition of time difference, t_5 enters the leaf node and is excluded.

When w_1 travels to (3.5, 1), t_4 appears, and $D_4(t, t_{now}) = -5 < 0$. Since t_4 does not meet the condition of time difference, t_4 enters the leaf node and is excluded.

When w_1 continues to (4, 1), t_2 and t_3 appear simultaneously. At this time $D_2(t, t_{now}) = 10 > 0$, $D_3(t, t_{now}) = 20 > 0$. Since t_2 and t_3 meet the condition of time difference, we enter the next feature node and calculate the travel cost ratio.

(2) Travel cost ratio feature node

We assume that the dynamic threshold $\lambda = 0.2$. The distance of t_1 is $d(e_{t1} - i_{t1}) = 10$, the distance between t_2 and w_1 is $d(l_w - i_{t2}) = 2$, the distance between t_3 and w_1 is $d(l_w - i_{t3}) = 1$. $L(t, t_{now}) = \frac{d(l_w - i_{t2})}{d(e_{t1} - i_{t1})} = \frac{2}{10} = 0.2 \le \lambda$, $L(t, t_{now}) = \frac{d(l_w - i_{t3})}{d(e_{t1} - i_{t1})} = \frac{1}{10} = 0.1 \le \lambda$. t_2 and t_3 meet the condition of the travel cost ratio, and we enter the next feature node and calculate the coincidence rate of the routes.

(3) Coincidence rate of routes feature node

The route points of t_2 and t_3 are the same, $f_2 = f_3 = 3$. $h(w_1, t_2) = \frac{2+3+4+3+2+2+3+2+3}{3} = 8$, $h(w_1, t_3) = 4$. Since w_1 has arrived at (4, 1), the angle between t_2 and w_1 is $0°$, and the angle between t_3 and w_1 is $14°$. If $\alpha = 1$, $R_{c2} = (1 - \frac{8}{10}) \times 1 = 20\%$, $R_{c3} = 58.2\%$.

From the R_{c2} and R_{c3} calculations, the variance of the tasks t_2 and t_3 is 0.0365, and we assume $C = 0.1$, $y = \lceil \frac{0.1}{0.0365} \rceil = 3$. There are only two alternative tasks at this time. We use t_2 and t_3 to meet the condition of the coincidence rate of route, enter the next feature node, and calculate the income growth rate of workers.

(4) Income growth rate of workers feature node

Assume $\delta = 2$, $per = 1.5$, $b_{t2} = 1 - \frac{5}{2 \times 10} = \frac{3}{4} = 0.75$ $b_{t3} = 0.8$. Assume that the cost rate of workers driving in idle state is $c_w = 0.2$, At this time, $I_2 = 0.75 \times 1.5 \times (10 + 7) - 1.5 \times 0.2 \times 5 - 1.5 \times 10 = 2.625 > 0$, $I_3 = 5.4 > 0$, can be found that $G_{I_2} = \frac{2.625}{15} = 17.5\%$, $G_{I_3} = \frac{5.4}{15} = 36\%$. Obviously $G_{I_3} > G_{I_2}$, so the algorithm makes the allocation $\langle t_3, w_1 \rangle$ and worker w_1 updates the route to continue driving until a new task appears for the algorithm to calculate.

4.4 Algorithm Complexity

In MTADT, the time complexity of crowdsourcing tasks that satisfy the time difference and cost ratio constraints is found to be $O(n)$ and $O(m)$, respectively. The time complexity of meeting the coincidence rate of the route and ranking the worker's income growth rate is $O(\log n)$. Therefore, the time complexity of the MTADT is $O(m+n)$.

5 Simulation Results and Analysis

Experiments were run on a machine with a 2.4-GHz Intel Core i5 processor and 4 GB of RAM. The operating system was Windows 10, and the programming language was Java.

5.1 Analysis of Results Based on Simulated Data

We analyzed the effects of different parameters on the Greedy, modified Hausdorff distance (MHD), and proposed MTADT algorithms in terms of running time, average travel cost, and memory overhead. The simulation parameters are listed in Table 2.

Table 2. Experimental parameters

Parameter	Value
Task size $|T|$	500, 1000, 1500, 2000, 2500
Work size $|W|$	500, 1000, 1500, 2000, 2500
Task radius R	1, 2, 3, 4, 5
Time difference D	5, 10, 15, 20, 25

(1) The experimental results on running time are shown in Fig. 2. Overall, the running time of the Greedy algorithm was the greatest, and MTADT ran slightly longer than MHD. When changing the number of tasks and workers from 500 to 2500, the running times of the three algorithms slowly increased within a reasonable range. But the Greedy algorithm took longest. When the radius of the selected area was changed, as the range of the area continued to expand, the number of workers and tasks also increased to a certain extent, and MTADT took longer than the MHD algorithm. As the task time difference increased, MHD and the Greedy algorithm were not affected, but MTADT took longer.

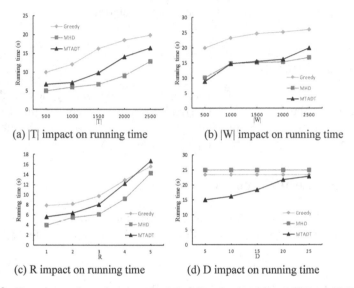

(a) |T| impact on running time   (b) |W| impact on running time

(c) R impact on running time   (d) D impact on running time

Fig. 2. Experimental running time results of Greedy algorithm, MHD, and MTADT

(2) The experimental results regarding travel costs are shown in Fig. 3. When the number of tasks and workers changed, the average travel costs of three algorithms increased rapidly with the increase of T and W, while the cost growth of MTADT was significantly lower than that of the Greedy algorithm and MHD; with the increase of radius, more workers and tasks appeared in the area. Therefore, the average travel cost of the algorithms slowly increased, and the cost of MHD grew fastest. As the task time difference increased, the cost of the Greedy algorithm and MHD was higher than that of MTADT.

(3) Results regarding memory consumption are shown in Fig. 4. When the number of tasks changed, memory overhead of the Greedy algorithm and MHD was slightly higher because they needed to continuously calculate the distance between routes. As the number of workers continued to increase, the Greedy algorithm consumed more memory. As the radius of the range increased, MHD consumed more memory, and MTADT consumed more when changing the task time difference.

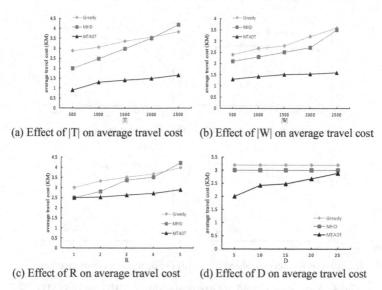

(a) Effect of |T| on average travel cost   (b) Effect of |W| on average travel cost

(c) Effect of R on average travel cost   (d) Effect of D on average travel cost

Fig. 3. Average travel cost of the Greedy algorithm, MHD, and MTADT

5.2 Analysis of Results Based on Real Data

The real data of the experiments in this chapter were downloaded and compiled from the JuShuLi network station (http://dataju.cn) [30]. The radius of the selected area of the experiment evenly increased from 1 to 5. The experimental results are shown in Fig. 5. It can be found that the real dataset is similar to the simulated dataset.

(1) The MHD algorithm had superior running time to the Greedy algorithm and MTADT, and MHD maintained a stable growth rate in various parameters.
(2) MTADT was significantly better than MHD and the Greedy algorithm regarding average trip cost, which greatly reduced the travel cost expense for workers and platforms.
(3) MTADT was better than MHD and the Greedy algorithm in terms of memory overhead.

5.3 Experimental Results

By experimenting with different simulated and real datasets, we can conclude the following.

(1) MTADT consumes less running time than the Greedy algorithm and MHD.
(2) MTADT has significantly lower average travel cost than MHD and the Greedy algorithm.
(3) MTADT consumes slightly more memory than the Greedy algorithm and MHD.

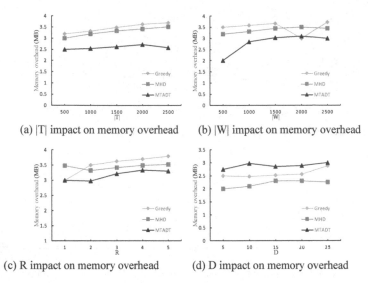

(a) |T| impact on memory overhead   (b) |W| impact on memory overhead

(c) R impact on memory overhead   (d) D impact on memory overhead

Fig. 4. Memory overhead of Greedy algorithm, MHD, and MTADT

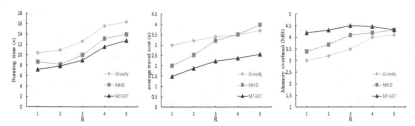

(a) R Impact on running time (b) Effect of R on average travel cost (c) R impact on memory overhead

Fig. 5. Real data results

In summary, our proposed algorithm has good scalability and can effectively solve the multitask assignment problem.

6   Research Outlook

We have proposed a method that assigns multiple tasks to single workers in online spatial crowdsourcing. However, some problems still require improvement and optimization, and this constitutes our future research direction.

(1) We did not combine spatial attributes to predict the occurrence of tasks. We can calculate the probability of occurrence of all unknown tasks using the SVM model and the Subsyn algorithm. And we use time, coordinates, and other information as the input of the SVM to train. We can calculate the probability of the final task by weighting [31].

(2) In future research, we will continue to explore the optimization problem introduced in this paper and improve the performance of the algorithm. We will consider the problem of assigning multitasks to multiple workers. We will comprehensively consider many factors to establish a task-assignment method that can adapt to various scenarios in spatial crowdsourcing.

7 Conclusion

We proposed a candidate workers selection algorithm based on minimizing location entropy for a single-task assignment. We designed a multitask assignment algorithm based on a decision tree, which builds the decision tree and calculates the time difference of tasks, travel cost, coincidence rate of the route, and income growth rate of workers to select the best new task to complete the multitask assignment to provide workers with the reward of low travel cost. The platform can effectively improve the utilization rate of resources. Finally, experiments prove that the proposed algorithm has a good approximation effect and can provide optimal distribution results while satisfying various constraints.

Acknowledgments. Here, I would like to express my gratitude to the teachers and students who helped me with the organization in the process of writing.

This work is partially supported by the National Key R&D Program of China (No. 2018YFB1003801), the National Natural Science Foundation of China (No. 61702378), the Technology Innovation Special Program of Hubei Province (No. 2018ACA13).

References

1. Chen, L., Shahabi, C.: Spatial crowdsourcing: challenges and opportunities. Bull. Tech. Comm. Data Eng. **39**(4), 14–25 (2016)
2. Tong, Y.X., She, J., Ding, B.L., et al.: Online mobile micro-task allocation in spatial crowdsourcing. In: IEEE International Conference on Data Engineering, pp. 49–60 (2016)
3. Tang, F., Zhang, H.: Spatial task assignment based on information gain in crowdsourcing. IEEE Trans. Netw. Sci. Eng. 1 (2019)
4. Yu, X.D., Liu, R., Chen, H.: Exploration of human resource management mode in the context of shared economy: a case study of DiDi. Human Resources Development of China, pp. 6–11 (2016)
5. Gautam, B., Koushik, G., Ananda, S.: Granger causality driven AHP for feature weighted KNN. Pattern Recogn. **66**, 425–436 (2017)
6. Yu, D.H., Zhang, L.L., Fu, C.: Online task allocation of spatial crowdsourcing based on dynamic utility. J. Electron. Inf. Technol. **40**(7), 1699–1706 (2018)
7. Saaty, T.L.: Analytic hierarchy process. In: Encyclopedia of Biostatistics, pp. 19–28. Wiley (2013)
8. Wu, P., Ngai, E.W.T., Wu, Y.: Toward a real-time and budget-aware task package allocation in spatial crowdsourcing. Decis. Support Syst. **110**, 107–117 (2018)
9. Zhao, Y.J., Han, Q.: Spatial crowdsourcing: current state and future directions. IEEE Commun. Mag. **54**(7), 102–107 (2016)

10. Howe, J.: The rise of crowdsourcing. Wired Mag. **14**(6), 1–4 (2016)
11. Gao, D., Tong, Y., She, J., Song, T., Chen, L., Xu, K.: Top-k team recommendation in spatial crowdsourcing. In: Cui, B., Zhang, N., Xu, J., Lian, X., Liu, D. (eds.) WAIM 2016. LNCS, vol. 9658, pp. 191–204. Springer, Cham (2016). https://doi.org/10.1007/978-3-319-39937-9_15
12. Deng, D.X., Shahabi, C., Demiryurek, U., et al.: Task selection in spatial crowdsourcing from worker's perspective. Geoinformatica **20**(3), 529–568 (2016)
13. Hassan, U.U., Curry, E.: Efficient task assignment for spatial crowdsourcing: a combinatorial fractional optimization approach with semi-bandit learning. Expert Syst. Appl. **58**, 36–56 (2016)
14. Song, T.S., Tong, Y.X., Wang, L.B., et al.: Trichromatic online matching in real-time spatial crowdsourcing. In: IEEE International Conference on Data Engineering, pp. 1009–1020. IEEE (2017)
15. Kazemi, L., Shahabi, C.: GeoCrowd: enabling query answering with spatial crowdsourcing. In: International Conference on Advances in Geographic Information Systems, pp. 189–198. ACM (2012)
16. Craw, S.: Manhattan distance. In: Sammut, C., Webb, G.I. (eds.) Encyclopedia of Machine Learning. Springer, Boston (2011). https://doi.org/10.1007/978-0-387-30164-8
17. Quinlan, J.R.: Induction on decision tree. Mach. Learn. **1**(1), 81–106 (1986)
18. Felzenszwalb, P.F., Girshick, R.B., Mcallester, D., et al.: Cascade object detection with deformable part models. Commun. ACM **56**(9), 97–105 (2013)
19. Guo, B., Yan, L., Wang, L., et al.: Task allocation in spatial crowdsourcing: current state and future directions. IEEE Internet Things J. **5**, 1749–1764 (2019)
20. Liang, Y., Lv, W.F., Wu, W.J., et al.: Mission planning based on friendship in mobile crowdsourcing environment. Front. Inf. Technol. Electron. Eng. **18**(1), 107–121 (2017)
21. Yuan, G., Sun, P., Zhao, J., et al.: A review of moving object trajectory clustering algorithms. Artif. Intell. Rev. **47**(1), 123–144 (2017)
22. Zhang, J., Pang, J.Z., Yu, J.F., et al.: An efficient assembly retrieval method based on Hausdorff distance. Robot. Comput.-Integr. Manuf. **51**, 103–111 (2018)
23. Zhang, X.B., Yang, D.S.: Hausdorff distance about spatial-temporal trajectory similarity based on time restriction. Appl. Res. Comput. **34**(7), 2077–2079 (2017)
24. Ji, P., Zhang, H.Y.: A subsethood measure with the Hausdorff distance for interval neutrosophic sets and its relations with similarity and entropy measures. In: Control and Decision Conference, pp. 4152–4157. IEEE (2017)
25. Song, T., Xu, K., Li, J., et al.: Multi-skill aware task assignment in real-time spatial crowdsourcing. GeoInformatica **2**, 1–21 (2018)
26. Xia, Z.Q., Hu, Z.Z., Luo, J.P., et al.: Adaptive trajectory prediction for moving objects in uncertain environment. J. Comput. Res. Dev. **54**(11), 2434–2444 (2017)
27. Sun, D., Ke, X., Hao, C., et al.: Online delivery route recommendation in spatial crowdsourcing. World Wide Web **11**, 1–22 (2018)
28. To, H., Fan, L.Y., Tran, L., et al.: Real-time task assignment in hyperlocal spatial crowdsourcing under budget constraints. In: IEEE International Conference on Pervasive Computing and Communications, pp. 1–8. IEEE (2016)
29. Shahabi, C.: Spatial crowdsourcing. In: The International Encyclopedia of Geography. Wiley (2017)
30. JuShuLi. Uber New York City by car data (2017). http://dataju.cn/Dataju/web/datasetInstanceDetail/210
31. Asghari, M., Shahabi, C.: On on-line task assignment in spatial crowdsourcing. In: IEEE International Conference on Big Data (2018)

TIMOM: A Novel Time Influence Multi-objective Optimization Cloud Data Storage Model for Business Process Management

Erzhou Zhu, Meng Li, Jia Xu, Xuejun Li, Feng Liu, and Futian Wang[✉]

School of Computer Science and Technology, Anhui University, Hefei 230601, China
{ezzhu,xjli,fengliu,wft}@ahu.edu.cn, meng77898@gmail.com,
xujia_ahu@qq.com

Abstract. In recent years, lots of the BPM (business process management) data storage strategies are proposed to utilize the computation and storage resources of the cloud. However, most of them are mainly focusing on the cost-effective methods and little effort is paid on the time influence of different datasets in the BPM. Storing the datasets with high time influence in the cloud is conducive to reduce the total cost (storage cost and computation cost) which is also important to the better performance of the BPM. Aiming at this problem, this paper proposes TIMOM, a novel time influence multi-objective optimization cloud data storage model for BPM. In the TIMOM model, the time influence model of the BPM is firstly constructed to calculate the time influence value for each dataset. Based on the time influence value, a new strategy, SHTD, for storing datasets in the cloud is designed. Then, by taking response time and total cost as two objectives, the multi-objective optimization method is designed to optimize the performance of the SHTD strategy. This method conducts the non-dominant sorting and calculates crowding-distance for all datasets. By doing this, important datasets are selected and stored in the cloud for repeatedly usage. Experimental results have demonstrated that the proposed TIMOM model can effectively improve the performance of the cloud BPM systems.

Keywords: Business process management · Multi-objective optimization · Cloud data storage

1 Introduction

With the growing data scale of applications, many business process management systems are seeking high performance cloud computing environments (such as Amazon AWS, Microsoft Azure and Google GCE) for data processing [1]. In the cloud computing environments, large volumes of datasets will be generated during the execution of the business processes. These datasets are also called as the intermediate datasets. Some of the intermediate datasets are important because they may be used repeatedly. However, huge storage cost will be paid if all these immediate datasets are stored in the cloud. On

© Springer Nature Switzerland AG 2020
S. Wen et al. (Eds.): ICA3PP 2019, LNCS 11944, pp. 315–329, 2020.
https://doi.org/10.1007/978-3-030-38991-8_21

the contrary, additional computation cost will be paid if all the intermediate datasets are deleted. In this case, whether some intermediate datasets to be stored or not will bring huge impact on the performance of the business processes.

Although many of immediate datasets are generated during the execution of the business processes, some of them are rarely repeatedly used. A good strategy for storing datasets for the business processes can improve the experience (less response time, small storage cost and small computation cost) of end users on using the cloud computing environments [2]. In order to get the high quality of service (QoS) to end users, strategies for storing immediate datasets must resolve two important issues: "what kind" and "how many" of datasets should be stored in the cloud?

At present, many dataset storage strategies are proposed to improve the performance of the business processes when they rent the cloud resources for task computing and data processing. The SAD (Store All Datasets) strategy stores all datasets during the execution of BPM. Because all the data are stored in the cloud servers, this strategy has the merits of zero response time and little computation cost. However, it also incurs huge storage cost. In the SND (Store None Dataset) strategy, except for the original datasets, none of immediate dataset will be stored. Since only the required datasets will be regenerate, there is little storage cost incurred in this strategy. However, it also consumes considerable computation cost and incurs high response time. The SHFD (Store High Frequency Datasets) [3] strategy stores the top 10% frequently immediate datasets. In the BPM, some datasets are more frequent used than the others. If these datasets are stored, the response time will decrease sharply. So, the SHFD is an effective strategy which can reduce total cost and response time simultaneously. The SHGTD (Store High Generation Time Datasets) [4] strategy stores 10% datasets with the highest generation times. In a BPM, some large-scale datasets will incur huge response time if they are deleted from the cloud servers. This strategy stores these datasets to reduce the response. However, it will incur high storage cost. The SID (Store Important Datasets) [5] strategy stores the 10% high important datasets. As can remarkably reduce the response time and total cost, the SID is an effective strategy.

Many of the existing dataset storage strategies are the cost-effective methods. However, little effort is paid on the time influence of different datasets in the business processes. Storing the dataset with high time influence in the cloud is conducive to improve the total cost which is also important to reduce the response time of the cloud BPM.

In this paper, we propose TIMOM, a novel time influence multi-objective optimization cloud data storage model for BPM. In the TIMOM model, the time influence model of the BPM is firstly constructed to obtain the time influence value for each dataset. Based on the calculated time influence values, a new SHTD strategy is designed to store datasets in the cloud. Then, by taking the response time and the total cost (storage cost and computation cost) as two objectives, the multi-objective optimization method is designed to optimize the performance of the SHTD strategy. Generally speaking, the contributions of this paper are listed as follows:

(1) *Builds a new time influence model.* Based on the time influence maximization method, the time influence model is constructed to effectively calculate the time influence value of each dataset.

(2) *Proposes the new SHTD dataset storage strategy.* Based on the time influence values calculated by the time influence model, the SHTD strategy is designed to select and store datasets for the cloud BPM under a given budget. The SHTD strategy stores the dataset with the highest time influence value at first. Then, the time influence values of its derived datasets are recalculated.

(3) *Proposes the new TIMOM model.* The TIMOM is built based on the SHTD strategy. In order to improve the performance of the SHTD strategy, the TIMOM model employs the multi-objective optimization method. By taking the total cost and the response time as two objectives, this method conducts the non-dominant sorting and calculates crowding-distance for all datasets.

2 Implementation of the TIMOM

The proposed TIMOM model is implemented by the following steps:

2.1 Mapping BPM to DDG

The DDG (Data Dependency Graph) is a directed acyclic graph. In the DDG, each node donates an intermediate dataset. By the DDG, all the intermediate datasets generated in the execution of business process can be connected. When some deleted datasets need to be regenerated, we can ask the help from their preorder datasets rather than regenerating them from the initial input datasets. If new datasets are generated, their references will be recorded by the DDG. During the running period of the business process, each dataset D_i has its own attributes, such as data size (marked as $Size(D_i)$), generation time (marked as $Generation(D_i)$) and usage frequency (marked as $Frequency(D_i)$). In the DDG, the generation time of a node is the time taken by utilizing the direct predecessors to generate this node. In the BPM, the generation time generally ranges from one hour to ten hours. Usage frequency refers to how many times of a dataset is used in the business process. The usage frequency of a dataset is usually acquired from the initial datasets. In this paper, the usage frequency is set one to ten times per ten days.

Definition 1 (*Preorder set*). The preorder set (marked as *Preorderset*) of a dataset is the set composed of all precursor datasets which directly or indirectly participate in the generation of this dataset.

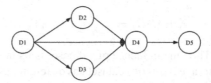

Fig. 1. An example of DDG for BPM.

E. Zhu et al.

As the DDG showing in Fig. 1, the D_1 node represents the original dataset. Since datasets D_2 and D_3 can be calculated from D_1, the preorder set of D_2 and D_3 is only composed of D_1, that is $Preorderset(D_2) = Preorderset(D_3) = \{D_1\}$. Furthermore, since D_5 is directly or indirectly derived from D_1, D_2, D_3 and D_4, the preorder set of D_5 is composed of D_1, D_2, D_3 and D_4 $(Preorderset(D_5) = \{D_1, D_2, D_3, D_4\})$.

Definition 2 (*Derived set* and *Derived datasets number*). For a given dataset D, the derived set (marked as $Derivedset(D)$) is the set composed of all datasets which can be directly or indirectly generated from this dataset. Correspondingly, the derived datasets number of D (marked as $I(D)$) is the number of elements in dataset $Derivedset(D)$.

For example, since D_5 in Fig. 1 has no successor node, the derived set of D_5 is empty $(Derivedset(D_5) = \emptyset, I(D_5) = 0)$. Since nodes D_2, D_3, D_4 and D_5 can be calculated from D_1, the derived set of D_1 is composed of D_2, D_3, D_4 and D_5 $(Derivedset(D_1) = \{D_2, D_3, D_4, D_5\}, I(D_1) = 4)$.

2.2 Dataset Storage Strategy Based on Time Influence Value

By constructing the time influence model (TIM), the time influence value of each dataset can be calculated. Then, based on the time influence value, the strategy for selecting and storing datasets in the cloud is designed.

Definition 3 (*Response time*). The response time (marked as Rt) of a dataset is the minimum time users spent to deal with this dataset in the entire running period of the business process.

The response time is an important index to measure the computational performance of the cloud environment. In the cloud environment, the respond time of a dataset D $(Rt(D))$ is determined by the generation time of its preorder datasets, the generation time of D $(Generation(D))$ and the usage frequency $(Frequency(D))$ of D. As showing in Fig. 1, if all the datasets are deleted except for the original dataset D_1, the response time of each dataset can be calculated as: $Rt(D_1) = 0$; $Rt(D_2) = Generation(D_2) \times Frequency(D_2)$; $Rt(D_3) = Generation(D_3) \times Frequency(D_3)$; $Rt(D_4) = (Generation(D_2) + Generation(D_3) + Generation(D_4)) \times Frequency(D_4)$; $Rt(D_5) = (Generation(D_2) + Generation(D_3) + Generation(D_4) + Generation(D_5)) \times Frequency(D_5)$.

Based on the definitions of datasets represented by the DDG, the response time of the corresponding DDG (suppose that there are n nodes in this DDG) can be calculated by formula (1).

$$Rt(DDG) = \sum_{i=1}^{n} Rt(D_i) \tag{1}$$

Time Influence Value. The time influence of a dataset is its influence on the response time of DDG. Different storage status of a dataset has different influences on the response time. Once a dataset is stored, users can get it without regeneration so that the response time of this dataset is 0. Furthermore, the response time of its derived datasets can also be decreased because the regeneration process of a dataset relies on its preorder datasets. Based on this observation we propose two characteristics to illustrate the time influence of datasets, the *Quantitative characteristic* and the *Quality characteristic*. In the DDG, we called a dataset has the *Quantitative characteristic* if it has more derived datasets. The dataset has the *Quantitative characteristic* will have great time influence on the calculation of the response time. In the DDG, we called a dataset has the *Quality characteristic* if it can pass all the time influence to its derived datasets.

The time influence transmission is to indicate that the time influence of a dataset can be transferred to its derived datasets. In a DDG, no dataset can be regenerated if all datasets in its preorder set are deleted. Based on this observation, the response time of a dataset includes the generation time of its preorder datasets. In order to get the time influence value of each dataset, the TIM is constructed by the following three steps:

(1) *Get the derived relationship.* The derived relationship among datasets is presented as the time influence flag δ. As showing in formula (2), the value (0 or 1) of the flag δ_{ij} depends on the relationship between dataset D_i and dataset D_j. If dataset D_i is in the preorder set of D_j, δ_{ij} is set to 1, otherwise, δ_{ij} is set to 0.

$$\delta_{ij} = \begin{cases} 1 & D_i \in PreorderSet(D_j) \\ 0 & D_i \notin PreorderSet(D_j) \end{cases} \tag{2}$$

(2) *Maximize the time influence.* In order to maximize the time influence, we define a social network $N(V, E)$. Where, V ($|V| = n$) is the nodes set; E ($|E| = m$) is the direct edges set. In the social network $N(V, E)$, each direct edge $e \in E$ is an associated with a propagation probability $p(e) \in [0,1]$. Meanwhile, the *timestamp* is defined as the time that an event that has already happened before this time. Given the social network $N(V, E)$, the process of the influence propagation is as follows:

(a) At timestamp 1 (initial stage), nodes in the seed node set A (a subset of V) are activated, while the remainder nodes in V are set to inactive.

(b) If a node u is activated at *timestamp i*, u is put into set A. Then, for each direct edge e ($e = <u, v>$), v will be activated by u at the *timestamp i* $+1$ with a probability value of $p(e)$. Where, u is the source node of edge e, v is the terminate node of e. After timestamp $i +1$, u cannot activate any other nodes.

(c) If a node is activated, it will remain active in the later *timestamps*.

Let $I(A)$ be the number of activated nodes when the above process terminates. At this time, no more nodes can be activated. Generally speaking, for a given social network N, a constant number k and a seed set A, the problem of influence maximization of N is to repeatedly expand A with the propagation probability until the number of elements in A reaches the expected number k (i.e. $I(A) = k$). The time influence model proposed in this paper is based on the above influence propagation process. Where node set D of DDG can be analogous to the node set V in social network N; the time influence value of a node in DDG is corresponding to the propagation probability value of $p(e)$ in the social network N; the derived dataset of D_i ($DerivedSet(D_i)$) is corresponding to activated nodes set of node A in the social network N.

The problem of time influence maximization is to find a seed set at first. Then, continuous expend the seed set with nodes that are directly or indirectly derived from the seeds until the expected maximum number of nodes is acquired. The final expended set is composed of immediate datasets with the highest time influence values. These datasets are the candidates for storing in the cloud to achieve smaller total cost while without sacrificing much response time.

Specifically, for a given dataset D_i, the initial number of its derived dataset ($I(D_i)$) is set to 1. Then, for any dataset D_j in the DDG, the propagation process is executed according to formula (3).

$$I(D_i) = \begin{cases} I(D_i) & \delta_{ij} = 0 \\ I(D_i) + 1 & \delta_{ij} = 1 \end{cases} \tag{3}$$

(3) *Calculate time influence value.* In the workflow of a business process, the respond time of a data set D_i ($Rt(D_i)$) is decided by generation time of datasets in its *PreorderSet(D_i)*, the generation time and the usage frequency of D_i itself. Based on the analysis of time influence transmission, we know that datasets with the quality characteristic can be transferred to its derived datasets completely. Specifically, the response time of dataset D_i is calculated by formula (4).

$$Rt(D_i) = \left(Generation(D_i) + \sum\nolimits_{D_j \in PreorderSet(D_i)} Generation(D_j) \right) \times Frequecy(D_i) \tag{4}$$

where, *Generation(D_i)* and *Frequency(D_i)* represent the generation time and usage frequency of dataset D_i.

To sum up, the TIM is built as a matrix time influence contribution matrix as follows:

$$\begin{bmatrix} 0 & \delta_{12} \times Rt(D_1) \times I(D_2) & \cdots & \delta_{1n} \times Rt(D_1) \times I(D_n) \\ \delta_{21} \times Rt(D_2) \times I(D_1) & 0 & \cdots & \vdots \\ \vdots & \vdots & \ddots & \ddot{a}_{2n} \times Rt(D_2) \times I(D_n) \\ \ddot{a}_{n1} \times Rt(D_n) \times I(D_1) & \ddot{a}_{n2} \times Rt(D_n) \times I(D_2) & \cdots & 0 \end{bmatrix}$$

where, $\delta_{ij} \times Rt(D_i) \times I(D_j)$ in the matrix can be interpreted as the time influence of D_i which contributes to D_{ij}. Based on the time influence contribution matrix, we can calculate the time influence value (*TIV*) of each dataset as defined in formula (5).

$$TIV(D_i) = \sum\nolimits_{D_j \in DDG} \left(\ddot{a}_{ij} * Rt(D_i) * I(D_j) \right) \tag{5}$$

Time Influence Dataset Storage Strategy (*SHTD*). As requiring no response time, the best strategy is to store all datasets. By this method, we can immediately get the required dataset without any data generation process. However, as showing in Fig. 2, it will incur huge total cost (including the storage cost and the computation cost). A good strategy that trade-offs between the response time and the total is urgently needed.

Based on the time influence value calculated above, the SHTD strategy is proposed to select and store datasets in the cloud. The SHTD strategy can be described as the following four steps:

(a) Calculate the time influence value for each dataset D ($TIV(D)$) in the execution of the BPM.

(b) Store the dataset with the highest time influence value. According to the change storage status of this dataset, its response time is set to 0. Meanwhile, the response time of its related datasets will also be changed.
(c) For the remaining datasets, recalculate the time influence values and store the dataset with the second highest time influence value.
(d) Repeatedly execute the above steps until the budget runs out.

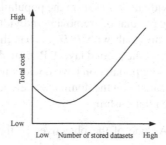

Fig. 2. Relationship between total cost and stored datasets

In the process of executing the BPM, the above time influence strategy is used to select the intermediate datasets. As described in the above steps, our time influence strategy always stores the dataset with the highest time influence value at first. Then, the time influence values of its derived datasets are recalculated. So, the dataset storage under this strategy is a dynamic process. As always selecting dataset with the highest time influence, the storage of intermediate datasets is more accurate than traditional methods in selection datasets.

2.3 Data Storage Strategy with Multi-objective Optimization

In this part, the multi-objective optimization method is used to optimize the above SHTD strategy. In this optimization method, the response time (defined by formula (1)) and the total cost (will be defined in the forthcoming of this section) of the dataset are taken as the two objectives.

Brief Introduction to the Multi-objective Problem. The multi-objective problem can be generally defined as the results of a set of objective functions under a set of constraints [6]. The general form of a multi-objective problem can be defined as follows:

$$
\begin{aligned}
&min[f_1(X), \ldots f_r(X)]/max[f_{r+1}(X), \ldots f_m(X)] \\
&\text{subject to} : g_i(X) \geq 0, h_j(X) = 0, i = 1, 2, \ldots, p, j = 1, 2, \ldots, q
\end{aligned}
\tag{6}
$$

where, $X = (x_1, x_2, \ldots, x_n)$ is the n-dimensional decision vector; $f_i(X)$ $(i = 1, 2, \ldots, m)$ is the objective function; inequalities $g_i(X) \geq 0$ $(i = 1, 2, \ldots, p)$ and equations $h_j(X) = 0$ $(j = 1, 2, \ldots, q)$ are the constraint functions; there are r minimizing objective functions and m-r maximizing objective functions.

In a multi-objective optimization algorithm, each solution is sorted according to the individual non-dominant attributes [7]. If all other solutions do not dominate the optimal solution, the solution is non-dominant. In the algorithm, for each individual D_i, two sets, $P(D_i)$ and $Q(D_i)$ are calculated. Where, $P(D_i)$ is the set composed of individuals dominate D_i in the population; $Q(D_i)$ is the set composed of individuals which are dominated by D_i. At the initial stage of the multi-objective optimization algorithm, individuals with $|P(D_i)| = 0$ in the population are selected and stored in the set M_1 (the first layer). Remove individuals of M_1 from the population. Consequently, for each individual D_j in the remaining population, removing individuals D_i from set of $P(D_i)$ if D_i is in the set of M_1. Then, individuals with $|P(D_j)| = 0$ in the remaining population are selected and stored in the set M_2 (the second layer). Repeatedly execute the above steps. If all individuals in the remaining population have no dominate individuals, individuals in the original population are satisfying the dominance relationship. Otherwise, they are satisfying the non-dominance relationship.

Calculation of Total Cost. As one of the objectives of the multi-objective, the total cost is defined as follow:

Definition 4 (*Total cost*). For a given dataset D, the total cost (marked as $Totalcost(D)$) is defined as the sum of the computation cost ($CC(D)$) and storage cost ($SC(D)$) when we want to regenerate data set D from the cloud. Specifically:

$$CC(D) = Rt(D) \times CP \tag{7}$$

$$SC(D) = Size(D) \times SP \tag{8}$$

$$TotalCost(D) = CC(D) + SC(D) \tag{9}$$

where, CP and SP are the computation price and storage price of dataset D specified by the cloud service providers.

As large generation time of datasets, the management of business processes in the cloud is an extremely complex task. Huge computation cost should be paid if we delete all datasets. The generated DDG is like a network. In this network, the propagation property and the derivation relationships of the storage of datasets can be easily presented. In the cloud environment, an important work is to select immediate datasets with maximum time influence and store them in the cloud to improve the performance (i.e. relatively short response time and small total cost) of the BPM.

Multi-objective Optimization Method for Dataset Storage. In our algorithm, the non-dominated sorting based on the crowding-distance for the time influence is performed at first. By doing this, datasets are divided hierarchically. Datasets with different priorities are put into different layers. Then, crowding-distances among datasets in the optimal layer are calculated and selected. Finally, we can balance the response time and total cost of datasets. As a result, datasets with little response time and small total cost are acquired for retrieving dataset in the cloud.

In order to perform the non-dominant sorting [8] for all datasets in the DDG, we must calculate two attributes, the time influence value and the data size, for all datasets at first. According to the constructed TIM, the time influence value of each dataset D_i can be calculated $(TIV(D_i))$ by formula (5). We take the dataset size as another attribute because it has the great influence on total cost (as showing in formula (8) and formula (9)).

Then, based on the two attributes, the non-dominant sorting is conducted for all datasets in the DDG. Algorithm 1 gives the main steps of the non-dominant sorting. By the non-dominant sorting, datasets are put into different layers. We can select the optimal number of layers to calculate the crowding-distances. The crowding-distance is defined as the density of the surrounding individuals at a given point in the population. The crowding-distance of an individual i forms the smallest rectangle of individuals around i. Take the 2-dimensional decision vector as an example, the crowding-distance of individual 2 in Fig. 3 can be calculated by formula (10).

$$CD_2 = (x_3 - x_1) + (y_1 - y_3) \tag{10}$$

Algorithm 1. Non-dominated sorting

Input: A general DDG representing relationship among dataset in the BPM.
Output: The ranked non-dominated solution.

Initialize the current front of each dataset to 0;
//The current front of a dataset is "0" means it has not been included in a layer yet.
For each dataset D_i with current front is 0 in the DDG do
  Calculate the dominant set $P(D_i)$ for it;
//In the BPM, if the $TIV(D_i)$ is large than the $TIV(D_j)$ and the size of D_i is smaller than the size of D_j, we called D_i dominants D_j.
  According to the dominant set $P(D_i)$ of dataset D_i, get the dominated number $|P(D_i)|$ of dataset D_i.
End-For.
$F \leftarrow 0$; //F specifies the number of layers of a dataset.
While not all datasets are layered do
  $F \leftarrow F+1$;
  For all datasets D_i with $|P(D_i)|=0$ do
    Set the current front of dataset D_i to F;// dataset D_i is put into the F-th layer.
    For each dataset D_j the remaining datasets do
      If it is dominated by dataset D_i
        Remove D_i from dominant set $P(D_j)$ of the dataset D_j;
        $|P(D_i)| \leftarrow |P(D_i)|$-1; // the dominated number of datasets D_j is reduced by 1.
      End-If.
    End-For.
  End-For.
End-While.

In order to calculate the crowing-distance of datasets, in Fig. 3, the X-axis is set as the total cost and the Y-axis is set as the response time. The total cost includes the computation cost and the storage cost. The computation cost is generated when the deleted datasets need to be regenerated; the storage cost is generated when these datasets are stored. In this paper, the total cost and response time are taken as the two objectives to calculate the crowding-distance of all datasets in the optimal layers. By doing this, datasets with the largest crowding-distances are selected and stored in the cloud. The remaining datasets are deleted for saving storage cost.

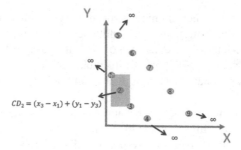

$$CD_2 = (x_3 - x_1) + (y_1 - y_3)$$

Fig. 3. Crowding-distance of individual 2.

3 Experiment Results

This part tests the performance of the SHTD strategy at first. Then, the performance of the entire TIMOM which integrates the SHTD strategy and the multi-objective optimization method is analyzed. In order to better verify the performance, five existing data storage strategies (SAD, SND, SHFD [3], SHGTD [4] and SID [5]) are compared with our methods. Among the five strategies, the SAD and SND, are the two classical strategies. The SHFD, SHGTD and SID are the newly proposed data storage strategies.

In this part, we use randomly generated acyclic DDGs to test the performance of different strategies. Each generated DDG should have only one root dataset and each dataset should have 1–2 successors. Meanwhile, generated DDGs should have the properties listed in Table 1.

Table 1. Properties settings of the DDG.

Dataset Size	10–100 GB
Generating time	1–10 h
Usage frequency	1–10 times per ten days
Computation price	$0.1 per hour unit per CPU
Storage price	$0.15 per Gigabyte per month

3.1 Performance of the SHTD Strategy

Total Cost of the SHTD Strategy. This experiment tests the role of the proposed SHTD strategy played in reducing the total cost when it is used to selected and stored datasets in the cloud. This strategy always selects datasets with high time influence values and stores them in the cloud. In this experiment, the DDG composed of 1000 datasets is randomly generated under the configurations listed in Table 1. Initially, the time influence values of all datasets are calculated. Based on the calculated time influence values, the dataset with the highest time influence value is selected and stored in the cloud. Meanwhile, the total cost and the response time of the DDG are also calculated. Since the dataset with the highest time influence is stored, time influence values of the remainder datasets are recalculated. Then, the dataset with the highest time influence in the remainder datasets is selected and stored in the cloud. Repeatedly execution the above steps until the required datasets (the number of datasets been stored in the cloud are determined by the budget of the users) are selected and stored in the cloud. As Fig. 4 shows, each time the dataset with the highest time influence value been stored in the cloud, the total cost (as shown in Fig. 5) and the response time (as shown in Fig. 6) of the DDG are also calculated.

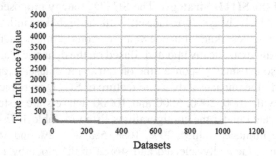

Fig. 4. Time influence values of datasets.

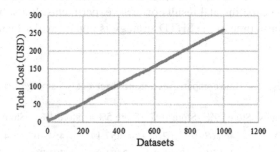

Fig. 5. Growth trend of the total cost.

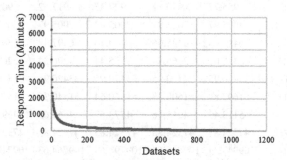

Fig. 6. Growth trend of response time.

As can be seen in Fig. 4, some datasets have much higher time influence values than the others. However, when all datasets are stored, the time influence values of the remainder datasets are quite small. Meanwhile, as shown in Fig. 5, the total cost is positive growing with the increasing number of datasets is stored in the cloud. So, it is impossible to store all datasets under a limited budget. Fortunately, as can be seen in Fig. 6, the decline of response time tends to be flat after datasets with high time influence values are stored. Based on the experimental results, it is only needed to store part of important datasets (with high time influence values) rather than store them all. So, the SHTD strategy proposed in this paper expends small total cost while without incurs high response time.

Response Time of the SHTD Strategy. The SHTD strategy proposed in this paper is a dynamic process which always store datasets with the high time influence value. Once the dataset with the highest time influence value is stored, time influence values of the remainder datasets are dynamically updated. Based on the dynamic feature of the SHTD strategy, we can also get small response time on retrieving datasets from the cloud.

In order to verify the impacts of dynamic feature of SHTD on reducing the response time, four DDGs with 1000, 2000, 5000 and 10000 datasets are tested. These DDGs are randomly generated under the configurations listed in Table 1. In this experiment, according the time influence values, 20%, 10%, 5%, 4%, 2% and 1% of datasets in different DDGs are sequentially selected and stored in the cloud by 1 time, 2 times, 4 times, 5 times, 10 times and 20 times respectively. Meanwhile, the experimental results on response time of three different strategies (SHTD, SHGTD and SHFD) are compared.

Table 2 lists the experimental results of these experiments. It can be seen from the table that the response time of the SHTD strategy is much smaller than the ones of the other two strategies.

Table 2. Response time comparisons among three dataset storage strategies (s).

DDGs	Strategies	Store 20%	Store 10%	Store 5%	Store 4%	Store 2%	Store 1%
DDG_1000	SHTD	468.68	462.21	457.19	454.10	451.38	449.48
	SHGTD	2491.86	2491.86	2491.86	2491.86	2491.86	2491.86
	SHFD	5086.50	5067.58	5051.87	5042.83	5033.84	5029.31
DDG_2000	SHTD	955.77	940.81	927.37	917.52	908.70	904.78
	SHGTD	7005.97	7005.97	7005.97	7005.97	7005.97	7005.97
	SHFD	6443.96	6425.66	6409.40	6397.42	6386.99	6378.75
DDG_5000	SHTD	2430.02	2365.74	2325.44	2296.54	2269.69	2254.58
	SHGTD	33420.95	33420.95	33420.95	33420.95	33420.95	33420.95
	SHFD	51697.89	51600.43	51560.64	51540.36	51527.37	51513.47
DDG_10000	SHTD	5113.49	4915.66	4773.67	4664.63	4590.91	4538.47
	SHGTD	47324.85	47324.85	47324.85	47324.85	47324.85	47324.85
	SHFD	109947.29	109846.64	109798.64	109764.63	109740.25	109720.76

3.2   Performance of the Entire TIMOM Model

This part tests the performance of the entire TIMOM model which integrates the SHTD data storage strategy with the multi-objective optimization method. Meanwhile, the performances of the other 5 existing data storage strategies are compared. According to the configurations listed in the Table 1, 6 DDGs which composed of 1000, 2000, 3000, 5000, 7000 and 8000 datasets are randomly generated. The total cost and response time of different data storage strategies are recorded in Fig. 7. As can be seen from Fig. 7, there are huge differences among the performance of different strategies. In order to better display the experimental results, the upper right corner of each sub-graph magnifies part of the experimental results.

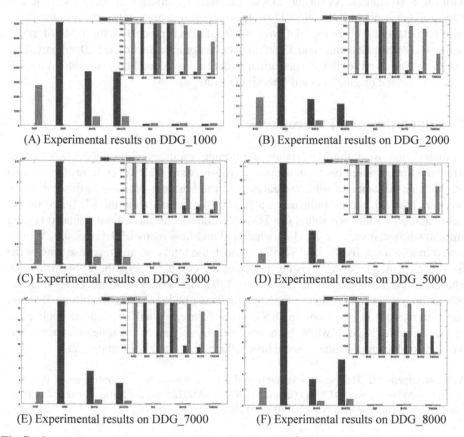

(A) Experimental results on DDG_1000   (B) Experimental results on DDG_2000

(C) Experimental results on DDG_3000   (D) Experimental results on DDG_5000

(E) Experimental results on DDG_7000   (F) Experimental results on DDG_8000

Fig. 7. Comparisons on response time (s) and total cost ($\$$) among 7 different data storage strategies.

It can be seen from the 6 sub-graphs in Fig. 7 that the SAD strategy requires no response time. However, due to all datasets are stored on the cloud, the total cost of this strategy is unacceptable. The SND strategy incurs 0 total cost as it doesn't store any dataset in the cloud. However, the response time of strategy is very high. Since only store datasets with high frequency and high generating time, the SHFD and SHGTD strategies exhibit better performances than the ones of the SAD and SND strategies. However, due to only consider single attribute, the SHFD and SHGTD strategies also incur high total cost and response time.

As for the SHTD strategy and the improved TIMOM model based on SHTD proposed in this paper, the response time and the total cost of them are much smaller than the other five existing strategies. Meanwhile, the larger scales of the tested DDGs the better performance of our strategies than the others. For example, our SHTD strategy reduces 1.4% response time and 2.4% total cost compared with strategy SID when they are used to process the DDG with 1000 datasets. However, when the DDG with 8000 datasets is processed, the SHTD strategy reduces 7.4% response time and 3.4% total costs compared with the SID strategy. As for the TIMOM model, it reduces 2.2% response time and 7.8% total cost compared with SID strategy when they are used to process the DDG with 1000 datasets. When the DDG with 8000 datasets is processed, the TIMOM model reduces 9.0% response time and 33.8% total costs compared with the SID strategy. This experiment also verifies that the performance our improved method (the TIMOM model) is better than the original method (the SHTD strategy).

4   Conclusion and Future Work

To make better use of the computation and storage resources of the cloud, many storage strategies are proposed based on the pay-as-you-go model. However, little effort is paid on the time influence of different datasets in the business process. In this paper, we propose TIMOM, a time influence multi-objective optimization model. Based on the calculated time influence value, the TIMOM model combines non-dominated sorting and crowding-distance for deciding what kind and how many of datasets should to be stored in the cloud. By the new TIMOM model, the total cost and response time of the BPM are balanced. Several existing strategies are compared with the TPMOM model. The experimental results have demonstrated that our model performs better than the other strategies in both response time and total cost under various conditions. Because our current work is based on the DDG, in the future, we will expand the application scope our methods. Meanwhile, because of heavily relying on the time influence value, we will conduct more studies on the time influence among different datasets.

Acknowledgement. This work was supported by the University Natural Science Research Project of Anhui Province (China) (Grant No. KJ2018A0022) and the National Natural Science Foundation of China (Grant No. 61972001, 61300169).

References

1. Pourmirza, S., Peters, S., Dijkman, R., et al.: BPMS-RA: a novel reference architecture for business process management systems. ACM Trans. Internet Technol. (TOIT) **19**(1) (2019). Article No.13

2. Wang, Y., Xu, R., Wang, F., et al.: Sliding-window based propagation-aware temporal verification for monitoring parallel cloud business workflows. In: IEEE 22nd International Conference on Computer Supported Cooperative Work in Design (CSCWD), pp. 449–454 (2018)
3. Yuan, D., Liu, X., Yang, Y.: Dynamic on-the-fly minimum cost benchmarking for storing generated scientific datasets in the cloud. IEEE Trans. Comput. **64**(10), 2781–2795 (2015)
4. Yuan, D., Yang, Y., Liu, X., et al.: A highly practical approach toward achieving minimum data sets storage cost in the cloud. IEEE Trans. Parallel Distrib. Syst. **24**(6), 1234–1244 (2013)
5. Xu, R., Zhao, K., Zhang, P., Yuan, D., Xie, Y., Yang, Y.: A novel data set importance based cost-effective and computation-efficient storage strategy in the cloud. In: IEEE International Conference on Web Services (ICWS), pp. 122–129 (2017)
6. Bahri, O., Talbi, E.-G.: Dealing with epistemic uncertainty in multi-objective optimization: a survey. In: Medina, J., Ojeda-Aciego, M., Verdegay, J.L., Perfilieva, I., Bouchon-Meunier, B., Yager, Ronald R. (eds.) IPMU 2018. CCIS, vol. 855, pp. 260–271. Springer, Cham (2018). https://doi.org/10.1007/978-3-319-91479-4_22
7. Paul, A., Shill, P.: New automatic fuzzy relational clustering algorithms using multi-objective NSGA-II. Inf. Sci. **448**, 112–133 (2018)
8. Singh, G.P., Thulasiram, R.K., Thulasiraman, P.: Non-dominant sorting Firefly algorithm for pricing American option. In: 2016 IEEE Symposium Series on Computational Intelligence (SSCI), pp. 1–10 (2016)

RTEF-PP: A Robust Trust Evaluation Framework with Privacy Protection for Cloud Services Providers

Hong Zhong[1,2] , JianZhong Zou[1,2] , Jie Cui[1,2] , and Yan Xu[1,2(✉)]

[1] School of Computer Science and Technology, Anhui University, Hefei 230039, China
xuyan@ahu.edu.cn
[2] Anhui Engineering Laboratory of IoT Security Technologies, Hefei 230039, China

Abstract. The trust problem of Cloud Services Providers (CSPs) has become one of the most challenging issues for cloud computing. To build trust between Cloud Clients (CCs) and CSPs, a large number of trust evaluation frameworks have been proposed. Most of these trust evaluation frameworks collect and process evidence data such as the feedback and the preferences from CCs. However, evidence data may reveal the CCs' privacy. So far there are very few trust frameworks study on the privacy protection of CCs. In addition, when the number of malicious CCs' feedback increases, the accuracy of existing frameworks is greatly reduced. This paper proposes a robust trust evaluation framework RTEF-PP, which uses differential privacy to protect CCs' privacy. Furthermore, RTEF-PP uses the Euclidean distances between the monitored QoS values and CCs' feedback to detect malicious CCs' feedback ratings, and is not affected by the number of malicious CCs' feedback rating. Experimental results show that RTEF-PP is reliable and will not be affected by the number of malicious CCs' feedback rating.

Keywords: Cloud computing · Trust evaluation · Robust · Differential privacy

1 Introduction

With the development of distributed intelligent computing and the rise of cloud computing, more and more companies and individuals are willing to deliver their work to CSPs. However, compared to the traditional service model, if the CSPs replace the CCs to exercise more capabilities, the CCs will lose the right to directly control the data [8]. According to a survey shown by Fujitsu, 88% potential CCs worried about the trustworthiness problem of CSPs [7]. Trust evaluation of CSPs has become an important research topic.

In recent years, a large number of trust evaluation frameworks have been proposed [2–4,9,12,13,17,18,20]. These trust evaluation frameworks can be divided into two types: (1) Objective trust evaluation frameworks [2–4,9,17,18,20] attempting to monitor the actual Quality of Service (QoS) performance of cloud

© Springer Nature Switzerland AG 2020
S. Wen et al. (Eds.): ICA3PP 2019, LNCS 11944, pp. 330–344, 2020.
https://doi.org/10.1007/978-3-030-38991-8_22

services in runtime, and compare the monitored QoS values with the claimed QoS values in cloud services' Service Level Agreement (SLA), (2) Subjective trust evaluation frameworks [12,13] attempting to employ CCs' feedback ratings to evaluate the trust of CSPs. However, in objective trust evaluation frameworks, the reliability of monitored QoS will be affected in the unpredictable Internet environment. Worse still, not all QoS attributes in SLA can be monitored. Therefore, objective trust evaluation frameworks may lead to incomplete evaluation. In subjective trust evaluation frameworks, the reliability of CCs' feedback is often affected by its own professional knowledge. What is worse, subjective trust evaluation frameworks are affected by malicious CCs' feedback ratings.

Subsequently, some papers [6,11,16,19] attempted to combine objective with subjective trust evaluation frameworks to improve the reliability of trust evaluation. But the above frameworks still do not effectively resist the impact of malicious CCs' feedback ratings. Although [19] proposed a detection method to find malicious CCs' feedback ratings, the accuracy of the framework will be greatly reduced when the number of malicious CCs' feedback ratings increases. Therefore, it is necessary to propose a robust trust evaluation framework when there are many malicious CCs' feedback ratings.

In addition, the privacy security issue in the trust evaluation process have also become a problem. The trust evaluation may collect some evidence data from CCs, such as feedback and personal preference, which may reveal CCs' sensitive information [21]. Privacy leakage may cause CCs to be reluctant to provide evidence data. However, the lack of sufficient evidence will influence the accuracy of trust evaluation [22]. The existing trust frameworks did not take into account the CCs' privacy protection.

In order to solve the above mentioned problems, this paper proposes a robust trust evaluation framework with privacy protection for CSPs. The main contributions can be summarized as follows:

(1) We propose a robust trust evaluation framework, which can detect malicious CCs' feedback ratings. Our framework firstly computes the Euclidean distances between the monitored QoS values and CCs' feedback ratings, and then identifies malicious CCs' feedback ratings according to the value of the Euclidean distances. Compared to the most common method [19], our framework is robust and not affected by the number of malicious CCs' feedback ratings.

(2) RTEF-PP is the first trust evaluation framework for CSPs with privacy protection. We use differential privacy to add noise to CCs' feedback ratings to protect the CCs' privacy. Furthermore, RTEF-PP is suitable for different application environments for the flexible adjustment of privacy parameters by the Laplace mechanism.

The rest of this paper is organized as follows. Section 2 reviews the related work on the trust evaluation for CSPs. Section 3 describes the system architecture and differential privacy. Section 4 details the specific process of RTEF-PP. Section 5 validates the effectiveness of RTEF-PP through an experiment. And Sect. 6 is conclusions.

2   Related Work

With the widespread use of cloud services, how to build trust between CCs and CSPs is a significant research issue. More and more trust evaluation frameworks have been proposed. These trust evaluation frameworks can be divided into two types: objective trust evaluation frameworks [2–4,9,17,18,20] and subjective trust evaluation frameworks [12,13].

Alhamad et al. [2] introduced the SLA validation mechanism into cloud computing. And then Alhamad et al. [3] proposed a trust management framework that used the SLA criteria and the experience of CCs to evaluate trustworthiness of cloud services, but the reliability of the CCs' experience may be affected by subjective factors. Muchahari et al. [9] proposed a novel trust model that employed a Dynamic Trust Monitor to get the real-time trust value of cloud services. However, there is no simulation and experimental results to prove the validity of the framework [9]. Chakraborty et al. proposed a quantitative model, and then extracted parameters from SLA and similar documents to compute trustworthiness as a fraction between 0 and 1 [4]. However, there is no case study and experiment either. Singh and Sidhu proposed two trust evaluation frameworks that used the SLA compliance to determine the trustworthiness of CSPs [17,18]. However, in their frameworks, QoS in the SLA is defined as a certain value and does not distinguish between positive and negative attributes, which is not suitable for real-world applications. Wang et al. proposed a dynamic cloud service trust evaluation framework based on SLA and privacy-awareness [20], which can dynamically update the trust value.

Since not all QoS attributes can be monitored or quantified, objective trust evaluation frameworks may lead to incomplete evaluation. Therefore some subjective trust evaluation frameworks based on CCs' feedback ratings are proposed. Noorian et al. evaluated CSPs' trust by combining both the service requester's and the other CCs' experience opinions, and proposed a preference-oriented QoS-based service selection framework [13]. However, there is no effective filtering of malicious CCs' opinions in this framework. Noor et al. proposed a reputation-based trust management framework employing trust feedback [12], which designed a novel protocol to prove the reliability of feedback information.

As the accuracy of the subjective trust evaluation framework is easily influenced by the CCs' own expertise and malicious CCs' feedback, there are some works which combined objective and subjective trust evaluation frameworks to evaluate the trustworthiness of CSPs. Ding et al. proposed a framework by combining QoS monitoring and CCs' feedback ratings, which can improve the accuracy and integrity of trust evaluation [16]. However, it does not consider the influence of dishonest ratings. Tang et al. [19] proposed an integrated trust evaluation framework TRUSS, which employed a Trust Evaluation Middleware to identify the dishonest CCs and unfair ratings. However the accuracy of TRUSS will decrease when there are many malicious feedback ratings.

3 Preliminaries

In this section, we firstly introduce the system architecture of RTEF-PP. We then introduce differential privacy that will be used in the trust evaluation process.

3.1 System Architecture

As shown in Fig. 1, the system architecture includes three participants: CSPs, CCs and an evaluation party. CSPs are cloud service providers. CCs are users who need to use cloud services of the CSPs. The evaluation party is an important component to evaluate the trust of CSPs. In our system architecture, the evaluation party has four modules as follows:

- Information storage module: Storing CSPs' SLA, the monitored QoS values and valid CCs' feedback ratings.
- QoS monitoring and malicious evaluation detection module: Monitoring the QoS data during the service operation. In addition, before performing the subjective trust evaluation, the module detects the feedback ratings of each CCs and deletes the malicious feedback ratings.
- Data protection module: Adding noise to the CCs' feedback ratings.
- Trust calculation module: Calculating the objective trust score, subjective trust score and integrate trust score.

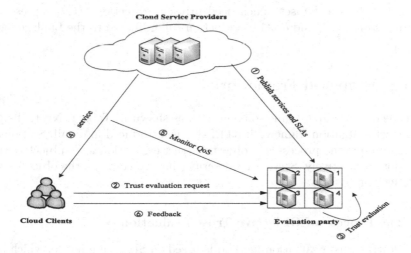

Fig. 1. The system model

The process of trust evaluation requires the concerted efforts of CSPs, CCs, and an evaluation party. As shown in Fig. 1, a complete trust evaluation process consists of six processes. At first, CSPs need to send registration information to the evaluation party before publishing cloud services. CCs then will send a trust evaluation request to the evaluation party. The evaluation party will conduct a trust evaluation for all the candidate CSPs according to CCs' request. After completing the trust evaluation, the CSP with the highest trust ranking will be recommended to provide services for CCs. When CCs use the cloud service, the evaluation party will monitor its actual performance and the monitored QoS values will be used for the next trust evaluation. Finally, CCs will give feedback ratings after using the cloud service.

3.2 Differential Privacy

Differential privacy is a privacy preserving mathematical model which was proposed by Dwork in 2006 [5], whose basic idea is to add noise to the original data, the conversion of the original data, or the statistical results to achieve privacy protection. In practice, the Laplace Mechanism and Exponential Mechanism are commonly used to implement differential privacy protection.

In this paper, the feedback ratings may reveal the CCs' privacy information. To preserve CCs' privacy, the Laplace mechanism is used to add random noise that satisfies the Laplace distribution. The magnitude of noise depends on the sensitivity of query function Δf and the privacy budget ε. In the Laplace mechanism, the output $N_f(D)$ after adding noise is calculated according to the formula (1):

$$N_f(D) = f(D) + lap(\frac{\Delta f}{\varepsilon}) \tag{1}$$

Where (D) is a dataset which need privacy protection, $f(D)$ represents a query function on D, and $lap()$ is a function subordinated to the Laplace distribution [5].

4   The Proposed Framework

In this section, we first define some notions as shown in Table 1. Next, the proposed trust evaluation framework RTEF-PP is introduced in detail. Sections 4.1 and 4.2 describe the processes of objective trust evaluation and subjective trust evaluation, respectively. Section 4.3 describes how to combine the objective and subjective trust scores.

4.1   The Process of Objective Trust Evaluation

The objective trust evaluation method is based on SLA compliance, which indicates the difference between the monitored QoS values and the claimed QoS in the SLA. The monitored QoS values can easily be acquired by using monitor software [23].

We firstly introduce the calculation method of single attribute compliance in Sect. 4.1.1. Next, we propose an objective trust evaluation method for multi-attribute compliance based on TOPSIS in Sect. 4.1.2.

Table 1. Definition of notations

Notation	Definition
$S = (S_1, S_2, S_3, \cdots S_m)$	A set of m cloud services with equivalent functionalities provided by different CSP, where $S_i(1 \leq i \leq m)$ represents the ith cloud service
$A = (A_1, A_2, A_3, \cdots A_n)$	A vector of QoS attributes of cloud services specified in their SLAs, where $A_j(1 \leq j \leq n)$ represents the jth attribute
$q_i = (q_{i1}, q_{i2}, q_{i3}, \cdots q_{in})$	A vector of QoS values of service S_i claimed in SLA, where q_{ij} represents the QoS value on A_j. $q_{ij} = \left[q_{ij}{}^s, q_{ij}{}^l\right]$, where $q_{ij}{}^s$ and $q_{ij}{}^l$ are respectively the lower and upper bound QoS values claimed by service S_i
$M_i^r = \{M_{i1}^r, M_{i2}^r, \cdots M_{ic}^r\}$	The rth monitored QoS value on service S_i, $c \leq n$, because not all attributes can be monitored. Where $M_{it}^r, (1 \leq i \leq m, 1 \leq t \leq c)$ represents the rth monitored value on the jth attribute of service S_i
$e_i^p = \{e_{i1}^p, e_{i2}^p, \cdots e_{is}^p\}$	The pth CCs' feedback rating on service S_i, $s \leq n$, because CCs don't need to give feedback on all attributes. Where $e_{iq}^p, (1 \leq i \leq m, 1 \leq q \leq s)$ represents the pth CCs' feedback rating on the qth attribute of service S_i

4.1.1 Single Attribute Compliance Calculation

The single attribute compliance is calculated by comparing the difference between the monitored QoS values and the claimed QoS in the SLA. In order to reduce the impact of network fluctuations on the monitored QoS values, RTEF-PP firstly computes the average of the multiple monitored QoS values. In addition, since each attribute has a different unit and range of values, it is necessary to normalize the QoS values into a unified range before calculating the compliance. The specific process is as follows:

(1) **Compute the average of the multiple monitored QoS values**

   While the CSP provides the service, the evaluation party will perform multiple monitoring on the services' monitorable attributes. Because not all QoS attributes are capable of being monitored, RTEF-PP monitors c attributes, $c \leq n$. $M_i^r = \{M_{i1}^r, M_{i2}^r, \cdots M_{ic}^r\}(1 \leq r \leq k)$ is the rth monitored QoS values for service S_i, where k represents the number of monitoring. Next, the average of the multiple monitored QoS values is computed according to the formula (2):

$$m_{ij} = \frac{1}{k} \sum_{r=1}^{k} M_{ij}^r, \qquad (j = 1, 2, 3 \cdots c) \qquad (2)$$

   Finally, we get the service S_i's average monitored QoS values $M_i = \{m_{i1}, m_{i2}, \cdots, m_{ic}\}$, m_{ij} represents the average of the jth attribute.

(2) **Normalization of the average of QoS attribute values**
As each attribute has a different unit and range of values, the average value m_{ij} of each QoS attribute should be normalized into a unified range $[0,1]$. If the QoS attribute A_j is a positive one, e.g., throughput, a larger value indicates better QoS and formula (3) will be employed. If the QoS attribute A_j is a negative one, e.g., response time, a smaller value indicates better QoS and formula (4) will be employed.

$$norm(m_{ij}) = \begin{cases} 0 & m_{ij} < q_{ij}^s \\ \frac{m_{ij} - q_{ij}^s}{q_{ij}^l - q_{ij}^s} & q_{ij}^s \le m_{ij} \le q_{ij}^l \\ 1 & m_{ij} > q_{ij}^l \end{cases} \tag{3}$$

$$norm(m_{ij}) = \begin{cases} 0 & m_{ij} > q_{ij}^l \\ \frac{q_{ij}^l - m_{ij}}{q_{ij}^l - q_{ij}^s} & q_{ij}^s \le m_{ij} \le q_{ij}^l \\ 1 & m_{ij} < q_{ij}^s \end{cases} \tag{4}$$

(3) **Calculate the compliance of a single attribute**
The normalized value $norm(m_{ij})$ is used as the compliance $SC_{(m_{ij})}$ for the jth QoS attribute of service S_i, as shown in the formula (5):

$$SC_{(m_{ij})} = norm(m_{ij}) \tag{5}$$

4.1.2 Multi-attribute Objective Trust Evaluation Based on TOPSIS

Because service S_i has multiple QoS attributes, which can be seen as a Multi-Criteria Decision Making (MCDM) problem. RTEF-PP employs an improved TOPSIS [10] method to evaluate the trust of CSPs. The improved TOPSIS method has five steps for objective trust evaluation with multiple QoS attributes as follows:

(1) **Step1: Construct a decision matrix**
We need to construct a decision matrix CSP_{m*c} consisting of m cloud services and c QoS attributes. As shown in formula (6), each item $SC_{(m_{ij})}$ in the matrix CSP_{m*c} means the compliance of the jth attribute of the ith cloud service, which is calculated in Sect. 4.1.1. Each row of the matrix represents a certain cloud service, and each column represents a QoS attribute of the cloud service.

$$CSP_{m*c} = \begin{bmatrix} SC_{(m_{11})} & SC_{(m_{12})} & \cdots & SC_{(m_{1c})} \\ SC_{(m_{21})} & SC_{(m_{22})} & \cdots & SC_{(m_{2c})} \\ SC_{(m_{31})} & SC_{(m_{32})} & \cdots & SC_{(m_{3c})} \\ \vdots & \vdots & \vdots & \vdots \\ SC_{(m_{m1})} & SC_{(m_{m2})} & \cdots & SC_{(m_{mc})} \end{bmatrix} \tag{6}$$

(2) **Step2: Determine the combined weight of each QoS attribute**
The impact of each attribute on the final decision is different. RTEF-PP uses information entropy to determine the combined weight of each QoS attribute.

Due to space limitations, the method of information entropy is not described in detail. According to the calculation method of Shannon [15], we get the jth QoS attribute's entropy weight λ_j.

(3) **Step3: Calculate the weighted decision matrix**

The decision matrix CSP_{m*c} will be multiplied by the weight λ_j to obtain the weighted decision matrix. Firstly, for each item $SC_{(m_{ij})}$ in decision matrix CSP_{m*c}, we get a new weighted value $NSC_{(m_{ij})}$ of jth QoS attribute of service S_i, according to the formula (7):

$$NSC_{(m_{ij})} = SC_{(m_{ij})} \cdot \lambda_j \tag{7}$$

Next, the weighted decision matrix $NCSP_{m*c}$ is obtained as shown in the formula (8):

$$NCSP_{m*c} = \begin{bmatrix} NSC_{(m_{11})} & NSC_{(m_{12})} & \cdots & NSC_{(m_{1c})} \\ NSC_{(m_{21})} & NSC_{(m_{22})} & \cdots & NSC_{(m_{2c})} \\ NSC_{(m_{31})} & NSC_{(m_{32})} & \cdots & NSC_{(m_{3c})} \\ \vdots & \vdots & \vdots & \vdots \\ NSC_{(m_{m1})} & NSC_{(m_{m2})} & \cdots & NSC_{(m_{mc})} \end{bmatrix} \tag{8}$$

(4) **Step4: Determine the Ideal alternative (A^+) and the Non Ideal alternative (A^-) for each QoS attribute:**

$$A^+ = \{A_j^+, j = 1, 2, \cdots, c\} \tag{9}$$

$$A^- = \{A_j^-, j = 1, 2, \cdots, c\} \tag{10}$$

Where $A_j^+ = \max\{(NSC_{(m_{ij})}), i = 1, 2, \cdots m\}$ is the best value of the jth QoS attribute in all cloud services, and $A_j^- = \min\{(NSC_{(m_{ij})}), i = 1, 2, \cdots m\}$ is the worst value of the jth QoS attribute in all cloud services.

(5) **Step5: Calculate the objective trust score of each service**

At first, we calculate the distance from every cloud service to the Ideal alternative (A^+) and the Non-Ideal alternative (A^-), respectively.

$$ds_i^+ = \sqrt{\sum_{j=1}^{c} (NSC_{(m_{ij})} - A_j^+)^2}, \qquad (j = 1, 2, \cdots, c) \tag{11}$$

$$ds_i^- = \sqrt{\sum_{j=1}^{c} (NSC_{(m_{ij})} - A_j^-)^2}, \qquad (j = 1, 2, \cdots, c) \tag{12}$$

Where ds_i^+ is the distance between the ith cloud service and the Ideal alternative (A^+), and ds_i^- is the distance between the ith cloud service and the Non-Ideal alternative (A^-). The objective trust score TOS_i of service S_i can be calculated according to the the formula (13):

$$TOS_i = \frac{ds_i^-}{ds_i^+ + ds_i^-}, \qquad (i = 1, 2, \cdots, m) \tag{13}$$

4.2   The Process of Subjective Trust Evaluation

In this section, we propose a subjective trust evaluation method based on CCs' feedback ratings. In addition, RTEF-PP introduces the Laplace mechanism to protect the privacy of CCs.

4.2.1   Malicious CCs' Feedback Identification

In a complex cloud environment, malicious CCs could provide malicious feedback ratings to affect the measurement results for commercial benefit. Therefore, it is necessary to recognize and remove these malicious feedback ratings before subjective trust evaluation. This paper proposes a detection method for malicious CCs' feedback based on the difference between CCs feedback ratings and the monitored QoS values.

(1) **CCs' feedback rating quantization**

At first, after using the service, the CCs evaluates the QoS attributes of the cloud service based on his own experience. Usually, CCs' feedback is a semantic description like "good", "normal", "bad" and so on. To facilitate subjective trust evaluation, RTEF-PP translates CCs' feedback into a numerical value.

As shown in Table 2, CCs' feedback is divided into five levels. For each level, a semantic description is translated into a numerical value. Let $e_i^p = \{e_{i1}^p, e_{i2}^p, \cdots, e_{is}^p\}$ be the numerical value of service S_i's the pth CCs' feedback rating, and $s \leq n$, for the reason that CCs don't need to evaluate all attributes.

Table 2. CCs' ratings on service

Level	Level description	Value
1	Pretty good	[0.8–1]
2	Good	[0.6–0.8)
3	Normal	[0.4–0.6)
4	Bad	[0.2–0.4)
5	Pretty bad	[0–0.2)

(2) **Detect malicious CCs' feedback ratings**

In RTEF-PP, the set of feedback QoS attributes may not be the same as the set of monitored QoS attributes. For example, reliability may appear in CCs' feedback data, but will not appear in the monitored data. So we firstly take out the same attributes' data from CCs' feedback ratings e_i^p and monitored QoS values M_i to construct $e_i^{p*} = \{e_{i1}^{p*}, e_{i2}^{p*}, \cdots, e_{ix}^{p*}\}$ and $M_i^* = \{m_{i1}^*, m_{i2}^*, \cdots, m_{ix}^*\}$. Next, we calculate the Euclidean distance $p(e_i^{p*}, M_i^*)$ between e_i^{p*} and M_i^* according to the formula (14):

$$p(e_i^{p*}, M_i^*) = \sqrt{\left(e_{i1}^{p*} - m_{i1}^*\right)^2 + \left(e_{i2}^{p*} - m_{i2}^*\right)^2 + \cdots + \left(e_{ix}^{p*} - m_{ix}^*\right)^2} \quad (14)$$

When $p(e_i^{p*}, M_i^*) \leq \varphi$, we think e_i^p is honest and store it in the evaluation party's information storage module. Where φ is a threshold, which is determined by expert's opinions and specific situations.

4.2.2   Privacy Protection for CCs' Feedback Ratings

RTEF-PP adopts the Laplace mechanism to protect the CCs' privacy. This process contains three steps which are shown as follows:

(1) **Extract CCs' feedback ratings data**

Extract CCs' feedback ratings from the evaluation party's data storage module to form the CCs' feedback ratings matrix $(CCs - CSP_i)_{p \times s}$:

$$(CCs - CSP_i)_{p \times s} = \begin{bmatrix} e_{i1}^1 & e_{i2}^1 & \cdots & e_{is}^1 \\ e_{i1}^2 & e_{i2}^2 & \cdots & e_{is}^2 \\ e_{i1}^3 & e_{i2}^3 & \cdots & e_{is}^3 \\ \vdots & \vdots & \vdots & \vdots \\ e_{i1}^p & e_{i2}^p & \cdots & e_{is}^p \end{bmatrix} \tag{15}$$

Each item e_{ij}^k of $(CCs - CSP_i)_{p \times s}$ represents the kth CCs' feedback ratings on the jth attribute of service S_i. Each row of the matrix represents a feedback ratings record of the CCs, and each column represents a QoS attribute of service S_i.

(2) **Calculate the average of CCs' feedback**

CCs' feedback data will be sent to the evaluation party's data protection module to add noise. In order to avoid that the other three modules of the evaluation party get the CCs' original feedback data, RTEF-PP calculates the average of CCs' feedback data, and then sends it to the evaluation party's data protection module.

For service S_i, the average of CCs' feedback ratings $(CCs - CSP_i)_{avg} = \{avg_{i1}, avg_{i2}, \cdots, avg_{is}\}$ of attribute of service S_i can get according to the formula (16):

$$avg_{iq} = \frac{1}{p} \sum_{j=1}^{p} e_{iq}^j, \qquad (j = 1, 2, 3 \cdots, p) \tag{16}$$

Where $avg_{iq}(1 \leq q \leq s)$ represents the average of the qth attribute's feedback ratings.

(3) **Add Laplace noise**

After the CCs' feedback data is obtained, the evaluation party's data protection module adapts to the Laplace mechanism to add noise to protect CCs' privacy. For service S_i, according to the formula (17), we add noise to the average of CCs' feedback data to get $(CCs - CSP_i)_{noise} = \{E_{i1}, E_{i2}, \cdots, E_{is}\}$. And $E_{ij}(1 \leq j \leq s)$ is the CCs' feedback rating of the jth attribute of service S_i after adding noise.

$$E_{ij} = avg_{ij} + lap(\frac{\Delta f}{\varepsilon}) \tag{17}$$

Finally, we send $(CCs - CSP_i)_{noise}$ to the evaluation party's trust evaluation module for subjective trust evaluation.

4.2.3 Calculate Subjective Trust Score

We use the CCs' feedback ratings data after adding noise $(CCs - CSP_i)_{noise} = \{E_{i1}, E_{i2}, \cdots, E_{is}\}$ obtained in Sect. 4.2.2 to calculate the subjective trust score TSS_i of each service S_i according to the formula (18):

$$TSS_i = \sum_{j=1}^{s} E_{ij} * \psi_j \tag{18}$$

where ψ_j is the weight of the jth attribute, which can be computed using the method of information entropy in Sect. 4.1.2.

4.3 Integrate Trust Score

This section calculates an integrate score TS_i for service S_i using the objective trust score and the subjective trust score according to the formula (19). An adjustable parameter β, $0 \le \beta \le 1$ is applied to various scenarios for improving the applicability of the framework.

$$TS_i = TOS_i * \beta + TSS_i * (1 - \beta) \tag{19}$$

The value of β is determined according to expert opinions and specific application scenarios, and it is usually set to 0.5 by default. However, if there are more malicious CCs in the application environment, the reliability of subjective trust evaluation may be relatively low, which means that β can be set greater than 0.5. Otherwise, β can be set smaller than 0.5.

After getting the integrated trust score of all candidate cloud services, RTEF-PP ranks these services according to the score. Top ranked cloud services will be recommended to CCs.

5 Experiments

To evaluate the performance of RTEF-PP, we conduct some experiments base on a real world dataset QWS [1]. The original dataset includes many services with different functionality. We extract 103 services with "email" functionalities from the 5825 services via a keyword-based method to construct a smaller QoS dataset which contains 339 * 103 records. The smaller QoS dataset will be used for trust evaluation. We evaluate the performance of RTEF-PP from accuracy, robustness, and accuracy after privacy protection. The experimental platform is MATLAB R2016a on Windows 7 system. The specific configuration is Intel Core i5-4430 CPU with 3.00 GHZ, 6 GB RAM.

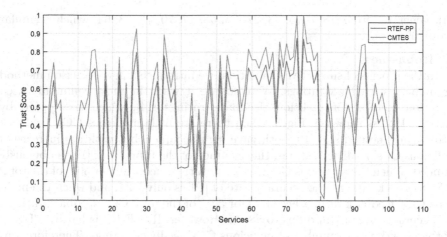

Fig. 2. The result comparison between RTEF-PP and CMTES

(1) Accuracy

As shown in Fig. 2, the trust scores of 103 cloud services are calculated through RTEF-PP. We can see that the top ten services number are in turn: 73, 75, 27, 35, 68, 70, 77, 76, 93, 64. Furthermore, we implement another existing trust evaluation method Compliance-based Multi-dimensional Trust Evaluation System, which is denoted as CMTES [18]. As shown in Fig. 2, the results calculated by the two frameworks do not coincide, since each framework has different scoring methods. But it can be observed that the overall trend of the two lines is almost the same. That means that the ranking results of the services calculated by the two evaluation framework are similar. And the top ten services number in

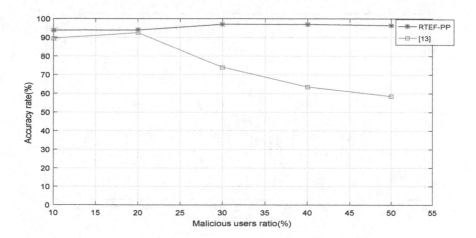

Fig. 3. The robustness comparison between RTEF-PP and [13]

turn generated from [18] is 75, 73, 27, 35, 68, 77, 70, 76, 93, 64, which is similar to our framework.

(2) Robustness

In the process of subjective evaluation, we introduce a new detection method for malicious feedback ratings. We compare it with the most commonly used detection method for malicious feedback ratings [14], which is adopted by TRUSS [19]. As shown in Fig. 3, when there are little malicious CCs' feedback ratings, our method and [19] both have high accuracy, but with the increase of malicious CCs' feedback ratings, the accuracy of the framework [19] is obviously reduced. When the malicious CCs' feedback ratings accounts for 30% of the total CCs' feedback ratings, the accuracy rate of [19] is only 74%, and when the malicious CCs' feedback ratings accounts for 50% of the total CCs' feedback ratings, the accuracy rate of [19] drops to 58.5%. However, RTEF-PP is hardly affected by the growth in the number of malicious CCs' feedback ratings. Therefore, our framework is robust for environments with more malicious feedback ratings.

(3) Accuracy after privacy protection

In RTEF-PP, we introduce differential privacy to protect CCs' privacy. However, the value of the privacy parameter ε in differential privacy can affect the accuracy and security of the trust evaluation framework. Usually, a larger means adding less noise, so the result of the frame will be more accurate, while a smaller ε means adding more noise, so the accuracy of the result of the frame will be reduced. In order to prove that our framework still has high accuracy after introducing differential privacy technology, we set ε to 0.1, 0.5 and 1 respectively. As shown in Fig. 4, it can be observed that the overall trend of the four lines is almost the same. This means that the trust evaluation results after adding noise still have high accuracy.

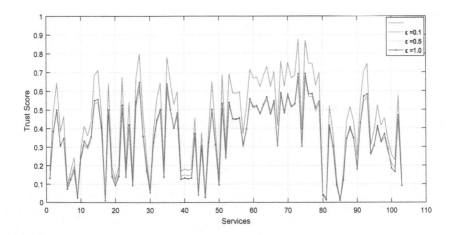

Fig. 4. Impact of ε on the accuracy of RTEF-PP

6 Conclusion

In this paper, we propose a robust trust evaluation framework with privacy protection for CSPs, which is denoted as RTEF-PP. In order to avoid the trust evaluation process revealing sensitive information of CCs, RTEF-PP firstly introduces differential privacy to protect the privacy of CCs. In addition, RTEF-PP uses the Euclidean distances between the monitored QoS values and CCs' feedback to detect malicious CCs' feedback ratings, which are not affected by the number of malicious CCs' feedback ratings. The experimental results prove that RTEF-PP is reliable and robust.

Acknowledgements. The work was supported by the National Natural Science Foundation of China (No. 61572001, No. 61702005), the Natural Science Foundation of Anhui Province (No. 1708085QF136), the Special Fund for Key Program of Science and Technology of Anhui Province, China (Grant No. 18030901027).

References

1. Al-Masri, E., Mahmoud, Q.H.: Discovering the best web service. In: Proceedings of the 16th International Conference on World Wide Web, WWW 2007, Banff, Alberta, Canada, 8–12 May 2007, pp. 1257–1258 (2007)
2. Alhamad, M., Dillon, T., Chang, E.: Conceptual SLA framework for cloud computing. In: IEEE International Conference on Digital Ecosystems and Technologies (2010)
3. Alhamad, M., Dillon, T.S., Chang, E.: SLA-based trust model for cloud computing. In: International Conference on Network- based Information Systems (2010)
4. Chakraborty, S., Roy, K.: An SLA-based framework for estimating trustworthiness of a cloud. In: IEEE International Conference on Trust (2012)
5. Dwork, C., McSherry, F., Nissim, K., Smith, A.: Calibrating noise to sensitivity in private data analysis. In: Halevi, S., Rabin, T. (eds.) TCC 2006. LNCS, vol. 3876, pp. 265–284. Springer, Heidelberg (2006). https://doi.org/10.1007/11681878_14
6. Huang, L., Deng, S., Li, Y., Wu, J., Yin, J., Li, G.: A trust evaluation mechanism for collaboration of data-intensive services in cloud. Appl. Math **7**(1L), 121–129 (2013)
7. Fujitsu Research Institute: Personal data in the cloud: a global survey of consumer attitudes. http://www.fujitsu.com/downloads/SOL/fai/reports/fujitsu_personal-data-in-the-cloud.pdf
8. Ma, Z., Jiang, R., Yang, M., Li, T., Zhang, Q.: Research on the measurement and evaluation of trusted cloud service. Soft. Comput. **22**(6), 1–16 (2016)
9. Muchahari, M.K., Sinha, S.K.: A new trust management architecture for cloud computing environment. In: International Symposium on Cloud and Services Computing (2013)
10. Nazemi, S., Vesal, H.: Multiple Attribute Decision Making (1981)
11. Nguyen, H.T., Zhao, W., Jian, Y.: A trust and reputation model based on Bayesian network for web services. In: IEEE International Conference on Web Services (2010)
12. Noor, T.H., Sheng, Q.Z., Yao, L., Dustdar, S., Ngu, A.H.H.: CloudArmor: supporting reputation- based trust management for cloud services. IEEE Trans. Parallel Distrib. Syst. **27**(2), 367–380 (2016)

13. Noorian, Z., Fleming, M., Marsh, S.: Preference-oriented QoS-based service discovery with dynamic trust and reputation management. In: Proceedings of the ACM Symposium on Applied Computing, SAC 2012, Riva, Trento, Italy, 26–30 March 2012, pp. 2014–2021 (2012)
14. Qiu, W., Zheng, Z., Wang, X., Yang, X., Lyu, M.R.: Reputation-aware QoS value prediction of web services. In: IEEE International Conference on Services Computing (2013)
15. Shannon, C.E., Weaver, W.: The mathematical theory of communication. Bell Labs Tech. J. **3**(9), 31–32 (1950)
16. Shuai, D., Yang, S., Zhang, Y., Liang, C., Xia, C.: Combining QoS prediction and customer satisfaction estimation to solve cloud service trustworthiness evaluation problems. Knowl.-Based Syst. **56**(3), 216–225 (2014)
17. Sidhu, J., Singh, S.: Improved topsis method based trust evaluation framework for determining trustworthiness of cloud service providers. J. Grid Comput. **15**(1), 1–25 (2016)
18. Singh, S., Sidhu, J.: Compliance-based multi-dimensional trust evaluation system for determining trustworthiness of cloud service providers. Future Gener. Comput. Syst. **67**, 109–132 (2017)
19. Tang, M., Dai, X., Liu, J., Chen, J.: Towards a trust evaluation middleware for cloud service selection. Future Gener. Comput. Syst. **74**(C), 302–312 (2017)
20. Wang, Y., Wen, J., Zhou, W., Luo, F.: A novel dynamic cloud service trust evaluation model in cloud computing, pp. 10–15 (2018)
21. Xu, K., Zhang, W., Zheng, Y.: A privacy-preserving mobile application recommender system based on trust evaluation. J. Comput. Sci. **26**, 87–107 (2018)
22. Zheng, Y., Ding, W., Niemi, V., Vasilakos, A.V.: Two schemes of privacy-preserving trust evaluation. Future Gener. Comput. Syst. **62**(C), 175–189 (2016)
23. Zheng, Z., Wu, X., Zhang, Y., Lyu, M.R., Wang, J.: QoS ranking prediction for cloud services. IEEE Trans. Parallel Distrib. Syst. **24**(6), 1213–1222 (2013)

A Privacy-Preserving Access Control Scheme with Verifiable and Outsourcing Capabilities in Fog-Cloud Computing

Zhen Cheng[1], Jiale Zhang[1], Hongyan Qian[1(✉)], Mingrong Xiang[2], and Di Wu[3]

[1] College of Computer Science and Technology, Nanjing University of Aeronautics and Astronautics, Nanjing 211106, China
{czheng,jlzhang,qhy98}@nuaa.edu.cn
[2] Faculty of Science, Engineering and Built Environment, Deakin University, Melbourne, VIC 3216, Australia
mxiang@deakin.edu.au
[3] School of Computer Science, Centre for Artifical Intelligence, University of Technology Sydney, Sydney, NSW 2007, Australia
Di.Wu@uts.edu.au

Abstract. Fog computing is a distribution system architecture which uses edge devices to provide computation, storage, and sharing at the edge of the network as an extension of cloud computing architecture, where the potential network traffic jams can be resolved. Whereas, the untrustworthy edge devices which contribute the computing resources may lead to data security and privacy-preserving issues. To address security issues and achieve fine-grained access control to protect privacy of users, ciphertext-policy attribute-based encryption (CP-ABE) mechanism has been well-explored, where data owners obtain flexible access policy to share data between users. However, the major drawback of CP-ABE system is heavy computational cost due to the complicated cryptographic operations. To tackle this problem, we propose a privacy-preserving access control (PPAC) scheme and the contributions are tri-folded: (1) we introduce outsourcing capability in fog-cloud computing (FCC) environment; (2) the outsource verification mechanism has been considered to guarantee the third party execute the algorithm correctly; (3) we design a partiality hidden method to protect the privacy information embedded in the access structures. The experimental results show that our proposed PPAC is efficient, economical and suitable for mobile devices with limited resources.

Keywords: Fog-cloud computing · Attribute-based encryption · Privacy-preserving · Access control

1 Introduction

With the high demand of data storage and sharing in the Internet of Things (IoT), cloud computing has become solution regard to its ability to make all of

© Springer Nature Switzerland AG 2020
S. Wen et al. (Eds.): ICA3PP 2019, LNCS 11944, pp. 345–358, 2020.
https://doi.org/10.1007/978-3-030-38991-8_23

the connected devices work together and store the unprecedented amount of data [1]. However, transferring huge amount of data from IoT devices to the cloud server is not only adding latency of the data transportation, but also consumes constrained network resources. For example, vehicles in the autonomous vehicle network might produce gigabytes data in one second, where shorter response time could help the vehicle to avoid the accident [2]. In order to transport large amount of data and reduce the response time, fog computing has been proposed as a promising solution to mitigate the limitation of IoT devices with constrained resources. As an extension of cloud computing, fog computing offloads the communication and computation burden of the network by processing data near the sources of data [3]. However, the security and privacy issues still present practical concerns for fog computing, where the cloud server can not be fully trusted and the edge devices are untrustworthy as well [4]. Therefore, the purpose of this work is to achieve lightweight and privacy-preserving access control in fog-cloud computing (FCC) by outsourcing partial cryptosystem computations to the fog and cloud servers.

The notion of attribute-based encryption (ABE) was first introduced by Sahai and Waters based on fuzzy identity encryption [5]. Furthermore, Goyal et al. proposed the key-policy attribute-based encryption (KP-ABE) scheme [6] and Bethencourt et al. present ciphertext-policy attribute-based encryption (CP-APE) construction methods [7]. Subsequently, a large number of ABE schemes were proposed [8–12], which add different functions on original ABE, such as ciphertext auditing and privacy protecting. Due to the heavy computation cost for the encryption and decryption phases, a new method of outsourcing decryption of ABE was proposed by Green et al. [13]. However, since the proxy server cannot be fully trusted, verifiable outsourced computation is required. The technique proposed in [14,15] can be used to outsource the operations of decryption in ABE systems, because the verification mechanism is considered along with outsourcing ABE schemes. Recently, [16,17] demonstrate that the ABE scheme can be applied to fog computing environment to solve the secure storage problem. However, these solutions still have security and privacy problems: preventing to the untrustworthy third party from learning private information and ensuring the correctness of results returned by the untrustworthy third party.

To tackle the above problems, we propose a verifiable and outsourced ABE mechanism, which not only protects data privacy, fine-grained access control and brings verifiability to the outsourced computation, but also reduces the computation cost. The contributions can be summarized as follows:

- **A lightweight access control model in FCC:** We first propose a novel lightweight ABE system to outsource the computation of both encryption and decryption procedure. As a result, most of the complicated cryptographic operations can be outsourced to the fog or cloud servers.
- **A verifiable mechanism for outsourcing capability:** In order to avoid the incorrect results from cloud servers, we present a concrete verification mechanism for outsourced computation capability.

- **A partially hidden access structure:** To protect the private information embedded in the access structures, we further design a partially hidden method along with the outsourced ABE system. In the proposed system, the attributes are hidden from the users whose private key does not satisfy the access structure.

The rest of this paper are organized as follows. In Sect. 2, we briefly introduce the basic knowledge of attribute-based encryption method. The system model is presented in Sect. 3, and the construction of proposed privacy-preserving access control scheme is detailed in Sect. 4. Extensive experimental evaluation is conducted in Sect. 5. Finally, Sect. 6 gives the conclusion.

2 Preliminaries

In this section, we briefly revisit the basic definition of bilinear groups, CP-ABE scheme, access structure, and the linear secret sharing schemes (LSSS).

2.1 Bilinear Maps

Let an algorithm that inputs a security parameter λ and outputs a tuple $(q, \mathbb{G}, \mathbb{G}_T, e)$, where $\mathbb{G}$ and $\mathbb{G}_T$ are two multiplicative cyclic groups of prime order q. The bilinear map $e : \mathbb{G} \times \mathbb{G} \to \mathbb{G}_T$ fulfills the following properties:

- Bilinearity: $e(g^a, g^b) = e(g, g)^{ab}$ for all $g \in \mathbb{G}$, $a, b \in \mathbb{Z}_q^*$.
- Non-degeneracy: There exist $e(g, g) \neq 1$ such that $g \in \mathbb{G}$.
- Computable: There is an efficient algorithm to compute e, for all $g \in \mathbb{G}$.

2.2 CP-ABE

CP-ABE scheme is consisted of the following four algorithms:

- **Setup** (λ, U): The setup algorithm inputs security parameter λ and universe description U. It outputs the public parameters PK and master key MSK.
- **KeyGen** (PK, MSK, S): The key generation algorithm takes as inputs the PK, the master key MSK, and a set of attributes S. It outputs the private key SK.
- **Encrypt** (PK, M, A): The encryption algorithm takes as input the PK, the message M, and an access structure A. It outputs the ciphertext CT.
- **Decrypt** (SK, CT): The decryption algorithm takes as inputs the private key SK with attribute set S. If S can satisfy the access structure A, it outputs the message M. Otherwise, it outputs $\perp$.

2.3 Access Structure

Let $\{P_1, P_2, \cdots, P_n\}$ is a collection of attributes, we say $A \in 2^{\{P_1, P_2, \cdots, P_n\}}$ is a monotone selection attribute set if $\forall B : B \in A$ and $B \subseteq C$ then $C \in A$. The monotone access structure A is a non-empty subsets of $\{P_1, P_2, \cdots, P_n\}$, i.e., $A \in 2^{\{P_1, P_2, \cdots, P_n\}} \setminus \{\emptyset\}$.

2.4   Linear Secret Sharing Schemes

A secret sharing scheme Π over a set of parties is called linear over $\mathbb{Z}_q$. The definitions are shown below:

- Assuming there are some shares among different parties from a vector over $\mathbb{Z}_q$.
- There exists a matrix M with ℓ rows and n columns called the share-generating matrix for Π. A function ρ which maps each row of the matrix to an associated party (for $i = 1, \cdots, \ell$), where $\rho(i)$ is the party associated with row i.
- When we consider the column vector $v = (s, r_2, \cdots, r_n)$, where $s \in \mathbb{Z}_q$ is the secret to be shared, and $(r_2, \cdots, r_n) \in \mathbb{Z}_q$ are randomly selected numbers, then M_v is the vector of ℓ shares of the secret s according to Π. Note that the share $(M_v)_i$ belongs to party $\rho(i)$.

3   System Model

3.1   Overview of Basic Outsourcing CP-ABE Scheme

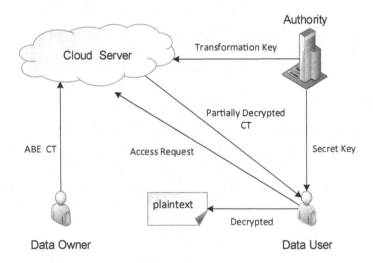

Fig. 1. The framework of ABE with outsourced decryption in cloud computing

In the conventional outsourced CP-ABE scheme [3], there exists a fully trusted authority to assign two types of keys for the data users and cloud server. One is the user's private key, called SK, which can be used to decrypt the ciphertext requested from the central server. Another is the corresponding transformation key TK used to construct the outsourcing part of ciphertexts. The whole work-flow of basic outsourced CP-ABE scheme in the cloud computing scenario is shown in Fig. 1. At first, the data owner encrypts his private data by using ABE scheme and sends the ciphertext to the cloud center for storage. Once receive the access request from the data user whose attributes satisfy the access policy, the cloud server converts the original ABE ciphertext into a simple ElGamal-form ciphertext by using TK. Then, this ElGamal-form ciphertext will be sent to the data user. Note that this ElGamal-form ciphertext is encrypted under the user's private key SK for the same message. At last, the data user can obtain the plaintext by a simple operation based on the secret key SK.

In our work, we mainly focus on the following issues over the basic outsourced CP-ABE scheme in the cloud computing scenario:

- Although conventional outsourced ABE scheme can shift part of decryption operations to the third party, the encryption step still brings the heavy computation costs, which will consume a lot of resources of IoT device and waste the network bandwidth.
- The correctness of the ciphertext which partially decrypted in the cloud computing center cannot be guaranteed. Some cloud servers may become "lazy" to save the overhead of bandwidth and resource, which will return "fake results" to the real ones.
- The access structures embedded in the ciphertext usually contain private information of data owner, that sensitive information could be disclosed by the untrusted data user.

3.2 Our Proposed PPAC Framework

To solve the aforementioned challenge, we take both verification and privacy into consideration and propose a privacy-preserving access control scheme with verifiable and outsourced capabilities in fog-cloud computing, named PPAC. Similar to the outsourced decryption mechanism, the data owner can also outsource the encryption to the fog server and verifies the correctness of the results returned by the fog server. Moreover, the embedded access structure is partially hidden from the data user whose private keys do not satisfy the access policy.

As depicted in Fig. 2, the whole access control system consists of five entities: authority, cloud server, fog servers, data owners, and data users.

- **Authority:** The authority is a fully trusted entity whose duty is to bootstrap the whole system and generates key materials associated with access control policies.
- **Cloud:** The cloud is the central server for data storage and sharing. Its duty is to collect all the users' data from fog servers and further make some analytics.
- **Fog servers:** It mainly provides the data users with the computing and storing function on the edge network to reduce the delay and improve the service quality.
- **Data owners:** In the fog-cloud model, IoT devices such as smart-phones, wearable devices, smart home appliances and etc., which gather data nearby and uploads them in an encrypted form. However, IoT devices with limited resources can only bear lightweight operations.

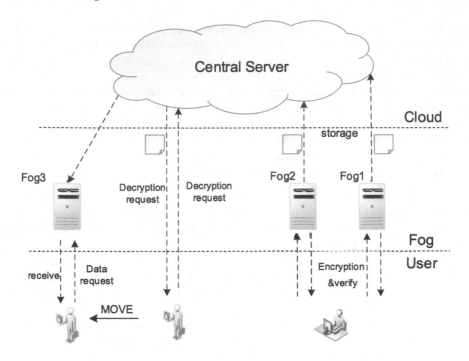

Fig. 2. Overview of proposed privacy-preserving access control system

- **Data users:** Data users play a role with data consumers who can request private data stored in the cloud. They also cannot accept the heavy computation costs and a long-delayed response due to its resources-constraint.

4   Construction of Proposed PPAC

In this section, we present the concrete PPAC scheme with verifiable and outsourcing capabilities which consists of eight algorithms. Note that the algorithms of Setup, KeyGen, and GenTK.out are performed on a certain trusted authority.

4.1   Algorithms in PPAC Scheme

1. **Setup** $(\lambda, U) \rightarrow (PK, MSK)$: On input a security parameter and an attribute universe description $U = 1, 2....\ell$. It outputs public parameters PK and a master key MSK. The setup algorithm works as follows:
   - Chooses a bilinear map $D = (p, \mathbb{G}, \mathbb{G}_T, e)$, where p is the prime order of the groups $\mathbb{G}$ and $\mathbb{G}_T$. Then, it chooses three random generators (g, h, u) and one exponent $\alpha \in \mathbb{Z}_p$.
   - Chooses a collision-resistant hash function $H : \{0,1\}^* \rightarrow \mathbb{Z}_p^*$ and adopts the key derivation function (KDF1) [18], where the length of function output is defined as $\mathcal{L} = |key| + |p|$ and p is the prime order.

- Publishes the public parameters as $PK = \{D, e(g,g)^\alpha, g, h, u, H, \mathcal{L},$ $KDF_1\}$, and the corresponding master key can be formed as $MSK = (PK, \alpha)$.

2. **KeyGen** $(PK, MSK, S) \rightarrow (SK)$: On input public parameters PK, the master secret key MSK and a set of attributes $S = \{A_1, \cdots, A_n\}$. It outputs a private key SK associated with S. The algorithm works as follows:
   - Generates a set of random values $(r, r_1, \cdots, r_k) \in \mathbb{Z}_p$.
   - Computes $k_0 = g^\alpha u^r$, and $k_1 = g^r$.
   - For all $i \in [1, k]$, calculates $k_{i,2} = g^{r_i}$ and $k_{i,3} = (uh^{A_i})^{r_i}$.
   - Then, it sets the private key as $SK = (S, PK, k_0, k_1, \{k_{i,2}, k_{i,3}\}_{i \in [1,k]})$.
   - Sends the private key SK to the data owners.

3. **GenTK.out** $(SK) \rightarrow (TK, RK)$: On input a private key SK, it outputs a transfer key TK and a corresponding retrieval key RK. The algorithm works as follows:
   - Randomly chooses a number $\tau \in \mathbb{Z}_p^*$.
   - Computes $k_0' = u^{\alpha/\tau} u^{r/\tau}$, $k_1' = g^{r/\tau}$, $k_{i,2}' = g^{r_i/\tau}$, and $k_{i,3}' = (uh^{A_i})^{r_i/\tau}$. Then, it sets the transfer key as $TK = (S, PK, k_0', k_1', \{k_{i,2}', k_{i,3}'\}_{i \in [1,k]})$ and the related retrieval key can be formed as $RK = (TK, \tau)$.
   - Sends the retrieval key RK to the data users.

Note that, the aforementioned three algorithms (i.e., Setup, KeyGen, and GenTK.out) are all executed at the authority side.

4. **Encrypt.out1** $(PK, N) \rightarrow (IT_1)$: This algorithm is performed by the *fog1*. On input public parameters PK and maximum bound of N rows in any LSSS access structure, it outputs a intermediate ciphertext IT_1. The algorithm works as follows:
   - Chooses three random numbers $(x_i', y_i', \lambda_i') \in \mathbb{Z}_p$, where $i \in [1, N]$ and computes $C_{i,1}' = u^{\lambda_i'}$ and $C_{i,2}' = (uh^{x_i'})^{y_i'}$.
   - Randomly picks $s' \in \mathbb{Z}_p$ and computes $C_0' = g^{s'}$. Then, the intermediate ciphertext can be formed as $IT_1 = (s', C_0', \{x_i', y_i', \lambda_i', C_{i,1}', C_{i,2}'\}_{i \in [1,k]})$.
   - Sends IT_1 to the data owner for consolidation and creates a replica for the storage in cloud center.

5. **Encrypt.out2** $(PK, N) \rightarrow (IT_2)$: This algorithm is executed by the *fog2*. On input public parameters PK and a maximum bound of N rows in any LSSS access structure, it outputs a intermediate ciphertext IT_2. The algorithm works as follows:
   - Picks three random exponent $(x_i'', y_i'', \lambda_i'') \in \mathbb{Z}_p$, where $i \in [1, n]$ and calculates $C_{i,1}'' = u^{\lambda_i''}$ and $C_{i,2}'' = (uh^{x_i''})^{y_i''}$.
   - Randomly picks $s'' \in \mathbb{Z}_p$ and computes $C_0'' = g^{s''}$. Then, the intermediate ciphertext is $IT_2 = (s'', C_0'', \{x_i'', y_i'', \lambda_i'', C_{i,1}'', C_{i,2}''\}_{i \in [1,k]})$.
   - Returns IT_2 to the data owner for consolidation and send a copy to the cloud center for storage.

6. **Encrypt.owner** $(PK, IT_1, IT_2, A, \rho, \tau) \rightarrow (CT, key)$: It is performed by the data owners. On inputs public parameters PK and two intermediate ciphertexts IT_1, IT_2 and an access structure (A, ρ, τ), where A is an $\lambda \times n$ matrix, ρ is a map from each row A_i of A to an attribute name, and τ is a set consists of $\{\tau_{\rho(1)}, \cdots, \tau_{\rho(\lambda)}\}$ and $\tau_{\rho(i)}$ is the value of attribute $\rho(i)$ calculated from the access policy. At last, the algorithm outputs the ciphertext CT and keeps key locally. The algorithm works as follows:

   - Adopts a collision-resistant hash function $H : \{0,1\}^* \rightarrow \mathbb{Z}_p^*$ and recomputes $s = s' + s''$ and $C_0 = C_0' \cdot C_0'' = g^{s'+s''}$, where $key = e(g,g)^{\alpha s}$.
   - Recomputes the parameters $\lambda_i = \lambda_i' + \lambda_i''$, $y_i = y_i' + y_i''$, and $x_i = x_i' + x_i''$. Then, calculates $C_{i,1} = C_{i,1}' \cdot C_{i,1}'' = u^{\lambda_i}$ and $C_{i,2} = C_{i,2}' \cdot C_{i,2}'' = (uh^{x_i})^{y_i}$.
   - Randomly chooses $v_2, \cdots, v_n \in \mathbb{Z}_p$, denotes a vector as $V = (s, v_2, \cdots, v_n)^T$ and computes another vector of shares of s as $(b_1, \cdots, b_\lambda) = M \cdot V$. For $i = 1$ to λ, it computes $C_{i,3} = b_i - \lambda_i$, $C_{i,4} = \rho(i)y_i - x_i y_i$, and $C_{i,5} = -y_i$.
   - Generates a new key SSK with the key derivation function and computes $KDF1(key, L) = ssk\|d$, $\hat{c} = u^{H(ssk)}v^{H(d)}$, where d is the length of SSK. Here, we add a Pedersen commitment $\hat{c}$ to achieve the verification of the outsourced computation.
   - Randomly chooses $t' \in \mathbb{Z}_p$ and computes $F = e(g,g)^{\alpha t'}$.
   - Picks a set of random values $v_2', \cdots, v_n' \in \mathbb{Z}_p$, denotes a vector $V' = (t', v_2', \cdots, v_n')$ and computes another vector of shares of t' as $b_1', \cdots, b_\lambda' = M \cdot V'$. Then, $C_{i,3}' = b_i' - \lambda_i$. At last, it sets the ciphertext as $CT_1 = (M, \rho, C_0, \hat{c}, \{C_{i,1}, C_{i,2}, C_{i,3}, C_{i,3}', C_{i,4}', C_{i,5}'\}_{i \in [1,\lambda]})$.

   Note that each attribute consists of an attribute value and an attribute name. If the attribute set does not satisfy the access structure, the attribute value in the access structure is hidden, while the attribute name can be broadcast. The role of CT_2 is to help the user who satisfies the access structure to decide which attribute set satisfies the access structure and further prevent the cloud from getting too much information during decryption. The scheme can partially hide the access structure to the cloud center and fog server.

7. **Transform.out** $(TK, CT) \rightarrow CT'$: This algorithm is performed by the cloud server. On inputs a conversion key TK and the ciphertext CT, it outputs a transformed ciphertext CT'. We define L to be the smallest set of subsets that satisfy the access structure (M, ρ). When received a ciphertext CT and a conversion key TK for attribute sets S from the data users, the cloud server first checks whether the user's attributes meet the smallest set of subsets L. The algorithm works as follows:

   - Checks whether there exists an $I \in L$ that satisfies:

$$F = \frac{e(C_0, k_0')}{e(u^{\sum_{i \in I} C_{i,3}' w_i}, k_1') \cdot \prod_{i \in I}(e(C_{i,1}, k_1')e(C_{i,2}h^{C_{i,4}}, k_1')e(g^{C_{i,5}}, k_{i,3}))^{w_i}} \tag{1}$$

where $\sum_{i \in I} w_i M_i = (1, 0, \cdots, 0)$. If there is no member of L that satisfies this equation, it outputs $\perp$. Otherwise, we calculate the encapsulated key as:

$$\frac{e(C_0, k_0')}{e(u^{\sum_{i \in I} C_{i,3} w_i}, k_1') \cdot \prod_{i \in I} (e(C_{i,1}, k_1') e(C_{i,2} h^{C_{i,4}}, k_1') e(g^{C_{i,5}}, k_{i,3}))^{w_i}}$$

$$= \frac{e(g^s, g^{\alpha/\tau}) e(g^s, u^{r/\tau})}{e(u^{\sum_{i \in I} (b_i - \lambda_i) w_i}, g^{r/\tau}) e(u^{\lambda_i}, g^{r/\tau})^{w_i}}$$

$$= \frac{e(g^s, g^{\alpha/\tau}) e(g^s, u^{r/\tau})}{e(u^s, g^{r/\tau})} = e(g, g)^{\alpha s/\tau}$$

$$(2)$$

- It sets the transformed ciphertext as $CT' = (e(g, g)^{\alpha s/\tau}, T' = \hat{c}, M, \rho)$ and sends the CT' to the data users.

8. **Decrypt.user** $(RK, CT, CT') \rightarrow m$: This algorithm is executed by the data users. On input a retrieval key RK, a ciphertext CT and a transformed ciphertext CT', it outputs $\perp$ when $T' \neq \hat{c}$. Otherwise, it outputs $key = (key')^\tau = e(g, g)^{\alpha s}$. In this situation, it is necessary to check the correctness of the decryption results returned by the cloud server. We take the method of adding a verification message (Pedersen commitment $\hat{c}$) in the encryption process. When the data user requests data from the cloud server, if the attribute set meets the access structure, the cloud will send ciphertext to the data user, because the cloud does not need to do any calculation in this process, so there is no reason for cloud server sends a false ciphertext. Then, the data user requests the cloud server for the partially decrypt ciphertext. The Pedersen commitment $\hat{c}$ of the decrypted ciphertext is compared with the verification information T' in the original ciphertext by the user. If $T' = \hat{c}$, it computes $m = \frac{m \cdot e(g,g)^{\alpha s}}{e(g,g)^{\alpha s}}$.

4.2 Correctness of Batch Verification for Outsourced Encryption

Due to the complexity of fog computing, we consider a scenario where a fog server might perform only a portion of the computation and return incorrect results to save computing resources. According to [19], we can use a naive way to verify the modular exponentiation in some group, i, e, m given (g, x, y), check that whether $g^x = y$ is correct or not. In our work, there exists two types of ciphertexts: the ciphertext with the same one generator $(C_0' = g^{s'}, C_0'' = g^{s''}, C_{i,1}' = u^{\lambda_i'}, C_{i,1}'' = u^{\lambda_i''})$ and ciphertext with two different generators $(C_{i,2}' = (u h^{x_i'})^{y_i'}, C_{i,2}'' = (u h^{x_i''})^{y_i''})$. Thus, we give the correctness of two-types of ciphertexts as follows:

[Type-1]:

- Given: Let $P = |\mathbb{G}|$ be the order of the group $\mathbb{G}$ and g is a primitive element of $\mathbb{G}$. Defining $\{(x_1, y_1), \cdots, (x_n, y_n)\}$ with $x_i \in \mathbb{Z}_p$ and $y_i \in \mathbb{Z}_p$, as well as a security parameter k.

- Check: $\forall i \in [1, \cdots, n]$, $y_i = g^{x_i}$.
- Picks a random subset $S = (b_1, \cdots, b_n) \in \{0,1\}^k$ and computes $x = \sum_{i=1}^{n} x_i b_i \mod p$, $y = \prod_{i=1}^{n} y^{b_i}$. If $g^x = y$, returns accept, else reject.

[Type-2]:

- Given: Let $P = |\mathbb{G}|$ be the order of the group $\mathbb{G}$ and g is a primitive element of $\mathbb{G}$. Defining $\{(x_1, y_1, z_1), \cdots, (x_n, y_n, z_n)\}$ with $(x_i, y_i) \in \mathbb{Z}_p$ and $z_i \in \mathbb{G}$, as well as a security parameter k.
- Check: $\forall i \in [1, \cdots, n]$, $z_i = u^{x_i} h^{y_i}$.
- Picks a random subset $S = (b_1, \cdots, b_n) \in \{0,1\}^k$ and computes $x = \sum_{i=1}^{n} x_i b_i \mod p$, $y = \prod_{i=1}^{n} y^{b_i}$, and $z = \prod_{i=1}^{n} z_i b_i$. If $u^x h^y = z$, returns accept, else reject.

4.3   Security Analysis

Here, we give a brief discussion about the security properties of our PPAC scheme. Note that the cloud and the fog servers are assumed to be semi-trusted in our security model. In particular, we describe the security problems of the scheme from the following aspects.

Data Confidentiality: First of all, we outsource the encryption to two fog servers, both of which are run by different operators. Note that there's no reason for these two fog servers to collude with each other and reveal the user's private key information. Besides, the cloud center cannot get any plaintext information about the user during the partial decryption process. Finally, when the data is requested, the data user cannot get any useless information about the plaintexts unless the attribute set satisfies the access structure.

Data Validation: For the outsource verification, we adopt the batch verification mechanism to check the correctness of encryption result, which has been widely used in cryptographic to verify modular exponentiation operations. In general, the expected error of this method is within our acceptable range. For the outsourced decryption validation, the key derived function KDF and the anti-collision hash function have been used to hide the final decryption results in the ciphertexts, which can protect the privacy of original data. For the user side decryption phases, the data user can obtain the ciphertexts from the cloud center only if the user's attribute set satisfies the access structure. In this way, the user can get the verification information contained in the correct ciphertext to verify the results and further ensure the correctness of outsourcing decryption.

Privacy Protection: In our proposed PPAC scheme, the attribute values of the access structure are hidden from the cloud and fog servers while only the attribute names of the access structure are public. Besides, when the cloud checks whether the attribute set of the data user satisfies the access structure, the data user is also unable to know the attribute value in this process, which means the data owner's private information embedded in the access structures can be protected.

5 Performance Evaluation

5.1 Theoretical Analysis

Computation Cost Comparison: The computation cost of our scheme depend on the number of modular exponentiation and pairing operations. In Table 1, we compare the efficiency of our scheme with the original ABE [5] and the other two outsourced ABE schemes [13] and [16]. Note that, E_p and P denote the modular exponentiation and pairing operation, respectively. ω denotes the attribute set and λ represents the number of rows of the matrix M for LSSS. As shown in Table 1, our scheme only leaves three exponentiations and a pairing computation (for hiding access policy) during the encryption phase. The computation cost of our schemes is less than [5] and [13]. The ABE technique [16] has only three exponentiations, but the outsource server in [16] know the access structure which may contain sensitive information. Thus, our scheme still behaves better. During the decryption phase, the computational costs for the data user are three exponentiations which less than ABE schemes in [5] and [13]. Therefore, our scheme requires less computation costs compared with above schemes.

Table 1. Comparison of computation costs

Schemes	KeyGen	Enc.user	Dec.user	Enc.verify	Dec.verify	Hidding policy
[3]	$2|\omega|E_p$	$2|\lambda|(E_p + P)$	$2|\lambda|(E_p + P)$	×	×	×
[4]	$2|\omega|E_p$	$(2 + 5|\lambda|)E_p$	E_p	×	✓	×
[19]	$2|\omega|E_p$	$3E_p$	$3E_p$	✓	✓	×
Our	$2|\omega|E_p$	$3E_p + P$	$3E_p$	✓	✓	✓

5.2 Experimental Evaluations

Here, we estimate the performance of our proposed PPAC scheme and compare the results with Green's work [13]. We implement the experiment by using pairing-based cryptography (PBC) library for a variety of cryptographic operations and Charm framework [20] for attribute-based encryption method. To achieve a high-security level, we choose the 224-bit MNT elliptic curve from the PBC library. The experiments are conducted on Intel(R) Core(TM) i5-8500 3.00 GHz CPU with Ubuntu 12.04.1 LTS 64-bit environment and using Python 3.4 as programming language.

In order to evaluate the influence of access policies on performance, we set 30 different policies in the form of $(a_1, \cdots, a_n)$, where a_i represents the attributes and $1 \leq n \leq 100$. As shown in Fig. 3, the running time for outsourced key generation is about 100–856 ms. Furthermore, from Fig. 4(a) and (b), we can see that the time for outsourced encryption and decryption are increased close to

linearly with the growth of policy attributes. In Fig. 5(a) and (b), we further
compared the complexity of encryption and decryption on user side. The results
demonstrate that the user side computation time of our proposed PPAC scheme
is a little higher than that of [13]. The reason for this situation is that the
verification function will bring some extra computation cost.

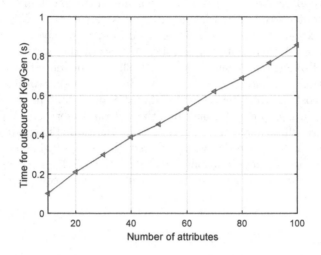

Fig. 3. Time for outsourced key generation

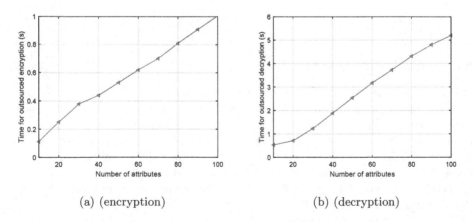

(a) (encryption)   (b) (decryption)

Fig. 4. Time for outsourced encryption and decryption

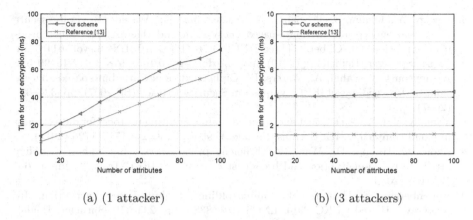

(a) (1 attacker)   (b) (3 attackers)

Fig. 5. Encryption and decryption time on user

6 Conclusion

Outsourced attributed-based encryption has performed its advantages on security data sharing in cloud computing. In this paper, we proposed a novel CP-ABE scheme in fog-cloud computing scenario, named PPAC, which contains the capabilities of verifiable outsourced encryption, decryption, and partially hidden access structure. The security analytics demonstrate that our presented PPAC scheme can achieve data confidentiality, data validation, and privacy protection simultaneously. Experimental results illustrate that PPAC can be effectively applied to the resource-constraint model devices. For the future, we will focus on integrating privacy-preserving techniques to machine learning services in the fog-cloud computing environment.

References

1. Shi, W., Cao, J., Zhang, Q., Li, Y., Xu, L.: Edge computing: vision and challenges. IEEE Internet Things J. **3**(5), 637–646 (2016)
2. Zhang, K., Mao, Y., Leng, S., He, Y., Zhang, Y.: Mobile-edge computing for vehicular networks: a promising network paradigm with predictive off-loading. IEEE Veh. Technol. Mag. **12**(2), 36–44 (2017)
3. Jalali, F., Hinton, K., Ayre, R., Alpcan, T., Tucker, R.S.: Fog computing may help to save energy in cloud computing. IEEE J. Sel. Areas Commun. **34**(5), 1728–1739 (2016)
4. Zhang, J., Chen, B., Zhao, Y., Cheng, X., Hu, F.: Data security and privacy-preserving in edge computing paradigm: survey and open issues. IEEE Access **6**, 18209–18237 (2018)
5. Sahai, A., Waters, B.: Fuzzy identity-based encryption. In: Cramer, R. (ed.) EUROCRYPT 2005. LNCS, vol. 3494, pp. 457–473. Springer, Heidelberg (2005). https://doi.org/10.1007/11426639_27

6. Lewko, A., Okamoto, T., Sahai, A., Takashima, K., Waters, B.: Fully secure functional encryption: attribute-based encryption and (hierarchical) inner product encryption. In: Gilbert, H. (ed.) EUROCRYPT 2010. LNCS, vol. 6110, pp. 62–91. Springer, Heidelberg (2010). https://doi.org/10.1007/978-3-642-13190-5_4

7. Bethencourt, J., Sahai, A., Waters, B.: Ciphertext-policy attribute-based encryption. In: Proceedings of IEEE Symposium Security Privacy (SP 2007), pp. 321–334 (2007)

8. Yang, K., Jia, X.: An efficient and secure dynamic auditing protocol for data storage in cloud computing. IEEE Trans. Parallel Distrib. Syst. **24**, 1717–1726 (2013)

9. Zhou, Z., Huang, D., Wang, Z.: Efficient privacy-preserving ciphertext-policy attribute based-encryption and broadcast encryption. IEEE Trans. Comput. **64**, 126–138 (2015)

10. Hohenberger, S., Waters, B.: Online/offline attribute-based encryption. In: Krawczyk, H. (ed.) PKC 2014. LNCS, vol. 8383, pp. 293–310. Springer, Heidelberg (2014). https://doi.org/10.1007/978-3-642-54631-0_17

11. Zhou, J., Cao, Z., Dong, X., Lin, X.: TR-MABE: white-box traceable and revocable multi-authority attribute-based encryption and its applications to multi-level privacy-preserving e-healthcare cloud computing systems. In: IEEE Conference on Computer Communications (INFOCOM 2015), pp. 2398–2406 (2015)

12. Rouselakis, Y., Waters, B.: Practical constructions and new proof methods for large universe attribute-based encryption. In: ACM SIGSAC Conference on Computer and Communications Security (CCS 2013), pp. 463–474 (2013)

13. Green, M., Hohenberger, S., Waters, B.: Outsourcing the decryption of ABE ciphertexts. In: Proceedings of USENIX Security Symposium (USENIX Security 2011) (2011)

14. Alderman, J., Janson, C., Cid, C., Crampton, J.: Access control in publicly verifiable outsourced computation. In: 10th ACM Symposium on Information, Computer and Communications Security (ASIACCS 2015), pp. 657–662 (2015)

15. Li, J., Huang, X., Li, J., Chen, X., Xiang, Y.: Securely outsourcing attribute-based encryption with checkability. IEEE Trans. Parallel Distrib. Syst. **25**(8), 2201–2210 (2014)

16. Ma, H., Zhang, R., Wan, Z., Lu, Y., Lin, S.: Verifiable and exculpable outsourced attribute-based encryption for access control in cloud computing. EEE Trans. Dependable Secure Comput. **14**(6), 679–692 (2017)

17. Xue, K., Hong, J., Ma, Y., Wei, D.S.L., Hong, P., Yu, N.: Fog-aided verifable privacy preserving access control for latency-sensitive data sharing in vehicular cloud computing. IEEE Network **32**, 7–13 (2018)

18. Krawczyk, H.: Cryptographic extraction and key derivation: the HKDF scheme. In: Rabin, T. (ed.) CRYPTO 2010. LNCS, vol. 6223, pp. 631–648. Springer, Heidelberg (2010). https://doi.org/10.1007/978-3-642-14623-7_34

19. Bellare, M., Garay, J.A., Rabin, T.: Fast batch verification for modular exponentiation and digital signatures. In: Proceedings of Advances in Cryptology (CRYPTO 2007), pp. 74–90 (2007)

20. Akinyele, J.A., et al.: Charm: a framework for rapidly prototyping cryptosystems. J. Cryptogr. Eng. **3**(2), 111–128 (2013)

Utility-Aware Edge Server Deployment in Mobile Edge Computing

Jianjun Qiu[1], Xin Li[1,2,3]($\boxtimes$), Xiaolin Qin[1], Haiyan Wang[4], and Yongbo Cheng[5]

[1] CCST, Nanjing University of Aeronautics and Astronautics, Nanjing, China
qiujj@nuaa.edu.cn, lics@nuaa.edu.cn
[2] State Key Laboratory for Novel Software Technology,
Nanjing University, Nanjing, China
[3] Collaborative Innovation Center of Novel Software Technology
and Industrialization, Nanjing, China
[4] Zhejiang Gongshang University, Hangzhou, China
[5] Nanjing University of Finance and Economics, Nanjing, China

Abstract. Traditional Mobile Cloud Computing (MCC) has gradually turned to Mobile Edge Computing (MEC) to meet the needs of low-latency scenarios. However, due to the unpredictability of user behaviors, how to arrange edge servers in suitable locations and rationally allocate the computing resources is not easy. Besides, the workload between the servers maybe unbalanced, which could lead to a shrinkage of system utility and waste of energy. So we analyze the workloads in a large MEC system and use one day to represent a workload cycle rotation. Combining the idea of differential workload changes with the local greedy method, we propose a new *Gradient* algorithm under the constraint of given limited computing capacity. We conduct extensive simulations and compared it with the algorithm based on the average workload as the Weight and the Greedy algorithm, which shows that the *Gradient* algorithm can reach the maximum utility compared with Weight and Greedy methods.

Keywords: Edge computing · MEC · Server deployment ·
Delay-sensitive

1 Introduction

It is an unprecedented challenge to guarantee the QoS while dealing with a massive amount of data for cloud computing, which has pushed the horizon of edge computing [1]. The environment of MEC is characterized by low latency, proximity, high bandwidth, and real-time insight into radio network information and location awareness [2]. By offloading computing tasks to physically close MEC servers, the quality of device energy consumption and execution latency, can be greatly improved [3,4]. In MEC systems, computing, network control, and storage of the cloud center are sunk to the edges that closer to the end user so as to reduce the transmission and corresponding delay, thereby achieving a higher

© Springer Nature Switzerland AG 2020
S. Wen et al. (Eds.): ICA3PP 2019, LNCS 11944, pp. 359–372, 2020.
https://doi.org/10.1007/978-3-030-38991-8_24

quality of service (QoS). MEC seems to solve the problem of delay perfectly, but due to cost constraints, it is impossible to place a large number of edge servers in many areas. Furthermore, small scale and decentralized deployment means that it cannot obtain the high equipment efficiency brought by scale advantages [5]. Besides the workload type and workload change of each area are also irregular. In MEC system, the first step to face is the deployment of the edge servers [6]. Meanwhile considering a large number of end users and the wide distribution of features, it is still not an easy work to deploy numerous physical servers under a limited cost constraint. Fortunately, thanks to the existing telecom network infrastructures [6], it is a feasible idea to deploy edge servers directly in existing telecom base stations such as macro BSs (Base Station) [7,8]. Meanwhile, it also provides a possibility for cooperation [9] and computation offloading [10,11] between edge servers. Undoubtedly, more cutting-edge MEC researches must be based on the premise that the edge server has been well installed. The problem of placement of the edge server is solved. The next step is how to properly allocate the computing capacity of the servers. Due to the unpredictability of user behaviors, how to arrange each edge server in a suitable location and rationally allocate the vacant computing resources after the user makes a request, thus ensuring that the request is processed within a certain period of time is not easy. As for the methods of deploying the edge servers, there has been some discussions in the academic field. For example, the two strategies described in the second section. However, none of the related studies have taken into account the characteristics of workload cyclical changes between edge servers, such as the workloads in different areas in the morning and evening. Based on this important characteristic, we analyzed the workload changes, comprehensively considered the maximum deployment utility, and proposed a *Gradient* algorithm to solve the deployment problem of edge servers.

In this paper, we focus on the MEC system, as shown in Fig. 1. The Edge servers can be placed in the BS. Then how to properly allocate the computing capacity of each server? EUs (End Users) can send requests directly to the edge servers that are closer to them so as to reduce latency, or they can request services from the remote cloud data center. Besides, we put together the base stations, which have similar workload change characteristics. For example, BSs near residential areas can be put together because the area have a much larger workload at night, and the workload will be small during the day due to going to work.

Based on the above scenario, how to reasonably allocate the computing capacity to maximize the overall utility is the main content of our research. And based on the characteristics of workload changes and local greedy methods, we propose a Gradient algorithm.

To represent the efficiency of the deployment of edge servers, we introduce utility to measure the results of task completion, which are executed under the servers that we installed, and we take utility maximization as a major objective for the server deployment problem. According to the delay requirement, we classify the tasks into two categories, delay-sensitive, and delay-tolerant. For the

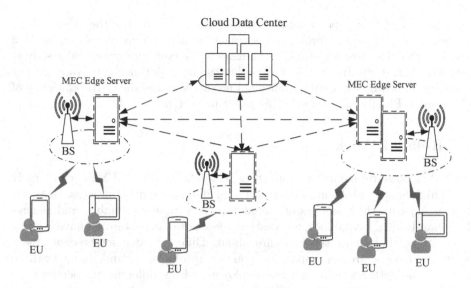

Fig. 1. System scenario

delay-sensitive tasks, we should return the feedback in time. That means we should try our best abilities to execute them in the local cloud so as to minimize the response time. On the other hand, the delay-tolerant tasks are treated differently. We can throw it directly into the remote cloud when the workload is merely tight. Hence, we define various utility functions based on task features. Naturally, for delay-sensitive tasks, we are eager to execute them in a short time, and we disgust it with a long time delay. So, it should get the most benefits can be obtained when they are forced to be sent to the remote cloud. The situation is not the same for the task of delay- tolerant. They can get the same utility whether they are executed in the local cloud or in the remote cloud since they don't care about their delays. Therefore, delay-sensitive tasks have a higher priority under some conditions, such as when the server's workload reaches a certain preset percentage value, which can be 80%, etc. When the workload on this local server reaches this percentage value, it refuses to accept the tasks of delay-tolerance and throws them directly into the remote cloud to ensure the benefits of delay-sensitive tasks.

In summary, we make the following contributions in this paper.

- We give a detailed description of the allocation of the edge server deployment, quantify some of the parameters and propose a specific method for measuring the effectiveness of deployment.
- We establish a model for computational resource-allocation strategy rationally. Meanwhile, a mathematical optimization method called Gradient method is used to give a way to maximize the total benefits.
- We conduct extensive experiments, the results show that our strategy has a significant performance compared with Weight and Greedy methods.

The rest of this paper is organized as follows. We review the related work in Sect. 2 and give the problem statement of server deployment in Sects. 3 and 4 introduces how to establish the model of server computational resource-allocation strategy. In Sect. 5, we conduct some experiments to validate our strategy and compare it with traditional methods and evaluate the efficiency of the results. Finally, we conclude this paper in Sect. 6.

2   Related Work

MCC has powerful computing and storage capability [12] and MEC can largely solve high latency problem in the traditional cloud computing process. However, how to allocate the resources of the edge servers more reasonably, and achieve high bandwidth, low latency, low cost, and low power consumption have become a hot topic that many scholars care about. Different edge server deployment solutions have a greater impact on computing efficiency, and many learn to conduct in-depth research on server deployment from different perspectives.

There is an optimal layout algorithm [13] based on the enumeration algorithm. The method proposes how to push services from the cloud center to the edge of the mesh to minimize the effect of average data transmission. Enumeration of K edge calculations in the edge mesh is enumerated by an enumeration algorithm. All deployment scenarios for the server, assessing the average data traffic for each placement case. In order to reduce the complexity, the author proposes to divide the deployment problems of K edge servers into K sub-problems, each of which will find the optimal location of an edge server.

The other method [14] is based on the limited server capacity, this paper studies how to allocate computing resources from the perspective of allocating mobile terminal requests to edge servers to achieve the ultimate goal of minimizing the average access latency of mobile terminals to service nodes, which is divided into two cases: When the number of edge servers is small, using integer programming to solve the problem, you can get a more accurate solution. When there are more edge servers, use the greedy algorithm. Besides, there are others methods for allocating resources [15–18].

However, none of the related studies have taken into account the characteristics of cyclical changes of the workloads, such as the workloads in different areas in the morning and evening. In this paper, we consider a cyclical change of workload and propose a new deployment method called Gradient method. Meanwhile, we measure the efficiency of the deployment algorithm by the total utility.

3   Scenario and Problem Analysis

3.1   Scenario and Preliminaries

We consider the edge server deployment problem with the scenario as shown in Fig. 1. The edge servers are deployed in the base stations, but the computing capacity of the specific deployment needs to be specified. We compare the efficiency of various deployment algorithms by the final utility. And we assume that the tasks that need to be executed are divided into two types, delay-sensitive and delay-tolerant tasks. They can be executed either in the edge cloud or in the remote cloud. Normally, we are eager to execute all of the tasks in a short time rather than with a long time waiting. However, Compared to the limited resources for computation in edge cloud, the remote cloud center is sufficient for providing users with rich enough resources. So, when the workload on this edge server reaches a percentage value, delay-sensitive tasks have higher priority than delay-tolerant tasks. The edge server refuses to accept the tasks of delay-tolerance and throws them directly into the remote cloud data center. In the traditional MEC system, we creatively divide a cycle (one day) into N intervals according to the characteristics of the workload cycle changes. And the workload of each interval may be different.

In order to evaluate the efficiency of computation capacity allocation, we calculate the overall utility. Next, we will give a description of the area, task, and utility function.

3.2 Area and Task Description

Our research is based on a large area with several sub-areas. A sub-area may contain some base stations. Moreover, among the M ($M >= 1$) sub-areas, the workload variation of each sub-areas may be different also. In order to distinguish the change of workload, we choose a day as the period of sub-area workload changes. Besides, we divide a period into N ($N >= 1$) intervals. Actually, this is reasonable enough, because there may be periodic changes in the local area in reality. To clearly describe the sub-areas, we use a tuple to characterize it as $\Lambda = <\Gamma, \psi>$, where

- Γ is the unique identifier for each sub-area;
- ψ is the set of workloads in N intervals of current sub-area;

As described in Sect. 1, all of the tasks can be divided into two categories, delay-sensitive and delay-tolerant. Delay-sensitive tasks must be executed in a short time, or it will get a terrible feedback. But for delay-tolerant tasks, there is no requirement for execution time. To clearly describe the task, we use a tuple to characterize it as $\Upsilon = <\xi, \psi, \Gamma, \chi>$, where

- ξ is the unique identifier for each task;
- ψ is the full utility of current task;
- Γ indicates which area current task belongs to;
- χ is used to identify the task is a delay-sensitive task or a delay-tolerant task. For the task, we have

$$\chi(\Upsilon_j) = \begin{cases} 1, & \text{if } \Upsilon_j \text{ is a delay-sensitive task;} \\ 0, & \text{if } \Upsilon_j \text{ is a delay-tolerant task.} \end{cases}$$

3.3 Utility Function

We do our best to maximize the utility for the deployment strategy, and also hope that all tasks can be done in the edge cloud so as to guarantee the shortest time. However, the local cloud server is just limited. So, when the workload on the edge server reaches a certain percentage value, we will abandon the tasks of delay-tolerance on the edge server so that delay-sensitive tasks occupy it only, thereby improving the overall utility. When different tasks are executed under different circumstances, naturally, different utilities will be obtained.

For the delay-sensitive tasks, the utility will drop a lot when they are executed in the remote cloud and they will get a full utility if they are executed in the local cloud. By this, we define the utility function for the delay-sensitive tasks as:

$$\eta(\Upsilon_j) = \begin{cases} \psi, & \text{if } \Upsilon_j \text{ is executed in local edge cloud;} \\ \alpha \cdot \psi, & \text{if } \Upsilon_j \text{ is executed in remote cloud.} (0 < \alpha < 1) \end{cases}$$

For the delay-tolerant tasks, it's very simple, there is no difference between the two cases because they don't care about their delays. So we can easily get the utility function for the delay-tolerant tasks as:

$$\eta(\Upsilon_j) = \psi.$$

3.4 Problem Statement

We aim to maximize the total utility in this whole system. First of all, we have to consider the problem of a sub-area utility. Because a sub-area contains N intervals in a period, and the cumulative sum of the utility of all intervals is the total utility of this sub-area. So, we can get the sub-area utility as:

$$h(\Lambda_k) = \sum_{j=1}^{N} \eta(\Upsilon_j).$$

Besides, the entire area contains M sub-areas, and the total utility of the entire area is accumulated by summing all the utility of each sub-area up. Moreover, maximizing this total utility is our goal. The objective can be represented as:

$$max. \quad \Omega = \sum_{k=1}^{M} h(\Lambda_k).$$

As described above, we should calculate the utility of each sub-area firstly. Assuming that the total computation capacity is limited to Φ. Besides, the computations capacity of the M areas have been assigned in order of $\lambda_1, \lambda_2, \lambda_3, ...\lambda_M$. Obviously the following equation exists.

$$\Phi = \lambda_1 + \lambda_2 + \lambda_3 + \cdots + \lambda_M$$

According to the description of the above problem, it can be preliminarily guessed that the edge server resource-allocation problem is an NP-hard problem. The specific proof is given below.

Theorem 1: The computational resource-allocation problem is NP-hard.

Proof: In the scenario we have stated above, we have M different sub-areas that need to be assigned suitable computing capacity under the limit of total computing capacity of Φ. Then we can show this case can be reduced from the knapsack problem.

For the knapsack problem, given a set of items $\cup B_i (1 \leq i \leq k)$, each with a weight ω_i and a value v_i, the problem is to select some of the items so that the total weight is less than or equal to Π and the total value is maximized. In our scenario, we constructed the set of areas $\cup \Lambda_k (1 \leq k \leq M)$, $\Pi = \Phi$, and the values generated by each sub-area change with the amount we allocate, which can be represented as $\eta(\Upsilon_j)$. And we aim to maximize the total utility by the strategy that we proposed. Then let $\eta(\Upsilon_j)$ be a constant value, then the problem degenerates into a normal knapsack problem. Since the general backpack is NP-hard, this promotion is also NP-hard. So, the computational resource-allocation problem is NP-hard.

4 Server Deployment Strategy

After giving the establishment of the scene model, parameterization and description of the problem, now we will consider the evaluation criteria and strategy of this problem. The first is the calculation of the individual task utility, and then the maximization of the overall utility.

4.1 Calculating Utility

When calculating the utility of a task, you should consider the type of the task firstly. According to the description of the utility functions above, the utilities of different tasks executed in different clouds are different naturally. In addition, as mentioned above, when the workload on this edge server reaches a percentage value $\varepsilon (0 < \varepsilon < 1)$, the edge server refuses to accept the tasks of delay-tolerance and throws them directly into the remote cloud center to ensure the utilities of delay-sensitive tasks. So, when considering whether the task is executed in the edge cloud or cloud center, we also need to think the workload of the current edge server.

In short, we should take the following three situations into consideration when calculating the utility of the task. The specific algorithm is shown in Algorithm 1 and explained as follows.

(1) When the edge server's workload ratio reaches ε and without overload. If it is a delay-sensitive task, the edge cloud accepts it (see Line 5). On the contrary, if it is a task of delay tolerance, the edge cloud rejects it and throws it directly to the cloud center in line 8. Although they are executed in different places, they can get the same full utilities at this time.

(2) When the edge server's workload ratio is less than ε. At this time, the edge server's workload is relatively light, so they are all accepted in line 11, and get the same full utilities.

(3) When the state of the edge server is overloaded. At this case, because the workload of the server is very tight, it will refuse all tasks to be executed on the edge cloud and throw them all to the center cloud in line 18. However, due to the different characteristics of the tasks, they get two completely different benefits. As for delay-tolerant tasks, they can get a full utility. However, it can only get α multiples of utility for delay-sensitive tasks.

Algorithm 1. calUtility()

Input: Υ_j: all tasks in entire area.

1: **for** *each* $\Lambda_k \in \Lambda$ **do**
2:    **for** *each* $\Upsilon_j \in \Lambda_k$ **do**
3:       **if** $workload(\Lambda_k) \geq 0.8$ **and** $workload(\Lambda_k) < 1$ **then**
4:          **if** $\chi(\Upsilon_j) = 1$ **then**
5:             $accept(\Upsilon_j)$;
6:             $\eta(\Upsilon_j) \leftarrow \psi$;
7:          **else**
8:             $refuse(\Upsilon_j)$;
9:             $\eta(\Upsilon_j) \leftarrow \psi$;
10:       **else if** $workload(\Lambda_k) < 0.8$ **then**
11:          $accept(\Upsilon_j)$;
12:          $\eta(\Upsilon_j) \leftarrow \psi$;
13:       **else**
14:          **if** $\chi(\Upsilon_j) = 1$ **then**
15:             $refuse(\Upsilon_j)$;
16:             $\eta(\Upsilon_j) \leftarrow \alpha \cdot \psi$;
17:          **else**
18:             $refuse(\Upsilon_j)$;
19:             $\eta(\Upsilon_j) \leftarrow \psi$;

4.2 Maximizing Utility

We have just completed the calculation of the utilities under the different conditions of two kinds of tasks. When calculating the total utility, we consider one of the sub-areas firstly. When we look at a sub-area, we will pay attention to the load of N intervals in one cycle. For example, supposing that current sub-area has been installed λ computing capacity. And we make the workload of this sub-area in order of $\delta_1, \delta_2, \delta_3, ...\delta_N$. Therefore, the benefit we can get is always the smaller one between the current workload and the computation capacity that we installed. Then, because a cycle is divided into N intervals, the result of adding N intervals is the total utility of this area. We should add the utilities of the M sub-areas to obtain the total utility of the entire area. Finally, We can measure the efficiency of the strategy of this computational resource-allocation by maximizing this total utility.

Based on the above descriptions, Let us quantify the utility of one of the sub-areas. So, for each sub-area, we can get the utility of each sub-area as:

$$h(\Lambda_k) = \sum_{i=1}^{N} min(\delta_i, \lambda_i).$$

Through the image of this function expression, it is not difficult to analyze that the utility of one sub-area is a piecewise function that varies with the amount of resources invested. And this function is divided into π segments, which represent the number of N intervals of workload values that are different in one cycle. Besides, with the amount of resources we invest, the utilities will gradually increase in π segments, but when the input reaches the maximum workload, the utilities will not change, which means that investing more resources is useless.

Bring the above function to the total utility function, we can easily get the total utility of the whole area as:

$$max. \quad \Omega = \sum_{k=1}^{M} \sum_{i=1}^{N} min(\delta_i, \lambda_i).$$

The specific algorithm description is shown in Algorithm 2. The basic idea is that we calculate the utilities of all the workloads of a sub-area in N intervals, and then add the utilities of the M areas to get the total utility of the whole system. Finally, maximize this value and get the final utility we need.

Algorithm 2. maxTotalUtility()

Input: $\cup\lambda$: the computations capacity of the M areas have been assigned in order of $\lambda_1, \lambda_2, ...\lambda_M$;

$\cup\delta$:the workload of current sub-area in N intervals in order of $\delta_1, \delta_2, ...\delta_N$.

1: **for** *each* $\Lambda_i \in \Lambda$ **do**

2:    **for** *each* $\delta_j \in \cup\delta$ **do**

3:        $h(\Lambda_k) \leftarrow h(\Lambda_k) + min(\lambda_i, \delta_j)$;

4: $max(h(\Lambda_k))$;

Obviously, the total utility function is composed of M piecewise functions. Therefore, under the limited total resource, there can be an allocation method that maximizes the value of this function. However, the computational resource-allocation problem is NP-hard as proved above. We use a heuristic algorithm called gradient algorithm to solve this problem.

4.3   Gradient Algorithm

To explain the gradient algorithm more clearly, let's simplify our scenario first. We divide the period of the sub-area into three intervals. Suppose that we already know the average workload data of a sub-area for these three intervals in order of 200, 300, 600. And The number of computing resources that we invested is x. We can get the utility as

$$h(\Lambda_k) = min(200, x) + min(300, x) + min(600, x).$$

So we can easily get the function expression image as Fig. 2.

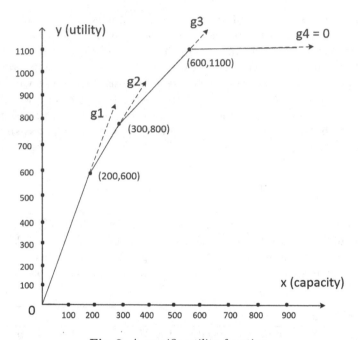

Fig. 2. A specific utility function

     Through the function expression, it is not difficult to analyze that the utility of one sub-area is a piecewise function that varies with the amount of resources invested. And this function is divided into π segments, which represent the number of N intervals of workload values that are different in one cycle. Besides, with the amount of resources we invest, the utilities will gradually increase in π segments, but when the input reaches the maximum workload, the utilities will not change, which means that investing more resources is useless.

     We can get the gradient of the function image of each segment as $g1$, $g2$, $g3$, $g4$. And it can be clearly seen that their gradients are decreasing in turn. When the resources that we invested exceed the maximum workload, the slope is 0. When the resource we invested is greater than or equal to the maximum load, the gradient is 0, which means that the utility does not change anymore. In our scenario, the entire area consists of M sub-areas. So the total utility is composed of M such functions which is shown in Fig. 3. In order to show the difference in utility growth at each segment, we used different colors to represent them. The

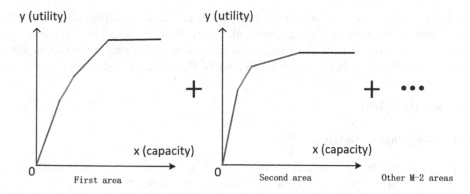

Fig. 3. Total utility composition (Color figure online)

red part indicates the fastest growth segment of the utility, followed by green, blue, and black. Therefore, we always invest in the red stage of each area first and then invest in green, blue and black parts.

Algorithm 3. calUtilityByGradient()

Input: $\cup \lambda$: the computations capacity of the M areas will be assigned in order of $\lambda_1, \lambda_2, ... \lambda_M$;
  $\cup \delta$: the workload of current sub-area in N intervals in order of $\delta_1, \delta_2, ... \delta_N$.
  $usedCapicity = 0$: indicates the currently allocated resource. The default is 0.
  Ω: Total computing capacity.
1: **for** *each* $\Lambda_i \in \Lambda$ **do**
2:   **for** *each* $\delta_j \in \cup \delta$ **do**
3:     **if** $usedCapicity < \Omega$ **then**
4:       $\lambda_i \leftarrow \delta_i$;
5:       $usedCapicity \leftarrow usedCapicity + \delta_i$;
6:       i++;
7:       **if** $i > M$ **then**
8:         $i \leftarrow 1$;
9:         j++;
10:      **else**
11:        break;
12:    **else**
13:      break;

Based on the above analysis, we propose a *Gradient* algorithm to allocate computation capacity. As the computation capacity invested increases, the gradient of the function curve gradually decreases, which means that the utility increases more and more slowly. According to this important feature, we prefer to invest resources to the part, which occupies the front smaller workload part that is more effective in increasing utility.

Specifically, how to calculate the computing resource algorithm allocated by each area by the *Gradient* algorithm is shown in Algorithm 3. The basic idea is that we invest the part of the computing resources, which can make the utility fastest growth. And then we invest the part of the gradient reduction.

5   Evaluation

5.1   Simulation Settings

There are other two algorithms for our approach: *Greedy* and *Weight* methods. The greedy algorithm refers to always making the best choice at the moment when solving the problem. That is to say, first consider the maximization of local utility, and do not consider the overall optimality. What it is doing is a local optimal solution in a certain sense. As for the *Weight* algorithm, there are three steps as follows:

- First, calculate the average workload of one cycle per sub-area and use it as the weight of the allocation computing capacity.
- Next, sum the average workload of the entire area up as the whole workload.
- Finally, the method of computing capacity is allocated according to the weight of the average workload of the sub-area to the whole workload of this area.

We randomly generated many different tasks in each area as the workload and selected three sets of representative experiments as three cases for analysis. In each of the three cases, the workload of each case changed dynamically with time. Meanwhile, we set $\alpha = 0.3$. Because execution in the remote cloud center will lead to higher response time, the utility will naturally be reduced a lot. We set $M = 10$, which means the whole system is divided into 10 sub-areas. $N = 3$, which roughly divides the workload of 24 h in one day into three stages. This is based on real-life scenes to divide one day into three stages: 0 h–8 h, 9 h–16 h, 17 h–24 h. It means morning, noon, and evening. Because the workload of different regions in these three time periods may change a lot relatively.

5.2   Result Analysis

The experimental results of the three cases are displayed in the Figs. 4, 5 and 6 respectively. And Fig. 4 shows the results on utilities with Φ different values. The x-axis is total computation capacity that we invested, while y-axis is the final utility. From the results we know that *Gradient* method achieves the maximum when there are more than 1700 capacities. Figure 5 shows that our algorithm always has better utilities than the other two algorithms. Although they all achieved the same utility in the end, this is because as the amount of computing resources invested increases, all the workload will be satisfied, and the allocation strategy is irrelevant. However there will be such a scenario just in the experiment, the cost constraint will definitely be considered in reality. The results of the third group are shown in Fig. 6. The *Gradient* algorithm has

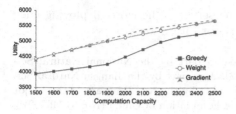

Fig. 4. Case I

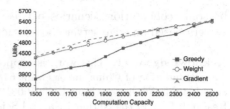

Fig. 5. Case II

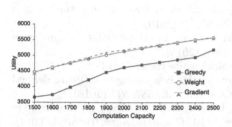

Fig. 6. Case III

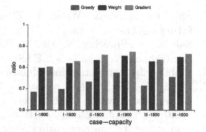

Fig. 7. Robustness of Algorithm

to be slightly hesitant to the Weight algorithm, both of which are much better than the *Greedy* algorithm. In general, in the three groups of experiments, the *Gradient* algorithm that we proposed always has the highest utility compared with the other two methods.

We also considered whether the *Gradient* algorithm has some robust properties, which means that whether the change of the workloads will affect the efficiency of the algorithm under a certain average workload condition. Figure 7 shows the results. This y-axis represents the ratio of the utility of an algorithm to the maximum utility that can be achieved in this case. The results show that our algorithm always has the best utility compared with the other two groups, and the utility ratio is stable above 0.8, which shows that our algorithm has a good robust feature.

6   Conclusion and Future Work

In this paper, we investigate the edge server deployment problem, quantify some of the parameters for measuring the efficiency of deployment and propose a *Gradient* algorithm. Meanwhile, we establish a model for computational resource-allocation strategy and conduct some extensive experiments, which shows that our algorithm contributes a lot on maximizing the total utility compared with other traditional algorithm and also has a good robust feature. The model we are based on is that there is no cooperation between the MEC edge servers. Actually, under the 4G/5G network architecture, there exists an X2 logical interface between the base stations for their communications. There may be some cooperation scenarios between the servers. Therefore, deploying edge servers based on edge server cooperation is our future work.

Acknowledgement. This work is supported in part by the National Natural Science Foundation of China under Grant 61802182, in part by the Jiangsu Natural Science Foundation under Grant BK20160813, in part by the Key Program of the Major Research Plan of the National Natural Science Foundation of China (No. 91746202).

References

1. Shi, W., Zhang, X., Wang, Y., Zhang, Q.: Edge computing: state of the art and future directions. J. Comput. Res. Dev. **56**, 69–89 (2019)
2. Sabella, D., Sprecher, N., Hu, Y., Patel, M., Young, V.: Mobile edge computing: a key technology towards 5G. ETSI White Pap. (2015)
3. Mobile-edge computing-introductory technical white paper (2014)
4. Chen, M., Hao, Y.: Task offloading for mobile edge computing in software defined ultra-dense network. IEEE J. Sel. Areas Commun. **36**, 587–597 (2018)
5. Mao, Y., You, C., Zhang, J., Huang, K., Letaief, K.B.: A survey on mobile edge computing the communication perspective. IEEE Commun. Surv. Tutorials **19**(4), 2339–2340 (2017). Fourth Quarter
6. Li, Y., Wang, S.: An energy-aware edge server placement algorithm in mobile edge computing. In: IEEE International Conference on Edge Computing, July 2018
7. Andrews, J.G., Claussen, H., Dohler, M., Rangan, S., Reed, M.C.: Femtocells: past, present, and future. IEEE Commun. Mag. **30**(3), 497–508 (2012)
8. Dhillon, H.S., Ganti, R.K., Baccelli, F., Andrews, J.G.: Modeling and analysis of k-tier downlink heterogeneous cellular networks. IEEE J. Sel. Areas Commun. **30**(3), 550–560 (2012)
9. Hu, X., Wong, K., Yang, K.: Wireless powered cooperation-assisted mobile edge computing. IEEE Trans. Wireless Commun. **17**(4), 2375–2388 (2018)
10. Mach, P., Becvar, Z.: Mobile edge computing: a survey on architecture and computation offloading. IEEE Commun. Surv. Tutorials **19**(3), 1628–1656 (2017). Third Quarter
11. Fan, W., Liu, Y., Tang, B.: Computation offloading based on cooperations of mobile edge computing-enabled base stations. IEEE Access **6**, 22622–22633 (2017)
12. Li, X., Lian, Z., Qin, X., Abawajy, J.: Delay-aware resource allocation for data analysis in cloud-edge system. In: IEEE International Symposium on Parallel and Distributed Processing with Applications (2018)
13. Zhao, L., Liu, J., Shi, Y., Sun, W., Guo, H.: Optimal placement of virtual machines in mobile edge computing. In: IEEE Global Communications Conference (2017)
14. Xu, Z., Liang, W., Xu, W., Jia, M., Guo, S.: Efficient algorithms for capacitated cloudlet placements. IEEEE Trans. Parallel Distrib. Syst. **27**, 2866–2880 (2016)
15. Wang, Z., Zhao, Z., Min, G., Huang, X., Ni, Q., Wang, R.: User mobility aware task assignment for mobile edge computing. Future Gener. Comput. Syst. **85**, 1–8 (2018)
16. Yin, H., Zhang, X., Liu, H., Luo, Y., Tian, C., Zhao, S., Li, F.: Edge provisioning with flexible server placement. IEEE Trans. Parallel Distrib. Syst. **28**, 1031–1045 (2016)
17. Zhang, H., Lin, X., Ji, H., Li, X., Wang, K.: MEC server selection algorithm based on fairness. IEEE Trans. Parallel Distrib. Syst. (2017)
18. Zeng, F., Ren, Y., Deng, X., Li, W.: Cost-effective edge server placement in wireless metropolitan area networks. Sensors **19** (2018)

Predicting Hard Drive Failures for Cloud Storage Systems

Dongshi Liu, Bo Wang, Peng Li, Rebecca J. Stones, Trent G. Marbach, Gang Wang, Xiaoguang Liu, and Zhongwei Li[✉]

Nankai-Baidu Joint Laboratory, College of Computer Science, Nankai University, Tianjin, China
{liudongshi,wangb,lipeng,rebecca.stones82,trent.marbach,wgzwp, liuxg,lizhongwei}@nbjl.nankai.edu.cn

Abstract. To improve reactive hard-drive fault-tolerance techniques, many statistical and machine learning methods have been proposed for failure prediction based on SMART attributes. However, disparate datasets and metrics have been used to experimentally evaluate these models, so a direct comparison between them cannot readily be made.

In this paper, we provide an improvement to the Recurrent Neural Network model, which experimentally achieves a 98.06% migration rate and a 0.0% mismigration rate, outperforming the state-of-the-art Gradient-Boosted Regression Tree model, and achieves 100.0% failure detection rate at a 0.02% false alarm rate, outperforming the unmodified Recurrent Neural Network model in terms of prediction accuracy. We also experimentally compare five families of prediction models (nine models in total), and simulate the practical use.

1 Introduction

Modern cloud storage systems and other large-scale data centers often host hundreds of thousands of hard drives as their primary data storage device. While the theoretical annual failure rate of a single hard drive is low, in such large numbers they are a primary source of failure in today's cloud storage systems [22,23]. Hard-drive failure leads to service unavailability, which negatively impacts the user experience, and can cause permanent data loss.

Modern hard drives incorporate Self-Monitoring, Analysis and Reporting Technology (SMART) [1], but SMART attributes cannot directly provide satisfactory hard-drive failure prediction performance [15]. As a result, many statistical and machine learning methods utilize SMART attributes to substantially improve upon hard-drive failure prediction performance [2,4–16,18–21,24–31]. However, differences in experimental setups make it hard to compare their respective performances and determine which model is most effective.

Difficulties comparing model performance hinder the realistic application in cloud storage systems. Two major differences are the choice of experimental dataset and the choice of experimental metric. A wide variety of datasets have

© Springer Nature Switzerland AG 2020
S. Wen et al. (Eds.): ICA3PP 2019, LNCS 11944, pp. 373–388, 2020.
https://doi.org/10.1007/978-3-030-38991-8_25

Table 1. Datasets used in previous work

Dataset		Reference(s)
No. drives	No. failed drives	
1,936	9	[8]
3,744	36	[9]
369	191	[14, 15, 20, 24–26, 30]
23,395	433	[10–12, 16, 28, 31]
38,989	170	[11, 12, 28]
10,157	147	
Backblaze data center datasets		[3, 4, 13]

been used; we tabulate them in Table 1 below[1]. Basic statistics of the datasets in this paper are listed in Table 3.

Most prior work [2, 4, 5, 7–10, 12–15, 18–21, 24–26, 30, 31] uniformly treated hard-drive failure prediction as a binary classification problem, and evaluated the model performance in terms of *failure detection rate* (FDR), defined as the proportion of failed drives that are correctly classified as failed, and *false alarm rate* (FAR), defined as the proportion of good drives that are incorrectly classified as failed. Some previous work [10, 12, 20, 24, 25] also incorporate *time in advance* (TIA), which is defined as the mean time between predicted failure and actual failure.

Instead of binary classification, some prediction models [16, 28] predict the residual life of hard drives, described by a drive's *health degree*. In this context, a drive's residual life is ordinarily predicted to fall into an interval, and prediction accuracy is measured by the number of predictions falling into the correct interval.

Li et al. [11] recently proposed two new performance metrics for hard drive failure prediction models: the *migration rate* (MR), defined as the proportion of data that is successfully migrated before its disk failed, and the *mismigration rate* (MMR), defined as the proportion of data on healthy disks that is migrated needlessly. Along with the traditional metrics (FDR, FAR, and TIA), we make use of these stricter metrics in this paper.

Four other significant issues that hinder model comparison are the following: (a) small datasets may be insufficient for adequately training the models, potentially leading to under-evaluated prediction performance results [8, 9, 14, 15, 24–26, 30] and contain too few failed drives, which negatively impacts both training and experimentation:

---

[1] Here and throughout the paper, despite the grammatical mismatch between "good" vs. "failed", for brevity we use "failed" as an adjective to describe hard drives which fail during data collection; all other hard drives are "good". This awkward nomenclature is consistent with many papers on this topic.

... detailed studies of very large populations are the only way to collect enough failure statistics to enable meaningful conclusions—Pinheiro et al., 2007 [17];

(b) different authors have chosen varying sets of SMART attributes to include and exclude in their experimental evaluations; (c) partitioning drives into test, training, and (possibly) validation sets has been done in various ways; and (d) some datasets [14,15,20,24–26,30] can be regarded as obsolete: their SMART information is inconsistent with the current SMART standard.

The main contributions of this paper are as follows:

- *Recurrent Neural Network Model Improvements.* We optimize the Recurrent Neural Network (RNN) model in terms of MR and MMR. We define a four-layer network model in which the output layer contains two additional nodes (MR and MMR). Further, during training, samples are instead considered over the entire life of a typical hard drive.
- *Nine models from five families.* We experimentally compare the performance of nine hard-drive failure prediction models on the same datasets collected from a real-world data center, using metrics MR and MMR, along with the traditional metrics FDR, FAR, and TIA.
- *Data center simulations.* We continue experimentation on six reasonable models by simulating their use on various drive families, in small-scale data centers and data centers with a mixture of drive models.

The paper is organized as follows: In Sect. 2, we survey related work on hard-drive failure prediction using SMART attributes. Section 3 presents the modified RNN model. Section 4 first gives a description of the datasets and how they are preprocessed, then presents the experimental results. We summarize the impact of this work in Sect. 5.

2   Related Work

SMART is a monitoring system which is widely used in modern hard drives. However, a simple SMART threshold-based method for hard-drive failure prediction results in an impractically poor FDR of around 3–10% when achieving a suitably low FAR around 0.1% [15]. Here we survey the methods used to overcome this problem.

The majority of prior work considered hard-drive failure prediction as a binary classification problem, including Bayesian approaches [4,8], the Wilcoxon rank-sum test [9,14,15], a support vector machine method [15,31], hidden Markov models [30], a method involving Mahalanobis distance [24–26], back-propagation artificial neural networks [31], a classification tree model [10], a regularized greedy forest model [2], a Gaussian mixture model [20], an online random forest model [27], and a method using the FastTree algorithm [29].

Realistically, hard drives deteriorate gradually, so some previous work instead studied "health degree" prediction. Li et al. [10] proposed a Regression Tree model and defined a hard drive's health degree as its failure probability. Pang et al. [16] and Xu et al. [28] treated health degrees as the remaining working time of a hard drive before an actual failure occurs. The accuracy of health degree prediction was used to test a combined Bayesian network model [16] and a recurrent neural network model [28]. Li et al. [11,12] improved the gradient-boosted regression tree (GBRT) method for hard-drive failure prediction.

From among these various methods, we select the most competitive to compare with our modified RNN model experimentally, which we describe in detail in the next section. Except for [11], all prior experimental evaluations did not incorporate data migration, which is a more critical factor for large-scale storage systems, such as cloud storage systems. One of the major motivations of this paper is thus to reevaluate these methods in an identical and up-to-date setting.

3   Extended RNN Model

We modify the network structure as depicted in Fig. 1. In [28], the previous time step's hidden layer is fed into the current hidden layer, whereas we feed in not only the previous time step's hidden layer but also the previous time step's output layer; this is a main distinction between the two methods.

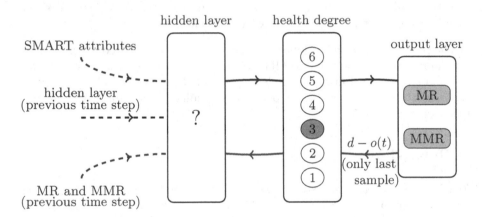

Fig. 1. The modified RNN model at time t; the SMART attributes at time t, hidden layer and output from time $t-1$ is fed into the hidden layer. It has four layers: an input layer, a hidden layer, a health-degree layer, and an output layer. We proceed drive by drive, then SMART-attribute sample by sample. MR is calculated from failed drives, and MMR is calculated from good drives.

When training RNN models, we incorporate the entire life of hard drives in the training set. The health-degree layer uses the softmax function to ensure the values of the six nodes form a valid probability distribution (i.e., all values are greater than 0 and their sum is 1). Each node's label represents the residual-life level a sample maps into. In the training process, for each sample we choose the

---

Algorithm 1. Modified RNN model training procedure

Input: samples for all hard drives, a four-layer RNN model with initial weight matrices
Output: network weight matrices
1: **for** hard drive D **do**
2:     **for** each SMART sample for hard drive D **do**
3:         compute the hidden layer and health degree probabilities (health-degree layer) as per an unmodified three-layer RNN as per [28], subjoining the previous time step's output layer as input
4:         find the node in the health-degree layer with the maximum value (i.e., probability), and calculate MR or MMR of the current SMART sample
5:         **if** not the last sample of a hard drive **then**
6:             update the MR and MMR output $o(t)$ at the current time step t
7:             feed back as unmodified three-layer RNN as per [28] (i.e., excluding $d-o(t)$), subjoining the previous time step's output layer
8:         **else**
9:             feed back using the four-layer RNN and reset MR and MMR (i.e., including $d - o(t)$)
10:         **end if**
11:     **end for**
12: **end for**
13: **return** network weight matrices

---

node with the maximum value (i.e., the residual-life level a sample most likely maps into) in the health-degree layer. Then we calculate MR or MMR at this time step assuming the migration rate in the corresponding level as in [11], and update the MR or MMR value until the last sample of a hard drive.

When feeding back into the network, for each sample, we feed back the network excluding the output layer as in [28], subjoining the previous time step's output layer. If it is the last sample of a hard drive, we also feed back the gradient of the output layer and then reset MR and MMR to 0. The gradient of the output layer is $d - o(t)$ where $o(t)$ is the assessed MR or MMR value at time t, and d is the target values of MR and MMR, namely 1 and 0, respectively. Algorithm 1 gives the details for training the modified RNN prediction model.

When testing, for each SMART record, we feed forward the network using weight matrices obtained from the training process. If the node with the maximum value in the health-degree layer falls into a level 1 through 5, the record is labeled as failed, otherwise good. Then we calculate FDR, FAR, and TIA like other binary classifiers. Metrics MR and MMR are calculated as in the training process.

4 Experimental Results

We test the modified RNN model and eight other hard-drive failure prediction methods, which divide into five categories, tabulated in Table 2. We evaluate the various hard-drive failure prediction methods in terms of FDR, FAR, TIA, MR, and MMR, on the datasets described in Sect. 4.1. When measuring FDR,

Table 2. Models we evaluate in this paper

Model family	Name	Year	Reference(s)
Probabilistic	Naive Bayes classifier	2001	[8]
	Bayesian network	2016	[4]
Support Vector Machine (SVM)		2005+	[15,31]
Decision Tree	CT (part of CART)	2014	[10]
	RT (part of CART)		
Boosting	GBRT	2016+	[11,12]
	XGBoost	New	
Time series	HMM	2010	[30]
	RNN	2016	[28]

FAR, and TIA, we apply a voting-based failure detection algorithm [31]. When measuring MR and MMR, we process the samples sequentially for each hard drive, like in [11].

4.1   Dataset Description and Preprocessing

Datasets and Preprocessing. Our datasets are from two real-world data centers. The data from the first data center, called dataset W, was released in [31]. The data from the second data center, called datasets M and S, were first used in [28]. The details of the three datasets are listed in Table 3.

Table 3. Dataset statistics

Dataset	Class	No. disks	No. samples
W	Good	22,962	3,837,568
	Failed	433	158,150
S	Good	38,819	5,822,850
	Failed	170	97,236
M	Good	10,010	1,681,680
	Failed	147	79,698

We use three non-parametric statistical methods—reverse arrangement test, rank-sum test, and z-scores [15] to select features as the SMART attributes in our datasets are non-parametrically distributed (which is consistent with the observations in [9,15]). We list the selected features for dataset W in Table 4 and for datasets M and S in Table 5. The differences between SMART attributes selected in Tables 4 and 5 are because drives of data sets W and M, S are from different data centers and drive models, with different collected SMART attributes. We divide the three datasets, taking 70% of data as training set, 15% as validation set, and 15% as test set.

When using the RNN model, we map all input data to $[0, 1]$, which we do by replacing the original value x of a feature by

$$x \mapsto \frac{x - x_{\min}}{x_{\max} - x_{\min}}$$

where $x_{\max}$ and $x_{\min}$ are the maximum and minimum values of this feature in the training set, respectively. To leverage the relatively long sequence of historical information of SMART attributes, we sample one SMART record in each 24-h period, consistent with [28].

Table 4. SMART features for dataset W

ID	Attribute name	Type
1	Raw Read Error Rate	Basic, change rate
2	Spin Up Time	Basic
3	Reallocate Sectors Count	Basic
4	Seek Error Rate	Basic
5	Power on Hours	Basic
6	Reported Uncorrectable Errors	Basic
7	High Fly Writes	Basic
8	Temperature Celsius	Basic
9	Hardware ECC Recovered	Basic, change rate
11	Reallocated Sectors Count (raw value)	Basic, change rate

Table 5. SMART features for datasets M and S

ID	Attribute name	Type
1	Raw Read Error Rate	Basic, change rate
2	Spin Up Time	Basic
3	Reallocate Sectors Count	Basic, change rate
4	Seek Error Rate	Basic
5	Power on Hours	Basic
8	Temperature Celsius	Basic
10	Current Pending Sector Count (raw value)	Basic

4.2   Dataset W

For dataset W, Fig. 2 plots the FDR and FAR using voting-based failure detection of each of the 9 methods. We see that increasing the number of voters generally reduces the FAR. Based on this experiment, in the subsequent experiments, we set the number of voters $N = 7$; there does not appear to be any significant benefit to choosing a greater N value. Not included in the figure are the TIA measurements; for $N = 7$, they ranged from 217 h (HMM) to 264 h (SVM), none of which would be problematic in practice. Excluding the naive Bayes classifier and the HMM model, all of the models give 24+ hours warning for 90%+ of the drives and 72+ hours warning for 84%+ of the drives.

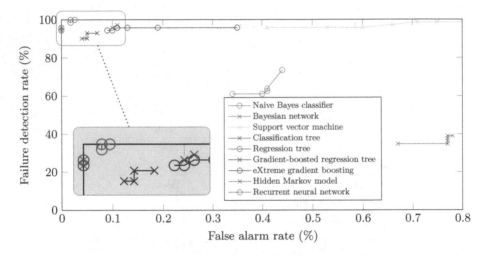

Fig. 2. Impact of the voting-based method on prediction accuracy (FDR vs. FAR) of the various prediction models, as the number of voters N varies in $\{1, 3, 5, 7, 11, 17, 27\}$ as in [10]; dataset W. The FAR generally decreases as N increases, so the plots are from right (small N) to left (large N). Some plots appear to have fewer than seven points because the FDR and FAR does not always vary with N in these experiments.

In Fig. 2, we see that the results of the Bayesian network model outperforms the naive Bayes classifier, indicating that SMART attributes not following Gaussian distributions, which agrees with the observations in [9,15].

Comparing the decision-tree and boosting models (CT, RT, GBRT, and XGBoost) using Fig. 2, we observe that all four consistently achieve high FDR (93%+ when $N = 7$). Moreover, CT, RT, and GBRT achieve a small false alarm rate (when $N = 7$, the maximum observed is in the CT model at 0.11%).

In Fig. 2, the SVM model achieves 97.22% FDR with 0.66% FAR (when $N = 7$), which is significantly better than the results observed in [15,31]. We attribute this discrepancy primarily to two factors: a different choice of SMART attributes than in [31], and a much larger dataset than in [15] (369 hard drives; 191 failed).

These results highlight the importance of re-testing the various methods on an "equal playing field".

The results for the HMM are the worst among those in Fig. 2; this level of prediction accuracy would render it impractical for use in a real-world setting. Curiously, if we only use the SMART attribute "Raw Read Error Rate" to build the HMM, the model achieves 90.27% FDR at 0% FAR, outstripping its results in Fig. 2. This result indicates that the HMM is better suited to a small set of attributes and performs poorly when using multi-dimensional attributes. When we use the SMART attribute "Reallocate Sectors Count", HMM only achieves 36.11% FDR at 0.02% FAR, which illustrates HMM does not always achieve high FDR when using a single SMART attribute, so its prediction accuracy varies according to different choice of SMART attributes.

The modified RNN model consistently achieves both a high FDR of 100.0% and a low FAR of 0.02% (when $N = 7$), and Fig. 2 indicates it outperforms all models except the Bayesian network and RT models, which achieve a fractionally better FAR of 0.00%, but a fractionally worse FDR of 96% (when $N = 7$).

We further compare the modified RNN model with the unmodified model in Table 6. The modified RNN model outperforms the unmodified RNN model in terms of both prediction accuracy and data migration. The MR increases by more than 18% points. When data fails to migrate, there is a heightened risk of data loss: it takes fewer additional failures to lose data permanently. Furthermore, in erasure-coded cloud storage systems, if migration is incomplete before a hard drive fails, we trigger reconstruction, consuming system resources.

Table 6. The unmodified vs. modified RNN model, along with the RT, GBRT and XGBoost models

Model	FDR (%)	FAR (%)	TIA (h)	MR (%)	MMR (%)
Unmodified RNN	95.83	0.03	255	79.92	0.01
Modified RNN	**100.0**	0.02	**263**	**98.06**	**0.00**
RT	95.83	**0.00**	262	91.31	0.15
GBRT	93.06	0.05	259	94.62	0.07
XGBoost	95.83	0.11	262	94.65	0.03

Experimental results for the RT, GBRT, XGBoost, and RNN models are included in Table 6. The GBRT and XGBoost models outperforms the RT model using the data migration metrics (MR and MMR), but the opposite is true for prediction accuracy metrics (FDR, FAR, and TIA). We also observe that the modified RNN model outperforms the RT, GBRT, and XGBoost models in terms of MR and MMR.

The motivation behind introducing MR and MMR in [11] was that FDR, FAR, and TIA are sometimes misleading in practice, and these observations further support this claim. High prediction accuracy does not necessarily imply more appropriate data migration is taking place, and thus does not necessarily imply a more reliable system.

4.3 Simulating Practical Use

We evaluate model performance by simulating their practical use in real-world data centers: (a) being used with different hard drive families, (b) being used in small-scale datasets, and (c) being used with a mixture of drive models. We exclude the naive Bayes classifier and the HMM, due to their poor performance. We also abandon the SVM model due to its higher FAR on dataset W compared with the remaining six models.

Table 7. Prediction and migration accuracy on datasets M and S

Model	M		S	
	FDR (%)	FAR (%)	FDR (%)	FAR (%)
Bayesian network	95.56	0.69	92.16	0.36
CT	95.45	0.57	96.15	0.52
RT	93.33	0.74	94.11	0.13
GBRT	86.36	0.15	84.62	0.06
XGBoost	95.45	0.55	96.15	0.14
RNN	**100.0**	**0.02**	**100.0**	**0.01**
Model	M		S	
	MR (%)	MMR (%)	MR (%)	MMR (%)
RT	95.45	0.23	91.00	0.04
GBRT	91.76	0.09	93.08	**0.02**
XGBoost	95.45	0.12	93.33	0.07
RNN	**98.75**	**0.03**	**98.58**	**0.02**

Datasets M and S. Hard drive models, manufacturers, and other environmental factors influence the statistical behavior of failures [22]. Even if made by the same manufacturers, different hard drive models have different characteristics, which may influence their reliability. Therefore, effectiveness on various hard drive models is an important factor in prediction models. With this motivation, we evaluate the six remaining models on datasets M and S, whose hard drive models are different to dataset W. Experimental results are tabulated in Table 7.

Comparing Table 7 to Fig. 2, we observe changes in the FAR for the Bayesian network model (from 0.00% to 0.36%+), the CT model (from 0.11% to 0.52%+), and the RT model (from 0.00% to 0.13%+). We also observe that the GBRT model has an FDR of around 93% for the W dataset (FAR 0.05%), whereas it is

Table 8. Prediction and migration accuracy on synthetic small-scale datasets

Model	A		B		C	
	FDR (%)	FAR (%)	FDR (%)	FAR (%)	FDR (%)	FAR (%)
Bayesian network	98.61	0.76	98.61	0.08	98.61	0.82
CT	98.61	0.38	97.22	0.20	98.61	0.13
RT	97.22	0.31	95.83	0.04	95.83	**0.00**
GBRT	76.39	**0.01**	84.72	**0.02**	90.28	0.04
XGBoost	100.0	0.08	97.22	0.08	95.83	0.12
RNN	**100.0**	**0.02**	**100.0**	0.04	**100.0**	0.03
Model	A		B		C	
	MR (%)	MMR (%)	MR (%)	MMR (%)	MR (%)	MMR (%)
RT	84.82	0.79	91.74	0.04	92.70	0.09
GBRT	93.18	0.66	92.99	0.07	94.31	0.06
XGBoost	89.37	0.12	90.40	**0.01**	91.14	**0.01**
RNN	**98.06**	**0.40**	**97.89**	0.24	**98.06**	0.30

around 86% for the M dataset (FAR 0.15%) and around 85% for the S dataset (FAR 0.06%). On the M and S datasets, the modified RNN model outperforms other models according to the metrics FDR, FAR, MR, and MMR.

Small-Scale Datasets. The datasets W, S, and M all contain a large number of hard drives. However, in real-world data centers, prediction models may be used on small or medium-sized datasets. We compare the remaining models on three "synthetic" datasets, like in [10, 11], named A, B, C, by randomly choosing 10%, 25%, and 50% of all the good and failed hard drives respectively from dataset W. Table 8 tabulates the experimental results on these small-scale datasets.

In Table 8, we observe only minor performance degradation as the size of the dataset decreases, although the FDR for GBRT drops from around 90% to around 76%. The modified RNN model outperforms the others in terms of FDR and MR, while the metrics FAR and MMR do not strongly favor a method.

Non-ideal Datasets. A real-world data center often has many engine rooms containing multiple hard-drive models. Though building a distinct prediction model for each hard drive model would achieve better results, this is impractical due to the time spent on data collection. To experimentally evaluate a non-ideal setup, we simulate two situations that might arise in a data center:

1. we have a large number of drives in the same drive family, together with different drive families whose data are insufficient for building models; and
2. we have multiple drive families with individually insufficient data, but together provide sufficient data.

For the first case, we build models using the dataset M and test model performance using dataset S (denoted M→S) or vice versa (denoted S→M). For the

Table 9. Prediction and migration accuracy for M→S and S→M

Model	S→M		M→S	
	FDR (%)	FAR (%)	FDR (%)	FAR (%)
Bayesian network	**100.0**	99.63	**100.0**	99.25
CT	77.27	92.79	69.23	67.99
RT	77.27	92.56	50.00	8.61
GBRT	77.27	76.55	61.54	23.72
XGBoost	77.27	92.78	73.08	92.03
RNN	**100.0**	**0.20**	**100.0**	**0.00**
Model	S→M		M→S	
	MR (%)	MMR (%)	MR (%)	MMR (%)
RT	77.29	19.13	56.01	5.94
GBRT	77.27	29.17	69.56	7.12
XGBoost	77.29	19.08	60.08	6.44
RNN	**98.75**	**0.16**	**99.78**	**0.00**

second case, we create a mixed dataset (denoted MS) by merging 25% of hard drives from the M and S datasets. We build models using the dataset MS and test model performance using datasets M, S, and MS. The results are denoted MS→S, MS→M, and MS→MS. We do not use dataset W here because the number of SMART attributes is inconsistent with the datasets M and S.

Table 9 tabulates the experimental results for M→S and S→M, and Table 10 tabulates the experimental results for MS→M, MS→S, and MS→MS.

In Table 9, the proposed RNN model is the only model which consistently achieves practicable performance for all metrics, other models all have impractically high FAR and high MMR. This may because RNN utilizes the long-term dependencies among SMART attributes and builds models according to the historical fluctuations of data, whereas the other five models are all based on numerical values. Though the data in datasets M and S have numerical differences, they may have similar historical fluctuations.

Since HMM also utilizes historical fluctuations of data, we perform an additional test for this model: we observe that HMM achieves 34.62% FDR at 6.31% FAR for M→S and 68.18% FDR at 0.00% FAR for S→M. We likewise test the unmodified RNN model, which achieves 76.47% FDR at 0.06% FAR for M→S and 37.78% FDR at 53.27% FAR for S→M. These observations are somewhat consistent with the hypothesis that long-term dependencies are responsible for the observations in Table 9, but there may be additional reasons for these results.

Importantly, the results in Table 9 strongly indicate how a hard-drive failure model trained for one drive model may be unusable for predicting failures in another model, and how an idealized experimental environment may exaggerate

Table 10. Prediction and migration accuracy for dataset MS

Model	MS→M		MS→S		MS→MS	
	FDR (%)	FAR (%)	FDR (%)	FAR (%)	FDR (%)	FAR (%)
Bayesian network	95.56	0.77	92.16	0.34	94.79	0.45
CT	86.36	0.65	92.31	0.42	89.58	0.49
RT	90.91	1.09	92.31	0.77	91.67	0.80
GBRT	86.36	**0.16**	84.62	0.21	83.33	0.16
XGBoost	86.36	0.43	92.31	0.20	89.58	0.22
RNN	**100.0**	0.44	**100.0**	**0.00**	**100.0**	**0.01**
Model	MS→M		MS→S		MS→MS	
	MR (%)	MMR (%)	MR (%)	MMR (%)	MR (%)	MMR (%)
RT	90.93	0.26	88.20	0.04	89.48	0.14
GBRT	88.91	**0.16**	88.33	0.09	88.60	0.11
XGBoost	90.92	0.24	92.92	0.05	91.39	0.13
RNN	**100.0**	0.44	**100.0**	**0.00**	**99.22**	**0.00**

the effectiveness of hard-drive failure prediction. In these results, the difference is extreme: going from nearly 0% FAR to nearly 100% FAR.

The results of all six models for the experiments MS→S and MS→M in Table 10 are similar to or slightly worse than the corresponding results in Table 7 (where we use training and test data from the same dataset). These results indicate simply creating a mixed training set is a practical method to overcome the problem arising in Table 9. All six models have practicable results for MS→MS, yet again we see the proposed RNN model outperforming the others.

5 Conclusion

In this paper, we implement a modified RNN model, and evaluate eight other models from five families on real-world datasets using various metrics experimentally. As a result, we give a fairer comparison between the models, which shows that the modified RNN model consistently achieves the best or nearly the best experimental results among the methods studied.

While experiments mostly give rise to comparable results to their original authors, for the support vector machine method, we observe a far higher prediction accuracy than presented by their authors in [15, 31], which we attribute to their small-size dataset and a different choice in SMART attributes.

The traditional metrics (FDR, FAR, and TIA) can sometimes suggest that one method outperforms another, while after incorporating migration (MR and MMR), the opposite is true. An example of this is the surprising observation that the RT method outperforms the GBRT and XGBoost methods on the traditional metrics, but in Table 6, we see that GBRT and XGBoost methods outperform the RT method in terms of MR and MMR.

In Sect. 4.2, we make the curious observation about how, for the hidden Markov model, prediction accuracy changes wildly depending on the selection of SMART attributes. In most work on this topic (including this paper), the authors make a selection of SMART attributes they consider most suitable. It would be interesting to expand this work to include the impact of the choice of SMART attributes, which we observe is significant in the example above.

We put forward the following advice when evaluating hard-drive failure prediction in cloud storage system:

– *Evaluation metric selection.* We observe that a high FDR does not necessarily imply a high MR. In a cloud storage system and other large-scale storage systems, we need to continuously migrate at-risk data, thereby consuming system resources. Thus migration-based metrics, such as MR and MMR, are better suited for evaluating model performance for cloud storage systems, than the less sophisticated metrics FDR and FAR.
– *Mixed drive models.* In Table 9, we make an observation that a model trained using data from one drive model may be useless at predicting hard-drive failure for a different drive model. An storage system operator should bear this in mind when training models for experimental evaluation. Further, in Table 10 we observe that this problem can be alleviated by using a training set that includes drives from both models. This is particularly relevant for cloud storage systems, which are likely to have multiple drive models.

Acknowledgments. This work is partially supported by NSFC (61872201, 61702521, 61602266, and U1833114); STDP of Tianjin (17JCYBJ15300, 16JCYBJ41900, 18ZXZNG00140, and 18ZXZNGX00200); and Fundamental Research Funds for Central Universities, KLMDASR.

References

1. Allen, B.: Monitoring hard disks with SMART. Linux J. (117), 74–77 (2004)
2. Botezatu, M.M., Giurgiu, I., Bogojeska, J., Wiesmann, D.: Predicting disk replacement towards reliable data centers. In: Proceedings of SIGKDD, pp. 39–48 (2016)
3. Chaves, I.C., de Paula, M.R.P., Leite, L.G.M., Gomes, J.P.P., Machado, J.C.: Hard disk drive failure prediction method based on a Bayesian network. In: Proceedings of IJCNN, pp. 1–7 (2018)
4. Chaves, I.C., de Paula, M.R.P., Leite, L.G., Queiroz, L.P., Gomes, J.P.P., Machado, J.C.: BaNHFaP: a Bayesian network based failure prediction approach for hard disk drives. In: Proceedings of BRACIS, pp. 427–432 (2016)
5. Ganguly, S., Consul, A., Khan, A., Bussone, B., Richards, J., Miguel, A.: A practical approach to hard disk failure prediction in cloud platforms: big data model for failure management in datacenters. In: Proceedings of BigDataService, pp. 105–116 (2016)
6. Garcia, M., et al.: Review of techniques for predicting hard drive failure with smart attributes. Int. J. Mach. Intell. Sens. Signal Process. **2**(2), 159–172 (2018)
7. Goldszmidt, M.: Finding soon-to-fail disks in a haystack. In: Proceedings of HotStorage (2012)

8. Hamerly, G., Elkan, C.: Bayesian approaches to failure prediction for disk drives. In: Proceedings of ICML, pp. 202–209 (2001)
9. Hughes, G.F., Murray, J.F., Kreutz-Delgado, K., Elkan, C.: Improved disk-drive failure warnings. IEEE Trans. Rel. 51(3), 350–357 (2002)
10. Li, J., et al.: Hard drive failure prediction using classification and regression trees. In: Proceedings of DSN, pp. 383–394 (2014)
11. Li, J., Stones, R.J., Wang, G., Li, Z., Liu, X., Xiao, K.: Being accurate is not enough: new metrics for disk failure prediction. In: Proceedings of SRDS, pp. 71–80 (2016)
12. Li, J., Stones, R.J., Wang, G., Liu, X., Li, Z., Xu, M.: Hard drive failure prediction using decision trees. Reliab. Eng. Syst. Saf. 164, 55–65 (2017)
13. Mahdisoltani, F., Stefanovici, I., Schroeder, B.: Proactive error prediction to improve storage system reliability. In: Proceedings of USENIX ATC, pp. 391–402 (2017)
14. Murray, J.F., Hughes, G.F., Kreutz-Delgado, K.: Hard drive failure prediction using non-parametric statistical methods. In: Proceedings of ICANN/ICONIP (2003)
15. Murray, J.F., Hughes, G.F., Kreutz-Delgado, K.: Machine learning methods for predicting failures in hard drives: a multiple-instance application. J. Mach. Learn. Res. 6, 783–816 (2005)
16. Pang, S., Jia, Y., Stones, R., Wang, G., Liu, X.: A combined Bayesian network method for predicting drive failure times from SMART attributes. In: Proceedings of IJCNN, pp. 4850–4856 (2016)
17. Pinheiro, E., Weber, W.D., Barroso, L.A.: Failure trends in a large disk drive population. In: Proceedings of FAST (2007)
18. Pitakrat, T., van Hoorn, A., Grunske, L.: A comparison of machine learning algorithms for proactive hard disk drive failure detection. In: Proceedings of SIGSoft Symposium on Architecting Critical Systems, pp. 1–10 (2013)
19. Qian, J., Skelton, S., Moore, J., Jiang, H.: P3: priority based proactive prediction for soon-to-fail disks. In: Proceedings of NAS, pp. 81–86 (2015)
20. Queiroz, L.P., Rodrigues, F.C.M., Gomes, J.P.P., et al.: A fault detection method for hard disk drives based on mixture of Gaussians and nonparametric statistics. IEEE Trans. Ind. Inform. 13(2), 542–550 (2017)
21. Rincón, C.C.A., Pâris, J.F., Vilalta, R., Cheng, A.M., Long, D.D.: Disk failure prediction in heterogeneous environments. In: Proceedings of SPECTS, pp. 1–7 (2017)
22. Schroeder, B., Gibson, G.A.: Disk failures in the real world: what does an MTTF of 1,000,000 hours mean to you? In: Proceedings of FAST, vol. 7, pp. 1–16 (2007)
23. Vishwanath, K.V., Nagappan, N.: Characterizing cloud computing hardware reliability. In: Proceedings of SoCC, pp. 193–204 (2010)
24. Wang, Y., Ma, E.W., Chow, T.W., Tsui, K.L.: A two-step parametric method for failure prediction in hard disk drives. IEEE Trans. Ind. Inform. 10, 419–430 (2014)
25. Wang, Y., Miao, Q., Ma, E.W., Tsui, K.L., Pecht, M.G.: Online anomaly detection for hard disk drives based on Mahalanobis distance. IEEE Trans. Rel. 62, 136–145 (2013)
26. Wang, Y., Miao, Q., Pecht, M.: Health monitoring of hard disk drive based on Mahalanobis distance. In: Proceedings of PHM-Shenzhen, pp. 1–8 (2011)
27. Xiao, J., Xiong, Z., Wu, S., Yi, Y., Jin, H., Hu, K.: Disk failure prediction in data centers via online learning. In: Proceedings of ICPP, p. 35 (2018)
28. Xu, C., Wang, G., Liu, X., Guo, D., Liu, T.Y.: Health status assessment and failure prediction for hard drives with recurrent neural networks. IEEE Trans. Comput. 65(11), 3502–3508 (2016)

29. Xu, Y., et al.: Improving service availability of cloud systems by predicting disk error. In: Proceedings of USENIX ATC, pp. 481–494 (2018)
30. Zhao, Y., Liu, X., Gan, S., Zheng, W.: Predicting disk failures with HMM- and HSMM-based approaches. In: Perner, P. (ed.) ICDM 2010. LNCS (LNAI), vol. 6171, pp. 390–404. Springer, Heidelberg (2010). https://doi.org/10.1007/978-3-642-14400-4_30
31. Zhu, B., Wang, G., Liu, X., Hu, D., Lin, S., Ma, J.: Proactive drive failure prediction for large scale storage systems. In: Proceedings of MSST, pp. 1–5 (2013)

Distributed and Parallel Algorithms

Efficient Pattern Matching on CPU-GPU Heterogeneous Systems

Victoria Sanz[1,2]([⊠]), Adrián Pousa[1], Marcelo Naiouf[1],
and Armando De Giusti[1,3]

[1] III-LIDI, School of Computer Sciences, National University of La Plata,
La Plata, Argentina
{vsanz,apousa,mnaiouf,degiusti}@lidi.info.unlp.edu.ar
[2] CIC, Buenos Aires, Argentina
[3] CONICET, Buenos Aires, Argentina

Abstract. Pattern matching algorithms are used in several areas such as network security, bioinformatics and text mining, where the volume of data is growing rapidly. In order to provide real-time response for large inputs, high-performance computing should be considered. In this paper, we present a novel hybrid pattern matching algorithm that efficiently exploits the computing power of a heterogeneous system composed of multicore processors and multiple graphics processing units (GPUs). We evaluate the performance of our algorithm on a machine with 36 CPU cores and 2 GPUs and study its behaviour as the data size and the number of processing resources increase. Finally, we compare the performance of our proposal with that of two other algorithms that use only the CPU cores and only the GPUs of the system respectively. The results reveal that our proposal outperforms the other approaches for data sets of considerable size.

Keywords: Pattern matching · CPU-GPU computing · CPU-GPU heterogeneous systems · Hybrid programming · Aho-Corasick

1 Introduction

Pattern matching algorithms locate some or all occurrences of a finite number of patterns (pattern set or dictionary) in a text (data set). These algorithms are key components of DNA analysis applications [1], antivirus [2], intrusion detection systems [3,4], among others. In this context, the Aho-Corasick (AC) algorithm [5] is widely used because it efficiently processes the text in linear time.

The ever-increasing amount of data to be processed, sometimes in real time, led several authors to investigate the acceleration of AC on emerging parallel architectures. In particular, researchers have proposed different approaches to parallelize AC on shared-memory architectures, distributed-memory architectures (clusters), GPUs and multiple GPUs [6–10].

© Springer Nature Switzerland AG 2020
S. Wen et al. (Eds.): ICA3PP 2019, LNCS 11944, pp. 391–403, 2020.
https://doi.org/10.1007/978-3-030-38991-8_26

Although modern computers include multiple CPU cores and at least one GPU, little work has been done to accelerate pattern matching on such systems. In [11] the authors present a hybrid parallelization of AC on CPU-GPU heterogeneous systems. Briefly, the algorithm consists of generating the data structures needed for pattern matching on the CPU cores and performing the matching process on the GPU. Similarly, in [12] the authors propose a hybrid CPU-GPU pattern-matching algorithm that uses the CPU to filter incoming data and the GPU to complete the matching process. The filter phase detects blocks of data suspected of containing patterns, thus it reduces the GPU workload and data transfers. In summary, previous work has focused on using only one type of processing unit (PU) for the matching process, i.e. CPUs or GPUs, leading to underutilization of system resources. To our best knowledge, no work has focused on using both PUs in a collaborative way to accelerate the matching process of pattern matching algorithms.

In this paper, we present a novel hybrid pattern matching algorithm that efficiently exploits the computing power of a heterogeneous system composed of multicore processors and multiple GPUs. Also we address the problem of load balancing among the processing resources of the system. We evaluate the performance of our algorithm on a machine with 36 CPU cores and 2 GPUs and study its behaviour as the data size and the number of processing resources increase. Finally, we compare the performance of our proposal with that of two other algorithms that use only the CPU cores and only the GPUs of the system respectively. The results reveal that our proposal outperforms the other approaches for data sets of considerable size.

This paper extends the work in [13] by presenting (1) a generalization of our algorithm to a wider range of heterogeneous systems, (2) an analysis of its behaviour as the problem size and the number of processing resources increase, and (3) a detailed comparison with previous approaches.

The rest of the paper is organized as follows. Section 2 introduces the AC algorithm. Section 3 summarizes two approaches to parallelize AC. Section 4 introduces a hybrid OpenMP-CUDA programming model and a workload distribution strategy, which are specific to CPU–GPU heterogeneous computing. Section 5 describes our parallel algorithm for pattern matching on CPU-GPU heterogeneous systems. Section 6 shows our experimental results. Finally, Sect. 7 presents the main conclusions and some ideas for future research.

2   The Aho-Corasick Algorithm

The AC algorithm [5] has been widely used since it is able to locate all occurrences of user-specified patterns in a single pass of the text. The algorithm consists of two steps: the first is to construct a finite state pattern matching machine; the second is to process the text using the state machine constructed in the previous step. The pattern matching machine has valid and failure transitions. The former are used to detect all user-specified patterns. The latter are used to backtrack the state machine, specifically to the state that represents the longest proper suffix, in order to recognize patterns starting at any location of the text. Certain states are designated as "output states" which indicate that a set of patterns has been found. The AC machine works as follows: given a current state and an input character, it tries to follow a valid transition; if such a transition does not exist, it jumps to the state pointed by the failure transition and processes the same character until it causes a valid transition. The machine emits the corresponding patterns whenever an output state is found. Figure 1 shows the state machine for the pattern set {he, she, his, hers}. Solid lines represent valid transitions and dotted lines represent failure transitions.

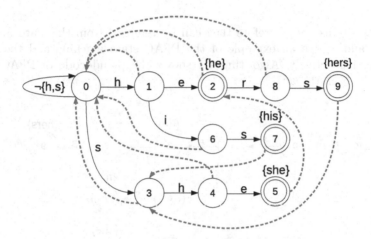

Fig. 1. Aho-Corasick state machine for the pattern set {he, she, his, hers}

3 Previous Approaches to Parallelize Aho-Corasick

The most straightforward way to parallelize AC [6] is based on dividing the input text into segments and making each processor responsible for a particular segment (i.e., each processor performs AC on its segment). All processors use the same state machine. The disadvantage of this strategy is that patterns can cross the boundary of two adjacent segments. This problem is known as the "boundary detection" problem. In order to detect these patterns, each processor has to compute an additional chunk known as "overlapping area", whose size is equal to the length of the longest pattern in the dictionary minus 1. However, this additional computation is an overhead that increases as the text is divided into more segments of smaller size. Figure 2 illustrates this strategy.

Another approach is the Parallel Failureless Aho-Corasick algorithm (PFAC) [7] that efficiently exploits the parallelism of AC and therefore is suitable for GPUs. PFAC assigns each character of the text to a particular thread. Each thread is responsible for identifying the pattern beginning at its assigned position and terminates immediately when it detects that such a pattern does not exist (i.e., when it cannot follow a valid transition). Note that PFAC does not use

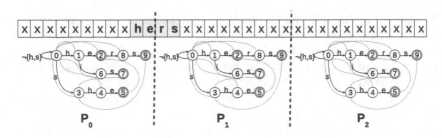

Fig. 2. Parallel AC

failure transitions and therefore they can be removed from the state machine. Figures 3 and 4 give an example of the PFAC state machine and the PFAC algorithm, respectively. Algorithm 1.1 shows the pseudocode of PFAC (code executed by each thread).

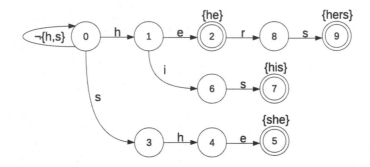

Fig. 3. PFAC state machine for the pattern set {he, she, his, hers}

Algorithm 1.1. Pseudocode of PFAC

```
pos = start
state = initial state
while ( pos < text size ){
    if (there is no transition for the current state and input character)
        break
    state = next state for the current state and input character
    if (state is an output state)
        register the pattern located at the position "start"
    pos = pos + 1
}
```

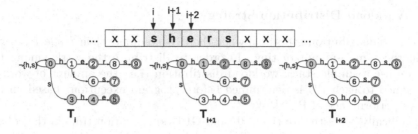

Fig. 4. Example of PFAC

4   A General CPU-GPU Computing Model

One of the reasons that motivate CPU-GPU heterogeneous computing is to improve the utilization of the processing units (PUs). The use of both types of PUs in a collaborative way may improve the performance of the application. In this section we introduce a hybrid OpenMP-CUDA programming model and a workload distribution strategy, which are specific to CPU-GPU heterogeneous computing.

4.1   Hybrid OpenMP-CUDA Programming Model

Assuming that the system is composed of N CPU cores and M GPUs, the proposed model creates two sets of threads. The first set has M threads, each one runs on a dedicated CPU core and controls one GPU, which involves the following steps: allocating memory on the GPU to store input and output data, transferring the data from CPU to GPU, calling the kernel function, transferring the results from GPU to CPU and freeing memory on the GPU. On the other hand, the second set has $N - M$ threads, which concurrently perform the corresponding calculations on the remaining CPU cores. Figure 5 depicts this hybrid programming model.

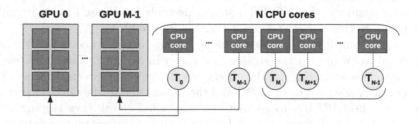

Fig. 5. Hybrid OpenMP-CUDA programming model

4.2   Workload Distribution Strategy

A workload distribution is optimal when the PUs complete their respective work within the same amount of time. In order to distribute the work among the PUs, we use a simple static workload distribution (i.e., the amount of work to be assigned to each PU is determined before program execution) based on the relative performance of PUs [14].

Specifically, we estimate the CPU and GPU(s) execution time in the collaborative implementation as $T'_{cpu} = T_{cpu} \cdot R$ and $T'_{gpu} = T_{gpu} \cdot (1 - R)$, respectively, where R is the proportion of work assigned to the CPU cores, T_{cpu} represents the execution time of the OpenMP algorithm on the available CPU cores and T_{gpu} is the execution time of the single-GPU or multi-GPU algorithm using CUDA, as appropriate. Clearly, the execution time of the collaborative implementation reaches its minimum when $T'_{cpu} = T'_{gpu}$, i.e. $T_{cpu} \cdot R = T_{gpu} \cdot (1 - R)$. From this equation we obtain $R = \frac{T_{gpu}}{T_{cpu}+T_{gpu}}$.

In our scenario, R has to be recalculated when the input data vary or the configuration of the system changes. According to R, the workload assigned to the CPU cores is $D_{cpu} = R \cdot D_{size}$ and the workload assigned to the GPU(s) is $D_{gpu} = D_{size} - D_{cpu}$, where D_{size} is the length of the text string.

Although it is impractical to run both OpenMP and CUDA applications in order to obtain R, we plan to use this first approach as a baseline to derive an estimation model for R.

5   Pattern Matching on CPU-GPU Heterogeneous Systems

Our implementation is based on the PFAC algorithm and uses the hybrid programming model proposed in Sect. 4.1.

Our algorithm generates the state machine on the CPU sequentially. The state machine is represented by a State Transition Table (STT) that has a row for each state and a column for each ASCII character (256). Each entry of the STT contains the next state information. Once generated, the STT is copied to the texture memory of the GPU(s) since this table is accessed in an irregular manner and, in this way, the access latency is reduced.

The algorithm distributes the workload (input text) between the CPU and the GPU(s) according to the strategy proposed in Sect. 4.2. Thus, the text is divided into two segments and the "boundary detection" problem appears. The first segment is assigned to the CPU and the second one to the GPU(s) (Fig. 6). In this way, the CPU has to compute an additional chunk (overlapping area) already residing in main memory and thus we reduce the amount of data to be transferred to the GPU(s).

When the heterogeneous system has several identical GPUs, the workload (segment) is distributed equally among them, taking into account the overlapping area. Each thread in charge of managing one GPU copies its segment into the global memory. Note that large segments may exceed the global memory

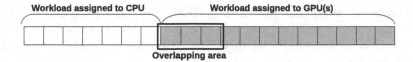

Fig. 6. Workload distribution between the CPU and the GPU(s)

capacity. In that case, the thread subdivides the segment into smaller segments and then it transfers and processes them one by one. It should be noted that each sub-segment must be transferred with the corresponding overlapping area. The implementation details of the PFAC algorithm on GPU (PFAC kernel) can be found in [7]. In summary, the kernel is launched with 256 threads per block. Each thread block handles 1024 positions of the input segment (i.e., each thread processes 4 positions). Each thread block loads the corresponding data into shared memory. Then, threads read input bytes from the shared memory in order to perform their work.

The threads that operate on the CPU cores distribute the workload (segment) equally among them via the OpenMP 'for' work-sharing directive.

6 Experimental Results

Our experimental platform is a machine composed of two Intel Xeon E5-2695 v4 processors and 128 GB RAM. Each processor has eighteen 2.10 GHz cores, thus the machine has thirty-six cores in total. Hyper-Threading and Turbo Boost were disabled. The machine is equipped with two Nvidia GeForce GTX 960; each one is composed of 1024 cores and 2 GB GDDR5 memory. Each CUDA core operates at 1127 MHz.

Test scenarios were generated by combining three English texts of different sizes with four English dictionaries with different number of patterns. All the texts were extracted from the British National Corpus [15]: text 1 is a 4-million-word sample (21 MB); text 2 is a 50-million-word sample (268 MB); text 3 is a 100-million-word sample (544 MB). The dictionaries include frequently used words: dictionary 1 with 3000 words; dictionary 2 with 100000 words; dictionary 3 with 178690 words; dictionary 4 with 263533 words.

To evaluate the effectiveness of our proposal, we compared the sequential version of PFAC (PFAC_SEQ) with the following parallel implementations:

- PFAC_CPU: implementation of PFAC on a multicore CPU using OpenMP.
- PFAC_GPU: implementation of PFAC on GPU using CUDA and executed with 256 threads per block.
- PFAC_MultiGPU: implementation of PFAC on multiple GPUs using CUDA and OpenMP, which is used only for managing the GPUs.
- PFAC_CPU-GPU: hybrid OpenMP-CUDA implementation of PFAC for heterogeneous systems composed of multicore processors and 1 GPU.
- PFAC_CPU-MultiGPU: hybrid OpenMP-CUDA implementation of PFAC for heterogeneous systems composed of multicore processors and multiple GPUs.

It should be noted that PFAC_SEQ, PFAC_CPU and PFAC_GPU are the original implementations provided by Lin et al. [7]. We developed the remaining versions described above, which are based on the aforementioned original implementations.

Our experiments focus on the matching step since it is the most significant part of pattern matching algorithms. For each test scenario, we ran each implementation 100 times and averaged the execution time. PFAC_CPU, PFAC_CPU-GPU and PFAC_CPU-MultiGPU were executed with the following system configurations: 6, 12, 18, 24, 30 and 36 threads/CPU cores. We considered the data transfer time (host-to-device and device-to-host, aka H2D and D2H) when evaluating the algorithms that use GPU(s), since it represents a significant portion of the total execution time [13] (i.e. it is not negligible).

First, we calculated the load balance of each run for the algorithms that use several processing resources (PFAC_CPU, PFAC_MultiGPU, PFAC_CPU-GPU and PFAC_CPU-MultiGPU). Load balance [16] can be defined as the ratio between the average time to finish all of the parallel tasks and the maximum time to finish any of the parallel tasks ($\frac{T_{avg}}{T_{max}}$). A load balance value near 1 means a better distribution of load.

In PFAC_CPU, each OpenMP thread represents a parallel task. On the other hand, in PFAC_MultiGPU, the work done by each GPU is a parallel task. Considering all tests for each algorithm, both achieve an average load balance of 0.99.

Table 1. Values of R used by PFAC_CPU-GPU and PFAC_CPU-MultiGPU

		No. of CPU cores					
		6	12	18	24	30	36
PFAC_CPU-GPU	Text 1	0.30	0.38	0.43	0.40	0.38	0.37
	Text 2	0.36	0.49	0.58	0.61	0.62	0.62
	Text 3	0.37	0.51	0.59	0.64	0.67	0.68
PFAC_CPU-MultiGPU	Text 1	0.21	0.28	0.33	0.30	0.28	0.28
	Text 2	0.24	0.35	0.43	0.47	0.48	0.48
	Text 3	0.24	0.36	0.44	0.49	0.52	0.54

PFAC_CPU-GPU and PFAC_CPU-MultiGPU consist of two parallel tasks: one is performed by the CPU cores and the other by the GPU(s). We distributed the input text among the PUs according to the strategy proposed in Sect. 4.2. Table 1 shows the value of R used by both algorithms, for each text and system configuration. Similarly, Table 2 presents the load balance achieved. The results reveal that our workload distribution strategy provides a good load-balance, which ranges between 0.75 and 0.96 for PFAC_CPU-GPU, and between 0.77 and 0.93 for PFAC_CPU-MultiGPU. Additionally, both algorithms follow a similar trend: the load balance improves as the size of the text increases.

Table 2. Load balance achieved by PFAC_CPU-GPU and PFAC_CPU-MultiGPU

		No. of CPU cores					
		6	12	18	24	30	36
PFAC_CPU-GPU	Text 1	0.82	0.86	0.84	0.80	0.77	0.75
	Text 2	0.89	0.92	0.95	0.91	0.91	0.92
	Text 3	0.90	0.94	0.96	0.95	0.94	0.94
PFAC_CPU-MultiGPU	Text 1	0.77	0.86	0.88	0.84	0.80	0.79
	Text 2	0.79	0.88	0.91	0.91	0.89	0.89
	Text 3	0.81	0.89	0.93	0.93	0.92	0.91

Next, we evaluated the performance (Speedup[1]) of the mentioned algorithms. For each algorithm and system configuration, the average speedup for each text is shown. This is because the speedup does not vary significantly with the dictionary.

Figure 7 illustrates the average speedup of PFAC_GPU and PFAC_MultiGPU, for different texts. In both cases, the speedup increases when going from Text 1 to Text 2, but then it plateaus. This is mainly due to the fact that the data transfer time (H2D and D2H) increases with the size of the text. In particular, for large texts, this overhead has a greater impact on performance. Note that the performance of PFAC_MultiGPU is less affected by data transfers, compared to PFAC_GPU. This is because PFAC_MultiGPU distributes the load equally among the GPUs of the system. Thus, independent small data transfers occur in parallel. Also, it can be observed that PFAC_MultiGPU achieves higher performance than PFAC_GPU.

Figure 8 shows the average speedup of PFAC_CPU, PFAC_CPU-GPU and PFAC_CPU-MultiGPU, for different texts and system configurations (the number of threads/CPU cores is indicated between parentheses). In each case, it can be seen that the system configuration that provides the best performance depends on the text. Moreover, in some cases different system configurations give the same performance for a given text. For this reason, fewer resources should be used to achieve an acceptable speedup with a low energy consumption. The analysis of energy consumption is out of the scope of this paper and is the subject of future work. Additionally, note that, for a fixed number of processing resources, the speedup of PFAC_CPU, PFAC_CPU-GPU and PFAC_CPU-MultiGPU tends to increase with the size of the text. Therefore, we conclude that these algorithms behave well as the workload increases.

Figure 9 compares the performance of the algorithms. In the case of PFAC_CPU, PFAC_CPU-GPU and PFAC_CPU-MultiGPU, we selected the system configuration that provides the best performance for each text. For example:

---

[1] Speedup is defined as $\frac{T_s}{T_p}$, where T_s is the execution time of the sequential algorithm and T_p is the execution time of the parallel algorithm.

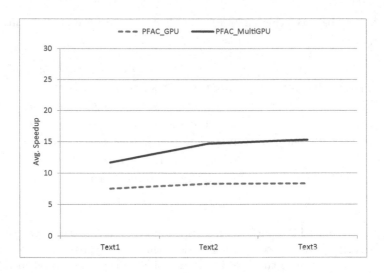

Fig. 7. Average speedup of PFAC_GPU and PFAC_MultiGPU

for PFAC_CPU and Text 1, we show the average speedup achieved with 18 CPU cores, whereas for Texts 2 and 3 we show the average speedup achieved with 36 CPU cores.

As it can be observed, PFAC_MultiGPU achieves the best performance for Text 1, followed by PFAC_CPU-MultiGPU (18 CPU cores + 2 GPUs), PFAC_CPU-GPU (18 CPU cores + 1 GPU), PFAC_GPU and PFAC_CPU (18 CPU cores). For this text, the algorithms achieve an average speedup of 11.69, 11.48, 8.14, 7.51, and 5.69 respectively.

Furthermore, PFAC_CPU-MultiGPU (24 CPU cores + 2 GPUs) achieves the best result for Text 2, with an average speedup of 20.85, followed by PFAC_CPU-GPU (18 CPU cores + 1 GPU) with 16.37, PFAC_MultiGPU with 14.70, PFAC_CPU (36 CPU cores) with 13.69 and PFAC_GPU with 8.31.

Finally, for Text 3, PFAC_CPU-MultiGPU achieves the best average speedup (25.41) by using all available resources, followed by PFAC_CPU-GPU (36 CPU cores + 1 GPU) with 21.23, PFAC_CPU (36 CPU cores) with 17.71, PFAC_MultiGPU with 15.31 and PFAC_GPU with 8.33.

Note that for Texts 2 and 3, the algorithms that use the CPU cores outperform those that use only 1 or 2 GPUs. In general, this is due to the impact of H2D/D2H transfers on PFAC_GPU and PFAC_MultiGPU. The former algorithm transfers the entire data between CPU and GPU, whereas the latter transfers an equal amount of data to each GPU in parallel. On the other hand, both PFAC_CPU-GPU and PFAC_CPU-MultiGPU are less affected by data transfers since they transfer a smaller portion of data, according to R.

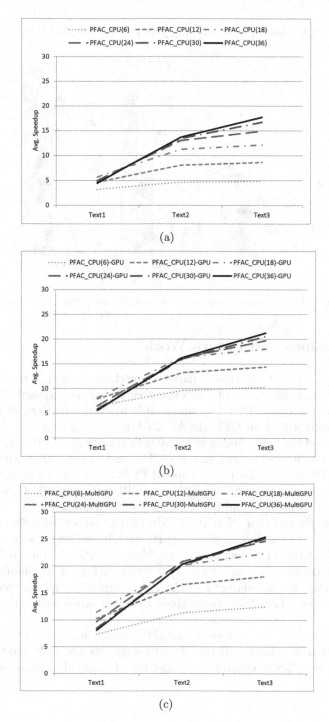

Fig. 8. Average speedup of (a) PFAC_CPU, (b) PFAC_CPU-GPU and (c) PFAC_CPU-MultiGPU

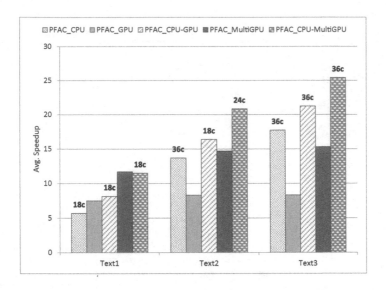

Fig. 9. Performance comparison of parallel matching algorithms

7   Conclusions and Future Work

In this paper we presented a novel pattern matching algorithm that efficiently exploits the computing power of a heterogeneous system composed of multicore processors and multiple GPUs. Our proposal is based on the Parallel Failureless Aho-Corasick algorithm for GPU (PFAC_GPU).

We evaluated the performance of our algorithm (PFAC_CPU-MultiGPU) on a machine with 36 CPU cores and 2 GPUs, and compared it with that of: PFAC_GPU; PFAC_MultiGPU, a version of PFAC that runs on multiple GPUs; PFAC_CPU, a version of PFAC for multicore CPUs; PFAC_CPU-GPU, a hybrid version of PFAC that uses multiple CPU cores and a single GPU.

The results showed that PFAC_CPU-MultiGPU outperforms the other algorithms, for texts of considerable size. In particular, it reaches an average speedup of 25.41 for the largest problem considered. However, PFAC_MultiGPU achieves the best performance for small texts. Furthermore, for a fixed number of processing resources, the speedup of PFAC_CPU-MultiGPU tends to increase with the size of the text. Therefore, we conclude that it behaves well as the workload increases.

As for future work, we plan to extend the experimental work to evaluate the PFAC algorithm on other parallel architectures such as Xeon Phi and CPU-GPU clusters. Also we plan to construct a model based load-balancing strategy.

References

1. Tumeo, A., Villa, O.: Accelerating DNA analysis applications on GPU clusters. In: IEEE 8th Symposium on Application Specific Processors (SASP), pp. 71–76. IEEE Computer Society, Washington D.C. (2010)
2. Clamav. http://www.clamav.net
3. Norton, M.: Optimizing pattern matching for intrusion detection. Sourcefire Inc., White Paper. https://www.snort.org/documents/optimization-of-pattern-matches-for-ids
4. Tumeo, A., et al.: Efficient pattern matching on GPUs for intrusion detection systems. In: Proceedings of the 7th ACM International Conference on Computing Frontiers, pp. 87–88. ACM, New York (2010)
5. Aho, A.V., Corasick, M.J.: Efficient string matching: an aid to bibliographic search. Commun. ACM **18**(6), 333–340 (1975)
6. Tumeo, A., et al.: Aho-Corasick string matching on shared and distributed-memory parallel architectures. IEEE Trans. Parallel Distrib. Syst. **23**(3), 436–443 (2012)
7. Lin, C.H., et al.: Accelerating pattern matching using a novel parallel algorithm on GPUs. IEEE Trans. Comput. **62**(10), 1906–1916 (2013)
8. Arudchutha, S., et al.: String matching with multicore CPUs: performing better with the Aho-Corasick algorithm. In: Proceedings of the IEEE 8th International Conference on Industrial and Information Systems, pp. 231–236. IEEE Computer Society, Washington D.C. (2013)
9. Herath, D., et al.: Accelerating string matching for bio-computing applications on multi-core CPUs. In: Proceedings of the IEEE 7th International Conference on Industrial and Information Systems (ICIIS), pp. 1–6. IEEE Computer Society, Washington D.C. (2012)
10. Lin, C.H., et al.: A novel hierarchical parallelism for accelerating NIDS using GPUs. In: Proceedings of the 2018 IEEE International Conference on Applied System Invention (ICASI), pp. 578–581. IEEE (2018)
11. Soroushnia, S., et al.: Heterogeneous parallelization of Aho-Corasick algorithm. In: Proceedings of the IEEE 7th International Conference on Industrial and Information Systems (ICIIS), pp. 1–6. IEEE Computer Society, Washington D.C. (2012)
12. Lee, C.L., et al.: A hybrid CPU/GPU pattern-matching algorithm for deep packet inspection. PLoS One **10**(10), 1–22 (2015)
13. Sanz, V., Pousa, A., Naiouf, M., De Giusti, A.: Accelerating pattern matching with CPU-GPU collaborative computing. In: Vaidya, J., Li, J. (eds.) ICA3PP 2018. LNCS, vol. 11334, pp. 310–322. Springer, Cham (2018). https://doi.org/10.1007/978-3-030-05051-1_22
14. Wan, L., et al.: Efficient CPU-GPU cooperative computing for solving the subset-sum problem. Concurr. Comput. Pract. Exp. **28**(2), 185–186 (2016)
15. The British National Corpus, version 3 (BNC XML Edition). Distributed by Bodleian Libraries, University of Oxford, on behalf of the BNC Consortium (2007). http://www.natcorp.ox.ac.uk/
16. Rahman, R.: Intel Xeon Phi Coprocessor Architecture and Tools: The Guide for Application Developers. Apress, Berkeley (2013)

Improving Performance of Batch Point-to-Point Communications by Active Contention Reduction Through Congestion-Avoiding Message Scheduling

Jintao Peng[1], Zhang Yang[1,2(✉)], and Qingkai Liu[1,2]

[1] Laboratory of Computational Physics, Institute of Applied Physics
and Computational Mathematics, Beijing 100088, China
{yang_zhang,liuqk}@iapcm.ac.cn
[2] CAEP Software Center for High Performance Numerical Simulation,
Beijing 100088, China

Abstract. Communication performance plays a crucial role in both the scalability and the time-to-solution of parallel applications. The share of links in modern high-performance computer networks inevitably introduces contention for communications involving multiple point-to-point messages, thus hinders their performance. Passive contention reduction such as the congestion control of the networks can mitigate network contention but with extra protocol cost, while application-level active contention reduction such as topology mapping techniques can only reduce contention of applications with static communication patterns. In this paper, we explore a different approach to actively reduce network contention through a congestion-avoiding message scheduling algorithm, namely CAMS. CAMS determines how to inject the messages in groups to reduce contention just in time before injecting them into the network, thus it is useful in applications with dynamic communication patterns. Experiments with a 2D halo-exchange benchmark on the Tianhe-2A supercomputer shows that it can improve communication performance up to 27% when messages get large. The proposed approach can be used in conjunction with topology mapping to further improve communication performance.

Keywords: Congestion-avoiding · Message scheduling ·
Communication optimization · Network contention · Fat-tree

1 Introduction

Communication performance has long been a critical part of the performance of parallel applications. It directly impacts the parallel scalability and sometimes even dominates the application running time [11]. With supercomputing approaching the exaflops era, due to the increase of architectural parallelism, parallel applications exhibit an increasing communication bottleneck, which restricts both their parallel scalability and their time-to-solution [23].

© Springer Nature Switzerland AG 2020
S. Wen et al. (Eds.): ICA3PP 2019, LNCS 11944, pp. 404–418, 2020.
https://doi.org/10.1007/978-3-030-38991-8_27

In an ideal supercomputer with a fully connected network, the only restriction on communication performance would be the link latency and bandwidth, together with how the application maps the communication to the nodes. However, this would require link count to be the square of the node count and is not viable to the supercomputers with tens of thousands of nodes. Thus modern high-performance computer networks share links between nodes with link count often proportional to the node count. This fact introduces contention on the links and causes network congestion [7], which significantly impacts the performance of parallel application communication [26]. Thus mitigating and reducing network contention is an important approach to improve communication performance.

In this paper, we focus on batch point-to-point communications which are widely used in parallel scientific and engineering applications. As indicated in [17], point-to-point communications can account for 90% of the total communication operations of typical scientific and engineering applications. Most applications of this type implement a bulk-synchronous-parallel (BSP) style of parallelization where communications are carried out in a batch manner. If not handled properly, batch point-to-point communications can flood the network with a large number of packets and cause network congestion, thus significantly downgrade the network performance and ultimately worsen the communication performance.

Modern high-performance networks such as Infiniband (IB) build in various congestion control mechanisms to mitigate network contention in the hope of better communication performance [3,10]. These mechanisms work passively at the packet level, detect congestion automatically, and deal with it by end-to-end traffic control. However, this comes with an extra protocol cost where extra congestion-control packets are generated and transmitted, which can make the congestion worse when the congestion is very high. On the other hand, applications use topology mapping techniques [12] distribute messages across spatial dimensions, thereby actively reduce network contention. These techniques view application communication as a static topology and are inadequate for applications with changing communication patterns, which is common in real-world parallel applications.

This paper explores a different approach to reduce network contention actively in the application just before a batch of point-to-point communications are carried out. Briefly, for a batch point-to-point communication, we propose a congestion-avoiding message scheduling algorithm, namely CAMS, to decide which messages to send and when to send, so the messages in-flight only introduce moderate contention which does not downgrade the network performance. The algorithm works online so that is capable of applications with dynamic communication patterns. Since the proposed algorithm does not change how the communicating processes map to computing nodes, it can be used in conjunction with topology mapping techniques to further improve communication performance. In summary, this paper makes the following contributions:

- Propose a congestion-avoiding message scheduling algorithm for batch point-to-point communications so that the network contention is reduced actively in the application just in time;
- Verify by experiments that the proposed message scheduling approach can eventually improve communication performance up to 27% with a typical halo-exchange communication pattern.

2   Background and Motivation

We briefly introduce in this section the batch point-to-point communications and their characteristics, and then analyze why active contention reduction is necessary and why current practice is not enough.

2.1   Batch Point-to-Point Communications

Parallel scientific and engineering applications mostly use a BSP-style parallelization [25]. The most significant characteristic of these applications is that they separate communication from the computation. After necessary initialization, these applications execute in communication-computation locksteps called supersteps. In each superstep, processes of an application would first acquire necessary remote data cooperatively (communication), and then computes independently with all data being local (computation). BSP makes parallel programming easy but can introduce network contention since communications are carried out in a burst manner.

The communication part of the supersteps can be modeled as batch point-to-point communications. Rigorously, a batch point-to-point communication is a set of independent point-to-point messages where the message order can be arbitrarily changed without impacting the correctness. Each message is uniquely determined by its source, destination and message tag. This type of communication is perhaps the most frequent in scientific and engineering applications. Halo-exchange [14], load migration [22], and particle communication [21] are such examples.

Although batch point-to-point communications may seem simple at a first glance, it usually gets complicated in real-world parallel applications. Firstly, the communication pattern can keep changing throughout the application since each superstep may require data from different partners. Secondly, the message sizes can keep changing throughout the application since each superstep can require different data. This highly dynamic nature of such applications makes communication optimization difficult.

2.2   Network Contention Mitigation and Reduction

Undoubtedly, modern high-performance computer networks share links to reduce architectural complexity. Thus network contention is always a concern of network design. High-end networks introduce credit-based flow-control mechanisms in the physical layer to ensure network contention does not cause network fault. It does not resolve contention, though. End-to-end congestion control protocols such as ECN and SRP are introduced in networks such as Infiniband [3] where the network will detect congestion and slow down the sender to mitigate the congestion. These protocols can reduce network contention but comes with a cost of extra protocol costs. Firstly, they may cause large overhead for small messages as shown in [16]. Secondly, they often lag behind the contention since the criteria of activating the slowing down are when the network resource is exhausted [20], which can often be too late. Last but not least, these protocols usually introduce extra congestion-control packets and re-send these packets when time is out, which can worsen the congestion. In the author's experience, the network can be stuck in such cases on the Tianhe-2 supercomputer.

The above mechanisms and protocols are passive in that they confront the contention after it is discovered. Job schedulers can actively reduce network contention ahead of time by allocating nodes with similar linking topology to the application communication topology. Runtimes and application themselves can actively reduce network contention by putting heavily-communicating processes in nearby nodes using topology mapping techniques. However, these existing active contention reduction approaches require the knowledge of application communication topology beforehand and view the communication as a static topology, thus do not fit in the real-world scenarios where the communication pattern is dynamic and hard to know beforehand.

Are there approaches to actively reduce network contention just in time and suitable for applications with dynamic communication patterns? To what extent can the application benefit from active contention reduction? These two questions motivate us to seek approaches to reduce contention from the origin of communication in applications, hence the work in this paper.

3   Congestion-Avoiding Message Scheduling for Batch Point-to-Point Communications

Given the fact that parallel scientific and engineering applications rely heavily on batch point-to-point communication, we propose in this section a just-in-time active contention-reduction approach through congestion-avoiding message scheduling, namely CAMS.

3.1   The Key Idea

One key characteristic of batch point-to-point communications is that the messages are independent of each other in a batch. Thus the order and time to send these messages do not impact the correctness of results. Then it is possible to schedule these messages so that the messages in flight do not introduce network contention, which eliminates congestion just in time. Messages scheduling can be invoked separately for every batch so it works in applications with dynamic communication patterns where the communication pattern is unknown beforehand and keeps changing during application running.

To illustrate the key idea and to show why this idea would improve communication performance, we introduce the following example as illustrated in Fig. 1: A batch of 6 messages $[F1, F2, \ldots, F6]$ transfers on a network of 4 switches and 5 links. The size of $F1$ and $F2$ is s bytes, and the size of other messages is $\frac{s}{3}$ bytes. All links have the same g byte/s bandwidth. And the network maintains a message-fair congestion control mechanism. When all messages are injected into the network at the same time (the left part), they will first contend on link $L4$ and $L5$ for the first $\frac{s}{3}$ bytes, then $F1$ and $F2$ will continue contending on $L1$ for their remaining $\frac{2s}{3}$ bytes. Given s is sufficiently large, it takes $\frac{s}{g} + \frac{4s}{3g} = \frac{7}{3}\frac{s}{g}$ s to finish. But we can schedule the messages as in the right part: $F2$, $F5$ and $F6$ are injected first, and $F1$, $F3$ and $F4$ are injected after $F2$, $F5$ and $F6$ go off the network. The contention on $L1$ is eliminated and the overall turn-around time is reduced to $2\frac{s}{g}$ s, which is a 16.7% performance improvement. This example illustrates a very counter-intuitive fact that *delaying message injection can improve instead of downgrade communication performance*. Thus the key idea of congestion-avoiding message scheduling is to partition the messages into such groups that finishing them one after another will improve the performance.

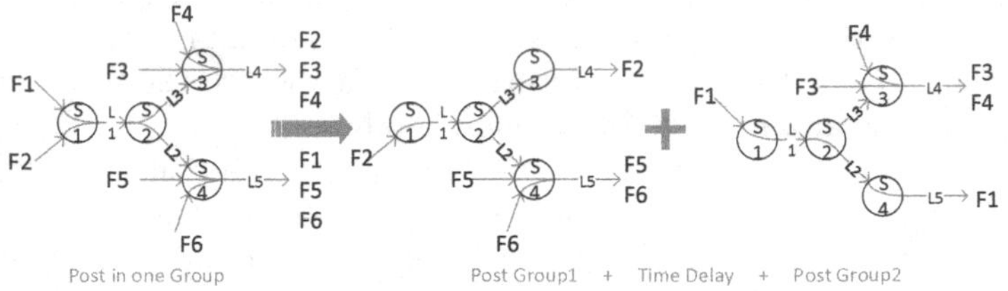

Fig. 1. An example of congestion-avoiding message scheduling: $F1$ and $F2$ are s bytes and others are $s/3$ bytes. The scheduling in the right part improves 16.7% the performance.

3.2   Overview of the Algorithm

The congestion-avoiding message scheduling algorithm, namely CAMS, is a simple three-step process as illustrated in Fig. 2:

1. **Initialization:** Map the messages on to the network to determine which links the message goes through.
2. **Group partition:** Partition the messages into disjoint groups so that intra-group contention is eliminated or reduced.
3. **Inter-group delay calculation:** Determine an inter-group delay so that the network is sufficiently utilized without introducing a performance penalty.

CAMS takes a batch of independent point-to-point messages, the network topology, and the routing table as the input, and produces a group partition of the messages together with inter-group delay as the output. The partitioning step generates the groups in the hope of reducing intra-group network contention

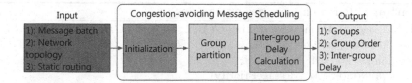

Fig. 2. Overview of the congestion-avoiding message scheduling algorithm

while maintaining sufficient network utilization. The delay-calculation step compacts the inter-group delay to improve network utilization while controlling the performance penalty of congestion. The algorithm produces a message schedule plan that is reusable.

To help to present the algorithm details, we define several notations here. We denote terminal nodes of the network as N, switches as R, and links as L. We denote the batch of messages as $M = \{m_i : (N_{s_i}, N_{d_i}, s_i), i = 1..n\}$, where N_{s_i} and N_{d_i} are the id of the source and destination nodes and s_i is the message size. The routes taken by the messages are then represented by a bipartite graph $G(M, L, E)$, where $e = (m_p, l_q), m_p \in M, l_q \in L$ indicates that message m_p goes through link l_q. The output of the CAMS algorithm can be represent as a sequence $[M_1, t_1, \ldots, M_2, t_2, \ldots, M_k]$, where M_i is the i-th group (a subset of M) and t_i is the delay after injecting the i-th group.

3.3   Partitioning Messages into Groups

The key part of CAMS is the partition step. In general, a partition of a set of n elements contains $B(n)$ alternatives, with $B(n)$ being the Bell number. The possible orders of the resultant sequence take at least another $n!$ alternatives. Thus we seek heuristics for a viable solution. The main idea of our heuristics is to search for groups that match a congestion-free communication patterns inherited in the network architecture. The routing algorithms of fat-tree networks exhibit a congestion-free pattern called "shift permutation" [27,28]: $\{(N_{s_i}, N_{d_j}) : j = (i + S) \bmod C, i = 0..C - 1\}$. Here S is a shift constant and C is the count of terminal nodes in the network. We make use of this fact in our heuristics.

The partition algorithm is presented in Algorithm 1. We define link contention degree $LCD(l)$ as the number of messages going through link l and $LCD_{max} = \max \{LCD_l : \forall l \in L\}$. We introduce a parameter LCD_{lim} to control the acceptable extent of contention and $LCD_{lim} = 1$ means no contention is acceptable. The algorithm goes through the following steps: it firstly sorts the messages to determine their priority in group construction, then traverses the sorted messages and constructs a maximal group M_i that $LCD_{max}(M_i) < LCD_{lim}$. It constructs the group in such a greedy manner to minimize the number of groups since more groups would introduce more inter-group delay and increase turn-around time.

The sort algorithm prioritizes messages as shown in Algorithm 2. It first aligns messages to MTU of the network so that the real packet to transmit is counted.

Algorithm 1. Partition messages into groups

Input: Message set M, message routing $G(M, L, E)$, and LCD_{lim}
Output: Message partition $[M_1, M_2, \ldots, M_k]$
1: partition := []; i := 1
2: DESCENDINGSORT(M, MessagePriority)
3: **while** $M \neq \phi$ **do**
4:     $M_i := \phi$
5:     **for** $m \in M$ **do**
6:         **if** $LCD_{max}(M_i \cup \{m\}) < LCD_{lim}$ **then**
7:             $M_i := M_i \cup \{m\}$; $M := M - \{m\}$
8:         **end if**
9:     **end for**
10:     partition := partition + $[M_i]$; i := i + 1
11: **end while**

Larger messages have a higher priority since they are more likely to be impacted by contention and to introduce contention. High priority makes them handled earlier to reduce the likelihood of contention. Given the same message size, messages with shorter node distance have higher priority since they are less likely to contend. Lastly, messages with a smaller node number have higher priority. This algorithm will put messages with the same node distance, i.e. messages matching the same shift-permutation pattern, in a continuous range. Thus it maximizes the possibility that a group will contain mostly the congestion-free shift-permutation pattern.

Algorithm 2. Determine the greater-than relationship of message priority

---

Input: Message m_i and m_j, message routing $G(M, L, E)$, LCD, MTU
Output: True for greater-than and False for less-than
1: **function** GREATERTHAN(m_i, m_j)
2:     s1 := $\lceil \frac{s_i}{MTU} \rceil$, s2 := $\lceil \frac{s_j}{MTU} \rceil$
3:     **if** s1 > s2 **then**
4:         **return** True
5:     **else if** s1 < s2 **then**
6:         **return** False
7:     **else if** $|N_{s_i} - N_{d_i}| < |N_{s_j} - N_{d_j}|$ **then**
8:         **return** True
9:     **else if** $|N_{s_i} - N_{d_i}| > |N_{s_j} - N_{d_j}|$ **then**
10:         **return** False
11:     **else**
12:         **return** $N_{s_i} < N_{s_j}$
13:     **end if**
14: **end function**

3.4   Determining the Inter-group Delay

The inter-group delay t_i lays in interval $[0, \max\{t_{m_j} : m_j \in M_i\}]$. It means group M_{i+1} is injected into the network immediately after M_i is injected when $t_i = 0$, while it means to wait until all messages in M_i go off the network when $t_i = \max\{t_{m_j} : m_j \in M_i\}$. The former can introduce heavy contention and the latter can waste too much time waiting and leave links idle. Good inter-group delay shall reduce contention while maintain high network utilization.

We introduce the concept of *critical link* to determine the inter-group delay. The critical link is a link $l \in L$ which have the smallest idle time given $\{t_i = 0 \ \forall \ i = 1..k-1\}$. It will become the most saturated link when messages are injected in the way defined by message partitioning. Thus to avoid congestion, the delay t_i shall not be shorter than the busy time of the critical link when transmitting M_i. However, moderate contention can help increase network utilization if it can be handled by the congestion control mechanisms without a performance penalty. Thus we introduce a scaling factor $\alpha \in [0, 1]$ so that one can control the acceptable extent of congestion. Algorithm 3 presents the details on how the critical link is selected and how the delay is computed.

Algorithm 3. Determine the inter-group delay

Input: Message partition $[M_1, M_2, \ldots, M_k]$, $G(M, L, E)$, link bandwidth g, and scaling factor $\alpha \in [0, 1]$.

Output: Inter-group delay $[t_1, t_2, \ldots, t_{k-1}]$.

1: **for** $i \in 1..k$ **do**
2:     $W_l^i := \sum\{s_j : \forall \ m_j \in M_i \text{ and } (m_j, l) \in E\}, \forall l \in L$ ▷ Link workload in M_i
3:     $W_{max}^i := \max\{W_l^i : \forall l \in L\}$ ▷ Max Link workload in M_i
4: **end for**
5: **for** $l \in L$ **do**
6:     $W_l^{idle} := \sum_{i=1}^{k}(W_{max}^i - W_l^i)$ ▷ The idle duration measured as workload
7: **end for**
8: $l_{critical} := \text{argmin}\{W_l^{idle} : \forall l \in L\}$
9: **for** $i \in 1..k-1$ **do**
10:     $t_i := \alpha \dfrac{W_{l_{critical}}^i}{g}$
11: **end for**

3.5   Complexity Analysis

Now we analyze the complexity of CAMS under the assumption of fat-tree networks. On fat-tree networks, there exists a constant d as the upper limit of path lengths. Given a batch of n point-to-point messages and the network topology and routing, we construct the message routing graph $G(M, L, E)$ which will take at most $O(nd)$ operations. In the partition process, the sorting algorithm will take $O(n \log n)$ operations. Forming group M_i will take $|M| - \sum_{j=1}^{i-1}|M_j|$ operations, then forming all groups will take at most $k|M| = O(kn)$ operations. As $k \leq n$, the partition step will take $O(n \log n) + O(n^2) = O(n^2)$ as the worst-case complexity. In the inter-group delay calculation process, computing per-group link load and their maximum take at most $O(k|L|)$ operations. Computing the idle duration also takes $O(k|L|)$ operations. Determining the critical link takes $|L|$ operations and the final computation of all t_is takes k operations. Thus CAMS takes $O(nd) + O(n \log n) + O(kn) + 2O(k|L|) + |L| + k$ total operations. Given $k \leq n$ and $|L| = O(n)$, the worst-case complexity of CAMS is $O(n^2)$ with respect to the total number of messages n. In real-world scenarios where $k \ll n$, the complexity is reduced to $O(n \log n)$. This indicates that CAMS enjoys a low algorithmic complexity.

4   Experimental Evaluation

In this section, we validate that scheduling batch point-to-point communications by partitioning can improve communication performance and the proposed CAMS algorithm can produce such partition, on the Tianhe-2A supercomputer with a typical 2D halo-exchange communication pattern.

4.1   Experimental Platform and the Benchmark

The Tianhe-2A supercomputer is currently one of the top ten most powerful supercomputers in the top500 list. It is equipped with a proprietary high-speed network named TH-Express. The network implements a multi-level fat-tree topology and uses static routing, thus makes a perfect platform for validating CAMS.

We choose a 2D halo-exchange pattern to validate CAMS. 2D halo-exchange is a famous communication pattern and represents the most frequent communication pattern in scientific and engineering applications. As illustrated in Fig. 3, the communication forms a 2D-mesh topology and each node sends a message of s bytes to its left, right, upper and lower neighbor. *This pattern is a typical case of neighborhood communications that are widely believed neither to cause congestion nor to be significantly impacted by congestion.* Therefore it is adequate for our validation purpose.

We implement our 2D halo-exchange benchmark directly upon the GLEX RDMA programming interface (the TH-Express alternative to Infiniband Verbs) so that we can inject the messages into the network directly. In our experiments, we first obtain a sequence of computing nodes and their inter-connection topology as part of the full fat-tree. We then schedule the messages onto these nodes using CAMS. After we get the scheduling result, we inject the messages into the network according to the order of groups and the inter-group delay as specified by the scheduling result and wait until messages finish transmitting. The inject-and-wait process is repeated to last for at least 1 s and the average time per iteration is computed to eliminate side-effects impacting data accuracy. The LCD_{lim} parameter is set to 1 so no intra-group contention is allowed. We compare CAMS with naive scheduling where all messages are injected into the network at once.

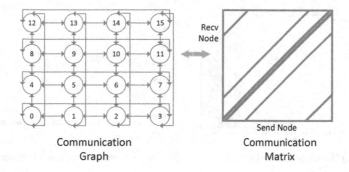

Communication   Communication
Graph   Matrix

Fig. 3. An example of 4×4 2D halo-exchange communication

4.2 Validation of Congestion-Avoiding Message Scheduling

We use 16 nodes to compare a 4×4 halo-exchange instance, where one process occupies one node so no intra-node communication is introduced. We vary the message size s from 4 KB to 1 MB so that it covers a representative range. We try the experiment twice on different sets of nodes on different sub-tree of the fat-tree and the results are shown in Figs. 4 and 5.

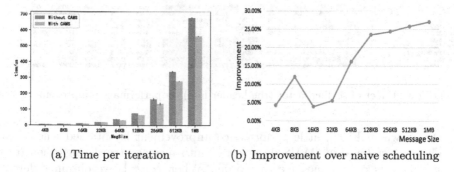

(a) Time per iteration   (b) Improvement over naive scheduling

Fig. 4. Turn-around time and improvement for 4×4 halo-exchange with scaling factor $\alpha = 0.2$ on node set A

The data in Figs. 4 and 5 shows that CAMS improves the performance for all message sizes from small to large. Improvement is not significant for small messages below 32 K, but it grows as the message size gets large. For messages over 64 KB, CAMS improves at least 15% the performance. And the maximum improvement of 27% is achieved in 1 MB message size. The improvement retains with two different node sets, although the numbers vary. This validates that even for neighborhood communications such as 2D halo-exchange, CAMS can improve their performance by actively reduce network contention just in time.

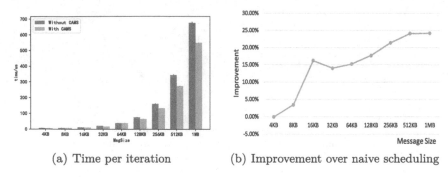

(a) Time per iteration   (b) Improvement over naive scheduling

Fig. 5. Turn-around time and improvement for 4×4 halo-exchange with scaling factor $\alpha = 0.2$ on node set B

4.3 Effect of the Scaling Factor α

The scaling factor α is introduced to improve network utilization while still maintaining acceptable network contention. We now analyze how α can impact the performance improvement of CAMS using the same 4×4 halo-exchange instance as in Sect. 4.2. We vary α from 0 to 1 and the message size s from 4 KB to 1 MB and present the improvement over naive scheduling in Fig. 6.

Improvement	0	1/15	2/15	1/5	0.266667	1/3	2/5	7/15	0.533333	3/5	2/3	11/15	0.8	13/15	14/15	1
size = 4KB	0.01%	0.44%	0.21%	-0.08%	-1.08%	0.84%	1.34%	3.52%	2.65%	3.20%	3.72%	3.17%	3.83%	2.30%	4.50%	-0.97%
size =8KB	-0.87%	1.73%	2.03%	3.42%	4.65%	3.08%	4.71%	6.14%	6.86%	5.02%	3.93%	2.47%	0.93%	-1.62%	-2.30%	-2.25%
size =16KB	-1.74%	2.13%	13.37%	16.21%	17.01%	16.34%	10.97%	6.92%	4.70%	4.30%	1.65%	2.34%	1.24%	1.83%	3.77%	6.26%
size =32KB	3.66%	8.10%	24.16%	14.02%	6.90%	3.36%	4.07%	7.17%	14.18%	16.92%	15.83%	16.53%	18.17%	17.11%	17.58%	15.50%
size =64KB	-2.65%	3.66%	1.19%	15.26%	2.74%	19.14%	18.30%	17.61%	16.00%	16.24%	14.33%	14.18%	14.03%	12.50%	11.34%	10.35%
size =128KB	1.83%	20.71%	16.18%	17.69%	17.34%	16.64%	15.57%	15.04%	12.37%	13.50%	13.98%	12.39%	12.84%	11.76%	10.30%	8.60%
size =256KB	5.63%	1.24%	21.85%	21.38%	21.34%	18.98%	18.92%	19.68%	18.03%	17.15%	16.76%	15.30%	14.55%	12.97%	10.41%	7.53%
size =512KB	2.34%	25.35%	25.95%	24.09%	23.49%	22.72%	20.99%	19.80%	20.84%	18.60%	16.47%	15.27%	17.42%	13.99%	11.06%	10.91%
size =1MB	1.59%	25.30%	26.74%	24.18%	23.07%	23.00%	20.79%	19.26%	20.05%	17.43%	13.64%	13.71%	14.40%	-19.99%	-24.22%	-27.57%

Fig. 6. Effect of scaling factor (the header line) on performance improvement

As shown in Fig. 6, little performance improvement is obtained when $\alpha = 0$. This is just because CAMS degenerates into message reordering when $\alpha = 0$ and there is only 1 message to reorder. When $\alpha = 1$, we obtain different improvement with different message sizes and a downgrade with 1 MB message. $\alpha \in [0.2, 0.8]$ produces stable performance improvements and it seems smaller α tends to produce higher improvements. This data shows that keeping moderate contention may still be beneficial.

5 Application Prospects

In this section, we discuss the application prospects of CAMS. In summary, the proposed algorithm can be embedded in the application to improve the performance of every batch point-to-point communication by reducing network contention actively just in time with often negligible extra cost, and can be used in conjunction with other communication optimization techniques to further improve communication performance.

Online Per-Pattern Message Scheduling: CAMS takes a description of a batch point-to-point communication as the input and produces a schedule plan as the output. Thus it can be embedded into most parallel scientific and engineering applications, as well as parallel programming frameworks. The application uses CAMS to schedule the messages before actually injecting them into the network. The algorithm itself introduces an overhead, which can out-weight the performance benefits of contention reduction. Luckily, since parallel applications usually exhibit repetitive communication patterns and each communication pattern repeats hundreds to thousands of times, the application can call CAMS once to generate a schedule plan and reuse it, thus reduces the overhead of CAMS to be negligible.

Pipelined Communication Optimization: CAMS fits well in the communication optimization pipeline. Job schedulers firstly minimize potential network contention by allocating computing nodes with minimal link sharing. Then application-level topology mapping techniques can be applied to place processes so that inter-node communication is minimized. After that, the application itself can apply techniques such as message aggregation to improve network utilization. CAMS can then kick in to reduce network contention before actual message transmission. Finally, runtime systems such as MPI and GASNet can optimize message transmission by making the best use of architectural features of the underlying network.

6 Related Works

Since network contention is a significant source of communication performance penalty, lots of efforts are devoted to either mitigate or reduce it.

The most obvious approach would be the congestion control mechanisms built in every modern high-performance computer network (with Infiniband [3] being the most well-known one), such as those in [10,15,16,24]. These methods are commonly referred to as end-to-end congestion control, where the sender will be asked to slow down packet injection when congestion is detected in the transmission path. Adaptive routing algorithms are proposed to reduce contention by taking less contended routes when congestion is detected [2,5,8]. These approaches can mitigate contention but with several limitations due to their unawareness of application-level traffic characteristics. For example, even adaptive routing leaves several cases unhandled [9]. Zahavi et al. proposed in [28] a static routing algorithm for fat-trees where an all-to-all shift communication pattern is congestion-free.

Topology mapping is another popular approach to improve communication performance [1,4,12,13]. These algorithms map processes to CPU cores to reduce contention ahead of time, but require knowledge of application communication topology and is not suitable for applications with dynamic communication patterns.

Luo et al. explore in [20] a way to limit the message rate of sending cores on multi-core systems to reduce congestion caused by concurrent message injection. The idea of message scheduling is explored by Lavrijsen et al. in [18,19], where a network-status-aware message reordering algorithm is used to reorder asynchronous point-to-point messages in the MPI runtime. However, these algorithms can not distinguish application communication patterns and can only reduce contention for messages in a limited queue and is passive in nature. Application-level activate contention reduction is employed to optimize FFT in [6] where the one-shot all-to-all communication in FFT is decomposed into many point-to-point communications. This technique can only be applied in special applications since it relies on specific algorithmic structure.

7   Conclusion and Future Works

This paper explores a new way to reduce network contention and improve communication performance by actively reducing contention just in time in the parallel application through congestion-avoiding message scheduling, namely CAMS. With the help of communication pattern and network topology and routing knowledge, CAMS partitions the messages into disjoint groups where intra-group contention is reduced and determines an inter-group delay so that the network utilization is improved without incurring performance penalty of congestion. In this way, the algorithm improves up to 27% the performance of a 2D halo-exchange pattern on the Tianhe-2A supercomputer. Since CAMS works per communication batch online, it applies to applications with dynamic communication patterns. Since it retains the rank order, it can be used in conjunction with topology mapping techniques to further improve communication performance.

A limitation of the current algorithm is that it relies on a congestion-free communication pattern existing on the fat-tree network. One of our future work would be to extend the algorithm to more network typologies. We also look forward to parallelizing the algorithm so that its performance is improved.

Acknowledgement. The authors would like to thank the National Supercomputer Center in Guangzhou for providing the experimental platform and tremendous help on usage of the Tianhe-2A supercomputer. This research was supported partially by Science Challenge Project (No. TZ2016002), National Key R&D Program (No. 2016YFB0201300) and Defense Industrial Technology Development Program (C1520110002). The authors also thank the reviewers for their helpful comments.

References

1. Agarwal, T., Sharma, A., Laxmikant, A., Kalé, L.V.: Topology-aware task mapping for reducing communication contention on large parallel machines. In: Proceedings 20th IEEE International Parallel and Distributed Processing Symposium, pp. 10-pp. IEEE (2006)
2. Alverson, R., Roweth, D., Kaplan, L.: The Gemini system interconnect. In: 2010 18th IEEE Symposium on High Performance Interconnects, pp. 83–87. IEEE (2010)

3. InniBand Trade Association: Inniband architecture specification, vol. 1 & 2, release 1.2, October 2004
4. Bhatelé, A., Bohm, E., Kalé, L.V.: Optimizing communication for Charm++ applications by reducing network contention. Concurr. Comput. Pract. Exp. **23**(2), 211–222 (2011)
5. Chen, D., et al.: The IBM Blue Gene/Q interconnection fabric. IEEE Micro **32**(1), 32–43 (2011)
6. Doi, J., Negishi, Y.: Overlapping methods of all-to-all communication and FFT algorithms for torus-connected massively parallel supercomputers. In: Proceedings of the 2010 ACM/IEEE International Conference for High Performance Computing, Networking, Storage and Analysis, SC 2010, pp. 1–9. IEEE (2010)
7. Escudero-Sahuquillo, J., et al.: A new proposal to deal with congestion in infiniband-based fat-trees. J. Parallel Distrib. Comput. **74**(1), 1802–1819 (2014)
8. Faanes, G., et al.: Cray cascade: a scalable HPC system based on a Dragonfly network. In: Proceedings of the International Conference on High Performance Computing, Networking, Storage and Analysis, p. 103. IEEE Computer Society Press (2012)
9. Gomez, C., Gilabert, F., Gomez, M.E., López, P., Duato, J.: Deterministic versus adaptive routing in fat-trees. In: 2007 IEEE International Parallel and Distributed Processing Symposium, pp. 1–8. IEEE (2007)
10. Gran, E.G., et al.: First experiences with congestion control in infiniband hardware. In: 2010 IEEE International Symposium on Parallel and Distributed Processing (IPDPS), pp. 1–12. IEEE (2010)
11. Gustafson, J.L.: Reevaluating Amdahl's law. Commun. ACM **31**(5), 532–533 (1988)
12. Hoefler, T., Jeannot, E., Mercier, G.: An overview of topology mapping algorithms and techniques in high-performance computing, chap. 5, pp. 73–94. Wiley, Hoboken (2014). https://doi.org/10.1002/9781118711897.ch5
13. Hoefler, T., Snir, M.: Generic topology mapping strategies for large-scale parallel architectures. In: Proceedings of the International Conference on Supercomputing, pp. 75–84. ACM (2011)
14. Jain, N., Bhatele, A., Robson, M.P., Gamblin, T., Kale, L.V.: Predicting application performance using supervised learning on communication features. In: Proceedings of the International Conference on High Performance Computing, Networking, Storage and Analysis, p. 95. ACM (2013)
15. Jiang, N., Becker, D.U., Michelogiannakis, G., Dally, W.J.: Network congestion avoidance through speculative reservation. In: IEEE International Symposium on High-Performance Computer Architecture, pp. 1–12. IEEE (2012)
16. Jiang, N., Dennison, L., Dally, W.J.: Network endpoint congestion control for fine-grained communication. In: Proceedings of the International Conference for High Performance Computing, Networking, Storage and Analysis, SC 2015, pp. 1–12. IEEE (2015)
17. Kamil, S., Oliker, L., Pinar, A., Shalf, J.: Communication requirements and interconnect optimization for high-end scientific applications. IEEE Trans. Parallel Distrib. Syst. **21**(2), 188–202 (2009)
18. Lavrijsen, W., Iancu, C.: Application level reordering of remote direct memory access operations. In: 2017 IEEE International Parallel and Distributed Processing Symposium (IPDPS), pp. 988–997. IEEE (2017)
19. Lavrijsen, W., Iancu, C., Pan, X.: Improving network throughput with global communication reordering. In: 2018 IEEE International Parallel and Distributed Processing Symposium (IPDPS), pp. 266–275. IEEE (2018)

20. Luo, M., Panda, D.K., Ibrahim, K.Z., Iancu, C.: Congestion avoidance on manycore high performance computing systems. In: Proceedings of the 26th ACM International Conference on Supercomputing, pp. 121–132. ACM (2012)

21. Madduri, K., et al.: Gyrokinetic toroidal simulations on leading multi- and many-core HPC systems. In: Proceedings of 2011 International Conference for High Performance Computing, Networking, Storage and Analysis, SC 2011, pp. 1–12. IEEE (2011)

22. Márquez, C., César, E., Sorribes, J.: A load balancing schema for agent-based SPMD applications. In: Proceedings of the International Conference on Parallel and Distributed Processing Techniques and Applications (PDPTA), p. 12. The Steering Committee of the World Congress in Computer Science, Computer Engineering and Applied Computing (WorldComp) (2013)

23. Oliker, L., et al.: Scientific application performance on candidate petascale platforms. In: 2007 IEEE International Parallel and Distributed Processing Symposium, pp. 1–12. IEEE (2007)

24. Pfister, G., et al.: Solving hot spot contention using infiniband architecture congestion control. In: Proceedings HP-IPC 2005, p. 6 (2005)

25. Valiant, L.: A bridging model for parallel computation. Commun. ACM **33**(8) (1990). https://doi.org/10.1145/79173.79181

26. Vetter, J.S., Mueller, F.: Communication characteristics of large-scale scientific applications for contemporary cluster architectures. In: Proceedings 16th International Parallel and Distributed Processing Symposium, pp. 10-pp. IEEE (2001)

27. Zahavi, E.: D-Mod-K routing providing non-blocking traffic for shift permutations on real life fat trees. CCIT Report 776, 840 (2010)

28. Zahavi, E., Johnson, G., Kerbyson, D.J., Lang, M.: Optimized infiniband TM fat-tree routing for shift all-to-all communication patterns. Concurr. Comput. Pract. Exp. **22**(2), 217–231 (2010)

Applications of Distributed and Parallel Computing

An Open Identity Authentication Scheme Based on Blockchain

Yuxiang Chen[1,2]($\boxtimes$), Guishan Dong[1], Yao Hao[1,2], Zhaolei Zhang[1],
Haiyang Peng[1], and Shui Yu[3]

[1] No. 30 Inst, China Electronics Technology Group Corporation,
Chengdu 610041, Sichuan, China
2392827595@qq.com
[2] Science and Technology on Communication Security Lab,
Chengdu 610041, Sichuan, China
[3] School of Software, University of Technology Sydney,
Sydney, Australia

Abstract. With the development of Public Key Infrastructure (PKI), there implements lots of identity management systems in enterprises, hospitals, government departments, etc. These systems based on PKI are typically centralized systems. Each of them has their own certificate authority (CA) as trust anchor and is designed according their own understanding, thus formalizing lots of trust domains isolated from each other and there is no unified business standards with regard to trust delivery of an identity system to another, which caused a lot of inconveniences to users who have cross-domain requirements, for example, repeatedly register same physical identity in different domains, hard to prove the validity of an attestation issued by a domain to another. Present PKI systems choose solutions such as Trust list, Bridge CA or Cross-authentication of CAs to break trust isolation, but practice shows that they all have obvious defects under existing PKI structure. We propose an open identity authentication structure based on blockchain and design 3 protocols including: Physical identity registration protocol, virtual identity binding protocol and Attribution attestation protocol. The tests and security analysis show that the scheme has better practice value compared to traditional ones.

Keywords: Open identity · Authentication · Trust delivery ·
Blockchain PKI

1 Introduction

1.1 Background

Identity management is one of the key technology of information security, which includes identity's establishment, description, definition, cancellation, etc.

Supported by National Key Research and Development Program of China (2017YFB0802300) and (2017YFB0802304). Science and technology projects in Sichuan Province (2017GZDZX0002) and Sichuan Science and Technology Program No. 2018JY0370.

© Springer Nature Switzerland AG 2020
S. Wen et al. (Eds.): ICA3PP 2019, LNCS 11944, pp. 421–438, 2020.
https://doi.org/10.1007/978-3-030-38991-8_28

In terms of application, it is related to the identification of people and equipment, access control and so on. The identification contents of access control includes MAC addresses, IP addresses, DNS (Domain Name System) analysis, etc. According to ITU (International Telecommunication Union)'s induction, identity management contains following aspects: protection of user's identity attestation and account privacy, security of operators and providers, economic requirements, management of government and enterprise, public service requirements, cyber security, public policy requirements, NGO (Non-Governmental Organization)'s privacy protection requirements, etc. [1].

Entering the era of big data, continuously increasing data is putting more and more pressure on the centralized identity management system, who has a high risk of being the target of adversary. At the same time, user cannot control their identity completely, moreover, they don't know whether their centralized stored data is leakage. Under the circumstances of such situation, Christopher Allen put forward self-sovereign identity of ten principles, which includes: Existence, Control, Access, Transparency, Persistence, Portability, Interoperability, Consent, Minimization, Protection, etc. These principles described that user entity should exist independently, control their own identity and data, visit their data directly, whereas identity itself should have the property of persistent existence, light portability, interoperability and being widely used, while the attestation of users should be minimized and user rights are protected [2–4].

1.2   Related Work in Traditional Authentication

Traditional identity management is typical centralized management, especially that massive authenticated user's data stores in third party's storage which makes it extremely dangerous for data management and trust delivery under the circumstance of widely used network.

Traditional PKI System: PKI (Public Key Infrastructure) is the foundation of current network security, which makes use of asymmetric cryptography system as infrastructure. It establishes security communication by using certificate to bind the owner and public key together. Using cryptographic method to verify the certificate to provide trust foundation for users on the platform. With the development of CA (Certificate Authority) in China and other countries, there are lots of PKI trust domains who are isolated from each other. Meanwhile there is no unified PKI standard or management which in turn makes it more difficult to transfer trust. Currently the trust transferring methods between domains usually includes: Bridge CA method, Cross-authentication between CAs, and CA trust list method. The choice of the methods usually depends on the practical environment, which in turn forms different structure systems. The main methods and their characteristics are shown in Table 1 [5–9].

We can see that CA's number and hierarchy are main facts that influence identity management's performance. Current PKI system usually chooses one or mix of several models to break isolation between domains or organizations according to actual conditions, but the trust isolation and uncontrollable factors are deepened among the trust domains at a higher level. Overall, bridge CA

Table 1. Comparison of typical cross authentication.

Concepts	Model				
	Strict hierarchical authentication trust model	Reticulation trust model	Bridge CA trust model	Web trust model	User-centric trust model
Trust anchor (centered)	Unique root CA	Local CA issuing certificate	Local CA issuing certificate	Previously installed trust CA	Users determine
Trust chain construction	Unique certificate path from top to bottom	Complicated, multiple path, deadlock	Easy, all cross-domain certificate pass through bridge CA	Simple, one path from top to bottom	Medium, decided by user
Trust domain expansion	Only trust root CA, cannot expand	Easy	Easy	Previously installed cannot expand	Easy
Trust domain expansion	Only trust root CA, cannot expand	Easy	Easy	Previously installed cannot expand	Easy
Trust establishment	Trust from top to bottom	Bilateral trust between CAs	Unilateral trust inside each domain, bilateral trust between domain and bridge CA	Several bilateral trust from top to bottom	Users decide
Applied environment	Within one organization	Within one organization or multiple organizations	One trusted bridge CA with multiple organizations	Several organizations supported by browser vender	Small domains recognized by users
Dependence on LDAP server	Low	High	Medium	Low	Low
Model expansion direction	From top to bottom	Equality between CAs	Growing through bridge CA	Previously installed in browser	Users decide

seems the best to extend trust domains and construct trust chain, but it's hard to find a trusted third party who acts as the bridge CA in practice [10,11].

OAuth Protocol: OAuth (Open Authentication) is an authorization layer set between user client and service provider, allowing third-party's website to access user's information stored in the service provider under the authorization of the user, user will not provide their own account passwords, but will provide third-party websites in the form of tokens, where the user defined the token's scope, permission and validity of time. OAuth system structure is shown in Fig. 1.

OAuth protocol provides an unified authentication solution to realize trust transferring and returns part of the data's management right to the user to some extent. The third party only accesses the user data after obtaining the permission of the user. However, the user's permission to access is too extensive, the third party can access all the data stored after obtaining the access right and get a lot of unnecessary data at the same time [12].

FIDO (Fast Identity Online): FIDO is committed to transforming today's large number of password-based online authentication technologies, eliminating user's reliance on passwords. It includes "no password experience" (biometric) UAF standard and "bi-factor experience" (passwords and specific devices) U2F standard, which can be used to solve problems of different authentication technologies that may cause the domains' isolation to a certain extent. FIDO's related parts are shown in Fig. 2.

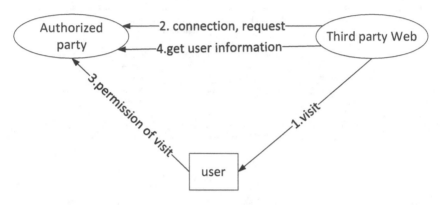

Fig. 1. Structure of OAuth.

FIDO's non password experience better solved the cumbersome problem of user authentication process and used bi-factor to strengthen security. However, from the perspective of identity management's security, user's authenticated data still stored in centralized databases, users don't know who use, modify or delete their data, with more data breaches appearing, FIDO is just focusing on a small part of the whole target [13–16].

Except for drawbacks in trust delivery of traditional solutions above, there is a lot of thinking about the future development of trust delivery [24–27]. The most representative scheme is put forward by Canetti et al. [23] who suggest Global PKI that can be used for universally composable authentication and key-exchange, which can solve trust transferability and non-repudiable proof problems. However, it is costly to subvert existing isolated PKI systems and facing a lot of obstacles. In addition to the certificate authentication method, a lot of countries such as US, India, Pakistan attempted to collect citizens' biometric identities in a unified way, none of them have been identified as successful because of privacy, cost, security and other considerations [28]. Although there are hundreds of improved methods in the field of authentication [29], they raised the threshold of identity association and tracking in different applications objectively, which caused lots of inconvenience, for example: users have to manage lots of accounts and it's hard for supervise department to correlate identities of different applications that may cheats users. We try to break the isolation by using blockchain to solve trust delivery problem.

1.3 Contribution

In this paper, to solve the trust delivery and user identity management problems discussed above, we propose a blockchain-based open identity authentication scheme, which can not only break the isolation of different identity domains, but can also balance the requirements of all kinds of participants. Under this structure, 3 protocols are designed, which includes: Physical identity registration protocol, Virtual identity binding protocol and Attribution attestation protocol, the first two protocols realize mapping from physical identity in reality to virtual identity in network, which makes it possible for supervision departments to trace back responsibility of illegal activities while still protects users' interests such as privacy. The rest attribution attestation protocol use blockchain instead of identity provider in attribute proof, which means any node who has a ledger in blockchain can provide validity verification.

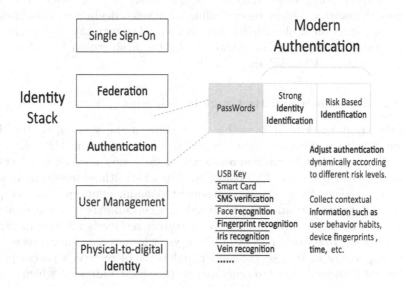

Fig. 2. Scope of FIDO.

Organization of the paper is shown as follows. In Sect. 2, we describe the key technologies' characteristics, its usage and goals in the scheme. In Sect. 3, we describe the details of open identity structure model and three protocols. We perform security analysis in Sect. 4 and evaluate its performance with centralized system (without blockchain) under the same experiment condition in Sect. 5. We conclude in Sect. 6 at last.

2 Current Technologies and Its Usage in Scheme

Traditional PKI systems mentioned above face a lot of troubles. Existed methods attempted to provide user better service, stronger data protection from perspective of system structure, user login method and authorization, etc. But absolutely centralized structure decides these attempts cannot thoroughly solve the problems, which mainly include: centralized databases's pressure of management and protection from rapidly growing data, cross-domain trust transferring, privacy and security of data. This paper realised a prototype system that combined blockchain and PKI, which have a better effect on the main problems above.

2.1   Blockchain and Its Usage in Scheme

Blockchain is the foundation of famous bitcoin, who has been running safely as a distributed, decentralized system for decades, which also subverts the centralized management system architecture. Blockchain has property of distribution, transparency, tamper-resistance, which are suitable to combine with identity system. The open identity authentication scheme uses blockchain to act as a trust transmitter, which objectively lowers the threshold for identity to be used in different domains and gives users flexible right when dealing with their data. With the introduction of blockchain, there is no need to maintain the centralized database, data's authenticity is guaranteed by the hash stored by CA or IDP (Identity Provider) on blockchain [2,17,19].

2.2   Stealth Address and Its Usage in Scheme

Considering that blockchain is visible to anyone in the system, so the data stored on it is its hash result. Similarly, Users care much about their data privacy in the open identity environment, where they may encounter adversary who analyzes their behavior. This paper introduced stealth address usually used in Crypto-currencies to identity management domain, which will better protect user's information from being analyzed to a certain extent. As is shown in Fig. 3, stealth address's typical usage is to protect receiver's identity in digital transaction area. Receiver's parent public keys (which is also known as stealth address) is publicly known, so the sender makes use of receiver's parent public key pairs and random number to generate one-time used addresses, which is publicly released with transactions. Only the true receiver who has corresponding secret key can decrypt it. This scheme chooses derived algorithm same to that used in CryptoNote [20], which can be briefly described as follows:

Sender Alice gets receiver Bob's public key pairs (A,B), She generates random number r and compute one-time used key pairs (P, R), where one-time used address is $P = H_S(rA)G + B$, one-time used public key is $R = rG$. Alice encrypts plain text like money amount with public key R to get cipher text and publicly releases it in the form of $(P, R, ciphertext)$. Anyone checks it with their private key (a, b) by computing $P' = H_S(aR)G + bG$ to check whether $P' = P$. Then compute secret key $x = H_S(aR) + b$ if the first checking result is right and finally decrypts the cipher text.

3 Model Structure and Protocol Design

3.1 Open Identity Authentication Structure

Understanding the key technologies introduced above, readers can better understand the scheme shown in Fig. 4. The scheme combines of user, Agent server, IdP(Identity Provider), blockchain and relying parties who maintain their own application servers. This scheme designed an identity management app which can be installed on user's phone. User operates app on the phone to register, get services from different relying parties, get authenticated identity attributions from IdP, who stores authenticated data's hash on the blockchain. Blockchain acts as the role of trust delivery, which guarantees information's authenticity in the network. Compared with traditional identity management in Table 1, new structure based on blockchain almost integrates all kinds of tradition PKI's advantages. Open identity structure gives users more initiative in the aspect of data control which is similar to OAuth's intent, releases Idp's pressure from storing massive data. With regard to trust delivery, this structure can be considered as replacing the bridge CA or trust list with a blockchain, which potentially makes it easier for users who have cross-domain requirements, Network participants no longer doubt the credibility of bridge CA maintained by third parties. As for authentication part, scheme absorbed thoughts of FIDO which can be tell from combination of password and face recognition when authenticate. The two-factor authentication used on App draws on FIDO's idea of "bi-factor experience".

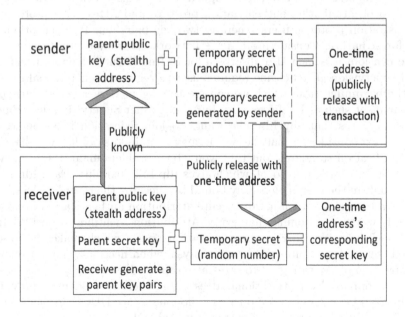

Fig. 3. Usage of stealth address

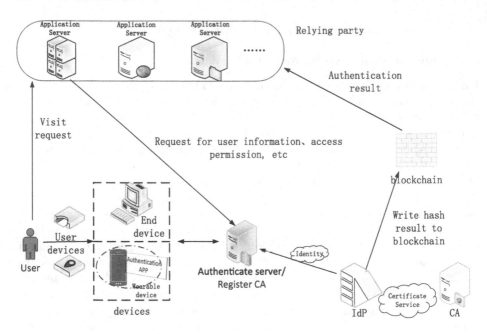

Fig. 4. Structure of open identity authentication scheme

Under the structure above, we designed 3 protocols which include registration protocol, attribution authentication protocol and Attestation protocol. In the registration protocol, users generates two key pairs, one for encryption, the other for signature. Considering today's network environment is so open that anyone can join it, which also absorbs a lot of fraudulent behaviors because of lack of supervision. The Scheme requires user to register their physical identity from the beginning, from which we can see the Agent Server better suits public department such as Public security bureau or other Supervision department. Scheme supposes that any Idp who wants to join federation must be strictly audited. Next considering that the system may contain lots of Idps, who have different verification standards, but that won't be a problem under the framework of Binding Attribution protocol, which gives Idp full flexibility. Each Idp firstly need to confirm the user is physically bound by checking the record of AS (Agent Server) on blockchain, then it can execute attribution authentication in its own way. It is worth mentioning that every Attribution identifier is derived from initial blockchain identifier in the registration phase, which makes it possible for agent server to execute correlation analysis when necessary while irrelevant adversary cannot. With regard to attestation, procedure actually becomes simpler when compared with traditional attestation, there is no identity provider's participation. Query and record operation no longer face DDOS problems at the same time. Signals in protocols are shown in Table 2.

3.2 Registration and Binding Protocol

Figure 5 shows the registration protocol of the scheme.

Step 1. User first downloads the identity app and registers the items of her account include nickname, passwords, fingerprints and so on, while the app automatically generates signing key pair (spk1, ssk1) and encryption key pair (epk1, esk1).

Step 2. App will then send out public key pairs to let agent server know.

Step 3, 4. AS checks whether the public key pair is already used, it would generate an UID and a certificate for the user if the answer is no. After that, AS returns user with the cert.

Step 5, 6. User verifies certificate and uploads her physical identity information such as her ID card, driver license and photo (face recognition), which are signed and encrypted.

Step 7. AS decrypts uploaded message, verifies signature, check the physical information, then derives key pairs (spk2, R2) from (spk1, epk1) as the users' blockchain identifier, which has no relation with first public key pairs (spk1, epk1) from the perspective of third parties except AS or someone who get random number r2 or secret key pairs (ssk1, esk1) can verify that relationship.

Step 8. Agent Server (AS) writes the hash of identity information to blockchain with user's identifier (spk2, R2).

Step 9, 10. Blockchain checks the signature, records it and return a result to AS to indicate that whether it is successfully recorded (step 9 and 10).

Table 2. The meaning of symbols used in protocols.

Symbol	Meaning
$(epki, eski), i = 1, 2...$	Encryption key pairs, epki is public key, eski is secret key
$(spki, sski), i = 1, 2...$	Signing key pairs, spki is public key, sski is secret key
$Nonce$	Number used once
$PKE_X(m)$	Use participant X's public key to encrypt message m
UID	Original identifier for user
$(spki, Ri), i = 1, 2...$	Derived one time used key pairs, used as blockchain identifier
$xi, i = 1, 2...$	Secret key corresponding to (spki, Ri)
$SIG_X(m)$	It means to sign all the message m before it by using sender X's private key. If the data has a hierarchy, it indicates the signature of all the data in parentheses
$Label$	It indicates the kinds of attribution requirements from Idp or relying party
$Ci, i = 1, 2...$	$Ci = E_{password}(spki, Ri)$, it stores in agent server to help user to find account back in case the user use a new device
AS/CA	Agent Server/Certificate Authority

Step 11. If AS get the successful result, it will return the derived key pair (spk2, R2) with result to user after signing and encryption.

Step 12, 13. User will decrypt, verify and compute corresponding secret key x2, use password to encrypt it to get result C2 and return it to AS.

Step 14. AS gets the returned message, store C2, and mark identity label as bound.

3.3   Binding Attribution Protocol

Figure 6 shows the attribute binding procedure.

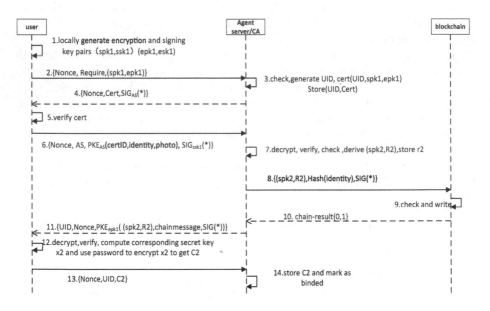

Fig. 5. Registration protocol

Step 1, 2. User sends AS her request for attribution binding, identity and attributes with her blockchain identifier (spk2, R2). Agent server checks user state and sends the idp with its related information including certs, blockchain identifier, identity and attribution.

Step 3, 4, 5. Idp (identity provider) decrypts and verifies the message, then query the blockchain with user's blockchain identifier (spk2, R2), get the identity's hash result in order to verify its validity.

Step 6, 7. After confirming the blockchain identity's validity, Idp will then check the submitted attribution, use random number r3 to generate corresponding identifier (spk3, R3) to mark attribution, of which r3 is designated by AS and has the meaning of supervision. Send the plain text's hash with its new identifier to blockchain to store.

Step 8, 9. Blockchain verifies the signature, records it and returns chain-result to indicate whether the writing is successful.

Step 10, 11, 12. Idp returns authenticated hash result AS with its identifier (spk3, R3), AS decrypts message, verifies signature, verifies (spk3, R3)'s validity by using r3, modifies the attribute states corresponding to (spk3, R3) as authenticated. AS then forwards the information to the user.

Step 13. Finally, the user can decrypt, verify and compute secret key x2, which can be used to sign in new Idp domain.

3.4   Attributes Attestation

Figure 7 shows the attestation of user to relying parties who also trust the same IDP.

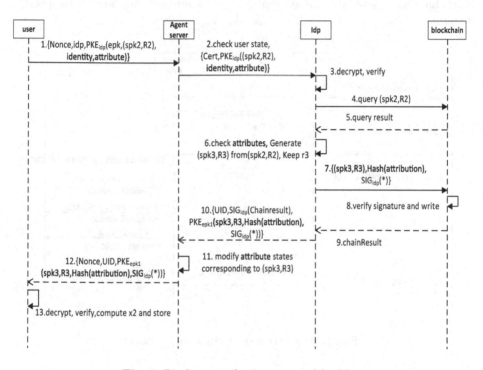

Fig. 6. Binding attribution protocol by Idp

Step 1, 2. When user wants to get relying party's service, she needs to satisfy its access control condition. RP (Relying Party) scans user's QR code to get its UID, which will redirect the RP to AS (Agent Server), which also means RP's request for specific attestation of user's attribution.

Step 3, 4. AS sends the user pre-settled data including the attestation's type "Lable". User will return back the corresponding data encrypted by RP, which will be forwards to RP by AS.

Step 5, 6, 7, 8. RP decrypts, verifies the signature and attribution, and queries the validity of the attribution on blockchain. RP then uploads authentication result to blockchain to record, and get its consequent.

Step 9, 10. Finally, RP returns the user its verifying result which is relayed at the AS.

4   Security Analysis of Scheme

4.1   User Privacy and Validity of Certificate

With the introduction of blockchain, user's data can be divided into three parts: such as identifiers $(spki, Ri), i = 1, 2...$ publicly known on blockchain, specific attribution that only the user knows which can be authenticated by the hash on blockchain and metadata like attribution label kind that only known to specific domains.

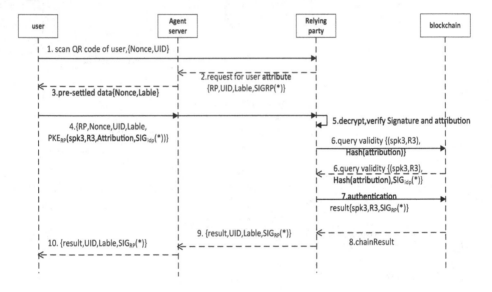

Fig. 7. Attestation of attribution protocol

There is no plain text of user's sensitive data stored in centralized database, which is different to all the traditional structure analyzed in Sect. 1.2. Also the trust transmit becomes much easier by using blockchain, there is no more need to worry about third parties' trustworthiness for participants.

4.2   Prevention of Replay Attack

Considering that replay attack is one of the most common attack methods used by adversaries in the network. Traditional challenge-response solution increases the communication overhead of the system, which needs at least 3 interactions between sender and receiver. This scheme uses nonce (Number used once) combined with time stamp instead to guarantee the same effect while still reduce the number of interactions from 3 to 2 [21].

Usually each HTTP request needs to be signed with time stamp, if the current receiving time exceeds the sending time by 60 s, the request would be judged to be illegal. Under the condition of most implementations currently, it takes hackers far more than 60 s to replay request from scratch. However, if the adversary is advanced enough to finish the attack within 60 s, time stamp used alone cannot guarantee that the request is valid.

Nonce can solve that extreme situation perfectly, it is a random string that is used only once in a specific context, if the returned message is within 60 s and the returned nonce matches the nonce previously used in the collection, this message would be classified as illegal. Because nonce only works within the range of 60 s, there is no need to worry about the increase of nonce collection maintained in storage, which can be cleared regularly while still keeps the verification function in extreme cases mentioned above.

4.3 Prevention of DDOS (Distributed Denial of Service)

Use blockchain to realize every certificate hash's distributed storage objectively makes it possible to relieve pressure of server's centralized database. Attestation protocol design gets rid of interaction with Idp, but use blockchain instead. Blockchain as a trust credential has the characteristics of multi-node backup information. Even if an attacker breaks a node, other nodes can still provide attestation services to exchange promises, which will effective resist DDoS and enhance system security.

5   Analysis of System Test and Feasibility

Blockchain itself has no plain text but stores hash of plain text instead. Data of any length is converted into length of 32 bytes by using hash functions. Lots of hash results will be packed as a block and dealt in the way of Merkle tree when its quantities reach a certain amount. The benefit of using Merkle tree is that if there exists no follow-up transaction for the current transaction. It can be deleted and replaced with its hash, which will not change its cryptographic security and integrity while minimised its storage.

The prototype of the scheme is realized based on Consortium blockchain Hyperledger Fabric, experiment parameters are shown in Table 3. The blockchain combines of 5 nodes, which includes one orderer and four peers, each of them contains same ledger. This paper analyzed system performance from perspective of registration, binding attribution and attestation which are related to protocols designed above and compared the scheme (circle line) with traditional scheme (without blockchain star line in following fig) under the same condition. In each phase, x-axis is transactions submitted per second (TPS) which means concurrent operations of registration, binding attribution and attestation per second, while the y-axis is the total consuming time, whole submitted transaction's success rate and throughput of the system, where throughput is the rate at which transactions are committed to the ledger.

Table 3. Parameters of experiment environment.

Items	Explanation
Consensus algorithm	SOLO
Smart contract	Node SDK
Size	One Orderer, four peers
Distribution	Single host
TPS	Transactions per seconds

5.1   System Performance of Registration Protocol

Figure 8 shows the system performance of registration. With the growing of TPS (Transactions per second), the time consumption rose rapidly after TPS reached 200. The success rate is always 100% and is not affected by the number

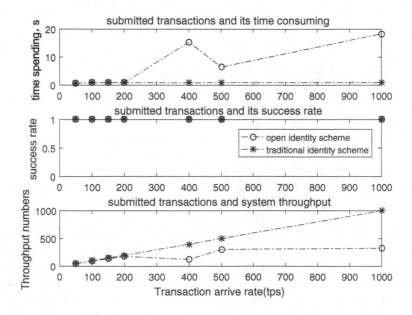

Fig. 8. System test of registration protocol

of concurrent registrations while the throughput is stable within 400 after TPS exceeds 500. As for traditional scheme (star line), it performs extremely good in aspects of consuming time and throughput under the same condition when tps exceeds 200, because submitted transactions don't need to be processed by blockchain system, such as cryptographic process, data synchronization of blockchain nodes, which saved a lot of resources.

5.2   System Performance of Binding Attribution Protocol

Figure 9 reflects the system performance of binding attribution phase. This phase includes more interactions with blockchain such as write and query operation, the consuming time and throughput is steadily growing up within 500 TPS, but the success rate dropped significantly after TPS exceeds 200. When compared with traditional scheme without blockchain, it performs similarly with the registration phase because of similar reasons.

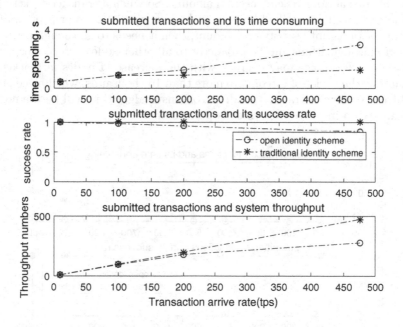

Fig. 9. System test of binding attribution protocol

5.3   System Performance of Attestation Protocol

With regard to attestation (Fig. 10) phase, the system performs better than first two phases. Consuming time have a significant rise after TPS exceeds 500. The success rate always keeps at a high level while the throughput is similar to the other two phases. But this time, new schemes performs better than traditional scheme, not only consuming time costs less, but also bigger in throughput. That's because the blockchain consists of 5 nodes, each one has full ledger that can provide attestation service, it is 5 versus 1 when compared with traditional scheme in this case, which means if more nodes join the prototype system, it can performs better in this phase.

5.4   Comprehensive Analysis of System

Overall, when the tps is within 200, the designed prototype system can guarantees the normal operation of first two protocols and performs even more effective than traditional scheme in attestation phase while solves the problems of traditional PKI analyzed above, such as trust transmitting, right of user data control, etc. However, the amount of water in the barrel depends on the shortest board, the shortest board in this case is the blockchain, especially when a transaction is written to it. When a transaction is submitted to blockchain, peer node would first verify its signature and transferred it to orderer peer who is in charge of the generation of block. However, a block is only generated when the number of transactions reach a certain amount. So when a transaction is sent to blockchain, it must be verified by peer node first, sent to orderer peer, wait for other transactions until satisfy the amount, then it needs to go though a series of cryptographic algorithms and be broadcast to all other nodes, which means that the system performance may degrade with the increase of nodes in blockchain. For example, the nodes of bitcoin are more than 10 thousands, which needs more than 10 h to confirm a transaction. The consensus algorithm will be the next to be researched to improve the system's performance.

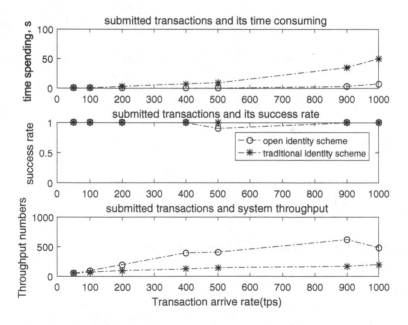

Fig. 10. System test of attestation protocol

In terms of deployment cost, the scheme don't change the original Idp-centric PKI (Public Key Infrastructure) operation mode, which would minimize costs as much as possible. Also the association between the user blockchain identity issued by Idp and original blockchain identifier issued by AS is provided, which makes it possible to trace back responsibility when necessary [22].

6   Conclusion

This paper puts forwards an open identity authentication scheme based on blockchain, introducing blockchain without changing the existing PKI authentication model. Using stealth addresses to derive blockchain identifiers for users to use in different Idp domains, which objectively prevent user's behaviors from analyzing by adversary, while still guarantees its traceability for supervision departments who maintains Agent Server in the scheme. Through the non-modifiable blockchain, it gives participants lots of conveniences with regard to trust delivery, which almost gets rid of DDOS problem. This scheme gives currently isolated identity management systems more flexibility to join the identity alliance. Each system is able to choose its own access control strategy when deal with request from other systems, which would even be potentially more effective in subsequent cross-domain research.

References

1. ITU-T: NGN Identity Management Framework. http://www.itu.int/rec/T-REC-Y.2720-200901-I. Accessed 4 July 2018
2. Allen, C.: The Path to Self-Sovereign Identity. http://www.coindesk.com/path-self-sovereign-identity/. Accessed 4 July 2018
3. UK Government: Identity Proofing and Verification of an Individual. https://www.gov.uk/government/uploads/system/uploads/attachment_data/file/370033/GPG_45_. Accessed 4 July 2018
4. Guel, M.D.: A framework for choosing your next generation authentication/authorization system. Information Security Technical Report 7.1, pp. 63–78 (2002). https://doi.org/10.1016/S1363-4127(02)00107-3
5. Adams, C., Lioyd, S.: Understanding PKI: Concepts, Standards and Deployment Considerations, 2nd edn. Addison Wesley Professional, Reading (2002)
6. Myers, M., Ankney, R., Malpani, A., et al.: X.509 Internet Public Key Infrastructure: Online Certificate Status Protocols. EITFRFC 2560. PKIX Working Group (1999)
7. Weimerskirch, A., Thonet, G.: A distributed light-weight authentication model for ad-hoc networks. In: Kim, K. (ed.) ICISC 2001. LNCS, vol. 2288, pp. 341–354. Springer, Heidelberg (2002). https://doi.org/10.1007/3-540-45861-1_26
8. Ma, M., Meinel, C.: A proposal for trust model: independent trust intermediary service (ITIS). In: Proceedings of the ICWI 2002, pp. 785–790 (2002)
9. Thompson, M.R., Olson, D., Cowles, R., et al.: CA-Based trust model for grid authentication and identity delegation. In: Proceedings of the GGF7 (2003)
10. Comodo: PKI-Public Key Infrastructure - What is it?. https://www.comodo.com/resources/small-business/digital-certificates1.php. Accessed 4 July 2018

11. Peng, B., Liu, J., Gong, Z., et al.: Cross-domain Trust Model Based on Bridge CA. Ship Electronic Engineering (2017)
12. Leiba, B.: OAuth web authorization protocol. IEEE Internet Comput. **16**(1), 74–77 (2012)
13. FIDO Alliance: The FIDO Alliance specification. http://fidoalliance.org. Accessed 4 July 2018
14. Lindemann, R., Baghdasaryan, D., Tiffany, E., et al.: FIDO UAF Protocol Specification v1.0: FIDO Alliance Proposed Standard. https://fidoalliance.org/spces/fido-uaf-v1.0-ps-20141208/fido-uaf-protocol-v1.0-ps-20141208.html. Accessed 4 July 2018
15. Lindemann, R., Baghdasaryan, D., Tiffany, E., et al.: FIDO UAF Protocol Specification v1.0: FIDO Alliance Proposed Standard. https://en.bitcoin.it/wiki/Block. Accessed 4 July 2018
16. Kexin, H.: Research on FIDO UAF Authentication Protocols Security. University of Science and Technology of China (2016)
17. Satoshi, N.: Bitcoin: A Peer-to-Peer Electronic Cash System (2009). http://bitcoin.org/bitcoin.pdf. Accessed 4 July 2018
18. Baars, D.: Towards Self-Sovereign Identity using Blockchain Technology. http://essay.utwente.nl/71274/1/Baars_MA_BMS.pdf. Accessed 4 July 2018
19. Antonopoulos, A.M.: Mastering Bitcoin: Unlocking Digital Crypto-Currencies. O'Reilly Media Inc, Sebastopol (2014)
20. Bergan, T., Anderson, O., Devietti, J., et al.: CryptoNote v 2.0. http://xueshu.baidu.com/s?wd=paperuri. Accessed 4 July 2018
21. Haber, S., Kaliski, B., Stornetta, S.: How do Digital Time-stamps Support Digital Signatures. http://www.x5.net/faqs/crypto/q108.html. Accessed 4 July 2018
22. Hyperledger: Project Charter. www.hyperledger.org/about/charter. Accessed 4 July 2018
23. Canetti, R., Shahaf, D., Vald, M.: Universally Composable Authentication and Key-Exchange with Global PKI. http://xueshu.baidu.com/s?wd=Universally+composable+authentication+and+key-exchange+with+global+PKI. Accessed 4 July 2018
24. Canetti, R., Krawczyk, H.: Universally composable notions of key exchange and secure channels. In: Knudsen, L.R. (ed.) EUROCRYPT 2002. LNCS, vol. 2332, pp. 337–351. Springer, Heidelberg (2002). https://doi.org/10.1007/3-540-46035-7_22
25. Canetti, R.: Universally composable signature, certification, and authentication. In: CSFW, p. 219. IEEE Computer society (2004)
26. Maurer, U., Tackmann, B., Coretti, S.: Key exchange with unilateral authentication: composable security definition and modular protocol design. IACR Cryptology ePrint Archive 2013, 555 (2013)
27. Kohlweiss, M., Maurer, U., Onete, C., Tackmann, B., Venturi, D.: (De-)Constructing TLS. IACR Cryptology ePrint Archive 2014, 20 (2014)
28. Weinberg, J.T.: Biometric identity. Soc. Sci. Electron. Publ. **59**(1), 30–32 (2016)
29. Ding, W., Ping, W.: Two birds with one stone: two-factor authentication with security beyond conventional bound. IEEE Trans. Dependable Secure Comput. **PP**(99), 1 (2016)

RBAC-GL: A Role-Based Access Control Gasless Architecture of Consortium Blockchain

Zhiyu Xu[1], Tengyun Jiao[1($\boxtimes$)], Lin Yang[1($\boxtimes$)], Donghai Liu[1], Sheng Wen[2], and Yang Xiang[3,4]

[1] Swinburne University of Technology, Hawthorn, VIC 3122, Australia
101982028@student.swin.edu.au, linyang@swin.edu.au
[2] Cyberspace Security Research Center, Peng Cheng Laboratory, Shenzhen, China
[3] State Key Laboratory of Integrated Service Networks,
Xidian University, Xi'an, China
[4] School of Software and Electrical Engineering, Swinburne University of Technology,
Melbourne, Australia

Abstract. Blockchain-based decentralized applications (DApps) have been used in various industrial areas. More are more companies are willing to participate in blockchain technology. Notheisen [18] proposed a framework based on Ethereum [3] to trade lemons on the blockchain platform. In this work, we discuss the application aims to digitize valued assets in the commercial area, such as real estate and watches. This article introduces a general DApp framework and some standard attack methods in the beginning. Our proposed novel role-based access control (RBAC) model improves the system permission control, and gasless mechanism enhances security and reliability. It allows users to publish their assets and use cryptocurrency to trade online. Meanwhile, this work can prevent several categories of attacks, such as gas-related attacks and malicious API invokes, according to our improvements. Besides, the efficiency performance of the system remains the same as before.

Keywords: Blockchain · Assets digitization · Role based access control

1 Introduction

Blockchain technology, recognized as an advanced version of distributed system, has attracted plenty of attention from various fields. As a representative of blockchain 2.0 era, Ethereum [3] has been widely used in the finance area to solve the problem of insufficient expansion of bitcoin. Through applying smart contracts, enterprises can customize smart contracts according to their business features in order to implement role-based access control. The core concept of Role-Based Access Control (RBAC) [10] is that users do not have full permissions in a complete system. In contrast, their privileges will be allocated according to

© Springer Nature Switzerland AG 2020
S. Wen et al. (Eds.): ICA3PP 2019, LNCS 11944, pp. 439–453, 2020.
https://doi.org/10.1007/978-3-030-38991-8_29

roles responsibilities. Authentication management will be simplified if a system has a similar scheme. Currently, many industrial fields like banking and agriculture have built their eco-system based on blockchain. Liquid assets management is a potentially valuable area that can combine with blockchain.

In this paper, we proposed a framework for the digitization of assets. We build our consortium chain based on Ethereum and implements our contracts on the EVM (Ethereum Virtual Machine). Currently, assets management is intricate work. It requires people from different fields to corporate jointly. We take the consortium chain as our ledger to store data. In consortium chain, members only have limited permissions to interact with the chain. Consortium chain provides member management, member authentication, and monitoring member. It provides several beneficial characteristics (e.g., reliability and transparency) for our platform. Besides, by using smart contracts, we can implement role-based access control in the system. For different categories of clients, we restrict their permissions in order to enable them to visit limited resources. In this way, we can not only manage our users hierarchically but also prevent malicious operations that can harm the system.

In our testing phase, a cyber attack hits our system. Attackers manipulated accounts to transfer Ethers from our address to their addresses. Fortunately, we were still in the testing phase and had the chance to address the security problems. Moreover, since we are using consortium chain, the attackers did not get any substantial benefit. By analyzing the attack model, we build an RBAC-GL: role-based access control gasless architecture to prevent further attacks. When the safety performance is improved, the transaction efficiency is not reduced. The major contributions of this article are as following:

- We proposed a consortium blockchain architecture that achieves role-based access control and gasless mechanism presented a solution for user asset trading.
- We discussed a realistic attack confronted in our research; As a proof-of-concept, we also studied the scheme to address security issues.
- We analyzed the security issues and the efficiency performance of the modified design.

The rest parts of this paper are presented as follows: Sect. 2 presents the general design of decentralized application (DApp) framework and three classic kinds of attacks in blockchain system. Section 3 illustrates our system design and how we prototype the system. Section 4 discusses the attack we confronted and our analysis. Section 5 provides the improvements of our new design in four aspects. Section 6 gives the evaluation of our framework based on security and efficiency aspect. Section 7 concludes the paper and outlines future work.

2   Background Overview

2.1   General DApp Framework

Due to the dramatic development of blockchain technology, DApp attracts attentions from the society in recent years [11]. For decentralized applications, the main difference is that data interactions are operated by smart contracts, while general applications process data by a centralized authority. A general DApp framework has the following features [16]:

– Applications are open-sourced and autonomous. No one can control more than 51% tokens in the system. Besides, applications should update themselves according to users' feedback. The update needs to be permitted after users' voting.
– Application data needs to be encrypted and stored on public blockchains.
– Incentive mechanism is necessary for the system to ensure miners and maintenance nodes can get rewards from their works.

Steemit [15] is a typical example of DApp. It rewards users with their tokens (steem) if they publish articles on their platform. Also, the blockchain would record your articles to help authors manage their essays.

Ethereum: Most DApps use Ethereum as the underlying structure. Typically, DApp providers have their own cryptocurrencies work as the exchange medium and use smart contracts to implement business logic. Smart contracts are valid codes that can be executed under preset conditions. Smart contracts can help exchange balance, property, or any other items in a transparent and conflict-free way. Besides, it can also perform some complex functions such as signature, managing agreements between users and storing application data. Ethereum, launched in July 2015, is the largest and most well-established underlying development platform that enables smart contracts to be built and run on the blockchain. Moreover, Ethereum is also considered to be the most popular development platform for DApp (Decentralized Application). Ethereum has on the following characteristics:

– Ethereum is a "Turing-complete" [5] system that can solve all the computational problems theoretically. The supporting of smart contracts can implement more complex logic in real-world application scenarios.
– Ethereum performs well while building consortium blockchain. Last year, the world's most extensive consortium system Hyperledger [13], announced that EEA (Enterprise Ethereum Alliance) and Hyperledger would collaborate to meet global demand for enterprise blockchain.
– Ethereum system is flexible and scalable and Ethereum platform is still evolving. Ethereum developer community proposed Ethereum Improvement Proposal (EIP) [1] as the standards of new features, protocols, contracts or APIs.

2.2 Attack Methods

We would introduce three representative attacks, including mining level, network level, and gas-related attacks. These three types of attacks can cause irreversible blockchain system damage.

Fig. 1. Eclipse attack

Selfish-Mining Attack. Proof of Work (PoW) is the most common consensus protocol used in blockchain technologies like Bitcoin and Ethereum. It is generally believed that the PoW is incentive-compatible, which means it can prevent conspiracy attacks and encourage miners to mine in the prescribed manner. However, scholars such as Eyal [9] proposed a deviant mining strategy that allows some miners to obtain more revenue than they honestly mine based on the consensus protocol, which names as the selfish mining.

Selfish mining attack is an attack on mining and incentive protocol. The core idea of selfish mining attack is that the selfish mining pool deliberately delays the broadcasting of its new validated blocks and constructs a self-controlled private branch. Honest miners will mine on the top of the public branch. Meanwhile, selfish miners will focus on the private branches controlled by themselves.

If the selfish pool validates plenty of blocks, the length of the private branch they maintain is naturally ahead of the public branch, but selfish mining pool will not release these new blocks at this time. When the length of the public branch is close to the private branch, selfish miners will broadcast the blocks they validated before. Malicious nodes can gain the revenue of these blocks. Honest miners would lose their validated blocks and waste mining power. Although in Selfish Mining attack, both honest and selfish pools will waste computing power, the honest pool will waste more power while selfish pool can obtain more profits than conducting honest mining. As a result, the miners are more willing to join Selfish Mining.

Eclipse Attack [12]. Usually, attackers need to have at least 51% mining power in the network so that they can modify data on the Blockchain. This assumption is based on all parties can visit full transactions. It is quite rough for a group that intends to raise a 51% attack. Taking Bitcoin as an example, it relies on Peer to Peer (P2P) networks to propagate information. For this reason, Eclipse takes advantage of P2P network to control all incoming and outgoing flow of nodes. When all flows are under control of attackers, data on the Blockchain can be modified to some extent.

In Bitcoin, each node has up to eight outgoing TCP connections and 117 incoming TCP connections. Those connections compose the P2P gossip network in order to broadcast blocks and transactions. However, not all nodes contain incoming connections. Eclipse attack focuses on nodes which support incoming connections.

According to Fig. 1, we can see that attack nodes take 40% mining power in the network. In Fig. 1, attackers partition miners into two groups with 30% mining power individually. Due to attack nodes have more considerable power, they can generate new blocks quicker than others. In this way, they can gain consensus from the whole network. This situation enables attackers to censor all blocks and transactions. Also, attackers could alter history in Blockchain.

Gas Related Attack. In the Ethereum blockchain system, Ethereum Virtual Machine (EVM) processes transactions through smart contracts. Moreover, there is a specified cost for every execution of the contract, which evaluates by a certain amount of gas. The more computation power is consumed, the higher gas will be charged. The gas mechanism aims to encourage developers to put only necessary computations on-chain and to incentivize the miners. However, it also brings security risks.

Block stuffing is a common gas related attack method. As the Ethereum computing resources are finite, every block has a block gas limit. By placing several transactions that consume the entire block gas limit, the attack can prevent other transactions from being included in the blockchain. Since miners prefer to choosing transactions with a higher gas price, the higher the attacker set the gas price, the higher is the chance of a successful attack. As a result, there is an opportunity for the attackers to pose a denial of service (DOS) attack on the blockchain system by filling up blocks with deviant transactions. As long as the attack specifies a generous enough gas price for transactions, rational miners will pack attackers' transactions on the blockchain. In the FOMO3D game [2], the winner used this block stuffing attack to obtain the final prize.

3   Prototyping

Customized Design: Our prototype is designed to meet digital assets business features. Users can publish their assets and use the cryptocurrency as the exchange medium. The cryptocurrency is based on ERC-20 standard [19]. We made some modifications in the smart contracts to fit our business scenes. ERC-20 (Ethereum Request for Comment 20) is an official protocol of EIP, which aims to provide a standard for the token implementation. It allows tokens on Ethereum to be reused by other applications. As of June 2019 [4], there are approximately 200,000 ERC 20-compliant tokens deployed on the Ethereum network.

Besides, we implemented a customized contract inherited from ERC-721 token standard [8]. ERC-721 is a standard for building unique or non-fungible tokens on the Ethereum blockchain, which could be used to record user assets in our system. ERC-721 standard brings a new perspective to the blockchain area. It can increase the emphasis on the uniqueness and the property rights of user assets. The ultimate goal of our solution is to truly digitize user assets and provide transparent, traceable, and non-repudiable transaction records. As for transaction management logic, we introduce the RBAC model. RBAC helps manage the identity of different users, so that fulfill the preset transaction rules.

The whole system includes client-side and server-side. The client-side is implemented as an online trading website that provides an intuitive and user-friendly interface. The server-side runs on another cloud server, which receives

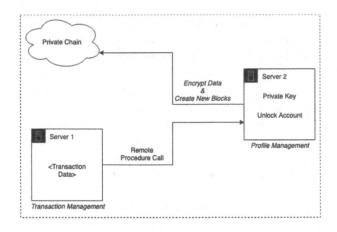

Fig. 2. Our framework

the RPC from the client side and then processes transactions to our private blockchain by the smart contracts.

Proof-of-Authority: In order to solve the Byzantine problem, Nakamoto [17] introduced a competitive mining mechanism in the Bitcoin system. A hash-based Proof-of-Work (PoW) consensus algorithm is adopted to ensure fairness. However, PoW has flaws like security issues, computation power waste, and efficiency limitation. Aside from PoW, other consensus protocols like Proof-of-Stake (PoS), Proof-of-Authority (PoA) [20], Delegated-Proof-of-Stake (DPoS) [14], all have their characteristics. Through our comparative analysis, we found that PoA is more suitable for enterprise solutions because each participant has the right to vote and verify transactions. As a result, we choose PoA as the consensus protocol. Besides, a PoA-based system has the following advantages:

- Low Power Cost: The computing power requirement of PoA based network is shallow, and it does not consume a large amount of electricity as PoW does [7].
- Higher Efficiency: The transaction confirmation time of PoA is significantly faster than PoS or PoW. PoA can address the limitations in terms of transactions per second (TPS).
- Better Scalability: It is quite easy to build and maintain a decentralized application (DApp) using PoA based network, which is much more scalable than PoW or PoS based networks.

Our Framework: Figure 2 briefly explains the architecture of our prototype. Our system runs on two separate cloud servers. One holds the client-side websites, and another configures the EVM platform for running smart contracts. Based on the typical online trading platform, the client-side implements the wallet interface and personal asset management system for each user. Meanwhile, there is an admin backend to manage accounts and other details of the whole system for the administrators. Our system aims to provide the users with an ultimate one-stop solution of assets digitizing, management, and trading. After the user creates an account in the system, he/she can conduct a series of operations, such as buying tokens, posting own assets, trading assets. We wrote our smart contracts to implement business logic. The following table shows some core APIs of our contracts and their details (Table 1).

Table 1. Core interfaces in the system

Name	Parameter(S)	Return(S)	Comment(S)
_addAccount	user_address	1: address, 0: "0 × 0"	Create a new account on the node
_unlockAccount	account_address, account_password, unlock_time, user_address	Promise<boolean> 1: True, 0: False	Normally, this function is used for unlocking the administrator account while there are transactions being sent.
_buyToken	buyer_address, amount	Promise<address> 1: hashed transaction value, 0: "0 × 0"	Purchase ERC20 tokens
_addNewAsset	User_address, object{asset details}	Promise<address> 1: asset address, 0: "0 × 0"	Create ERC721 tokens, represent unique physical assets
_trade	from_address, to_address, asset_address	Promise<address> 1: hashed transaction value 0: "0 × 0"	Normal trade API transfer user asset from one user to another
_checkTransaction	transaction_address	Promise<object>	Get the transaction detailed information
_lockAccount	account_address	Promise<boolean> 1: True, 0: False	Lock the administrator account

4 Potential Issues

Recently, our prototyping system encounters an attack from others. The issue was first exposed when our testing engineers published some sample assets on the system. Fortunately, as our system is still in the testing phase, we immediately terminated the services to avoid further attacks. By checking system log files, we found that all the Ethers in our admin account were transferred to an invalid address. Since there was no left Ether to pay for the gas, transactions cannot be sent, and the system crashed. After analysis and discussion, we believe that the attacker used remote RPC API to send transactions to our admin account and transferred all the remaining Ethers to his/her private account. We drew a possible flow chart of this attack, as shown in Fig. 3.

At first, the attacker used some scanning tools to obtain our IP address and found that the RPC port of our system was open. Then the attacker called some standard RPC APIs to query the balance of our admin account. At this time,

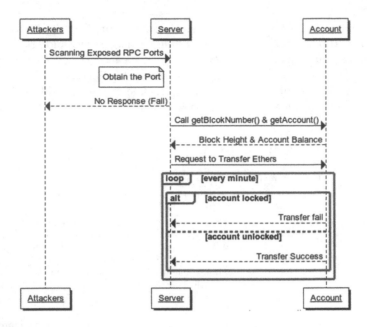

Fig. 3. Posible attack flow

the attacker was aware that there are many Ethers in our account and started to send transactions to transfer the Ethers to his/her address. The submitted transactions need the admin account's signature first. Typically, the signature is only available while the account is unlocked, which means the attacker needs to wait for the processing of an ordinary transaction. When the admin account signs for the victim's transaction, the account is unlocked, so that the attacker's transactions can be signed and included in the new block. However, in our system, the administrator account is never locked. As a result, the attacker's transactions were easily signed, processed, and then the transfer was successful.

Since we deployed our system on a consortium chain, there is no valid account in the attacker's address, so that the Ethers that the attacker transferred is theoretically burned. Though the attacker did not gain any benefits from this attack, the attack did stop our smart contracts from being able to function, which resulted in the crash of our system. After this attack, we realized that our previous design has security flaws. In the next section, we will point out our problems and introduce the improvements of our new design.

5   Proposed Improvements

To address the security flaws and upgrade our system design, we provide our improvements in the four following aspects:

Role-Based Access Control: In our previous design, the administrator account is in charge of signing transactions and also creating new blocks.

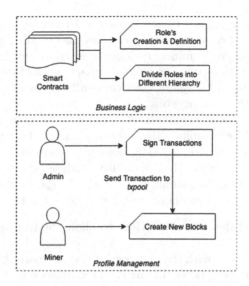

Fig. 4. Improved RBAC structure

We made a mix of roles that can bring potential issues. Through analyzing our programs, we found this kind of role integration is flawed for the framework. Maintaining consortium chain should not let the same account sign transactions and create blocks together.

Figure 4 illustrates the structure of the improved RBAC model. We would discuss this model in business logic and system security area.

- Business Logic: We restrict granted permission for different categories of clients on platform, which called transaction management. Our framework includes various user account types. Those accounts would not be granted to operate remote interfaces individually. However, user roles have associations with operations. The advantage of this concept is simplifying privileges management and understanding. If business requirements changed in the future, we could update the role's privileges instead of modifying all users permissions. For system maintainers, through constructing roles' definition, roles' hierarchical structure, and relationships to enable them control system access in an abstract level [6]. As a result, the RBAC model makes our system more adaptable for industrial business.
- System Security: We divide the previous account into two separate accounts, which called profile management. The admin account only focuses on signing transactions. It would accept transactions from the client-side. Then admin account gives signature with the private key and sends transactions to miner account with a signature. The miner account would fetch transactions from the transaction pool (would be discussed in the following part). After validating transactions and the signature, it would create new blocks on the consortium chain. With this separate accounts method, we can improve the security of the system.

Gasless Mechanism: As mentioned in part 2, Ethereum network uses gas to measure the number of computational resources that it will consume to execute certain operations. When the miners include transactions inside new blocks, they will use their computation power to validate smart contracts. Gas allows miners to charge a certain fee. This gas mechanism can help incentive miners to work and maintain the Ethereum network.

However, in our encountered attack, the attacker transferred the admin account's Ethers. No remaining Ethers for paying for the gas resulted in the system crash. In order to address this problem, we introduced a gasless mechanism in our new architecture. After feasibility analysis, we think that removing the gas in our consortium blockchain will not affect logic validation and transaction processing. The improvements are mainly reflected in the following aspects:

- We modified the Ethereum client Geth to allow the transactions with 0 gas being processed.
- Typically, transactions with the higher gas price are more attractive to miners and will be confirmed quicker. In our customized node protocol, since all the transactions are with 0 gas, miners will pack transactions in the chronological order.
- In our improved design, although we removed the gas system, the transactions are still generated based on the standard Ethereum transaction format.

Through these improvements, our new architecture can fundamentally prevent all the gas-related attacks and meanwhile, satisfy our business logic and keep a measure of system compatibility.

Restrict RPC Access: In Sect. 4, we discussed the attack we encountered and the possible attack methods. There is an important question of how the attacker successfully transferred all the Ethers of our admin account. In this part, we will give detailed answers and the security upgrade we have made to address this issue.

UnlockAccount Code Sample

```
func (s *PrivateAccountAPI) UnlockAccount(ctx context.Context, addr
    common.Address, password string, duration *uint64) (bool, error) {
    ......
    // Use a series if-else statements to validate the input unlock time
    // If 'duration' is valid: call TimeUnlock() to unlock account as
       follows
    ......
    err := fetchKeystore(s.am).TimedUnlock(accounts.Account{Address:
        addr}, password, d)
    return err == nil, err
}

func (ks *KeyStore) TimedUnlock(a accounts.Account, passphrase string,
    timeout time.Duration) error {
```

```
    a, key, err := ks.getDecryptedKey(a, passphrase) // Decrypt private
        key from '/keystore'
    ......
    ......
    if timeout > 0 {
        u = &unlocked{Key: key, abort: make(chan struct{})}
        ks.unlocked[a.Address] = u // If unlock duration is set,
            expire ks.unlocked when timeout
    }
    ......
}
```

Firstly, let us figure out how *UnlockAccount* works: In *UnlockAccount* code, we can see that the function will check if the input unlock-time is valid. The function *UnlockAccount* will check if the input unlock-time is valid. If yes, it will invoke another function *TimedUnlock*. This function will extract and decrypt the account's private key and then stores the private key in a field *ks.unlocked*. To be mentioned, in our previous design, the administrator account is set to be unlocked indefinitely.

The transaction method is fulfilled by three functions: *SendTransaction()* to *wallet.SignTx()* to *w.keystore.SignTx()*.

As we can see from the *SendTransaction* code, function *wkeystoreSignTx* loads the private key stored in ks.unlocked and signs the transaction. As a result, if the unlock-time doesn not run out, the signature process will always succeed, and transactions can be sent through RPC APIs. In the default configuration of Go-ethereum, there is no authentication requirement for RPC APIs, and the *unlockAccount* process does not require user's identity either, which means anyone can use RPC APIs to send transactions when the admin account is unlocked. Since our admin account was not lock in the origin design, the attacker was able to transfer all remained Ethers.

To solve this problem, we restricted the RPC port only to serve the intranet network. The new setting assures that only our admin account is allowed to send transactions. Besides, we no longer set the admin account to be always unlocked. In the modified design, the admin account will be unlocked for 30 seconds while signing a new transaction.

SendTransaction Code Sample

```
func (s *PublicTransactionPoolAPI) SendTransaction(ctx context.Context,
    args SendTxArgs) (common.Hash, error) {
    ......
    ......
    signed, err := wallet.SignTx(account, tx, s.b.ChainConfig().ChainID)
    ......
    return SubmitTransaction(ctx, s.b, signed)
}
```

```
func (w *keystoreWallet) SignTx(account accounts.Account, tx
    *types.Transaction, chainID *big.Int) (*types.Transaction, error) {
    ......
    return w.keystore.SignTx(account, tx, chainID)
}

func (w *keystoreWallet) SignTx(account accounts.Account, tx
    *types.Transaction, chainID *big.Int) (*types.Transaction, error) {
    ......  // Look up the key to sign with and abort if it cannot be
        found
    unlockedKey, found := ks.unlocked[a.Address]
    if !found {
        return nil, ErrLocked
    }
    if chainID != nil {
        return types.SignTx(tx, types.NewEIP155Signer(chainID),
            unlockedKey.PrivateKey)
    }
    return types.SignTx(tx, types.HomesteadSigner{},
        unlockedKey.PrivateKey)
}
```

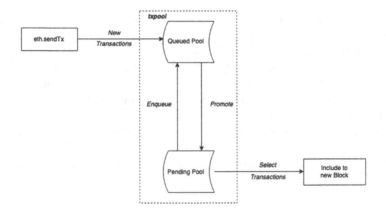

Fig. 5. Transaction pool structure

Expand Pending Pool

In Ethereum, the current submitted transactions are stored in a transaction pool (txpool) before being included in the new block. Figure 4 briefly describes inter-actions in txpool. The latest submitted transactions will be added to the txpool first. Then the miner will select transactions from this pool and include them to new blocks. There are two sub-pools in txpool. Pending pool stores transactions that are ready to be processed and included in the block. Queued pool stores transactions that are currently waiting for execution. The role of the queued pool is to cover out of order transactions arriving from the network. Ethereum uses a field called nonce to keep tracking the total number of transactions executed in an account. If the transaction nonce value does not match the address nonce value, the incoming transaction is not in the right order and would be pushed into the queued pool (Fig. 5).

There are several parameters in txpool configuration files that determine the maximum number of transactions. In default, pending pool's storage capability is 4,096 transactions for all accounts and 16 for the individual account. Numbers of queued pool are 1,024 and 64 respectively. In our improved design, we expand these numbers to 65,536 (except for the queued size for one account, which is hardcoded and unchangeable). This customized expansion of txpool storage capacity aims to meet requirements of our assets trading system. Based on our estimation, this expanded space is theoretically enough for high load conditions. In this way, we can keep user transactions from being discarded due to the buffer size limitation.

6   Analysis of RBAC-GL

In this section, we analyze our system after improvements. We focus on three aspects: model feasibility, attacks defending and efficiency performance. Our framework is constructed on two consortium chain servers and one client server.

- consortium chain servers configurations:
  2 Amazon web services (AWS) servers 8x 2.3 GHz CPUs, Intel Broadwell E5-2686v4, 32 GiB memory, 64 GiB hard disc, Geth version 1.8.17
- client server configuration:
  1 Amazon web services (AWS) servers 4x 2.3 GHz CPUs, Intel Broadwell E5-2686v4, 16 GiB memory, 128 GiB hard disc, Geth version 1.8.17

Business Model: In traditional assets management area, agents play a role to help users publish and propagate their assets. Some agents are not qualified, which may damage the interests of users. These agents might put their own profits first. This kind of fragmented management model can bring some potential issues, which could harm the market environment. Applying blockchain technology can bring following benefits:

- One of the most significant features in assets liquidity is transparency. Users can view assets' owner and any detailed information on the platform. When publishing commodities, the system requires sellers to fill some detailed information in order to help buyers have a better acquaintance of items.
- Traceability is an indispensable factor in the trading model. We record provenances of assets on blocks, which can ensure data is non-repudiable.

Security Analysis: Through restricting RPC access, we can prevent the system from malicious API invokes. Through applying role-based access control, we allocate different roles with separate functionalities. The miners and the admin account will do their jobs separately. The gasless mechanism can thoroughly address all the gas-related issues. Besides, we expand the size of the transaction pool in order to enhance the reliability of confirming transactions. Moreover, since there is no mining process in a PoA network, and our prototype uses a consortium Blockchain, the attack methods in mining and network levels we mentioned before will not work for our system.

Efficiency Performance: Due to the platform encountered an attack from others, we made some improvements in the framework to enhance the security. Meanwhile, we need to ensure that the efficiency performance of our new design does not decrease. We test five main Blockchain interaction operations to analyze response speed as the efficiency performance. As we can see from Table 2, there is no performance degradation in the improved framework.

Table 2. Interfaces average response time

Interface name	Response time (1)	Response time (2)	Response time (3)	Response time (4)	Response time (5)	Average time	Previous response time
_buyToken	1508.773 ms	1518.100 ms	1513.130 ms	1531.792 ms	1515.668 ms	1517.493 ms	1602.348 ms
_addAccount	3558.379 ms	3445.547 ms	3281.870 ms	3601.529 ms	3434.446 ms	3464.354 ms	3397.265 ms
_addNewAsset	3200.452 ms	3108.472 ms	3122.200 ms	3012.268 ms	3107.753 ms	3110.229 ms	5423.996 ms
_updateAsset	1318.262 ms	1285.962 ms	1227.463 ms	1318.929 ms	1226.489 ms	1275.421 ms	1762.203 ms
_trade	8498.235 ms	8561.315 ms	8758.722 ms	8752.807 ms	8214.863 ms	8557.188 ms	8487.143 ms

Limitation and Future Work: Although we have discussed the business model, security, and efficiency performance, there are still intrinsic flaws brought by PoA. For example, if validators on the consortium chain have associations offline, elections of those authorities would be meaningless. In the future, there are two points we need to fix. The first point is to ensure that validators on the platform do not have cross-correlations. The second point is to improve the peak throughput of the consortium chain. Due to our system is still at an initial stage, the concurrency value won't be high. We need to consider how to raise it in the future.

7   Conclusion

The blockchain application has attracted many attentions from society. With the development of Ethereum, it provides more opportunities for companies to build decentralized applications based on smart contracts. Smart contracts help companies to build the system according to business features. More conventional industries (e.g., agriculture) embrace blockchain technology to build a transparent and secure environment. Assets management is a significant area that can apply blockchain technology. Traditionally, some agents would help users to broadcast and manage assets. We propose a prototype that can help investors publish, manage and trade their assets easily. Blockchain records all related data in the prototype. We improved the framework by restricting RPC and applying gasless mechanism, which upgrades our security settings. In the future, we need to consider the authenticated nodes allocation. Under the PoA consensus algorithm, if some nodes pose malicious operations on the blockchain, the framework will crash. As a result, we need to consider these problems in future work.

References

1. Ethereum Improvement Proposals (2015). http://eips.ethereum.org/all
2. Fomo3D (2018). https://fomo3d.hostedwiki.co/
3. Ethereum White Paper (2019). https://github.com/ethereum/wiki/wiki/White-Paper
4. Ethescan Token Tracker (2019). https://etherscan.io/tokens
5. Atzei, N., Bartoletti, M., Cimoli, T.: A survey of attacks on ethereum smart contracts (SoK). In: Maffei, M., Ryan, M. (eds.) POST 2017. LNCS, vol. 10204, pp. 164–186. Springer, Heidelberg (2017). https://doi.org/10.1007/978-3-662-54455-6_8
6. Cruz, J.P., Kaji, Y., Yanai, N.: Rbac-sc: role-based access control using smart contract. IEEE Access 6, 12240–12251 (2018)
7. De Vries, A.: Bitcoin's growing energy problem. Joule 2(5), 801–805 (2018)
8. Entriken, W., Shirley, D., Evans, J., Sachs, N.: ERC-721 Non-fungible Token Standard. Ethereum Foundation (2018)
9. Eyal, I., Sirer, E.G.: Majority is not enough: Bitcoin mining is vulnerable. Commun. ACM 61(7), 95–102 (2018)
10. Ferraiolo, D., Cugini, J., Kuhn, D.R.: Role-based access control (rbac): features and motivations. In: Proceedings of 11th Annual Computer Security Application Conference, pp. 241–48 (1995)
11. Foroglou, G., Tsilidou, A.L.: Further applications of the blockchain. In: 12th Student Conference on Managerial Science and Technology (2015)
12. Heilman, E., Kendler, A., Zohar, A., Goldberg, S.: Eclipse attacks on Bitcoin's peer-to-peer network. In: 24th {USENIX} Security Symposium ({USENIX} Security 15), pp. 129–144 (2015)
13. Hyperledger: Enterprise ethereum alliance and hyperledger to advance the global blockchain business ecosystem. https://cn.hyperledger.org/announcements/2018/10/01/enterprise-ethereum-alliance-and-hyperledger-to-advance-the-global-blockchain-business-ecosystem (2018)
14. Larimer, D.: Delegated proof-of-stake (dpos). Bitshare whitepaper (2014)
15. Larimer, D., Scott, N., Zavgorodnev, V., Johnson, B., Calfee, J., Vandeberg, M.: Steem: an incentivized, blockchain-based social media platform. March. Self-published (2016)
16. Miraz, M.H., Ali, M.: Applications of blockchain technology beyond cryptocurrency. arXiv preprint arXiv:1801.03528 (2018)
17. Nakamoto, S., et al.: Bitcoin: a peer-to-peer electronic cash system (2008)
18. Notheisen, B., Cholewa, J.B., Shanmugam, A.P.: Trading real-world assets on blockchain. Bus. Inf. Syst. Eng. 59(6), 425–440 (2017)
19. Vogelsteller, F., Buterin, V.: ERC-20 token standard, pp. 04–13 (2015). https://github.com/ethereum/EIPs/blob/master/EIPS/eip-20.md GitHub Site. Accessed 2018
20. Wood, G.: PoA Private Chains. https://github.com/ethereum/guide/blob/master/poa.md GitHub Site. Accessed 2015

Developing Patrol Strategies for the Cooperative Opportunistic Criminals

Yanan Zhao[1,2]🆔, Mingchu Li[1,2](✉)🆔, and Cheng Guo[1,2]🆔

[1] Dalian University of Technology School of Software, Dalian, China
yananzhao@mail.dlut.edu.cn, {mingchul,guocheng}@dlut.edu.cn
[2] Key Laboratory for Ubiquitous Network and Service Software of Liaoning Province, Dalian, China

Abstract. Stackelberg security game (SSG) has been widely used in counter-terrorism, but SSG is not suitable for modeling opportunistic crime because the criminals in opportunistic crime focus on real-time information. Hence, the opportunistic security game (OSG) model is proposed and applied in crime diffusion in recent years. However, previous OSG models do not consider that a criminal can cooperate with other criminals and this situation is very common in real life. Criminals can agree to attack the selected multiple targets simultaneously and share the utility. The police may be unable to decide which target to protect because multiple targets are attacked at the same time, so criminals can gain more utility through cooperation and interfere with police decisions. To overcome this limitation of previous OSG model, this paper makes the following contributions. Firstly, we propose a new security game framework COSG (Cooperative Opportunistic Security Game) which can capture bounded rationality of the adversaries in the cooperative opportunistic crime. Secondly, we use a compact form to solve the problem of crime diffusion in the cooperative opportunistic crime. Finally, extensive experiments to demonstrate the scalability and feasibility of our proposed approach.

Keywords: Game theory · Opportunistic crime · Cooperation mechanism · Human behavior

1 Introduction

Stackelberg security game (SSG) is a security game framework that describes the interaction between the security agents and terrorists. There are usually two roles of leader and follower in this model, and each participant has his own set of strategies. In each round of the game, the leader always makes decisions first, and the follower makes decisions after observing the leader's strategy. The combination of their decisions affects their ultimate interests.

Supported by National Nature Science foundation of China under grant Nos: 61572095, 61877007.

© Springer Nature Switzerland AG 2020
S. Wen et al. (Eds.): ICA3PP 2019, LNCS 11944, pp. 454–467, 2020.
https://doi.org/10.1007/978-3-030-38991-8_30

In recent years, with the growing threat of terrorist organizations, preventing terrorist incidents has become an important and challengeable task for security agencies. Many agencies use limited resources to protect the targets and produce the strategies about how to allocate these resources [22]. However, the terrorist organizations' strategies are various and it is difficult to predict the target which will be attacked to the security agencies. Fortunately, SSG has made a huge contribution to helping the security agencies in allocating resources reasonably. Many decision systems are designed based on the security game theory. The Trusts system [24] assigns police resources in the subway network to prevent railway crimes, such as fare evasion and theft. The PROTECTION system [17] protects the coast by combating the criminal activities. The PAWS system [25] helps the rangers to find poachers' traps in Queen Elizabeth National Park (QENP) in Uganda. Most applications use the SSG theory, where defenders use limited resources to cover some targets firstly and the attackers observe the defenders' actions to make their most profitable strategies.

Many criminals do not need to take long time to observe the defenders' actions in real life. They only care about the real-time information and find the opportunity to commit crimes. We call this type of crime opportunistic crime. Opportunistic crime theory is widely used in transportation networks. Attacker seeks the target by transportation and when attacker arrives at a station, criminal can choose to commit crime if there is no police or continue to search another target if the attacker observes the police at the same station. The attacker can continue starting the next round of crime after completing a crime or stop. The SSG is not suitable for the opportunistic crimes, because the attacker needs long time observation (weeks or even months) in SSG. Therefore, a new opportunistic security game model is presented in [27]. This work has proposed the concept of OSG, and defined three characteristics of opportunistic criminals who (1) opportunistically and continually seek for targets and diffuse by transportation; (2) have real-time observation rather than long-term latency; and (3) know limited knowledge of the defender.

Traditional SSG assumes human behaviors are completely rational, and this is only fit for modeling terrorist attacks, which require long-term plans. In most cases, attackers are boundedly rational. Just like opportunistic crimes, attackers do not need long-term plan. Many human behavior models have been proposed to consider the bounded rationality of the attackers. Three well-known models are Prospect Theory (PT) and Quantal Response (QR) and Subjective Utility Quantal Response (SUQR). PT states that attackers make decisions based on the potential value of losses and gains rather than the final outcome [10]. QR indicates that the attackers are more likely to choose the targets with higher expected utility [14]. SUQR uses a linear combination of features to replace the expected utility in QR [15]. By experiments in [1], SUQR has the best performance in predicting the human behavior. In this paper, COSG uses the SUQR model to calculate the probability of a target being chosen by the player.

In the OSG criminals can cooperate with each other, for example the criminals agree to attack different targets simultaneously to increase the probability of successful crime and the excepted utility. Recent work on security game has pointed out the application of cooperative attack in wildlife protection [23]. They have built a multi-round Stackelberg game and proposed a new human behavior model based on it.

We summarize the previous work and propose a novel model COSG (Cooperative Opportunistic Security Game). The contributions of this paper are as follows: Firstly, we combine the cooperation mechanism in SSG with the OSG and establish the COSG model, and COSG better describes the attacker's behavior who can cooperate with other attackers and continuously commit the crime. Secondly, we modify the resource allocation algorithm in traditional opportunistic crime, so that it can be quickly applied in cooperative opportunistic crime resource allocation problem. Finally, we conduct experiments to demonstrate the scalability and feasibility of our proposed approach by inviting 50 volunteers to provide us with data. Experimental results show that our model can effectively help defenders to deal with the cooperative opportunistic criminals.

2   Motivating Scenario

An example of the typical cooperative opportunistic crime is the free market in China. Free markets are trading markets that are spontaneously generated in certain places. Some small free markets generally focus on selling food, clothes, and daily supplies. In the free markets, prices are not regulated by the government. Vendors are free to set prices, and buyers can bargain with sellers. Vendors in the free market can transfer their booths according to the number of customers at different times to maximize their earnings. Although the free markets provide convenience to people, free markets generate many garbage which seriously pollutes the environment during business hours (see Fig. 1(a)). In addition, free markets affect the surrounding traffic conditions and the noise can also interfere people's daily life. Therefore, the security department has set up patrols to combat the vendors in the free markets (see Fig. 1(b)).

To facilitate the understanding, we call the vendor in the free market as attacker and the ranger as defender. The model of free market has two following features: (1) The attackers are opportunists whose behaviors are accorded with the features of opportunistic crime and they attack the targets where they can gain high expected utility. (2) An attacker can choose to cooperate or not cooperate with other attackers. If attackers agree to cooperate, they will attack different targets at the same time and their gains will be divided equally, and if not they will get the pay-offs for individual attacks. Whether to cooperate depends on the utility that an attacker can gain in cooperation and non-cooperation. As far as we know, previous models do not take into account the cooperation mechanism in opportunistic crime, but the cooperation between attackers is indeed a very common phenomenon in reality.

(a)   **(b)**

Fig. 1. (a): A vendor generated smoke in free market. (b): The rangers combated vendor and confiscated their tools.

3 Related Work

Crime in transportation network is a very important and challengeable to the security agencies, because the criminals can opportunistically seek targets and transfer by bus or metro train [22]. For example, an attacker arrives a station by metro train and will commit a crime if the attacker does not observe police at the same station. If the criminal finds police at the current station, the attacker will move to another station until gives up or finds a new target. We assume that the crime occurs at the station where there are many people and the probability of successful crime is high. SSG theory is not suitable for this kind of situation due to the long-term planning of the attacker. [27] presents a more flexible model OSG based on the Markov strategy and gives the algorithm EOSG to allocate defender's resource. p_{ij} in Markov transition matrix is the probability of attacker is at station i and defender is at station j at the same time. Obviously if the number of stations grows exponentially, the Markov transition matrix can be very complicated. They also use the COPS algorithm which simplifies the transition matrix to solve the large-scale OSG problems, but the scalability of COPS algorithm is still limited. Previous studies have shown that the performance of OSG with Markov models can be affected by the size of the problem, and we can find another abstract method to simulate the transition with reducing the size of the transition matrix.

Actually, the machine learning methods such as decision tree and cluster have been used in crime prediction [11]. To some extent, these methods are viable, because criminals have certain regularity in committing crimes, and different criminals have their own delinquency proneness. We can collect these crime data and train model to predict the crime hotspot. However, generalization of the model in machine learning is closely related to the data set. We must consider the difficulty of collecting criminal data. [8] points out that in wildlife protection there is a large amount of unlabeled data, and very little labeled data. If we train the model with a small amount of labeled data, it may lead to overfitting.

[23] have provided the basis for our study of the attacker's cooperation mechanism, and they use SUQA to model the behavior of boundedly rational adversaries. In SUQA, a new utility function called subjective utility is defined, which is a linear combination of key features. Experiments show that the SUQA model performs better than the QA model. In this paper, considering the impact of the boundedly rational adversaries, we apply the SUQA model and the cooperative strategy in the OSG to propose a new framework COSG.

4   COSG Model

In this section, we discuss the novel cooperative opportunistic security game model. For convenience, we call the vendor in the free market as attacker and the ranger as defender. We assume that: (1) The entire area is divided into grids of the same size and each grid is called zone. Time is divided into time steps of the same size. (2) Each zone is a target and an attacker can commit opportunistic crime in specific zones. If the attacker finds that there is no defender in the current target, the attacker commits a crime (**S**uccess). Otherwise, the attacker does not commit crime (**F**ailure) and utility is zero. (3) Multiple attackers will share their total utility fifty-fifty if attackers agree to cooperate. (4) COSG is zero-sum game. The defender's utility is non-positive, and the attacker's utility is non-negative. For simplicity, we explain the model with two attackers and one defender.

4.1   Utility

The SSG has two players, an attacker and a defender, and in the COSG we have multiple attackers $\Psi_1, \Psi_2, ..., \Psi_N$ and one defender Θ with M resources. The defender can cover M targets and each attacker can attack a target at the same time. The attacker chooses which target to move to at the next time step based on the expected utility of cooperation and non-cooperation and the probability that the defender will appear in this position. When the target is already protected by a defender's resource, the attacker does not attack and the attacker's utility is zero. Attackers and defender's resources can move to the adjacent zone of the current position at the next time step or stay in the current zone. Attackers cannot observe the coverage distribution of the defender and only when an attacker and a defender's resource move to the same position, the attacker will observe the defender. Similarly, defender cannot observe the positions of attackers, only know the attackers are opportunists. Attackers have a possible initial distribution of defender based on their historical experience and they can use this distribution to measure the attractiveness of a target, but this distribution is not the true distribution of the defender. Defender's transition strategy is the common knowledge of all the players.

To discuss this model more specifically, let T be a set of targets and T_1, T_2 are two subsets of all the targets T, where $T_2 = T - T_1$. T_1 is available to the first attacker Ψ_1 and T_2 is available to the second attacker Ψ_2. At any time step, the positions of the two criminals are t_1 and t_2 respectively. The two attackers determine their targets through the pay-off of cooperation and non-cooperation and Table 1 summarizes the players' pay-off in all cases. $U_{\Psi_1}^u(t_1)$ indicate the pay-off of Ψ_1 at uncovered target t_1 and $U_{\Psi_2}^u(t_2)$ is the pay-off of Ψ_2 at uncovered target t_2 respectively. Similarly, $U_{\Psi_1}^c(t_1)$ indicate the pay-off of Ψ_1 at covered target t_1 and $U_{\Psi_2}^c(t_2)$ is the pay-off of Ψ_2 at covered target t_2. The defender's pay-offs in uncovered targets t_1, t_2 are $U_\Theta^u(t_1)$ and $U_\Theta^u(t_2)$ respectively. $U_\Theta^c(t_1)$ and $U_\Theta^c(t_2)$ are the pay-offs of defender in covered targets. In order to introduce a cooperative mechanism, we use ϵ to represent the reward factor, and the reward factor will motivate two attackers to cooperate. If attackers successfully cooperate with each other they will share all of their pay-offs fifty-fifty and they will receive the reward ϵ.

Table 1. Pay-offs for attacks.

Attackers: Ψ_1, Ψ_2	Crime success status	Cooperation status
$U_{\Psi_1}^u(t_1)$, $U_{\Psi_2}^u(t_2)$	Ψ_1 S, Ψ_2 S	Noncooperation
$U_{\Psi_1}^u(t_1)$, $U_{\Psi_2}^c(t_2)$	Ψ_1 S, Ψ_2 F	Noncooperation
$U_{\Psi_1}^c(t_1)$, $U_{\Psi_2}^u(t_2)$	Ψ_1 F, Ψ_2 S	Noncooperation
$U_{\Psi_1}^c(t_1)$, $U_{\Psi_2}^c(t_2)$	Ψ_1 F, Ψ_2 F	Noncooperation
$(U_{\Psi_1}^u(t_1) + U_{\Psi_2}^u(t_2) + 2\epsilon)/2$	Ψ_1 S, Ψ_2 S	Cooperation
$(U_{\Psi_1}^u(t_1) + U_{\Psi_2}^c(t_2) + \epsilon)/2$	Ψ_1 S, Ψ_2 F	Cooperation
$(U_{\Psi_1}^c(t_1) + U_{\Psi_2}^u(t_2) + \epsilon)/2$	Ψ_1 F, Ψ_2 S	Cooperation
$(U_{\Psi_1}^c(t_1) + U_{\Psi_2}^c(t_2))/2$	Ψ_1 F, Ψ_2 F	Cooperation

Ψ_i is opportunistic criminal, and does not attack when defender's resource cover the current target, so the pay-off of Ψ_i is zero in this case. The COSG is zero-sum game, and the pay-off of defender Θ is $-(U_{\Psi_1}(t_1) + U_{\Psi_2}(t_2))$. In our COSG model, the two factors that affect the probability of the attacker committing a crime are the utility $U_{\Psi_i}(i = 1, 2)$ and the probability that the police will protect a specific target. The two attackers compare their expected utility in case of non-cooperation or cooperation and choose the optimal strategy. If the best choices for both attackers are cooperation, they will attack cooperatively and share the pay-offs. Otherwise, one of attackers is not willing to cooperate, they will commit crime individually.

4.2   Transition and Diffusion

In order to simulate the scenario of opportunistic crime in real life, we introduce the transition and diffusion of the opportunistic crime. The whole area where crime may occur is divided into zones of the same size. The criminals Ψ_1, Ψ_2 and defender Θ can move in the specific zones. We have a more compact division of time steps so that the players can only move to the adjacent zones or stay in the same zone at each time step, e.g. moving from one zone to a neighboring zone, is assumed to take one time step. At each time step, Θ and Ψ_1, Ψ_2 firstly move at the same time, and then Ψ_1 and Ψ_2 commit crimes at their current targets (or not commit) at this time step, the defender Θ protects the target where they are currently simultaneously. At the next time step, the three players develop their own optimal strategy and repeat the previous process. We explain the transition and diffusion mechanism of players by two 4×4 zones T_1, T_2 (see Fig. 2).

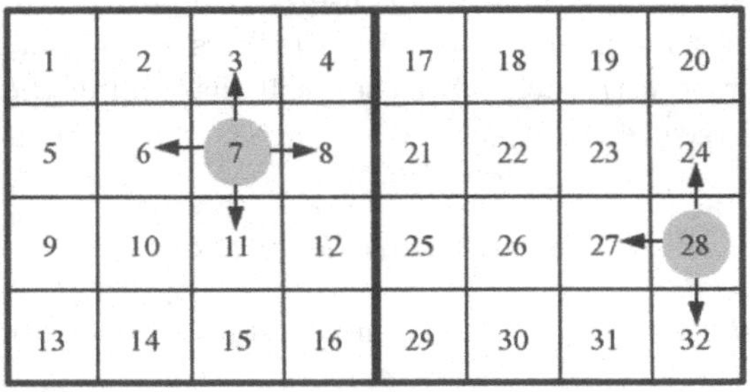

Fig. 2. Distribution of two attackers at time unit t, and they can move to a neighboring zone at next time step $t + 1$.

The attackers Ψ_1, Ψ_2 and defender Θ move within their specific zones. We define that Ψ_1 moves in the left area T_1, and Ψ_2 moves in the right area T_2 and Θ can move in the $T = T_1 \cup T_2$. At this time step t, Ψ_1 is in target 7 of T_1, and Ψ_2 is in target 28 of T_2. If the two attackers commit crimes in their current zones based on if they observe defender Θ at time step t. At the next time unit $t + 1$, all the players can move to an adjacent zone to commit opportunistic crime or protect the target. For example, Ψ_1 only can move to the target $3, 6, 8, 11$ or stay in the target 7 at time step $t + 1$. For an attacker, which target to be chosen depends on the utility of the target when cooperation and non-cooperation and the probability of this target is protected, so we can give the transition matrix of Ψ_1, Ψ_2 and Θ. To the attacker Ψ_1, transition matrix of Ψ_1 is T_{Ψ_1}, and T_{Ψ_1} is a $5 \times (4 \times 4)$ matrix. The (4×4) represents each target number in T_1, and the 5 represents a strategy choice to move up, down, left, right or stay. So each element $x_{i,j}$ in row i and column j is the probability of moving to a neighboring zone when Ψ_1 is in zone j.

In the transition matrix (see Fig. 3), the vector in column 2 is $(0, \frac{1}{4}, \frac{1}{3}, \frac{1}{6}, \frac{1}{4})^T$, so when attacker Ψ_1 is in target 2, the probability of moving down to target 6 is $\frac{1}{4}$, and the probability of moving left to target 1 is $\frac{1}{3}$, and the probability of moving right to target 3 is $\frac{1}{6}$, and the probability of staying in target 2 is $\frac{1}{4}$. There is no target for attacker to move up when in target 2, the probability of moving up is 0. The attacker can move from target 1 to target 2 and also move from target 2 to target 1, So these two probabilities are equal in T_{Ψ_1} and they are $\frac{1}{3}$.

$$
T_{\Psi_1} = \begin{array}{cccc} 1 & 2 & \dots & 16 \end{array}
\begin{pmatrix}
0 & 0 & \dots & \frac{3}{7} \\
\frac{1}{3} & \frac{1}{4} & \dots & 0 \\
0 & \frac{1}{3} & \dots & \frac{1}{7} \\
\frac{1}{3} & \frac{1}{6} & \dots & 0 \\
\frac{1}{3} & \frac{1}{4} & \dots & \frac{2}{7}
\end{pmatrix}
\begin{array}{l}
\text{move up} \\
\text{move down} \\
\text{move left} \\
\text{move right} \\
\text{not move (stay)}
\end{array}
$$

Fig. 3. Transition matrix of attacker Ψ_1.

The probability that Ψ_1 moves from zone i to zone j and Ψ_2 moves from zone m to zone n at next time step is

$$
\begin{aligned}
p(j,n) &= p_{\Psi_1}(i,j) \cdot p_{\Psi_2}(m,n) \\
s.t. \quad & i,j \in T_1, \quad j \in Adj(i) \\
& m,n \in T_2, \quad n \in Adj(m)
\end{aligned} \tag{1}
$$

where Adj is the set of all adjacent zones of a specified zone. Let $p_{\Psi_1}(i,j)$ denotes the probability of Ψ_1 in target i and choose to attack target j at next time step. Similarly, $p_{\Psi_2}(i',j')$ denotes the probability of Ψ_2 in target i' and choose to attack target j' at next time step. $p_{\Psi_1}(i,j)$, $p_{\Psi_2}(m,n)$ can be obtained from attackers' transition matrix T_{Ψ_1} and T_{Ψ_2} respectively. We give the equation of $p_{\Psi_1}(i,j)$, and $p_{\Psi_2}(m,n)$ is calculated similarly to Eq. (2).

$$
p_{\Psi_1}(i,j) = \max\{(1 - \overrightarrow{c_{b,t}(j)}) \cdot Att(j)\} \tag{2}
$$

In Eq. (2) $\overrightarrow{c_{b,t}}$ represents the attacker's belief states of defender's place at next time step, so $\overrightarrow{c_{b,t}(j)}$ is the probability distribution that the police Θ may appear in the target j to the attacker Ψ_1. Att is the attractiveness of the neighboring zones to the attacker. The attractiveness of targets Att is attacker's probability distributions of choosing target at next time step based on the subjective utility and Att is calculated as Eq. (3).

$$
Att(j) = \max\{p_{\Psi_1}^{nc}(j), p_{\Psi_1}^{c}(j)\} \tag{3}
$$

We have known that Ψ_1 and Ψ_2 both are bounded rational, so we use the SUQR model to describe the probability that they choose their own targets and whether they prefer to cooperate. The SUQR extends the classic quantal response model by replacing the excepted utility with a subjective utility function. In the case of cooperation and non-cooperation, the probabilities that Ψ_1

will choose the zone j at next time step are shown in Eq. (4), and the equation of Ψ_2 can be generated likewise.

$$p_{\Psi_1}^{nc}(j) = \frac{e^{SU^{nc}(j)}}{\sum\limits_{I \in T_1} e^{SU^{nc}(I)}}$$

$$p_{\Psi_1}^{c}(j) = \frac{e^{SU^{c}(j)}}{\sum\limits_{I \in T_1} e^{SU^{c}(I)}} \tag{4}$$

where the SU^c and SU^{nc} are the subjective utility functions of an attacker in an zone when the two sides cooperate successfully (c) or cooperate unsuccessfully (uc). If one's best choice is to cooperate but another's optimal choice is not to cooperate or both attackers choose not to cooperate, we refer to these situations as cooperation failed. In this case, Ψ_1 and Ψ_2 commit crimes individually. Only when the best choices for criminals both are cooperation, they will attack cooperatively and share their pay-offs.

For the defender Θ, we do not need to consider whether to cooperate, and the defender arranges resources based on the subjective utility, so the probability of Θ will protect target y in next time step is

$$p_\Theta(y) = \frac{e^{SU(y)}}{\sum_{X \in T} e^{SU(X)}} \tag{5}$$

where $y \in T$. Let $p_\Theta(x, y)$ denotes the probability that a defender's resource in zone x and move to zone y at next time step, and $p_\Theta(x, y)$ is an element in the transition matrix T_Θ.

$$p_\Theta(x, y) = p_\Theta(y) \tag{6}$$

The two opportunistic criminals can attack target j and target n at next time step when the target is not covered. Thus, we only pay attention to if the two targets which will be attacked are protected by the defender's resources. The probability distribution of Θ covers the two targets after t time steps is shown as Eq. (7) where $\vec{c_0}$ is the initial state distribution of the resources.

$$\vec{c_t}(j, n') = T_\Theta^t \cdot \vec{c_0}(j, n) \tag{7}$$

4.3   Optimal Resource Allocation Strategy

In our model, we can obtain the combination of the attackers' locations and the defender's locations at each time step. Attackers make choices on account of defender's place and the attractiveness of the neighboring zones. Defender dispatches m resources based the initial state distribution and the transition matrix. We have described the behavior of the attackers and defender in COSG model, and in this section we give the formulas to find the optimal patrol strategy. We focus on the interaction between attackers Ψ_1, Ψ_2 and defender Θ.

Our optimal resource allocation strategy is to minimize the loss of defender in each state. We consider the locations where the attackers appeared, and whether the defender can protect these targets timely. The attackers are opportunists, and they do not take action if they notice the defender in the same zone, so the defender's loss will be reduced. We have

$$U_p(k) = V_p \cdot X_k \tag{8}$$

where $U_p(k)$ is the defender's expected utility at kth time step. V_p represents the utility for the two targets where the attackers in and we can get the utility based on the probability of the attackers launch crimes in the two targets accordingly. X_k is the probability of defender cover the targets at kth time step.

The defender's goal is to minimize the total expected utility of all time steps. The more time steps we set, the more we can simulate opportunistic crimes in reality. Therefore, the objective of defender is

$$
\begin{aligned}
Obj &= \lim_{K \to \infty} \sum_{k=0}^{K} U_p(k) \\
&= \lim_{K \to \infty} \sum_{k=0}^{K} V_p \cdot ((1 - \alpha) \cdot T_\Theta)^k \cdot X_1
\end{aligned}
\tag{9}
$$

Only unknown variable in Eq. (9) is T_Θ, and it can be denoted as the defender's decision.

5   Experiments

We evaluate the performance of our approach based on extensive experiments. We use Jupyter Notebook (version 5.7.6) and all results are performed on a 64-bit PC with a 3.30 GHz CPU and a 16.0 GB RAM. Each data point we report is an average of 50 different samples.

5.1   Data Sets

In our experiments, we simulate the cooperative opportunistic crime in real life, and simplify the model of real-world. In our model, the whole area is divided into zones of the same size, and we set the entire area to two $N \times N$ zones. Time is divided into continuous and equal time steps, and player moves from one zone to a neighboring zone at one time step. Defender has m resources and the initial distribution of these resources which the two attackers do not know is set based on the importance of targets. The two attackers know a possible distribution of police based on their historical experience. The defender's transition matrix is the common knowledge of all players. We randomly set the attractiveness function Att of each target i, so the attacker's route is more random and difficult to predict. The defender's utility of not covering at least one target is $U_\Theta < 0$ and

$U_\Theta = 0$ is covering both two targets in our experiments. Similarly, the criminal's utility of attacking a target successfully is $U_{\Psi_i} > 0$ and $U_{\Psi_i} = 0$ is giving up committing crime. The attacker decides whether to cooperate according to the utility in case of noncooperation and cooperation. We set the exit rate of the attacker $\alpha = 0.1$.

5.2   Results

The experiment involves four models and there are COSG model, Markov chain model, random patrol model, and no patrol model. Players in our model decide how to move based on the subjective utility. In the random patrol model, defender chooses the next target to protect randomly and we set the value of each element in the transition matrix to $\frac{1}{5}$. To the no patrol model, the defender does not cover any targets, so the attackers commit the crime arbitrarily. In the Markov chain model, the players can move to any zones at a time step.

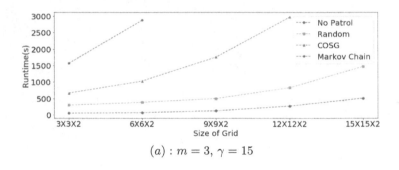

(a) : $m = 3$, $\gamma = 15$

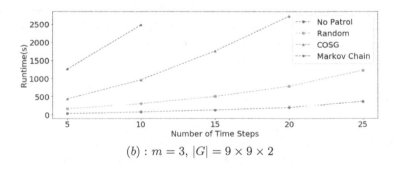

(b) : $m = 3$, $|G| = 9 \times 9 \times 2$

Fig. 4. Runtimes analysis.

We compare the scalability of our model. The result is shown as Fig. 4(a) where the x-axis represents the size of grid and the y-axis indicates runtime. γ denotes the number of time step. Our model is greatly affected by the size of grid and cannot scale up to the size of grid larger than $12 \times 12 \times 2$ with the runtime cap of 3000 s. Random strategy and no patrol strategy are always faster than our model, because they do not need to consider the subjective utility of the defender. The Markov chain model requires the maximum runtime, because the transition matrix of it is more complicated than the other three models. In our experiments, we also compare the runtime of our strategy with different time length. The result is shown in Fig. 4(b) and $|G|$ denotes the size of grid. Runtime rises faster as the number of time steps increase, because more boundary constraints are considered. We deal with small-scale issues in this paper, and we can find that scalability is still a major challenge.

Figure 5 shows that the number of defender's resources m can influence the defender's utility. When we set the number of time steps $\gamma = 15$ and the size of grid $|G| = 9 \times 9 \times 2$, the defender's utility rises as the number of resources increases. The random patrol model and no patrol model cannot give a more satisfactory patrol strategy, because the two models do not consider too much bounded rationality of the players. Our COSG model and Markov chain model perform better than the previous two models.

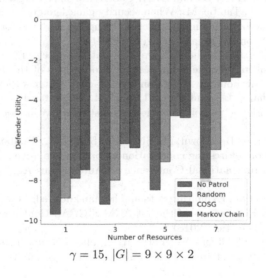

$$\gamma = 15, \ |G| = 9 \times 9 \times 2$$

Fig. 5. Different resources of problem.

6 Summary

In this paper, we propose an innovative model for cooperative opportunistic and boundedly rational criminals. Furthermore, we introduce a compact form of transition matrix representation unlike previous opportunistic security game which used Markov chain to describe players' diffusion. Traditional Stackelberg game models require attacker to make decision in advance, but the attacker in our model reacts to real-time information. As shown in our experimental results, the runtime of our model is better than the Markov chain model. However, scalability is still a major challenge, and the current model is only sui table for small-scale problems. In future work, we can use the abstract method or constraint generation to compress the scale of the problem. In addition, it is also feasible to construct the opportunistic security game model by using neural networks.

References

1. Abbasi, Y.D., Short, M., Sinha, A., Sintov, N., Zhang, C., Tambe, M.: Human numes in opportunistic crime security games: evaluating competing bounded rationality models. In: Proceedings of the Third Annual Conference on Advances in Cognitive Systems ACS (2015)
2. Abbasi, Y., et al.: Know your adversary: insights for a better adversarial behavioral model. In: CogSci (2016)
3. Bondi, E., Fang, F., Hamilton, M., Kar, D., Dmello, D., Choi, J., Nevatia, R.: Spot poachers in action: augmenting conservation drones with automatic detection in near real time. In: Thirty-Second AAAI Conference on Artificial Intelligence (2018)
4. Fang, F., Stone, P., Tambe, M.: When security games go green: designing defender strategies to prevent poaching and illegal fishing. In: Twenty-Fourth International Joint Conference on Artificial Intelligence (2015)
5. Fang, F., Jiang, A.X., Tambe, M.: Optimal patrol strategy for protecting moving targets with multiple mobile resources. In: Proceedings of the 12th International Conference on Autonomous Agents and Multiagent Systems (2013)
6. Gholami, S., Zhang, C., Sinha, A., Tambe, M.: An extensive study of Dynamic Bayesian Network for patrol allocation against adaptive opportunistic criminals (2015)
7. Gholami, S., Wilder, B., Brown, M., Sinha, A., Sintov, N., Tambe, M.: A game theoretic approach on addressing cooperation among human adversaries. In: Proceedings of the 15th International Conference on Autonomous Agents and Multiagent Systems (2016)
8. Gurumurthy, S., et al.: Exploiting data and human knowledge for predicting wildlife poaching. In: In Proceedings of the 1st ACM SIGCAS Conference on Computing and Sustainable Societies (2018)
9. Halvorson, E.D., Conitzer, V., Parr, R.: Multi-step multi-sensor hider-seeker games. In: Twenty-First International Joint Conference on Artificial Intelligence (2009)
10. Kahneman, D., Tversky, A.: Prospect theory: an analysis of decision under risk. Econometrica **47**(2), 263–292 (1979)
11. Kar, D., et al.: Cloudy with a chance of poaching: adversary behavior modeling and forecasting with real-world poaching data. In: Proceedings of the 16th International Conference on Autonomous Agents and Multiagent Systems (2017)
12. Li, M., Cao, Y., Qiu, T.: Optimal patrol strategies against attacker's persistent attack with multiple resources. In: IEEE SmartWorld, Ubiquitous Intelligence & Computing, Advanced & Trusted Computed, Scalable Computing & Communications, Cloud & Big Data Computing, Internet of People and Smart City Innovation (2017)
13. Laan, C.M., Barros, A.I., Boucherie, R.J., Monsuur, H., Timmer, J.: Solving partially observable agent-intruder games with an application to border security problems. Naval Res. Logist. **66**(2), 174–190 (2019)

14. McKelvey, R.D., Palfrey, T.R.: Quantal response equilibria for normal form games. Games Econ. Behav. **10**(1), 6–38 (1995)
15. Nguyen, T.H., Yang, R., Azaria, A., Kraus, S., Tambe, M.: Analyzing the effectiveness of adversary modeling in security games. In: Twenty-Seventh AAAI Conference on Artificial Intelligence (2013)
16. Pita, J., Jain, M., Ordóñez, F., Tambe, M., Kraus, S., Magori-Cohen, R.: Effective solutions for real-world stackelberg games: when agents must deal with human uncertainties. In: Proceedings of the 8th International Conference on Autonomous Agents and Multiagent Systems (2009)
17. Shieh, E., et al.: Protect: a deployed game theoretic system to protect the ports of the united states. In: Proceedings of the 11th International Conference on Autonomous Agents and Multiagent Systems (2012)
18. Sinha, A., Fang, F., An, B., Kiekintveld, C., Tambe, M.: Stackelberg security games: looking beyond a decade of success. In: Twenty-Seventh International Joint Conference on Artificial Intelligence (2018)
19. Tsai, J., Kiekintveld, C., Ordonez, F., Tambe, M., Rathi, S.: IRIS-a tool for strategic security allocation in transportation networks (2009)
20. Tayebi, M. A., Ester, M., Glässer, U., Brantingham, P. L.: Crimetracer: activity space based crime location prediction. In: Proceedings of the 2014 IEEE/ACM International Conference on Advances in Social Networks Analysis and Mining (2014)
21. Varakantham, P., Lau, H. C., Yuan, Z.: Scalable randomized patrolling for securing rapid transit networks. In: Twenty-Fifth IAAI Conference (2013)
22. Wang, X., An, B., Strobel, M., Kong, F.: Catching Captain Jack: Efficient time and space dependent patrols to combat oil-siphoning in international waters. In: Thirty-Second AAAI Conference on Artificial Intelligence (2018)
23. Wang, B., Zhang, Y., Zhou, Z.H., Zhong, S.: On repeated stackelberg security game with the cooperative human behavior model for wildlife protection. Appl. Intell. **49**(3), 1002–1015 (2019)
24. Yin, Z., et al.: Trusts: scheduling randomized patrols for fare inspection in transit systems. In: Twenty-Fourth IAAI Conference (2012)
25. Yang, R., Ford, B., Tambe, M., Lemieux, A.: Adaptive resource allocation for wildlife protection against illegal poachers. In: Twenty-Fourth International Joint Conference on Artificial Intelligence (2015)
26. Yang, Z., Zhu, J., Teng, L., Xu, J., Zhu, Z.: A double oracle algorithm for allocating resources on nodes in graph-based security games. Multimedia Tools Appl. **77**(9), 10961–10977 (2018)
27. Zhang, C., Jiang, A.X., Short, M., Brantingham, P.J., Tambe, M.: Modeling Crime diffusion and crime suppression on transportation networks: an initial report. In: 2013 AAAI Fall Symposium Series (2013)
28. Zhang, C., Sinha, A., Tambe, M.: Keeping pace with criminals: Designing patrol allocation against adaptive opportunistic criminals. In: Proceedings of the 14th International Conference on Autonomous Agents and Multiagent Systems (2015)

Deep Learning vs. Traditional Probabilistic Models: Case Study on Short Inputs for Password Guessing

Yuan Linghu[1], Xiangxue Li[1,2(✉)], and Zhenlong Zhang[1]

[1] School of Computer Science and Technology/Software Engineering,
East China Normal University, Shanghai, China
xxli@cs.ecnu.edu.cn
[2] Westone Cryptologic Research Center, Beijing, China

Abstract. The paper focuses on the comparative analysis of deep learning algorithms and traditional probabilistic models on strings of short lengths (typically, passwords). The password is one of the dominant methods used in user authentication. Compared to the traditional brute-force attack and dictionary attack, password guessing models use the leaked password datasets to generate password guesses, expecting to cover as many accounts as possible while minimizing the number of guesses. In this paper, we analyze the password pattern of leaked datasets and further present a comparative study on two dominant probabilistic models (i.e., Markov-based model and Probabilistic Context-Free Grammars (PCFG) based model) and the PassGAN model (which is a representative deep-learning-based method).

We use Laplace smoothing for the Markov model and introduce particular semantic patterns to the PCFG model. Our output shows that the Markov-based models can cover the vast majority of the passwords in the test set and PassGAN demonstrates surprisingly the worst effect. Nevertheless, considering the threat that an attacker may adjust the training set, the PCFG model is better than the Markov model. Using Passcode with high-frequency passwords can increase the coverage while reducing the number of guesses. Brute-force attack can also work better when used in conjunction with probabilistic models. For the same billion guesses, brute-force attack can be used to crack pure digital passwords of 4 to 8 lengths, and original-PCFG and modified-PCFG could increase by 11.16% and 8.69%, respectively.

Keywords: Password guessing · Deep learning · PassGAN · Markov model · PCFG

1 Introduction

In most computer systems and websites, using textual passwords is the primary method for user authentication. Textual passwords are easy to implement and

© Springer Nature Switzerland AG 2020
S. Wen et al. (Eds.): ICA3PP 2019, LNCS 11944, pp. 468–483, 2020.
https://doi.org/10.1007/978-3-030-38991-8_31

require less storage space. Although some alternative methods [1] have been proposed, textual passwords are unlikely to be replaced by them in the near future [2].

However, passwords of random strings are hard to remember, users then tend to choose easy-to-remember passwords, which puts their passwords in a high-risk situation. In order to guide users to create stronger passwords, most websites implement some strict password creation policies that the users must obey and the strength of the password is fed back through password strength meter when users input their passwords. The metric of password strength recommended by NIST [3] is based on estimating password's information entropy. Nevertheless, for the modern password guessing attack the NIST's metric is ineffective [4]. Therefore, the key to the study of password security lies in the password guessing attack. Password guessing attack is designed to crack a target password. More detailed, the attacker first builds a candidate set containing possible passwords, then the attacker uses these passwords as a guess one by one until a guess is same as the target password or the candidate set is exhausted.

Brute-force attack is the most preliminary attack strategy, namely, try all possible guess on a specific password space until get the right one. While ensuring guessing of any possible passwords, this approach is impractical in most cases because the cost of time and storage space that an attacker can afford increases with the password space. The number of guesses required to completely cover a given password space using the brute-force attack is: $Guess\_Number = \sum_{k=i}^{n} x^k$, where x is the size of the character set, and n and i represent the maximum length and minimum length of the target password, respectively.

Dictionary attack is space-reduction compared to brute-force attack. As guessing a password, the attacker tries the possible ones from a pre-defined dictionary one by one. The entries that make up the dictionary are usually considered to be strings that users are more likely to use to create passwords. The attacker can also produce more modified guesses throughout mangling-rules. For example, if there is a mangling-rule that converts the first two characters into lowercase letter, the attacker can make the guess *PAssword123*, if *password123* in the dictionary.

Several recent seminal work introduces to password guessing probabilistic models and deep learning models. The coming section reviews the current state of the art. Now, the effective of password guessing attack depends highly on the password guessing model that attackers can use. By analyzing the leaked datasets that can be accessed by the attacker, this paper evaluates the password patterns in 8 leaked datasets and utilizes the probabilistic models and PassGAN (which is a deep learning-based model) to conduct the guessing attack.

2   Related Work

In [5], Weir et al. create a probabilistic context-free grammar model which trains a disclosed password dataset, and fills mangling-rules generated from PCFG with dictionary vocabulary as password guesses. Each password is divided into substrings by letters, digits, and symbols. Later, Houshmand et al. [6] introduce keyboard mode and multi-word mode into probabilistic context-free language to enhance the attack effect.

Wang et al. [7] analyze the disclosed datasets leaked from Chinese websites and systematically investigate the characteristics of Chinese users' passwords. They improve PCFG-based password guessing model by adding pinyin, date and other patterns. One essential finding of them is that attackers can increase the password coverage by evaluating the characteristics of the datasets.

Unlike other research that focus on the shallow segmentation pattern of password, Veras et al. [8] explore the semantics of passwords (the meaning of passwords). They use Natural Language Processing techniques to extract the understanding of semantic patterns in passwords and add these patterns to PCFG-based password model.

In [9], Narayanan and Shmatikov first propose a Markov chain-based password guessing model. Their main observation is that the passwords that are easy to remember have a similar distribution to the language used by the users, so they generate passwords by low-order Markov chains. Ma et al. [10] find that Markov models can cover more passwords than PCFG models when added by normalization and smoothing tricks.

Deep learning-based model is used as another type of password guessing. Hitaj et al. [11] propose PassGAN, which is a Generative Adversarial Network (GAN). Unlike probabilistic password guessing models, PassGAN does not need to manually analyze datasets. In [12], Xu et al. introduce Long Short-Term Memory recurrent neural network (LSTM) into password guessing. When producing the next character of a password, the LSTM-based model considers all the characters in the prefix.

3   Analysis of Datasets

3.1   Data Cleaning

In recent years, frequent password leakage incidents provide attackers with natural training data, so that passwords generated by attacker models can be quite close to the choice of the real user. Meanwhile, the website administrators analyzing the leaked passwords can help their management. We collect 8 password datasets and analyze them in this section.

We first perform data-cleaning on these datasets. We remove the passwords of a length less than 4 or greater than 18, and reserve the legal passwords that contain only 95 printable characters. Table 1 shows basic information of these data sets, as well as the amount of raw data, the size of dataset after data-cleaning and the number of unique passwords (no duplicate). We mention that gmail data set is the only one for which user language is English. The difference in language may make the dataset show different characteristics in some aspects from other data sets. In addition, 7k7k and 17173 are the same in the user language and website service, and the sizes of the dataset and unique password distribution are quite close. Thus we can choose 17173 as the training set and

Table 1. 8 leaked datasets

Dataset	User language	Web service	Original size	After cleansing	Unique size
Tianya	Chinese	Social forum	30,763,347	30,700,778	12,873,816
7k7k	Chinese	Gaming community	18,891,570	18,873,452	4,926,404
17173	Chinese	Gaming community	18,295,141	18,294,212	5,210,172
Duduniu	Chinese	Gaming community	16,176,598	16,154,501	10,076,299
CSDN	Chinese	IT forum	6,422,902	6,411,479	4,022,960
Zhenai	Chinese	Dating site	5,241,488	5,231,992	3,495,701
Gmail	English	E-mail	4,900,359	4,899,798	3,100,789
Renren	Chinese	Social platform	4,657,575	4,645,144	2,806,814

7k7k as the test set for some particular experiments[1]. Other six datasets are also available as test set in our experiment.

3.2 Password Pattern

Table 2. The percentage of top 5 password's simple pattern in each dataset

Dataset	D	LD	L	DL	LDL
17173	51.84%	28.79%	10.23%	5.40%	1.26%
7k7k	59.49%	18.31%	11.09%	4.57%	2.15%
CSDN	45.11%	28.53%	12.38%	6.48%	1.94%
Duduniu	30.76%	46.05%	10.92%	7.98%	1.59%
Gmail	15.79%	31.49%	40.09%	2.94%	3.56%
Renren	53.38%	19.08%	20.75%	2.97%	1.25%
Tianya	63.78%	15.82%	10.12%	4.42%	1.73%
Zhenai	59.81%	21.17%	6.78%	5.71%	1.47%

Table 2 shows the password's simple patterns of the top five in each data set. D stands for digit sequence and L stands for letter sequence. For example, D means that the password is composed entirely of digits, and LD means that the password pattern is a letter sequence followed by a digit sequence. The percentage is obtained by dividing the number of occurrences of a particular password pattern in a dataset by the number of passwords in that dataset. The password pattern D occupies the largest proportion in Chinese datasets,

---

[1] The interchange of the 17173 and 7k7k datasets is feasible. Whereas, the choice of the two datasets is not arbitrary, but stems from their substantive comparability (same language, similar data sizes, same web services (see Table 1)) so that the resulting comparison is meaningful.

Table 3. Classification of password's semantic patterns:Strings that match the first seven semantic patterns are digit sequence, others are letter sequence.

Pattern	Description
Birthday1	YMD format
Birthday2	MDY format
Birthday3	MMDD
Birthday4	YYYY
Birthday5	YYYYMM
Birthday6	YYMMDD
Mobilephone_number	Chinese Mobile phone number
Chinese_surname	Pinyin of Chinese surname
Chinese_fullname	Pinyin of Chinese fullname
Chinese_place	Pinyin of Chinese place name
English_name	Common English names
English_word	Common English words
Keystroke	Continuous keyboard strokes

especially in tianya, whose percentage of pattern D reaches 63.78%. The two password patterns D and LD associated with digit sequence add up to more than 70% in every Chinese datasets. These results show that Chinese users are prefer digits to letters. Although the gmail users select letters in their passwords, the pure letter sequence is the most password pattern L (accounts for 40.09%) in gmail dataset.

In order to remember the password, users prefer to add their own birthday, name and other information. We refer to some previous work [7,8,13,14], and classify the password's semantic patterns as shown in Table 3. Before we count the percentage of semantic patterns in 8 datasets, we create dictionary for the semantic patterns of letter sequence and implement regular expressions for the semantic patterns of digit sequence. For the three Chinese semantic patterns, we first collect Chinese vocabulary, then use a python package [15] to convert them into Chinese Pinyin. We collect 568 surnames from Hundred Family Surnames and Chinese provincial and municipal district names as the Chinese vocabulary of Chinese_surname and Chinese_place respectively. To get Chinese vocabulary for Chinese fullname, we collect a 20 million leaked reservation dataset from the work in [7] and three leaked online purchase datasets, all of them includes buyer's name (Dangdang contains 1 million data, Vancl contains 200,000 data and the size of Amazon(China) is 200,000), then we select all Chinese fullname in these 4 datasets. According to [8], English_name dictionary is from [16] which includes popular baby names in American since 1879. English_word is from [17] which includes 58,110 lowercase English words [7].

Table 4 shows the percentage of password's semantic patterns in each dataset. For each password, we determine whether a certain substring of them conforms

Table 4. The percentage of password's semantic pattern in each dataset

Patterns	17173	7k7k	CSDN	Duduniu	Gmail	Renren	Tianya	Zhenai
Birthday1	8.56%	9.45%	11.86%	7.30%	0.55%	6.71%	9.93%	10.58%
Birthday2	0.83%	0.82%	0.93%	0.78%	1.02%	0.76%	0.83%	0.94%
Birthday3	6.77%	7.11%	7.62%	6.05%	3.12%	5.54%	7.61%	6.64%
Birthday4	2.52%	2.31%	2.99%	3.26%	2.59%	2.41%	2.41%	2.75%
Birthday5	0.14%	0.19%	0.20%	0.26%	0.03%	0.18%	0.30%	0.37%
Birthday6	7.70%	9.85%	8.97%	6.72%	1.43%	6.99%	12.65%	10.93%
Mobilephone_number	2.82%	2.01%	3.52%	2.43%	0.03%	3.43%	2.76%	9.23%
Chinese_surname	3.49%	2.30%	3.95%	5.13%	1.13%	1.80%	2.32%	3.56%
Chinese_name	8.66%	7.02%	9.52%	9.55%	0.72%	5.64%	6.16%	8.27%
Chinese_place	0.09%	0.08%	0.13%	0.09%	0.02%	0.06%	0.10%	0.09%
English_name	0.73%	1.12%	1.45%	2.32%	12.48%	4.45%	1.09%	0.50%
English_word	1.55%	2.21%	4.95%	3.51%	21.90%	6.38%	2.08%	0.93%
Keystroke	3.21%	1.96%	2.88%	2.21%	2.03%	1.46%	1.84%	2.21%

to the definition of a semantic pattern. The percentage in Table 4 is obtained by dividing the number of passwords that match a certain semantic pattern by the size of the dataset. Any substring in one password cannot be used to form a semantic pattern in multiple times. E.g., *zhangsan* is a Chinese fullname, and the substring *zhang* is a Chinese surname at the same time. In this case, we only count Chinese_surname, which is based on the principle of length first.

When we observe the statistical results, some user behavior can be found. The date of birth plays a key role in the memory of password, although the semantic patterns about birthday in Gmail is much less than the other 7 datasets, we mention that there are up to 40.01% (see Table 2) pure letter sequence in this dataset. Birthday2 is more than Birthday1 in Gmail, which can be attributed to Western countries accustomed to recording the year after the date. Name is also a group of semantic pattern that is often used, in addition to the names of the respective languages, English names appear in Chinese datasets, and Chinese fullname and surname appear in the Gmail. As expected, Gmail users prefer to English words and English names, and Renren has the most English semantic patterns among Chinese datasets, which can attribute to the user community based on young students who receive better English education in China. The Mobilephone_number pattern also has a proportion in Chinese datasets (from 2.01% to 9.23%), which can help us guess long passwords. The percentage of Keystroke pattern in 8 datasets ranges from 1.46% to 3.21%, which also means that continuous keyboard words are easy to remember.

4   Attack Models

4.1   Original-PCFG

Probabilistic Context-Free Grammar is first used by Weir et al. for password guessing attack in [5]. The model can split passsword into many substrings, which are the sequence of digit, letters or symbols. For example, given a password **abcd1234**, the model parses it into L_4D_4 template, which means that the string consists of first 4 letters and 4 digits. The probability of **abcd1234** assigned by PCFG is computed as $P(abcd1234) = P(L_4D_4) \times P(L_4 \rightarrow abcd) \times P(D_4 \rightarrow 1234)$, where $P(L_4D_4)$ is the frequency that the L_4D_4 template appears in training set, $P(L_4 \rightarrow abcd)$ is the frequency that **abcd** appears in the "4-letters" substring set, and $P(D_4 \rightarrow 1234)$ is the frequency that **1234** appears in the "4-digits" substring set.

Weir et al. use the external dictionary to fill the substring of letters in the template when PCFG generates passwords. For example, if **!** and **123** are in the sets of "1-symbol" and "3-letters" respectively, the $S_1L_3D_3$ template could be evolved to **!$L_3$123**. Weir et al. call it pre-terminal structure, which is very similar to the mangling-rule in dictionary attack. The $S_1L_3D_3$ pre-terminal structure means character **!** followed by 3 letters and a substring **123**, then letter sequence of length 3 in the external dictionary will be selected to replace L_3. In their work, PCFG model automatically derives mangling-rule from the training set, but the effective will depend on the quality of the external dictionary. Unlike them, we will also use the letter sequences from the training set to fill the L-segments in the template.

4.2   M-Order Markov Model

In 2005, Narayanan and Shmatikov first use Markov-based model in password guessing attack [9]. An m-order Markov model corresponds to an N-gram, where $N = m + 1$, and the next character will be determined by its first m characters. Under an m-order Markov model, the probabilty of a string of $c_1c_2...c_l$ is computed as $P(c_1c_2...c_l) = \prod_{i=1}^{l} P(c_i|c_{i-m}c_{i-m+1}...c_{i-1})$, each probability on the right is computed from training set, and the calculation formula is

$$P(c_i|c_{i-m}c_{i-m+1}...c_{i-1}) = \frac{O(c_{i-m}c_{i-m+1}...c_{i-1}c_i)}{\sum_{c \in \sum} O(c_{i-m}c_{i-m+1}...c_{i-1}c)} \tag{1}$$

where $O(c_{i-m}c_{i-m+1}...c_{i-1}c_i)$ means the occurrence of $c_{i-m}c_{i-m+1}...c_{i-1}c_i$ in the training set, $\sum$ stands for character set, which includes 95 printable characters and a special filling-symbol $\perp$. From the equation, we can see that when $i \leq m$, the length of the prefix is not enough to count, so each password should be pre-processed before our training. We add m filling-symbols before the first character of the password, and add a filling-symbol $\perp$ at the end of the password. The special symbol $\perp$ also acts as end-symbol, which stops generating more guesses in password generation.

When the order of the Markov model is too high, it will cause many strings to appear as 0, Ma et al. [10] introduce smoothing technics to solve data sparsity. In this paper, we consider Laplace smoothing, which adds a count δ to $O(c_{i-m}c_{i-m+1}...c_{i-1}c)$. In [10], the value of δ is 0.01, considering the size of $\sum$ is 96.

Ma et al. model the process of generating a password as a search tree. The nodes in the tree represent a string $c_1c_2...c_l$, whose parent node is for $c_1c_2...c_{l-1}$. The connection between parent node and child node is labeled with the transition probability which is calculated by the Markov model. The nodes in the search tree are stored in the priority queue, sorted in descending order of probability. At each iteration, the node with the highest probability is dequeued, and more child nodes are pushed into the priority queue. If the dequeue node's last character is $\perp$, the string corresponding to the node is a password guess. However, priority queue is not desirable in practice. Even if we use the probability threshold, which can prevent nodes with too small probability from entering the priority queue, we cannot generate a large number of passwords in this way. Therefore, we deprecate priority queue and output password guesses with their probability. After generating all passwords above the given probability threshold, we divide them into blocks according to the probability interval and sort each block's passwords.

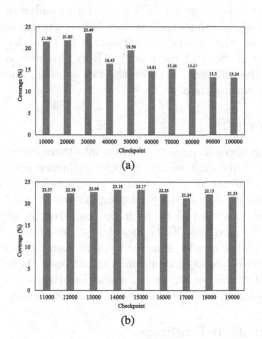

Fig. 1. The coverage for 7k7k on various number of iteration (on PassGAN)

4.3 PassGAN

PassGAN is a deep learning-based model, proposed by Hitaj et al. [11]. Different from probabilistic models, PassGAN leverages Generative Adversarial Networks (GAN) and can automatically obtain the distribution of passwords from the training set. GAN consists of a generative deep neural network G (generator) and a discriminative deep neural network D (discriminator). At each iteration, it will be given an input I, generator G generates a fake sample, and the goal of discriminator D is to distinguish between the generated samples from G and the real samples from I. When D cannot distinguish between the fake sample generated by G and the real sample, the training is considered as complete.

PassGAN can be instantiated by the Improved Wasserstein GAN (IWGAN) [18] which is used for text generation. Since IWGAN requires the input strings to be of uniform length, we use filling-symbol to extend the length of passwords in training set to 19, this is because the maximum length of passwords is 18 after data-cleaning. Before using PassGAN to generate password guesses, we need a preliminary experiment to determine the number of iterations of the model. We set 100,000 iterations and other parameters follow [11]. The 17173 data set is used as the training set. In this process, some checkpoints are set. When the number of iterations reaches the number in the checkpoints, the model generates 5 million password guesses for attacking 7k7k.

Figure 1 shows the coverage for 7k7k on various number of iteration. As can be seen from Fig. 1(a), coverages are less than 20% when the number of iterations exceeds 30,000. From the results of the reaction in Fig. 1(b), we choose 14000 as the number of iterations of the model. Although the coverage is higher at 30,000 iterations, the difference between the two is small, and 14,000 iterations consume fewer computing time.

4.4 Modified-PCFG

Original-PCFG can segment password strings into three types of substrings, and their meanings are only digit sequence, letter sequence and symbol sequence, but the passwords chosen by users often have semantics (see Sect. 3.2). In our modified-PCFG model, we add the semantic patterns in the Table 3, which segment the string of password along with the patterns in original-PCFG. The segmentation of the modified-PCFG model first determines whether the substring is a digit sequence or a letter sequence, and then determines whether they conform to the definition of a certain semantic pattern. The substring that does not conform is still defined as D_n or L_n. Algorithm 1 presents the segmentation process of the modified-PCFG model for the password string.

5   Implementation Findings

5.1 Markov Model vs. PCFG

We set the probability threshold to 10^{-10}, the 17173 data set as the training set, the order of the Markov model between 1 and 5, and also add Laplace smoothing techniques to models of all orders. Since the minimum password production is 220,301,565, we set the total number of guessing attack to 200 million for comparison. Table 5 shows the coverage of m-order Markov model for the 7k7k. In our experiments, the Laplace smoothing technique has no effect on the results of the 1st and 2nd order Markov models and could work better when the order is higher.

Table 5. The coverage of m-order Markov model for the 7k7k dataset given guessing time = 200 million.

Order	No Laplace	Laplace
1	53.10%	53.10%
2	57.86%	57.86%
3	61.85%	61.86%
4	65.34%	65.39%
5	67.16%	67.39%

The GuessNumber-Coverage is a type of graph which shows the relationship between the number of guesses and the coverage. In this graph, the x-axis represents the number of guesses and the y-axis represents the corresponding coverage. For the probabilistic models, the number of passwords generated by them is output in order of probability from large to small, so the Guess Number N also represents the use of the first N passwords for guessing.

We select the two best performing models in Table 5, 5-Markov-Laplace and 4-Markov-Laplace and compare them with original-PCFG and modified-PCFG. Similarly, 17173 is still a training set, the test set is 7k7k, and we set the number of guesses to 1 billion. When the probability threshold is 2.5×10^{-11}, 4-Markov-Laplace produces 1,000,109,995 passwords and 5-Markov-Laplace generates 1,070,641,997 passwords. Figure 2 is the GuessNumber-Coverage graph of four different probabilistic password guessing models[2]. From the trend of the

---

[2] When looking into password guessing attacks, one may observe that it is pretty hard to obtain even a small leap for the coverage (especially with the increasing of the number of guesses). For example, one can exploit modified-PCFG model to produce 1 billion passwords which are then used to crack the 17173 training set. Next, take the difference set (i.e., all the passwords not covered by the collection of the 1 billion passwords) as new training set and generate 200 million more passwords. Now the coverage gain is only 0.55% (contrary to expectations). This kind of striking contrast (in experiments) may not be thoroughly experienced from comparison exhibition (in a figure). Whereas, as can be seen from Fig. 2, it is very clear that the Markov-based model reports significantly better performance than the PCFG-based model when the number of guesses reaches 1 billion.

The models themselves are suitable for different datasets/languages (which will affect the resulting coverage and lead to different outputs). Generally, a cut-and-dried dictionary would be exploited for the dataset of a specific user language to capture the semantic pattern of the user group (for PCFG-based models). However, this does not mean that the models are not applicable to other languages.

Algorithm 1. The segmentation of Modified-PCFG

Input: A string of password
Output: A list of Substring after segmetion

Procedure Segmentation(Password):
1: SubStrs ← DLS_Segmentation(Password)
2: List ← [ ]
3: **for** SubStr **in** SubStrs **do**
4:     **if** SubStr is digit sequence **then**
5:         **if** SubStr has sementic pattern **then**
6:             Left, Mid, Right ← DigitsSplit(Substr)
7:             List ← Segmentation(Left) + List
8:             List ← List + Mid
9:             List ← List + Segmentation(Right)
10:        **else**
11:            List ← List + Substr
12:        **end if**
13:    **end if**
14:    **if** SubStr is letter sequence **then**
15:        **if** SubStr has sementic pattern **then**
16:            Left, Mid, Right ← LettersSplit(Substr)
17:            List ← Segmentation(Left) + List
18:            List ← List + Mid
19:            List ← List + Segmentation(Right)
20:        **else**
21:            List ← List + Substr
22:        **end if**
23:    **end if**
24:    **if** SubStr is symbol sequence **then**
25:        List ← List + SubStr
26:    **end if**
27: **end for**
28: **return** List

---

curve in the figure, we can see that when the Guess Number increases, the increasing rate of coverage becomes lower and lower, which means that for those passwords that are difficult to be cracked, more guesses need to be tried. From the figure, we can also easily find that the four models can cover 10% of the data in the test set when the Guess Number reaches only 10^3. The curves of modified-PCFG and original-PCFG are much close when the number of guesses is less than 10^8, but after that, the coverage of the modified-PCFG model seems higher. When the Guess Number is 10^9, the coverages of two Markov Models are higher than that of the two PCFG models, and 5-Markov-Laplace is the highest. However, PCFG-based models have higher coverage than the Markov-based models when the Guess Number is small (such as 10^7 times). This provides us a reasonable strategy for the attacker: when the number of guessing attack is

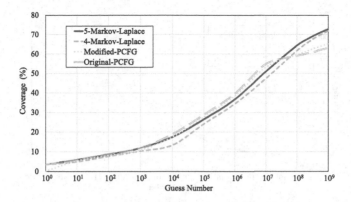

Fig. 2. The guessing attack on 7k7k dataset

limited, the password guessing model that performs best under the corresponding Guess Number can be selected.

In order to explore the impact of the size of the training set, we randomly select 1%, 10% and 50% of the data in the 17173 dataset as three datasets for training the modified-PCFG model and producing one billion passwords. Figure 3 shows guessing attack on 7k7k when using different percentages of the training set[3]. At the Guess Number $= 10^4$, coverages do not show much different. While the coverage of the 10% training set is less than the number of guesses of 10^6, it will be lower than the coverage of using more training set data. After the number of guesses reaches 10^7, the coverage of the full training set gradually widens with the coverage of the 50% training set. This result also shows that the model can learn complete information when using a more complete training set, but this advantage can only be reflected when the number of guesses reaches a certain number. The coverage with the 10% training set is lower than the coverage with more data when the number of guesses is 10^6. When the number of guesses reaches 10^7, the advantage of using the full training set is gradually reflected. This result also shows that the model can learn completed information of password distribution when using the full training set, but this advantage can only be reflected when the Guess Number reaches a certain number.

---

[3] This experiment package manifests that small datasets may convey incomplete information, and this incompleteness could expose inherent defects especially when the number of guesses is large. According to the rationale of these models, even with 1% of the training set, the models could generate 1 billion password guesses (used to conduct comparative experiments). Yet what's more concerning is the coverage in the context of password guessing. For the probabilistic model, the password used for the N-th guessing is of the N-th largest probability produced by the model.

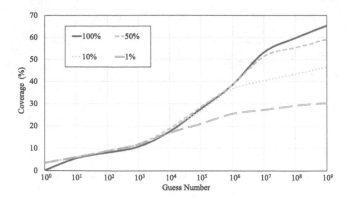

Fig. 3. Modified-PCFG model attacks on 7k7k dataset using different percentages of training set.

Table 6. Five password guessing models attack on all datasets when the Guess Number is 10^9.

Models	17173	7k7k	Tianya	Renren	Zhenai	Duduniu	CSDN	Gmail
4-Markov-Laplace	77.57%	71.94%	69.68%	64.66%	59.58%	58.41%	53.23%	35.28%
5-Markov-Laplace	86.10%	72.81%	67.87%	60.64%	58.03%	57.85%	51.40%	29.96%
Modified-PCFG	87.43%	65.28%	56.80%	53.29%	50.01%	57.75%	49.02%	34.31%
Original-PCFG	91.32%	62.97%	51.56%	48.36%	45.55%	57.11%	42.89%	30.52%
PassGAN	41.36%	44.65%	47.97%	38.48%	40.78%	24.93%	24.65%	9.84%

5.2   PassGAN vs. Probabilistic Models

In Sect. 4.3, we determine the number of iterations of the PassGAN model, i.e., 14,000. The 17173 dataset is still used as a training set. We set the number of password generations to 3 billion. Since GAN generates duplicate passwords, we de-duplicate 3 billion passwords and get 1,183,886,617 unique passwords (i.e., no duplicate). In order to compare with the four probability models, we randomly select one billion of them for guessing attacks on different datasets.

We first implement experiments with five password guessing models on the other six datasets. In our experimental results showed in Table 6, the 4-Markov-Laplace model has the highest coverage for the six test sets, range from 35.28% to 69.68%. But the effect of this model is not always the best when the number of guesses does not reach 10^9, which is similar to 5-Markov-Laplace (in Fig. 2). Similarly, Modified-PCFG performed better on the other six datasets than the Original-PCFG, which increases by 0.64% to 6.13%. PassGAN performs the worst in all datasets. However, as a deep learning-based approach, PassGAN does not require any prior knowledge of passwords. The effects of the four probabilistic models in the experiments for Duduniu and Gmail are not much different. Especially for the attack on gmail dataset, the coverage of the four models is the lowest, no more than 40%. As can be seen from Table 1, the user language of the Gmail dataset is English. The difference in language makes the two datasets have different distributions, so the experimental results are reasonable. The different coverage of the same model on different training sets demonstrates that the degree of difference between the training set and the test set will affect the attack effect of the model.

The coverage for 17173 dataset is also showed in the Table 6, which means we perform the guessing attacks on the training set. The starting point for designing this experiment is: assuming that the attacker can adjust his training set according to the characteristics of the dataset to be attacked, and the risk of this approach cannot be ignored. Considering an extreme situation, attackers get a training set which is nearly no different from the test set, then the password guess generated in this case will have a very good effect. The training set and the test set are identical, the coverage will be high, and we should not underestimate it in the face of this risk. As shown in Table 6, the experimental results for the 17173 dataset are also quite different from those previously obtained in other test sets. For the four probability models, the two PCFG-based models have better attack effects on the 17173 dataset than the two Markov-based models. The Original-PCFG with the worst performance in other data sets can achieve 91.32% coverage, the 5-Markov-Laplace model is slightly worse than the Modified-PCFG model, and the coverage of the 4-Markov-Laplace model is lowest among probability models. PassGAN still does not achieve the desired attack effect, even if the training set is the same as the test set. Our understanding is that PassGAN is suitable for training high frequency length passwords in datasets, as in [11].

We want to take out the passwords of lengths 4 to 9 in the 17173 dataset, and input all the passwords of each length separately into PassGAN model for training. In the 17173 dataset, there are more passwords from 6 to 9 in length, and we set the number of generated passwords to one billion. There are fewer passwords with lengths of 4 and 5, so we set the number of generations to 100 million. We compare the Modified-PCFG model with PassGAN and Table 7 shows the results of the guessing attack on the 7k7k dataset according to the length of the password. The column of Size in the table records the number of guesses after deduplication. PassGAN has a higher coverage rate for passwords of lengths 5 and 7, and Guess Number tried for length 7 is smaller, which means that when using the Modified-PCFG model for guessing attacks, using the PassGAN model to crack passwords of length 7 can increase the number of passwords covered while reducing the total number of guesses.

5.3   Combined Attacks

Another set of experiments we design is to use a combination of brute force and probabilistic models to attack passwords. In particular, we use brute-force attack to crack digit sequence of length 4 to 8. According to Eq. 1, the Guess Number is 111,110,000. We then select the first 88,890,000 passwords generated by the four probabilistic models to form a billion guesses, and the training set is 7k7k. Table 8 shows the increasing coverage of the probabilistic models after being

Table 7. The coverage of the guessing attack on the 7k7k dataset according to the length of the password

Length	PassGAN		Modified-PCFG	
	Size	Coverage	Size	Coverage
4	128,585	82.53%	367,225	98.22%
5	1,251,496	83.68%	1,088,252	81.03%
6	11,576,964	82.41%	38,659,965	88.79%
7	46,336,785	78.79%	74,463,399	72.59%
8	262,203,066	53.91%	206,026,231	72.78%
9	498,585,564	22.23%	301,921,929	58.71%

Table 8. Four probabilistic models combined with Brute-force attack

Models	No brute-force	With brute-force
4-Markov-Laplace	71.94%	72.04%
5-Markov-Laplace	72.81%	73.21%
Modified-PCFG	65.28%	73.97%
Original-PCFG	62.97%	74.13%

used in combination with brute-force attack. Brute-force attack is not obvious for the performance growth of two Markov-based models. In the case of the same Guess Number, Modified-PCFG increased the coverage of 7k7k from 65.28% to 73.97%, and the coverage of Original-PCFG increased by 11.16% to 74.13%.

Acknowledgement. The paper is supported by the National Natural Science Foundation of China (Grant Nos. 61572192, 61971192) and the National Cryptography Development Fund (Grant No. MMJJ20180106).

References

1. Zhu, B., Yan, J., Bao, G., Yang, M., Xu, N.: Captcha as graphical passwords- a new security primitive based on hard AI problems. IEEE Trans. Inf. Forensics Secur. **9**, 234–240 (2014)
2. van Herley, C., Oorschot, P.: A research agenda acknowledging the persistence of passwords. IEEE Secur. Priv. **10**, 28–36 (2012)
3. Burr, W.E., et al.: NIST SP800-63-2: Electronic authentication guideline. National Institute of Standards and Technology, Reston, VA, Technical report, Special Publication (NIST SP) - 800-63-1 (2011)
4. Matt, W., Sudhir, A., et al.: Testing metrics for password creation policies by attacking large sets of revealed passwords. In: ACM CCS 2010, pp. 162–175 (2010)
5. Matt, W., Sudhir, A., et al.: Password cracking using probabilistic context-free grammars. In: IEEE Symposium on Security and Privacy, pp. 391–405 (2009)

6. Houshmand, S., Aggarwal, S., Flood, R.: Next gen PCFG password cracking. IEEE Trans. Inf. Forensics Secur. **10**, 1776–1791 (2015)
7. Wang, D., Cheng, H., et al.: Understanding Passwords of Chinese Users: Characteristics, Security and Implications, July 2014. https://www.researchgate.net/profile/Ding_Wang12/publication/269101022_Understanding_Passwords_of_Chinese_Users_Characteristics_Security_and_Implications/links/5544e2700cf23ff7168696a8.pdf
8. Veras, R., Collins, C., Thorpe, J.: On the semantic patterns of passwords and their security impact. In: Network and Distributed System Security Symposium (2014)
9. Arvind, N., Vitaly, S.: Fast dictionary attacks on passwords using time-space trade-off. In: ACM CCS 2005, pp. 364–372 (2005)
10. Ma, J., Yang, W., Luo, M., Li, N.: A study of probabilistic password models. In: Proceedings of IEEE Symposium on Security and Privacy, pp. 689–704 (2014)
11. Hitaj, B., Gasti, P., Ateniese, G., Pérez-Cruz, F.: PassGAN: a deep learning approach for password guessing. CoRR abs/1709.00440 (2017)
12. Xu, L., Ge, C., et al.: Password guessing based on LSTM recurrent neural networks. In: IEEE International Conference on Computational Science and Engineering, pp. 785–788 (2017)
13. Li, Y., Wang, H., Sun, K.: A study of personal information in human-chosen passwords and its security implications. In: IEEE INFOCOM, pp. 1-9 (2016)
14. Wang, D., Zhang, Z., Wang, P., Yan, J., Huang, X.: An underestimated threat. In: ACM CCS, Targeted Online Password Guessing (2016)
15. ChineseTone. https://github.com/letiantian/ChineseTone
16. ssa.gov. https://www.ssa.gov/oact/babynames/limits.html
17. English words. http://www.mieliestronk.com/wordlist.html
18. Gulrajani, I., Ahmed, F., et al.: Improved Training of Wasserstein GANs. CoRR abs/1704.00028 (2017). http://arxiv.org/abs/1704.00028

A Fully Anonymous Authentication Scheme Based on Medical Environment

Jing Lv$^{(\boxtimes)}$, Ning Xi$^{(\boxtimes)}$, and Xue Rao$^{(\boxtimes)}$

Xidian University, Xian, China
{lvj,xrao}@stu.xidian.edu.cn, nxi@xidian.edu.cn

Abstract. The RFID (Radio Frequency IDentification) technology have been widely applied in the medical environment. Meanwhile, the demand of protecting the patients' identity information and private medical data is becoming more and more urgent. In addition, many existing schemes only preserve privacy against the external attackers but without considering the threats from servers and readers. In this paper, we propose a fully anonymous medical mutual authentication scheme, which allows identifiers to authenticate to servers without revealing the identity. The proposed scheme applies the anonymous credential and ECC (Elliptic Curve Cryptography) to achieve anonymousness and mutual authentication. Moreover, through the security analysis, our scheme can provide privacy preserving against both the outsider and insider attackers during the authentication compared with typical ECC-based schemes. And the experimental results indicate that the cost on our novel scheme is affordable.

Keywords: Medical information systems · Anonymous authentication · Elliptic Curve Cryptography · RFID authentication

1 Introduction

With the rapid development of cloud computing and Internet of Things (IOT) technologies [1], smart healthcare has become one of the most important way to provide healthcare services [2], e.g. using RFID technology to store and query patient information, manage hospital equipments and track drug information. However, due to a wide range of attacks [3] on the security of medical data, such as man in middle, spoofing, counterfeiting, traceability, eavesdropping, desynchronization etc, a great challenge is posed on medical privacy. These attacks lead to a leakage on patients' privacy and even cause severe threats on their physical and mental health. Therefore, more and more researchers have concentrated on the topic of protecting patients' medical privacy.

In the medical environment, one of the most important way to protect privacy is the authentication of patients' tags. And the privacy protection is especially essential during the authentication. In order to protect the patients' privacy in this phase, there is a vast amount of literature on RFID privacy-preserving authentication schemes. Due to the limited resources, most of identifiers (e.g. RFID tags) can not execute complex cryptographic operations.

© Springer Nature Switzerland AG 2020
S. Wen et al. (Eds.): ICA3PP 2019, LNCS 11944, pp. 484–498, 2020.
https://doi.org/10.1007/978-3-030-38991-8_32

Thus, many schemes adopt hash-based cryptography [4–8] instead of using the traditional public key cryptography. These schemes assure a high quality on security with low cost, which can provide ID privacy, replay attack resistance, forward security etc. However, most hash-based schemes suffer from scalability problems [9–12], in which performing a linear search for every identification can be time consuming. In order to overcome these security flaws, lightweight ECC-based authentication schemes [14,15,17,18] have been proposed for the better scalability with high security assurance. And nowadays, they are widely applied in medical environment as well as hash-based schemes.

But, the majority of the existing approaches assume RFID readers and the back-end server to be fully trusted, which often cannot be guaranteed in practical environment. The identifier doesn't preserve anonymity towards HIS server during the mutual authentication. Most medical systems store the identity and medical information in a plaintext form, in which insider attackers such as HIS administrator can easily steal the identity of patients during their authentication. Besides, if the server is compromised, attackers can also access to their identity.

In order to ensure the privacy of patients towards HIS and readers during the authentication, we have made the following contributions. Firstly, we present a group signature algorithm to support the anonymous authentication, which can preserve ID privacy towards HIS. Secondly, we propose a privacy-preserving mutual authentication scheme, which meets the security requirements of being fully anonymous. Finally, we design a ID privacy-preserving authentication system for smart healthcare in a practical medical environment.

The remainder of this paper is structured as follows: In Sect. 2, we introduce some related work about RFID authentication schemes applied in medical environment. In Sect. 3, we propose a fully anonymous RFID authentication scheme based on ECC cryptography and anonymous credential, and the details are given. In Sect. 4, we analysis and evaluate the security as well as performance of the proposed approach. Finally, we conclude this paper.

2   Related Work

In order to protect the medical data from being illegally collected, there are several authentication methods applied in medical environment, including hash-based schemes, ECC-based schemes and schemes based on anonymous credentials.

The existing Hash-based schemes achieve tag's authentication by sending the identity's hash value to the server for verification. Most of these schemes are partitioned into two major types: hash-lock schemes and hash-chain schemes. Hash-lock [4,5] schemes use a random number generator while hash-chain schemes [6–8] don't. The hash-based schemes [4–8] provide unlinkability by using one-way hash functions, which are lightweight for RFID systems. However, hash-based approaches suffer from scalability problems. In a large-scale RFID system, the server has to perform an exhaustive search among all tags in the system to identify the interrogated one. Besides, attackers can launch denial-of-service (DoS) attacks through submitting fake identity to the server. It would be a time and resource consuming work to search one fake tag in the system before realizing its invalidity [11].

In order to solve these problems, the ECC-based RFID authentication scheme is proposed by Tuyls and Batina [13]. ECC-based schemes use public key encryption instead of searching ID to validate the tag, which are more flexible and scalable. Besides, the lightweight elliptic curve implementations [14,15] require less computation than some cryptographic hash algorithms [16]. Zhang et al. [17] propose an RFID authentication scheme using ECC for healthcare systems, which provides high security assurance and good scalability with low cost. However, Zhang's scheme can not provide forward untraceability. When an adversary get the tag identifier, he could easily track the tag. Farash et al. [18] proposed a similar medical ECC-based scheme, which can provide reply attack resistance, forward privacy, man-in-middle resistance etc.

Most of the existing approaches assume RFID readers and HIS to be fully trusted. However, tags in these schemes are not anonymous towards readers and servers. If the server is compromised, attackers can also access to private data during the authentication. In order to overcome this security flaw, anonymous authentication of RFID tags to readers has been discussed in [19–22]. These anonymous RFID schemes trade identification efficiency for the privacy. Gope et al. [22] propose a lightweight anonymous RFID authentication scheme. However, it doesn't provide tag's anonymity towards the server. Xie's scheme [19] reduces the computational load on tags by storing the pre-computed data into tags. But the cost is still too high for an electric identifier because of the complex of the identifier's signature calculation. Armknecht et al. [20] presents a fully anonymous scheme which does not require tags to perform public key cryptography. A third party anonymizer is introduced in this scheme, which plays a role anonymizing the identity of the electronic identifier by distributing an anonymous credential. However, it is difficult to assure the anonymizer to be fully trusted in practice. Besides, this scheme requires the electronic identifier interacting with the anonymizer at the beginning of each session, which bring an extra communication cost.

To overcome above security weakness, an anonymous medical authentication scheme is presented in Sect. 3, which uses the ECC cryptography and anonymous credentials to achieve mutual authentication. Anonymizers are needless in our scheme, which has a better performance on security.

3   Anonymous RFID Authentication Scheme

3.1   System Model

The proposed fully anonymous RFID authentication model consists of four entities, i.e. the Certification Authority (CA), the Identifier (e.g. RFID tag and so on), the Identifier Reader (IR) and the Hospital Information System (HIS). The architecture of our scheme is shown in the Fig. 1.

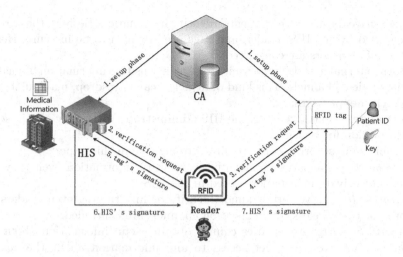

Fig. 1. The fully anonymous medical authentication model

-CA. CA works as a trusted third party which initializes the entire system, including generating and managing cryptographic parameters, the private/public keys and so on. The CA also performs pre-calculations to reduce the calculation executed by the tag.

-Electronic Identifier. The identifier is a kind of resource-constrained data carrier storing the patient's information. When a patient enters the hospital, an RFID wristband (Tag) issued by the hospital will be assigned to indicate the identity of the patient. Moreover, some medical RFID tags is capable of performing basic cryptographic operations for security consideration, such as random number generating and hashing. But public-key encryption is still not supported.

-Identifier Reader. The RFID reader is used to deliver messages between tags and HIS. It is capable of reading and writing the tag.

-HIS. The HIS is a database which stores all medical information (e.g. patient name, drug list, medical history and so on) running as a server as well as a verifier in the RFID system.

Our scheme is aimed to implement the mutual authentication between tags and HIS while maintaining the anonymity of the tag to both reader and HIS. Each tag is initialized by CA, and it stores a unique private key TSK_s and a public key PK_H of HIS. During the authentication phase, we use the ECC cryptography and group signature to achieve mutual authentication. Anonymous credentials are also used to keep tag's anonymity towards HIS administrators. HIS authenticates the tag without knowing the identity of it.

3.2 Threat Model

In our system, HIS is a "honest-but-curious" server. That means HIS follows our proposed scheme, and responds faithfully for queries, but it may aggressively collect sensitive data, e.g. the patients' identity information. Besides, the wireless channels used by the HIS, reader, and tags are assumed to be insecure. Hence, two types of attackers are considered:

(1) An external attacker who wants to obtain sensitive information of medical data via wireless channels. This kind of attacker can eavesdrop, manipulate and discard scheme messages;

(2) An internal attacker (e.g. the HIS administrator) who may obtain sensitive information for benefits.

The following security goals are aimed to achieve in this paper:

Anonymity. Adversaries cannot obtain identity information even they can get access to private information stored in the HIS.

Untraceability. Adversaries cannot identify or link the tag even readers are corrupted or the wireless communication channels are controlled.

Forward Security. Adversaries cannot obtain useful information about the previous session even they get access to any information about the current session.

Mutual Authentication. It should not be authenticated successfully when an adversary disguised as a legal tag or the HIS.

3.3 Preliminaries and Notations

In this section, we provide some preliminaries to introduce the required primitives and notations used in our scheme.

Preliminaries. ECC cryptography and bilinear maps are used in our scheme to maintain the identity anonymity to reader and HIS. They are defined as follows by referring to [20]:

Definition 1. Elliptic Curve: *The elliptic curve E defined on the finite field Z_q is expressed as $E = (q, a, b, G, n)$, all points on the curve satisfy the equations:*

$$y^2 = x^3 + ax + b(mod\ q) \tag{1}$$

$$\forall a, b \in Z_q, 4a^3 + 27b^2 \neq 0(mod\ q) \tag{2}$$

Where G is a point on the curve E with order n, called the generation point.

Definition 2. Admissible Pairing: *Let G_1, G_2, and G_T be three groups of large prime exponent q for security parameter $l_q \in N$. The groups G_1, G_2 are written additively with identity element 0 and the group G_T multiplicatively with identity element 1. A pairing is a mapping $e : G_1 \times G_2 \to G_T$ that is*
Bilinear: *For all $Q, Q' \in G_2$ it holds that*

$$e(P + P', Q + Q') = e(P, Q) \cdot e(P, Q') \cdot e(P', Q) \cdot e(P', Q') \tag{3}$$

Non-degenerate: *For all $P \in G_1^*$ there is a $Q \in G_2^*$, and for all $Q \in G_2^*$ there is a $P \in G_1^*$, $e(P, Q) \neq 1$.*

Computable: *There is a probabilistic polynomial time algorithm that computes $e(P, Q)$ for all $(P, Q) \in G_1 \times G_2$.*

Notations. For a clear description on the proposed authentication scheme, some of the notations involved in the scheme are shown in Table 1.

Table 1. Notations

Notation	Description
r_k, f_k, x, y	Random numbers generated by CA
A_k, B_k, C_k	The credential of tag
IPK	The tags' public key
TSK_s	The tags' private key
P_1, P_2	The generator of cyclic groups
N_d	A random number generated by HIS
E_H	The generator of a cyclic group
PK_H	The HIS's public key
k_H	The HIS's private key
r_1, z_1, z', t^*	Random numbers generated by CA
h', h	The computed hash value
$\sigma = (A_k^*, B_k^*, C_k^*, h', s)$	The signature of tag

3.4 Fully Anonymous RFID Authentication Scheme

In order to overcome the existing security flaws, i.e., the exposure of identity towards HIS and readers, we propose fully anonymous RFID mutual authentication scheme to authenticate tag and HIS anonymously.

The proposed scheme includes three phases: setup phase, authentication phase and revocation phase. In the setup phase, the system parameters including IPK, TSK_s are generated by CA. In the authentication phase, the HIS and tags authenticate each other. In the revocation phase, outdated tags can be flexibly revoked.

Setup Phase. This phase is the preparation of the authentication phase. The keys of the HIS and tags used in the authentication phase are generated by CA first. The patients's medical data are stored in the HIS, e.g. the patient ID, the medical list, the EPC code of all the drugs involved. The patient's client id, specifically expressed as C , is stored in the patient's RFID wristband. The server id is expressed as S, and it is stored in the HIS server.

There are six steps included in the setup phase. The HIS and tag can obtain the secret parameters through the security channel between the HIS, tag and the CA in this phase.

Step 1: CA generates the secret parameters $l = (l_q, l_h, l_\epsilon, l_n), \{l_q, l_h, l_\epsilon, l_n \in N\}$. After that, three cyclic group G_1, G_2, G_T of a large prime exponent q are generated according to the parameter l_q. In addition, a bilinear map $e : G_1 \times G_2 \to G_T$ is a map with both $\langle P_1 \rangle = G_1$ and $\langle P_2 \rangle = G_2$.

Step 2: CA chooses two secret parameters $x, y \in Z_q$, and computes the result of X and Y to satisfy $X = x \cdot P_2$, $Y = y \cdot P_2$. And then CA chooses an ideal collision-resistant hash function $H : 0, 1^* \to 0, 1^{l_h}$.

Step 3: CA generates a series of random numbers $f_1, f_2, f_3 \cdots, f_k \in Z_q$, $r_1, r_2, r_3 \cdots, r_k \in Z_q$ where k is the number of tags. Then we compute $A_k = r_k P_1$, $B_k = y A_k$, $C_k = (x + xy f_k)A$, $h = H(A_k, B_k, C_k)$ and $\beta = e(B_k, X)$. Here we denote β^{2^i} as β_i.

Step 4: CA generates the tag's private key $TSK_s(f_k, A_k, B_k, C_k, h)$ while the public key of the tag is denoted as $IPK(l, q, G_1, G_2, G_T, P_1, P_2, e, X, Y, H)$. After that, CA initializes the revocation list RL and the secret database DB, with the tag stored in the form of TSK_s in the DB.

Step 5: CA generates an elliptic curve parameter E_H of HIS and generates a public key PK_H and a private key k_H of HIS.

Step 6: After all the steps above, we store the server private key k_H, system parameter E_H and RL in HIS, i.e., $S = IPK, k_H, E_H$. Then we store the tag private key TSK_s, the server public key PK_H , system parameter E_H and the pre-computed data in the patient tag, i.e., $C = TSK_s, PK_H, E_H, \gamma$.

Authentication Phase. In this phase, the patient tag and HIS authenticate each other anonymously based on ECC cryptography and anonymous credentials. The details are shown as follows as shown in Fig. 2: *Step 1*: HIS generates N_d randomly and send it to tag who is requested to sign it later for anonymous authentication.

Step 2: Upon the receipt of N_d, the tag chooses random numbers $Z' \in Z_q, t^* \in Z_q, z_1 \in Z_q, r_1 \in Z_q$ and then computes the result of $z = t^* \cdot z' \mod q$. Then the tag initializes τ as 1. the tag will update $\tau = \tau \cdot \beta_k (0 \leqslant k \leqslant l_q - 1)$ when the kth bit of z is 1. After that, the tag calculates $A_k^* = A_k \cdot t^*$, $B_k^* = A_k \cdot t^*$, $C_k^* = A_k \cdot t^*$, $R_1 = r_1 \cdot E_H$, $h' = H(A_k^*, B_k^*, C_k^*, \tau, N_d)$, $s = z' + h' \cdot f_k \mod q$.

Based on all these calculations, the tag generates its signature

$$\sigma = (A_k^*, B_k^*, C_k^*, h', s)$$

and sends it along with R_1 to HIS.

Step 3: HIS uses the public key to verify the signature base on following equation

$$e(A_k^*, Y) = e(B_k^*, P_2) \tag{4}$$

Step 4: If the verification is successful, HIS explores the revoked list RL to validate the tag. If there is for $f \in RL$ which satisfies the equation $e(A_k^* + f \cdot B_k^*, X) = e(C_k^*, P_2)$, the tag can be proved to be invalid and the authentication will be fail.

Step 5: HIS computes

$$\tau' = e(A_k^*, X)^{h_1} \cdot e(B_k^*, X)^s \cdot e(C_k^*, P_2)^{h_1} \tag{5}$$

If there is $h' = H(A_k^*, B_k^*, C_k^*, \tau', N_d)$, it is proved that the tag is legal.

Step 6: If the tag is legal, then HIS computes $TK_s = R_1 \cdot k_H$ and get the parameter $r_1 = TK_s / E_H$. After that, the tag calculates $TK_{s_1} = R_1 \cdot r_1$ and

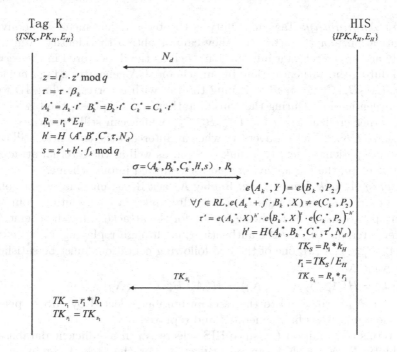

Fig. 2. Overview of our RFID authentication scheme

sends it to the patient tag to authenticate. Upon the reception of TK_{s_1}, the tag computes $TK_{r_1} = R_1 \cdot r_1$. If there is $TK_{s_1} = TK_{r_1}$, it turns out HIS is legal.

Initialization and Revocation Phase. In practice, it is necessary to generate a new tag for the new patients who just enter the hospital. In the same way, when a patient leaves the hospital, his tag should be revoked. Therefore, the tag's generation and revocation are included in our scheme. The procedure is detailed as follows:

For the revocation of expire tags, CA first checks whether the revoked tag $(T_k, f_k, A_k, B_k, C_k)$ exists in the secret database DB. If it exists, CA adds (T_k, f_k) into the RL, and sends the RL through the secure channel to HIS. After that, the tag's records will be removed from DB;

For the generation of fresh tag T_k, CA randomly chooses parameters $r_k, f_k \in Z_q$ to generate A_k, B_k, C_k, and stores $(T_k, f_k, A_k, B_k, C_k)$ to the secret database DB, as well as the private key, system parameters and pre-computed data.

4 Analysis and Evaluation

4.1 Security and Privacy Analysis

In this section, we give the security analysis of our scheme against some possible attacks in a medical authentication system.

Tag Anonymity: Tag anonymity is the basis requirement on anonymous authentication. For the attackers, they cannot obtain the identity even if they illegally access the relevant information, such as the data stored in the server or passed during the authentication. In our scheme, CA generates a tag's credential $cred = (A_k, B_k, C_k)$ instead of issuing the tag with a corresponding ID for the later authentication. During the mutual authentication phase, the tag refreshes it to another credential $cred = (A_k^*, B_k^*, C_k^*)$ which can still be verified by the public key IPK. Even the adversary who can intercept the credential still cannot infer to the patient's identify or link the tag, as well as the internal attackers in HIS. Therefore, the tag anonymity can be achieved in our scheme.

Replay Attack Resistance: Replay Attack Resistance means that obtaining the current session's information hardly makes a contribution doing some further operations in the following session for the attackers. In our scheme, if the attacker attempts to pass the authentication through replaying an old message $\sigma = (A_k^*, B_k^*, C_k^*, h', s)$, one of the two following conditions must be satisfied:

(i) $N_d = N_d'$

(ii) $h' = H(A_k^*, B_k^*, C_k^*, \tau, N_d) = H(A_k^*, B_k^*, C_k^*, \tau, N_d)'$,

where N_d, N_d' correspond to the random number selected by HIS respectively when the signature is first generated and replayed.

In terms of condition (i), since HIS selects N_d in a sufficient database, the probability $Pr[N_d = N_d']$ can be neglected. For the second condition, since the hash function chosen by CA is considered to be ideal collision-resistant, the adversary cannot generate the same hash value without N_d. Therefore, our scheme can withstand replay attacks.

Forward Security: Getting access to any secret information about the current session will not make any contribution to obtain useful information about the previous session. In our proposed scheme, the random numbers N_d, r_1, t^*, z', z_1 are used once for current session, and their previous value is hard to obtain for the attacker. In addition, the random numbers change in each session, which can assure that any secret information containing those numbers is fresh in the current session. So it is difficult to obtain any previous information according to the messages in current session. Therefore, our proposed scheme satisfies forward secrecy.

Mutual Authentication: Mutual authentication means that it should not be successfully authenticated when an adversary disguised as a legal tag or the HIS. To prove this, we need some intractability assumptions, which we introduce in the followings by referring to [20].

Definition 3. ***Bilinear LRSW Assumption:*** *Let $O_{X,Y}$ be an oracle that on input, it outputs $(A, yA, (x + fxy)A)$. Let S be the set of oracle queries. The bilinear LRSW assumption is that for all p.p.t. adversary, it holds that*

$$Pr[f \notin S \wedge f \in Z_q \wedge B = yA \wedge C = (x + fxy)A] \qquad (6)$$

is neglected.

Definition 4. *Elliptic Curve Discrete Logarithm Problem (ECDLP):*
Given an elliptic curve defined over a finite field F_q, a point $P \in E(F_q)$ of order
n, and a point $Q = LP$ where $0 \leq l \leq n - 1$, determine l is difficult.

Given an access to oracle $O_{X,Y}$ and the public key IPK, the adversary do not
know the secret parameters x, y. Hence, according to the bilinear LRSW assumption, The probability is negligible for adversary to generate a valid signature
$\sigma = (A_k^*, B_k^*, C_k^*, h', s)$. So the tag authentication is assured.

Based on the ECDLP, the probability is negligible for insider attackers to
reduce the private key k_H based on the public key PK_H and the elliptic curve
parameter E_H parameter. So the HIS authentication is assured.

Security Comparisons: Security performance comparisons are made
between some related typical schemes and our scheme, which is shown in Table 2.
In Table 2, '$\sqrt{}$' means the corresponding property is satisfied, while '$\times$' means
the corresponding property is not satisfied.

Table 2. Security performance comparison

Authentication	Replay Attack Resistance	Forward Secrecy	Mutual Authentication	Anonymity against Outside Attacker	Anonymity against Server
Liao's scheme [26]	$\sqrt{}$	$\times$	$\times$	$\sqrt{}$	$\times$
Chou's scheme [25]	$\sqrt{}$	$\times$	$\times$	$\times$	$\times$
Zhao's scheme [23]	$\sqrt{}$	$\times$	$\sqrt{}$	$\sqrt{}$	$\times$
Xie's scheme [19]	$\sqrt{}$	$\sqrt{}$	$\times$	$\sqrt{}$	$\sqrt{}$
Our scheme	$\sqrt{}$	$\sqrt{}$	$\sqrt{}$	$\sqrt{}$	$\sqrt{}$

In Table 2, the result indicates that our scheme has additional security features, which make it provides the better security than other schemes. Most of
the schemes do not provide anonymity and untraceability for server and readers except Xie's [19] and ours. But Xie's and Liao's [26] scheme fails to provide
mutual authentication. Besides, Zhao's [23] , Liao's and Chou's [25] schemes
fail to provide forward secrecy. He's [24] scheme does not provide anonymity
and untraceability against outside attacker, which will cause a severe threat to
patient's privacy protection. It is vital for the medical environment to keep the
tag's anonymity, and the absence of this security feature makes it easy to obtain
the identity information illegally. The attacker can also get access to the secret
information of the previous sessions.

4.2 Performance Analysis

In this section, we focus on the resources required by the tag to execute the tag
authentication scheme. In particular, we consider the computational, communication and memory effort of the tag.

Computation. The proposed authentication scheme requires the tag to do
some computations. During the authentication, the tag generates l_q random bits,
and performs six multiplications, one addition in Z_q, $(l_q - 1)/2$ multiplications in

G_T, and one hash digest. *Communication.* In the proposed scheme, server sends a random value of l_n bits and one element of G_1 to tag. The tag sends four elements of G_1, one G_1 bit hash digest and one element of Z_q to server. Hence, the total communication complexity of the authentication scheme is $l_n + 3l_{G_1} + l_h + l_q$ bits. In particular, l_{G_1} is the size of an element of G_1.

Memory. Each tag must store two elements of Z_q, one key of l_e bits, l_q elements of G_T, four elements of G_1, and one hash digest of l_h bits. This means that each tag must store $2l_q + l_e + l_q \cdot l_{G_T} + 4l_{G_1} + l_h$ bits in total, where l_{G_T} is the size of an element of G_T.

4.3   Experiments and Evaluations

In this section, we investigate the performance of our scheme and make comparisons with the other two typical ECC-based RFID authentication schemes. There are two different phases, i.e, setup and mutual authentication phase, we respectively evaluate the time cost and memory cost in the two phases. Also we consider the situation that several servers execute the scheme concurrently. The basic configuration is shown as Table 3.

Table 3. Configuration

General	
Virtual machine number	5
Cores and RAM	2(2.09 GHz), 4G
Operation system	Ubuntu 14.04 LTS
Development environment	eclipse
Compiler	gcc
Library	PBC

As it is shown in Fig. 3, we tested the time cost of setup phase in our scheme where the number of tags varies from 1000 to 10000. It shows that even if there is a large enough number of patients in the hospital, and the overhead is acceptable in practice. It is similar to the case of the new tag's generation phase.

Figure 3 also shows the the memory usage of setup phase where the number of tags is 1000 to 10000. Here we ran the test case 100 times. It shows that the memory usage varies little with the increase of the number of tags, and it doesn't result in a heavy cost when there are lots of tags.

Figure 4 shows the time cost when there are multiple servers concurrently running the tag's setup phase of our scheme. It is shown that the time cost effectively decreases with the increase of the number of concurrent servers, which indicates that multiple servers running concurrently will result in a lower cost, we can adapt this measure when there are lots of tags to be setup.

Figures 5 and 6 respectively shows the time cost and memory utilization of the mutual authentication phase on different schemes. In practice, the probability of large amounts of tags being authenticated at the same time is low,

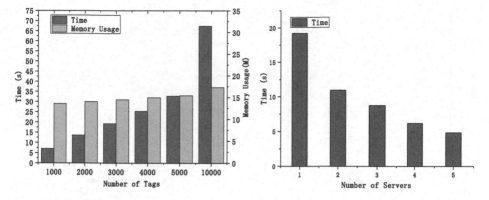

Fig. 3. Time cost and memory usage in setup phase

Fig. 4. Time cost by servers

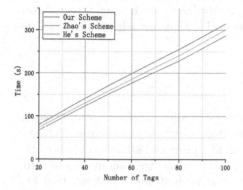

Fig. 5. Time cost of authentication phase

Fig. 6. Memory usage of authentication phase

so we test our scheme when there are 20 to 100 tags to be authenticated. The time cost varieties linearly with the increase of tags during the authentication phase.

Compared to He's [24] and Zhao's scheme [23], our scheme provides tag anonymity for readers and server, which is absent in He's and Zhao's. The time cost and memory utilization on our scheme is slightly higher than He's and Zhao's scheme. But, it is visible that there is not much difference between ours, He's and Zhao's. With the increase of the numbers of tags, both time and memory utilization in all schemes increase.

Figure 7 shows the time cost when there are multiple servers concurrently running the authentication phase of our scheme. It is shown that the time cost of our scheme, He's and Zhao's will decrease effectively with the number of servers increases. In the case of lots of tags to be authenticated, it is a good choice to run multiple servers at the same time for a lower time cost.

Figure 8 shows time cost of the authentication phase on our scheme and Armknecht et al.'s scheme. We test two schemes when there are 20 to 100 tags to be authenticated. With the increase of the numbers of tags, the time cost in two schemes increase linearly. And Armknecht et al.'s scheme requires the electronic

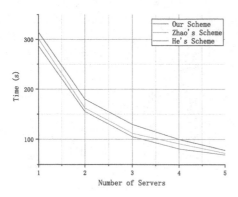 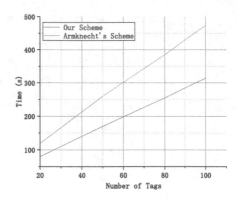

Fig. 7. Time cost of multiple servers of authentication phase

Fig. 8. Time cost of authentication phase

identifier interacting with a third party (anonymizer) at the beginning of each session, which bring an extra cost. So it is visible that our scheme has less time cost than Armknecht et al.'s scheme.

5   Conclusion

In this paper, we propose a fully anonymous medical authentication system, along with a mutual authentication scheme based on medical environment, and the analysis as well as evaluation are given. The proposed schemes enhances the security of healthcare environments, achieving the identifier's anonymity towards the hospital server. Through experiments and evaluations, we shows that our scheme not only offers security against well-known cryptographic attacks, but also provides fully anonymity. And we slightly increased the affordable overhead for additional security features.

References

1. Jia, X., Feng, O., Fan, T., Lei, Q.: RFID technology and its applications in Internet of Things (IoT). In: 2012 2nd International Conference on Consumer Electronics, Communications and Networks (CECNet), 1282–1285 (2012)
2. Turner, G.S., Tjaden, K.: United nations, department of economic and social affairs, population division, world fertility report 2013: fertility at the extremes. Popul. Develop. Rev. **41**(3), R6987–R6989 (2015)
3. Gope, P., Lee, J., Quek, T.Q.S.: Resilience of DoS attack in designing anonymous user authentication protocol for wireless sensor networks. IEEE Sens. J. **17**(2), 498–503 (2017)
4. Sun, H., Li, P., Xu, H., Zhu, F.: An improvement RFID security authentication protocol based on hash function. In: Barolli, L., Xhafa, F., Javaid, N., Enokido, T. (eds.) IMIS 2018. AISC, vol. 773, pp. 375–384. Springer, Cham (2019). https://doi.org/10.1007/978-3-319-93554-6_35

5. Yu, W., Jiang, Y.: Mobile RFID mutual authentication protocol based on hash function. In: International Conference on Cyber-Enabled Distributed Computing and Knowledge Discovery (CyberC), October 2017
6. Ohkubo, M., Suzuki, K., Kinoshita, S.: Cryptographic approach to a privacy friendly Tag. In: RFID Privacy Workshop@MIT, November 2003
7. Ohkubo, M., Suzuki, K., Kinoshita, S.: Hash-chain based forward-secure privacy protection scheme for low-cost RFID. In: 2004 Symposium on Cryptography and Information Security, vol. 1, pp. 719–724, January 2004
8. Nohara, Y., Inoue, S.: A secure and scalable identification for hash-based RFID systems using updatable pre-computation. In: WiSec 2010, Hoboken, New Jersey, USA, 22–24 March 2010 (2010)
9. Ning, H., Liu, H., Mao, J., Zhang, Y.: Scalable and distributed key array authentication protocol in radio frequency identification based sensor systems. IET Commun. $5(12)$, 1755–1768 (2011)
10. Song, B., Mitchell, C.J.: Scalable RFID security protocols supporting tag ownership transfer. Comput. Commun. $34(4)$, 556–566 (2011)
11. Alomair, B., Clark, A., Cuellar, J., Poovendran, R.: Scalable RFID systems: a privacy-preserving protocol with constant-time identification. IEEE Trans. Parallel Distrib. Syst. $23(8)$, 1536–1550 (2012)
12. Alomair, B., Poovendran, R.: Privacy versus scalability in radio frequency identification systems. Comput. Commun. $33(18)$, 2155–2163 (2010)
13. Tuyls, P., Batina, L.: RFID-tags for anti-counterfeiting. In: Pointcheval, D. (ed.) CT-RSA 2006. LNCS, vol. 3860, pp. 115–131. Springer, Heidelberg (2006). https://doi.org/10.1007/11605805_8
14. Hein, D., Wolkerstorfer, J., Felber, N.: ECC is ready for RFID – a proof in silicon. In: Avanzi, R.M., Keliher, L., Sica, F. (eds.) SAC 2008. LNCS, vol. 5381, pp. 401–413. Springer, Heidelberg (2009). https://doi.org/10.1007/978-3-642-04159-4_26
15. Lee, Y.K., Sakiyama, K., Batina, L., Verbauwhede, I.: Elliptic curve based security processor for RFID. IEEE Trans. Comput. $57(11)$, 1514–1527 (2008)
16. N.N.I. of Standards.: Technology: Cryptographic Hash Algorithm Competition. http://csrc.nist.gov/groups/ST/hash/sha-3/index. html
17. Zhang, Z., Qi, Q.: An efficient RFID authentication protocol to enhance patient medication safety using elliptic curve cryptography. J. Med. Syst. $38(5)$, 47 (2014). https://doi.org/10.1007/s10916-014-0047-8
18. Farash, M., Nawaz, O.: A provably secure RFID authentication protocol based on elliptic curve for healthcare environments. J. Med. Syst. 40, 165 (2016). https://doi.org/10.1007/s10916-016-0521-6
19. Xie, R., Xu, C., Chen, W., Li, W.: An RFID authentication protocol anonymous against readers. J. Electron. Inf. Technol. 37, 1241–1247 (2015). https://doi.org/10.11999/JEIT140902
20. Armknecht, F., Chen, L., Sadeghi, A.-R., Wachsmann, C.: Anonymous authentication for RFID systems. In: Ors Yalcin, S.B. (ed.) RFIDSec 2010. LNCS, vol. 6370, pp. 158–175. Springer, Heidelberg (2010). https://doi.org/10.1007/978-3-642-16822-2_14
21. Wu, F., Xu, L., Kumari, S., Li, X., Das, A.K., Shen, J.: A lightweight and anonymous RFID tag authentication protocol with cloud assistance for e-healthcare applications. J. Ambient Intel. Humanized Comput. $9(4)$, 919–930 (2018)
22. Gope, P., Lee, J., Quek, T.: Lightweight and practical anonymous authentication protocol for RFID systems using physically unclonable functions. IEEE Trans. Inf. Forensics Secur. $13(11)$, 2831–2843 (2018)

23. Zhao, Z.: A secure RFID authentication protocol for healthcare environments using elliptic curve cryptosystem. J. Med. Syst. **38**(5), 46 (2014). https://doi.org/10.1007/s10916-014-0046-9

24. He, D., Kumar, N., Chilamkurti, N., Lee, J.: Lightweight ECC based RFID authentication integrated with an ID verifier transfer protocol. J. Med. Syst. **38**, 116 (2014). https://doi.org/10.1007/s10916-014-0116-z

25. Chou, J.: An efficient mutual authentication RFID scheme based on elliptic curve cryptography. J. Supercomput. **70**(1), 75–94 (2013). https://doi.org/10.1007/s11227-013-1073-x

26. Liao, Y., Hsiao, C.: A secure ECC-based RFID authentication scheme integrated with ID-verifier transfer protocol. Ad. Hoc. Netw. **18**, 133–146 (2013). https://doi.org/10.1016/j.adhoc.2013.02.004

Service Dependability and Security

RaNetMalDozer: A Novel NN-Based Model for Android Malware Detection Over Task Kernel Structures

Xinning Wang[1,2] and Chong Li[1(✉)]

[1] School of Engineering, Ocean University of China, Qingdao 266100, China
{xzw0033,czl0047}@auburn.edu
[2] Department of Computer Science and Software Engineering, Auburn University, Auburn, AL 36849, USA

Abstract. The traditional neural network can maintain the performance of identifying Android malware by learning Android characteristics. But owing to the rapid growth of Android information, it cannot deal with the massive data efficiently for millions of Android applications. In this paper, we propose a novel NN-based model (RaNetMalDozer) for Android malware detection to improve the accuracy rate of classification and the training speed. The RaNetMalDozer (RNMD) can dynamically select hidden centers by a heuristic approach and expeditiously gather the large-scale datasets of 12,750 Android applications. To systematically analyze Android kernel behaviors, we investigate 112 Android kernel features of *task_struct* and offer forensic analysis of key kernel features. Furthermore, compared to the traditional neural network, EBPN method which achieves an accuracy rate of 81%, our RNMD model achieves an accuracy rate of 94% with half of training and evaluation time. Our experiments also demonstrate the RNMD model can be used as a better technique of Android malware detection.

Keywords: Analysis system · Android system · Kernel features · Malware detection · Machine learning · Neural network

1 Introduction

Currently, the growing market share of Android devices has been accompanied by the unprecedented rise of malicious threats, e.g., web threats and application threats [3]. The Android web threats exploit vulnerable websites, e.g., phishing, spam, and other fraudulent methods, to mislead customers to browser a malicious website and then they collect users' important information. The application threats masquerade Android malicious applications (apps) as the legitimate for users to install and execute. As discussed in a survey of Android security in 2015 [26], there were numerous shortcomings in its system security when facing web threats and application threats, in part because of its open-source framework, install-time permission, and because of the lack of isolation with third-part

© Springer Nature Switzerland AG 2020
S. Wen et al. (Eds.): ICA3PP 2019, LNCS 11944, pp. 501–517, 2020.
https://doi.org/10.1007/978-3-030-38991-8_33

applications. As a result, countless Android smartphones have become routinely susceptible to malicious apps. Accordingly a number of Android malware detection techniques [6,15,23,24] have been proposed to protect Android systems. Among them, Android malware detection based on the kernel structure [15,24] has become prominent due to its fine-grained data collection. Moreover, the Android kernel structure is able to audit all the active processes on Android phones and obtain detailed log information from Linux 14.04.1-Ubuntu kernel layer. In [24], kernel-based malware detection approximately achieves the accuracy rate of 90% by collecting and analyzing Android kernel features.

In terms of kernel-based malware detection techniques, the number of kernel features and the size of data records decide the correctness and scalability of malware detection. A short list of kernel features (32 features) in [24] not only decrease the classification performance, but also cause the overfitting issue as well as in [7] where few features lead to overfitting. Hundreds of thousands of Android features also influence the effectiveness of classification. To accurately predict Android malware in [27], 56,354 features are gathered from Android apps. But because of the high dimensional data input, learning a precise model becomes impossible while processing their original dataset. We find that high dimensional kernel features in Android can preserve the accuracy rate of Android malware detection since 112 kernel features in **task_struct** elaborate the activities of each process. Hence in our experiment we collect and analyze 112 kernel features to leverage kernel-based malware detection.

Neural Network (NN) approaches [12] can detect the malware based on the techniques of machine learning through linear or nonlinear classification models. The main advantage of NN-based detection is that the approaches can capture more characteristics from the undisciplined data samples and achieve a good classification result. However, its main drawback is that NN-based methods cannot train precise models for the large-scale dataset, even reduce the classification performance. The traditional method, Error Back Propagation Network (EBPN) has been mostly utilized [25] to improve the performance of malware classification. However, when the size of datasets is growing exponentially, massive numerical computation introduces the rise of computation overhead. Due to the characteristics of massive data, training the EBPN model is time-consuming and the classification performance is not always global optimal. To more accurately detect Android malware, we propose a RaNetMalDozer (RNMD) model to dynamically monitor Android apps in kernel layer. Furthermore, RNMD speeds up the training of malware detection twofold and obtains a higher accuracy performance of training the large-scale data samples.

In this work, our RNMD model automatically collects a large-scale dataset of Android apps and efficiently detects Android malware. Moreover, it examines 112 Android kernel features from **task_struct** in physical Android phones. In summary, our contributions are shown as below:

1. We propose the new NN-based model (RNMD) to systematically analyze kernel features and precisely detect Android malware, and design the heuristic approach of clustering to select and calculate the initial clustering centers.

2. Our RNMD model collects 112 kernel features from 6375 malware apps and 6375 benign apps. It processes massive data, preserves the classification performance and conducts a comprehensive analysis of kernel features.
3. Compared to EBPN model with an average accuracy rate of 81%, our RNMD achieves an accuracy rate of 94% on average. Furthermore, RNMD requires less resource allocation and less execution time. Our experiment also shows the comparisons of weight vectors of RNMD and EBPN.

The rest of this paper is organized as follows. Section 2 introduces related work on Android malware detection. Then Sect. 3 discusses the design and implementation of RNMD model. Section 4 evaluates our RNMD model and conclusions are discussed in Sect. 5.

2 Related Work

In general, malware detection falls into a plethora of categories based on different classification methodology. Kim *et al.* [16] propose power-aware malware detection by collecting power consumption samples and calculating the Chi-distance. Behavior-based analysis for malware detection is proposed by Shabtai *et al.* [23], where they offer a high level framework of malware detection, including feature selection and the number of the top features. Rastogi *et al.* [21] study the behaviors of anti-malware software and provide a method named DroidChameleon, which listens to the system-wide message broadcast and compares their footprints with a single rule. In addition, Lanzi *et al.* [18] propose a system-centric model to predict Android malware performing a large-scale data collection of call sequence and training the data with n-gram model. Demme *et al.* [9] propose a machine learning based detection technique with performance counter. They analyze the feasibility of online malware detector and come up with tentative plans of hardware implementation of malware detector.

The static analysis of Android malware in [19] is based on permissions and source code analysis, which can run on Android devices without root access. But in [19] 387 Android apps are classified by machine learning algorithms. In order to reduce the manpower of feature extraction, a CNN-based Android malware detection (R2-D2) system is proposed to convert bytecodes of source files to RGB color codes in [14]. Although the technique analyzes the Android archive files from a new perspective, the special way of declaring data variables makes Android malware detection more complicated. In [13] Hou *et al.* propose a deep brief network method named DroidDelver as an alternative of Android malware detection techniques. But they only use the existing data collection of API calls from Comodo Cloud Security Center. Ghorbanzadeh *et al.* [10] devise a neural network approach to category validation by learning permissions and likelihoods of Android apps categories. In [29] a deep learning based approach (DeepFlow) to malware detection is proposed for the analysis of data flows. But DeepFlow achieves a lower accuracy rate than other machine learning algorithms. Grosse *et al.* [11] show how to investigate the viability of adversarial crafting against neural networks based on DREBIN dataset.

Table 1. Feature Categoof Kernel Structure **task_struct**: **task_struct** contains 6 **task_state** features, 48 **mem_info** features, 15 **sche_info** features, 30 **signal_info** features, and 13 **others** features.

Categories (#.)	112 Android Kernel Feature Names & Their Number Signs
task_state (6)	[3]exit_state, [4]exit_code, [5]exit_signal, [6]pdeath_signal, [7]jobctl, [8]personality
mem_info (48)	[9]maj_flt, [10]min_flt, [11]arg_end, [12]arg_start, [13]end_brk, [14]start_brk, [15]cache_hole_size, [16]def_flags, [17]start_code, [18]end_code, [19]start_data, [20]end_data, [21]env_start, [22]env_end, [23]exec_vm, [24]faultstamp, [25]mm_flags, [26]free_area_cache, [27]hiwater_rss, [28]hiwater_vm, [29]last_interval, [30]locked_vm, [31]map_count, [32]mm_count, [33]mm_users, [34]mmap_vmoff, [35]mmap_base, [36]nr_ptes, [37]pinned_vm, [38]reserved_vm, [39]shared_vm, [40]stack_vm, [41]total_vm, [42]task_size, [43]token_priority, [44]nivcsw, [45]nvcsw, [46]start_stack, [47]rss_stat_events, [48]usage_counter, [49]nr_dirtied, [50]nr_dirtied_pause, [51]dirty_paused_when, [52]normal_prio, [53]utime, [54]stime, [55]utimescaled, [56]stimescaled
sche_info (15)	[57]last_queue, [58]pcount, [59]run_delay, [60]state, [61]on_cpu, [62]on_rq, [63]prio, [64]static_prio, [65]rt_priority, [66]policy, [67]rcu_read_lock_nesting, [68]stack_canary, [69]last_arrival, [70]flags, [71]ptrace
signal_info (30)	[72]group_exit, [73]signal_nr_threads, [74]signal_notify_count, [75]signal_flags, [76]signal_leader, [77]signal_utime, [78]signal_cutime, [79]signal_stime, [80]signal_cstime, [81]signal_gtime, [82]signal_cgtime, [83]signal_nvcsw, [84]signal_nivcsw, [85]signal_cnvcsw, [86]signal_cnivcsw, [87]signal_maj_flt, [88]signal_cmaj_flt, [89]signal_cmin_flt, [90]signal_inblock, [91]signal_oublock, [92]signal_cinblock, [93]signal_coublock, [94]signal_maxrss, [95]signal_cmaxrss, [96]signal_sum_sched_runtime, [97]signal_audit_tty, [98]signal_oom_score_adj, [99]signal_oom_score _adj_min, [100]sas_ss_sp, [101]sas_ss_size
others (13)	[102]gtime, [103]link_count, [104]total_link_count, [105]sessionid, [106]parent_exec_id, [107]self_exec_id, [108]ptrace_message, [109]timer_slack_ns, [110]default_timer_slack_ns, [111]curr_ret_stack, [112]trace, [113]trace_recursion, [114]plist_node_prio

3   A Novel NN-based Android Malware Detection Model

This section introduces the hierarchical architecture of RNMD model and the design and implementation of RNMD which focuses on function approximation and prediction as well.

3.1   Android Task Kernel Structures task_struct

The kernel data structure, **task_struct** [24], describes the elemental information of executing programs in Android kernel layer. As a useful descriptor of progress interaction, **task_struct** includes 112 features to manage the resource utilization while a process is running. It hence elaborates the information of process state, process priority, scheduling methods, signal interaction, etc. Moreover, each Android application frequently invokes the 112 kernel features to request or release its resources. The noticeable features of **task_struct** are necessary to predict the trend of system resources. Due to the full access to system resources, Android kernel features are used to delimit Android malicious behaviors [17,23,24].

A detailed footprint of kernel features is crucial to Android malware detection due to the decent coverage of process behaviors. We hence design a software agent to gather Android kernel data in **task_struct**. In order to systematically manage data samples, these 112 features are grouped into 5 categories in Table 1 with the following form: $\langle hash\_key, classifier, task\_state, mem\_info, sched\_info,$ $signal\_info, others \rangle$, where **hash_key** and **classifier** are not listed in Table 1.

3.2   Modeling of Feed-Forward RNMD

In the feed-forward neural network, to minimize the errors between outputs and targets, error back propagation is generally considered as the efficient method in the traditional neural network [25]. The neural network contains a layer of inputs, a hidden layer of training neurons, and a layer of outputs. There are the similar layers with the simple NN architecture in RNMD. However, RNMD requires to return errors between outputs and targets to the previous step of numerical calculation. The errors are propagated back to all the former layers. The error vector, err, is a format of the difference between known target value $target \rightarrow t$ and calculated value $output \rightarrow o$ given by the activation function (1)

$$o_j = f(net_j) = \frac{1}{1+e^{-\beta net_j}} \tag{1}$$

where β is the constant of activation function. A smaller value of β leads to a soft transition and a larger value of β causes a hard transition in activation levels. RNMD provides a canonical derivation using the squared errors as Eq. (2):

$$err = \sum_{j=1}^{np}(t_j - o_j)^2 = \sum_{j=1}^{np}(t_j - f(net_j))^2 \tag{2}$$

where j is the number of patterns from 1 to np, t_j, o_j are the target value and the output value of the j-th pattern and the function f is the same function as Eq. (1).

To reduce the error rate of the feed-forward computation, RNMD backward passes the error signals for each neuron and updates the weights in hidden layers. The weight updating rule is demonstrated as the following equation:

$$W_{k+1} = W_k + c \times \Delta \times Input \tag{3}$$

where W_{k+1}, W_k represent the new weight vector and the former weight vector, respectively. The parameter c is the learning constant to control the step size of correcting the errors and Δ denotes the inner product of the errors of the present layer and the derivatives of the output function. $Input$ is the input value of each layer and middle hidden layers receive their input value from the previous layer unlike the first hidden layer which directly obtains its input value from the original or reduced input without the calculation of intermediate NN layers.

Equation (3) indicates the procedure of updating weights for neurons in the same hidden layer. The value of Δ_2 in the output layer is computed by multiplying err and the derivative of the output function g'. In hidden layers, the errors have to be calculated by summating the product of weights W_2 and Δ_2. Similarly, the value of Δ_1 of the hidden layers is given by the Eq. (4):

$$\Delta_1 = f' \times \sum W_2 \times \Delta_2 = f' \times \sum W_2 \times g' \times err \tag{4}$$

Following the weight updating rules of (3), the weights of hidden layers, W_1, are updated by adding the product of learning rate c, Δ_1 and the input value X. With thousands of iterations of updating weights, the overall error rate can be reduced to a lower level.

3.3   Two Strategies to Reduce Data Size in RNMD

RNMD Kernel Function. Radial basis function [8,20] is proposed to solve the problems of nonlinear classifications or nonlinear approximation. However, the traditional neural network algorithm such as EBPN [28] is not suitable for the exascale computing of big data. In order to satisfy category validation of Android apps, radial basis function is designed and implemented in RNMD methodology. Hence our RNMD model utilizes a Gaussian kernel in the hidden layers to accomplish the nonlinear transformation of input samples, which enhances the training performance of exascale datasets. Specifically, its hidden layers perform the nonlinear transformation and map the original space to the new space using the following Eq. (5):

$$net_j = exp(- \|X - C_j\|^2 / \sigma_j^2) \tag{5}$$

where net_j, X, C_j, σ represents the j-th neuron's net value, the input vector, the j-th neuron's center position, and the j-th neuron's standard deviation, respectively. $\|.\|$ denotes the Euclidean norm and $\|X - C_j\|$ stands for the Euclidean distance between the pattern and the center. The RNMD utilizes the center's information C and its standard deviation between input layers and hidden layers.

The output layers of RNMD model need to combine each output from hidden layers as the given function (6):

$$o = \sum_{j=1}^{k} W_j \times net_j \tag{6}$$

where o, j, W_j, net_j represent the output value of output layers, the number of neurons from 1 to k, the vector of the weights and the j-th neuron's net value, respectively. The selection of clustering centers is demanded to be finished by the clustering algorithms before applying the RNMD transformation. Furthermore, the transformation given by the Eq. (5) can calculate the net values and output results also can be obtained according to the output function (6).

Algorithm 1. K-means clustering

1: **Input:** Training dataset D, number of clusters k
2: **if** *the first iteration* **then**
3:     Initialize the k clusters randomly
4: **else**
5:     Read the k clusters c_j from the last step
6: **end if**
7: Set $sum_j = 0$ and $n_j = 0$ for $j = \{1, ..., k\}$
8: **while** $TRUE$ **do**
9:     **for** $x_i \in D$ **do**
10:         **for** $j \in \{1, ..., k\}$ **do**
11:             $j_{min} = \arg\min \|x_i - c_j\|$
12:             $sum_{j_{min}} = sum_{j_{min}} + x_i$
13:             $n_{j_{min}} = n_{j_{min}} + 1$
14:             $D_j \leftarrow x_i$
15:         **end for**
16:     **end for**
17:     **for** $j \in \{1, ..., k\}$ **do**
18:         $c_j = sum_j/n_j$
19:     **end for**
20: **end while**

Algorithm 2. Gradient Descent with Constant Learning Rate

1: **Input:** Training dataset D, α, TE_{min}, clustering centers set C, kernel width set σ
2: Randomly choose the weight vector W, initialize the target output vector TP and the input vector X with dataset D
3: **while** $TE > TE_{min}$ **do**
4:     $NET = EXP(-\|X - C\|^2/\sigma^2)$
5:     $OP = W * NET$
6:     $\Delta w = -\alpha * NET * (TP - OP)$
7:     $\Delta c = -\alpha * NET * (X - C)/\sigma^2 * W * (TP - OP)$
8:     $\Delta\sigma = -\alpha * NET * (X - C)/\sigma^3 * W * (TP - OP)$
9:     $W = W + \Delta w, C = C + \Delta c, \sigma = \sigma + \Delta\sigma$
    Compute the new total errors TE
10:         $OP2 = W * NET$
11:         $ERR = TP - OP2$
12:         $TE = sum(sum(ERR \cdot ERR))$
13: **end while**

K-means to Calculate RNMD Centers. K-means algorithm can separate a clustering of data points into K regions. It selects the K centers randomly before its first iteration and then iteratively performs the following steps:

1. Calculate the centroid which is the closest to the current data point and assign the current data point to this cluster.
2. Update the selected cluster which is the closest to the current data point with the mean of the new dataset including the current data point.

Algorithm 1 explains the K-means clustering methods widely used to find a locally optimal partition of datasets. Firstly, we initialize them by randomly choosing k data samples from the dataset (Lines 2–6). S_j is the sum of all the data points belonging to the j-th center, n_j is the total number of all the data points belonging to the j-th center. During the procedure of iterative computation of k clustering regions, additional variables, sum_j and n_j, are required to temporally save the intermediate results (Line 7). The K-means algorithm assigns each data point x_i to a region D_j that is the closest centroid to x_i, and calculates the relevant cluster statistics (Lines 9–16). Meanwhile, the centroids of the k clusters are updated with the mean of the data set of these clusters (Lines 17–19). The execution of this algorithm can be terminated until all the centroids of the k clusters rarely change or the number of iterations exceeds the threshold.

3.4   Two Steps to Minimize Errors in RNMD

Select the Kernel Width (σ). The kernel width can be determined by different setting schemes [22]. In this study, we investigate a popular method for the setting of the kernel width. In this case, the K-means method is utilized to calculate the centroid c_j. The kernel width σ_j is set to the mean of Euclidean distances between data points and their cluster centroid as Eq. (7)

$$\sigma_j = \frac{1}{n_j} \sum_{x \in D_j} \|x - c_j\| = \frac{1}{n_j} \sum_{x \in D_j} (x_i - c_j)^2 \qquad (7)$$

In this situation of kernel width, the values of the parameters n_j, D_j, c_j, which represent the number of data points belonging to the j-th cluster, the data collection of the j-th cluster, and the j-th clustering center, respectively, are retrieved from the Algorithm 1.

Gradient Descent to Reduce Error Rate. The RNMD model iteratively reduces the error rate by gradient descent [4] to obtain the minimal error in Eq. (8)

$$TE = \sum_{i=1}^{n} \sum_{j=1}^{k} (t_{i,j} - o_{i,j})^2 \qquad (8)$$

where $t_{i,j}$ is the target response of the i-th output from the j-th neurons and $o_{i,j}$ is the actual response of the i-th output from the j-th neuron. Actually, the value of $t_{i,j}$ is known and the value of $o_{i,j}$ is achieved by the Eq. (6). The minimal error is that the derivatives of the parameters clustering center c_j, kernel width σ_j and the output weight w_j vanish. Therefore, an iterative computation of the gradient descent with the direction of the negative gradient $-\frac{\partial TE}{\partial w}, -\frac{\partial TE}{\partial c}, -\frac{\partial TE}{\partial \sigma}$ can solve this issue.

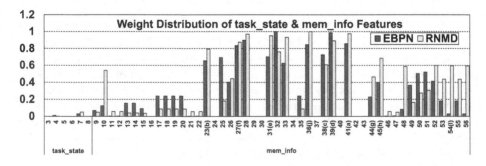

Fig. 1. Weight Distribution of **task_state** and **mem_info** Features: x-axis denotes the number and category of kernel features in Table 1 from Number 3 to 56 and y-axis denotes the weight value of these kernel features. The letters (a)–(l) represent the number of kernel features in Table 4.

Combining the Gaussian basis function with the error reduction of the gradient descent, the updating rules of the RNMD can be obtained as the following Eqs. (9), (10), and (11):

$$\Delta w_j = -\alpha \sum_{i=1}^{n} net_j(x_i)(t_{i,j} - o_{i,j}) \tag{9}$$

$$\Delta c_j = -\alpha \sum_{i=1}^{n} net_j(x_i) \frac{x_i - c_j}{\sigma_j^2} \sum_{j=1}^{k} w_j(t_{i,j} - o_{i,j}) \tag{10}$$

$$\Delta \sigma_j = -\alpha \sum_{i=1}^{n} net_j(x_i) \frac{(x_i - c_j)^2}{\sigma_j^3} \sum_{j=1}^{k} w_j(t_{i,j} - o_{i,j}) \tag{11}$$

where $\Delta w_j, \Delta c_j, \Delta \sigma_j$ refer to the elements in vectors $\Delta w, \Delta c, \Delta \sigma$ and α is the learning rate constant which is significant to control the convergence to a minimum [5]. Algorithm 2 shows the procedure of the gradient descent with the constant learning rate. The input values of Line 1 are obtained from Algorithm 1. In the iterative computation, the three vectors, $\Delta w, \Delta c, \Delta \sigma$, are used to update the previous vectors of W, C, σ (Lines 6–9). Then the new total errors can be triggered (Lines 10–12).

4   Performance Evaluation

We evaluate the RNMD framework and compare its results with EBPN method in this section. Firstly, we introduce our experimental configuration of the exascale computation and our datasets. Then, we analyze the forensics of Android kernel task structures and demonstrate their characteristics along with the changes of execution time. Finally, we discuss the resource usage and the comparisons of the accuracy rate.

4.1   Experiment Configuration

We evaluate the performance on our IBM super computer **Cirrascale**. The super computer supports parallel computing with 48 CPU cores and 260 GB

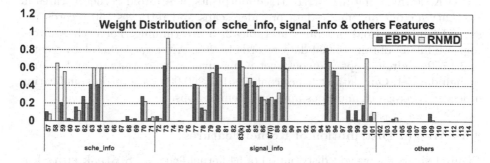

Fig. 2. Weight Distribution of **sche_info**, **signal_info** and **others** Features: x-axis denotes the number and category of kernel features in Table 1 from Number 57 to 114 and y-axis denotes the weight value of these kernel features. The letters (a)–(l) represent the number of kernel features in Table 4.

Table 2. Android malware families

Malware	Samples	Malware	Samples
ADRD	261	IconoSys	246
Adwo	140	jSMSHider	161
Agent	120	KMin	124
AnserverBot	120	LoveTrap	112
Asroot	110	NickySpy	119
BaseBridge	140	Plankton	136
BigServ	109	Pjapps	110
Boxer	110	SMSKey	107
Dowgin	200	SMSreg	131
DroidDreamLight	256	SMSReplicator	156
Droidkungfu 1–4	688	SndApps	134
FakeDoc	111	OpFake	130
FakeInst	132	TapSnake	126
FakePlayer	210	Waspx	114
Geinimi	281	YZHC	113
GingerMaster	210	Youmi	116
GoldDream	207	Zhash	111
Gone60	206	ZitMo	107
HippoSMS	204	Zsone	207

Table 3. Android benign categories

Benign	Samples	Benign	Samples
AR Apps	260	Music	260
Books	260	Navigation	260
Business	260	News	260
Education	260	Photo	260
Entertainment	260	Productivity	260
Finance	260	Reference	260
Food	260	Shopping	260
Health	260	Social	260
Kids	260	Sports	300
Lifestyle	260	Travel	300
Magazines	260	Utilities	300
Medical	260	Weather	275

memory, which attributes to the in-memory calculation of the exascale data. Furthermore, in order to improve the performance NVIDIA Tesla K80 graphic cards are configured in our super computer. To simplify the evaluation process, we implement these algorithms with MATLAB R2016a and collect the Android application samples with Python programming language remotely.

4.2   Malware and Benign Datasets

As shown in Tables 2 and 3, we collect 6375 representative Android malicious apps by VirusTotal from 38 Android malware families and 6375 Android benign apps in 24 popular benign categories from Google Play Store where there is an APK downloading mirror. To trace footprints of Android kernel features in **task_struct** our automatic data provider scans and retrieves 750 records per second while Android programs in Tables 2, 3 are running. For each Android app, we finally obtain 15,000 data records in 20 s. After gathering data records of 12,750 Android apps, we construct a massive malware dataset and a large benign dataset. Our dataset hence contains 191,250,000 data records, indicating a decent coverage of malicious apps and benign apps.

4.3   Weight Distribution of Kernel Features

Figure 1 shows the weight distribution of 54 kernel features. Six **task_state** features achieve small weight values which are less than 0.05. The results of EBPN and RNMD are similar for **task_state** kernel features, indicating the **task_state** features are less relevant to malware detection than others. For **mem_info** features, we can see according to our RNMD method, 16 weight values are between

Table 4. Key Features of Active Processes

#	Feature	Description
(a)	total_vm	total number of memory pages over all VMAs
(b)	exec_vm	number of executable memory pages
(c)	reserved_vm	number of reserved memory pages
(d)	shared_vm	number of shared memory pages
(e)	map_count	number of memory mapping areas
(f)	hiwater_rss	high water mark of resident set size
(g)	nivcsw	number of in-volunteer context switches
(h)	nvcsw	number of volunteer context switches
(i)	maj_flt	number of major page faults
(j)	nr_ptes	number of page table entries
(k)	signal_nvcsw	number of signals of volunteer context switches
(l)	stime	time eclapsed in system mode

0.5 and 1.0 and 24 weight values are between 0 and 0.5. Obviously, 13 weight values are more than 0.5 and less than 1.0 and 27 weight values are more than 0 and less than 0.5 by EBPN method. The letters from (a) to (l) stand for the descriptors of key variables of active processes in Table 4. Take 41(a) as an example, the number 41 in Table 1 illustrates total_vm and the letter (a) in Table 4 also represents total_vm.

Figure 2 describes the weight distribution of the remaining kernel features, **sche_info**, **signal_info** and **others**. EBPN achieves 7 weight values which are more than 0.5 and less than 1.0 and 26 weight values which are between 0 and 0.5. Compared to EBPN, RNMD obtains 12 weight values between 0.5 and 1.0 and 22 weight values between 0 and 0.5. The non-zero weight values are principal unit of neural networks. They decide how much influence of input values works on output values. Therefore, the weight values of kernel features show the relative importance to the input values.

4.4   Forensic Analysis of Task Kernel Variables

The dataset of 112 kernel features has more challenges in mathematical statistics and analysis in terms of the advances of data collection and storage capability [28]. As a matter of fact, not all the collected features are identically helpful for distinguishing the malicious software. Hence we measure the key kernel features and analyze their phenomena illustrated in Table 4 and Fig. 3.

Table 4 shows the names and descriptions of 12 key features of **tast_struct** in Android kernel. The benign app which is a safe and stable finance apk app implemented by Microsoft Corporation, MSN Money-Stock Quotes & News Free [1], is installed into the Android platform for analyzing the feature trend. A Trojan virus, Porns.apk [2], accesses the information of telephone services, manipulates

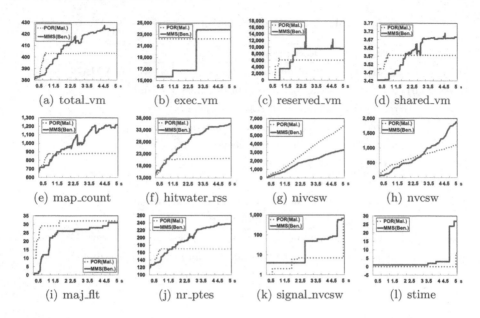

Fig. 3. Variation of **task_struct** variables in Table 4 along with time (ms): beneficial parameters for malware detection bring discriminating thresholds or different trends while programs reach balance status

the SMS operation, and steals the account privacy. The interesting traits of the two apk apps are depicted in Fig. 3. As we can see from 12 figures (a)–(l), none of the features of process execution converge at some steady state. In contrast, the malware lines of selected features approach a smooth and steady status after a sharp growth. Take Fig. 3(d) shared_vm as an example, the vertical value of the red line representing benign application exceeds the blue dash line of malware applications at the 2nd second. The vertical unit of shared_vm parameter is 10^5. The points behind the crossing point (1.7, 3.57) fall into two clusters instead of aggregating at some points again as the value of shared_vm, but there are few noise points at point (0.75, 3.58) which disturbs the feature extraction after benign and malware processes become stable.

Unlike non-malware apps which request reasonable resources discretionarily, malware apps require resources at the beginning, thereafter maintaining a steady status over a long time duration. The parameters listed in Table 4 exhibit this distinct trends of Fig. 3 between malware and non-malware, e.g., the number of different types of memory pages in Fig. 3(a), (b), (c), and (d), mapping counts of Fig. 3(e), the maximum of RSS from Fig. 3(f), major page faults in Fig. 3(i), page table numbers in Fig. 3(j), signal information of volunteer context switches of Fig. 3(k), and system time from Fig. 3(l). As shown in Fig. 3(g) and (h), it seems that the two parameters, the number of in-volunteer context switches (nivcsw) and the number of volunteer context switches (nvcsw), follow the linear distribution as the time goes up.

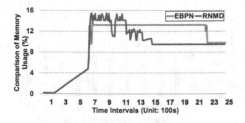

Fig. 4. Memory usage

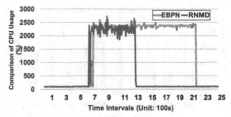

Fig. 5. CPU usage

4.5 Resource Allocation Results

Figure 4 shows the comparison of memory usage of EBPN and RNMD methods. We can see that the EBPN method uses 2200 s to finish the training job and the RNMD framework finishes its job within less time (1400 s). In detail, both of RNMD and EBPN methods request the similar memory resources to load the whole dataset to the memory from the disk in 700 s due to the same data sizes. The RNMD framework continues to calculate the cluster centers of the dataset with larger memory resources than the EBPN method between 700 s and 1100 s. When the RNMD framework finishes the computation of data centroids, its memory usage decreases by 6% approximately and its training procedure is performed from 1100 s to 1500 s. In contrast, the EBPN method requires less memory than the RNMD framework during the process of cluster centers' calculation from 700 s to 1100 s. Then its need of memory exceeds the RNMD framework before accomplishing the training procedure.

Figure 5 shows the comparison of their CPU usage. The EBPN method spends the less execution time (1400 s) to train the classification model than the EBPN method (2200 s). In 700 s, the RNMD technique and EBPN method load the whole dataset to the memory for the numerical computation. Therefore, their CPU demands are reduced to 100%. When finishing the loading job, the RNMD framework and the EBPN method begin to train the model from 700 s to 2200 s. We can see that the RNMD framework calculates the neuron information with a lower CPU usage (1960%) than the EBPN method (2160%). When the RNMD framework accomplishes the computation in 1300 s, its CPU usage decreases to 100% dramatically, which is similar with the EBPN method in 2100 s.

4.6 Comparison of Accuracy Rate

In our study, recognizing an Android app as benign is accounted to be positive and classifying a malicious app as malware is considered as negative. The classification performance is quantified by **T**rue **P**ositive (**TP**: the proportion of identifying benign apps as being nonmalware), **T**rue **N**egative (**TN**: the proportion of recognizing malicious apps as being malware), and Accuracy Rate (**Acc.**: the portion of all the benign and malware which are correctly categorized as nonmalware and malware).

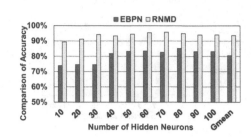

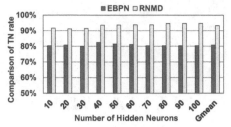

Fig. 6. Accuracy rate   **Fig. 7.** True negative rate

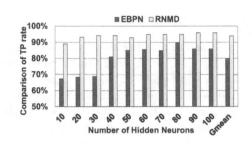

Fig. 8. True positive rate

Figure 6 shows the accuracy rates of RNMD and EBPN methods. We can see that RNMD can lead to a higher accuracy rate (93.65%) than EBPN (80.57%) on average. In the experiment, 6375 malware and 6375 benign applications have been used to train the precise model. In contrast with the EBPN which iteratively reduces the error for all data samples, the RNMD method can preserve the accuracy rate of 94% since its clustering centers gather the similar application samples into the same regions. The RNMD method chooses the clustering centers as its hidden neurons, however the EBPN method retrieves hidden neurons through the iterative computation of the entire dataset. Additionally, training results of the EBPN method alters from 74% to 83% along with the changes of the number of hidden neurons. However, our RNMD method can avoid the situation due to its clustering computation and gradient descent method.

Figure 7 illustrates the comparison of the true negative rate of RNMD method and EBPN method. RNMD achieves about 94% true negative rate on average compared to EBPN with 81% true negative rate. RNMD improves 13% of true negative rate than EBPN method. As shown in Fig. 7, the number of hidden neurons (comprising in clustering centers in RNMD) does not lead to the change of classification performance. When the number of hidden neurons increases from 30 to 40, the true negative rate of RNMD and EBPN increases by 2%.

Figure 8 shows the true positive rate of RNMD method and EBPN method. On average RNMD can achieve 94% true positive rate, while EBPN obtains 80%

true positive rate. Due to the clustering centers in RNMD, the classification performance is increased by more than 10%. The increase of the number of hidden neurons (clustering centers) in RNMD does not lead to the decrease of classification performance because RNMD has grouped the similar data records into the same clustering center. According to the rule of thumb, the number of hidden neurons approximately equals to $\sqrt{\#.\ of\ Android\ apps}$, the square root of the number of Android apps.

5   Conclusions

In this paper, we propose a novel NN-based model (RNMD) for Android malware detection. To measure the similarity of massive data, our RNMD model calculates the clustering centers and assigns each data record into different regions. We further provide the forensic analysis of 112 kernel features, the resource usage of EBPN and RNMD methods as well as their performances. Compared to EBPN, our RNMD model preserves a higher TP rate, a higher TN rate, and a higher accuracy rate with less execution cost and time. Hence RNMD improves the prediction performance for the exascale computation.

Acknowledgement. This research was funded by Fundamental Research Funds for the Central Universities under Grant 201962012.

References

1. http://apkpc.com/apps/com.microsoft.amp.apps.bingfinance.apk
2. https://forums.tomsguide.com/threads/porn-virus-on-smartphone.185058/
3. Android malicious threats. http://usa.kaspersky.com/internet-security-center
4. Andrychowicz, M., et al.: Learning to learn by gradient descent by gradient descent. In: Advances in Neural Information Processing Systems, pp. 3981–3989 (2016)
5. Armijo, L.: Minimization of functions having lipschitz continuous first partial derivatives. Pac. J. Math. **16**(1), 1–3 (1966)
6. Bugiel, S., Davi, L., Dmitrienko, A., Fischer, T., Sadeghi, A.R.: Xmandroid: a new android evolution to mitigate privilege escalation attacks. Technische Universität Darmstadt, Technical report TR-2011-04 (2011)
7. Cawley, G.C., Talbot, N.L.: On over-fitting in model selection and subsequent selection bias in performance evaluation. J. Mach. Learn. Res. **11**(Jul), 2079–2107 (2010)
8. Chen, S., Cowan, C.F., Grant, P.M.: Orthogonal least squares learning algorithm for radial basis function networks. IEEE Trans. Neural Networks **2**(2), 302–309 (1991)
9. Demme, J., et al.: On the feasibility of online malware detection with performance counters. ACM SIGARCH Comput. Archit. News **41**(3), 559–570 (2013)
10. Ghorbanzadeh, M., Chen, Y., Ma, Z., Clancy, T.C., McGwier, R.: A neural network approach to category validation of android applications. In: 2013 International Conference on Computing, Networking and Communications (ICNC), pp. 740–744. IEEE (2013)

11. Grosse, K., Papernot, N., Manoharan, P., Backes, M., McDaniel, P.: Adversarial perturbations against deep neural networks for malware classification. arXiv preprint arXiv:1606.04435 (2016)
12. Hagan, M.T., Demuth, H.B., Beale, M.H., De Jesús, O.: Neural Network Design, vol. 20. PWS Publishing Company, Boston (1996)
13. Hou, S., Saas, A., Ye, Y., Chen, L.: DroidDelver: an android malware detection system using deep belief network based on API call blocks. In: Song, S., Tong, Y. (eds.) WAIM 2016. LNCS, vol. 9998, pp. 54–66. Springer, Cham (2016). https://doi.org/10.1007/978-3-319-47121-1_5
14. Hsien-De Huang, T., Kao, H.Y.: R2–d2: color-inspired convolutional neural network (CNN)-based android malware detections. In: 2018 IEEE International Conference on Big Data (Big Data), pp. 2633–2642. IEEE (2018)
15. Isohara, T., Takemori, K., Kubota, A.: Kernel-based behavior analysis for android malware detection. In: 2011 Seventh International Conference on Computational Intelligence and Security (CIS), pp. 1011–1015. IEEE (2011)
16. Kim, H., Smith, J., Shin, K.G.: Detecting energy-greedy anomalies and mobile malware variants. In: Proceedings of the 6th International Conference on Mobile Systems, Applications, and Services, pp. 239–252. ACM (2008)
17. Kim, H.H., Choi, M.J.: Linux kernel-based feature selection for android malware detection. In: The 16th Asia-Pacific Network Operations and Management Symposium, pp. 1–4. IEEE (2014)
18. Lanzi, A., Balzarotti, D., Kruegel, C., Christodorescu, M., Kirda, E.: Accessminer: using system-centric models for malware protection. In: Proceedings of the 17th ACM Conference on Computer and Communications Security. pp. 399–412. ACM (2010)
19. Milosevic, N., Dehghantanha, A., Choo, K.K.R.: Machine learning aided android malware classification. Comput. Electr. Eng. **61**, 266–274 (2017)
20. Que, Q., Belkin, M.: Back to the future: radial basis function networks revisited. In: AISTATS, pp. 1375–1383 (2016)
21. Rastogi, V., Chen, Y., Jiang, X.: Droidchameleon: evaluating android anti-malware against transformation attacks. In: Proceedings of the 8th ACM SIGSAC Symposium on Information, Computer and Communications Security, pp. 329–334. ACM (2013)
22. Schwenker, F., Kestler, H.A., Palm, G.: Three learning phases for radial-basis-function networks. Neural Networks **14**(4), 439–458 (2001)
23. Shabtai, A., Kanonov, U., Elovici, Y., Glezer, C., Weiss, Y.: "andromaly": a behavioral malware detection framework for android devices. J. Intell. Inf. Syst. **38**(1), 161–190 (2012)
24. Shahzad, F., Akbar, M., Khan, S., Farooq, M.: Tstructdroid: realtime malware detection using in-execution dynamic analysis of kernel process control blocks on android. National University of Computer & Emerging Sciences, Islamabad, Pakistan, Technical Report (2013)
25. Socher, R., Manning, C.D., Ng, A.Y.: Learning continuous phrase representations and syntactic parsing with recursive neural networks. In: Proceedings of the NIPS-2010 Deep Learning and Unsupervised Feature Learning Workshop, vol. 2010, pp. 1–9 (2010)
26. Tan, D.J., Chua, T.W., Thing, V.L., et al.: Securing android: a survey, taxonomy, and challenges. ACM Comput. Surv. (CSUR) **47**(4), 58 (2015)
27. Wu, W.C., Hung, S.H.: Droiddolphin: a dynamic android malware detection framework using big data and machine learning. In: Proceedings of the 2014 Conference on Research in Adaptive and Convergent Systems, pp. 247–252. ACM (2014)

28. Yao, J., Zheng, S., Bai, Z.: Sample Covariance Matrices and High-Dimensional Data Analysis, vol. 2. Cambridge University Press, Cambridge (2015)
29. Zhu, D., Jin, H., Yang, Y., Wu, D., Chen, W.: Deepflow: deep learning-based malware detection by mining android application for abnormal usage of sensitive data. In: 2017 IEEE Symposium on Computers and Communications (ISCC), pp. 438–443. IEEE (2017)

Moving Target Defense
Against Injection Attacks

Huan Zhang⊚, Kangfeng Zheng(✉)⊚, Xiaodan Yan(✉)⊚, Shoushan Luo,
and Bin Wu

School of Cyberspace Security, Beijing University of Posts and Telecommunications,
Beijing 100876, China
huan5699@163.com, zkf_bupt@163.com, {xdyan,binwu}@bupt.edu.cn,
buptlou@263.net

Abstract. With the development of network technology, web services become more convenient and popular. However, web services are also facing serious security threats, especially SQL injection attack(SQLIA). Due to the diversity of attack techniques and the static of defense configurations, it is difficult for existing passive defence methods to effectively defend against all SQLIAs. To reduce the risk of successful SQLIAs and increase the difficulty of the attacker, an effective defence technique based on moving target defence (MTD) called dynamic defence to SQLIA (DTSA) was presented in this article. DTSA diversifies the types of databases and implementation languages dynamically, turns the Web server into an untraceable and unpredictable moving target and slows down SQLIAs. Moreover, the period of mutation was determined by the concept of dynamic programming so as to reduce the hazards caused by SQLIAs and minimize the impact on normal users as much as possible. Final, the experimental results showed that the proposed defence method can effectively defend against injection attacks in relational databases.

Keywords: Moving target defense · SQL injection attack · WEB service · Mutation period · Network security

1 Introduction

With the rapid development of computer network technology, the range of network applications has been extended from traditional computer applications to mobile smart phones [1] and Internet of Things applications [2]. The convenience of the network applications leads to the growing dependency of people on web applications and the storage of almost all information about visitors and servers in databases. These databases usually become suitable attack targets because of containing sensitive information. OWASP [3] released a report on the ten most critical web application security risks, which showed that SQLIA had the highest frequency among web application attacks. With SQL injection, the attacker attempted to exploit data providers at the back end through the injection of malicious SQL queries, which usually resulted in the leakage of information [4].

© Springer Nature Switzerland AG 2020
S. Wen et al. (Eds.): ICA3PP 2019, LNCS 11944, pp. 518–532, 2020.
https://doi.org/10.1007/978-3-030-38991-8_34

In order to protect web applications from SQLIAs, researchers have proposed a lot of techniques to assist webmasters, such as IDS, taint-tracking, static and dynamic analysis depending on predefined untrusted patterns. However, the configuration environments for these technologies are static. In real offensive and defensive confrontation, attackers have sufficient time to study the configuration information and vulnerabilities of the target. In contrast, the defenders are limited to passive defense.

In general, the attacker needs to determine the targets system attributes and then launches an attack based on the attributes obtained. By analyzing the samples of SQLIAs, it could be found that SQLIAs relied on the type of database and programming language. The attacker is in need of different attack payloads when web applications database and implementation language were different.

In order to reduce the asymmetry of the cost of attack and defense, MTD [5] was developed. By actively changing the system state and configuration, MTD can transform the attack surface exposed to the attacker, thereby reducing the ability of the attacker to effectively launch an attack. This paper proposed a novel and proactive approach (namely DTSA) to web applications by using the ideas of MTD. If the defender dynamically changed the attributes of web applications, the attacker would spend time in determining the attributes repeatedly and have to continuously reassess his attack strategy. It is worth noting that DTSA is not a substitute for existing security techniques, but a complement to them. It was necessary to make two major decisions to protect web applications with MTD ideas. First, the attributes mutated in web applications needed to be determined. Furthermore, the decision on the mutation period is also one of the keys to MTD. For the first decision, several attributes of web applications can be select to mutate, including system, database, IP address, port number, etc. However, there is no detailed analysis of the effects of different attributes on SQL injection defense. Therefore, the qualitative analysis of the defense effects of different attributes will become the key to the selection of mutation attributes. For the second factor, most existing studies adopt fixed period variation, but ignore the impact of mutation on service and ordinary clients. Thus this scenario motivated us to select the efficient attributes under real-time network service environment. The main contributions of this study are as follows:

(1) The defense effects of different attributes of web applications on different attack types were summarized and analyzed. In order to improve the defense efficiency while minimizing the consumption of resources available to the system, implementation language and databases were chosen for mutation as the most important attributes against SQLIAs.
(2) By weighing the hazard of SQLIAs and the impact of mutations on services, the problem of optimal mutate period was formulated as an optimal stopping problem which was solved by using the concept of dynamic programming.
(3) We conducted comprehensive experiments to evaluate the defense effect of DTSA. The experimental results showed that DTSA could effectively reduce the success rate of SQLIAs, thereby increasing the attack cost of the attacker.

The rest of this paper was organized as follows. Section 2 described the relate works about SQLIA detection and prevention. Section 3 described the architecture of DTSA. Section 4 contained an analytical approach to randomize the database and background language to minimize the action cost and attack damage. Section 5 contained a simulation evaluation. Section 6 concluded the paper.

2 Related Work

A number of researches and approaches have been devoted for detection and prevention of SQLIA over the past decade [6,7]. Basically, the detection and prevention techniques could be classified into four broad categories. First approach was static analysis [8,9]. Static analysis protects against intrusion attacks by discovering security vulnerabilities in the source code. However, defenders need to manually discover possible vulnerabilities in the system, which is time and energy consuming.

Different from static analysis, dynamic analysis analyzed the response and SQL queries from a web application [10]. First, it sent every kind of input to the target, then it received the response and collect SQL queries which are generated in the application. Second, it launched SQL injection attack to collect the response and SQL queries. Then the normal SQL queries and responses were compared and analyzed. If there was a difference, then there was an attack. In addition, taint analysis [11,12] was also a dynamic analysis. Different from static analysis, dynamic analysis could locate vulnerabilities without making any modifications to web applications. However, dynamic analysis could not find all vulnerabilities in the web application pages because the predefined attack codes are fixed.

In addition, combining the advantages of static analysis and dynamic analysis, the researchers proposed a defence technology that combines the static analysis and dynamic analysis. For instance, Lee and Jeong [13] proposed an effective detection method that extracts the attribute value in the SQL request statement in the submit parameters, then compares this attribute value with a predetermined one.

Some researchers detect and prevent SQLIA by machine learning. Chen et al. [14] analyzed the characteristics of SQLIA and used SVM algorithm to identify SQLIA. Wu et al. [15] proposed a method to prevent SQLIA in the cloud environment. They judged whether the user's input is an attack by computing the execution time of the SQL query statement and comparing it with the parse tree. Li et al. [16] proposed a LSTM-based SQL injection attack detection method and they also proposed an injection samples generation method based on communication attack behaviors analysis. Yan et al. [17] extracted the features from abstract syntax tree of JavaScript and constructed a deep learning network to detect vulnerable applications from normal ones.

Similar to traditional SQL injection, NoSQL databases were still vulnerable to injection attacks [18–20]. Ron et al. [21] introduced the several SQL injection technologies for the NoSQL database and some methods to mitigate SQL injection. Hou et al. [22] analyzed the malicious injection in NoSQL databases and propose defense approaches to prevent the malicious injection attacks. However, they are all static defenses that can be bypassed by attackers. Son et al. [23] designed a new run time tool that precisely and efficiently detects code injection attacks. Eassa et al. [24] presented a testing tool to detect NoSQL injection attacks in web application. And it can be applied to a variety of NoSQL databases. Eassa et al. [25] detected the NoSQL injection attacks by compare the reserved keywords and operators which may be used in injection attack.

The last category was mutation. Appelt and Nguyen [26] proposed a blackbox automated testing approach that generate adequate test data sets by constantly mutation the query to detect and find the SQL injection vulnerabilities. Taguinod and Doupe [27] applied MTD concepts to web applications, but it is only a theoretical explanation. Vikram and Yang [28] randomized HTML form pages on which users could submit data to defend against web vulnerabilities, however, their main defense was the Web bot attack, which could not effectively defend against SQLIAs. Keromytis and Angelos [29] diversified the SQL statement by appending a random tag in the SQL execution statement. Boyd et al. [30] rewrote all keywords with the random key appended. Simon et al. [31] combines various kinds of software diversification together. These diversification strategies include natural software diversity and automatic diversity. Natural software diversity achieves the same application function through different software modules and automatic diversity generate diverse versions of some application components. Although these methods could prevent injection attacks, the authors did not study the time of mutation. If the frequency of mutation was too low, there was a possibility that the attacker will bypass the defense.

3   Architecture

In order to improve attack efficiency, the attacker tended to first determine the type of database and high-level implementation language and then launched a series of attacks to obtain sensitive information. Figure 1 shows the overall architecture of DTSA composed of four components, including Web server, Core Unit and logic processing scripts and databases.

The Web server provided the online services needing to be protected. Core Unit was the core of DTSA that decided the process to be used and the time to change. Responsible for handling requests, logic processing scripts could be implemented in different languages, such as Python, PHP and Perl. The defence model of DTSA contains relational and NoSQL databases.

Different processing scripts had the same processing logic in order to reduce the impact on the Web server. Similar to logic processing scripts, the same data was stored in different databases. Notably, it was not necessary for DTSA to implement all logic processing scripts by different languages. Instead, it was only necessary for DTSA to handle a small percent of scripts where data could be submitted by users for database execution. Similarly, only sensitive information was stored in different databases. Therefore, DTSA had relatively lightweight workload.

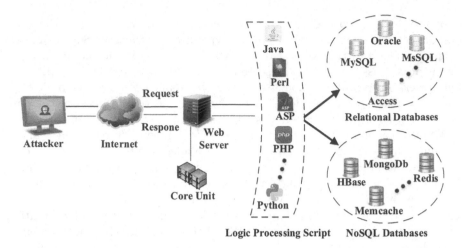

Fig. 1. Overall DTSA architecture.

In the defence model, the script handling the request was dynamic. Core Unit evaluated the damage of possible SQLIAs and the cost of mutation to determine the time of mutation, and then selected one of the execution scripts to handle user requests randomly. For example, PHP was the current script language handling user requests and MySQL was the database. After assessed by Core Unit, the script was expected to change in 5 min. After 5 min, Core Unit randomly selected another script which might be written by Perl. The database was Oracle. In this model, DTSA implemented at the server-side. Hence, the mutation will be transparent to normal clients and malicious attackers.

4   Mutation Strategy

The primary purpose of DTSA was to increase the difficulty of the attacker and deter or thwart the progress of the attack. Although there are many attributes in the web service that can be used for diversification, we should choose the most effective mutation attributes for the SQLIAs to reduce the impact on the service. Based on our previous research [32], we can analyze the defense effect of different attributes against different attacks in detail.

As can be seen in Table 1, the defense efficiency of different attributes was divided into three levels, high (H), medium (M) and low (L). Each level was be divided into two sub-levels. Level 1 indicates that the system attribute can defend against all attacks of a type of attack, and level 2 indicates that the system attribute can defend against some of the attack modes of a class of attacks. In this table, the most effective system attributes for the SQL injection attack are database, implementation language, and instruction set. In consideration of the high expertise of instruction set mutation required, this article had determined the mutation of implementation language and database.

Table 1. Defensive effects of different mutant elements against Web attacks

System Attribute	Network Element	Token	IP	Port	Topology	Protocol	MAC	Proxy	WEB Server	Operation System	Database	Implementation language	File Information	Database Information	Software Diversity	VM	Instruction Set	Service Version	Address Space Layout
Attack Style	Scan	L1	L1	H2	H2	H2	H1	H1	H2	H2	H2	H2	M2	M1	M1	L1	H1	H2	L1
	DDoS	L1	L1	H2	H1	L2	L1	L1	H1	L1	L1	L1	L1	L1	L1	L1	L1	L1	L1
	CSRF	L1	H2	L1	L1	L1	L1	L1	L1	L1	L1	L1	L1	L1	L1	L1	L1	L1	L1
	Redirect	L1	H2	L1	L1	L1	L1	L1	L1	L1	L1	L1	L1	L1	L1	L1	L1	L1	L1
	XSS	L1	L1	L1	L1	L1	L1	L1	L1	L1	L1	L1	L1	L1	L1	L1	H2	M1	L1
	SQL Injection	M1	L1	L1	L1	L1	L1	L1	L1	L1	H2	H2	M1	H1	L1	L1	H2	M1	L1

Then the frequency of mutation turned into the focus of this study. The lower the frequency of mutation was, the longer static attributes would be exposed, which provided enough time for the attacker to discover and exploit SQL injection vulnerabilities and led to the insignificance of MTD mutation. In addition, too high frequency of would raise network overhead accordingly, which would inevitably reduce service quality. A random or fixed value was often used as the mutation period in existing research. However, an appropriate mutation period should be adjusted according to different environments and systems.

In this case, the damage caused by SQLIAs covered the disclosure of sensitive information about users and Web services while that caused by the attacker got the root permissions of the web application. In addition to the damage caused by the attack, the cost arising from mutation should also be considered, mainly including network latency. The following equation was the evaluation function of the total cost, the sum of two costs:

$$T_S = T_A + T_M \tag{1}$$

Wherein, T_S represented the sum of costs; T_M referred to the cost arising from mutation; T_A meant the sum of the damage caused by SQLIAs before mutation. Let T_A be a linear function of N. N indicated the time of exposing current configuration to the attacker. D_A stood for the average loss arising from the attack in one unit of time. Then, the evaluation function (1) could be expressed as:

$$T_S = D_A * N + T_M \tag{2}$$

As described previously, D_A could be classified as hazardous to sensitive information and visitors. N should be less than S if the successful time of SQLIAs was S. Otherwise, the mutation would be meaningless. Therefore, D_A could be defined as:

$$D_A = \frac{\lambda_1 * I + \lambda_2 * C}{S} \tag{3}$$

Wherein, I and C denoted the number of sensitive data in the database and the average number of visitors during a certain period of time respectively, both of which contributed differently to D_A when the number of visitors was different. λ_1 and λ_2 represented the weights of I and C respectively, and $\lambda_1 + \lambda_2 = 1$. λ_1 would be larger than λ_2 if C was larger than I, which meant that the damage to visitors would be greater than that caused by the leakage of sensitive data when the number of visitors was greater than the amount of sensitive data. Similarly, λ_1 would be smaller than λ_2 and the loss caused by leakage would be greater than the damage to visitors when C was smaller than I. In addition, λ_1 would be equal to λ_2 when C was equal to I.

This paper aimed to find the time N minimizing the expectation value of the total cost T_S. The concept of dynamic programming was adopted to introduce the function $T(m, N)$ obtaining the number of time m to minimize the expectation value of the evaluation function 1. Let N be the optimal time of mutation. The function 1 indicated the loss of the web application when the attack occurred at $m\,(m \leq N)$ unit times. As a result, function 1 could be defined as follows:

$$T(N, N) = D_A * N + T_M \qquad (m = N) \tag{4}$$

$$T(m, N) = P(\bar{F}_{m+1}|O_m)*m*D_A + P(F_{m+1}|O_m)*T(m+1, N) \qquad (m < N) \tag{5}$$

In this process, each attack was independent, whose process could be treated as a binomial distribution. An event where attack occurred at m times was defined as O_m. The probability of no attack in time $m+1$ was defined as $P(\bar{F}_{m+1})$ while the probability of attack was defined as $P(F_{m+1})$. Thus, $P(\bar{F}_{m+1}|O_m)$ meant the condition probability that the attack occurred at m times and the attack did not occur at the $m + 1$ time. $P(F_{m+1}|O_m)$ represented the condition probability that the attack occurred at $m + 1$ time. The relationship between these two conditional probabilities could be expressed as:

$$P(\bar{F}_{m+1}|O_m) = P(O_m) - P(F_{m+1}|O_m) \tag{6}$$

Starting from the definition of the function $T(m, N)$, the goal was to find the N minimizing $T(0, N)$ because N was the same as m minimizing the expectation value of the function (1).

By iterating (5), the general presentation of $T(0, N)$ could be presented as:

$$
\begin{aligned}
T(0, N) = C * \prod_{j=0}^{N-1} (P(O_j) - P(\bar{F}_{j+1}|O_j)) + I * \sum_{j=1}^{N-1} j * P(\bar{F}_{j+1}|O_j)) * \prod_{i=0}^{j} (P(O_i) \\
- P(\bar{F}_{i+1}|O_i))) + N * D_A * \prod_{i=0}^{N-1} (P(O_i) - P(\bar{F}_{i+1}|O_i)))
\end{aligned}
\tag{7}
$$

By calculation and comparison, the smallest N in (7) would be found. Different web application environments were simulated to illustrate the effectiveness of this method. In the simulated environment, the database contained 10,000, which meant the value of I would be 10,000. At three different time periods, the number of visitors was 5,000, 10,000, and 20,000 respectively. It took 60 seconds for the attacker to attack. In these three different scenarios, the parameters λ_1 and λ_2 in (3) were assigned to (0.7, 0.3), (0.5, 0.5), (0.3, 0.7) respectively. The period of mutation depended on the number of visitors and the average damage of the attack. Therefore, the relationship between them could be seen from Fig. 2.

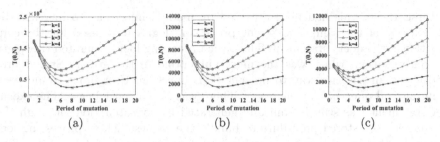

Fig. 2. Mutation period for different circumstances, (a) $I = 10,000$, $C = 20,000$, (b) $I = 10,000$, $C = 10,000$, (c) $I = 10,000$, $C = 5,000$

In these experimental, the value of k represented the multiple of average attack damage. When k was equal to 1, the average damage would be D_A. When k was equal to 2, it meant that the average damage of the attack to the web application was twice that of D_A, and so on. As displayed from the figure, the period of mutation would be shorter when k was larger, which was intuitively clear. It was necessary to shorten the period of mutation when the damage caused by the attack increased, thereby increasing the difficulty of the attacker.

5 Evaluation

In this study, the experimental environment includes Python2.7 and Virtual Machine with Ubuntu 16.04 and 8 GB memory, MATLAB R2017b, Apache2.4.9 and a server with Windows 10 and Intel Xeon E5-2620 CPU and 32 GB memory.

First of all, a sample with SQL injection vulnerability was constructed. A user could enter the username and password into a simple web page to log on the system. The main source code was:

$< td >< inputname = "login\_n" type = "text" id = "login\_nm" >< /td >$

$< td >< inputname = "login\_pw" type = "text" id = "login\_pw" >< /td >$

After receiving the username and password entered by the client, the web application would query the database to confirm the existence of the user. If the user existed, the web application would return a prompt for a successful login. If the user did not exist or the password was incorrect, the web application would return the information of failed login. It was a typical SQL injection vulnerability due to the lack of filtering for the construction of a query statement. e.g.

$\$query = "select * from logins where username='\$user' and password ='\$pwd' ";$

$\$result = mysqli\_query(\$mysqlconn, \$query);$

$\$user$ and $\$pwd$ were corresponding to the username and password entered by the client. Then, the malicious attacker could attack this web page by various attack types.

First, the sample had been attacked by manual injection. To increase the authenticity of the sample, different processing logics were used for the construction of the query statement. The basic processing logic generated the query by concatenation, which was the most vulnerable to the attack. Stored procedure was an effective way to prevent SQL injection. User input would be a string instead of executing the query. There was also a special case. The execution statement was still dynamically generated by concatenation although the webmaster used stored procedure to handle the request. These processing logics were implemented in different implementation languages and databases. The results of manual injection in relational databases are shown from Tables 2, 3 and 4.

Table 2. Attack in basic logic

Mutate elements		SQLIA type							
Database	Language	Tautology	Piggy backed	Union	Logically incorrect	Boolean injection	Timing attacks	False injection	Alternate encoding
MySQL	PHP	Y	N	Y	N	Y	Y	N	Y
	Perl	Y	N	Y	N	Y	Y	N	Y
	Python	Y	Y	Y	Y	Y	Y	Y	Y
Oracle	PHP	Y	N	Y^1	Y	Y^1	N	N	Y^1
	Perl	Y	N	Y^1	N	Y^1	N	N	Y^1
	Python	Y	N	Y^1	Y	Y^1	N	N	Y^1
SQL Server	PHP	Y	Y	Y	Y	Y^2	Y^2	Y^2	Y
	Perl	Y	Y	Y	Y	Y^2	Y	Y^2	Y
	Python	Y	×	Y	Y	Y^2	Y^2	Y^2	Y

Table 3. Attack in store procedure

Mutate elements		SQLIA type							
Database	Language	Tautology	Piggy backed	Union	Logically incorrect	Boolean injection	Timing attacks	False injection	Alternate encoding
MySQL	PHP	N	N	N	N	N	N	N	N
	Perl	N	N	N	N	N	N	N	N
	Python	N	N	N	N	N	N	N	N
Oracle	PHP	N	N	N	Y	N	N	N	N
	Perl	N	N	N	N	N	N	N	N
	Python	N	N	N	N	N	N	N	N
SQL Server	PHP	N	N	N	Y	N	N	N	N
	Perl	N	N	N	Y	N	N	N	N
	Python	N	N	N	N	N	N	N	N

Table 4. Attack in dynamic store procedure

Mutate elements		SQLIA Type							
Database	Language	Tautology	Piggy backed	Union	Logically incorrect	Boolean injection	Timing attacks	False injection	Alternate encoding
MySQL	PHP	Y^1	N	Y^2	N	Y^3	Y^2	N	Y^2
	Perl	Y^1	N	Y^1	N	Y^3	N	Y^3	Y^2
	Python	Y^1	Y^1	Y^2	Y	Y^3	Y^2	Y^2	Y^2
Oracle	PHP	N	N	N	N	N	N	N	N
	Perl	N	N	N	N	N	N	N	N
	Python	N	N	N	Y	N	N	N	N
SQL Server	PHP	N	Y^2	Y^3	Y	Y^2	Y^2	Y^2	Y^2
	Perl	N	N	N	Y	N	N	N	N
	Python	N	N	N	Y	N	N	N	N

In these tables, 'Y' indicated that the attack was successful while 'N' indicated the attack failed. Different 'Y' suggested that the attacker needed different attack loads to attack successfully in different environments. For example, 'SELECT * FROM products WHERE id=1-SLEEP (15)' would be used when the database was MySQL if the attacker wanted to perform a time-based attack. 'SELECT * FROM products WHERE id=1; WAIT FOR DELAY '00:00:15'' was required to achieve the same purpose since SQL Server did not support the sleep function. The main reason was that different databases had specific tables and supported functions, and the functions of executing query statements in different implementation languages were also different. For instance, the function 'execute' in the python language could execute multiple statements simultaneously, which might result in piggy-backed attack. However, the function 'mysqli_query' in PHP just executed one statement.

It could be seen from the tables that an attack could be successful in some circumstances, but fail in other circumstances. Thus, the attacker would need to constantly adjust the attack payload for a successful attack if constant changes occurred in the database and language.

Then, this sample was scanned by Acunetix Web Vulnerability Scanner (AWVS). As a commonly-used tool for scanning and exploiting SQL injection vulnerabilities, AWVS automatically found a series of vulnerabilities for each page. Containing high, medium, low and informational alerts, the scanning results listed the number of each alert and the attack load for each vulnerability. Therefore, the details of SQL injection vulnerabilities which could be successfully attacked could be presented. The scanning results obtained by use of AWVS are shown in Fig. 3.

In this experiment, was assigned as a value of one minute, which meant that it would take one minute for the attacker to attack successfully. In addition, I was 10000 and C was 10000. The experiment was repeated ten times.

The blue column in the Fig. 3 represented the number of SQL injection vulnerabilities scanned by AWVS under the static condition. The number of vulnerabilities scanned could not change because the database and implementation

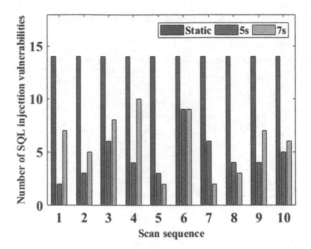

Fig. 3. Number of SQL injection vulnerability between Static and mutation.(Color figure online)

language were static. Red and yellow columns indicated the number of SQL injection vulnerabilities scanned by AWVS when the periods of mutation were 5s and 7s respectively. According to the results, the number of vulnerabilities scanned when the database and implementation language were mutated was less than that when attributes were static. The reason for this phenomenon was that the attacker would determine the type of database first and proceed to the next attack when detecting an injection point. However, the payload of attack for the previous database might be no longer applicable to the new database if the database was changed in the process of attacking, giving rise to the failure of the attack.

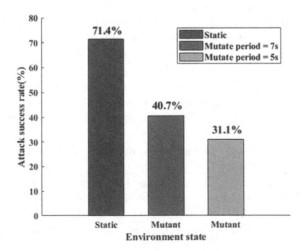

Fig. 4. Comparison of attack success rate under different mutation period.

The attack load provided by AWVS was applied to exploit these vulnerabilities. The success rates of exploitation in different mutation periods are recorded in Fig. 4. It could be seen that the exploitation of the vulnerability in the case of dynamic mutations in the database and implementation language was lower than that in static state. The reason was that the attack payload might no longer be available to the new environment even if available vulnerabilities could be scanned by the attacker. The attack success rate when the period of mutation was 7 s was larger than that when the period of mutation was 5 s, which resulted from the randomization of mutations. The same attack payload produced different effects in different environments, which thus resulted in the certain randomness of the final experimental results.

The time cost of DTSA was examined through the measurement of page loading time (PLT) which referred to the time interval between the timestamps of the browsers sending the request and completely loading a webpage. The developer tools in Microsoft Edge were used to calculate the time.

Compared with the PLT of static properties, PLT would also change irregularly when a mutation occurred. In this experiment, the period of mutation was 5 s. As shown in Fig. 5, the highest overhead was nearly 200% and the smallest overhead was only 34% of the average in a static state. The change in PLT was believed to be acceptable to increase security. In addition, the cache would be empty first when PLT was calculated. Thus, the overhead could even see more decrease in a communication session with cache storage in reality. Additionally, the program occupied less than 1% of CPU and less than 0.002% of the memory during operation.

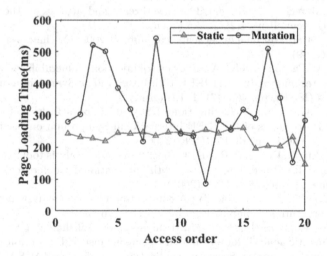

Fig. 5. Comparison of PLT between static state and mutation.

6 Conclusion

In this paper, DTSA was presented, it was a framework that employs dynamic. For the two key factors of MTD, we first analyzed the defense effect of different system attributes against different attacks, and selected database and implementation language as the mutated elements of DTSA. In order to minimize the damage caused by the attack and the impact of mutation, the optimal mutation time problem in time series data has been formulated as an optimal stopping. With DTSA, we could effectively prevent the protected Web services from SQLIA. Our simulation showed that DTSA can effectively reduce the success rate of SQLIA with a reasonably low performance overhead.

In the future study, we will explore ways to implement the DTSA prototype system and apply DTSA to NoSQL database, such as MongoDB and Redis. In addition, we intend to conduct research on the rapid perception and prediction of attacks to select the most effective defense strategy.

References

1. Xu, Y., Wang, G., Ren, J., Zhang, Y.: An adaptive and configurable protection framework against android privilege escalation threats. Future Gener. Comput. Syst. **92**, 210–224 (2019)
2. Xu, Y., Ren, J., Wang, G., Zhang, C., Yang, J., Zhang, Y.: A blockchain-basednon-repudiation network computing service scheme for industrial IoT. IEEE Trans. Ind. Inf. (2019)
3. OWASP Top 10 Application Security Risks. https://www.owasp.org/index.php/Top_10-2017_Top_10. Accessed 27 Mar 2018
4. Holz, T., Marechal, S., Raynal, F.: New threats and attacks on the world wide web. IEEE Secur. Priv. **4**(2), 72–75 (2006)
5. Chong, F., et al.: National cyber leap year summit 2009: Co-chairs report. NITRD Program (2009)
6. Kindy, D.A., Pathan, A.S.K.: A survey on sql injection: vulnerabilities, attacks, and prevention techniques. In: 2011 IEEE 15th International Symposium on Consumer Electronics (ISCE), pp. 468–471. IEEE (2011)
7. Halfond, W.G., et al.: A classification of SQL-injection attacks and countermeasures. In: Proceedings of the IEEE International Symposium on Secure Software Engineering, vol. 1, pp. 13–15. IEEE (2006)
8. Gould, C., Su, Z., Devanbu, P.: JDBC checker: a static analysis tool for SQL/JDBC applications. In: Proceedings of the 26th International Conference on Software Engineering, pp. 697–698. IEEE (2004)
9. Chunmei, W.U., Xia, N., Mao, B.: A comparison of static analysis technology for intrusion prevention. Comput. Eng. **32**(3), 174–176 (2006)
10. Kosuga, Y., Kono, K., Hanaoka, M., Hishiyama, M., Takahama, Y.: Sania: syntactic and semantic analysis for automated testing against SQL injection. In: Twenty-Third Annual Computer Security Applications Conference (ACSAC 2007), pp. 107–117. IEEE (2007)
11. Naderi-Afooshteh, A., Nguyen-Tuong, A., Bagheri-Marzijarani, M., Hiser, J.D., Davidson, J.W.: Joza: hybrid taint inference for defeating web application SQL injection attacks. In: 2015 45th Annual IEEE/IFIP International Conference on Dependable Systems and Networks, pp. 172–183. IEEE (2015)

12. Nguyen-Tuong, A., Guarnieri, S., Greene, D., Shirley, J., Evans, D.: Automatically hardening web applications using precise tainting. In: Sasaki, R., Qing, S., Okamoto, E., Yoshiura, H. (eds.) SEC 2005. IAICT, vol. 181, pp. 295–307. Springer, Boston, MA (2005). https://doi.org/10.1007/0-387-25660-1_20

13. Lee, I., Jeong, S., Yeo, S., Moon, J.: A novel method for SQL injection attack detection based on removing SQL query attribute values. Math. Comput. Modell. **55**(1–2), 58–68 (2012)

14. Chen, Z., Guo, M., et al.: Research on SQL injection detection technology based on SVM. In: MATEC Web of Conferences. vol. 173, p. 01004. EDP Sciences (2018)

15. Wu, T.Y., Chen, C.M., Sun, X., Liu, S., Lin, J.C.W.: A countermeasure to SQL injection attack for cloud environment. Wireless Pers. Commun. **96**(4), 5279–5293 (2017)

16. Li, Q., Wang, F., Wang, J., Li, W.: LSTM-based SQL injection detection method for intelligent transportation system. IEEE Trans. Veh. Technol. **68**(5), 4182–4191 (2019)

17. Yan, R., Xiao, X., Hu, G., Peng, S., Jiang, Y.: New deep learning method to detect code injection attacks on hybrid applications. J. Syst. Softw. **137**, 67–77 (2018)

18. Hou, B., Shi, Y., Qian, K., Tao, L.: Towards analyzing mongodb nosql security and designing injection defense solution. In: 2017 IEEE 3rd International Conference on Big Data Security on Cloud (bigdatasecurity), IEEE International Conference on High Performance and Smart Computing (HPSC), and IEEE International Conference on Intelligent Data and Security (IDS), pp. 90–95. IEEE (2017)

19. Patil, P., Gaikwad, A., Badekar, R., Urade, A., Hirve, R.: Analysis and diminution of NoSQL injection attacks. IJRDO - J. Comput. Sci. Eng. **3**(3), 01–07 (2017). ISSN 2456-1843, https://www.ijrdo.org/index.php/cse/article/view/481

20. Algarni, A., Alsolami, F., Eassa, F., Alsubhi, K., Jambi, K., Khemakhem, M.: An open tool architecture for security testing of NoSQL-based applications. In: 2017 IEEE/ACS 14th International Conference on Computer Systems and Applications (AICCSA), pp. 220–225. IEEE (2017)

21. Ron, A., Shulman-Peleg, A., Puzanov, A.: Analysis and mitigation of NoSQL injections. IEEE Secur. Priv. **14**(2), 30–39 (2016)

22. Hou, B., Qian, K., Li, L., Shi, Y., Tao, L., Liu, J.: Mongodb NoSQL injection analysis and detection. In: 2016 IEEE 3rd International Conference on Cyber Security and Cloud Computing (CSCloud), pp. 75–78. IEEE (2016)

23. Son, S., McKinley, K.S., Shmatikov, V.: Diglossia: detecting code injection attacks with precision and efficiency. In: Proceedings of the 2013 ACM SIGSAC Conference on Computer & Communications Security, pp. 1181–1192. ACM (2013)

24. Eassa, A.M., Al-Tarawneh, O.H., El-Bakry, H.M., Salama, A.S.: NoSQL racket: a testing tool for detecting NoSQL injection attacks in web applicationss. Int. J. Adv. Comput. Sci. Appl **8**(11), 614–622 (2017)

25. Eassa, A.M., Elhoseny, M., El-Bakry, H.M., Salama, A.S.: NoSQL injection attack detection in web applications using restful service. Program. Comput. Softw. **44**(6), 435–444 (2018)

26. Appelt, D., Nguyen, C.D., Briand, L.C., Alshahwan, N.: Automated testing for SQL injection vulnerabilities: an input mutation approach. In: Proceedings of the 2014 International Symposium on Software Testing and Analysis. pp. 259–269. ACM (2014)

27. Taguinod, M., Doupé, A., Zhao, Z., Ahn, G.J.: Toward a moving target defense for web applications. In: 2015 IEEE International Conference on Information Reuse and Integration, pp. 510–517. IEEE (2015)

28. Vikram, S., Yang, C., Gu, G.: Nomad: towards non-intrusive moving-target defense against web bots. In: 2013 IEEE Conference on Communications and Network Security (CNS), pp. 55–63. IEEE (2013)
29. Keromytis, A.D.: Randomized instruction sets and runtime environments past research and future directions. IEEE Secur. Priv. **7**(1), 18–25 (2009)
30. Boyd, S.W., Keromytis, A.D.: SQLrand: preventing SQL injection attacks. In: Jakobsson, M., Yung, M., Zhou, J. (eds.) ACNS 2004. LNCS, vol. 3089, pp. 292–302. Springer, Heidelberg (2004). https://doi.org/10.1007/978-3-540-24852-1_21
31. Allier, S., Barais, O., Baudry, B., Bourcier, J., Daubert, E., Fleurey, F., Monperrus, M., Song, H., Tricoire, M.: Multitier diversification in web-based software applications. IEEE Softw. **32**(1), 83–90 (2014)
32. Zhang, H., Zheng, K., Wang, X., Luo, S., Wu, B.: Efficient strategy selection for moving target defense under multiple attacks. IEEE Access (2019)

Tiger Tally: Cross-Domain Scheme for Different Authentication Mechanism

Guishan Dong[1,3], Yuxiang Chen[1,2,3](✉), Yao Hao[1,2,3], Zhaolei Zhang[1,3], Peng Zhang[1,3], and Shui Yu[2,3]

[1] No.30 Inst, China Electronics Technology Group Corporation, Chengdu 610041, Sichuan, China
2392827595@qq.com
[2] Science and Technology on Communication Security Lab, Chengdu 610041, Sichuan, China
[3] School of Software, University of Technology Sydney, Sydney, Australia

Abstract. As the most effective way to improve the efficiency of government work, e-government has been built at all levels of China, accompanied by the construction of hundreds of authentication centers, which cause serious isolation of different systems, waste of resources, inconveniences for users who have business requirements across departments and districts. Currently, users need to repeatedly register and manage multiple different accounts, or even multiple different authentication methods. In the context of population migration, cross-departmental and regional business operations are growing, it is of great significance to find trust transfer methods for different government applications.

To overcome the issue, in this paper, we first explicitly put forward a trust delivery model named "Tiger tally" that can use consensus of all the participants instead of traditional centralized structure by using blockchain. Then design a cross-domain authentication protocol that is compatible with different authentication mechanism. As our main contribution, our scheme is advanced to resolve the trust delivery issues and it is strictly considered from perspective of security, low cost and unified regulatory requirements. In particular, by integrating "HMAC", traditionally the purview of message security with token standard, our scheme realized the idea of "Tiger tally" in ancient. It not only overcomes the long-standing trust delivery obstacles in e-government, but also achieves the traceability of responsibility and security guarantee beyond the isolated systems' security bound.

Keywords: Trust delivery · Consortium blockchain · Network security · E-government · Cross-domain authentication

Supported by National Key Research and Development Program of China (2017YFB0802300) and (2017YFB0802304), Science and technology projects in Sichuan Province (2017GZDZX0002) and Sichuan Science and Technology Program No.2018JY0370.

© Springer Nature Switzerland AG 2020
S. Wen et al. (Eds.): ICA3PP 2019, LNCS 11944, pp. 533–549, 2020.
https://doi.org/10.1007/978-3-030-38991-8_35

1   Introduction

1.1   Background and Meaning

Internet plus government service uses the Internet as an innovation platform to provide new services for public. The open environment of Internet is a high-risk area for data theft, unauthorized modification, fraud and privacy disclosure. Therefore, ensuring the sustainable service capability of government services and preventing users' privacy data disclosure is the basic security goal of the national government service system. As the basis of security, identity authentication is the primary requirement of the security guarantee system for government services. A convenient,efficient and secure identity authentication system can effectively enhance people's trust in online government services and willingness to participate, and can also effectively support government departments to put some services that originally had security concerns on the Internet to provide services.

In recent years,ministries, provincial and municipal governments of China have actively explored providing online services through the Internet, built many regional identity authentication platforms and thus accumulated valuable experience. For the public, if they want to use the services provided by the government, they must register, authenticate. However, they maintain multiple different user names and passwords because of different systems built by different departments and districts, which becomes a heavy burden. For legal entities, when conducting inter-provincial electronic bidding and other businesses, they need to register and use multiple certificates and media, the lack of uniform identification of the legal identity of the enterprises also increases the costs. With the convenience of transportation and economic development, population movement is generally on the rise. Therefore, identity authentication system needs to solve the inconvenience caused by repeated registration and authentication. State council of PRC explicitly proposed "Building a Unified Identity Authentication System for Mass Affairs" in [2016] No.23 Document [1]. The core purpose is to make full use of the established authentication services, strengthen multi-party cooperation, resonably distributing authentication pressure, quickly build a unified national authentication system to support "one registration, network-wide trustworthiness, one authentication and multi-point interconnection", explore and practicing the implementation of the national credible identity strategy [2,3].

1.2   Authentication Situation in E-Government

All current authentication methods can be classified into three categories: what the user has (USB key), what the user knows (password) and what the user is (one or more biometric characteristics) [4–6]. Identity authentication methods can be divided into account/password [7–9], OTP (One time password)[10–12], USB key [13–15], biometric [3,16–18], digital certificate [19–21], etc. Among these authentication methods, account/password is the least secure, biometrics and other methods are considered to have higher security. In practical applications, one or a combination of several authentication will be used according to security requirements [22,23]. However, in the e-government construction of various departments in China, due to lack of security awareness, saving costs, etc. most of them adopted account/password mechanism, only a small number of them adopted authentication mechanisms with higher security levels. This isolated systems have caused problems such as repeated registration and difficult in account management. How to realize trust delivery between different isolated systems becomes urgent, that is, user only need to register once and always use the registered authentication method to access other e-government applications (belong to different departments, different districts) with same or lower security level authentication mechanism.

1.3 Related Work in Cross-Domain Design

Traditionally, all the authentication methods above can be divided into two kinds with regard to trust delivery when cross domain authentication, one use certificate that must be indexed and traced back to superior issuing authorities, the other use password type such as account and password, two-factor authentication, finger print, etc, which must have a backup locally at the authentication server.

For certificate systems, there are mainly three ways for cross-domain authentication, cross authentication between CAs (Certificate Authorities), trust list of anchor CAs and Bridge CA method [20,21]. The first method's application range is too small, because it is easy to form complex topological structures when the application range is large which makes it hard to manage and search trust path. The last two need authoritative third parties, however, it faced with difficult to find trusted third parties that convinced all participants when practice. At the same time, the cost of maintaining a third party is high when the number of participants is large.

In addition to the certificate system, another bottleneck of trust delivery is authentication method of the client, for example, some use account and password, some use two-factor authentication, some others use fingerprint, iris, etc. Such kinds of authentication methods need to keep credential locally at authentication server [20,21]. Unlike public key certificate that can be publicly known, sensitive information cannot be shared with unrelated parties. In social networks, some organizations use single sign-on or OAuth. The theory of SSO (Single-Sign-On) [24] is that many applicat ions rely on a third-party authentication center, but its essence is still central authentication and the its scalability is poor. OAuth [25,26] can be used for cross-platform login. Its principle is to set up an authorization layer between client and service provider, allowing third party websites to access various information stored in the service provider under the premise of user authentication. User will not provide their own account passwords, but provide third-party websites in the form of tokens. Considering the low security of social network environment, these two methods are not adopted in China's government applications.

So when there are cross-domain requirements, there are usually two solutions, if one AS (authentication server) don't recognize a user, it redirects the user to its original AS, who is also trusted by the current AS. The user accomplishes its regular authentication at its original AS, The original AS will transfer the authentication evidence to current AS. After that, current AS will allow the user get the services. Besides, the rest way is to copy and rebuild the same identity management system in the current business system, obviously, this cumbersome and resource-consuming method cannot be popularized. In reality, different service departments don't know or just ignore their users' requirements and have no power to dock and reform their systems. Therefore, causing a lot of inconveniences to users.

1.4   Contribution and Article Structure

To address the trust delivery issues for E-government applications, in this paper, we propose an blockchain-based scheme which can realize trust delivery between different information system that use different authentication mechanism. Our original contribution can be summarized as follows:

- We put forward a trust transferring model without trusted third parties based on blockchain for E-government authentication systems. In this way, our scheme can realizes trust enhancement by using its anti-tampering and public transparency characteristics.
- Based on the designed model without trusted third party, we explore trust delivery method by using unified blockchain-federate token and designed a protocol for users belong to different authentication systems, that is, once authenticated by his original authentication method , the user can visit all the other federated applications with the same or lower security level.
- we proved the scheme's security and feasibility theoretically. Besides, our scheme will be of great value to the public and save cost when compared with other schemes.

The remainder of the paper is organized as follows. In Sect. 2, we outline the Trust delivery system model, analyze the threat and security requirements. In Sect. 3, we describe the preliminaries and key technologies of the trust delivery scheme. Then we design the "Tiger Tally" protocol in details in Sect. 4 and analyze the scheme's security and feasibility in Sect. 5, respectively. Finally, Sect. 6 concludes the paper.

2   Trust Delivery System Model, Threat Analysis and Security Requirements

The idea comes from ancient China, The emperor divided the "tiger tally" into two halves, one is placed in the barracks and the other is kept by the emperor himself. During the war, the emperor handed over his half to the general, general who needs to dispatch troops must present his identity to the barracks and join the two halves (one from barracks) together, only after all authentication have been passed can he dispatch the troops.

2.1 System Model

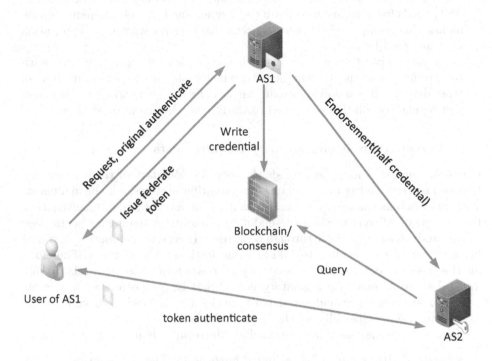

Fig. 1. Trust delivery system model: tiger tally

As is shown in Fig. 1, the trust delivery system model include three basic entities and blockchain.

- *AS1 (Authentication Server 1)*: The duty of AS1 is to prevent unauthorized users from obtaining e-government services and data. It backs up the identity information of each user and authenticate each user it manages. In our scheme it will authenticate users in the original way, but issue tokens at the federate level according to users' cross-domain needs. At the same time, it will send the authentication credential to the authentication server (AS2) of target domain through the secure channel
- *AS2 (Authentication Server 2)*: AS2 has the same duty as AS1, but the protected objects are different, that is AS1 and AS2 belongs to different E-government applications, and each backup, manage the identity data of their own users, which is also the root cause of the difficulty of trust delivery between different e-government applications. In the designed model, when AS2 needs to certify an unfamiliar user (belongs to AS1) that it does not back up, at the federate level, it must obtain the endorsement from the authentication server (AS1) of the user's domain previously and can use it to verify whether the user's credential matches the voucher recorded by AS1 on blockchain.

– *User of AS1*: The role of the user in the model is defined as cross-domain users based on actual requirements. The user (register on AS1 and belongs to AS1) needs federate token issued by AS1 when she has cross-domain requirements (for example, she wants to access another e-government application that managed by AS2).

– *Blockchain*: Blockchain has the characteristics of transparency, anti-tampering, consensus, traceability, etc, which makes it the perfect medium of trust delivery. Based on blochchain, we actually design a consensus, low cost, and regulatory solution compared to traditional ones mentioned above.

2.2 Threat Analysis and Security Requirements

Cross-domain authentication exceeds the security boundary of the original system and reaches the federate level including multiple systems, who have different authentication mechanism. So consider that at the level of federate, authentication servers of different mechanisms endorse each other by issuing authentication tokens and transferring trust through consensus. However, the issued token may be intercepted, thus leading to MITM (Man-in-the-middle Attack). Therefore, on the basis of consensus, it is necessary to restrict the tripartite relationship and take into account the regulatory needs, that is, for visitors with federate cross-domain tokens, an enhanced trust mechanism is needed to verify them again and ensure traceability of the access behavior.

Based on the analysis above, we conclude the security requirements as follows:

– *security of extraterritorial government application*: The key to ensuring security lies in the authentication server. The extra-domain authentication server needs to get the endorsement from the original domain server and be able to re-authenticate the authorized cross-domain users.

– *credential in federate has consensus*: Record credentials in blockchain to lay the foundation for "user authenticate once, the whole network is available" by using blockchain's characteristics of anti-tampering, consensus and public maintenance.

– *traceability of user's cross domain behavior*: At the federate level, all the authentication servers control users' cross-domain behavior and are responsible for cross-domain security. At the federate level, whole cross-domain behaviors need to be presented to departments with regulatory requirements.

3   Preliminaries and Key Technology

In this section, some theories will be introduced, which also play an important role in the protocol design.

3.1   HMAC (Hash Message Authentication Code)

HMAC [28–30] is an authentication mechanism that uses a hash function together with a shared key and can be expressed as:$MAC_k(m)$, where m is the message that delivers, k is the shared key of both parties. HMAC is used to authenticate short messages, guarantee data integrity and data source authentication. It can be used in conjunction with any cryptographic hash function such as(MD5,SHA-1) and has strong operability in blockchain-based cross-domain authentication.

However, the usage of HMAC in our scheme is slightly different. Based on the trust relationship model, similar authentication systems have similar security levels. An identity management system can trust other applications with same or higher security levels. That is, the account/password system can trust the authentication result of other account/password systems, and can also trust the authentication result with high security level, such as biometrics, two-factors key and so on.

The effect in the actual application is that the user selects the desired service (belongs to other identity authentication systems) in its original application interface and logs in according to the login method in the original domain. After the authentication is successful, the server issues a federate token to the user and returns the authentication credential to the server of targeted domain (i.e. when cross domain, the authentication server in the domain endorses the authentication server outside the domain), and the credential content includes the authentication result and credential itself. When initiating cross-domain access, the out-of-domain server can make a secondary challenge to prevent the attack from intercepting the token and logging in.

3.2 Authentication Token

Token refers specifically to crypto-currencies in blockchain, but refers specifically to certificates [27] in identity authentication, which is practically used to solve the problem of bank card information disclosure in network transmission. Similarly, since password authentication, biometrics, passwords and other sensitive information cannot be transferred over the network, authentication token has become the best way to transfer trust. International standards have already defined authentication token, taking JWT (JSON Web Token) as an example (as shown in the figure), a token contains three types of fields:

1. Header: including using algorithm (alg), whether to sign and encrypt. If not encrypted, the corresponding attribute is marked "none".
2. Payload: includes fields such as Issuer, Subject, Audience, Expiration time, Not Before, Issuer at, JWT ID, etc. In addition to required fields, other custom declarations can be added, Custom fields are skipped if they are not recognized by the recipient.
3. Signature.

In order to make the scheme more practical, we have borrowed the international standard and focused on the payload field properties, as shown below, similar to the X.509 standard, it includes required fields and optional items (custom items).

Table 1. Payload settings and meanings

Payload	Meaning
iss	The identifier of the token's issuer, contains the scheme, host, port number
sub	The subject user identifier, unique within a single information system
aud	The access target of the token (which can contain multiple access targets), such as the URL of the relying party
nonce	Session ID, used to prevent replay attack
exp	Expiration date, usually shorter than a few minutes, will not be allowed to access beyond a certain time
iat	The date of the token issuance, according to the UTC standard, accurate to the year, month, day, hour, minute and second
jti	A unique identifier for issued tokens, prevent tokens from being reused
auth-time (optional)	Authentication time
acr (optional)	Authentication text type, for example, when specified as 0, it does not meet the requirements of level 1
amr (optional)	Authentication method reference, is a JSON character array form, using the authentication method identifier such as:dynamic password, password

3.3   Challenge/Response

The use of authentication token is to solve the problem that sensitive information such as passwords cannot be disclosed. After the user's original domain is passed, he or she will obtain the federate token issue by the original authentication server. Considering the security of cross-domain authentication, that is, the token may be illegally intercepted, copied, use to log on by the attacker at the authentication server. Therefore, when users holding tokens access the extraterritorial authentication server, they need to use the trust enhancement method to authenticate visitors again. Our scheme will use challenge/response method [10–12] to re-verify visitors at the federate level. Currently, there are mainly asynchronous technologies based on time synchronization and challenge/response.

1. The time-based authentication technology takes the user's login time as a random factor and the time flow as a changeable factor, generally taking 60 s as a change unit and requires high time accuracy. The password generated within a certain period of time is usually recognized as synchronization, which may require the support of special hardware such as a time synchronization token.
2. Challenge/response (as is shown in Fig. 2) is that when the user logs in, the system generates a random number to the user (the challenge may also be

generated by the client), the user processes the challenge and his own password key with a built-in algorithm to get a response to the server. The authentication server can verifies the user's response with the password key (secretly shared between user and server) and cryptographic algorithm. The changeable factor at the time of challenge is the random sequence generated by the server. Because each random number is unique and not reused, it is generated and used in the same place, regardless of time and without synchronization problems.

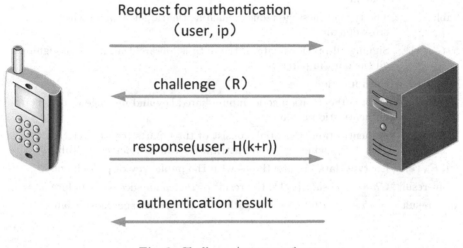

Fig. 2. Challenge/response theory

Considering that the purpose of the trust transfer scheme is "authenticate once, whole network available". It is better to make use of blockchain's characteristics of consensus, decentralization and smart contract. The use of time synchronization token is not only short in validity period, but also the hardware is not conducive to the system scalability. Therefore, we adopted challenge/response asynchronous technology to let the token have a longer validity period, reliable security and strong scalability.

4   Protocol Design

This protocol aims at the cross-domain authentication requirements of users of independent systems in the federate that may have different authentication mechanisms. Users log in by using the account/password, fingerprint, key and other methods in their own original system. After passing their original domain authentication, the local domain authentication server issues a federate token to realize cross-domain access. When accessing the server outside the domain, the user will be given a second challenge/response and cross-domain authentication will be realized by checking the blockchain.

According to the requirements of the trust delivery federate and referring to international standards, the token payloads contains the attributes fields shown in Table 1. And the meanings of protocol symbols are shown in Table 2.

Table 2. Symbol meanings of the protocol

Symbol	Meaning
noncex	x = 1,2,3... session identifier
UID	User's whole network identifier, managed by smart contract when cross domain
Lable	The type of message request, such as the target domain when cross-domain
$SIG_X(*)$	Signing all previous data, if the data is hierarchical, it means signing all the data in parentheses
Hash(*)	Hash function
$Hash_K(m)$	Hash authentication code, input shared key and message m to generate authentication result
Token	authentication credential, consist of three parts (can be referred to Sect. 3.2), including several attributes (can be referred to Table 1)
$PKE_X(*)$	Encrypt data in parentheses with the public key of participant X
chain-result	$Chain - result\epsilon\{0,1\}$, the result of the chain success or failure
auth-result	$Auth - result\epsilon\{0,1\}$, the authentication result succeeds or fails

The cross-domain authentication process of different authentication mechanisms is as follows (Fig. 3):

1. The user selects the application (extraterritorial application) to be accessed in the terminal interface and performs registration according to the registered authentication method in the original application interface such as: entering account password, Key, etc.
2. The terminal initiates authentication to the local domain server (AS1 Authentication Server (1), and the data structure is the same as the original authentication method except that "Lable" marks the cross-domain access request identifier.
3. The authentication server AS1 in this domain executes the original authentication method. If the authentication is passed, then step (4) is executed. Step (1) (2) (3) is the original identity authentication method in different government applications, user can select the same or lower authentication security level application in the original application interface. After passing the original intra-domain authentication, the user will obtain the federate token with cross-domain capability.
4. $AS1 \rightarrow blockchain : ID_{AS1}\|Hash_K(N_A)\| SIG_{AS1}(*)$. After AS1 verifies the identity of the user, according to the cross-domain identification

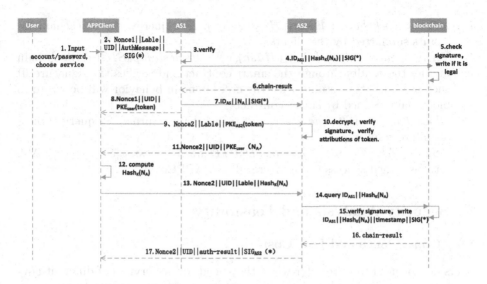

Fig. 3. Password type cross-domain authentication protocol

indicated by the user, it chooses to submit the authentication voucher $Hash_K(N_A)$ to the blochchain, where the symmetric key K is shared between AS1 and user's client, random number N_A is generated by AS1 itself and not shared with the cross-domain user. AS1 enters the two parts into the function $Hash_K(N_A)$ and write the result to the blockchain.

5. check signature, then write. Blockchain verifies the signature of AS1 cross-domain credential, track of it if it is legal.

6. *blockchain* → *AS*1: chain-result. Blockchain returns the chain record success message "chain-result".

7. $AS1 \to AS2 : ID_{AS1} \| N_A \| SIG_{AS1}(*)$. According to the application service required by the user, N_A is sent to the user's cross-domain target (out-of-domain authentication server) through the secure channel (HTTPS).

8. $AS1 \to APPClient : Nonce \| UID \| PKE_{user}(token)$. AS1 (authentication server) issues token to user.

Step (5) (6) (7) (8) is the federate token issuing stage, after which the user can initiate cross-domain access at any time within a certain period of time.

9. $APPClient \to AS2 : Nonce \| lable \| PKE_{AS2}(token)$. The user initiates a cross-domain request to AS2 through the client.

10. AS2 decrypts the token, verifies the signature and token attribute fields.

11. $AS2 \to APPClient : Nonce \| UID \| PKE_{user}(N_A)$. AS2 sends the endorsement of AS1 (secretly shared random number N_A) to the user for identity challenge.

12. APPClient then locally calculates $Hash_K(N_A)$ with symmetric key K and challenge value N_A.

13. $APPClient \to AS2 : Nonce \| UID \| Lable \| Hash_K(N_A)$. APPClient sends the calculation results to AS2 through a secure channel.

14. $AS2 \rightarrow blockchain : ID_{AS1}\|Hash_K(N_A)$. AS2 queries the blockchain for results submitted by the clients.
15. Verify signature, write $ID_{AS1}\|Hash_K(N_A)\|timestamp\|SIG(*)$ Blockchain queries the results through the smart contract and verifies the signature. If successful, the time stamp with the cross-domain behavior will be wrote to blockchain (Signed by smart contracts).
16. $Blockchain \rightarrow AS2 : chain - result$. Blockchain returns AS2 query success ot failure results.
17. $AS2 \rightarrow APPClient : SessionID\|UID\|auth - result\|SIG_{AS2}(*)$. AS2 returns signed success or failure results to APPClient.

5   Security Analysis and Feasibility

5.1   Convenience and Low Cost

Users do not need to re-register when they need to use services of different government applications with the same security level or lower security level, not do they need to face the problem of multi-account management difficulties. It only needs to register once in any government application in the blockchain consensus federate. If you need to obtain services, you can log in once, the whole network will be available. From the perspective of the construction of various departments benefit the public, there is no longer a need to face the problems of resource consumption, repeated construction and temporary solution rather than permanent cure. Any government application can log into other government applications of the same security level and there is no problem of high construction cost and high authentication pressure, because the federate center pressure disperses to the decentralized authentication system of each application.

5.2   Prevent Man-in-the-middle Attack (MITM) at Federate Level

MITM (Man-in-the-Middle attack) is the most common means adopted by attackers in the network. Our Password type cross-domain authentication protocol can be subdivided into two stages, we don't interfere with the internal authentication of original system too much (steps 1 to 3 in Fig. 3), from issuing tokens to initiating access with tokens (steps 4 to 17 in Fig. 3), we use the secondary challenge considering asynchronous requirements. In both stage interaction, a combination of timestamp and nonce is used to prevent MITM attacks.

During the communication interaction between issuing token and using token to initiate access, using secure channel http, each http request usually needs to be signed with a timestamp. If the message reception time exceeds 60 s, the request is determined to be illegal. In most cases, hackers need more than 60 s to complete the MITM attack. However if the hacker is advanced enough to complete within 60 s, the timestamp will be invalid, so we designed nonce to deal with this extreme case. Nonce is a random string that is used only once in a specific context. If the returned message is within 60 s and the returned nonce matches the used one saved locally, the message is classified as illegal, thus ensuring safe communication in extreme cases.

5.3 Regulatory

In the above security considerations, the authentication server in each government application's original domain bears the primary responsibility for alliance security. The signature of random numbers makes cross-domain authentication non-repudiation, the use of random numbers is one-time, every time a user logs in , smart contract will detect and record it. The trust enhancement is realized with the help of blockchain consensus and the feature of anti-tampering. At the same time, out scheme adopts the Hyperledger Fabric, which provides regulatory interface to present the situation of cross-domain authentication to the regulatory authorities, which can effective tracing the responsibilities.

5.4 System Test

Compared with the central authentication scheme, the bottleneck of the trust transfer scheme is limited by the efficiency of consortium blockchain platform. The TPS (Transactions per second) is hard to improve significantly because the overhead of algorithm is too high. However, it is possible to strike a balance between application requirements, security and efficiency. We simulated the blockchain record efficiency of credential submission to show the scheme's feasibility. Experiment environment is shown in Table 3.

Table 3. Experimental environment

Operation System	Ubuntu 16.04
Network simulation tool	TC (Traffic Control), realize network environment such as network delay and packet loss
Hyperledger Fabric edition	1.1.0
blockchain node status	4 virtual machines,4 node in common, Kafaka consensus
Test smart contract	chain code-example02.go (official example)

Figure 4 shows the pressure test of designed trust delivery system, it reflects the transaction success time of the first 200 test under different network delays (from 0 to 1400 ms). As the delay increases (100 ms each time), the time consuming fluctuations of successful transactions will gradually increase, which is consistent with the blockchain record mechanism. Unlike central system, which record success directly when a transaction is received, blockchain record success is defined as a transaction from submission to block release and can be queried.

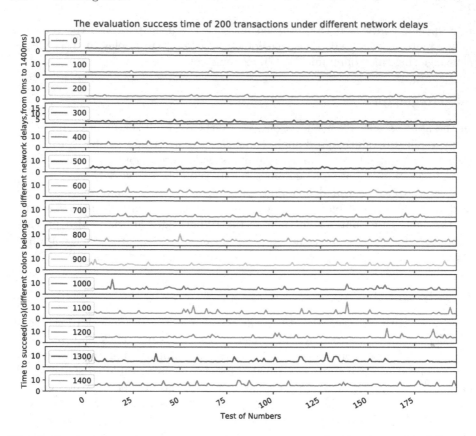

Fig. 4. Pressure test: Average transaction success time of 200 tests under different network delays

According to the blocking strategy designed by fabric, if any of the parameter conditions of batch-time-out or max-message are met, a new block will be triggered. In our simulation, stress tests were carried out, we don't need to consider batch-time-out because of the impact of large trading volume. As the network latency increases, the reason for more fluctuation is that each block needs to wait for the number of transactions to reach the set before it can be generated, that's why the fluctuation emerges.

Table 4 shows average time consuming under different network delay while Fig. 4 shows the linear relationship of network delay and average time consuming. Its linear fit is y = 0.00225x+2.72, we can see that within the 1400 ms delay, the average time taken for a successful transaction is basically linearly related to the network delay. From Fig. 4 we can see the average time consuming fluctuate a lot when network latency is beyond 300 ms. However, considering time consuming is within 10 ms when network delay is at a certain level, the number of cross-domain users is also smaller than the number of authenticated users within each system, our trust delivery system is valuable in use (Fig. 5).

Table 4. Network delay and average transaction success time consuming

Network delay (ms)	Average success time (ms)	Network delay (ms)	Average success time (ms)	Network delay (ms)	Average success time (ms)
0	2.699	500	3.78	1000	4.963
100	2.956	600	4.309	1100	5.171
200	3.127	700	4.371	1200	5.334
300	3.403	800	4.497	1300	5.675
400	3.507	900	4.679	1400	5.856

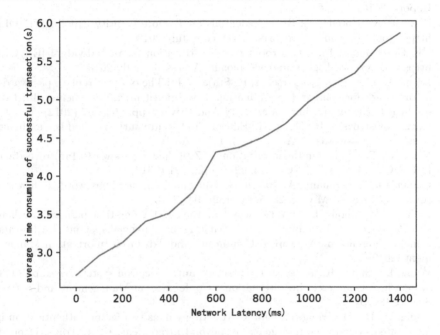

Fig. 5. Relation of network latency and average time consuming

6   Conclusion

In this paper, we take a first step towards breaking the isolation between lots of authentication system that have been built and used in e-government area of China. Beyond the proposal of the Tiger Tally Scheme which meets convenience, low cost, practicability and security, the trust delivery model based blockchain consensus lays a foundation for unified identity authentication, management and convenient service for benefiting the public in the future. To the best of our knowledge, we introduce "Tiger Tally" model, traditionally the purview of information security, into the trust delivery field for the first time. Based "Tiger Tally" model with the designed protocol, our scheme can provide users with a convenient experience and well address problems of repeated construction and waste of resources. In particular, the efficiency of the trust transfer scheme is simulated to determine the practicability and feasibility.

References

1. State Council of PRC, Notice of the General Office of the State Council on Forwarding the Implementation Plan of the Ministry of National Development and Reform Commission and other departments to promote the "Internet + Government Affairs Service". http://www.ndrc.gov.cn/zcfb/zcfbqt/201604/t20160426_799767.html. Accessed 4 July 2018
2. GOV.UK Verify. https://www.gov.uk/government. Accessed 4 July 2018
3. Weinberg, J.T.: Biometric identity. Soc. Sci. Electron. Publish. **59**(1), 30–32 (2016)
4. Lewison, K., Corella, F.: Backing rich credentials with a blockchain PKI. Technical Report (2016)
5. Lewison, K., Corella, F.: Rich Credentials for Remote Identity Proofing[EB/OL]. https://pomcor.com/techreports. Accessed 4 July 2018
6. UK Government. Identity Proofing and Verification of an Individual [EB/OL]. https://www.gov.uk/government/uploads. Accessed 4 July 2018
7. Bonneau, J., Herley, C., Oorschot, P., Stajano, F.: The quest to replace passwords: a framework for comparative evaluation of web authentication schemes. In: Proceedings IEEE Symposium on Security And Privacy, pp. 553–567 (2012)
8. Katz, J., Ostrovsky, R., Yung, M.: Efficient and secure authenticated key exchange using weak passwords. J. ACM **57**(1), 1–41 (2009)
9. Wang, D., Wang, P.: On the implications of Zipfs law in passwords. In: Proceedings ESORICS 2016, series LNCS, vol. 9878, pp. 1–21 (2016)
10. Camenisch, J., Lehmann, A., Neven, G.: Optimal distributed password verification. In: Proceedings ACM CCS 2015, pp. 182–194 (2015)
11. Li, Y., Li, X., Zhong, L., et al.: Research on the S/KEY one-time password authentication system and its application in banking and financial systems. In: International Conference on Networked Computing and Advanced Information Management (2010)
12. Wang, L., Zhang, R.: An security-enhanced authentication system based on OTP system in E-commerce. In: International Conference on Management and Service Science (2010)
13. Wang, D., He, D., Wang, P., Chu, C.-H.: Anonymous two-factor authentication in distributed systems: certain goals are beyond attainment. IEEE Trans. Depend. Secur. Comput. **12**(4), 428–442 (2015)
14. Huang, X., Chen, X., Li, J., Xiang, Y., Xu, L.: Further observations on smart-card-based password-authenticated key agreement in distributed systems. IEEE Trans. Para. Distrib. Syst. **25**(7), 1767–1775 (2014)
15. Wang, D., Gu, Q., Cheng, H., Wang, P.: The request for better measurement: a comparative evaluation of two-factor authentication schemes. In: Proceedings ACM ASIACCS 2016, pp. 475–486 (2016)
16. Wimberly, H., Liebrock, L.: Using fingerprint authentication to reduce security: an empirical study. In: IEEE IEEE Symposium on Security and Privacy 2011, pp. 32–46 (2011)
17. Zhang, F., Feng, D.G.: Fuzzy extractor based remote mutual biometric authentication. J. Comput. Res. Dev. **46**(5), 850–856 (2009)
18. Dodis, Y., Reyzin, L., Smith, A.: Fuzzy extractors: how to generate strong keys from biometrics and other noisy data. In: Cachin, C., Camenisch, J.L. (eds.) EUROCRYPT 2004. LNCS, vol. 3027, pp. 523–540. Springer, Heidelberg (2004). https://doi.org/10.1007/978-3-540-24676-3_31

19. Myers, M., Ankney, R., Malpani, A., et al.: X.509 Internet Public Key Infrastructure: Online Certificate Status Protocol. EITFRFC2560. PKIX Working Group, p. 6 (1999)
20. Perlman, R.: Overview of PKI trust models. IEEE Netw. 13(6), 38–43 (2002)
21. Lambrinoudakis, C., Gritzalis, S., Dridi, F., et al.: Security requirements for e-government services: a methodological approach for developing a common PKI-based security policy. Comput. Commun. 26(16), 1873–1883 (2003)
22. Yu, J., Wang, G., Mu, Y., Gao, W.: An efficient generic framework for three-factor authentication with provably secure instantiation. IEEE Trans. Inform. Foren. Secur. 9(12), 2302–2313 (2014)
23. Odelu, V., Das, A., Goswami, A.: A secure biometrics-based multi-server authentication protocol using smart cards. IEEE Trans. Inform. Foren. Secur. 10(9), 1953–1966 (2015)
24. Pashalidis, A., Mitchell, C.J.: A taxonomy of single sign 2 on systems. In: Proceedings of 8th Australasian Conference on the Information Security and Privacy, ACISP, pp. 249–264 (2003)
25. The OAuth 2.0 Authorization Protocol, IETF OAuth Working Group draft, work in progress, September 2011
26. Leiba, B.: OAuth web authorization protocol. IEEE Internet Comput. 16(1), 74–77 (2012)
27. The OAuth 2.0 Authorization Protocol: Bearer Tokens, IETF OAuth Working Group draft, work in progress, October 2011
28. Bellare, M., Canetti, R., Krawczyk, H.: Keying hash functions for message authentication. In: Koblitz, N. (ed.) CRYPTO 1996. LNCS, vol. 1109, pp. 1–15. Springer, Heidelberg (1996). https://doi.org/10.1007/3-540-68697-5_1
29. Chinas office of security commercial code administration: specification of sm3 cryptographic hash function [EB/OL] (2010). http://www.oscca.gov.cn/UpFile/20101222141857786.pdf. Accessed Apr 2010
30. Krawczyk, H., Bellare, M., Canetti, R.: HMAC: Keyed-Hashing for Message Authentication. Rfc (1997)

Simultaneously Advising via Differential Privacy in Cloud Servers Environment

Sheng Shen[1(✉)], Tianqing Zhu[1(✉)], Dayong Ye[1(✉)], Mengmeng Yang[2(✉)],
Tingting Liao[3(✉)], and Wanlei Zhou[1(✉)]

[1] School of Computer Science, Center for Cyber Security and Privacy,
University of Technology Sydney, Ultimo, Australia
Sheng.shen-1@student.uts.edu.au,
{Tianqing.zhu,Dayong.ye,Wanlei.zhou}@uts.edu.au
[2] Nanyang Technological University, Singapore, Singapore
melody.yang@ntu.edu.sg
[3] Wuhan Polytechnic University, Wuhan, China
tingtingliao000@gmail.com

Abstract. Due to the rapid development of the cloud computing environment, it is widely accepted that cloud servers are important for users to improve work efficiency. Users need to know servers' capabilities and make optimal decisions on selecting the best available servers for users' tasks. We consider the process that users learn servers' capabilities as a multi-agent Reinforcement learning process. The learning speed and efficiency in Reinforcement learning can be improved by transferring the learning experience among learning agents which is defined as advising. However, existing advising frameworks are limited by a requirement during experience transfer, which all learning agents in a Reinforcement learning environment must have the completely same available choices, also called actions. To address the above limit, this paper proposes a novel differential privacy agent advising approach in Reinforcement learning. Our proposed approach can significantly improve the conventional advising frameworks' application when agents' choices are not the completely same. The approach can also speed up the Reinforcement learning by the increase of possibility of experience transfer among agents with different available choices.

Keywords: Cloud computing · Cloud server · Reinforcement Learning · Advising

1 Introduction

Reinforcement learning (RL) [8] is widely used today to autonomously learn how to solve sequential decision-making problems interacting with the system environment [18]. Regular RL approaches need a large number of interactions with the system environment to learn a policy. Advising frameworks address the above problem by reusing previous knowledge in current repeated state from other agents [15]. Simultaneously learning agents advising has been come up with to accelerate learning when all agents start learning in a multiple-state system at the same time [3, 17]. Some of learning agents who have explored available choices, also called actions, in some states more times than

© Springer Nature Switzerland AG 2020
S. Wen et al. (Eds.): ICA3PP 2019, LNCS 11944, pp. 550–563, 2020.
https://doi.org/10.1007/978-3-030-38991-8_36

others, can be seen as more experienced to provide advice. A single agent can play both roles of teacher who can provide advice, and student who will ask for advice. The Fig. 1 shows a real-world example, one assignment in a course which related to cloud computing in a university. Both students and teaching staff need to execute this assignment on the university's cloud servers, but students can only access to public servers, and staff can access either public servers or private servers. In this example, the experienced students who have already executed their assignments more times than others can provide advice to other students to choose the optimal server with the best capability to run their assignments within traditional advising frameworks, and same to experienced teaching staff to transfer their experience to other staff.

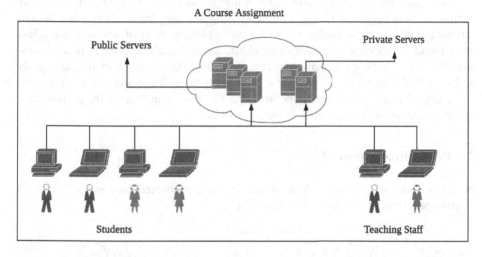

Fig. 1. An example in real-world: all students and teaching staff of a course need to execute their assignments on a cloud computing servers, but they have different available choices on them.

However, traditional advising frameworks are only appropriate for the situation which all agents have the same actions, but cannot be applied when agents' actions are not the completely same. In real-world, it's commonly that agents have different available actions even though they are solving the same problem. Back to Fig. 1, considering traditional frameworks, experienced teaching staff cannot advise students which server has a better capability and a better performance due to different available choices for servers even though they are working on the same assignment. Therefore, it is crucial to solve the problem on advising among agents with different available actions. Little previous work has thought about experience transfer between two agents with different available actions, and the challenge is how to transfer the experience for the actions which are different among teachers and students'.

We find that differential privacy mechanism can address the above challenge with the property of randomization. We consider all agents' available actions and their experience on each state as independent datasets. The two datasets can be considered as neighboring datasets in terms of differential privacy. In Fig. 1, if teaching staff and students record the execution time of tasks as their experience on selecting servers, these records for servers can be seen as datasets. The less time means the better capability of servers. When there is only one teaching-staff-access-only server, teachers datasets have one more records than students. In this case, the datasets of teachers and students can be considered as neighboring datasets. By adding randomization, these datasets can be considered as the same in a defined scale, and teaching staff and students can share experience with each other. The differential privacy property in our work is to guarantee that the experience for each action can still be used as advice if the two actions differ in, at most, one record. This approach can expand the applicable field of advising frameworks and increase the probability of the occurrence of advising.

There are two key contributions in this paper. Firstly, we propose a differentially private advising approach to improve the existing advising frameworks when agents' actions are different. Secondly, we widen the applicable field of advising and allow more possible experience transfer to accelerate agents' learning process in a RL system. We design a comprehensive experiment to show our approach in simultaneously multi-agent RL system. We provide a brief analysis of the convergence of agents' Q-learning. We also present our experiment results and the analysis of the performance of differential privacy mechanism in our approach.

2   Problem Statement

In this section, we provide a detailed scenario as a motivated case study to explain existing problems to which we are addressing.

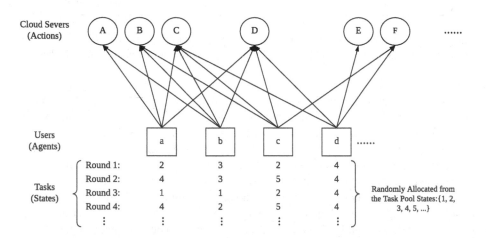

Fig. 2. Overview of scenario setting

2.1 Scenario Setting

The scenario shown in Fig. 2 is an extension of above example of Cloud computing environment, but there are more than four users, $a, b, c, d, \cdots$, and these users can be considered as agents participating in a RL process. Some tasks, $1, 2, 3, 4, \cdots$, in the task pool, can be perceived as states, and they form the set of states, S. Each agent will be allocated to a random task from the task pool, and we call a single cycle for an agent a round, from being allocated to a task to the time before being allocated to the next task. Cloud servers $A, B, C, D, \cdots$, are able to solve agents' tasks. Servers are called actions which are the available actions taken by agents to solve the tasks. All of one agent's available cloud servers compose a complete set of actions, A, for this agent. Servers can solve all tasks but behave differently on solving different tasks. In other words, if the capability of a server for a task can be considered as the time of executing tasks, this server may cost different time to solve different tasks. Therefore, the aim of all users is to solve the task with the optimal cloud server each round to minimizing task-solving time.

2.2 Scenario Discussion

In this subsection, we discuss how above scenario is working and how users achieve their aim. By solving a random task each round, an agent gains a feedback, we called it reward r, to describe the capability of server on solving this task. The higher capability the server has on executing this task, the higher reward is fed back to the agent, and vice versa. Each agent requires to know cloud servers' capabilities based on their previous sequence of learning experience, in order to make the best decision next time being allocated to the same tasks. Therefore, users should expand their own knowledge by receiving the rewards from previous rounds, and use the knowledge to make the next decision. The knowledge is probabilities distribution of selections for each agents' available actions in each state s. The agent needs to 'understand' this reward and update their probabilities distribution for available actions at states by a transition function T, in order to make a better decision in next time while being allocated to this task.

Agents can transfer experience to or ask experience from others to accelerate the learning process. In above scenario, due to the randomness of the task allocation to simultaneously learning agents, it is very likely to generate the gap of knowledge at different state. The knowledge gap is the gap on experience, based on difference of times that an agent is allocated to a task with others. An agent can be considered as more experienced than the other on a task if this agent has met this task enough times more than the other. A threshold is used to define 'experienced' with a minimum number of knowledge gap. An agent can ask for advice from the experienced agent while being allocated to this task again. However, it seems that agents meet a difficult problem on advising due to the limit we mentioned above, i.e. traditional advising frameworks only focus on same actions among agents. Therefore, our proposed approach in this paper is to address above problem. This approach allows agents to advise to accelerate their learning in the above scenario when RL agents have different available actions.

3   Preliminary

3.1   RL

RL techniques are usually used to solve sequential decision-making problems, which model the learning environment as a Markov decision processes (MDP). An MDP model can be depicted as a tuple $\langle S, A, T, R \rangle$, where S is the set of possible environment states, A is the set of agents' available actions, T is the probability of transition to next state s' while performing action a in state s, and R is the reward function for agents defining the reinforcement received from performing actions. At each learning step, an agent chooses an action after observing the state s, and then the next state is defined by T and a reward signal r can be observed. The aim of RL task for the agent is to find an optimal policy $\pi : S \to A$ which maps the best action for each possible state.

We focus on RL algorithms with Q-value estimates to determine the agent's policy, which is used to map every combination of state and action to an estimate of the long-term reward starting from the current state-action pair: $Q : S \times A \to R$. The policy, $\pi^*(s) = argmax_a Q^*(s, a)$, is optimal if the Q-value is accurate for this policy. The update of the Q-value during RL process based on experiences with the MDP. An *experience* can be described by a quadruple $\langle s, a, r, s' \rangle$ where action a is taken in state s, resulting in reinforcement reward r and a transition to next state s'. The update of Q-value is following the rule as: $Q(s, a) \leftarrow Q(s, a) + \alpha[R(s, a) + \gamma Q(s', a') - Q(s, a)]$, where α is learning rate, γ is a discount factor for next states, a and $a' \in A$ are the current and next action, respectively, and s and $s' \in S$ are the current and next state, respectively.

3.2   Advising in RL

Supposed that an agent has learned an optimal policy π for a specific task and become experienced, it can teach another agent which is beginning to learn the same task using this fixed policy. As the student learns, the teacher will observe each state s the student encounters and each action a taken by the student. In n of these states, the teacher can advise the student to take the 'correct' action $\pi(s)$ with its own learning experience or policy. The teacher-student framework aims at accelerating a student training process. However, a budget b can limit advising process. The teacher cannot provide further advice when the budget b is spent. Therefore, it is critical to define when to give advice to accelerate student's learning.

3.3   Differential Privacy

Differential privacy is a provable privacy notation that provided by Dwork et al. [6]. It provides a strong privacy guarantee that the outputs of the queries on the neighbouring datasets will be statistically similar. The formal definition of differential privacy is presented as follows:

Definition 1 (ϵ-**Differential Privacy**). *A randomized algorithm $\mathcal{M}$ gives ϵ-differential privacy for any pair of neighbouring datasets D and D', and for every set of outcomes Ω, $\mathcal{M}$ satisfies:*

$$Pr[\mathcal{M}(D) \in \Omega] \leq \exp(\epsilon) \cdot Pr[\mathcal{M}(D') \in \Omega]. \tag{1}$$

ϵ is the privacy parameter, which is defined as the *privacy budget* [5]. It controls the privacy preservation level. A smaller ϵ represents a greater privacy.

Definition 2 (Neighbouring Dataset). *The datasets D and D' are neighbouring datasets if and only if they differ in one record. This is denoted as $D \oplus D' = 1$.*

'$\oplus$' indicates the difference between two datasets.

Definition 3 (Sensitivity). *For a query $Q : D \rightarrow \mathbb{R}$, the global sensitivity of Q is defined as follow:*

$$S = max_{D,D'} \| Q(D) - Q(D') \|_1 \tag{2}$$

Definition 4 (*Laplace* mechanism). *Given a function $f : D \rightarrow \mathbb{R}$ over a dataset D, Eq. 4 provides ϵ-differential privacy.*

$$\widehat{f}(D) = f(D) + Laplace(\frac{s}{\epsilon}) . \tag{3}$$

A *Laplace* mechanism is used for numeric output. Differential privacy is achieved by adding *Laplace* noise to the true answer [12, 19, 20].

4 Algorithms for Advising with Differential Privacy

Our approach is developed in the teacher-student framework, where multiple agents simultaneously learn in a shared environment. This framework is independent of specific RL algorithms. Agents can accelerate the learning by asking for advice from another agent that already has more experience for the current state. In this framework, each agent can be either an advisor or an advisee or even both. Moreover, each agent has communication budget b_{ask} and b_{give} to control its communication overhead in both asking and giving advice separately. In this paper, an agent i asks for advice from other agents, it will cost 1 each time from communication budget b_{ask}, and cost 1 each time from communication budget b_{give} while providing advice to others. When agent i runs out of its communication budget in both, agent i can continue its RL process but without further advice process.

4.1 RL and Normalization

RL algorithm is given in Algorithm 1, which is the base of our further application of differential privacy mechanism. In the RL learning loop, from Line 6 to Line 8, agent j selects an action a_k, based on the probability distribution over the available actions in state s, and the agent receives a reward r by executing action a_k. This reward updates the Q-value of a_k in state s, which this update is based on: (1) $Q(s, a_k)$, current Q-value of a_k in state s, (2) $maxQ(s', a)$, the maximum Q-value in the new state s', (3) a learning rate α and a discount rate β and (4) the current reward r.

An agent should to adjust its probability distribution by the updated Q-value, which is shown from Line 9 to 12. $A(s)$ is the set of available actions in state s. After computing

---

Algorithm 1. RL

---

1 /* Taken agent j in state s as an example */
2 **Initialize probability distribution**
3 **Initialize Q-value for each actions**
4 **Initialize $count = 0$ for each actions**
5 **while** $count < allowed\ times$ **do**
6   Agent j selects an action, a_k, based on the probability distribution: $\pi(s) = \langle \pi(s, a_1), \cdots, \pi(s, a_n)\rangle$;
7      $r \leftarrow R\langle s, a_k \rangle$;
8      $Q(s, a_k) \leftarrow (1 - \alpha) Q(s, a_k) + \alpha [r + \gamma \max_a Q(s', a)]$;
9      $\bar{r} \leftarrow \Sigma_{a \in A(s)} \pi(s, a) Q(s, a)$;
10   **for** *each action* $a \in A_{(s)}$ **do**
11         $\pi(s, a) \leftarrow \pi(s, a) + \xi(Q(s, a) - \bar{r})$;
12     $\pi(s) \leftarrow Normalize(\pi(s))$;
13     $count(a_k, s) \leftarrow count(a_k, s) + 1$;
14     $s \leftarrow s'$;

Algorithm 2. Normalize()

---

1 For agent j, there are n available actions
2 Let $d = min_{1 \le k \le n} p(k)$, mapping centre $c_o = 0.5$ and mapping lower bound $\Delta = 0.001$;
3 **if** $d < \Delta$ **then**
4      $\rho \leftarrow \frac{c_o - \Delta}{c_o - d}$;
5   **for** $k = 1$ *to* n **do**
6          $p(k) \leftarrow c_o - \rho(c_o - p(k))$;
7 $\Pi \leftarrow \Sigma_{1 \le k \le n} p(k)$;
8 **for** $k = 1$ *to* n **do**
9      $p(k) \leftarrow \frac{p(k)}{\Pi}$;
10 **return** p;

---

the average reward $\bar{r}$ with the updated Q-value and the probability distribution (Line 9), the probability of each action $a \in A(s)$ is updated based on (1) the current selection probability of each action $\pi(s, a)$, (2) the current Q-value of each action $Q(s, a)$, (3) the average reward $\bar{r}$ and (4) a learning rate ϵ. Line 12 is a normalization algorithm to adjust new probability distribution for each action $a \in A(s)$, which must satisfy $\Sigma_{1 \le k \le n} p(k) = 1$. In the end of the loop, the current state is updated to the next state $s' \in S$ and restarting the loop.

The purpose of Algorithm 2 is to use proportion-based mapping to normalize an invalid probability distribution to a valid one. Proportion-based mapping can adjust invalid probabilities into the range $(0, 1)$ and keep the relative magnitude of the probabilities. It is noted that Algorithm 1 with Algorithm 2 does not run only once but runs a pre-determined number of times until all agents perform an adequate RL for optimal decisions on actions.

Algorithm 3. Advising with DP

---

1 /* Taken agent j, k in state s as an example */
2 **Initialize** $t(s)$ for each state s
3 **Initialize** $b_{give} = c$ for each agent, c is a constant
4 **Initialize** $b_{ask} = c$ for each agent, c is a constant
5 **Input** $n(s, a)$, times of an agent at state s
6 **Initialize** knowledge gap K, K is a constant
7 **while** *execution times $<$ allowed time* **do**
8   **if** $b_{ask}(a_j) > 0$ **then**
9   **if** $b_{give}(a_k) > 0$ **then**
10         $g = max_{1 \leq k \leq |Neig|}(n(s, a_k) - n(s, a_j))$
11   **if** $g \geq K$ **then**
12   **for** *each action $a \in A(s, a_j) \cap A(s, a_k)$* **do**
13            $Q(s, a_j, a) \leftarrow Q(s, a_k, a) + Lap(\frac{\Delta S}{\epsilon})$
14            $\pi(s) \leftarrow Normalize(\pi(s))$;
15   **if** *action $a \in A(s, a_j)$ and $a \notin A(s, a_k)$, and action $a_0 \in A(s, a_k)$ and $a_0 \notin A(s, a_j)$* **then**
16            $Q(s, a_j, a) \leftarrow Q(s, a_k, a_0) + Lap(\frac{\Delta S}{\epsilon})$
17            $\pi(s) \leftarrow Normalize(\pi(s))$;
18         $b_{ask}(a_j) \leftarrow b_{ask}(a_j) - 1 \ b_{give}(a_k) \leftarrow b_{give}(a_k) - 1$

4.2 Advising and Knowledge Transfer with Differential Privacy

Firstly, an agent j decides whether it is able to ask for advice in one learning step by checking its asking budget $b_{ask}(a_j) > 0$. It also needs to check whether there is one neighbouring agent can provide advise by checking its giving budget b_{give} and knowledge gap. If budget $b_{give}(a_k) > 0$ and knowledge gap $g > K$, K is a pre-determined constant which means the minimum requirement of knowledge gap to execute advising stage. Assume that agent a_k is able to ask agent a_j for advice, it sends an advice request message which is the current state of agent a_j, to agent a_i. Once agent a_i receives this request message, agent a_i needs to directly send its advice back to agent a_j. The advice includes agent a_i's knowledge and learning experience, Q-value, which assists agent a_j to adjust its own Q-value and make further decision on actions.

The Algorithm 3 shows the algorithm of advising strategy in our proposed approach. The Laplace noise, $Lap(\frac{\Delta S}{\epsilon})$ is added on the advisor's Q-value in Line 13 and Line 15. ΔS is the sensitivity of the maximum reward. As the current reward has a direct impact on the selection probability. The sensitivity ΔS is determined by the maximum reward. Advisee accepts advisor's Q-value with noisy on actions which are owned by both the advisor and the advisee, with discount rate α. According to Definition 1, the application of differential privacy here is to mitigate the impact of the difference on one action. Advisee can accept the Q-value for the action differed in advisor's. The advisor obtains an extra reward for each action in this state in advising stage and then accelerates learning speed.

4.3 Discussion

It is obvious that $Q(s, a)$ is convergent to the maximum reward $r(s, a)$ of from the action a under this state s. In our motivated case study, the next state s' is not necessary to be considered, therefore, the equation of Q-value is $Q(s, a_k) \leftarrow (1 - \alpha) Q(s, a_k) + \alpha r$, which is only correlated to current $Q(s, a)$, fixed discount rate α and reward r. Because the reward for an action at a state is fixed for all agents in the system, all agents are able to gain the same reward to adjust its Q-value. We assume that when $Q(s, a) = r$ with a fixed discount rate α, the value of $Q(s, a)$ cannot be changed according to above equation because all related variables' value are fixed. The same logic can also be used while considering the next state s' due to the transitivity of convergence. Ideally, after enough times of learning process, the Q-value can represent the interaction of the best action's reward and the best choice of the next state.

Secondly, the convergence of Q-value determines the satisfaction of differential privacy mechanism. We consider that all $Q(s)$ for each agent is an individual dataset, $Q(s) : \{Q(s, a_0), Q(s, a_1), \cdots, Q(s, a_n)\}$. As we discussed above, the value of $Q(s, a)$ should be same for each agent after convergence theoretically. In our case, when agents have different actions with differing in one record, their datasets can be seen as neighbouring datasets, and the actions can be randomized and be transferred between agents, according to Definition 1.

5 Experiment

5.1 Experimental Settings

We perform an experiment with the scenario introduced in Sect. 2. Our experiment consists in a setting of random numbers of agents, servers and tasks in range $[8, 12]$ which is better setting to demonstrate results. Agents adopt RL to explore and identify abilities of severs simultaneously. An agent will be randomly allocated to a task each round and will choose one server to address this task. Agents aims are to select the optimal servers with strongest abilities for specific tasks through self-learning experience. Agents can ask advice from others who are more experienced on the task. Meanwhile, they can also provide their learning experience to the others who lack of experience on the task. The agent state is composed of the tasks which are available in the task pool. The agent action is composed of the servers which are available for agents to solve tasks. However, agents may be not allowed to access all servers which results in the fact that agents' action options cannot completely same. Therefore, the difference in actions between agents are significantly considered in this experiment, and then satisfy the environment for our hypothesis.

In order to demonstrate our performance intuitively, especially to illustrate the performance of differential privacy mechanism in the traditional advising framework, we involve our approach and other works together in the experiment via control variable method. We compare the results including the rate of convergence and the ratio of convergence (accuracy). We investigate the following four strategies:

1. **Regular learning (No Advice)** – As reference, we evaluate the SARSA learning algorithm without advising;
2. **Normal advising (Agents with Same Actions)** – Advising process only occur between two agents with same available action options;
3. **Advising without differential privacy** – Advising process can occur between two agents with same action options or with maximum one difference in their action options. This strategy does not apply differential privacy, which means that advisee directly accept advisors' knowledge to adjust their own experience affecting actions;
4. **Advising with differential privacy** – Advising process can occurs between two agents with same action options or with maximum one difference in their action options. This strategy applies differential privacy, which means that advisee accept advisors' knowledge with additional noise to adjust their own knowledge influencing action selections.

The results we describe in the next section are the average of 1000 executions on the above procedure for each strategy. We generate reward with random numbers in the range $[0, 3]$ for each task for each server, to represent abilities of servers for each task. The minimum requirement of knowledge gap in this experiment is set as $K = 10$, which means the advisor must learn the same task 10 times more than the advisee to give advice. The communication budget in this experiment is set as $b_{ask} = 10$ and $b_{give} = 10$ for each agent, which means each agent can provide advice and ask advice 10 times separately.

5.2 Experimental Results

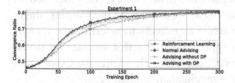

Fig. 3. Comparison between four strategies (Color figure online)

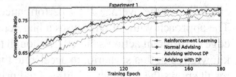

Fig. 4. Comparison between four strategies in learning and advising stages (Color figure online)

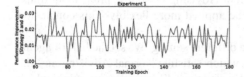

Fig. 5. Performance comparison on differential privacy mechanism in experiment 1

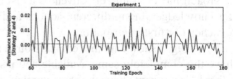

Fig. 6. Performance comparison on normal advising approach and differential advising approach (Color figure online)

In this section, we demonstrate experimental results and provide the explanation and analysis. Figure 3 shows the result from the first experiment for the performance comparison between four strategies mentioned above. The x-axis represents training epoch, and the y-axis means the ratio of the sum of actual rewards gained by all agents and the sum of theoretically maximum reward that all agents can obtain each training epoch. This ratio is driven by two factors, the value of actions' rewards and agents' probability distributions. The ratio ra for each agent can be described as follow:

$$ra = \frac{\sum_{k \in A}^{n} \pi(s, a_k) * R(s, a_k)}{\sum_{m \in A}^{n} \pi(s, a_m) * R(s, a_m)} \tag{4}$$

where n is each agent, k is each action of an agent, $\pi(s, a_k)$ is the probability of this action, $R(s, a_k)$ is the reward of this action, m is the action with theoretically maximum reward. Due to the fixed reward $r(s, a)$, the change of the ratio is affected by the change of probability. The greater convergence ratio means the higher probability for the greater reward action. The increase of the ratio with the increasing training epoch means the agents' effect of RL, and increasing speed for each strategy means the speed of learning, which we propose to compare in this experiment.

From the global perspective of Fig. 3, the convergence ratio for all four strategies start with the same point, about 35%, and converge to 80%, which proves that all strategies learn within the same environment setting. When we focus on the learning epoch, as shown in Fig. 4, RL locates in the lowest position which means RL approach has the lowest learning speed among all four strategies. It is clear that Advising without DP approach with yellow color is under our approach, and the comparison between both approaches shows the importance of differential privacy mechanism. Figure 5 proves the effectiveness of differential privacy mechanism, and indicates that differential privacy mechanism can improve the average 1.5% performance in this approach during the learning stage. The approach without DP causes a huge error while negatively transferring the knowledge of different action's directly, although the approach still transfer accurate 7 actions' knowledge in this scenario, to accelerate the process of RL. Therefore, the experiment result shows the accuracy of differential privacy mechanism applied here.

This experiment also demonstrates the feasibility of our proposed approach. Back to the Fig. 3, our proposed approach shows the almost same performance as the normal advising approach and two approaches are almost overlapped. In Fig. 6, the red line converges to 0, which is the average of the difference of two approaches' convergence ratio during the learning stage from epoch 60 to epoch 180. In other words, our proposed approach has the same effectiveness as Normal Advising but our approach can transfer knowledge among different actions and be applied more flexibly. This conclusion matches our first contribution.

The Fig. 7 shows another comparison in a different setting. We set the half of agents have the completely same available actions which satisfies Normal Advising and the other half of agents have the same number of actions but differing in one action, which can apply our approach. Same to the first experiment, the x-axis represents training epoch for both strategies, and the y-axis means the ratio of the sum of actual reward gained by all agents and the sum of theoretical maximum reward that all agents can obtain. The increase of the ratio is the speed of convergence which represents the learning speed. The two strategies run in this same environment. In a global view, because

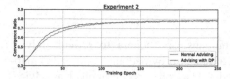

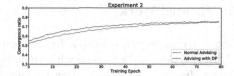

Fig. 7. Comparison between normal advising and advising with differential privacy

Fig. 8. Comparison between normal advising and advising with differential privacy in learning and advising stages (Color figure online)

Fig. 9. Performance comparison on differential privacy mechanism in experiment 2

our approach can only accelerate the speed but cannot improve the learning result, after training epoch 150, both curves are convergent to about 80%. In the result of this experiment, our approach has the quicker learning speed and reaches the maximum convergent ratio faster than normal advising approach, and has almost the same learning accuracy. It is shown more clearly in the Fig. 8. It shows the comparison between Normal Advising and Advising with differential privacy in learning and advising stages in training epochs from 60 to 180. Our approach as blue curve is always higher than the red curve during the whole learning stage.

The Fig. 9 demonstrates the ratio of performance improvement. Similar to the Fig. 5 in Experiment 1, our approach can accelerate the learning speed in learning and advising stage with the ratio about 8% as maximum during the whole training epoch. Both strategies converge to stable situations after epoch 125 but our approach seems perform 1% worse than Normal Advising. Here are two reasons that why our approach seems converge to a lower percentage than the other approach. Firstly, due to the setting of random reward, the convergent ratio is not fixed and floating in a reasonable range each time of running the experiment. Secondly, because both strategies converge in 150 epoch and we have many actions in this scenario, there is not easy to generate knowledge gap during learning stage, which means advising can still happen for two strategies after maximum level of convergence. Our approach by transferring knowledge with noise can not perform better than Normal Advising transferring accurate knowledge. In order to show the all advising process within the accurate experiment result, we did not set the threshold to stop the occurrence of advising after fully converging. However, in real-world, we can improve our approach by setting a threshold to stop advising if an agent is experienced and confident enough to choose action alone. Therefore, with the above two reasons, the experiment result adequately proves that our approach can widen the application field in advising frameworks which matches our second contribution.

6 Related Work

The early approaches mostly focused on humans as teachers to provide advice [2,9]. [13] proposed a method which requires a human-provided and hand-coded mapping to link the two tasks, to transfer knowledge from one task to another as advice. [7] proposed to train an agent using autonomous teacher that is assumed to perform at a moderate level of expertise for the task. However, this approach may restrict the performance of advising due to receiving too much advice. *Teacher-student* framework is firstly proposed by [14] and further improved by [10], which introduces a numeric communication budget to limit the number of times of advising during the learning steps, which becomes an essential part of the advising model to imitate humans' availability and attention capability in real-world [4]. [21] proposed that an agent should learn when to give advice with building a sequential decision-making problem. [11] proposed that multiple agents can broadcast their average reward at the end of each learning epoch while learning in the same system, and agents whose average reward is lower than the best one can ask for device from the best agent.

In above approaches, either advisor and advisee can trigger the process of advising in RL stage. [1] proposed a jointly-initiated framework that both advisor and advisee need to agree to receive and provide advice simultaneously. [3] proposed advising framework among simultaneously learning agents based on *teacher-student* framework, which agents start learning together without an experienced agent. Their method is also based on jointly-initiated advisor-advisee relations which are established on demand when an advisee is not confident enough to make choice alone, and an advisor has enough confidence to provide advice at the same time, which is highly related to our proposed method. [16,17] proposed to use differential privacy to protect benign agents against malicious agents in the advising stage.

7 Conclusion

In summary, we have presented a new differentially private advising approach for simultaneously multi-agent Reinforcement Learning in this paper, which can be used in server recommendation problem in real-world scenario based on the cloud server environment. This is also the first work to incorporate differential privacy on advising framework. We used differential privacy mechanism in the simultaneously multi-agent advising framework to allow two agents with different available actions to transfer their knowledge. There are two key contributions, firstly, we improved the advising frameworks when agents' actions are not the completely same. Secondly, we widen the applicable field of advising and allow more possible knowledge transfer in a RL system. We also demonstrated the experimental results based on the scenario we described in the introduction and scenario setting sections, regarding a multiple user shared cloud server environment.

Ackowledgement. This work is supported by an ARC Linkage Project (DP190100981) from Australian Research Council, Australia.

References

1. Amir, O., Kamar, E., Kolobov, A., Grosz, B.: Interactive teaching strategies for agent training (2016)
2. Clouse, J.A., Utgoff, P.E.: A teaching method for Reinforcement learning, pp. 92–110 (1992)
3. da Silva, F., Glatt, R., Costa, A.: Simultaneously learning and advising in multiagent Reinforcement learning. In: Proceedings of the 16th Conference on Autonomous Agents and MultiAgent Systems, pp. 1100–1108 (2017)
4. David, M., et al.: Distraction becomes engagement in automated driving. Proc. Hum. Factors Ergon. Soc. Annu. Meet. **59**, 1676–1680 (2015)
5. Dwork, C.: A firm foundation for private data analysis. Commun. ACM **54**, 86–95 (2011)
6. Dwork, C., McSherry, F., Nissim, K., Smith, A.: Calibrating noise to sensitivity in private data analysis. In: Halevi, S., Rabin, T. (eds.) TCC 2006. LNCS, vol. 3876, pp. 265–284. Springer, Heidelberg (2006). https://doi.org/10.1007/11681878_14
7. Clouse, J.A.: Learning from an automated training agent. In: Adaptation and Learning in Multiagent Systems (1996)
8. Littman, M.: Reinforcement learning improves behaviour from evaluative feedback. Nature **521**, 445–451 (2015)
9. Maclin, R., Shavlik, J.W.: Creating advice-taking reinforcement learners. Mach. Learn. **22**, 251–281 (1996)
10. Matthew, E.T., Nicholas, C., Anestis, F., Ioannis, V., Lisa, T.: Reinforcement learning agents providing advice in complex video games. Connect. Sci. **26**, 45–63 (2014)
11. Nunes, L., Oliveira, E.: On learning by exchanging advice. arXiv preprint cs/0203010 (2002)
12. Sun, N., Zhang, J., Rimba, P., Gao, S., Zhang, Y., Xiang, Y.: Data-driven cybersecurity incident prediction: a survey. IEEE Commun. Surv. Tutor. **21**, 1744–1772 (2018)
13. Torrey, L., Walker, T., Shavlik, J., Maclin, R.: Using advice to transfer knowledge acquired in one Reinforcement learning task to another. In: Gama, J., Camacho, R., Brazdil, P.B., Jorge, A.M., Torgo, L. (eds.) ECML 2005. LNCS (LNAI), vol. 3720, pp. 412–424. Springer, Heidelberg (2005). https://doi.org/10.1007/11564096_40
14. Torrey, L., Taylor, M.: Teaching on a budget: agents advising agents in Reinforcement learning. In: Proceedings of the 2013 International Conference on Autonomous Agents and Multiagent Systems, pp. 1053–1060 (2013)
15. Ye, D., He, Q., Wang, Y., Yang, Y.: An agent-based integrated self-evolving service composition approach in networked environments. IEEE Trans. Serv. Comput. **12**(6) (2019)
16. Ye, D., Zhang, M., Vasilakos, A.V.: A survey of self-organization mechanisms in multiagent systems. IEEE Trans. Syst. Man Cybern. Syst. **47**(3), 441–461 (2016)
17. Ye, D., Zhu, T., Zhou, W., Yu, P.: Differentially private malicious agent avoidance in multiagent advising learning. IEEE Trans. Cybern. (2019)
18. Ye, D., Zhang, M., Sutanto, D.: Cloning, resource exchange, and relationadaptation: an integrative self-organisation mechanism in a distributed agent network. IEEE Trans. Parallel Distrib. Syst. **25**(4), 887–897 (2013)
19. Zhu, T., Li, G., Zhou, W., Yu, P.: Differentially private data publishing and analysis: a survey. IEEE Trans. Knowl. Data Eng. **29**, 1619–1638 (2017)
20. Zhu, T., Xiong, P., Li, G., Zhou, W., Yu, P.: Differentially private model publishing in cyber physical systems. Future Gener. Comput. Syst. (2018)
21. Zimmer, M., Viappiani, P., Weng, P.: Teacher-student framework: a Reinforcement learning approach. In: AAMAS Workshop Autonomous Robots and Multirobot Systems (2014)

Feature Generation: A Novel Intrusion Detection Model Based on Prototypical Network

Shizhao Wang[1], Chunhe Xia[1], and Tianbo Wang[1,2(✉)]

[1] Beijing Key Laboratory of Network Technology, Beihang University, Beijing, China
{wangsz,xch,wangtb}@buaa.edu.cn
[2] School of Cyber Science and Technology, Beihang University, Beijing, China

Abstract. Intrusion detection becomes more and more essential to ensure cyberspace security. In fact, the detection is a process of classifying traffic data. However, attacks usually try to cover up themselves to be as similar as normal traffic to avoid being detected. This will cause a high degree of overlap among different classes in the input data, and affect the detection rate. In this paper, we propose a feature generation based prototypical network (**FGPNetwork**) model to solve overlapping data classification problem in intrusion detection. By analyzing the characteristics of data transmission in the network, we select the basic package characteristics and roughly divide them into several parts. Then, a contribution rate is used to calculate the specific contribution of basic features to classification. We order the features by rate descending in each part and generate the new features by Convolutional Neural Networks (CNN) with different kernels. The new features can obtain the intrinsic connection of original features and add more nonlinearity to the model. Finally, the combination of new features and original features will be input into the prototypical network. In prototypical network, data is mapped to a high-dimensional space, and separated by narrowing the distance of data and their respective cluster centers. Because of the uneven distribution of the intrusion detection dataset, we use undersampling method in each batch. The experimental result on NSL-KDD test dataset also shows that our model is better than other deep learning intrusion detection methods.

Keywords: Intrusion detection · Deep learning · Feature generation · Prototypical network

1 Introduction

At present, the security of cyberspace is increasingly severe, and network security measures are becoming more and more indispensable. Intrusion detection with the ability of active defense has an irresistible position. Intrusion Detection System (IDS) [1] is an active defense technology that analyzes real-time captured network packets or host logs to determine whether an intrusion has occurred. It is of great significance to network security. The essence of intrusion detection is to analyze the traffic data and identify whether it is an attack one [2]. As a result, many typical classification algorithms can be applied to intrusion detection.

© Springer Nature Switzerland AG 2020
S. Wen et al. (Eds.): ICA3PP 2019, LNCS 11944, pp. 564–577, 2020.
https://doi.org/10.1007/978-3-030-38991-8_37

In recent years, traditional machine learning methods have developed rapidly. It is free from the inefficiency of manual statistical analysis, and has the advantages of high classification accuracy and strong generalization ability. The representative machine learning algorithms such as Naive Bayes, Support Vector Machine (SVM), Decision Trees, etc. are widely used in intrusion detection. But machine learning methods usually need a lot of effort in features selection. And there is still room for the improvement of classification accuracy. With the development of hardware and the rise of big data, deep learning technology has received more attention. It composes multiple nonlinear layers to transform the input from lower levels to higher ones to discover the representations needed for classification [3]. And it has feature extraction and stronger generalization ability comparing with the traditional machine learning. Due to the excellent classification ability of deep learning, it has been widely used in intrusion detection [4–7].

In network, in order not to be identified, various attacks try to disguise themselves as much as possible to make them closer to the normal network traffic. So, it is difficult to classify attacks and normal traffic exactly. This serious issue hinders the performance of intrusion detection. The prototypical network [8] believes that data will gather around the representation in the embedded space. And the classification can be achieved by shortening the distance between data and its representation. The prototypical network is proved to have a state-of-the-art performance in few-shot learning. And we find that its clustering method is also beneficial to the classification of overlapping samples. So, we use prototypical network as classification method to make the data more differentiated in high dimensional space.

Although the prototypical network does a better classification of overlapping data, and it improves the result of intrusion detection to some extent, the final accuracy still need to be improved. We consider optimizing the model from features. We pick up and combine the basic package features of samples by analyzing the data transmission process in network. We believe more information can be mined from them. According to the analyzing result, we can pre-divide the features into several major categories according to different layers and functions. Next, we define a contribution rate calculated by variance, and determine the specific order of features in contribution descending in each major category. Then, different kernels are used to generate new features which can obtain intrinsic relationship. The new features are combined with the original features as a new description of the samples.

Using feature generation based prototypical network, the various types of intrusion traffic and normal traffic can be better separated. And the rest of the paper is organized as follows: Sect. 2 reviews the application of deep learning technology in the field of intrusion detection. Section 3 describes the theory and specific method of feature generation and prototypical network. Section 4 shows the experimental results of FGPNetwork and compares it with other basic methods. Section 5 concludes FGPNetwork and proposes the future work.

2   Related Work

Machine learning based intrusion detection can distinguish the attacks by learning the difference of categories automatically. Traditional misuse detection fails to detect unknown attacks, but machine learning methods can identify the new attacks by generalizing the representative characteristics [9]. There are numerous typical machine learning methods have been used in intrusion detection including Naïve Bayes [10, 11], Decision Tree [12, 13], Support Vector Machine [14, 15]. Ensemble learning, which can reduce the over-fitting in training stage and improve the generalization ability. It is also widely used in intrusion detection task such as Random Forest [16, 17].

The accuracy of classification and strong generalization ability is the important properties of intrusion detection. Deep learning usually does better in these aspects and it can also extract the representative characteristics from raw data which is important for classification. Therefore, many prior researches have applied deep learning methods to the intrusion detection tasks and get satisfactory results. Tang et al. [18] have used a 3 layers Deep Neural Network (DNN) in the intrusion detection with six basic features chosen from original features. The accuracy of detection in test dataset is 75.75% which is better than machine learning methods but still have room to be improved. The result of simple DNN without feature preprocessing is not ideal in complex classifications. Jin Kim [19] applied the CNN in intrusion detection. Preprocessed and normalized data are input into CNN to finish classification. Because of feature extraction ability of CNN, the intermediate results produced by it will be more conducive to the classification. The effect of this method is improved compared to DNN. In [20], authors use Text-CNN as the information extraction part, then use random forest to classify. This method has a superior performance in detection rate and false alarm rate, but the false negative rate is a little high. Recurrent Neural Network (RNN) has the ability of memory that can capture the relationship in the input data. Some papers also use the RNN in intrusion detection. In [21], RNN is used to detect Botnet with undersampling method to reduce the impact on classification results from data inequality. The attack detection rate is 0.809 on test dataset. [6] and [22] use long short-term memory (LSTM) in KDD 1999 intrusion dataset to do the intrusion detection. But these two papers just do the cross-validation and don't give the accuracy in KDD 1999 test dataset. In [5], auto-encoder based method is used, which can collect features representation from unlabeled data. The accuracy on NSL-KDD test set is 79.10%.

The results of above machine learning and deep learning methods are better than traditional ones, but still can be improved. As a result, we propose our new method according to the characteristics of intrusion data.

3 Theory and Methodology

Our model is mainly designed for data overlapping problems. It consists of three key parts, including features extraction part, features generation part and prototype network classification part.

In the first stage, we analyze the package transmission process, and obtain important features of protocols. Then, we reorder these features by the relationships among them. To order these basic features specifically, a contribute rate is proposed, which is a ratio of between-classes variance and in-class variance. We arrange the basic features by contribution rate descending in each group. Next, CNN is applied to complete the feature generation work considering its ability of feature extraction and nonlinear transformation. Different sizes of kernels including 1×2 1-filter and 1×3 1-filter are used.

These new generated features, with nonlinearity and the intrinsic relationship of basic features, are connected to the original features. The features generation part is trained separately so that the new features can perform best in intrusion classification. And then combine them with original features to form new detection inputs.

In feature generation stage, the prototypical network [8] is used. It believes that data will gather around the representation in the embedded space. The input will be mapped into an embedding space to get more nonlinearity. Then, the algorithm calculates the representative element and shorten the distance between data and its representation. In this way, different classes of data will be separated from each other.

The overall structure of the model is shown as Fig. 1. Where FCN is the fully connected network. Normal, DoS, Probe, U2R, R2L are the different types of the traffics. We use undersampling method in each batch in prototypical network training stage because the distribution of different classes in intrusion data is not balance.

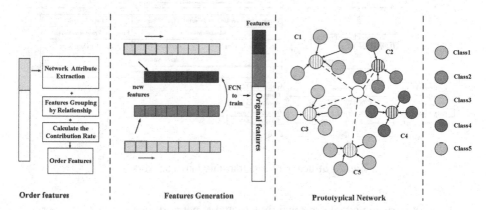

Fig. 1. The overall model

3.1 Convolutional Neural Network

Convolutional Neural Network (CNN) uses convolution operations to enhance features [3]. Usually, CNN contains convolution part and pooling part. Firstly, the kernels will do convolution with the input data. The size of kernel can be different such as 3×3 or 5×5 to extract different scale features. Then, the feature maps will be put into pooling layer which makes the next layer be able to pick up larger-scale details than just edges and curves [23]. Each layer of CNN can get many feature maps, so it is in fact much deeper than a normal neural network [23]. These properties make CNN more beneficial for feature learning and classification.

3.2 Feature Generation

The transmission of data in the network can be described by the OSI five-layer model. The five layers are physical layer, data link layer, network layer, transport layer and application layer. During the transmission, the lower layer will provide services for the upper layer. So, the lower layer can determine the function of the upper layer. From Fig. 2 we can see, the hosts have all the 5 layers, but in the data transmission, there are only the lower 3 layers.

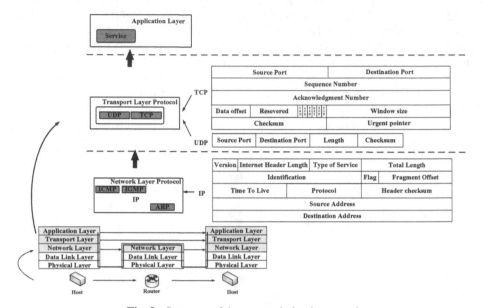

Fig. 2. Structure of data transmission in network

Network intrusion attacks are mainly targeted at network layer, transport layer and application layer. We extract the content of these three layers. The network layer has an IP protocol family, including IP and its affiliate protocols such as ICMP. Network layer will provide services for transport layer which mainly do the transmission of segments. There are link oriented TCP protocol and connectionless UDP protocol. Transport layer provides different services for the above application layer to accomplish specific tasks whose essence is the transmission of byte data. This is the vertical relationship in the network. We also give the header of protocols properties many protocol functions related to.

According to above protocol header information, we obtain the relevant attributes (see Fig. 3). The content of the circle is the main attribute of the protocol it points to. These attributes determine how the protocols process the different situations in data transmission. Attacks precisely perform by making use of the different response methods of the protocol. For example, in DoS attack, adversary usually sends lots of TCP SYN packages and enters in SYN_SEND state. And after the server return an ACK, the adversary does nothing. In this way, the connection becomes a semi-connected state and consume network resources. We can see that the link sate is important for detection. Therefore, using the relationship of protocol attributes to model is beneficial for intrusion detection.

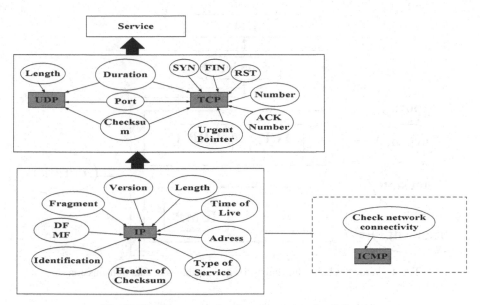

Fig. 3. Attribute extraction based on network relationship

From the whole data transfer processing we can see that the protocol and its attributes are very important in data transmission. The implementation of the attack is also considering the protocol rule in this process. Thus, it is meaningful to analyze the basic package characteristics related to the protocols. We can extract the relationship between these basic features and obtain new features.

Next, we abstract the vertical relationship and horizontal attributes in data transmission. It can be display as Fig. 4. Where the arrow indicates the service providing to upwards and the line indicates the attribute. The attributes in each layer have greater correlation. Besides, the attributes between layers also affect each other. For example, different protocols provide different services. We finally divide the package features into three parts like in the following picture.

After separating the basic features and getting the three major parts, we define a contribution rate as standard of the importance to classification to specifically order features in each group. It is calculated by ratio of variances between-classes and variances in-class. When the variance of a feature is small within a class and large among classes, the feature is more conducive to classification. Thus, we give the contribution rate (CR) of each feature as follow.

$$\text{CR} = \frac{(\sum_{i=0}^{4}(m_i - \overline{m})^2)/5}{\left(\sum_{i=0}^{4} V_i\right)/5} = \frac{\sum_{i=0}^{4}(m_i - \overline{m})^2}{\sum_{i=0}^{4} 1/n \sum_{j=1}^{n}(s_j - m_i)^2} \tag{1}$$

Where V_i is the variance of a feature in the i-th class, and there are totally 5 different classes. n is the number of samples in each class. s_j represents the j-th sample, and m_i is the mean of samples in the i-th class. $\overline{m}$ is the mean of all the data. When the larger the CR is, the greater the contribution of the feature is.

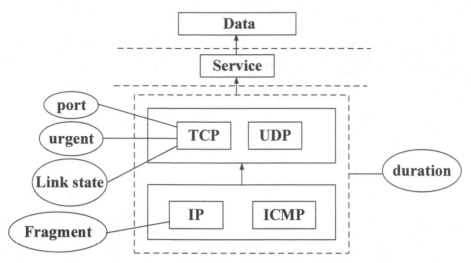

Fig. 4. Vertical relationship and horizontal attributes in data transmission

Using the formula above, we can calculate the contribution rate of basic package characteristics. Then we arranged the features in each group according to the contributions descending order. In this way, the high contribution features will have chance to combine by kernels and the new features will more helpful for classification. Finally, these features make up a 1×9 arrangement.

CNN have the ability of feature extraction, so we use CNN to generate new features. We apply different size of kernels in it including 1×2 1-filter and 1×3 1-filter. Then, we splice these two results. We think these kernels can extract the details of the in-group information, and can also combine the within-groups characteristics to obtain useful new features. We also do the experiments to check the optimal kernels, and the results show that when using 1×2 1-filter and 1×3 1-filter kernels, the detection results will be better. So, these two sizes of kernels are chosen. Because we just use CNN as a means of synthesizing features, we omit the pooling layer.

After extraction, we put the results of CNN to a fully connected network and use the gradient descent method to train. In this way, the new features can be more useful for classification.

3.3 Theory of Prototypical Network

The prototypical network proposed in [8] is to solve the problem of few-shot. It can extract the representative characteristics of the data. The prototypical network model maps input into an embedding space where each sample will cluster around their respective prototype representation. The map function is a CNN which can learn the non-linearity. and the classes' prototype is the mean center of each category. In the classification, a point belonging to which category depends on which distance from the class prototype is the nearest. The author computes the distance by Euclidean distance in Bregman divergence and prove that the performance of Euclidean distance is outstanding. In order to shorten the distance between the point and its prototype representation, a softmax loss is used. The loss formula in [8] is as follow.

$$J = -log \frac{exp(-d(f(x), c_k))}{\sum_{i=0}^{m} exp(-d(f(x), c_i))} \tag{2}$$

where x is the input, and $f(x)$ is the mapping neural network. c represents the mean center of each class, and c_k is the exact center of each category of the data. d is the Euclidean distance between data and representation. In original paper, author use CNN in $f(x)$ with 4 3 × 3 64-filter kernels. In order to adapt our intrusion data, we use a 3 × 3 32-filter and a 3 × 3 64-filter instead.

4 Experiments

4.1 Introduction of Detection Process

The Benchmark Data Set. We use intrusion detection dataset NSL-KDD as the benchmark to verify the performance of our FGPNetwork model. There are 125973 records in train set and 22544 records in test set. And some test set records are not in the train set, we count that only 83.3% of records in the test set are included in the training set. Each record has 41 features including continues features and discrete features [24]. And there are 5 types of the class including a normal one and 4 attacks in the dataset. The types of attacks include denial of service (DoS), probing (Probe), user to root (U2R), root to local (R2L). The number of different categories in the data set is shown in Table 1.

Table 1. Different categories in NSL-KDD

	Normal	DoS	Probe	U2R	R2L
NSL-KDD Train	67343	11656	45927	52	995
NSL-KDD Test	9711	7458	2421	200	2754

Preprocess of Dataset. In the NSL-KDD, most of the features are numeric except the protocol_type, service and flag which are character. So, we digitize them first. We respectively use integers from 0 to 2, from 0 to 69 and from 0 to 10 to replace 3 types of protocol_type, 70 types of services, and 11 kinds of flags. We use the one-hot code to process the 5 labels. After, we get rid of 20th feature because the values of it are found all be 0. So, there are 40 original features for training.

Next, we also use max-min method to normalize the original features to deal with the dimension problem in features. Formula of normalization is shown as follow.

$$x' = \frac{x - min}{max - min} \tag{3}$$

In this way, the values of each feature will be limited in [0, 1], and the impact of dimension problem in features will be solved.

Algorithm Model. We find that the 9 basic features are directly from the packets, and other features are statistical ones processed according to different characteristics of attacks. In other word, these 9 features are the basic attributes in data transmission. Therefore, we can extract more distinguishing features from these basic features to describe the samples better. We display the 9 basic features which are shown in Table 2.

Table 2. Basic package features in NSL-KDD

Attributes	Name	Type
Basic features	duration, src_bytes, dst_bytes, wrong_fragment, urgent	Numerical
	protocol_type, service, flag, land	Discrete

According to the grouping method in Sect. 3.3, we first make a rough division of features. The first group is the data to be transferred which are Src_bytes and Dst_bytes in dataset. Next is the services in application layer. The last one is the protocols and their attributes which are protocol_type, flags (is the link state), duration, land (is about the port), urgent and wrong_fragment in the dataset. Then, we calculate the contribution rate (CR) of each basic feature by formula (2). And order the features by CR descending in each group. The CR results are shown as Table 3.

Table 3. The contribution rate of each basic feature

Feature	Src_bytes	Dst_bytes	service	flag	Protocol_type	duration	Wrong_fragment	urgent	land
CR	0.4e−3	0.1e−3	0.17	0.23	0.12	0.05	0.02	0.02	0.3e−3

Finally, we order the features as "**src_bytes, dst_bytes, service, flag, protocol_type, duration, wrong_fragment, urgent, land**".

Then, we use two different sizes of kernels to extract new features, one is 1×2 1-filter, another is 1×3 1-filter. We don't use the pooling layer. After the convolution process, we stitch the results from two scales to get a 15-dimension features. To make the new features more advantageous for classification, we use a one-layer full connection network to train the parameters of these two kernels. And we use the parameters of CNN kernels when the accuracy is the best.

After getting the new 15 features, we concatenate them with original 40 features. These 55 features will be the input of next part. In classification stage, we use 2 kernels with sizes of 3×3 32-filter, 3×3 64-filer and a full connected network to map the input data to a 200-dimension embedding space. And in this space, we use representation and Euclidean distance to calculate the loss whose concrete formula is shown as formula (1) in 3.3. Considering the unbalanced distribution of different labels, we use batch-undersampling method. It makes the model not only focus on the normal traffics with large numbers, but also on various attack traffics and get a better classification result. The least category is u2r, only including 52 samples. According to it, we set the size of the batch to 260, and each category contains 52 samples. In this way, we can improve the detection rate of attack samples to some extent. And we use the Adam to train the model. The Adam adds a first and a second moment estimate to the general gradient descent to make the training and convergence faster.

Finally, we use our model on test dataset to verify our algorithm's effectiveness. The overall flow of it is shown in Fig. 5.

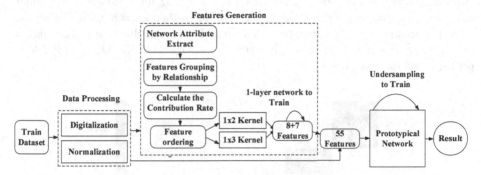

Fig. 5. Overall flow of the model

Evaluation Metrics. For evaluation indicators, we mainly focus on accuracy which represents the detection rate. But the imbalance of intrusion data in different categories leads to the accuracy cannot evaluate the result well, so we also calculate the false negative rate (FNR), false positive rate (FPR) and F1-Measure to judge the model. These indicators usually be used in intrusion detection to measure the results fully. The FNR is the ratio of samples that are misidentified as negative. The FPR is the ratio of samples that are misidentified as positive. And F1 combines the results of precision and recall. The formulas are shown as follows.

$$Acc = \frac{N_{nor} + N_{dos} + N_{probe} + N_{u2r} + N_{r2l}}{N_{total}} \tag{4}$$

$$FNR = \frac{FN}{TP + FN} \tag{5}$$

$$FPR = \frac{FP}{TN + FP} \tag{6}$$

$$F1 = \frac{2 \times \frac{TP}{TP+FP} \times \frac{TP}{TP+FN}}{\frac{TP}{TP+FP} + \frac{TP}{TP+FN}} \tag{7}$$

Where N_{nor}, N_{dos}, N_{probe}, N_{u2r} and N_{r2l} are the number of the data correctly assigned to class of Normal, DoS, Probe, U2R and R2L. The N_{total} is the total number of the samples. *TP*, *TN*, *FP*, *FN* are true negative, true negative, false positive and false negative.

4.2 Experiments and Discussion

We use the CNN with 1×2 1-filter and 1×3 1-filter kernels on the basic features. The CNN is trained separately with cross entropy loss and Adam in this stage for 20 epochs to get the best parameters, and the highest accuracy is 0.697. So, we pick the parameters of feature generation CNN this epoch.

After generate features, we put the combination of new ones and original features into the prototypical network. The network has a 3×3 32-filter and a 3×3 64-filter kernels in nonlinearity mapping with undersampling method in each batch. We get the confusion matrix in the best train result shown as Fig. 6. The diagonal is the samples that correctly classified. And from this matrix we can calculate the *TP, TN, FP, FN* to get the detection indicators.

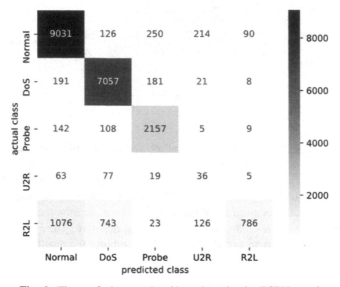

Fig. 6. The confusion matrix of best detection by FGPNetwork

Besides, we also compare the accuracy curve in training of our model and other basic models. The result is shown as Fig. 7. From the curve we can see that the accuracy of our model is better than other models. And it also shows that the accuracy of model with feature generation is higher than no feature generation.

Finally, we calculate the values of best detection indicators including Accuracy (ACC), FNR, FPR, F1, and compare the 4 detection indicators with other basic deep learning methods. Considering the unbalance distribution of data, we use undersampling method in all these algorithms. Where FG stands for feature generation, PN is the prototypical network. The results are shown in Table 4.

From above we can see, when we add the process of feature generation, the detection indicators are better than the basic models. It shows that our feature generation method is beneficial for intrusion classification. Besides, the clustering method in embedding space of the prototypical network also further promotes the detection effectiveness.

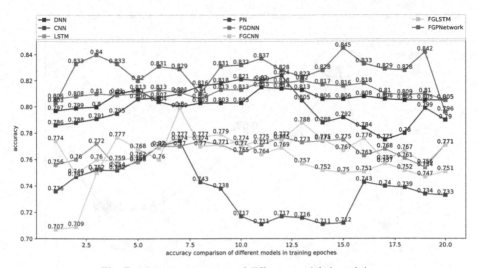

Fig. 7. The accuracy curve of different models in training

Table 4. Evaluation metrics on FGPNetwork

Method	DNN	CNN	LSTM	PN	FGDNN	FGCNN	FGLSTM	FGPNetwork
ACC	0.815	0.774	0.777	0.824	0.820	0.799	0.788	**0.846** ↑
FNR	0.215	0.229	0.233	0.172	0.176	0.185	0.236	**0.114** ↓
FPR	0.072	0.074	0.087	0.079	0.072	0.074	0.078	**0.070** ↓
F1	0.857	0.847	0.836	0.880	0.876	0.874	0.841	**0.913** ↑

Our FGPNetwork model combining these two improvements has the highest accuracy. The FNR, FPR of our model are lower than other models. And the F1 is larger than other basic methods. These all prove that our model is optimal in intrusion detection. So, the FGPNetwork is effective on intrusion detection.

To prove that the kernels we used in the feature generation stage is the best, we test different combinations of kernels on the 9 basic features including 1×2, 1×2 & 1×3, up to 1×2 to 1×9. The best accuracies with these kernels are shown in the Fig. 8. From the line chart we can see, when two kernels are 1×2 1-filter and 1×3 1-filter, the result is the best. So, we choose this set of parameters. (We use 1 for 1×2, 2 for 1×2 & 1×3, 3 for 1×2 & 1×3 & 1×4, until 8 for 1×2 & 1×3 & 1×4 & 1×5 & 1×6 & 1×7 & 1×8 & 1×9, and they are all 1-filter).

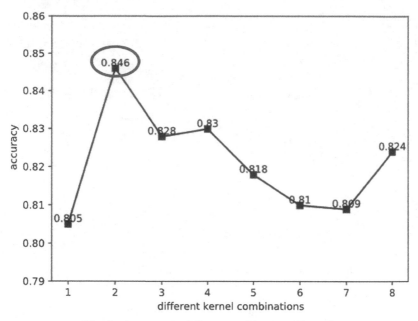

Fig. 8. Accuracy in different combination of kernels

5 Conclusion

In this paper, we propose the FGPNetwork model for data overlap in intrusion detection. First, we analyze the process of data transform in network, and extract the package basic features. Next, we roughly class them into some groups by relationship among them. We also propose a contribution rate to specifically measure the distinction of the basic features to classification and then order these features by CR descending in each group. Then, we use the CNN to generate new features from these basic package features. In this way, we can obtain the intrinsic relationship of features, so that newly generated nonlinearity features can classify data more accurately. Finally, we use the batch undersampling based prototypical network in the classification stage to make the data separate from each other better. The detection results show that FGPNetwork has superior performance in the metric indicators. In the future, we will pay more attention to the small attacks with fewer samples, and try to significantly increase their detection rate to make the intrusion detector more perfect.

Acknowledgment. This work is supported by National Natural Science Foundation of China (U1636208, F020605) and the National Natural Science Foundation of China (Grant No. 61902013).

References

1. Roesch, M.: Snort - lightweight intrusion detection for networks. In: Proceedings of the 13th USENIX Conference on System Administration. Lisa 1999, vol. 99, pp. 229 (1999)

2. Yin, C., et al.: A deep learning approach for intrusion detection using recurrent neural networks. IEEE Access **5**, 21954–21961 (2017)
3. LeCun, Y., Bengio, Y., Hinton, G.: Deep learning. Nature **521**(7553), 436 (2015)
4. Javaid, A., et al.: A deep learning approach for network intrusion detection system. In: ICST, pp. 21–26 (2016)
5. Shone, N., Ngoc, T.N., Phai, V.D., Shi, Q.: A deep learning approach to network intrusion detection. IEEE Trans. Emerg. Top. Comput. Intell. **2**, 41 (2018)
6. Staudemeyer, R.C.: Applying long short-term memory recurrent neural networks to intrusion detection. S. Afr. Comput. J. **1**(56), 136–154 (2015)
7. Lin, W., et al.: Using convolutional neural networks to network intrusion detection for cyber threats. In: 2018 IEEE International Conference on Applied System Invention (ICASI), pp. 1107–1110. IEEE (2018)
8. Snell, J., Swersky, K., Zemel, R.: Prototypical networks for few-shot learning, pp. 4077–4087 (2017)
9. Mishra, P., et al.: A detailed investigation and analysis of using machine learning techniques for intrusion detection. IEEE Commun. Surv. Tutorials **21**(1), 686–728 (2018)
10. Barbara, D., Wu, N., Jajodia, S.: Detecting novel network intrusions using Bayes estimators, pp. 1–17 (2001)
11. Panda, M., Abraham, A., Patra, P.M, Discriminative multinomial naive Bayes for network intrusion detection, pp. 5–10 (2010)
12. Rajeswari, L.P., Kannan, A.: An intrusion detection system based on multiple level hybrid classifier using enhanced C4. 5, pp. 75–79. IEEE (2008)
13. Stein, G., et al.: Decision tree classifier for network intrusion detection with GA-based feature selection, pp. 136–141. ACM (2005)
14. Kabir, E., et al.: A novel statistical technique for intrusion detection systems. Future Gener. Comput. Syst. **79**, 303–318 (2018)
15. Nskh, P., Varma, M.N., Naik, R.R.: Principle component analysis based intrusion detection system using support vector machine, pp. 1344–1350. IEEE (2016)
16. Chauhan, H., et al.: A Comparative study of classification techniques for intrusion detection, pp. 40–43 (2013)
17. Zhang, J., Zulkernine, M., Haque, A.: Random-forests-based network intrusion detection systems. IEEE Trans. Syst. Man Cybern. Part C (Appl. Rev.) **38**(5), 649–659 (2008)
18. Tang, T.A., et al.: Deep learning approach for network intrusion detection in software defined networking, pp. 258–263. IEEE (2016)
19. Kim, J., et al.: Method of intrusion detection using deep neural network, pp. 313–316 (2017)
20. Min, E., et al.: TR-IDS: anomaly-based intrusion detection through text-convolutional neural network and random forest. Secur. Commun. Netw. **2018**, 1–9 (2018)
21. Torres, P., et al.: An analysis of recurrent neural networks for botnet detection behavior, pp. 1–6 (2016)
22. Kim, J., et al.: Long short term memory recurrent neural network classifier for intrusion detection, pp. 1–5 (2016)
23. Convolutional Neural Networks – Basics (2017). https://mlnotebook.github.io/post/CNN1/
24. Derived Features. University Of California, I. (1999)

Topic Reconstruction: A Novel Method Based on LDA Oriented to Intrusion Detection

Shengwei Lei[1], Chunhe Xia[1], Tianbo Wang[1,2(✉)], and Shizhao Wang[1]

[1] Key Laboratory of Beijing Network Technology, Beihang University, Beijing, China
{leisw,xch,wangtb,wangsz}@buaa.edu.cn
[2] School of Cyber Science and Technology, Beihang University, Beijing, China

Abstract. Traditional intrusion detection methods are facing the problems of distinguishing different types of intrusion with high similarity. The methods use a single value to characterize each attribute and mine the relationship of each attribute at the feature extraction stage. However, this granularity of features extraction is not sufficient to distinguish different intrusions whose network flow characteristics are similar. Facing the problem, we establish an intrusion detection model based on Latent Dirichlet Allocation (ID-LDA) and propose a novel topic reconstruction method to extract the distinctive features. We mine the value distribution of each attribute and the association of multiple attributes to extract the more implicit semantic features. These features are more useful for identifying slight differences in different kinds of intrusions. However, the current LDA models are difficult in determining the most optimal topic number. Meanwhile, the recent methods ignore the multiple topics selection. These above problems result in difficulty in generating the perfect Document-Topic Distribution (DTD) and lower detection accuracy. So we propose a topic overlap degree and a dispersion degree to quantitatively assess the quality of the DTD. Finally, we get the most optimal topic number and select the best topics. Experiments on the public NSL-KDD dataset have verified the validity of the ID-LDA. These results outperform many state-of-the-art intrusion detection methods in terms of accuracy.

Keywords: Intrusion detection · LDA topic reconstruction · Optimal topic number determination · Topic overlap and dispersion degree · Multi-topic selection

1 Introduction

Existing intrusion detection methods are very effective in distinguishing different intrusions in high level. However, when the difference is weak, the effect is not satisfied. While many intrusions exhibit similar network flow characteristics. The mainstream intrusion detection methods [25–28] typically use weighting parameters to characterize the relationship of individual attributes. For a single attribute, it is usually represented by a single value (e.g., arithmetic mean, highest value, etc.). As a result, the features extracted from the above methods are not fine-grained to distinguish different intrusions with high similarity. In order to improve the detection accuracy, we need to extract more detailed features to identify the slight differences in different intrusions.

© Springer Nature Switzerland AG 2020
S. Wen et al. (Eds.): ICA3PP 2019, LNCS 11944, pp. 578–594, 2020.
https://doi.org/10.1007/978-3-030-38991-8_38

The topic model is a statistical model for clustering the implicit semantic structure of an essay. It can be used for classification by topic [6]. Fortunately, the LDA topic model [1] can not only describe the relationship of different attributes, but also extract the distribution of values for each attribute. Compared with the traditional methods (represent an attribute by a single value), the LDA model is able to extract features from multiple dimensions. The LDA model extracts some of the more granular and implicit semantic features, and helps to distinguish slight differences in different intrusions. At the same time, the LDA model can extract unique feature in a relatively intuitive way, which is very suitable for intrusion detection. So we propose a novel topic reconstruction method based on LDA to extract the distinctive intrusion features. Applying the LDA topic model to intrusion detection is to draw analogies between intrusion data and text document. The monitored intrusion data is considered as a text document. Each document (a type of intrusion data) is a mixture of topics (intrusion behavior) and each topic is a mixture of words (intrusion activities). Intrusion activity is reflected by the actual data monitored on network (i.e., the activity is the actual data). We need to adjust the LDA model to mine the unique behavior of each type of intrusion. **In the ideal situation,** the unique behavior must frequently occur on one type of intrusion. However, the unique behavior must be less common in other types of intrusion. In the LDA topic model, each document corresponds to a topic distribution (i.e., the unique behavior is the selected topic for each document). In Fig. 1(d), the ideal situation corresponds to a perfect DTD.

However, some different intrusions exhibit similar network flow characteristics (e.g., R2L and PROBE); some intrusions that mimic normal network flow are similar to normal (e.g., low-speed DOS and NORMAL) [24]. Therefore, the ideal situation is difficult to achieve in practice and construct a perfect DTD is a challenging work. As the DTD is strongly related to the topic number, determining the most optimal topic number is a large challenge. Furthermore, the majority researches are to choose the maximum weight topic as the feature when the DTD is generated, ignoring other lower weight topics. The selection of a single topic ignores a portion of semantic information that would result in poor effort. Above all, the current researches on LDA topic model are facing the challenges of determining the optimal topic number as well as ignoring the multiple topics selection.

In order to solve the above problems, we establish an ID-LDA model and propose a novel topic reconstruction method to extract feature. The main contributions are:

- We propose a method that formatting the intrusion data to satisfy the LDA topic model (i.e., convert the raw data into words).
- We design two indicators (topic *Overlap Degree* and topic *Dispersion Degree*) to assess the quality of the DTD. The indicators can help to determine the most optimal topic number. At the same time, we design a multi-topic selection algorithm to reconstruction a new topic as feature identifier and the corresponding matching mechanism.
- Experiments on the public NSL-KDD dataset have verified the effectiveness and superiority of the ID-LDA model.

The structure of the paper is as follows: Sect. 2 introduces the related work. Section 3 briefly introduces the knowledge of the LDA topic model and the Jensen Shannon divergence. It also states the problems faced by the LDA model. Section 4 gives a detailed introduction of the ID-LDA model and key technologies. Section 5 presents the experiments and assessments. Section 6 concludes this paper and points to the direction of future work.

2 Related Work

Some researches apply the LDA topic model to feature extraction [7–12], classification, certification [2–5], and anomaly recognition [13–15]. In [7, 8], user behavior features are extracted by using the LDA topic model from the mobile data. While in [8], the authors focus on extracting the behavior patterns of different users independent of each other. Each user selects a single topic with the highest weight as the feature. In [10], the LDA model is applied to process driving data with time series. The driving mode, relationship and user' driving behavior has been extracted. In [11, 12], the authors take into account the subject matter of the television program being watched and the time stamp of the viewing behavior. User behavior is analyzed in order to capture the inherent viewing mode and unique interest. In [13], the researchers propose a LDA-based computing framework for group spamming detection in product review data. Experiments show that the model can detect high quality spammer groups. The above researches apply the LDA model to extract features. However, most of them just select the single topic with the maximum weight as the feature and ignore other topics with less weight. In particular, selecting the single topic loses a significant portion of the semantics when the maximum weight is small, resulting in a poor detection effect. Consequently, applying the LDA topic models straightforwardly cannot solve our problem. A flexible multi-topic selection mechanism is urgently needed.

On the other hand, the topic number is strongly related to the DTD during the LDA model training phase. There are two cases of topic overlap or dispersion in different DTDs. Therefore, choosing the most optimal topic number to generate a suitable DTD is critical to classification. Some indicators are proposed to assess the DTD, including topic perplexity [21, 22], topic stability [16] and topic consistency [17–19], et al. The topic perplexity degree describes the uncertainty degree in which topic the document belongs to. But it does not distinguish the unique topic. In [16] a method based on repeated LDA operation and clustering to measure the stability is proposed. However, the topic instability may lead to systematic errors. In [18, 19], the authors select popular topic words according to the topic-word co-occurrence, and use a set of top-level keywords [20] to evaluate topic quality. Nevertheless, the above indicators are all from the word perspective and are not suitable for classification. It is more important to evaluate the DTD from the topical perspective. The existing methods are difficult to cope with the challenge, and a new assessment indicator is urgently needed.

3 Problem Statement

In this section, we introduce the basics of LDA topic model and Jensen-Shannon (JS) divergence. Then we state the problems and challenges of applying LDA model to intrusion detection in detail.

3.1 Background Knowledge

Table 1. Parameter description of LDA topic model

Parameter	Description
M	The number of documents
K	The number of topics
V	The number of words (non-repeated)
N_m	The number of words (repeated) in document m
$Z_{m,n}$	The topic assignment for the nth word in document m
$w_{m,n}$	The nth word in document m
$\overrightarrow{\theta_m}$	The topic distribution for document m
$\overrightarrow{\varphi_k}$	The word distribution for topic k
$\vec{\alpha}$	The hyper parameter which controls the sparsity of a topic that is assigned to each document
$\vec{\beta}$	The hyper parameter which controls the sparsity of per-document word distribution

LDA Topic Model. The topic modeling is a technique that uncovers the hidden thematic structure of document collections. The LDA topic model describes the process of generating a document. Each word in a document is obtained by a process of "choosing a topic with a certain probability and selecting a word with a certain probability from the topic" (see Formula 1, Table 1). The topic of the document follows a polynomial distribution, and the word of the topic follows a polynomial distribution. Each document represents a probability distribution of topics, and each topic represents a probability distribution of words. The topic is a linear combination of semantically related and differently weighted words.

$$P\left(w_m, z_m, \overrightarrow{\theta_m}, \overrightarrow{\varphi_k} \middle| \vec{\alpha}, \vec{\beta}\right) = \prod_{n=1}^{N_m} P\left(w_{m,n} \middle| \overrightarrow{\varphi_k}, z_{m,n}\right) P\left(z_{m,n} \middle| \overrightarrow{\theta_m}\right) P\left(\overrightarrow{\theta_m} \middle| \vec{\alpha}\right) P\left(\overrightarrow{\varphi_k} \middle| \vec{\beta}\right)$$

$$(1)$$

Jensen-Shannon Divergence. KL divergence (D_{KL}) and JS divergence (D_{JS}) [23] are common ways to measure the similarity of two distributions. $D_{KL}(P||Q) = \sum P(x) log \frac{P(x)}{Q(x)}, 0 \le D_{KL}(P||Q) \le 1, D_{KL}(P||Q) \ne D_{KL}(Q||P)$.

$$D_{JS}(P||Q) = \frac{1}{2}D_{KL}\left(P(x) \middle|| \frac{P(x) + Q(x)}{2}\right) + \frac{1}{2}D_{KL}\left(Q(x) \middle|| \frac{P(x) + Q(x)}{2}\right) \quad (2)$$

JS divergence (see Formula 2) is a variant of *KL* divergence. The range of D_{JS} value is [0, 1]. D_{JS} is 0 when the two distributions are equal; and the opposite is 1. The more similar the two distributions, the smaller the D_{JS}. $D_{JS}(P\|Q) = D_{JS}(Q\|P)$. Compared with *KL* divergence, *JS* divergence is more accurate in discriminating the similarity of distribution. Its characteristic is more suitable for the requirement of this paper. It is applied to determining the most optimal topic number and the matching mechanism.

3.2   Problem Statement

At the feature extraction stage, the selected topic is the feature of each type of intrusion in the LDA topic model. When the number of documents is *M* (*M*-Doc) and number of topics is *K* (*K*-Topic), a $M \times K$ DTD matrix is generated. In the DTD, $\sum_{k=1}^{K} \theta_i^k = 1.0$, where θ_i^k indicates the weight of topic *k* for document *i*. The ultimate goal of intrusion detection is to distinguish different types of intrusion. Therefore, **the perfect DTD must satisfy the following situation**: each document corresponds to a unique topic with 100% weight. The front $M \times M$ matrix presents an identity matrix. See Fig. 1(d), the green box denotes the unique topic.

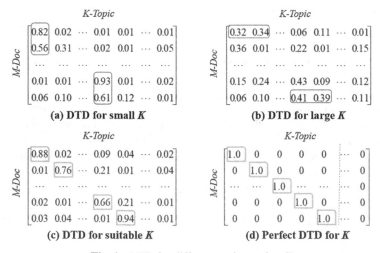

Fig. 1. DTD for different topic number *K*

As each document corresponds to a *K*-dimensions topic distribution, the sum of the *K* topics weights is 1.0. Therefore, the topic number *K* is closely related to the distribution of topic weights. The smaller the value *K*, the more concentrated the topic weights (see Fig. 1(a)). The larger the value *K*, the more scattered the topic weights (see Fig. 1(b)). There are two situations in reality:

(1) **Topic Overlap**: some topics with high weight appear in different documents at the same time in the DTD (see Formula 3)). In Fig. 1(a), the red box shows the topic overlap. As a consequence, the $\left(N_i^{test} * \theta_i^{k1} + N_j^{test} * \theta_j^{k2}\right)$ samples in document i, j may be misclassified due to the overlapping of topic $k1$. τ_{high} denotes a high weight. N_i^{test} denotes the sample number in test document i. $\mathcal{F}_{i \to j}^{classify}(num)$ denotes the num samples in document i will be misclassified to document j in a certain probability.

$$\because \exists k1 = k2 \leq K, i, j \leq M, i \neq j, \quad \theta_i^{k1} \geq \tau_{high} \wedge \theta_j^{k2} \geq \tau_{hign}$$
$$\therefore \mathcal{F}_{i \to j}^{classify}\left(N_i^{test} * \theta_i^{k1}\right) \vee \mathcal{F}_{j \to i}^{classify}\left(N_j^{test} * \theta_j^{k2}\right) \tag{3}$$

(2) **Topic Disperse**: the topic with the maximum weight is small or the difference between the maximum weight and the secondary weight is small in one document in the DTD (see Formula 4). In Fig. 1(b), the blue box shows the topic dispersion. As a consequence, neither single topic $k1$ nor topic $k2$ is suitable to be a feature. Otherwise, the $\left(N_i^{test} * \left(1 - \theta_i^{k1}\right)\right)$ samples in document i may be misclassified due to the single topic selection. $\tau_{interval}$ denotes a threshold of weight difference. τ_{max} denotes a maximum threshold of weight.

$$\because \exists i \leq M, \max_{k1 \leq K} \theta_i^{k1} < \tau_{max} \vee \max_{k1 \leq K} \theta_i^{k1} - \max_{k2 \leq K, k2 \neq k1} \theta_i^{k2} < \tau_{interval}$$
$$\therefore \mathcal{F}_{i \to j, j \neq i}^{classify}\left(N_i^{test} * \left(1 - \theta_i^{k1}\right)\right) \tag{4}$$

Based on the above analysis, a suitable topic number K is needed to generate a suitable DTD (see Fig. 1(c)). The suitable DTD must be close to the perfect DTD as possible. The problem of applying the LDA model can be formulated as Formula 5.

$$minimize \quad \|PDTD_K - ADTD_K\|$$
$$subject\ to \begin{cases} M \leq K \leq Num_{topic} \\ \max_{k1 \leq K} \theta_i^{k1} - \max_{k2 \leq K, k2 \neq k1} \theta_i^{k2} \geq \tau_{interval} \end{cases} \tag{5}$$

$ADTD_K$ denotes the actual DTD of topic number K, and $PDTD_K$ indicates the perfect DTD of topic number K. Num_{topic} denotes the maximum threshold of topic number. We need an indicator to quantitatively evaluate the different of $\|PDTD_K - ADTD_K\|$. In the Sect. 4.3, we present two indicators (*Overlap Degree* and *Dispersion Degree*) to specify. After the optimal topic number K is obtained, we need to make a multi-topic selection to make up for the missing semantics. It is specified in Sect. 4.3.

4 Intrusion Detection Model Based on LDA

4.1 Framework of ID-LDA

The overall framework of ID-LDA model is shown in Fig. 2. It consists of four modules: preprocessing, training, feature extraction and matching.

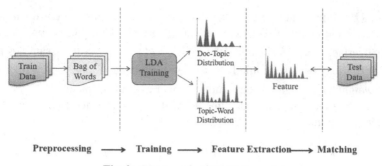

Fig. 2. Framework of ID-LDA model

4.2 Preprocessing Module

The preprocessing module is to convert the raw characteristic data into limited words for the LDA topic model. There are two types of raw intrusion data:

(1) **Limited discrete characteristic value.** E.g., the network protocol type has three types: *TCP*, *UDP*, and *ICMP*. This type of intrusion data is directly converted to words. E.g., $\{TCP, UDP, ICMP\} \rightarrow \{TCP_{protocol}, UDP_{protocol}, ICMP_{protocol}\}$.

(2) **Almost unlimited continuous characteristic value.** E.g., the number of bytes from source host to target host with the range of [0, 1379963888]. For such continuous values, it is not necessary to convert each value into a word because the physical meaning of some adjacent values is consistent. And too many words in LDA topic model will delay the training time and reduce the effect. These continuous characteristic values will be segmented according to its distribution. We should ensure that the value in each segment can distinguish different types of intrusion as possible (see Formula 6).

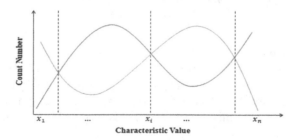

Fig. 3. Distribution of continuous value

As shown in Fig. 3, the X-axis is the continuous value, and the Y-axis is the count of each continuous value. The values in the range $[x_i, x_j](i < j \leq n)$ will be converted into one same value when satisfying Formula 6. N_{x_k} and M_{x_k} are the counts of two different types of intrusion. E.g., the red curve and the blue curve represent two different types of intrusion in Fig. 3. According to the segmentation algorithm (see Algorithm 1), the continuous values are divided into 4 segments: $\{x'_1, x'_2, x'_3, x'_4\}$.

$$\begin{cases} Sum_{i,j} = \left| \sum_{i \le k \le j} N_{x_k} - \sum_{i \le k \le j} M_{x_k} \right| \\ Sum_{i,j} > Sum_{i,j+1} \\ Sum_{i,j} > Sum_{i-1,j} \end{cases} \tag{6}$$

The preprocessing involves transforming characteristic values into words and generating the bag-of-words for each type of intrusion data. After completing the first step, the count of each word in each type of intrusion will be calculated to generate the bag-of-words. The bag-of-words is the input to the LDA training module.

Algorithm 1. The pseudo code of segmentation of raw continuous values

---

Input: $X, Y1, Y2, X', \mathcal{T}_{count}$ // $\mathcal{T}_{count}$ is the count of segments threshold
Output: X' // X' is the value vector after segmentation
1. $X = \{x_1, x_2, \cdots, x_n\}$, $Y_1 = \{N_{x_1}, N_{x_2}, \cdots, N_{x_n}\}$, $Y_2 = \{M_{x_1}, M_{x_2}, \cdots, M_{x_n}\}$;
2. **for** $i \leftarrow 1\,to\,n$, $j \leftarrow i+1\,to\,n$, $\tau \leftarrow 1$
3.   **while** $\tau < \mathcal{T}_{count}$ **do**
4. $Sum_{i,j} = |\sum_{i \le k \le j} N_{x_k} - \sum_{i \le k \le j} M_{x_k}|$;
5.   **if** $Sum_{i,j} > Sum_{i,j+1}$ **then**
6. $x'_\tau \leftarrow \{x_i, x_{i+1}, \cdots, x_j\}$;
7. $X' \leftarrow X' \cup \{x'_\tau\}$;
8. $\tau ++$;
9. $i \leftarrow j+1$;
10. $j \leftarrow i+1$;
11.   **end if**
12.   **end while**
13. **end for**

4.3 Training and Feature Extraction

Overlap Degree and Dispersion Degree. After the training stage, we define two indicator of topic *Overlap Degree* and *Dispersion Degree* to quantitatively assess the different of $\| PDT D_K - ADT D_K \|$ (see Formula 5).

(1) *Overlap Degree*: the degree to which topics appear in different documents at the same time in the DTD.

$$D_{overlap} = \sum_{k=1}^{K} \sum_{1 \le i < j \le M} \frac{N_i^{test} * \theta_i^k + N_j^{test} * \theta_j^k}{\sum_{m=1}^{M} N_m^{test}} * \frac{min\left(\theta_i^k, \theta_j^k\right)}{max\left(\theta_i^k, \theta_j^k\right)} \tag{7}$$

$N_i^{test} * \theta_i^k + N_j^{test} * \theta_j^k$ denotes the number of classification deviations that may occur due to the topic k overlap in test document i, j. $\sum_{m=1}^{M} N_m^{test}$ denotes the total sample number in test document. $\dfrac{min\left(\theta_i^k, \theta_j^k\right)}{max\left(\theta_i^k, \theta_j^k\right)}$ represents the occurrence probability of classification deviation due to the topic k overlap. $\displaystyle\sum_{1 \le i < j \le M} \dfrac{N_i^{test} * \theta_i^k + N_j^{test} * \theta_j^k}{\sum_{m=1}^{M} N_m^{test}} *$ $\dfrac{min\left(\theta_i^k, \theta_j^k\right)}{max\left(\theta_i^k, \theta_j^k\right)}$ represents the overlap degree of single topic k. The smaller the $D_{overlap}$ value, the smaller the different of $\| PDTD_K - ADTD_K \|$. E.g., the $D_{overlap}$ value of the perfect DTD in Fig. 1(d) is 0.

(2) **Dispersion Degree**: the degree to which topics scatter in the same document in the DTD.

$$D_{dispersion} = \sum_{1 \le i < j \le M} D_{JS}\left(\overrightarrow{\theta_i}, \overrightarrow{\theta_j}\right) * \frac{N_i^{test} + N_j^{test}}{\sum_{m=1}^{M} N_m^{test}} \tag{8}$$

$D_{JS}\left(\overrightarrow{\theta_i}, \overrightarrow{\theta_j}\right)$ indicates the dispersion degree of two topic distributions corresponding to the document i, j. $N_i^{test} + N_j^{test}$ denotes the number of samples affected in test document i, j. The larger the $D_{dispersion}$ value, the smaller the different of $\| PDTD_K - ADTD_K \|$. E.g., the $D_{dispersion}$ value of the perfect DTD in Fig. 1(d) is 1.

Multi-topic Selection. Existing researches almost select the single topic of maximum weight in feature selection, ignoring other topics with a slightly lower weight. It reduces the detection accuracy. While in the actual DTD, the sum of Top-K weights in one document nearly has accounted for the majority. Choosing too many topics will increase the topic *Overlap Degree*, and it results in lower accuracy. We must select the suitable Top-K (K-unfixed) topics for each document. In this paper, we design a flexible multi-topic selection algorithm for each document with the topic count unfixed. According to Algorithm 2, we select several topics for one document when the maximum weight is less than τ_{max} or the difference between the maximum weight and the secondary weight is less than $\tau_{interval}$. We select the topic whose weight is not less than τ_{min}. It ends until the sum of the weights is not less than τ_{sum}.

Algorithm 2. The pseudo code of multi- topic selection for document m

---

Input: $\overrightarrow{\theta_m}$, K, τ_{sum}, τ_{max}, τ_{min}, $\tau_{interval}$, $Topic_{select}$
Output: $Topic_{select}$
//$\overrightarrow{\theta_m}$: topic distribution for document m
//τ_{sum}: sum of the selected topic weights
//τ_{max}: maximum threshold of topic weight
//τ_{min}: minimum threshold of topic weight
//$\tau_{interval}$: the threshold of weight difference
1. **if** $max_{k \leq K}\ \theta_m^k \geq \tau_{max}$ $\vee$ $max_{k1 \leq K}\ \theta_m^{k1} - max_{k2 \leq K, k2 \neq k1}\ \theta_m^{k2} \geq \tau_{interval}$ **then**
2. $Topic_{select} \leftarrow Topic_{select} \cup \left\{ topic'_{max_{k \leq K}\ \theta_m^k} \right\}$;
3.   **return** $Topic_{select}$
4. **else**
5. $\langle P_{topic_1}, P_{topic_2}, \cdots, P_{topic_K} \rangle \leftarrow Sort(\overrightarrow{\theta_m})$ //sort from large to small
6.   **while** $i \leftarrow 1\ to\ K$, $P_{topic_i} \geq \tau_{min}$ **do**
7. $Topic_{select} \leftarrow Topic_{select} \cup \left\{ topic'_{P_{topic_i}} \right\}$;
8.   **if** $\sum_{j=1}^{i} P_{topic_j} \geq \tau_{max}$ **then**
9.   **return** $Topic_{select}$
10.   **end if**
11.   **end while**
12. **end if**

4.4 Matching Mechanism

We propose a matching mechanism for multiple topics. Assume the selected topics are: $topic_{k1}, topic_{k2}, \cdots, topic_{kt}$. $P_{topic_{ki}}$ denotes the weight of topic ki. We reconstruct a new topic ($topic_{new}^m$) as the feature of document m.

$$topic_{new}^m = \frac{\sum_{i=1}^{t} P_{topic_{ki}} * topic_{ki}}{\sum_{i=1}^{t} P_{topic_{ki}}} \tag{9}$$

The practical meaning of a topic is a linear combination of semantically related and differently weighted words. Each $topic_k$ corresponds to a V-dimensional word distribution $\overrightarrow{\varphi_k}$. So the feature of document m is the word distribution: $\overrightarrow{\varphi_{topic_{new}^m}}$.

$$\overrightarrow{\varphi_{topic_{new}^m}} = \frac{\sum_{i=1}^{t} P_{topic_{ki}} * \overrightarrow{\varphi_{ki}}}{\sum_{i=1}^{t} P_{topic_{ki}}} \tag{10}$$

We convert the test data (*Sample*) into words according to the preprocessing algorithm. Count the number of each word, then a V-dimensional word distribution ($\overrightarrow{\varphi_{Sample}}$) is generated. *JS* divergence [23] is used to calculate the similarity between the *Sample* and the M test documents: $D_{JS}(Doc_m || Sample)$. Finally, we classify this *Sample* in the document m with the smallest $D_{JS}(Doc_m || Sample)$ value.

$$D_{JS}(Doc_m || Sample) = D_{JS}(\overrightarrow{\varphi_{topic_{new}^m}}, \overrightarrow{\varphi_{sample}}) \tag{11}$$

5 Experiment and Evaluation

This section aims to demonstrate the performance of the ID-LDA model. We evaluate the model from the following aspects: detect accuracy, determination of optimal topic number, multi-topic selection, and preprocessing method.

We use NSL-KDD Dataset [24] for our testing database, which contains one type of normal network flow data (NORMAL) and 4 types of intrusion (DOS, R2L, U2R and PROBE) data (see Table 2). Each sample contains a 41-dimensional attribute and a class label.

Table 2. Description of NSL-KDD Dataset

	NORMAL	DOS	R2L	U2R	PROBE
Train Data	67343	45927	11656	52	995
Test Data	9711	7458	2421	200	2750

At the preprocessing stage, each attribute is divided to several (less to 10) segments (see Sect. 4.2). Finally, we get 151 segments for each sample (both Train Data and Test Data). At the training stage of generating the DTD, the hyper parameters $\vec{\alpha}$, $\vec{\beta}$ (see Table 2) in the Dirichlet distribution control the shape of the distribution. They are set to 0.01 to prefer the distinctive topic in each document, as more weight will be assigned to key topic when the hyper parameters are small. The number of iterations indicates the number of rounds of Gibbs sampling, which is the main technique used in LDA to approximate topic and word distributions. The number of iterations is set to 20000. As there are 5 types of label, we generate diffident DTDs according to the topic number (from 5 to 50). Then we get the most optimal topic number and choose the best topics to reconstruct new topic for each type of intrusion. Finally we use the test data to get the detection results.

We test different intrusion detection methods to compare the detection accuracy and other metrics. In order to verify the validity of the indicators (topic *Overlap Degree* and *Dispersion Degree*), we plot the actual curves of different topic numbers. We find the actual most optimal topic number and then verify the accuracy whether it reaches the highest at the point (the most optimal number). In order to validate the multi-topic selection mechanism, we conduct three sets of experiments (Top-1 topic selection, Top-K topics selection (K-fixed), multi-topic selection (K-unfixed)) to compare the accuracy. We also test the detection performance of different preprocessing methods.

5.1 Accuracy of Different Intrusion Detection Algorithms

We compare the performance of ID-LDA with the existing methods from the perspective of accuracy, false negative rate (FNR), false positive rate (FPR) and F1-score (F1). The mainstream methods include support vector machine (SVM), Random Forest (RF), Bayes Network (BN), Convolutional Neural Networks (CNN), Long Short-Term Memory (LSTM) [25–28]. The metrics are shown in Formula 12. Num'_i denotes the number of i-type intrusion successfully detected. Num_{total} denotes the number of all the samples. TP, TN, FP, FN denote the true positive, true negative, false positive and false negative.

$$
\left\{
\begin{array}{l}
Accuracy = \dfrac{Num'_{NORMAL}+Num'_{DOS}+Num'_{R2L}+Num'_{U2R}+Num'_{PROBE}}{Num_{total}} \\[2mm]
FNR = \dfrac{FN}{TP+FN}, \ FPR = \dfrac{FP}{TN+FP} \\[2mm]
F1 = \dfrac{2\times\frac{TP}{TP+FP}\times\frac{TP}{TP+FN}}{\frac{TP}{TP+FP}+\frac{TP}{TP+FN}}
\end{array}
\right.
\tag{12}
$$

Table 3. Detect effects of different methods

	SVM	BN	RF	CNN	LSTM	ID-LDA
Accuracy	0.569	0.618	0.761	0.774	0.777	**0.834**
F1	0.656	0.802	0.787	0.828	0.836	**0.904**
FNR	0.268	0.024	0.028	0.229	0.233	**0.101**
FPR	0.391	0.347	0.377	0.069	0.087	**0.068**

As shown in Table 3, our ID-LDA model is obvious superior to the mainstream methods in terms of accuracy, F1 and FPR. Though the FNR in BN and RF is less than the value in ID-LDA, but the accuracy is much lower than the ID-LDA model. On the premise of improving the accuracy and reducing the FPR, the FNR in ID-LDA still remains at a low level. The superiority of the ID-LDA model has been proved.

5.2 Effectiveness of Topic Overlap Degree and Dispersion Degree

We test the validity of the two indicators. We determine the most optimal topic number based on the curves of *Overlap Degree* and *Divergence Degree*. Since there are only 5 types of network flow data and the topic weight is more scattered when the topic number is too large. So we display the *Overlap Degree* and *Dispersion Degree* curves (see Fig. 4) with the topic number from 5 to 50. When the topic number exceeds 40, the topic weight is over scattered in practice. The maximum weight of some document is less than 0.3 in actual, resulting in lower accuracy. Therefore, the most optimal topic number must be no more than 40. On the other hand, the lower the topic *Overlap Degree* and the higher the *Dispersion Degree*, the better the test effect. From the *Overlap Degree* curve, the vector of Top-5 optimal topic number is 8, 39, 25, 20 and 12 respectively. According to the *Dispersion Degree* curve, the vector of Top-5 optimal topic number is 8, 20, 39, 26 and 23. Lastly, the most optimal topic number is determined to be "8".

The current method [1] is to calculate the most optimal topic number based on the *Perplexity* degree (see Formula 13). w_m denotes the word in test document m, and N_m denotes the count of words in test document m. The smaller the *Perplexity* degree, the better the effect. According to the $log(Perplexity)$ curse (see Fig. 5), the vector of Top-5 optimal topic number is 13, 6, 8, 5 and 7. Therefore, the most optimal topic number is

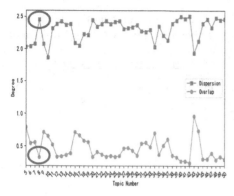

Fig. 4. *Overlap* and *Dispersion* degree curse

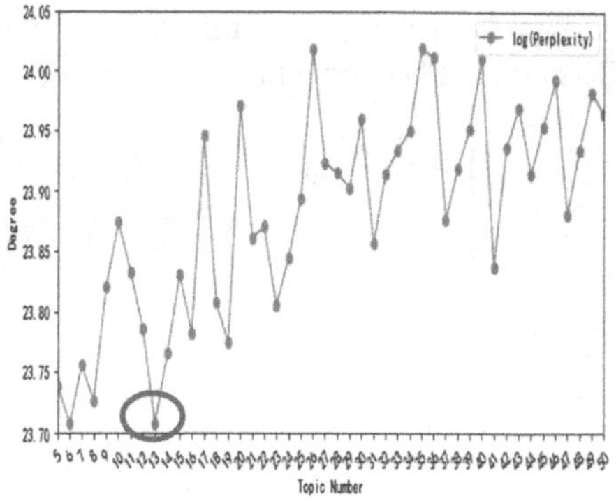

Fig. 5. *log(Perplexity)* curse

determined to be "13".

$$Perplexity = exp\left\{ -\frac{\sum_m^M \log p(w_m)}{\sum_{m=1}^M N_m} \right\}$$
(13)

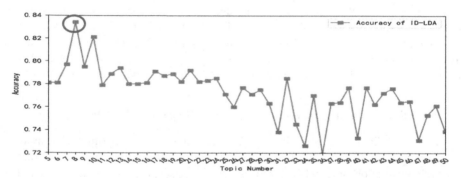

Fig. 6. Accuracy of different topic numbers

Experiments are carried out to calculate the accuracy of different topic numbers. In Fig. 6, the accuracy reaches the maximum when the topic number is "8". It is consistent with the most optimal number ("8") determined by the topic *Overlap Degree* and *Dispersion Degree* curses. However, the number ("13") calculated by the traditional *Perplexity* degree has a certain deviation from the actual most optimal topic number. Experiments have proved that the *Overlap Degree* and *Dispersion Degree* can effectively determine the most optimal topic number. This can effectively improve the application ability of the LDA topic model.

5.3   Accuracy of Different Topic Selection Methods

We compare the accuracy of different topic selection methods. In practice test, nearly the sum of top-4 weight takes up a large proportion. So we test the accuracy of the Top-1

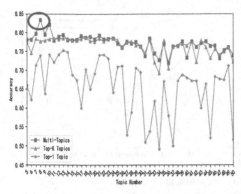 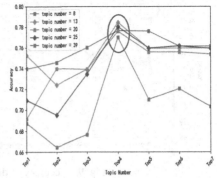

Fig. 7. Accuracy of different topic selections   **Fig. 8.** Accuracy of K-fixed topic selections

topic selection and Top-4 topics selection (see Fig. 7). According to the literature [8], we choose 0.6 as the maximum weight threshold in Top-1 topic selection. We discard the topics with the weight below 0.1. Selecting these topics may increase the *overlap degree* and result in lower accuracy. It is verified in Fig. 8. In our work, when the Top-1 topic weight is less than 0.6, multiple topics (each topic weight is not less than 0.1) are selected. It ends when the sum weight is not less than 0.8. Figure 7 shows the impact of Top-1, Top-K ($K = 4$) and multi-topic selection on accuracy. Figure 8 shows the accuracy when we choose the Top-K topics (K ranges from 1 to 7) for each document.

In Fig. 7, the accuracy of traditional Top-1 topic selection method is significantly lower than our multi-topic selection method. Meanwhile, our multi-topic method is also superior to the Top-K (K fixed) topics selection method in most different topic numbers. In particular, the advantage is obvious when the topic number is small. However, the advantage is not obvious when the topic number is larger. This may be related to the topic number. When the topic number is large, the topic is over-dispersed. In this circumstance, the multi-topic selection is more similar to the top-K selection.

In Fig. 8, the accuracy reaches the maximum when the K value is 4. Too many or too few topics selection will reduce the accuracy. In this situation, our multi-topic selection method is still superior to the Top-k ($k = 4$) topics selection method. Therefore, our multi-topic selection is more flexible and versatile. The experiments have proved that reconstructing a new topic from multiple topics can compensate for some of the missing semantics and has better generalization capabilities.

5.4 Effect of Different Preprocessing Methods

We compare the performance of our preprocessing method (segment algorithm) with the traditional method in term of detect accuracy. The traditional method is to segment the continuous value according to the total value distribution instead of the value distribution of different intrusions. In Fig. 9, our segment method is obvious superior to the traditional method in terms of accuracy. Experiments have proved the effectiveness of our preprocessing method.

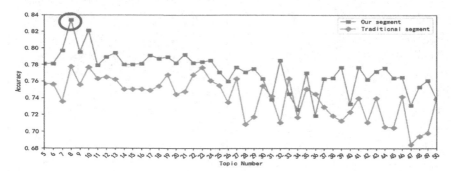

Fig. 9. Accuracy of different segment algorithm

6   Conclusion and Future Work

In this paper, we propose a novel topic reconstruction method based on LDA and establish an ID-LDA model. At the feature extraction stage, we propose the topic *Overlap Degree* and *Dispersion Degree* to evaluate the DTD quality and determine the most optimal topic number. In addition, we design a more flexible and scalable multi-topic selection, as well as the topic reconstruction mechanism. Experiments on the public NSL-KDD dataset validate the effectiveness of the two indicators. In terms of detection accuracy, the multi-topic selection is obviously superior to the traditional single topic selection. Our ID-LDA model is also superior to the mainstream detection models. In the future, we plan to apply the LDA model to dynamic data, adjusting the extracted topics in real time to achieve more accurate detection.

Acknowledgment.   This work is supported by the National Natural Science Foundation of China (U1636208, F020605, No. 61902013).

References

1. Blei, D.M., Ng, A.Y., Jordan, M.I.: Latent Dirichlet allocation. Mach. Learn. Res. **3**, 993–1022 (2003)
2. Zhang, Y., Chen, W., Zha, H., et al.: A time-topic coupled LDA model for IPTV user behaviors. IEEE Trans. Broadcast. **61**(1), 56–65 (2015)
3. Farrahi, K., Gatica-Perez, D.: Discovering routines from large scale human locations using probabilistic topic models. ACM Trans. Intell. Syst. Technol. **2**(1), (2011)
4. Huynh, T., Fritz, M., Schiele, B.: Discovery of activity patterns using topic models. In: Proceedings of the 10th International Conference on Ubiquitous Computing, Seoul, Korea, pp. 10–19. ACM (2008)
5. Guixian, X., Xu, W., Yao, H., et al.: Research on topic recognition of network sensitive information based on SW-LDA model. IEEE Access **7**, 21527–21538 (2019)
6. Zhang, Y., Wang, Z., Yongtao, Yu., et al.: LF-LDA: a supervised topic model for multi-label documents classification. IJDWM **14**(2), 18–36 (2018)
7. Casale, P., Pujol, O., Radeva, P., et al.: A first approach to activity recognition using topic models. In: Artificial Intelligence Research & Development, International Conference of the Catalan Association for Artificial Intelligence, CCIA, Vilar Rural De Cardona, October. DBLP (2009)

8. Yang, Y., Sun, J., Guo, L.: PersonaIA: a lightweight implicit authentication system based on customized user behavior selection. IEEE Trans. Dependable Secure Comput. 16(1), 113–126 (2019)

9. Wilson, J., Chaudhury, S., Lall, B.: Clustering short temporal behaviour sequences for customer segmentation using LDA. Expert Syst. e12250 (2009)

10. Xie, L., Shi, Y., Li, Z.: Driving pattern recognition based on improved LDA model. In: 5th IEEE International Conference on Cloud Computing and Intelligence Systems (CCIS), Nanjing, China, pp. 320–324 (2018)

11. Gao, Y., Wei, X., Zhang, X., et al.: A combinational LDA-based topic model for user interest inference of energy efficient IPTV service in smart building. IEEE Access 6, 48921–48933 (2018)

12. Chen, W., Zhang, Y., Zha, H.: Mining IPTV user behaviors with a coupled LDA model. In: IEEE International Symposium on Broadband Multimedia Systems & Broadcasting, London, pp. 1–6. IEEE (2013)

13. Wang, Z., Gu, S., Xu, X.: GSLDA: LDA-based group spamming detection in product reviews. Appl. Intell. 1, 1–14 (2018)

14. Budhiraja, A., Reddy, R., Shrivastava, M.: Poster: LWE: LDA refined word embeddings for duplicate bug report detection. In: 2018 IEEE/ACM 40th International Conference on Software Engineering: Companion Proceedings, Gothenburg, pp. 165–166. IEEE Computer Society (2018)

15. Andrzejewski, D., Mulhern, A., Liblit, B., Zhu, X.: Statistical debugging using latent topic models. In: Kok, J.N., Koronacki, J., Mantaras, R.L., Matwin, S., Mladenič, D., Skowron, A. (eds.) ECML 2007. LNCS (LNAI), vol. 4701, pp. 6–17. Springer, Heidelberg (2007). https://doi.org/10.1007/978-3-540-74958-5_5

16. Mäntylä, M., Claes, M., Farooq, U.: Measuring LDA topic stability from clusters of replicated runs. In: ESEM 2018 ACM, Oulu, Finland (2018)

17. Gollapalli, S.D., Li, X.-l.: Using PageRank for characterizing topic quality in LDA. In: 2018 ACM SIGIR International Conference on the Theory of Information Retrieval (ICTIR 2018), Tianjin, China, pp. 115–122 (2018)

18. Morstatter, F., Liu, H.: A novel measure for coherence in statistical topic models. In: Proceedings of the 54th Annual Meeting of the Association for Computational Linguistics, Berlin, Germany, pp. 543–548 (2016)

19. Newman, D., Lau, J.H., Grieser, K., et al.: Automatic evaluation of topic coherence. In: Human Language Technologies: Conference of the North American Chapter of the Association of the ACL, Los Angeles, California, pp. 100–108 (2010)

20. Jonathan, C., Boyd-Graber, J., et al.: Reading tea leaves: how humans interpret topic models. In: NIPS, Vancouver, British Columbia, Canada (2009)

21. Zhao, W., Chen, J.J., Perkins, R., et al.: A heuristic approach to determine an appropriate number of topics in topic modeling. BMC Bioinformatics 16(Suppl 13), S8 (2015)

22. Grant, S., Cordy, J.R., Skillicorn, D.B.: Using heuristics to estimate an appropriate number of latent topics in source code analysis. Sci. Comput. Program. 78(9), 1663–1678 (2013)

23. Lin, J.: Divergence measures based on the Shannon entropy. IEEE Trans. Inf. Theory 37(1), 145–151 (1991)

24. McHugh, J., Brugger, S.T. (1999). http://kdd.ics.uci.edu/databases/kddcup99.thml

25. Zhihua, C., Lei, D., et al.: Malicious code detection based on CNNs and multi-objective algorithm. Parallel Distrib. Comput. 129, 50–58 (2019)

26. Xiaoyu, G., Hui, Z., et al.: A single attention-based combination of CNN and RNN for relation classification. IEEE Access **7**, 12467–12475 (2019)
27. Yao, H., Sun, X., et al.: An enhanced LSTM for trend following of time series. IEEE Access **7**, 34020–34030 (2019)
28. Alguliyev, R.M., Aliguliyev, R.M., et al.: The improved LSTM and CNN models for DDoS attacks prediction in social media. IJCWT **9**(1), 1–18 (2019)

PDGAN: A Novel Poisoning Defense Method in Federated Learning Using Generative Adversarial Network

Ying Zhao[1], Junjun Chen[1(✉)], Jiale Zhang[2], Di Wu[3,4], Jian Teng[1], and Shui Yu[3]

[1] College of Information Science and Technology,
Beijing University of Chemical Technology, Beijing 100029, China
{zhaoy,chenjj,tengj}@buct.edu.cn
[2] College of Computer Science and Technology,
Nanjing University of Aeronautics and Astronautics, Nanjing 211106, China
jlzhang@nuaa.edu.cn
[3] School of Computer Science, University of Technology Sydney,
Sydney, NSW 2007, Australia
[4] Centre for Artificial Intelligence, University of Technology Sydney,
Sydney, NSW 2007, Australia
{Di.Wu,Shui.Yu}@uts.edu.au

Abstract. Federated learning can complete an enormous training task efficiently by inviting participants to train a deep learning model collaboratively, and the user privacy will be well preserved for the users only upload model parameters to the centralized server. However, the attackers can initiate poisoning attacks by uploading malicious updates in federated learning. Therefore, the accuracy of the global model will be impacted significantly after the attack. To address this vulnerability, we propose a novel poisoning defense generative adversarial network (PDGAN) to defend the poising attack. The PDGAN can reconstruct training data from model updates and audit the accuracy for each participant model by using the generated data. Precisely, the participant whose accuracy is lower than a predefined threshold will be identified as an attacker and model parameters of the attacker will be removed from the training procedure in this iteration. Experiments conducted on MNIST and Fashion-MNIST datasets demonstrate that our approach can indeed defend the poisoning attacks in federated learning.

Keywords: Federated learning · Poisoning defense · Generative adversarial network

1 Introduction

The traditional machine learning architecture [1] provides intelligent data analysis and automatic decision making that is known as machine-learning-as-a-service

© Springer Nature Switzerland AG 2020
S. Wen et al. (Eds.): ICA3PP 2019, LNCS 11944, pp. 595–609, 2020.
https://doi.org/10.1007/978-3-030-38991-8_39

[2], such as recommendation systems, keyboard input prediction, smart transportation, and health monitoring. However, such learning scenario requires its participants to outsource their private raw data to an unknown third party, causing a significant privacy concern where the sensitive data may be exposed to attackers [3]. Considering the above privacy issues, *federated learning* [4,5] has been explored recently, which has the natural ability to preserve the user data privacy by its unique distributed machine learning framework. The federated learning framework trains a global model across multiple participants in a distributed manner. Each participant can download the global model from the central server and train the model on their training datasets locally instead of outsourcing the sensitive training data in the traditional centralized training methods. Participants in federated learning only need to upload the model parameters (i.e., parameters of gradients and weights) generated from local training procedure, which provides a basic privacy guarantee. After receiving all the participant model updates, the central server will execute the federated average algorithm to update the global model further. The processes mentioned above will be performed iteratively until the global model tends to convergence.

However, the unique training scheme of the federated learning could be a double edge sword, where the central server cannot access participants private training data. This characteristic may be leveraged by the attackers to launch attacks such as poisoning attack. Poisoning attack [6] is a common attack method in the traditional centralized distributed system. The attackers in the poisoning attack [7,8] can change the learning model parameters by compromising partiality training data from other participants. Label-flipping [9] is a classic method to launch the poisoning attack, and it also can poison the federated learning. It requires the attacker to change the parameters of the target learning model in the training phase, and the poisoned model will be marked with some attacker targeted attributes. Then, the poisoned model will misclassify (take a classification task as an example) the attacker chosen inputs at the prediction stage. In federated learning, we notice that the poisoning attack can be easily initiated due to the following reasons. (1) Federated learning algorithm does not contain any authentication or identity verification mechanisms, so the trustworthy of participants cannot be guaranteed. (2) The participants' training data and training procedure are invisible to the server, so it is impossible to audit the accuracy of users' models from their updates. (3) Since the distributions of participants' training datasets are independent, which brings enormous difficulties for the anomaly detection of user's updates.

In this paper, we focus on the poisoning attack launched by the malicious participants in federated learning and try to defend such active attacks. To defend poison attacks, we deployed a generative adversarial network in the server to reconstruct user local training data and audit accuracy for each participant model using the generated data. The participant whose accuracy is lower than a predefined threshold is identified as the attacker.

1.1   Related Work

Known poisoning attack defense methods in centralized learning scenarios have been well explored. Secure multiparty computation and homomorphic additive cryptosystem [10–12] are two efficient tools to build training models while protecting training data privacy. However, these schemes introduce huge computation overhead to the participants and may bring a negative effect on model accuracy. Byzantine-tolerant machine learning methods [13,14] have been explored recently to guarantee the privacy of only Byzantine participants, which imitating the applicant in the scenario of federated learning. Besides, anomaly detection techniques have shown the significant advantages of detecting abnormal participants' behaviors. Aiming at detecting poisoning attacks in distributed learning models, mechanisms in [9,15,16] apply several algorithms (e.g., k-means and clustering) to check the participants' updates across communication rounds and remove the outliers.

Furthermore, some other defense methods [17,18], such as cosine similarity and gradients norm detection, were proposed to detect the gradients anomalies. However, the effectiveness of above-mentioned anomaly detection methods is quite low in the context of federated learning. That is mainly because the distributions of participants' training data in federated learning are considered as Non-IID (not independent and identically distributed), which means the participants' model updates are obviously different from each other.

1.2   Our Contributions

Aiming at solving the above problems, we propose a novel poisoning defense method. Briefly speaking, the contributions in our paper are threefold.

- We propose a detection scheme based on accuracy auditing, named poisoning defense generative adversarial network (PDGAN), to defend poisoning attacks in federated learning. The proposed scheme can be easily deployed in the real scenario.
- The PDGAN can use partial classes data to reconstruct the prototypical samples of participants' training data for auditing the accuracy of each participant's model.
- Experiments conducted on MNIST and Fashion-MNIST datasets demonstrate that our approach can indeed defend the poisoning attacks in federated learning.

1.3   Organization

The rest of this paper is organized as follows. In Sect. 2, we briefly introduce the basic knowledge of federated learning and generative adversarial nets. The overview of the poisoning attack in federated learning is presented in Sect. 3, and the construction of the proposed defense algorithm is detailed in Sect. 4. Extensive experimental evaluation is conducted in Sect. 5. Finally, Sect. 6 gives the conclusion and future work.

2 Preliminaries

In this section, we briefly review the preliminary knowledge of federated learning and the introduction of Generative adversarial nets (GAN) to facilitate understanding of our defense mechanism.

2.1 Federated Learning

Federated learning [4] is a centralized training framework which can preserve user privacy by its unique distribution learning mechanism. Unlike other collaborative learning methods, the participants in federated learning upload the model updates which generated by training learning model on participant private training data to the central server, and the central server will distribute the global models which share the same structure with participants models. In federated learning, all participants share the same learning objective and model structure, where the central server sends the current global model parameters to the selected participants m_t in each communication round t. Then, all the selected participants update their model and apply the model to train the local data. Each participant uploads the model updates after the local training procedure, where the uploaded model updates will be averaged and accumulated to the current central model. Equation 1 shows the updating procedure on the central model.

$$M_{t+1} = M_t + \frac{1}{m_t} \sum_{k=1}^{m_t} u_t^k, \tag{1}$$

where M_t is the current global model at the t-th iteration, and u_t^k represents the model updates uploaded by the k-th participant. A federated learning framework can achieve high satisfaction when the participants download the same model with the same initialization, which is averaged by the central model with all the valid uploads.

2.2 Generative Adversarial Nets

Generative adversarial networks [19] have shown the promised performance in computer vision as well as other research areas [20], which can generate high-quality fake images by training on the original images. The generator ($\mathcal{G}$) and discriminator ($\mathcal{D}$) of GAN play an adversarial game, where the $\mathcal{G}$ firstly generates and $\mathcal{D}$ discriminate if the image is from the generator or original image set where the output of $\mathcal{D}$ can be represented as fake (0) or real (1). The performance of both $\mathcal{G}$ and $\mathcal{D}$ can be improved by several training epochs. Equation 2 shows the training target of GAN.

$$\min_{\mathcal{G}} \max_{\mathcal{D}} V(\mathcal{D}, \mathcal{G}) = \mathbb{E}_{x \sim p_{data}(x)}[\log \mathcal{D}(x)]$$
$$+ \mathbb{E}_{z \sim p_z(z)}[\log(1 - \mathcal{D}(\mathcal{G}(z)))], \tag{2}$$

where $x \sim p_{data}(x)$ represents the original data distribution and $z \sim p_z(z)$ is the distribution of the random vector z. $\mathcal{D}$ and $\mathcal{G}$ will be trained by several epochs until the training procedure achieves the Nash equilibrium, while the $\mathcal{D}$ cannot discriminate the fake data from the real data.

3   Overview of Poisoning Attacks

In this section, we first introduce the threat model, including the learning scenario, the attacker's goals, and the attacker's capabilities. Then, we detailed discuss the poising attacks against federated learning and demonstrate the effectiveness of poisoning attacks through experimental evaluations.

3.1   Threat Model

Learning Scenario: As described in Sect. 2.1, we consider a federated learning scenario where multiples participants agree on a common learning objective and jointly train a global model on their localized training datasets. Besides, we assume that the distributions of all the participants' datasets are independent with each other, and the participant model updates will be averaged on the server side to achieve federated learning property. Without loss of generality, the main purpose of training a global model is to perform an image classification task in this paper.

Attacker's Goal: The attacker in the poisoning attack wants to conduct a specific global model that performs high accuracy on his chosen inputs while having less impact on overall accuracy. The attacker replaces the global model with a poison model that has the following two properties:

- Poisoning accuracy: the global model should behave good performance on the attacker-chosen poisoned inputs after the attacker's model updates were uploaded to the central server.
- Overall accuracy: the poisoned updates should have a less negative impact on overall accuracy, which means the poisoned global model cannot be discarded due to the attack behaviors.

Attacker's Capability: In a federated learning scenario, the attacker first pretends to be an honest user to participate in the learning system, while the main purpose of this attacker is to compromise the global model. Specifically, the attacker has the following capabilities:

- Knowledgeable: the attacker has a white-box access privilege to the global model, meaning that the model structure and parameters can be obtained.
- Active: the attacker is considered as an active insider because he can fully control the local training procedure and modify the model hyper parameters (e,g., epochs, learning rate).

3.2   Poisoning Attacks

Existing literatures [6, 9, 15] demonstrate that the poisoning attack can be easily launched by malicious participants in the context of federated learning. Here, we give a brief introduction about this poisoning attack and show how one attacker can successfully compromise the global model.

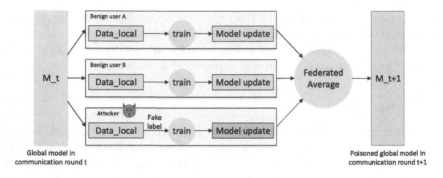

Fig. 1. Poisoning attack in federated learning

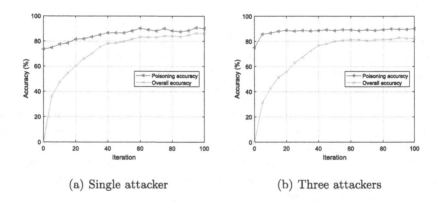

(a) Single attacker   (b) Three attackers

Fig. 2. Poisoning accuracy and overall accuracy under attacks

Figure 1 illustrates the detailed processes of the poisoning attack in federated learning. Considering there are three participants, and one of them is attacker, where all the participants agree on a common learning objective and model structure as depicted in the federated learning algorithm. The attacker's purpose is to compromise the global model by uploading the poisoned local model updates. To achieve this goal, he first changes the category label of his target class on his training dataset. Then, he feeds the modified datasets to the local model and computes the poisoned local updates. Note that these poisoned local updates can be scaled to speed up the attack process. Last, these local updates are sent to the central server. After model averaging, the global model can be contaminated by the poisoned model.

Figure 2(a) and (b) illustrate the poisoning accuracy and overall accuracy of image classification task on MNIST dataset. According to the results, we can see that the increasing rate of poisoning task accuracy on three attackers scenarios is faster than the single attacker scenario.

4   Defense Algorithm

In this section, we present the proposed defense approach PDGAN, which is specifically designed for poisoning defense in federated learning. To detect attackers and reduce the impact of the poisoning attack, the PDGAN reconstructs participants training data from their updates in the server and audits accuracy for each participant model using the generated data. The participant whose accuracy is lower than a predefined threshold will be identified as the attacker.

4.1   Overview of PDGAN

In poisoning attacks, attackers try to poison the global model by uploading malicious updates, where misclassification happens after the model has been poisoned on the server side. Anomaly detection methods of federated learning can be categorized into two classes: gradients distance detection and model accuracy auditing. The gradients distance detection method [9] detects anomaly by comparing the distances between gradients uploaded by different participants. For evading this detection, attackers [21] use the optimization algorithm to carry out stealthy poisoning attacks. The model accuracy auditing methods use an auxiliary dataset to identify attackers by checking the accuracy for each participant model.

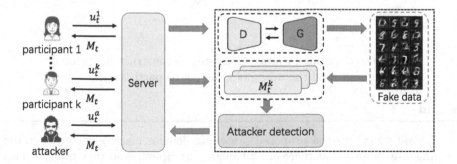

Fig. 3. Overview of the proposed PDGAN method in the federated learning.

Algorithm 1. PDGAN in Federated Learning

Data: Parameters updates u_t^k from participant k at iteration t; global model M_{t-1} at iteration $t-1$; Auxiliary data X_{aux}; Labellist L; Accuracy threshold η

1   Initialize Generator $\mathcal{G}$

2   **for** *Iteration* t **do**

3   Receive updates from the selected participants, $\{u_t^1, u_t^2,..., u_t^k\}$

4   Generate X_{fake} from $\mathcal{G}$

5   Update the Discriminator by the participant updates,
$\mathcal{D}_t = \mathcal{D}_{t-1} + \frac{1}{N}\sum_{k=1}^{N} u_t^k$

6   Train $\mathcal{D}_t$ by X_{aux} and X_{fake}, and Train $\mathcal{G}$

7   **if** $t \geq d\_iter$ **then**

8   **for** $k=1$ *to* N **do**

9   Initialize participant classification model, $M_t^k = M_{t-1} + u_t^k$

10   **foreach** x *in* X_{fake} **do**

11                  $L[k][x] = M_t^k(x)$

12   **end**

13   **end**

14   Assign labels for X_{fake} based on L

15   Calculate accuracy a^k of each participant classification model on X_{fake}

16   Initialize the sum of benign updates $S = 0$ and the number of benign participants $NC = 0$

17   **for** $k=1$ *to* N **do**

18   **if** $a^k \geq \eta$ **then**

19                  $S = S + u_t^k$

20                  $NC = NC + 1$

21   **end**

22   **end**

23           $M_t = M_{t-1} + \frac{S}{NC}$

24   Sent M_t to all participants

25   **end**

26   **else**

27   Federated learning averages updates to construct new global model and send the new global model to participants

28   **end**

29 **end**

Figure 3 overviews the proposed poisoning defense mechanism in federated learning. We assume that there are k benign participants and that the $(k+1)$th participant is the attacker who uploads malicious updates u_t^a to the server for poisoning the global model. In federated learning, there is usually an auxiliary data to audit participant model accuracy. However, it is hard that the auxiliary data includes all classes data in the real scenario because participants train models locally and do not share their private training data. To solve this problem, we implement a GAN on the server to reconstruct participant training data. The PDGAN does not detect attackers at the beginning of the training but after some iterations, for the generative adversarial network needs to take some iterations to train itself. We set the proposed method to begin to detect attackers at iteration d_iter. After obtaining the generated data, the server builds a classification model M_t^k for each participant by using the updates u_t^k uploaded by each participant and the global model M_{t-1} of the previous iteration. We can only reconstruct the training data with GAN, but we can not access the data labels. Therefore, we feed the generated data to each participant model and then get the predicted results. We specify that the label with the most occurrences is the true label for each data. After obtaining the generated data and its' labels, the accuracy of participant model can be calculated. Therefore, the participants can be divided into two clusters, benign participants and attackers, by a predefined accuracy threshold η. If one participant is judged to be an attacker, the PDGAN will ignore its updates in this iteration. The Algorithm 1 shows the procedures of federated learning under the PDGAN.

4.2   Structure of PDGAN

The PDGAN involves two components, discriminator and generator, which are deployed on the server side. The target of PDGAN is to generate samples closed to participants private training data by alternately optimizing the discriminator and the generator. In the training phase, the generator generates fake data, and the discriminator discriminates these generated data from real data. According to [21] and [22], there is an auxiliary dataset in the server for evaluating the learning process in federated learning. In the real scenario, the auxiliary dataset is hard to include all classes data, so we assume that the auxiliary dataset only consists of some classes data, which is used by discriminator to achieve the real-fake task. In federated learning, the server averages gradients uploaded by participants to construct a new global model. Therefore, the global model contains information about real data. According to [22] and [23], we use the global model as the discriminator in our proposed method.

Table 1. Network structure of PDGAN

Discriminator	$32^2 \times 1 \xrightarrow{Conv\ (stride\ =\ 2),\ LeakyReLU,\ Dropout}$
	$16^2 \times 64 \xrightarrow{Conv\ (stride\ =\ 2),\ BatchNorm,\ LeakyReLU}$
	$8^2 \times 64 \xrightarrow{Conv\ (stride\ =\ 2),\ BatchNorm,\ LeakyReLU}$
	$4^2 \times 64 \xrightarrow{Conv\ (stride\ =\ 1),\ BatchNorm,\ LeakyReLU}$
	$2^2 \times 128 \xrightarrow{Conv\ (stride\ =\ 1),\ BatchNorm,\ LeakyReLU}$
	$4^2 \times 128 \xrightarrow{Conv\ (stride\ =\ 1),\ LeakyReLU} 2^2 \times 128 \xrightarrow{AvgPool2d,\ FC,\ Softmax} 11$
Generator	$100 \xrightarrow{Deconv,\ BatchNorm,\ LeakyReLU}$
	$4^2 \times 256 \xrightarrow{Deconv,\ BatchNorm,\ LeakyReLU}$
	$8^2 \times 128 \xrightarrow{Deconv,\ BatchNorm,\ LeakyReLU} 32^2 \times 1 \xrightarrow{Tanh} 32^2 \times 1$

Table 1 shows the network structure of the PDGAN. For discriminator, the kernel size of the first three convolutional layers and the last three convolutional layers are 4×4 and 3×3, respectively. Additionally, we use *BatchNorm* layers and *Dropout* layers to achieve good model performance. For the generator, the kernel size is 4×4, and the input is a random vector of length 100. The *LeakyReLU* and *Tanh* are used as activation functions in different layers.

5   Experiments

5.1   Datasets

We used two public datasets in experiments to evaluate the effectiveness of the proposed defense method. The details of the two datasets are shown as follow.

- *MNIST:* The MNIST dataset [24] is a gray-scale handwritten digits dataset that consists of 70,000 images with the size of 28×28. There are totally 10 classes digits from 0 to 9 in the MNIST. The dataset is divided into two subsets, 60,000 training images and 10,000 test images, which are commonly used for training and evaluating image classification systems.
- *Fashion-MNIST:* The Fashion-MNIST dataset [25] is another benchmark dataset for machine learning evaluation. The dataset consists of 10 classes fashion product images which are divided into 60,000 training samples and 10,000 test samples.

We resized the images from the two datasets to 32×32 in the experiments. All experiments are done by using PyTorch framework on a server with Intel Xeon W-2133 3.6 GHz CPU, Nvidia Quadro P5000 GPU with 16 G RAM and RHEL7.5 OS.

5.2   Experimental Setting

To evaluate the effectiveness of the proposed method, we set two scenarios in our experiments.

1. **Single attacker:** There are 10 participants in the federated learning where single attacker uploads malicious updates to the server and other 9 participants are benign.
2. **Multiple attackers:** In the real scenario, the attackers are fewer than the benign participants; Therefore, we set the number of attackers to 3 in the experimental setting of multiple attackers. For each iteration, there are 3 attackers and 7 benign participants in federated learning.

In the above experimental setting, training data is randomly assigned to each participant. For the *MNIST*, the target of attackers is to misclassify digit 1 to digit 7. For the *Fashion-MNIST*, the target of attackers is to misclassify T-shirt to Pullover.

We train the federated learning model as a classification task and define two metrics to evaluate the proposed method. One is the poisoning accuracy, which represents the success rate of classifying the poisoned samples to the attacker chosen classes. The second is the overall accuracy, which means the success rate of getting correct classification results with all samples.

5.3   Data Reconstruction

In this section, we evaluate the proposed method's performance on reconstructing participants training data. We implement a GAN on the server side under the single attacker scenario, where 10 participants attend the training procedure in each iteration. Figure 4 shows the reconstruction results of MNIST dataset in different iterations. In this experimental setting, the auxiliary data in the server includes two classes data (digit 0 and digit 4), which is as the real data to feed the discriminator. After 400 iterations, the generated images are not blurring and can be used as the auditing data for detecting the attacker. Figure 5 shows the reconstruction results of Fashion-MNIST dataset. The auxiliary data includes 3 classed data (dress, coat, sandal). Because the images of Fashion-MNIST are more complex than the MNIST, it takes more iterations to generated not blurring images.

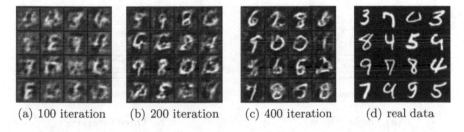

(a) 100 iteration   (b) 200 iteration   (c) 400 iteration   (d) real data

Fig. 4. MNIST reconstruction performance

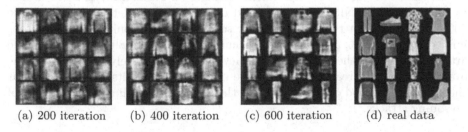

(a) 200 iteration   (b) 400 iteration   (c) 600 iteration   (d) real data

Fig. 5. Fashion-MNIST reconstruction performance

5.4   Poisoning Defense

Figure 6 illustrates the poisoning accuracy and the overall accuracy of image classification on the MNIST dataset. According to the results, we can see that the poisoning task accuracy has achieved a high level immediately once the attacker starts to upload the poisoning updates. That is mainly because the attackers well train their local model for the poisoning target. Besides, the increasing rate of poisoning accuracy on three attackers scenarios is faster than the accuracy of the single attacker scenario, and the overall accuracy is almost similar in both scenarios. We set the d_iter to 400, which means that the PD-GAN begin to detect attacks at the 400th iteration. From Fig. 6(a), poisoning accuracy immediately drops to 3.1% at the 400th iteration, and the overall accuracy increases to 90.1%. The reason behind this result is that the PDGAN removes the malicious updates of the attacker, and only benign participants can contribute to the global model. From Fig. 6(b), poisoning accuracy drops to 3.5% at the 400th iteration, and the overall accuracy increases to 89.2%.

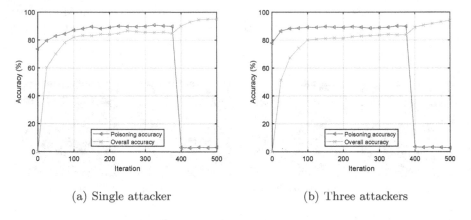

(a) Single attacker   (b) Three attackers

Fig. 6. Detection mechanism on MNIST dataset

As shown in Fig. 7, the experimental results on Fashion-MNIST are similar to the results of MNIST. The proposed defense method detect attacks at the 600-th iteration. For the single attacker scenario, the poisoning accuracy drops immediately to 4.5%, and the overall accuracy increases to 89.7%. For the three attackers scenario, the poisoning accuracy drops immediately to 5.1%, and the overall accuracy increases to 88.6%. The experimental results demonstrate that the proposed method can indeed detect the attackers and eliminate the effect of poisoning attacks.

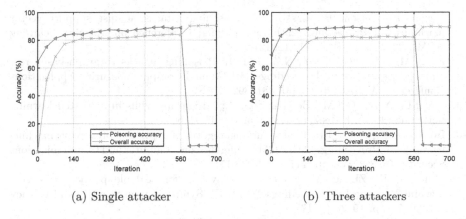

(a) Single attacker   (b) Three attackers

Fig. 7. Detection mechanism on Fashion-MNIST dataset

6   Conclusion and Future Work

In this paper, we propose a novel poisoning defense method PDGAN in federated learning. The proposed method is based on the generative adversarial network, which is implemented on the server side and can reconstruct participants training data. By using the generated data, the proposed method audits the accuracy of each participant's model and then identify attackers. Experiment results demonstrate that the PDGAN can effectively reconstruct the training data and successfully defend the poisoning attack by auditing the accuracy of the participant model. In future work, we plan to explore the poisoning defense for federated learning with device, class, or user-level differential privacy.

References

1. Ribeiro, M., Grolinger, K., Capretz, M.A.M.: MLaaS: machine learning as a service. In: Proceedings 14th International Conference on Machine Learning and Applications (ICMLA 2015), pp. 896–902 (2015)
2. Hesamifard, E., Takabi, H., Ghasemi, M., Wright, R.N.: Privacy-preserving machine learning as a service. In: Proceedings 19th Privacy Enhancing Technologies Symposium (PETS 2018), pp. 123–142 (2018)
3. Yu, S.: Big privacy: challenges and opportunities of privacy study in the age of big data. IEEE Access **4**, 2751–2763 (2016)
4. McMahan, H.B., Moore, E., Ramage, D., Hampson, S., Arcas, B.A.: Communication-efficient learning of deep networks from decentralized data. In: Proceedings 20th International Conference on Artificial Intelligence and Statistics (AISTATS 2017), pp. 1–10 (2017)
5. Yang, Q., Liu, Y., Chen, T., Tong, Y.: Federated machine learning: concept and applications. ACM Trans. Intell. Syst. Technol. **10**(2), 1–19 (2019)
6. Jagielski, M., Oprea, A., Biggio, B., Liu, C., N-Rotaru, C., Li, B.: Manipulating machine learning: poisoning attacks and countermeasures for regression learning. In: Proceedings 39th IEEE Symposium on Security and Privacy (SP 2018), pp. 19–35 (2018)

7. Biggio, B., Nelson, B., Laskov, P.: Poisoning attacks against support vector machines. In: Proceedings 29th International Conference on Machine Learning (ICML 2012), pp. 1807–1814 (2012)
8. Fang, M., Yang, G., Gong, N.Z., Liu, J.: Poisoning attacks to graph-based recommender systems. In: Proceedings 34th Annual Computer Security Applications Conference (ACSAC 2018), pp. 381–392 (2018)
9. Fung, C., Yoon, C.J.M., Beschastnikh, I.: Mitigating sybils in federated learning poisoning (2018). https://arxiv.org/abs/1808.04866
10. Bonawitz, K., et al.: Practical secure aggregation for privacy-preserving machine learning. In: Proceedings 24th ACM Conference on Computer and Communications Security (CCS 2017), pp. 1175–1191 (2017)
11. Mohassel, P., Zhang, Y.: SecureML: a system for scalable privacy-preserving machine learning. In: Proceedings 38th IEEE Symposium on Security and Privacy (SP 2017), pp. 19–38 (2017)
12. Phong, L.T., Aono, Y., Hayashi, T., Moriai, S.: Privacy-preserving deep learning via additively homomorphic encryption. IEEE Trans. Inf. Forensics Secur. **13**(5), 1333–1645 (2018)
13. Blanchard, P., Mhanmdi, E.M.E., Guerraoui, R., Stainer, J.: Machine learning with adversaries: byzantine tolerant gradient descent. In: Proceedings 32th Annual Conference on Neural Information Processing Systems (NIPS 2017), pp. 119–129 (2017)
14. Yin, D., Chen, Y., Ramchandran, K., Bartlett, P.: Byzantine-robust distributed learning: towards optimal statistical rates. In: Proceedings 35th International Conference on Machine Learning (ICML 2018) (2018)
15. Shen, S., Tople, S., Saxena, P.: AUROR: defending against poisoning attacks in collaborative deep learning systems. In: Proceedings 32nd Annual Computer Security Applications Conference (ACSAC 2016), pp. 508–519 (2016)
16. Baracaldo, N., Chen, B., Ludwig, H., Safavi, J.A.: Mitigating poisoning attacks on machine learning models: a data provenance based approach. In: Proceedings 10th ACM Workshop on Artificial Intelligence and Security (AISec 2017), pp. 103–110 (2017)
17. Shokri, R., Shmatikov, V.: Privacy-preserving deep learning. In: Proceedings 22nd ACM Conference on Computer and Communications Security (CCS 2015), pp. 1310–1321 (2015)
18. Zhang, X., Felix, X.Y., Kumar, S., Chang, S.-F.: Learning spread-out local feature descriptors. In: Proceedings IEEE International Conference on Computer Vision (ICCV 2017), pp. 4595–4603 (2017)
19. Goodfellow, I., et al.: Generative adversarial nets. In: Proceedings 29th Annual Conference on Neural Information Processing Systems (NIPS 2014), pp. 2672–2680 (2014)
20. Pan, S., Hu, R., Long, G., Jiang, J., Yao, L., Zhang, C.: Adversarially regularized graph autoencoder for graph embedding. In: Proceedings 27th International Joint Conference on Artificial Intelligence (IJCAI 2018) (2018)
21. Bhagoji, A.N., Chakraborty, S., Mittal, P., Prateek, M., Calo, S.: Analyzing federated learning through an adversarial lens (2018). https://arxiv.org/abs/1811.12470
22. Wang, Z., Song, M., Zhang, Z., Song, Y., Wang, Q., Qi, H.: Beyond inferring class representatives: user-level privacy leakage from federated learning. In: Proceedings 38th Annual IEEE International Conference on Computer Communications (INFOCOM 2019) (2018)

23. Hitaj, B., Ateniese, G., Perez-Cruz, F.: Deep models under the GAN: information leakage from collaborative deep learning. In: Proceedings 2017 ACM SIGSAC Conference on Computer and Communications Security (CCS 2017), pp. 603–618 (2017)
24. LeCun, Y., Bottou, L., Bengio, Y., Haffner, P.: Gradient-based learning applied to document recognition. In: Proceedings of the IEEE, pp. 2278–2324 (1998)
25. Xiao, H., Rasul, K., Vollgraf, R.: Fashion-MNIST: a novel image dataset for benchmarking machine learning algorithms (2017). https://arxiv.org/abs/1708.07747

A Geo-indistinguishable Location Privacy Preservation Scheme for Location-Based Services in Vehicular Networks

Li Luo, Zhenzhen Han, Chuan Xu$^{(\boxtimes)}$, and Guofeng Zhao

College of Communication and Information Technology,
Chongqing University of Posts and Telecommunications, Chongqing, China
1573756270@qq.com, 952139237@qq.com, {xuchuan,zhaogf}@cqupt.edu.cn

Abstract. In vehicular networks, the location-based services (LBSs) are very popular and essential for most vehicular applications. However, large number of location information sharing may raise location privacy leakage of in-vehicle users. Since the existing privacy protection mechanisms ignore the trajectory information, so that the location privacy of in-vehicle users and the trade-off between privacy and quality of service (QoS) cannot be effectively solved. In order to provide satisfactory privacy and QoS for in-vehicle users, in this paper, we propose an improved geo-indistinguishable location privacy protection scheme (GLPPS). Specifically, we first select an area of service retrieval (ASR) instead of user's real location to send to LBS, which can protect location privacy while avoiding the leakage of trajectory. Secondly, we establish an income model based on Stackelberg between in-vehicle users and attackers, and design an IM-ASR algorithm based on the Iterative method and Maximin theorem, to achieve optimal trade-off between privacy and QoS. Finally, we prove that GLPPS satisfies α-differential privacy. Moreover, the simulation results demonstrate that GLPPS can reduce loss of QoS while improving location privacy compared to other methods.

Keywords: Vehicular network · Location-based services · Location privacy · Differential privacy · Geo-indistinguishable

1 Introduction

The vehicular network is one of the most typical applications in the field of Internet of Things (IoT). The rapid development of vehicular networks has expedited the development of a variety LBSs [14]. When a vehicle needs LBS [13], such as checking the traffic status of a place or searching for the surrounding parking lots, a vehicle needs to submit real location to service provider for information exchange. Although LBSs provide convenience for in-vehicle users, location information of users are exposed in the network, which may trigger personal hobbies, salary income, political inclination, and even life-threatening safety is speculated by attackers [3,9]. The location privacy issue is resolved to alleviate the worries

© Springer Nature Switzerland AG 2020
S. Wen et al. (Eds.): ICA3PP 2019, LNCS 11944, pp. 610–623, 2020.
https://doi.org/10.1007/978-3-030-38991-8_40

of in-vehicle users. Therefore, location privacy has become an urgent problem to be solved in vehicular networks.

Differential privacy is increasingly used to protect location privacy in vehicular networks [4,5,16], since it does not rely on any background knowledge of attackers and has strong privacy guarantees [7]. In these studies, the researchers suggest uploading the actual data of vehicle to a trusted third-party server, then uniformly adding noises to the original data to achieve the location privacy protection. The authors propose adding noises on some frequently accessed sensitive locations in the driving trajectory, thereby protecting location privacy [11,12,15]. Han et al. [8] propose an algorithm to achieve privacy protection by perturbing the original data with additive noise. The authors in [12] construct location search tree (LST) for time-series location data publication, in order to ensure the inherent relation among location data and add Laplace noise to the node of LST. Andréset et al. [2,10] propose a regional geo-indistinguishability model based on differential privacy. Minor changes to user's location will have little effect on the results of query. Therefore, the corresponding privacy protection level can be established according to the change of the user's location. The geo-indistinguishable mechanism indicates that when the distance between the actual positions $l1$ and $l2$ is within the range of the user's required distance, it means that $l1$ and $l2$ are indistinguishable.

Existing studies mainly have following flaws. According to the road constraint and user's movement mode, attackers can accurately infer the trajectory of user and then deduce the location information. Meanwhile, the trade-off between privacy and QoS does not provide satisfactory service to users. In terms of current issues, we design a GLPPS based on geo-indistinguishability for in-vehicle users, to improve the privacy level while ensuring QoS. Specifically, users select an appropriate ASR instead of real location to request service, which can avoid trajectory information leakage while protecting location privacy and satisfying QoS. Meanwhile, due to the prioritization of behavior between attackers and users, we establish a Stackelberg privacy game model and design an IM-ASR algorithm to obtain an ASR of the best utility.

The contributions of this paper are summarized as follows:

1. For the location privacy issue of LBS in vehicular networks, we propose an improved location perturbation scheme (GLPPS). Specifically, an ASR that satisfies service accuracy to replace actual location is selected by in-vehicle user to request service. Even if attackers gain the ASR, location of in-vehicle user cannot be found.
2. In order to protect location privacy while avoiding leakage of trajectory, ASR is sent by user to LBS for service request. Meanwhile, the number of vehicles and roads in the ASR are quantified as trajectory influence factors, which enhances privacy level of in-vehicle users.
3. To obtain the best trade-off between privacy and QoS, we introduce Stackerberg to establish a privacy game model, and design IM-ASR algorithm to improve privacy level while ensuring QoS.

The rest of the paper is organized as follows. Section 2 introduces the problem statement and the system model in vehicular networks. Section 3 introduces we propose a GLPPS to ensure QoS while improving privacy. Section 4 illustrates the simulations, along with the parameter settings, followed by the result analysis and discussions. Finally, Sect. 5 conclude this paper and outlines the next steps.

2 Problem Statement and System Model

In this section, at first, we focus on overview of LBS system and then analyze the problem in vehicular networks. After that, we establish system model of LBS in vehicular network.

2.1 Problem Statement

As shown in Fig. 1, our system primarily consists of in-vehicle users, the network infrastructure and the LBS providers.

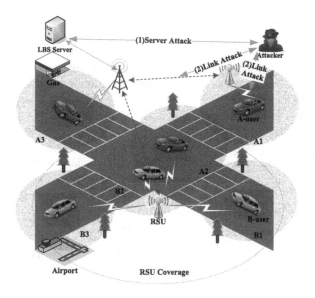

Fig. 1. LBS system of vehicular network.

If a vehicle is within the road side unit (RSU) coverage, the service request is sent to LBS through network infrastructure, such as RSU, base station (BS); otherwise, service request be sent directly to LBS provider over mobile network. There are two common attacks in this process, which are curious servers or RSUs and attackers who eavesdrop or intercept traffic channel information. After attackers obtain information of users, the location information is inferred by attackers, which can threaten personal privacy and even life safety.

Existing privacy mechanisms cannot avoid trajectory information leakage. Specifically, in combination with service request and road restrictions, attackers quickly infer trajectory of in-vehicle user and then decuce location information. As shown in Fig. 1, the destination of A and B is gas station and airport, respectively. Firstly, A and B upload $A_1 \rightarrow A_2 \rightarrow A_3$ and $B_1 \rightarrow B_2 \rightarrow B_3$ to LBS respectively for service request. Then, attackers can combine background knowledge and location information uploaded by user to infer the user trajectory information. Although A and B may appear at intersection at the same time to interfere attackers, attackers can still obtain the trajectory information of A by excluding B, since the destination of B is airport. Finally, location information of A is decuced.

2.2 System Model

In LBS system, due to the behaviors between in-vehicle users and attackers are orderly, we establish a Stackelberg privacy game model. In this model, in-vehicle users are regarded as leaders in our model, and attackers act as followers. Then, we add noise of Laplacian distribution in real location P_0 of user to generate a wrong location q. This privacy protection mechanism satisfies differential privacy [7] definition as shown the following.

Definition 1 (ε-Differential Privacy). *A randomized mechanism M is differentially private if and only if any two database D and D' contain at most one different record, and for any possible anonymized output $O \in Rang\,(M)$,*

$$Pr[M(D) = O] \le e^\varepsilon \times Pr[M(D\prime) = O] \tag{1}$$

where the probability is taken over randomness of M. ε is privacy budget and indicating the amount of noise added.

The game model requires leaders to determine the location F_i of service request. The probability that user selects location F_i at P_0 is defined as

$$L\,(F_i|p_0) = L\,(F_i = (x_i, y_i)\,|p_0 = (x_0, y_0)) \tag{2}$$

And followers optimize attack strategy based on background knowledge $\varphi\,(\cdot)$ after observing leader decision, ultimately determining the probability of speculative location $\tilde{p}$ is defined as follows

$$h(\tilde{p}|F_i) = \frac{f(F_i|\tilde{p})\varphi(\tilde{p})}{\sum_{p_0} f(F_i|p_0)\varphi(p_0)} \tag{3}$$

where $f\,(\cdot)$ is probability density function of Laplace.

We use the distance $d_{\tilde{p} \rightarrow p_0}\,(\tilde{p}, p_0)$ between the location guessed by attackers and the real location of users to measure the income of both players. When $d_{\tilde{p} \rightarrow p_0}\,(\tilde{p}, p_0)$ is larger, it indicates that income of users is larger; otherwise, the income of attackers is larger.

3   GLPPS: Geo-indistinguishable Location Privacy Protection Scheme

3.1   Overview of GLPPS

The distance distortion between the actual location and false location in many LBSs applications can not influence utility of service, as long as reported location meets accuracy requirements of users (i.e., the area of service interest (ASI) is included in the ASR in a certain proportion). For example, a user at A wants to know traffic condition within 2 km. If the user reports B that is 1 km away from A. At this time, LBS only needs to retrieve traffic conditions within 3 km around B can provide satisfactory service for user. Finally, the user can obtain desired information by filtering these results from LBS.

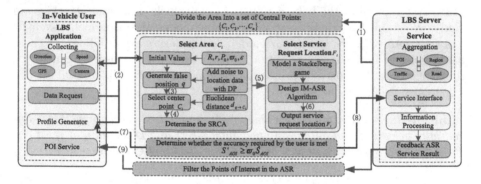

Fig. 2. Privacy protection scheme workflow of GLPPS.

Based on the above thought, we propose a GLPPS, and its workflow is shown in Fig. 2. In order to determine an optimal service request F_i, users first determine a service request candidate area (SRCA), where all points in the area can serve users. Then we establish a Stackelberg privacy model and design an IM-ASR algorithm to obtain a best solution. Finally, we use accuracy ϖ to ensure users can get satisfactory service. In addition, we use an example of Fig. 3 to explain in detail the workflow of Fig. 2. Specifically, our program includes the following steps:

- Firstly, LBS server releases a set of regional center points $C_1, C_2, \cdots, C_n$ for in-vehicle users (Step (1)), where $C_i = (x_i, y_i)$ denotes coordinates of center point of the i-th area are indicated. This example assumes that the area is divided into 3 cells by LBS (Fig. 2). The coordinates of these centers are $C_1 = (1,1)$, $C_2 = (3,1)$ and $C_3 = (5,1)$.
- Then, ASR is sent to LBS for making a service request. The ASR is determined to be mainly divided into the following 3 steps: (1) generating false location (Step (2)); (2) choosing a cell and determining SRCA (Step (4));

(3) selecting a center point and defining an ASR (Step (5)). In the Fig. 3, we suppose that location of user is $p_0 = (1.5, 1)$, $\varepsilon = 1$, the radius of user ASI is set to $r = 0.2$. By using Planar Laplacian, generating a false location is represented as $q = (2.3, 1.7)$. Since d_2 is the minimum, we use C_2 as the center point of SRCA and set $R = 1$. In SRCA, in order to improve privacy level while ensuring QoS, IM-ASR algorithm is used to obtain a center of ASR $F_i = (2.1, 1.2)$, and the radius of ASR is set to $R' = 1.5$. Thus, the F_i-centric ASR is reported to LBS for service.

- After selecting the ASR, user calculates in Step 6 whether service accuracy approximately satisfies the user requirement. If $\varpi \geq \varpi_0$, user performs Step 7 to re-adjust parameter to determine ASR; otherwise, it means that reported location can provide satisfactory service for user, so Step 8 is performed to upload result obtained at this time to LBS for service request.
- Finally, after the user reporting ASR, LBS retrieves information based on location and sends it to the user. The user filters all returned information and selects points of interests (POIs) (Step (9)).

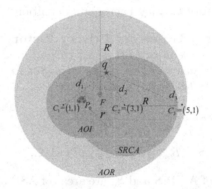

Fig. 3. Description of in-vehicle user's decision.

3.2   Problem Formulation

Our scheme GLPPS is proposed based on geo-indistinguishability, and geo-indistinguishability [2] definition is as follows:

Definition 2 (ε-Geo-Indistinguishability). *A mechanism M satisfies ε-geo-indistinguishability if and only if for all q, q' that satisfy $d(q, q') \leq r$:*

$$d_P(M(x), M(x')) \leq \varepsilon d_\chi(x, x')   x, x' \in \chi \qquad (4)$$

According to the geo-indistinguishability mechanism, the occurrence probability of obfuscated result obtained for a actrual location p_0 should decrease exponentially along the radius r around p_0. In linear space, obfuscated points can be gained with Laplace distribution. For the location based problem, dimension of location is two-dimensional, whose probability density function (pdf) is

Theorem 1 (Planar Laplacian [2]**).** *Given the parameter* $\varepsilon \in \Re^+$, *and the actual location* $x_0 \in \Re^2$, *the pdf of our noise mechanism, on any other point* $x \in \Re^2$, *is:*

$$f\left(x|x_0\right) = \frac{\varepsilon^2}{2\pi}e^{-\varepsilon d(x_0,x)} \tag{5}$$

where $\varepsilon^2/2\pi$ is a normalization factor. We call this function planar laplacian centered at x_0.

Privacy Degree of GLPPS. According to the Theorem 1, GLPPS defines probability that user selects F_i is an ASR center in p_0 as follows:

$$L(F_i|p_0) = \frac{\iint\limits_{F_i} \frac{\varepsilon^2}{2\pi}e^{-\varepsilon d(p_0,q)}\,dxdy}{\sum\limits_{i=1}^{N} \iint\limits_{F_i} \frac{\varepsilon^2}{2\pi}e^{-\varepsilon d(p_0,q)}\,dxdy} \quad (N=1,2,\ldots,N) \tag{6}$$

Location of in-vehicle user is replaced by the ASR. The trajectory privacy in ASR is γ_i-Area Trajectory Privacy Factor. Its definition is as follows:

Definition 3 (γ_i **- Area Trajectory Privacy Factor**). *User selects an ASR instead of real location to send to LBS, when current location of in-vehicle user is* p_0. *The number of vehicles and roads are* m *and* n *in the ASR, respectively. The trajectory impact factor of the ASR is defined as*

$$\gamma_i = \frac{-S\cdot n\cdot\sum_{\tilde{p}} h(\tilde{p}|F_i)\log h(\tilde{p}|F_i)}{S_0\cdot\log m} \tag{7}$$

$$S \in R^+, (n,m) \in Z^+, i = 1,2...N$$

where F_i is the center of ASR; S and S_0 are areas of ASR and LBS initialization domain, respectively. According to the conservation of traffic [6], the number of vehicles is basically constant within a certain time and area.

Based on the above analysis, user's privacy expectation can be defined as

$$\sum\nolimits_{p_0} \gamma_i\varphi(p_0)L(F_i|p_0)d_{\tilde{p}\to p_0}\left(\tilde{p},p_0\right) \tag{8}$$

According to the location privacy protection mechanism, the probability that F_i is selected by user is $L\left(F_i\right) = \sum_{p_0} \varphi\left(p_0\right)f\left(F_i|p_0\right)$. For simplicity, we can express privacy model that maximizes income of attackers as

$$x = \min_{\tilde{p}}\sum\nolimits_{p_0} \gamma_i\varphi(p_0)L(F_i|p_0)d_{\tilde{p}\to p_0}\left(\tilde{p},p_0\right) \tag{9}$$

To facilitate the calculation, the above formula is changed to a linear condition constraint formula as shown below,

$$x \leq \sum\nolimits_{p_0} \gamma_i\varphi(p_0)L(F_i|p_0)d_{\tilde{p}\to p_0}\left(\tilde{p},p_0\right) \tag{10}$$

When attacker adopts an optimal strategy, the best privacy expectation of user is defined as

$$\max \sum_{F_i} x \tag{11}$$

Based on the maximin Theorem [1], we can get the optimal trade-off of privacy between users and attackers.

Service Satisfaction of GLPPS. The service satisfaction of GLPPS is essentially QoS that user receives after requesting service.

- **Service Accuracy.** LBS works as follows: (1) vehicles submit their current locations to LBS servers; (2) servers return these results retrieved in ASR; (3) users filter the information and get a POI in ASI. The ideal service accuracy should be AOI are completely included in ASR, that is, accuracy value should be $\varpi = 1$. However, it is very difficult to achieve when the reported locations are perturbed. Therefore, we define user's service accuracy as follows:

Definition 4 (ϖ-Accuracy). *In-vehicle user obtains ϖ-accuracy is determined by the area that ASR is occupied by user's AOI. Service accuracy of user is defined as*

$$\varpi = \frac{S'_{ASI}}{S_{ASI}} \quad \varpi \in [0, 1] \tag{12}$$

where S_{ASI} is an area occupied by real location when user sends a request to LBS (the default initial ASI is completely included in ASR); S'_{ASI} is an area, which is ASI occupies ASR when user receives a service feedback from LBS.

In order to ensure that submitted ASR can provide users with satisfactory service accuracy, the following relationship will be satisfied.

$$\varpi_0 \leq \varpi \tag{13}$$

- **Quality of Service Loss.** If user selects ASR as service retrieval region, QoS depends on the false location F_i of GLPPS output. More similar actual and observed location are, the higher QoS is. Therefore, the mathematical definition of quality of service loss Q_{loss} is

$$Q_{loss} = \sum_{p_0, F_i} \varphi(p_0) L(F_i | p_0) d_{\tilde{p} \rightarrow p_0}(\tilde{p}, p_0) \tag{14}$$

According to the regional distribution, when Q_{loss} exceeds certain threshold Q_{loss}^{max}, user can get useless services. Therefore, in order to provide effective services for user, the relationship between Q_{loss}^{max} and Q_{loss} is as follows:

$$Q_{loss}(\varphi, L, d_{F_i \rightarrow p_0}) \leq Q_{loss}^{max} \tag{15}$$

We note the influence of threshold Q_{loss}^{max} that mainly depends on $d_{F_i \rightarrow p_0}$ and ϖ, hence it is also dependent on type of LBS that user is querying.

Privacy and QoS Trade-Offs. In summary, based on the provided symbols and assumptions, user's income is formulated as a constrained optimization problem and formally given as follows:

$$\max \sum_{F_i} x/Q_{loss} \tag{16}$$

$$
\begin{aligned}
s.t. \quad & d_{P_0 \rightarrow P}(p_0, p) \leq r \\
& \varpi_0 \leq \varpi \\
& x \leq \sum_{p_0} \gamma_i \varphi(p_0) L(F_i|p_0) d_{\tilde{p} \rightarrow p_0}(\tilde{p}, p_0) \\
& \sum_{p_0, F_i} \varpi \varphi(p_0) L(F_i|p_0) d_{F_i \rightarrow p_0}(F_i, p_0) \leq Q_{loss}^{\max} \\
& h(\tilde{p}|F_i) \geq 0, \ \forall F_i \\
& \sum_{\tilde{p}} h(\tilde{p}|F_i) = 1 \ \forall F_i
\end{aligned}
\tag{17}
$$

Equation (15) represents user's satisfaction while ensuring privacy and QoS; in Eq. (16), the constraint 1 indicates user's privacy level and interest areas; constraints 4 and 5 ensure that attackers can infer location by receiving service request location to maximize their income.

3.3   IM-ASR Algorithm

As we can see in Eq. (11), the privacy expectation is satisfied the maximin theorem. Therefore, we design an IM-ASR algorithm based on Iterative method and Maximin theory. The specific idea of IM-ASR algorithm is as follows.

Algorithm 1. IM-ASR Algorithm

Require: $R, R', r, P_0, q, \varepsilon, \varpi_0, \Delta = \{C_1, C_2, \ldots, C_n\}$
Ensure: F_x, F_y
1: $SRCA = (C_i, R) \leftarrow min\ d(q, \Delta)$
2: $ASR = (F_i, R') \leftarrow F_i = (F_x, F_y) \in SRCA$
3: **for** $F_i = (F_x, F_y) \in SRCA$ **do**
4:     Calculate the quality of service loss value Q_{loss}
5:     **if** $Q_{loss} \leq Q_{loss}^{max}$ **then**
6:         **for** $\hat{P} = (\hat{x}, \hat{y}) \in ASR$ **do**
7:             Calculate all x in ASR, choose the smallest
8:         **end for**
9:     **end if**
10:    Save all x in array A, $privacy = max\ A$
11:    Solve Equation (15) and save all incomes in array $B[\cdot]$
12:    Choose the largest user benefit $income(P_0) = max\ B$
13: **end for**
14: Output ASR corresponding to $income(P_0) = max\ B$
15: **if** $\varpi_0 \leq \varpi$ **then**
16:    output F_x, F_y;
17: **else**   The user resets the parameters
18: **end if**

In the lines 3–14, some points in SRCA are iterated to find the ASR center that improves privacy while ensuring QoS. In line 5, a threshold is set for obtaining a satisfactory QoS. From the lines 6th–8th, attackers set strategy to get an optimal income, after gets user request information. Based on the maximin theorem, we get user's maximum privacy expectations on the line 10. On the line 12, we find service point F_i that maximizes user's income. On the line 15, the accuracy is used to determine if the obtained F_i can provide satisfactory service to user. And if so, the location is reported; otherwise, user resets those parameters.

4   Evaluation

In this section, effectiveness of our proposed location protection mechanism is experimentally evaluated by simulations. The simulations are performed using MATLABr2018b in an ASUS computer with Intel Core i5-4590 (3.3 GHz) processor, 8 GB RAM, and Windows 10 operating system. We utilize the T-drive dataset [15] containing GPS trajectories of 10,357 taxis for simulation evaluation. In the dataset, there are approximately 15 million points, each sampled at 170 s intervals (average distance is approximately 620 m). The data set records a broad range of outdoor movements of users, which include shopping, sports, going to work, and going home, etc. Each trajectory is marked by a series of GPS points that contain users' ID, timestamp, and users' locations (latitude and longitude). They are the event-based simulations for our experiments.

We select the user data trajectory from February 2nd, 2008 to February 7th, 2008 as the training dataset to generate the attacker's background knowledge. The user data trajectory of February 8th, 2008 is used as the experimental data. Since there is no user query in this dataset, we simulate the actual conditions to generate a query for each point in the trajectory. In the experimental simulation, we set the distance between regional centers to $d_{C_i \to C_{i+1}} = 1$ km, $\varepsilon = 0.2$/km, user setting $r = 0.5$ km, $R = 1$ km, $R' = 2.5$ km. Moreover, we select 300 trajectories of in-vehicle users as training data set to generate background knowledge $\varphi(\cdot)$.

4.1   Proof on Privacy

Theorem 2. *The GLPPS can provide guaranteed protection for the location privacy of in-vehicle users, which means the scheme is to satisfy α-differential privacy, i.e.,*

$$L(F_i|p_0) \le e^\alpha L(F_i|p)$$

Proof. Combined with the Definition 1 and formula (7), we have that

$$\frac{L(F_i|p_0)}{L(F_i|p)} = \frac{\iint\limits_{ASR} \frac{\varepsilon^2}{2\pi} e^{-\varepsilon d(p_0,q)} dx dy}{\iint\limits_{ASR} \frac{\varepsilon^2}{2\pi} e^{-\varepsilon d(p,q)} dx dy}$$

$$\Rightarrow e^{-\varepsilon(d(p_0,q)-d(p,q))}$$

According to the triangle inequality theorem, their relationship is as follows

$$e^{-\varepsilon(d(p_0,q)-d(p,q))} \leq e^{-\varepsilon d(p_0,p)}$$

Since $d(p_0,p) \leq r$, it is easy to get the following inequality

$$\frac{L(F_i|p_0)}{L(F_i|p)} \leq e^{\alpha} \quad (\alpha = \varepsilon r)$$

According to the Definition 1, we have that GLPPS satisfies α-differential privacy. Therefore, it can provide privacy guarantee for users.

4.2   Trade-Off Between Privacy and QoS Loss

In order to guarantee QoS of in-vehicle user, we determine a threshold Q_{loss}^{max} of service quality loss by testing the relationship between the distance (i.e., user's real location and service request location) and the user's QoS.

As shown in Fig. 4, when $d(F_i, P_0) = 2.25\,\text{km}$, loss of user's QoS is maximized and the probability that user selects F_i as the service request location center is 1. When $d(F_i, P_0) = 2.5\,\text{km}$, user's location is on the ASR boundary and probability that for the area is selected as service request area is 0.5. As the $d_{F_i \to p_0}$ becomes larger, probability that F_i is selected to provide service to user gradually becomes smaller, and finally tends to 0. Therefore, when $d_{F_i \to p_0} \leq 2.25\,\text{km}$, i.e., $d_{F_i \to p_0}{}^{max}(F_i, P_0) = 0.9R'$, the QoS of user can be guaranteed.

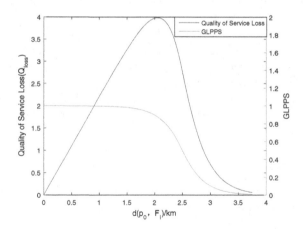

Fig. 4. The impact of the distance between reported location and real location of in-vehicle user on QoS

To prove the efficiency of GLPPS, we compare this scheme with existing privacy protection schemes [2,4] in terms of privacy and income of users. The assessment between privacy and loss of service quality is shown in Fig. 5. As the privacy of location increases, the QoS declines, which is consistent with actual

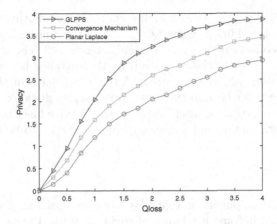

Fig. 5. Q_{loss} vs. *Privacy*: Relationship between QoS and privacy

situation. At the same time, since GLPPS considers trajectory privacy of in-vehicle users, it can be seen from comparison of the Fig. 5 under the same Q_{loss}, GLPPS has better privacy than the latest Convergence Mechanism and the most typical Planar Laplace.

The Fig. 6 indicates comparison results on trade-off between the privacy level and the loss of Q_{loss}. It can be seen from the Fig. 6 that GLPPS is significantly more profitable for users than Convergence Mechanism and Planar Laplace.

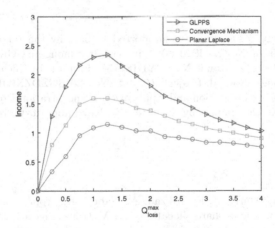

Fig. 6. Q_{loss}^{max} vs. *income*: Relationship between QoS thresholds and income of in-vehicle user

In the Fig. 6, the income of user is 0 at the beginning, because user's privacy is 0 at this time. With the increase of Q_{loss}^{max}, income's value of the in-vehicle user also rises. When Q_{loss}^{max} is close to 1.25, user makes use of GLPPS to obtain income is about 2.35, which is much higher than income of other privacy protection mechanisms. In addition, when $Q_{loss}^{max} > 1.25$, the income of the in-vehicle user gradually becomes smaller, and eventually tends to 0. There are two mainly reasons for the results. On the one hand, as the distance becomes larger, the probability that user selects the current ASR is smaller; On the other hand, when the distance is far beyond the user's best service distance, although ASR can ensure that the privacy of user is stable, this service has completely lost its meaning. Therefore, the in-vehicle user's income can tend to be 0.

5   Conclusion

With the rapid development of vehicular network systems, the location privacy protection has become hot issues of social concern. In this paper, we propose a GLPPS based on the geo-indistinguishable to protect location privacy and improve QoS. In addition, we utilize Stackelberg game to balance privacy between users and attackers, and design an IM-ASR algorithm based on Iterative method and maximin theory. Through compare with Convergence Mechanism and Planar Laplace, the performance of proposed location privacy protecting GLPPS is evaluated. Finally, we prove that GLPPS satisfies α-differential privacy. And effectiveness of proposed GLPPS are also verified through comparison experiments.

In the future, we will predict the relationship between the user's current location and the next moment based on the high mobility of the in-vehicle network users, and measure the amount of noise of the user to further protect the user's location privacy.

Acknowledgement. This work was supported in part by the Innovation Funds of Graduate PhD (Grant No. BYJS201803), in part by Chongqing Graduate Research and Innovation Project (Grant No. CYB18175), in part by Sichuan Major Science and Technology Special Project (Grant No. 2018GZDZX0014), in part by Chongqing Technology Innovation and Application Demonstration Project (Grant No. cstc2018jszx-cyzdX0120), in part by the 13th Five Key Laboratory Project (Grant No. 61422090301).

References

1. Amanatidis, G., Markakis, E., Nikzad, A., Saberi, A.: Approximation algorithms for computing maximin share allocations. ACM Trans. Algorithms (TALG) **13**(4), 52 (2017)
2. Andrés, M.E., Bordenabe, N.E., Chatzikokolakis, K., Palamidessi, C.: Geo-indistinguishability: differential privacy for location-based systems. arXiv preprint arXiv:1212.1984 (2012)
3. Chen, J., He, K., Yuan, Q., Chen, M., Du, R., Xiang, Y.: Blind filtering at third parties: an efficient privacy-preserving framework for location-based services. IEEE Trans. Mob. Comput. **17**(11), 2524–2535 (2018)

4. Chen, Z., Bao, X., Ying, Z., Liu, X., Zhong, H.: Differentially private location protection with continuous time stamps for VANETs. In: Vaidya, J., Li, J. (eds.) ICA3PP 2018. LNCS, vol. 11337, pp. 204–219. Springer, Cham (2018). https://doi.org/10.1007/978-3-030-05063-4_17
5. Chi, Z., Wang, Y., Huang, Y., Tong, X.: The novel location privacy-preserving CKD for mobile crowdsourcing systems. IEEE Access 6, 5678–5687 (2017)
6. Coclite, G.M., Garavello, M., Piccoli, B.: Traffic flow on a road network. SIAM J. Math. Anal. 36(6), 1862–1886 (2005)
7. Dwork, C.: Differential privacy. In: van Tilborg, H.C.A., Jajodia, S. (eds.) Encyclopedia of Cryptography and Security. Springer, Boston (2011). https://doi.org/10.1007/978-1-4419-5906-5
8. Han, S., Topcu, U., Pappas, G.J.: Differentially private distributed constrained optimization. IEEE Trans. Autom. Control 62(1), 50–64 (2016)
9. Hasrouny, H., Samhat, A.E., Bassil, C., Laouiti, A.: VANET security challenges and solutions: a survey. Veh. Commun. 7, 7–20 (2017)
10. Hua, J., Tong, W., Xu, F., Zhong, S.: A geo-indistinguishable location perturbation mechanism for location-based services supporting frequent queries. IEEE Trans. Inf. Forensics Secur. 13(5), 1155–1168 (2017)
11. Jiang, K., Shao, D., Bressan, S., Kister, T., Tan, K.L.: Publishing trajectories with differential privacy guarantees. In: Proceedings of the 25th International Conference on Scientific and Statistical Database Management, p. 12. ACM (2013)
12. Kang, H., Zhang, S., Jia, Q.: A method for time-series location data publication based on differential privacy. Wuhan Univ. J. Nat. Sci. 24(2), 107–115 (2019)
13. Wang, S., Hu, Q., Sun, Y., Huang, J.: Privacy preservation in location-based services. IEEE Commun. Mag. 56(3), 134–140 (2018)
14. Wang, X., et al.: A city-wide real-time traffic management system: enabling crowdsensing in social internet of vehicles. IEEE Commun. Mag. 56(9), 19–25 (2018)
15. Yuan, J., et al.: T-drive: driving directions based on taxi trajectories. In: Proceedings of the 18th SIGSPATIAL International Conference on Advances in Geographic Information Systems, pp. 99–108. ACM (2010)
16. Zhou, L., Yu, L., Du, S., Zhu, H., Chen, C.: Achieving differentially private location privacy in edge-assistant connected vehicles. IEEE Int. Things J. 6(3), 4472–4481 (2018)

A Behavior-Based Method for Distinguishing the Type of C&C Channel

Jianguo Jiang[1,2], Qilei Yin[1,2], Zhixin Shi[1,2(✉)], Guokun Xu[1,2], and Xiaoyu Kang[1,2]

[1] Institute of Information Engineering, Chinese Academy of Sciences, Beijing, China
{jiangjianguo,yinqilei,shizhixin,xuguokun,kangxiaoyu}@iie.ac.cn
[2] School of Cyber Security, University of Chinese Academy of Sciences, Beijing, China

Abstract. The botnet is one of the most dangerous threats to the Internet. The C&C channel is an essential characteristic of the botnet and has been evolved from the C&S type to the P2P type. Distinguishing the type of C&C channel correctly and timely, and then taking appropriate countermeasures are of great importance to eliminate botnet threats.

In this paper, we raise a behavior-based method to classify the type of C&C channel. In our method, we put forward a series of features relevant to C&C behavior, apply a feature selection approach to choose the most significant features, use the Random Forest algorithm to build an inference model, and make the final type determination based on the time slot results and their temporal relationship. The experimental result shows not merely that our method can distinguish the type of C&C channel with an accuracy of 100%, but also our feature selection approach can effectively reduce the number of required features and model training time while ensuring the highest accuracy, and then bring about significant improvements in method efficiency. Moreover, the comparison with another method further manifests the advantages of our work.

Keywords: Botnet analysis · Network security · C&C channel · Malware classification · Random Forest

1 Introduction

The botnet [22] is a network composed by many devices controlled by specific malware code (bot). It can be used to launch a series of destructive attacks such as the DDOS [17]. The Command and Control (C&C) channel is a unique and indispensable characteristic of botnet [9]. Either the botmaster or the bot requires it to keep the botnet alive.

Traditionally, the C&C channel is of the Client-Server (C&S) type, which means that the bots establish the C&C channel with one or several remote serves to get the latest control commands. However, this kind of C&C channel

© Springer Nature Switzerland AG 2020
S. Wen et al. (Eds.): ICA3PP 2019, LNCS 11944, pp. 624–636, 2020.
https://doi.org/10.1007/978-3-030-38991-8_41

may exposure the addresses of C&C servers and can be blocked by the common botnet countermeasure like blacklist or firewall easily. Currently, a certain number of new botnets adopt the Peer to Peer (P2P) based C&C channel [6,15,16], which delivers the control commands among bots through P2P communication topologies. It is robust to single point failure and can impair the effectiveness of traditional countermeasures. Thus, many researches put forward targeted countermeasures. For instance, Holz et al. [10] disable P2P botnet by sending fake commands made by benign content and the command keys. Besides, the attack against P2P network [7] can also be used to against the P2P based C&C channel. However, these targeted countermeasures typically consume more system and network resources. Also, they may interfere with the benign P2P traffic and applications. Hence, it is inappropriate and impractical to deploy these targeted countermeasures over a long time and use them to against any botnet. To eliminate the botnet threats, we should classify the type of C&C channels correctly and timely at first, and then adopt appropriate countermeasures.

To handle this problem, we propose a behavior-based method in this paper. It mainly adopts a series of behavior features and a Machine Learning method to infer the type of C&C channel accurately. As distinguishing the type of C&C channel is a binary classification problem (C&S or P2P), the P2P botnet detection method could also be utilized to solve it to some extent. However, compared with them, our work has three significant differences and merits. Firstly, it only needs the traffic of the individual bot and does not need us to monitor a collection of infected P2P bots or even a whole network simultaneously. Secondly, it can distinguish the C&C channel type of a long duration bot with complex activities, rather than just the maliciousness of single network flow or connection. Lastly and most importantly, we apply a feature selection approach to assess the behavior features and choose much smaller feature sets, which can also bring about the highest accuracy in different conditions. When deploying our work and other similar methods in the real world, the botnet traffic in large volume has to be converted into behavior feature vectors at first, and it is necessary to retrain or adjust the inference model regularly to learn the behavior patterns of latest training traffic. Hence, the number of behavior features and the model training time are essential to the method efficiency and the timeliness of deploying countermeasures. Also, there is no doubt that our selected small feature sets can consume much less time and resources for traffic profiling and inference model retraining, which means significant improvements in method efficiency.

To summarize, our main contributions are as follows:

- We propose a behavior-based method to distinguish the type of C&C channel from the aspect of the individual bot.
- In our method, we come up with 25 behavior features to profile the time slots of the individual bot, apply a feature selection approach to choose the most significant features, build a C&C channel type inference model using the Random Forest algorithm, and then determine the C&C channel type of each bot based on the slot results and their temporal relationship.

- We evaluated our method based on public datasets. The experimental result suggests that our method can distinguish the type of C&C channel with an accuracy of 100%. Also, it indicates that our feature selection approach is capable of obviously reducing the number of required features and model training time while ensuring the highest accuracy, and then brings about significant improvements in method efficiency.
- We also compared our work with another behavior-based P2P botnet detection method. The result shows that our method requires much less training time and behavior features to distinguish the type of C&C channel accurately and then manifests the efficiency advantage of our work.

The rest of the paper is organized as follows. We discuss related work, describe the details and perform the evaluation in Sects. 2, 3, and 4 respectively. Lastly, we conclude and plan the future works in Sect. 5.

2   Related Work

Based on the detection granularity, the traditional P2P botnet detection methods can be coarsely classified into two categories.

Group Bots based Method: This kind of method usually monitors a group of infected P2P bots simultaneously or even the whole network, and then utilizes the relationship between P2P bots to make the detection. For instance, Coskun et al. [5] show how to detect P2P bots in a network after one bot has been found. Also, it uses the pairwise mutual-contact relationships between pairs of bot peers. Nagaraja et al. [12] raise BotGrep, a content-agnostic algorithm to locate the bots in a network. It relies on the fast-mixing nature of the structured P2P botnet C&C graph and exploits the spatial relationships in communication traffic. Zhang et al. [18,19] extract statistical fingerprints from suspicious P2P hosts and utilize them to make the detection.

Dissimilarly, our method makes the C&C channel type classification based on the behavior of the individual bot, so that it does not need us to monitor many bots simultaneously or even a whole network at all and has better applicability.

Single Bot based Method: This kind of method directly profiles the behavior related to the individual bot (e.g., time slot, network flow, host) and uses their behavior patterns to make the judgment. For example, PeerRush [14] extracts a series of statistical features for each host and uses a binary classifier to check whether the traffic is generated by P2P applications, and uses many one-class classifiers to detect the real P2P bots. Alauthaman et al. [3] develop a P2P botnet detection approach based on the statistical features, and it adopts classical machine learning algorithms to detect the P2P botnet connections. BotSuer [11] filters out non-P2P traffic, aggregates P2P flows and then extracts statistical features from clusters to identify the P2P flows. Similarly, Zhao et al. [20,21] utilize 12 features and machine learning algorithms to detect P2P botnet traffic.

Compared with them, our work is not limited to check whether a single testing object belongs to P2P botnet. Since a long duration bot usually executed a series of complex activities, we divide the bot traffic into time slots, infer the

C&C channel type of each slot, and then determine the C&C channel type of each bot based on the slot results and their temporal relationship. Moreover, we utilize a feature selection approach to significantly reduce the training time of the type inference model as well as the number of required features.

3   Methodology

The main goal of our method is to distinguish the type of C&C channel from the aspect of the individual bot, and Fig. 1 provides an overview of it. As a long duration bot may perform a series of activities, there is a possibility that some behavior features related to specific activities are accumulated and then lead to an incorrect C&C type result. Hence, we divide the bot traffic into consecutive slots, infer the C&C channel type of each slot individually, and then determine the C&C channel type of the whole bot. Based on it, our method is consists of four steps, including **Behavior Profiling, Feature Selection, Slot Type Inference** and **C&C Channel Type Determination**. The first two steps are used to profile the bot traffic in each time slot, while the last two steps are responsible for inferring the slots type and determining the C&C channel type of testing bots.

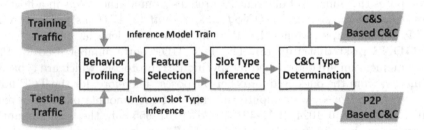

Fig. 1. The overview of our method.

3.1   Behavior Profiling

This step is responsible for profiling the bot traffic in each time slot. Since the bots may perform many diverse activities, our behavior features should be only related to C&C activity as much as possible. It means that only the bi-directional traffic should be analyzed as they can exchange the control commands and the status information between the botmaster and bot. Also, as the P2P protocols can generate the addresses of P2P peers, P2P communication does not need prior DNS activities in most cases. Thus, our behavior features are from the bi-directional traffic whose addresses do not appear in successful DNS responses to capture the difference between C&S and P2P based C&C activities. Specifically, assume T is the whole traffic of a bot, we split it into n consecutive time slots based on a time window W and describe it as:

$$T = (Class, W_1, W_2, \cdots, W_n) \tag{1}$$

where *Class* means the C&C channel type of T and W_i represents the traffic in the *i-th* time slot. For each W_i, we assemble its traffic into a network flows set called $Allflows_i$. Each flow can be identified by a 5-tuple (*srcip, dstip, protocol, srcport, dstport*), where the *srcip* is the IP address of the bot generating this flow and the *dstip* is that of the outside host communicating with this bot. If a flow from W_i meets the following conditions (1) Its *dstip* does not appear in any DNS response before this flow, (2) It contains bi-directional packets, we put it into a new flow set called $NODNSflows_i$. Then we use $NODNSflows_i$ to generate a feature vector V_i to profile W_i, which includes 25 behavior features in five categories:

- **NODNS communication ratio.** In each time slot, the more outside hosts interact with a bot through no-DNS and bidirectional flows, the bot is more likely to have a P2P based C&C activity. Hence, assume *count_ip* is a function that calculates the number of distinct *dstips* in a flow set, we compute $count\_ip(NODNSflows_i)/count\_ip(Allflows_i)$ as feature *F1* of V_i.
- **NODNS IP diversity.** The peers in the same P2P network usually scattered widely in the address space, which means that their IP prefixes may be diverse and then a P2P based C&C activity may contain relatively more different IP prefixes of outside hosts. Thus, assume *count_prefix* is a function that computes the number of different IP prefixes from all the *dstips* in a flow set, we regard $count\_prefix(NODNSflows_i)/count\_ip(Allflows_i)$ as feature *F2* of V_i. In this paper, we use the IP prefix with 16-bit length.
- **NODNS port distribution.** The TCP/UDP ports number used for P2P communication are determined by specific P2P protocols, which are typically different from the ones used for other purposes. Therefore, for the TCP flows in $NODNSflows_i$, we compute the ratios of the amounts of their *srcports* in three scopes: 0–1024, 1024–49151 and 49151–65535 to the total amount of *srcports* as three features *F3, F4, F5* of V_i. Then we calculated three features *F6, F7, F8* based on their *dstports* and another six features *F9, ..., F14* for all the UDP flows in the same way.
- **NODNS size distribution.** As the C&C traffic is generated by specific C&C protocols, the sizes of C&C flows may also follow particular patterns. Thus, we profile the flow size of $NODNSflows_i$ by computing the ratios of the amounts of flow sizes in five scopes: 1–512 Bytes, 512 Bytes-1 KB, 1–512 KB, 512 KB–1 MB and above 1 MB to the amount of all flows. For either the TCP or UDP flows in $NODNSflows_i$, we compute 5 features through this method respectively. Then, we represent these features as *F15, ..., F24* of V_i.
- **NODNS communication similarity.** As the bots in the same botnet send similar responses at a similar time [9], the traffic between P2P botnet peers also exhibit similarities to some extent. Thus, firstly we regard any two flows in $NODNSflows_i$ to be *similar* if they have the same pair of *protocol* and *srcport* attributes. Then, for each pair of attributes, we aggregate all its *similar* flows and count the number of their different *dstips*. Lastly, we compute the total amount of *dstips* for all attribute pairs containing more than one

different *dstip*, divide it by $count\_prefix(NODNSflows_i)$ and regard the result as a feature $F25$ of V_i.

With these features, we describe W_i as:

$$W_i = (Type, F_1, F_2, \cdots, F_{25}) \qquad (2)$$

where *Type* means the C&C channel type of this time slot. When W_i belongs to a bot traffic T whose C&C channel type is known, its *Type* is the same as the *Class* of T.

3.2 Feature Selection

Our 25 behavior features are based on several characteristics of C&C activity. However, extracting all the 25 features from the real world botnet traffic in large volume may consume a great many of time and system resources, so that it is a non-negligible shortage when we want to distinguish the C&C channel type of new bots efficiently and timely. Moreover, an inference model utilizing all the features may also require much longer training time. Thus, a smaller feature set is urgently needed, and we apply a feature selection approach to solve it. Specifically, we use the mutual information [2], a commonly used metric in feature evaluation, to assess the importance of each feature to the different C&C channel types based on the training traffic. Given two jointly discrete random variables X and Y, their mutual information is defined by Eq. 3:

$$I(X,Y) = \sum_{y \in Y} \sum_{x \in X} p(x,y) \, log(\frac{p(x,y)}{p(x)\,p(y)}) \qquad (3)$$

where $p(x,y)$ is the joint probability function of X and Y, while $p(x)$ and $p(y)$ are the marginal probability functions of X and Y respectively. Then, with the importance of each feature, we choose the most significant K features to assemble the new feature set.

3.3 Slot Type Inference

To infer the C&C channel type of the unknown time slots, we adopt a Machine Learning method, which can learn the behavior patterns of training time slots automatically and then use them to make the prediction. Specifically, we choose the Random Forest (RF) [4] algorithm. RF is a kind of ensemble classifier consisting of a collection of tree-structured base classifiers. When building an RF-based model composed by L base classifiers and the total number of training instances is V, we randomly sample V instances with replacement from the original training set to train a base classifier, and repeat this step for L times until we have trained each base classifier. The building of the base classifier is also different from the original growing of Decision Tree. When splitting a node of a Decision Tree model, we randomly select c features from the full d features belong to this node and then choose the best splitting feature from the new c features. Lastly, the label of a testing instance is a combination of the prediction results of base classifiers (e.g., The majority voting). Due to the random instances and features selection, the RF algorithm has a strong generalization ability and excellent performance. Moreover, it has the advantages of ease of implementation and low computational overhead. With a certain number of training slots whose C&C channel type is known, we profile them by our selected features, use the features vectors as the input of the RF to train an inference model. Then, it can be used to infer the C&C channel type of testing slots.

3.4   C&C Channel Type Determination

After inferring the C&C channel types of testing slots, we determine the C&C channel type of the whole bot, and it is based on two common characteristics of C&C activity including (1) The bots and C&C servers have to communicate constantly to keep botnet alive, which means the C&C activities should cover most of the life cycle of a bot. (2) The temporal interval between two consecutive C&C activities should be relatively small to ensure the timeliness of botnet commands. Thus, we search for the time slots in the same C&C channel type while having small temporal intervals, which may represent the consecutive C&C activities. Then we regard the sum of their temporal intervals as their coverage rate to the bot life cycle. Lastly, we label the C&C channel type of bot to be the type having a higher coverage rate.

Formally, assume Tarray = (Type_1, Type_2, ..., Type_n) is an array composed by the *Type* attributes of all the time slots of a bot trace, we create a sub-array P2Parray = (P2Pslot_1, P2Pslot_2, ..., P2Pslot_m) where P2Pslot_j ($1 \leq j \leq m$) is the temporal number of the *j-th Type* equalling P2P, to represent the time slots exhibiting P2P behavior. Also, we make a sub-array CSarray = (CSslot_1, CSslot_2, ..., CSslot_o) where CSslot_k ($1 \leq k \leq o$) is the temporal number of the *k-th Type* equalling C&S, to represent the time slots showing C&S behavior. Then we define two values *PSUM* and *CSUM* to calculate their coverage rates to the bot life cycle:

$$PSUM = \sum_{1 \leq p < q \leq m} interval\_check(P2Pslot_q - P2Pslot_p, c\_th) \qquad (4)$$

$$CSUM = \sum_{1 \leq r < u \leq o} interval\_check(CSslot_u - CSslot_r, c\_th) \qquad (5)$$

$$interval\_check(a, b) = \begin{cases} a, a < b \\ 0, else \end{cases} \qquad (6)$$

where *c_th* is a threshold that checking whether two adjacent temporal numbers are consecutive in C&C activity and we empirically set it to be 2. If $PSUM > CSUM$, we label the C&C channel type of this bot as P2P, otherwise we label it as C&S. If two values are equal, we label it to the type of the sub-array containing more temporal numbers.

4 Evaluation

4.1 Dataset

We created our evaluation dataset from several public botnet datasets including:

- **CTU-13 Dataset** [8]. It was captured in the CTU University, Czech Republic, in 2011, and has been widely used in many researches. We chose the CTU-13-1, CTU-13-2, CTU-13-3, CTU-13-9, CTU-13-12, and CTU-13-13 scenarios as their long duration time and the various network activities. We also split the scenarios (CTU-13-9 and CTU-13-12) mixed by multiple bots into sub-scenarios based on the address of each bot.
- **Stratosphere IPS Project Dataset** [1]. It was an online dataset containing more than 300 botnet scenarios. We chose the CTU-126-1, CTU-140-1, CTU-140-2, and CTU-66 scenarios as they had detailed C&C information.
- **PeerRush Dataset.** We also chose the P2P botnet dataset raised in [14]. And we used the Zeus scenarios in it as they preserved complete DNS records, which are indispensable to our behavior features. Besides, we removed the Zeus-2_3 scenario as its temporal duration was too short.

To create a training set and a testing set for evaluation, we split the scenarios having only one bot into two sub-scenarios with equal duration and put one in each set. For the scenarios containing sub-scenarios, we put half of the sub-scenarios into the training set and the others into the testing set. We show their information in Table 1. Besides, we filtered out some traffic for all traces such as the internal, DNS, DHCP traffic as they were harmless.

Table 1. The information of our evaluation dataset

C&C	Training set	Testing set	Size	Scenario	Family
C&S	CTU-13-1_train	CTU-13-1_test	56 MB	CTU-13-1	Neris
	CTU-13-2_train	CTU-13-2_test	35 MB	CTU-13-2	
	CTU-13-9_165, CTU-13-9_191 CTU-13-9_192, CTU-13-9_193 CTU-13-9_204	CTU-13-9_205, CTU-13-9_206 CTU-13-9_207, CTU-13-9_208 CTU-13-9_209	1.1 GB	CTU-13-9	
	CTU-13-3_train	CTU-13-3_test	123 MB	CTU-13-3	Rbot
	CTU-13-13_train	CTU-13-13_test	110 MB	CTU-13-13	Virut
	CTU-126-1_train	CTU-126-1_test	1.2 GB	CTU-126-1	Geodo
	CTU-140-1	CTU-140-2	503 MB	CTU-140	Bunitu
P2P	CTU-13-12_165, CTU-13-12_191	CTU-13-12_192	282 MB	CTU-13-12	NSIS.ay
	CTU-66_train	CTU-66_test	395 MB	CTU-66	Sality
	Zeus-1_1, Zeus-1_2, Zeus-1_3 Zeus-1_4, Zeus-1_5, Zeus-1_6	Zeus-1_7, Zeus-1_8, Zeus-1_9 Zeus-1_10, Zeus-1_11, Zeus-1_12	51 MB	Zeus-1	Zeus
	Zeus-2_1, Zeus-2_2, Zeus-2_4 Zeus-2_5, Zeus-2_6, Zeus-2_7 Zeus-2_8	Zeus-2_9, Zeus-2_10, Zeus-2_11 Zeus-2_12, Zeus-2_13, Zeus-2_14 Zeus-2_15	54 MB	Zeus-2	

4.2   Experiments Method

We implemented our method based on Scikit-learn [13], and we defined the accuracy of our approach as the number the testing bot traces whose C&C channel type was correctly predicted over the total amount of testing bot traces. As the time window W was a pre-defined parameter, we tested four different values including 2000, 3000, 4000, and 5000 s to assess the ability of our approach. For each time window, we changed K, which determined the number of significant features selected, from 1 to 25 at an interval of 1, built the corresponding type inference model based on the training set and then made the C&C channel type determination on testing set individually. Therefore, on the premise of the highest accuracy, we could achieve the smallest feature set and the set having the fastest model training time.

4.3   Experimental Result and Analysis

Features Ranking. Before choosing the significant features, it was necessary to rank the importance of the total 25 behavior features based on their mutual information. We listed the features importance ranking in Table 2 under different time window values, while the most important features were in the left side. Although their internal orders were different, it could be seen that the features: F1 (The NODNS communication ratio), F2 (The NODNS IP diversity), F13 (The ratio of UDP dstport in the range 1024–49151), F15 (The ratio of TCP size in the range 1–512 Bytes), and F20 (The ratio of UDP size in the range 1–512 Bytes) were always the five most important features for all the conditions. These five critical features were from four different feature categories and indicated that the characteristics behind them were necessary to distinguish different types of C&C Channel. The F25 (The NODNS communication similarity), which represented the fifth feature category, had relatively less importance under small time window values. However, its importance order also showed an increasing tendency as the time window becoming larger.

Table 2. The importance ranking for our features.

Time window	The ranking list
2000 s	F2,F1,F13,F20,F15,F16,F7,F11,F17,F25,F6,F10,F21,F14,F22,F19,F8,F18,F4,F3,F5,F24,F23,F12,F9
3000 s	F13,F2,F1,F20,F15,F16,F17,F11,F7,F25,F6,F10,F21,F14,F22,F19,F18,F8,F5,F4,F3,F24,F23,F12,F9
4000 s	F2,F1,F13,F20,F15,F16,F17,F7,F25,F6,F11,F14,F21,F10,F22,F19,F5,F18,F8,F4,F3,F24,F23,F12,F9
5000 s	F20,F13,F1,F2,F15,F25,F6,F16,F17,F7,F11,F14,F21,F10,F22,F5,F19,F18,F8,F4,F3,F24,F23,F12,F9

C&C Channel Type Distinction. With these importance ranking lists, we evaluated our method under different time window and K values and showed the results in Fig. 2. From it, we observed and concluded the following issues.

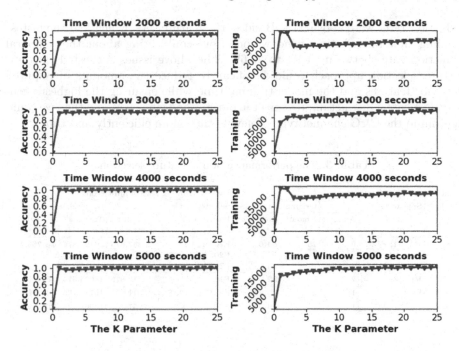

Fig. 2. The overall performance of our method.

Firstly, our method achieved the C&C channel type classification accuracy of 100% for all the time window values and most of the K values, which not only proved the effectiveness of our method, but also indicated that our features were able to mine the behavior difference between two types of C&C channels in detail and then made an accurate distinction.

Secondly, in some cases, a small number of chosen features may not bring about a short training time (e.g., K is 1 and 2 for time window 2000 s and 4000 s). It was mainly led by a fact that sometimes a Random Forest based inference model utilizing only a small amount of features has to build more complicated decision trees to split the training instances from different categories well so that it required more training time. Except for these cases, the training time generally raised as the number of chosen features increased.

Lastly, our feature selection approach was able to select smaller and quicker feature sets than the full behavior feature set. On the premise of achieving the highest accuracy (100%), we listed the information about the smallest feature sets and the ones having the fastest model training time in Table 3. Under the smallest time window (2000 s), the traffic volume and the C&C activities in each time slot were relatively less, and then we required more features to profile their behavior difference. Hence, we still needed 9 most significant features to achieve the highest accuracy, and the reduction rate of the feature amount was 64%. By contrast, under larger time windows, only 1 or 2 most important features were enough to classify the type of C&C channel accurately, and the corresponding reduction rates were above 90%. Besides, the feature sets, which had the fastest training time in each time window condition, could bring about the temporal reduction rates between 14.1% to 20.6%. The above issues suggested that our feature selection approach could significantly reduce the amounts of required behavior features and the model training time while ensuring the highest accuracy. It meant a distinct improvement in our method efficiency, and then we could determine the C&C channel type of new botnet more efficiently and timely.

Table 3. The performance of our feature selection

Time window	Smallest feature set				Fastest training set (in millisecond)				
	Selected set		Original set		Selected set		Original set		
	Features	Amount	Amount	Reduction rate	Features	Training time	Training time	Reduction rate
2000 s	F2, F1, F13 F20, F15, F16 F7, F11, F17	9	25	64.0%	F2, F1, F13 F20, F15, F16 F7, F11, F17	22347.875	26192.875	14.7%
3000 s	F13, F2	2	25	92.0%	F13, F2	14713.875	18025.375	18.4%
4000 s	F2	1	25	96.0%	F2, F1, F13, F20	13511.625	15731.125	14.1%
5000 s	F20	1	25	96.0%	F20	11957.75	15060.125	20.6%

4.4   Comparison with PeerRush

To further manifest the advantage of our work, we compared it with another P2P botnet detection method PeerRush [14]. We chose it because it also relied on the traffic of the individual bot, and we used its evaluation dataset. PeerRush utilized 13 behavior features in three categories to profile the behavior of each time window and detected P2P hosts through a traditional machine learning algorithm. To compare it with our method in the same conditions, we also changed its machine learning algorithm to be the Random Forest, trained it through our training set, and evaluated its performance on our testing set. Then we show the comparison result in Table 4 (Training time is in Millisecond).

Table 4. The comparison result

Time window	Our method (smallest feature set)			Our method (fastest feature set)			PeerRush		
	Accuracy	Feature amount	Training time	Accuracy	Feature amount	Training time	Accuracy	Feature amount	Training time
2000 s	100%	9	22347.875	100%	9	22347.875	100%	13	38462.75
3000 s	100%	2	14713.875	100%	2	14713.875	100%	13	27656.125
4000 s	100%	1	17777.125	100%	4	13511.625	100%	13	23045.375
5000 s	100%	1	11957.75	100%	1	11957.75	100%	13	20870.25

From the result, we could see that PeerRush was able to achieve the accuracy of 100% in each time window values. Thus, the PeerRush had the same ability in the C&C channel type classification with our method. However, our approach was able to achieve the highest accuracy with much fewer behavior features.

Moreover, the model training time based on our selected feature sets was much shorter than that of PeerRush. When the time window is 3000 s, the PeerRush required about 27656.125ms to train its Random Forest Model while our selected feature sets only consumed about 14713.875ms, and then we got the highest temporal reduction rate of about 46.8%. Besides, the temporal reduction rates under other time window values were in the range of about 22.9% to 42.7%. To sum up, compared with PeerRush, our method successfully chose the essential features in classifying C&C channel types and then significantly reduced the model training time as well as the number of required features. Thus, our method has better efficiency than PeerRush.

5   Conclusion

In this paper, to distinguish the type of C&C channel accurately, we propose a behavior-based method. In our method, we put forward 25 behavior features in 5 categories to profile the time slots of the individual bot, utilize a feature selection approach to choose the critical features, build a slot type inference model using the Random Forest algorithm, and then determine the C&C channel type of each bot based on the slot results and their temporal relationship. The evaluation result shows that our work can achieve an accuracy of 100%, and our feature selection process is capable of significantly reducing the model training time as well as the number of required features while ensuring the highest accuracy. Moreover, the comparison result further suggests the efficiency advantage of our work to the PeerRush approach.

In the near future, we are going to adopt the online machine learning algorithms. We believe they will enable our method to learn the behavior patterns of new training traffic automatically and then bring about better applicability in large-scale networks.

Acknowledgement. We thank the anonymous reviewers for their invaluable feedback. Our work was supported by the National Key R&D Program of China grant no. 2017YFB0801900.

References

1. Malware capture facility project. https://stratosphereips.org/category/dataset.html
2. Mutual information. https://en.wikipedia.org/wiki/Mutual_information
3. Alauthaman, M., Aslam, N., Zhang, L., Alasem, R., Hossain, M.A.: A P2P botnet detection scheme based on decision tree and adaptive multilayer neural networks. Neural Comput. Appl. **29**, 1–14 (2016)
4. Breiman, L.: Random forests. Mach. Learn. **45**(1), 5–32 (2001)
5. Coskun, B., Dietrich, S., Memon, N.D.: Friends of an enemy: identifying local members of peer-to-peer botnets using mutual contacts, pp. 131–140 (2010)
6. Dittrich, D., Dietrich, S.: P2P as botnet command and control: a deeper insight, pp. 41–48 (2008)

7. Douceur, J.R.: The sybil attack. In: Druschel, P., Kaashoek, F., Rowstron, A. (eds.) IPTPS 2002. LNCS, vol. 2429, pp. 251–260. Springer, Heidelberg (2002). https://doi.org/10.1007/3-540-45748-8_24
8. Grill, M., Stiborek, J., Zunino, A.: An empirical comparison of botnet detection methods. Comput. Secur. **45**, 100–123 (2014)
9. Gu, G., Zhang, J., Lee, W.: BotSniffer: detecting botnet command and control channels in network traffic. In: Network and Distributed System Security Symposium (2008)
10. Holz, T., Steiner, M., Dahl, F., Biersack, E.W., Freiling, F.C.: Measurements and mitigation of peer-to-peer-based botnets: a case study on storm worm, p. 9 (2008)
11. Kheir, N., Wolley, C.: BotSuer: suing stealthy P2P bots in network traffic through netflow analysis. In: Abdalla, M., Nita-Rotaru, C., Dahab, R. (eds.) CANS 2013. LNCS, vol. 8257, pp. 162–178. Springer, Cham (2013). https://doi.org/10.1007/978-3-319-02937-5_9
12. Nagaraja, S., Mittal, P., Hong, C.Y., Caesar, M., Borisov, N.: BotGrep: finding P2P bots with structured graph analysis. In: USENIX Conference on Security, p. 7 (2010)
13. Pedregosa, F., et al.: Scikit-learn: machine learning in python. J. Mach. Learn. Res. **12**(10), 2825–2830 (2013)
14. Rahbarinia, B., Perdisci, R., Lanzi, A., Li, K.: PeerRush: mining for unwanted P2P traffic. In: Rieck, K., Stewin, P., Seifert, J.-P. (eds.) DIMVA 2013. LNCS, vol. 7967, pp. 62–82. Springer, Heidelberg (2013). https://doi.org/10.1007/978-3-642-39235-1_4
15. Sinclair, G., Nunnery, C., Kang, B.H.: The waledac protocol: the how and why. In: International Conference on Malicious and Unwanted Software, pp. 69–77 (2009)
16. Wang, P., Sparks, S., Zou, C.C.: An advanced hybrid peer-to-peer botnet. IEEE Trans. Dependable Secure Comput. **7**(2), 113–127 (2010)
17. Zargar, S.T., Joshi, J.B.D., Tipper, D.: A survey of defense mechanisms against distributed denial of service (DDOS) flooding attacks. IEEE Commun. Surv. Tutor. **15**(4), 2046–2069 (2013)
18. Zhang, J., Perdisci, R., Lee, W., Luo, X., Sarfraz, U.: Building a scalable system for stealthy P2P-botnet detection. IEEE Trans. Inf. Forensics Secur. **9**(1), 27–38 (2014)
19. Zhang, J., Perdisci, R., Lee, W., Sarfraz, U., Luo, X.: Detecting stealthy P2P botnets using statistical traffic fingerprints. In: IEEE/IFIP International Conference on Dependable Systems & Networks, pp. 121–132 (2011)
20. Zhao, D., Traore, I., Ghorbani, A., Sayed, B., Saad, S., Lu, W.: Peer to peer botnet detection based on flow intervals. In: Gritzalis, D., Furnell, S., Theoharidou, M. (eds.) SEC 2012. IAICT, vol. 376, pp. 87–102. Springer, Heidelberg (2012). https://doi.org/10.1007/978-3-642-30436-1_8
21. Zhao, D., et al.: Botnet detection based on traffic behavior analysis and flow intervals. Comput. Secur. **39**(4), 2–16 (2013)
22. Zhu, Z., Lu, G., Chen, Y., Fu, Z.J., Roberts, P., Han, K.: Botnet research survey, pp. 967–972 (2008)

IoT and CPS Computing

Sparse Representation for Device-Free Human Detection and Localization with COTS RFID

Weiqing Huang[1,2,3], Shaoyi Zhu[2,3(✉)], Siye Wang[1,2,3(✉)], Jinxing Xie[2,3], and Yanfang Zhang[2,3]

[1] School of Computer and Information Technology, Beijing Jiaotong University, Beijing, China
[2] Institute of Information Engineering, Chinese Academy of Sciences, Beijing, China
{huangweiqing,zhushaoyi,wangsiye,xiejinxing,zhangyanfang}@iie.ac.cn
[3] School of Cyber Security, University of Chinese Academy of Sciences, Beijing, China

Abstract. Passive human detection and localization is the basis for a broad range of intelligent scenarios including unmanned supermarket, health monitoring, etc. Existing computer vision or wearable sensor based methods though can obtain high precision, they still face some inherent defects, such as privacy issues, battery power limitations. Based on the human movement induced backscattered signal changes, we propose a device-free human detection and localization system on radio-frequency identification (RFID) devices. The system extracts environment-independent features from both RSSI and phase for dynamic monitoring in the first stage, then the target is further located if the moving human is detected. In particular, an overcomplete dictionary is learned when creating the fingerprint library, which helps to make the representation of the location more compact and computationally simple. Moreover, PCA based dimensionality reduction method is then adopted to acquire valid features to determine the final position. Extensive experiments conducted in real-life office and bedroom demonstrate that the proposed system provides high accuracy for human detection and achieves the average distance error of less than 1 m.

Keywords: Device-free · Human detection · Indoor localization · Sparse representation · RFID

1 Introduction

Recently, the popularity of ubiquitous applications such as retail analytics and remote health monitoring has made the intelligent sensing technology a hot topic, the foundation of which is the device-free passive human detection and localization [15]. Take Amazon's first conceptual unmanned supermarket "Amazon Go" as an example, customers can complete the payment silently at the moment of

© Springer Nature Switzerland AG 2020
S. Wen et al. (Eds.): ICA3PP 2019, LNCS 11944, pp. 639–654, 2020.
https://doi.org/10.1007/978-3-030-38991-8_42

leaving without any device or assistance due to the radio frequency identification (RFID) technology. In this case, each product is attached with a unique RFID tag and can be registered automatically when passing through the RFID portal deployed at the exit [30]. Moreover, the real-time interest analysis of products can be performed through the localization of human in this RFID-based "smart space", which is beneficial for the supermarket to adjust marketing strategies.

Many efforts have been made to perceive the existence of a human and then sense his location or activity with cameras, wearable sensors or RF-signals. Computer vision based approaches [16,28] have made significant progress due to the great improvement of computer performance and the development of image processing techniques. However, several challenges [25] such as reliability issue [light and line-of -sight (LOS) restrictions], and the privacy issue (even more serious in sensitive areas) are still left to be solve. Location awareness and activity recognition from wearable or wireless sensors are relatively mature through past few years. This type of method requires human to carry at least one sensor (e.g. smart phone [21,26], Bluetooth device [5,13], RFID tag [9,20]) with him. Obviously, the disadvantages of device-based methods are mainly reflected in the two aspects. Firstly, it is difficult to guarantee that the human can wear the device all the time, which is uncomfortable and considered intrusive in terms of body, especially for the elderly. Secondly, battery power is another restriction for the widespread use of such methods, as most devices require continuous power supply except passive RFID tags. In view of the above defects, researchers began to study the possibility of sensing human's location and activity through a device-free approach, which is achieved by monitoring signal changes in the surrounding environment induced by the human. With the ubiquitous deployment of wireless networks, most of aforementioned works concentrated on working with Channel State Information (CSI) in WiFi [24]. Though providing subcarrier-level channel measurements for fine-grained gesture recognition [3,7], positioning [6,14] and crowd counting [8,22], this kind of parameter can only be obtained by modifying NIC, which is not universal and convenient enough. In the use of RFID technology for device-free perception, existing research focuses on activity and gesture recognition [17], few of them have explored its potentials in passive indoor localization.

In this paper, we attempt to propose a device-free system from human detection to localization in practical indoor environment on Commercial-Off-The-Shelf (COTS) RFID devices, exploring full signal information including Received Signal Strength Indication (RSSI) and phase. To achieve this, we stick the RFID tag array on the wall and place the antenna array in the appropriate place to form multiple antenna-tag links. After the RFID-based "smart space" is established, four novel features both from RSSI and phase are first extracted from adjacent group readings to characterize the stability of the signal, and then a cutting line can be found for distinguishing different situations (with human or without human) through support Vector Machine (SVM). When the moving human is detected and needs to be further located, we adopt sparse representation based method to predigest the links that less affected by human occlusion and reflection and make the signal matrix more compact and computationally simple. Particularly, an overcomplete dictionary is learned when creating the fingerprint library in pre-defined reference points, the sparse coefficients of the signal at a specific location can be presented via the dictionary. Additionally, to further reduce the computational complexity, we use Principal Component Analysis (PCA) algorithm to extract the most discriminative features and compare several Nearest Neighbor Classifier (NN) based algorithms for final localization. To validate our design, we prototype the system on COTS RFID devices in two scenarios, the results demonstrate that the system achieves nearly 100% accuracy of human presence and achieves the average distance error of less than 1 m.

The contributions of our work are summarized as follows:

- We build a framework from human detection to indoor localization in a passive way via RFID signals. In particular, sensitive and effective features of multiple antenna-tag links are selected to characterize the shadowing and reflection effect of human appearance, which are highly robust regardless of environmental changes.
- We propose a dictionary-based sparse representation method together with PCA dimensionality reduction for the compressive sensing of the signals in specific locations. Compared with the original one, the signal matrix expressed in this way is more compact, highly focused and faster to process.
- We design and implement the system with COTS RFID devices in real-life office and bedroom settings. Benefit from the novel features extracted from RSSI and phase, the system is capable of detection moving humans with high accuracy. Moreover, experiments on the different environments and different tag densities show that the system maintains robustness and effectiveness of detection and localization after the number of tags reaches a certain level.

The rest of this paper is organized as follows. The previous research is reviewed in Sect. 2, followed by the system overview in Sect. 3. The detailed methodology of the system is described in Sect. 4. We report the experimental results in Sect. 5 and conclude the paper in Sect. 6.

2   Related Work

Ubiquitous human detection, localization and activity recognition have drawn much attention owing to the rapid development of emerging smart applications. The existing solutions can be roughly divided into the following two categories according to the necessity of wearing additional devices: device-based and device-free.

Device-Based. Additional device such as smart phone, Bluetooth beacon, RFID tag need to be carried on the body to complete real-time monitoring in this kind of approaches. In [26], cell phone based pedestrian dead reckoning (PDR) was combined with WiFi localization through unscented Kalman filters (UKF) to estimate the final location, which improves performance a lot in terms of reliability and localization accuracy. AcMu [21] was proposed to self-update the radio map automatically and continuously for wireless indoor localization with cell phones carried with human. In addition, accelerometer and gyroscope could used to support activity recognition in daily life or remote healthy monitoring [11,27]. Since the dense deployment of IoT devices, [5,13] complete indoor localization by means of RSSI fingerprint reported by Bluetooth Low Energy 4.0 (BLE4.0) beacon. Conducive to RFID technology, a multi-direction weight position Kalman Filter [29] was proposed to combine the RSSI data from four different directions for accurate localization. RF-Brush [9], RF-IDraw [20], RF-Dial [4], RF-Kinect [18], RF-ECG [19] were designed to transform a pen, finger or even the whole body to an intelligent HCI device for movement tracking or heartbeat sensing. However, all of the above methods require human to carry an extra device anytime and anywhere, which is a cumbersome task, especially for the elderly and children.

Device-Free. Considering the inherent defects of the device-based schemes, more researchers have begun to seek solutions in a non-intrusive way. Recently, fine-grained CSI in WiFi has been highly studied since it can provide subcarrier-level channel information for target tracking and subtle gesture recognition. For example, Widar [6] and Widar 2.0 [14] were proposed in succession for passive human tracking, which achieves decimeter-level localization with AoA, TF and Doppler shift. Gesture recognition [3,7] was also studied using machine learning or radio image based methods. Although the information provided by CSI is sufficiently fine-grained, it can not be obtained directly from commercial WiFi device since the NIC needs to be modified. As the first step of using RFID for device-free AR/VR system, [17] extracted the hand reflection part from the mixed signal and complete finger tracking and gesture recognition by WKNN and CNN.

3   Overview

3.1   Smart Space Creation

As shown in Fig. 1, an array of the RFID tags is pasted onto three walls at a certain height from the ground with an antenna array places relatively to it. Thus, an RFID-equipped "smart space" is created with multiple antenna-tag links. Suppose there are A antennas with T tags communicate normally with each other, then $L = A * T$ links are formed. Since the reader cyclically activates the antenna in the order of 1-A, the RSSI measurements R_m and phase measurements P_m in the m-th round will be generated as:

$$R_m = \left\{ R_m^1, ..., R_m^A \right\}, P_m = \left\{ P_m^1, ..., P_m^A \right\} \tag{1}$$

Each R_m^i and P_m^i represents the RSSI values and phase values of all tags read by the antenna i in the m-th round:

$$R_m^i = \left\{ R_m^{i,1}, ..., R_m^{i,T} \right\}, P_m^i = \left\{ P_m^{i,1}, ..., P_m^{i,T} \right\} \tag{2}$$

When the background is static, that is, no one exists in the network, each link remains stable with R_m and P_m fluctuating in a very small range. However, when there is a human moving in this network, some links are affected by human occlusion or reflection and resulting in large changes in RSSI and phase. Particularly, the specific affected links depend on where the human is, which in turn can be used to locate the human in the network.

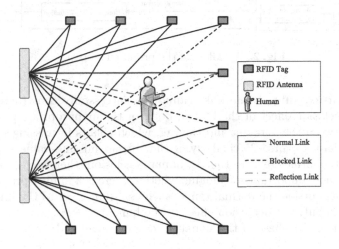

Fig. 1. System deploy.

3.2   System Architecture

The overall architecture of the proposed system is shown in Fig. 2, which mainly contains the following two components: *Human detection Module, Indoor localization module*. Specifically, *Human detection Module* extracts features from adjacent segments of RSSI and phase to describe the steady state of the links and facilitates SVM algorithm for discrimination. *Indoor localization module* first learns an overcomplete dictionary of all offline fingerprint points, thereby sparse representation of each point can be performed. PCA is then applied to extract the most discriminative feature for the further positioning algorithm training. When the system detects the presence of a human and needs to be further located, the data segment is first sparsely represented using the dictionary, and after the dimension is reduced, the final location can be calculated via the trained positioning model.

4   Methodology

4.1   Human Detection

Since the raw data may be missing or fluctuating, it is necessary to perform preprocessing and then extract effective features from the processed signal matrix for human detection.

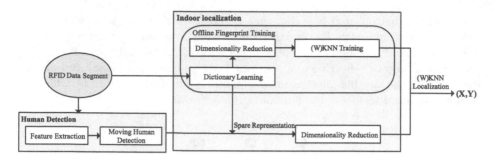

Fig. 2. Workflow of the proposed system.

Data Preprocessing. Some links cannot be successfully constructed considering the occlusion effect of the human body or the inappropriate position and distance between the antenna and the tag, which results in missing reading for some antenna-tag pairs. Particularly, if the current missing value is generated because of the former reason, that is, human occlusion, then the corresponding link is extremely important for judging whether there is a human present or not. Thus, a value outside the normal range, such as −100, is set to the link without reading to amplify the occlusion effect of the human.

In addition, the phase of the signal ϕ_T can be calculated as:

$$\phi_T = 2\pi \left(\frac{2d}{\lambda} + \phi_c \right) \bmod 2\pi \tag{3}$$

where d is the distance between the antenna and tag, λ is the wave length of RFID signal, and ϕ_c is the constant phase deviation due to device itself or the background. Therefore, as the distance changes linearly, the phase changes from 0 to 2π, and then immediately falls from 2π to 0, which causes the wrapped phase. For example, although the phase of 0.2 and 6.3 differ greatly in value, they are highly close in fact. One-Dimensional Phase Unwrapping method [10] can unwrap the phase effectively but will lead the phase value fall outside the range of $[0, 2\pi]$. In order to eliminate wrapped situation while maintaining a normal phase reading range, let $P_m^{i,j}$ and $P_{m+1}^{i,j}$ represent the phase of antenna i with tag j in round m and m+1, respectively, we calculate the 2π-unrelated phase difference $diff$ of the adjacent rounds as follows:

$$diff = \left| \left| P_m^{i,j} - 2\pi \right| - P_{m+1}^{i,j} \right| \tag{4}$$

If $diff \leq \varepsilon$, where ε is the threshold, here take 0.7, it indicates that the two phases belong to the above situation of volatility, we thus set the adjusted phase in (5). Figure 3 illustrates an example of the phase after 2π-unrelated preprocessing, which eliminates the volatility as expected compared to the origin version.

$$P_{m+1}^{i,j}{}' = \left| P_{m+1}^{i,j} - 2\pi \right| \tag{5}$$

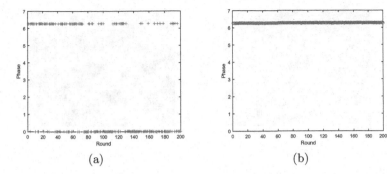

Fig. 3. Effect of phase preprocessing. (a) Phase before preprocessing, (b) Phase after preprocessing.

Feature Extraction and Dynamic Detection. In order to make the system robust and stable, environment-independent features need to be extracted for training, which means that the features related to absolute RSSI or phase values, e.g., median, mean, variance cannot be selected.

It can be seen from Fig. 4(a) and (b) that both RSSI and phase remain relatively stable in the static environment, while both have experienced large fluctuations with time passing throughout all links in dynamic case, as shown in Fig. 4(c) and (d). Motivated by this observation, we consider using the relationship between the values of adjacent rounds to describe the stability of the link, and further determine whether there is a moving human exists. Given the two adjacent group of RSSI and phase as $[R_m, R_{m+1}]$ and $[P_m, P_{m+1}]$, their corresponding Pearson Coefficient ρ_R and ρ_P can be calculated as:

$$\rho_R = \frac{\mathrm{cov}\,(R_m, R_{m+1})}{\sigma_{R_m}\sigma_{R_{m+1}}} \tag{6}$$

$$\rho_P = \frac{\mathrm{cov}\,(P_m, P_{m+1})}{\sigma_{P_m}\sigma_{P_{m+1}}} \tag{7}$$

where $\mathrm{cov}\,(R_m, R_{m+1})$ and $\mathrm{cov}\,(P_m, P_{m+1})$ are the covariance of $[R_m, R_{m+1}]$ and $[P_m, P_{m+1}]$, respectively. For both metrics, with higher correlation coefficient values, the network is more stable and free of human, yet low correlation coefficient indicates a high probability of the human presence in the network. Figure 5 demonstrates the validity of the selected two features, which can be clearly seen from the figure that the RSSI and phase are very stable due to the unaffected link, resulting in two correlation coefficients close to 1 and concentrated in the upper right corner. Conversely, the dynamic samples with moving human are not so well centralized compared to the static case and are scattered in areas with relatively small correlation coefficients.

For better distinguish between the two cases, in addition to the correlation coefficient, we also capture several other features from the difference between the two rounds of data $R_{diff} = [R_m - R_{m+1}]_{L \times 1}$, $P_{diff} = [P_m - P_{m+1}]_{L \times 1}$, all

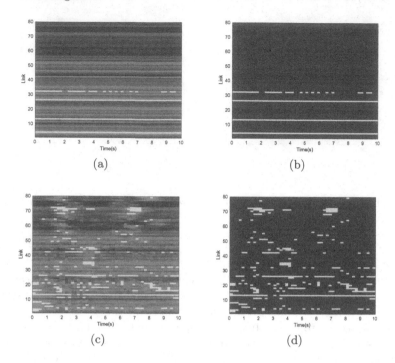

Fig. 4. Variances of both RSSI and phase in static and dynamic cases. (a) RSSI of static case, (b) Phase of static case, (c) RSSI of dynamic case, (d) Phase of dynamic case.

of them are listed in Table 1. As a consequence, we adopt the effective classifiers SVM to separate the two types of samples by training a cutting line, which is easy to find according to the above analysis. Noticing that although the absolute values of RSSI and phase vary with the size and layout of the environment, the effect of the discriminant model trained in the initial environment does not change much. The reason behind this is that the features we extracted are relative features of adjacent readings, which are only relate to the temporal changes.

4.2   Indoor Localization

Mainstream solutions for indoor localization including path loss model based, interval-based, trilateration-based, and fingerprint-based algorithms, where fingerprint-based method is superior than other methods to some extent since it requires almost no reference measurement points and the positioning accuracy is relatively high. Thus, we first need to perform RSSI and phase fingerprint collection at each point and sparsely represent it offline, and then PCA is applied to reduce its dimension. Through the above compressive sensing of the fingerprint vector, we can get the most representative features for further positioning on the one hand, and also can greatly reduce computation and storage overhead on the other hand.

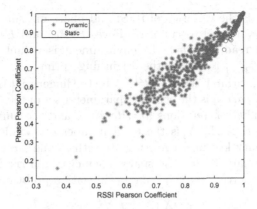

Fig. 5. Correlation Coefficient of RSSI and phase in static and dynamic cases.

Sparse Representation Dictionary Learning. As explained in Sect. 3, the location of the human in the network can have a negative or positive impact on the normal communication of a certain number of the links. In either case, the RSSI and phase values of the corresponding links will fluctuate greatly. Therefore, we attempt to highlight the changes caused by these affected links and weaken the normal links to better divide human's locations into the correct areas, which also simplifies the subsequent calculations and reduces the storage burden. To this end, we use a dictionary-based approach to get the sparse representation of the signal vector.

Assume the positioning segment contains 6 rounds of complete polling between all the antennas and tags, which means that the location of the human is determined almost every 1 s. Let $\mathbb{R} = \{R_m, ..., R_{m+5}\}$ and $\mathbb{P} = \{P_m, ..., P_{m+5}\}$ be the RSSI and phase in a specific time window, and the difference between the current and static signal vector is defined as:

$$\Delta\mathbb{R} = [E(\mathbb{R}) - E(\mathbb{R}_0)]_{L \times 1} \tag{8}$$

$$\Delta\mathbb{P} = [E(\mathbb{P}) - E(\mathbb{P}_0)]_{L \times 1} \tag{9}$$

Table 1. Features and brief descriptions

No	Feature	Description
1	ρ_R	Pearson Coefficient of $[R_m, R_{m+1}]$
2	ρ_P	Pearson Coefficient of $[P_m, P_{m+1}]$
3	R_{diff_Mean}	Mean RSSI of R_{diff}
4	P_{diff_Mean}	Mean phase of P_{diff}
5	R_{diff_Max}	Max RSSI of R_{diff}
6	P_{diff_Max}	Max phase of P_{diff}
7	R_{diff_Var}	RSSI variance of R_{diff}
8	P_{diff_Var}	phase variance of P_{diff}

where $\Delta\mathbb{R}$ and $\Delta\mathbb{P}$ is the variation of RSSI and phase measurement, $E(\mathbb{R})$ and $E(\mathbb{P})$ is the mean values of 6 sets of readings, $E(\mathbb{R}_0)$ and $E(\mathbb{P}_0)$ is the mean values of 6 sets of readings when the environment is vacant without human. Then $\mathbb{F} = [\Delta\mathbb{R}, \Delta\mathbb{P}]_{2L \times 1}$ is used as the origin fingerprint.

As shown in Algorithm 1, let $\mathcal{F} \in R^{2L \times N}$ be the fingerprint training set for all reference locations, *param* is the specified parameter set of the algorithm mainly containing the number of iterations *numIteration* and the number of atoms in the dictionary K, $K > 2L$, N is the total number of samples. We adopt K-SVD based dictionary learning algorithm [1] with $\mathcal{F}$ and *param* as inputs. An overcomplete dictionary D and the sparse coefficient matrix X of all samples can be obtained by solving the following optimization problem:

$$\min_{D,X} \|\mathcal{F} - DX\|_F^2, \ s.t. \ \forall i, \ \|x_i\|_0 \leqslant T_0 \tag{10}$$

The specific process is as follows, **Sparse representation** and **Dictionary update** are iteratively performed until the predetermined *numIteration* is reached.

- **Initialization** K column vectors are taken from the original sample matrix $\mathcal{F}$ as the atoms of the initial dictionary D_0.
- **Sparse representation.** Using Orthogonal Matching Pursuit (OMP) [12] to get the sparse coefficient matrix X according to the dictionary D.
- **Dictionary update.** Update $d_j, j \in [1, K]$ of D column by column. Specifically, the k-th row vector of X is extracted as x_T^k, and the error matrix E_k is calculated when updating d_k. Then the value of E_k corresponding to the position where x_T^k is not equal to 0 is extracted to form a new version as E_k'. Finally, SVD is applied on the E_k' to update d_k to the first column of U and x_T^k to $\Sigma(1,1) V(:,1)^T$.

Algorithm 1: K-SVD based Dictionary Learning

Input: Training fingerprint set $\mathcal{F} \in R^{2L \times N}$, *param*
Output: Dictionary $D \in R^{2L \times K}$, sparse coefficient matrix $X \in R^{K \times N}$
1 Initialize $D = D_0$;
2 **for** *i=1:numIteration* **do**
3   Calculation X by D using OMP algorithm;
4   **for** *k=1:K* **do**
5          $E_k = \mathcal{F} - \sum_{j \neq k} d_j x_T^j$;
6   Find $S_k = \{s|1 \leqslant s \leqslant N, x_T^k(s) \neq 0\}$;
7          $E_k' = E_k(S_k)$;
8   Apply SVD on E_k', $E_k' = U\Sigma V^T$;
9   Update $d_k = U(:,1)$ and $x_T^k = \Sigma(1,1) V(:,1)^T$;
10   **end**
11 **end**

Location Acquisition. Since the sparse coefficient matrix X of all samples is obtained during the dictionary learning phase, it is necessary to carry out further dimensional reduction processing to reduce computational complexity and acquire the most discriminative features. PCA (Principal Component Analysis) [23] is one of the most widely used data dimensionality reduction algorithms, whose main idea is to map m-dimensional features to n-dimensional new orthogonal features. Recall that $\mathbb{F} \in R^{2L \times 1}$ becomes $x_T \in R^{K \times 1}$, $K > 2L$ after sparse representation, which further becomes a $Q \times 1$ feature vector after PCA. Q is determined by the contribution rate, that is, the extent to which the reduced space can represent the original space, which is greater than 90% generally.

For online localization, let $\widetilde{\mathbb{F}} = \left[\widetilde{\Delta\mathbb{R}}, \widetilde{\Delta\mathbb{P}}\right]_{2L \times 1}$ be the original test signal vector, which is first sparsely represented as $\widetilde{x_T} \in R^{K \times 1}$ by the OMP algorithm according to the training dictionary D, and then reduced to the same dimension of Q by PCA. Among a large number of fingerprint localization algorithms, Nearest Neighbor (NN) based algorithms such as K-Nearest Neighbor (KNN) and Weighted K-Nearest Neighbor (WKNN) are simple and easy to perform, which can always select the reference point closest to the target effectively [2]. Take WKNN as an example, for an online test vector, calculate its distance from each fingerprint in the fingerprint library (such as Euclidean distance) and select the nearest K fingerprints, the final (X, Y) can be acquired by weighted averaging the coordinates of the reference points.

5 Experiment and Analysis

5.1 Experiment Settings

We implement a prototype of the system based on COTS RFID devices working between 860 MHz to 960 MHz with Impinj Speedway R420 reader, Laird S9028PCL antenna and Impinj H47 tags and evaluate it in two different environments as shown in Fig. 6(a): (a) an office room covering about $4 \times 10\,\mathrm{m}^2$ with one sofa and one desk; (b) a bedroom room covering about $8 \times 10\,\mathrm{m}^2$ with one bed and one desk. Take the office room as an example, shown in Fig. 6(b), 4 antennas are hung on the wall to form an antenna array, and the remaining three walls are pasted with two rows of passive tags to form a tag array, ensuring that the antennas can read tags as much as possible. The distance between the antenna (tag) and the ground should be within the height of the human, usually less than 1.8 m.

In order to verify the human detection effect of the system, we separately collect the signal of the empty room and the room with randomly moving human, each for 30 min. For the fingerprint library establishment phase, we choose the four corners of 0.6 * 0.6 floor tiles as the reference positions and collect the RSSI and phase data in each point for 1 min. When evaluating the tracking accuracy of the proposed system, the placement of human is vary across 45 different locations in two scenarios.

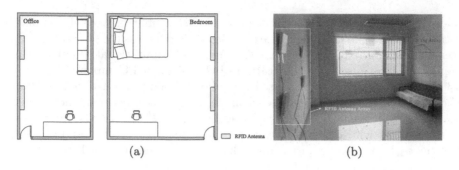

Fig. 6. Experimental setup. (a) floor plan of two rooms, (b) office setup.

5.2 Performance Evaluation

Metrics. We use the average localization error (ALE) combined with the Cumulative Distribution Function (CDF) curve to evaluate the positioning effect of the system. As for the human detection performance, the two metrics are defined as follows:

$$sensitivity = \frac{N_{D\_c}}{N_D} \tag{11}$$

$$specificity = \frac{N_{S\_c}}{N_S} \tag{12}$$

where N_{D_c} and N_{S_c} is the number of samples correctly classified as dynamic and static, respectively. N_D and N_S is the total number of dynamic and static samples, respectively.

Overall Performance. The best experimental result with total 32 tags on the wall indicates that sensitivity and specificity are both equal to 100% in two different environments, which means that our feature extraction method is effective, making the features highly representative.

In order to analyze the difference in performance of the three algorithms NN, KNN, WKNN, and study the effect of different K values on the positioning accuracy, we compare these three algorithms and take the value of K from 1 to 4 in Fig. 7. Obviously, KNN and WKNN perform better than the NN with smoother CDF curves and higher average accuracy. Specifically, the average localization error is listed in Table 2, from which we can see that NN performs relatively poor with ALE of 1.137 m. KNN and WKNN behave similarly, and ALE gradually decreases with the value of K increases in both case, which achieves lowest of 0.94 m when K = 4.

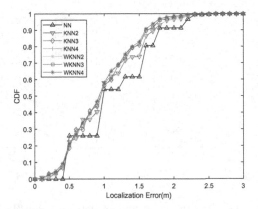

Fig. 7. Performance comparison of NN, KNN and WKNN.

Table 2. Average localization error in NN, KNN and WKNN

	NN	KNN			WKNN		
		KNN2	*KNN3*	*KNN4*	*WKNN2*	*WKNN3*	*WKNN4*
Average localization error (m)	1.137	1.0097	0.9689	0.9479	1.0096	0.9684	0.9481

5.3 Impact of Tag Density

It is intuitive to assume that the density of the tag has a decisive influence on the performance of the system. Too few tags may make the system less sensitive, and too many tags may greatly increase the processing burden of the system. Thus, we first give the effect of different tag density on human detection performance, as shown in Fig. 8(a). Not surprisingly, the system is strongly able to detect the presence of human as the number of tags increases, which is reflected in the improvement of sensitivity. Specifically, though sensitivity is the lowest when there is only 1 tag for detection, it also reached 85.9%. When the number of tags increases from 1 to 8, the performance improves significantly with sensitivity increasing directly to 97.33% and reached 100% when there are 32 tags on the wall. However, sensitivity remains the same or even drops slightly as the number of tags continues to increase. Since specificity does not change much with the tag density, 32 tags are suitable and sufficient for moving human detection.

The effect of different tag density on localization performance is shown in Fig. 8(b). It can be seen that when the number of tags is less than 20, the localization error of the system is large, but the performance improves with the increase of tags. Best positioning effect achieved when the number of tags is 32, and the localization error of all test points are within 2.13 m. In line with the above human detection situation, the positioning effect is slightly reduced as the number of tags continues to increase. This is likely because that introducing too many unnecessary tags means that the distance between tags is too small, which causes the preemption of read resources and affects the communication between the tag and the antenna, thereby increasing the probability of systematic misjudgment.

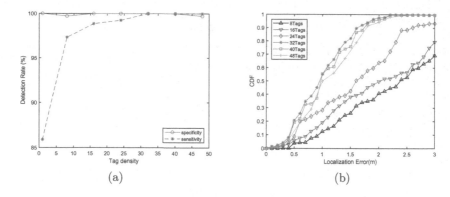

Fig. 8. Impact of tag density on (a) human detection and (b) localization.

Real-time performance is one of the important indicators to measure the quality of positioning systems. In many previous works, researchers are blindly pursuing the improvement of positioning accuracy [29], ignoring the real-time requirement, which leads to the impracticality of the system. Since the position of human is determined every 6 rounds of complete polling between all the antennas and tags, which is about 1s, the data processing and positioning time needs to be less than 1s to ensure real-time performance. It can be seen intuitively that as more tags are involved, the required processing time will be gradually increasing. When the number of tags is specified as 32 here, the corresponding time is 0.7s, which is less than 1s of data acquisition and is sufficient for real-time use.

6   Conclusion

In this paper, we present the design details and implementation methods for a device-free system that with human detection and localization capabilities on COTS RFID devices. In order to create a space with intelligent awareness, we attach a tag array on the wall of the room and place the antenna array in the appropriate place to form multiple antenna-tag links. To detect the presence of human, total 8 environment-independent features are extracted from RSSI and phase to characterize the stability of the link. In the positioning phase, the dictionary-based sparse representation is combined with the dimensionality reduction method to obtain the compressed perception of the signal vector at a specific location, which is further used as input for the subsequent localization algorithm. Experimental results acquired in real-life scenario demonstrate that our system has outstanding performance on human detection and can achieve an average localization error of less than 1 m.

In the future, we will expand our vision to dynamic and static combination of human detection and explore the possibility of detecting stationary human with RFID in a passive way, which is closer to the actual situation. Moreover, some trajectory optimization strategies such as Kalman filter will be considered to improve the accuracy of localization and better track the human.

Acknowledgement. This work was supported by the National Natural Science Foundation of China (61601459).

References

1. Aharon, M., Elad, M., Bruckstein, A.: K-SVD: an algorithm for designing overcomplete dictionaries for sparse representation. IEEE Trans. Signal Process. **54**(11), 4311–4322 (2006)
2. Ali-Loytty, S., Perala, T., Honkavirta, V., Piche, R.: Fingerprint Kalman filter in indoor positioning applications. In: 2009 IEEE Control Applications, (CCA) Intelligent Control, (ISIC), pp. 1678–1683 (2009)
3. Arshad, S., et al.: Wi-chase: a WiFi based human activity recognition system for sensorless environments. In: IEEE International Symposium on A World of Wireless (2017)
4. Bu, Y., et al.: RF-dial: an RFID-based 2D human-computer interaction via tag array. In: IEEE INFOCOM 2018 - IEEE Conference on Computer Communications, pp. 837–845, April 2018
5. Castillo-Cara, M., Lovón-Melgarejo, J., Bravo-Rocca, G., Orozco-Barbosa, L., García-Varea, I.: An empirical study of the transmission power setting for bluetooth-based indoor localization mechanisms. Sensors **17**(6), 1318 (2017)
6. Dan, W., Zhang, D., Xu, C., Wang, Y., Hao, W.: WiDir: walking direction estimation using wireless signals. In: ACM International Joint Conference on Pervasive & Ubiquitous Computing (2016)
7. Gao, Q., et al.: CSI-based device-free wireless localization and activity recognition using radio image features. IEEE Trans. Veh. Technol. **66**(11), 10346–10356 (2017)
8. Georgievska, S., et al.: Detecting high indoor crowd density with Wi-Fi localization: a statistical mechanics approach. J. Big Data **6**(1), 31 (2019)
9. Gong, Y., Xie, L., Wang, C., Bu, Y., Lu, S.: RF-brush: 3D human-computer interaction via linear tag array. In: 2018 IEEE 15th International Conference on Mobile Ad Hoc and Sensor Systems (MASS) (2018)
10. Itoh, K.: Analysis of the phase unwrapping algorithm. Appl. Opt. **21**(14), 2470 (1982)
11. Lau, S.L., König, I., David, K., Parandian, B., Carius-Düssel, C., Schultz, M.: Supporting patient monitoring using activity recognition with a smartphone. In: International Symposium on Wireless Communication Systems (2010)
12. Pati, Y., Rezaiifar, R., Krishnaprasad, P.: Orthogonal matching pursuit: recursive function approximation with applications to wavelet decomposition, pp. 40–44 (2009)
13. Peng, Y., Fan, W., Xin, D., Xing, Z.: An iterative weighted KNN (IW-KNN) based indoor localization method in bluetooth low energy (BLE) environment. In: Ubiquitous Intelligence & Computing, Advanced & Trusted Computing, Scalable Computing & Communications, Cloud & Big Data Computing, Internet of People, & Smart World Congress (2017)

14. Qian, K., Wu, C., Zhang, Y., Zhang, G., Yang, Z., Liu, Y.: Widar2.0: passive human tracking with a single Wi-Fi link. In: Proceedings of the 16th Annual International Conference on Mobile Systems, Applications, and Services, MobiSys 2018, pp. 350–361 (2018)

15. Qian, K., Wu, C., Zheng, Y., Liu, Y., Zhou, Z.: PADS: passive detection of moving targets with dynamic speed using phy layer information. In: IEEE International Conference on Parallel & Distributed Systems (2015)

16. Sample, A.P., Macomber, C., Jiang, L.T., Smith, J.R.: Optical localization of passive uhf RFID tags with integrated leds. In: IEEE International Conference on RFID (2016)

17. Wang, C., et al.: Multi - touch in the air: device-free finger tracking and gesture recognition via cots RFID. In: IEEE INFOCOM 2018 - IEEE Conference on Computer Communications, pp. 1691–1699, April 2018

18. Wang, C., Liu, J., Chen, Y., Xie, L., Liu, H.B., Lu, S.: RF-kinect: a wearable RFID-based approach towards 3D body movement tracking. Proc. ACM Interact. Mob. Wearable Ubiquit. Technol. **2**(1), 41:1–41:28 (2018)

19. Wang, C., Xie, L., Wang, W., Chen, Y., Bu, Y., Lu, S.: RF-ECG: heart rate variability assessment based on cots RFID tag array. Proc. ACM Interact. Mob. Wearable Ubiquit. Technol. **2**(2), 85:1–85:26 (2018)

20. Wang, J., Vasisht, D., Katabi, D.: RF-IDRAW: virtual touch screen in the air using RF signals. In: ACM Conference on Sigcomm (2014)

21. Wu, C., Zheng, Y., Xiao, C.: Automatic radio map adaptation for indoor localization using smartphones. IEEE Trans. Mob. Comput. **PP**(99), 1 (2018)

22. Xi, W., et al.: Electronic frog eye: counting crowd using WiFi. In: IEEE Conference on Computer Communications, IEEE INFOCOM 2014, pp. 361–369, April 2014

23. Xu, H., Caramanis, C., Sanghavi, S.: Robust PCA via outlier pursuit. IEEE Trans. Inf. Theory **58**(5), 3047–3064 (2012)

24. Yang, Z., Zhou, Z., Liu, Y.: From RSSI to CSI: indoor localization via channel response. ACM Comput. Surv. **46**(2), 25:1–25:32 (2013)

25. Yu, G., Zhan, J., Ji, Y., Jie, L., Gao, S.: MoSense: a RF-based motion detection system via off-the-shelf wifi devices. IEEE Internet of Things J. **PP**(99), 1 (2017)

26. Yu, J., Na, Z., Liu, X., Deng, Z.: WiFi/PDR-integrated indoor localizationusing unconstrained smartphones. EURASIP J. Wirel. Commun. Network. **2019**(1), 41 (2019)

27. Zhang, S., Mccullagh, P., Nugent, C., Zheng, H.: Activity monitoring using a smart phone's accelerometer with hierarchical classification. In: Sixth International Conference on Intelligent Environments (2010)

28. Zhongqin, W., Min, X., Ning, Y., Ruchuan, W., Haiping, H.: Computer vision-assisted region-of-interest rfid tag recognition and localization in multipath-prevalent environments. Proc. ACM Interact. Mob. Wearable Ubiquit. Technol. **3**(29), 1–30 (2019)

29. Zhu, D., Zhao, B., Wang, S.: Mobile target indoor tracking based on multi-direction weight position Kalman filter. Comput. Netw. **141**, 115–127 (2018)

30. Zhu, S., Wang, S., Zhang, F., Zhang, Y., Feng, Y., Huang, W.: Environmentally adaptive real-time detection of RFID false readings in a new practical scenario. In: 2018 IEEE SmartWorld, Ubiquitous Intelligence Computing, Advanced Trusted Computing, Scalable Computing Communications, Cloud Big Data Computing, Internet of People and Smart City Innovation (SmartWorld/SCALCOM/UIC/ATC/CBDCom/IOP/SCI), pp. 338–345, October 2018

A Novel Approach to Cost-Efficient Scheduling of Multi-workflows in the Edge Computing Environment with the Proximity Constraint

Yuyin Ma[1], Junyang Zhang[1], Shu Wang[5], Yunni Xia[1(✉)], Peng Chen[2], Lei Wu[3], and Wanbo Zheng[4(✉)]

[1] Software Theory and Technology Chongqing Key Lab, Chongqing University, Chongqing, China
xiayunni@hotmail.com
[2] School of Computer and Software Engineering, Xihua University, Chengdu, China
[3] School of Mathematics, University of Electronic Science and Technology of China, Chengdu, China
[4] School of Mathematics, Kunming University of Science and Technology, Kunming 650500, Yunnan, China
zwanbo2001@163.com
[5] School of Information, Liaoning University, Shenyang 110036, China

Abstract. The edge computing paradigm is featured by the ability to offload computing tasks from mobile devices to edge clouds and provide high cost-efficient computing resources, storage and network services closer to the edge. A key question for workflow scheduling in the edge computing environment is how to reduce the monetary cost while fulfilling Service-Level-Agreement in terms of performance and quality-of-service requirements. However, it's still a challenge to guarantee user-perceived quality of service of applications deployed upon edge infrastructures due to the fact that such applications are constantly subject to negative impacts, *e.g.*, network congestions, unexpected long message delays, shrinking coverage range of edge servers due to battery depletion. In this paper, we study the multi-workflow scheduling problem and propose a novel approach to Cost-Efficient Scheduling of Multi-Workflows in the Edge Computing Environment With Proximity Constraint. The proposed approach aims at minimizing edge computing costs while meeting user-specified workflow completion deadlines and leverages a discrete firefly algorithm for yielding the scheduling plan. We conduct experimental case studies based on multiple well-known scientific workflow templates and a real-world dataset of edge resource locations as well. Experimental results clearly suggest that our proposed approach outperforms traditional ones in terms of cost and makespan.

Keywords: Edge computing · Workflow scheduling · Cost efficiency · Proximity constraint

© Springer Nature Switzerland AG 2020
S. Wen et al. (Eds.): ICA3PP 2019, LNCS 11944, pp. 655–668, 2020.
https://doi.org/10.1007/978-3-030-38991-8_43

1   Introduction

In the past decades, the cloud computing paradigm has evolved as the key force to provide computing, storage and network services, which has been widely applied in various fields, *e.g.*, scientific workflow execution [3,5,16]. However, with the increasing popularity of Internet of Things (IoT), the architecture of cloud computing framework is facing great challenge [7,20]. Nowadays, we live in such a world where smart IoT devices are ubiquitous. These massive smart devices located at the edge of network require data processing with low-latency, location-awareness and mobility requirements. Edge computing is derived to satisfy these demanding requirements. This emerging computing paradigm is viewed as network edge cloud, which can effectively compensate for the disadvantages of cloud computing such as communication latency. In practice, the location of edge devices is close to end-user applications so that it can process delay-sensitive tasks quickly. Therefore, edge computing is more efficient than traditional cloud by taking some factors into account such as service latency, power consumption, network traffic, operating expenses, content distribution and so on.

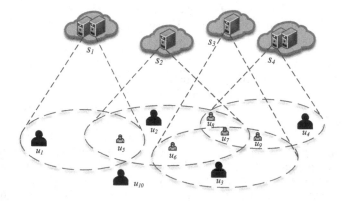

Fig. 1. Edge computing deployment example

Although the problem of execution workflow upon traditional cloud infrastructures with multiple objectives and constraints, which is known to be an NP-complete problem, attracted considerable research attentions recently [12,13]. The problem of scheduling workflows upon edge infrastructures can be fundamentally different and it remains a challenge how to reduce the cost of scheduling multi-workflows requested by edge users with the proximity constraint. Figure 1 shows a good example of deploying and offloading tasks among edge nodes with proximity constraint. Assuming that there are four edge servers in a particular area, each server covering a specific area, the user can offload computing tasks to any server within the coverage. Users who are not in coverage cannot establish a connection with the server for application offloading (proximity constraint). As shown in Fig. 1, user u_6 can offload computing tasks to servers s_2 and s_3, and user u_7 can offload applications to servers s_2, s_3, and s_4. User u_8 can be assigned to servers s_2 and s_4. User u_1 can only offload tasks to server s_1. Since user u_{10} is not within the coverage of any server, the task cannot be offloaded to the server. Each user makes a service request, which can be thought of as a workflow, and the workflow consists of tasks. For example, some tasks of the service request issued by the user u_7 may be assigned to the server s_2, some to the server s_3, and others to the server s_4. Tasks assigned to the server are executed in the VM. Because the server price is different, the task is divided and dispatched to different servers for execution, which can reduce the cost while solving the delay and network congestion.

In this paper, we study the problem of cost-efficient multi-workflow scheduling in the edge computing environment with the proximity constraint, *i.e.*, a computing task of a workflow can be offloaded to a nearby edge node only if the corresponding user are within the communication range of such edge node. We consider a multi-edge-user, multi-workflow, cost-reduction, proximity-limited, makespan-constraint formulation and leverage, a discrete firefly algorithm (DFA) to solve it. We show through simulation as well that our proposed method clearly outperforms traditional ones in terms of cost and makespan.

Table 1. Notations and Description.

Notation	Description	Notation	Description
n	The total number of workflows	n_i	The total number of tasks in workflow
m	The total number of edge servers	M	The total number of tasks in all workflow
m_p	The total number of virtual machines in server p	T_i	The set of tasks in workflow i, $T_i = \{t_{i1}, t_{i1}, \ldots, t_{in_i}\}$
t_{ij}	The jth task of workflow i	VM_{pk}	The kth virtual machine of server p
t_{ijpk}	The execute time of t_{ij} on VM_{pk}	$pred(t_{ij})$	All predecessor node tasks of t_{ij}
x_{ijpk}	A boolean variable indicating whether VM_{pk} is selected for t_{ij}	c_{pk}	The unit-price-time of VM_{pk}
G	The directed acyclic graph (DAG)	E	The set of dependencies in a workflow
$D(W_i)$	User defined deadline of the workflow i	$T(W_i)$	The finish time of the workflow i
d_{ip}	The distance between server p and workflow i	cov_p	The coverage area of server p
$ST(t_{ijpk})$	The start time of t_{ij} on VM_{pk}	$FT(t_{ijpk})$	The completion time of t_{ij} on VM_{pk}

2   Related Work

2.1   Workflow Scheduling

It is widely acknowledged that to schedule multi-tasks workflow on distributed platforms, *e.g.*, clouds or edge nodes, is an NP-hard problem. It is there for extremely time-consuming to yield optimal schedule through traversal-based algorithms. Fortunately, heuristic and meta-heuristic algorithms with polynomial complexity are able to produce approximate or near optimal solutions at the cost of acceptable optimality loss. For example, Habak *et al.* [2] proposed and designed the femtocloud system which provided cloud services through mobile devices at the edge of network. They consider task scheduling is the most important design of a system, and developed an optimization framework that yields scalable heuristic solutions to the problem. Mao *et al.* [14] developed a multi-users mobile devices resource management online algorithm, with the objective of minimizing the power consumption of the mobile devices and the mobile edge computing server, subject to a task buffer stability constraint. They also developed a delay improved mechanism to reduce the execution delay.

To ensure efficient utilization of cloud resources, Tong *et al.* [9] proposed a workload placement algorithm that decides the destination edge server and assigned computational capacity for a given task. Their basic idea is to opportunistically aggregate and serve the peak loads that exceed the capacities of lower tiers of edge cloud servers to other servers at higher tiers in the edge cloud hierarchy. Zhang *et al.* [25] studied the cost optimization of the task scheduling problem in the edge computing system, while satisfying Qos requirements of users. TTSCO (Two-stage Task Scheduling Cost Optimization) algorithm was proposed to solve the optimization problem, which consists of two parts are get the initial task scheduling strategy by BF algorithm optimize the scheduling scheme to get the final strategy. Zhao *et al.* [26] considered the offloading of delay-bounded tasks in heterogeneous cloud environment, which objective is to allocate resources reasonably in order to maximize the probability that tasks can have the delay requirements met. Sahni *et al.* [18] proposed a scheduling algorithm to deal scientific workflow with deadline-constrained in the cloud environment, and minimize cost efficiency, while taking into account the virtual machine (VM) performance variability and heterogeneous nature of cloud resources.

2.2   Firefly Algorithm

The Firefly algorithm (FA), proposed by Yuan X.S. [23,24], have shown its high potent in dealing with complex combinatorial optimization problems and versatile types of engineering and industrial optimization scenarios. It's a swarm based metaheuristic algorithm inspired by the flashing behavior of fireflies. Sanaei *et al.* [19] used the FA algorithm in Resource Constrained Project Scheduling Problem (RCPSP). The algorithm starts by generating a set of random schedules. After that, the initial schedules are improved iteratively using the flying approach proposed by the FA. By termination of algorithm, the best schedule found by the method is returned as the final result. Marichelvam *et al.* [15] proposed a discrete firefly algorithm to minimize makespan for M-stage hybrid flowshop scheduling problems with the objective function of makespan and mean flow time.

FA contains two elements, namely brightness and attractiveness, which are being repeatedly updated. Brightness refers to the position of the firefly and determines its moving direction. Attractiveness determines the distance that the firefly moves through. FA is with the following assumptions: (1) A firefly will be attracted to each other regardless of their sex because they are unisexual; (2) Attractiveness is proportional to their brightness whereas the less bright firefly will be attracted to the brighter firefly. However, the attractiveness decreased when the distance of the two fireflies increased; and (3) If the brightness of both fireflies is the same, the fireflies will move randomly. The generations of new solutions are by random walk and attraction of the fireflies. The brightness of the fireflies should be associated with the objective function of the related problem. Their attractiveness makes them capable to subdivide themselves into smaller groups and each subgroup swarm around the local models.

The relative brightness of fireflies can be calculated as:

$$I(r) = I_0 \cdot e^{-\gamma r_{ij}} \tag{1}$$

where, I_0 represents the fluorescent brightness of the firefly, *i.e.*, the fluorescence brightness when $r = 0$. the better the objective function value is, the higher the brightness is. γ is the light intensity absorption coefficient, r_{ij} is the distance between the fireflies i and j.

The attraction of fireflies can be calculated as:

$$\beta(r) = \beta_0 \cdot e^{-\gamma r_{ij}^2} \tag{2}$$

where β_0 is the maximum attraction when $r = 0$.

Firefly location is updated according to:

$$x_i = x_i + \beta(r) \cdot (x_j - x_i) + \alpha(rand - 1/2) \tag{3}$$

where x_i and x_j are the spatial locations of fireflies i and j. α is the step factor, $rand$ is a random factor that is uniformly distributed over (0,1). $(rand - 1/2)$ is a random disturbance to avoid prematurely falling into local optimum during the position update process.

FA According to the calculation, the firefly with low brightness moves to that with high brightness. The moving distance is decided by according to (2) and (3).

3   Model and System

3.1   System Architecture

The edge computing system is composed of an edge computing agent (ECA) and multiple edge servers. The edge computing agent manages all resources. An edge server owns several virtual machines (VMs) and each of which can handle a workflow task that a user offloads at a time. An edge server usually has limited capacity for storage and computation. For the energy saving purpose, offloading and execution of workflow tasks are subject to the proximity constraint, *i.e.*, an edge server can only support tasks of its nearby edge users that fall in with its communication range.

As can be seen in Fig. 2, an edge user organizes its tasks in the form of a workflow. Therefore, there exist multiple workflows in the edge environment to be handled and executed.

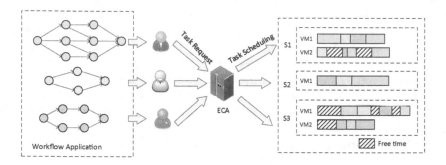

Fig. 2. Edge computing system model

A workflow refers to a directed acyclic graph DAG, $G = (T, E)$. T denotes the task set $T = \{t_1, t_2, \ldots, t_n\}$, E the set of edges between tasks, and $e_{ij} = (t_i, t_j)$ is a priority constraint, indicating that t_i is the precedent task of t_j (Table 1).

3.2   Problem Formulation

Based on the described system configuration, the problem that we are interested in is thus, for given proximity constraints of server-user communications, how to schedule workflows with as low as possible cost and completion time. The resulting formulation is thus:

$$Min \quad f = Cost = \sum_{p=1}^{m} \sum_{k=1}^{m_p} \sum_{i=1}^{n} \sum_{j=1}^{n_i} c_{pk} \cdot x_{ijpk} \cdot [FT(T_{ijpk}) - ST(T_{ijpk})] \quad (4)$$

subject to:

$$ST(T_{ij}) \geq \max[FT(T_{il})], T_{il} \in pred(T_{ij}) \ and \ l \in \{1, \ldots, n_i\} \quad (5)$$

$$T(W_i) \leq D(W_i) \quad (6)$$

$$d_{ij} \leq cov_p \quad (7)$$

$$x_{ijpk} \leq 1 \quad (8)$$

where

$$x_{ijpk} = \begin{cases} 1, & if \ VM_{pk} \ is \ selected \ for \ task \ T_{ij} \\ 0, & otherwise \end{cases} \quad (9)$$

4 Discrete Firefly Algorithm for Multi-workflows Scheduling

In this section, we present novel schemes of coding rules, decoding rules, and updating rules that deal with the location-aware and proximity-constrained multi-workflows scheduling problem.

4.1 Encoding

The encoding scheme includes two parts:

(1) Virtual machine assignment: The encoding of the VM selection part comprises a set of integer values. The encoding length is the same as the number of tasks of all workflows. We use a **VM-string** to represent its encoding. An example of such encoding scheme is shown in Fig. 3, where it is assumed that there are three workflows w_1, w_2, and w_3. w_1 can be assigned to s_2 and s_3, w_2 can be assigned to s_2, s_3, s_4, and s_2, s_4 cover w_3.

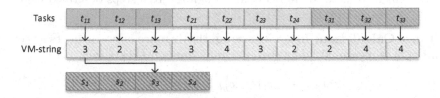

Fig. 3. VM-string representations

(2) Task scheduling: Tasks are scheduled according to the topological constraints of workflows. The encoding of the task scheduling part is similarly composed of a set of integer values in terms of a **Task-string**. Figure 4 shows an example of its encoding. As can be seen, from left to right, the task-string shows the first task of the third workflow, the first task of the second workflow, *etc.*

Fig. 4. Task-string representations

4.2 Initialization and Firefly Evaluation

Firefly population initialization consists of two parts: the **VM-string** initialization and the **Task-string** one. The former is related to proximity constraint, while the latter is related to the topological structure of the workflow. Populations generated follow the proximity and topological constraints to avoid the existence of infeasible solutions.

The fitness value of firefly is closely related to the light intensity. In this paper, we employ (10) represent the fitness, in terms of the light intensity of individuals.

$$
F = \begin{cases} f + \alpha \sum_{i=1}^{n} [T(W_i) - D(W_i)]^m, & T(W_i) > D(W_i) \ and \ m \geq 1 \\ \\ f, & T(W_i) \leq D(W_i) \end{cases} \tag{10}
$$

4.3 Firefly Update

The direction of movement of fireflies is determined by the light intensity and the fitness value. For any two fireflies, the darker one will move to the brighter one. The movement of fireflies are realized through discrete steps described below.

Distance Calculation: In a **VM-string**, the distance between any two fireflies p and p_{best} is measured by its corresponding hamming distance:

$$
r(x_i, x_j) = x_{id} \oplus x_{jd}, d \in \{1, 2, \ldots, M\} \tag{11}
$$

where an exclusive OR operations is employed to decide the hamming distance [11].

The distance between two fireflies in the **Task-string** is decided by the swap distance, which is the minimum number of exchanges required to exchange one string to another [11]. As shown by an example in Table 2 the swap distance between p and p_{best} is 4.

β-step Update: The firefly position update requires two consecutive steps: β-step and α-step. β-step is a process of moving fireflies towards brighter fireflies according to the following steps.

Step 1: Calculate the hamming distance of the **VM-string** and the swap distance of the **Task-string** as r_1 and r_2;

Step 2: Calculate the attractiveness, $\beta(r)$, according to (12);

Step 3: Generate $|r_1|$ and $|r_2|$ random numbers between 0 and 1. If the random number is lower than $\beta(r)$, the element of the p is replaced by the corresponding element of the brightest firefly;

Step 4: Fireflies move to the brightest one.

$$
\beta(r) = \frac{\beta_0}{(1 + \gamma r^2)} \tag{12}
$$

$$
x_i = x_i + \alpha(rand_{int}) \tag{13}
$$

Table 2. Solution updation

Updation	VM-string	Task-string
Current firefly p	$\{2, 3, 2, 2, 3, 3, 4, 2, 2, 4\}$	$\{2, 3, 1, 1, 3, 2, 1, 3, 2, 2\}$
Best firefly p_{best}	$\{3, 2, 2, 3, 4, 3, 2, 2, 4, 4\}$	$\{3, 2, 1, 2, 3, 1, 2, 1, 3, 2\}$
Distance r_1 and r_2	6	4
Attractiveness $\beta(r)$	0.22	0.38
$rand(\,)between(0, 1)$	$\{\boxed{0.17}, 0.59, 0.83, \boxed{0.06}, \boxed{0.11}, 0.51\}$	$\{\boxed{0.29}, 0.75, \boxed{0.33}, 0.42\}$
Update position after β-step	$\{3, 3, 2, 2, 4, 3, 2, 2, 2, 4\}$	$\{3, 2, 1, 1, 3, 2, 2, 1, 3, 2\}$
Update position after α-step	$\{3, 3, 2, \boxed{4}, \boxed{2}, 3, 2, 2, 2, 4\}$	$\{3, 2, 1, \boxed{3}, \boxed{1}, 2, 2, 1, 3, 2\}$

Algorithm 1: Firefly Algorithm

Objective function $F(x)$ Algorithm related parameters: α, β_0, γ, $iter_{max}$
Generate initial population of fireflies: $X_1, X_2, \ldots, X_n$
Light intensity I_i and x_i is determined by $F(X_i)$
while $t < iter_{max}$ **do**
    for $i = 1{:}n$ **do**
        for $j = 1{:}i$ **do**
            if $F(x_i) > F(x_j)$ **then**
             | move firefly i towards firefly j;
            else
             | move firefly j towards firefly i;
            end
            Update the attractiveness of all fireflies;
            Evaluate new solution and update light intensity;
        end
        Rank the fireflies and find the current best;
    end
end

α-**step Update:** It carries out random disturbance to avoid local optimum. Its details can be found in [15] and thus are not elaborated due to the page limit.

Based on the designs described above, the overall algorithm is illustrated as Algorithm 1.

5   Performance Evaluation

In this section, we evaluate the performance of our proposed method by comparing it with three existing methods, namely, MinMin+ [21], GAMEC [22], and

WaterDrop [1]. All experiments were performed on a Windows PC with an Intel Core i5-4210U processor and 12 GB RAM. The proposed algorithm is evaluated by two indexes, cost and workflow completion time.

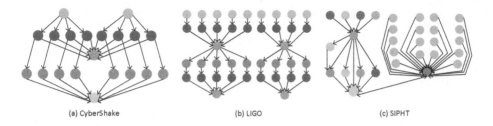

(a) CyberShake   (b) LIGO   (c) SIPHT

Fig. 5. The case templates of workflows

We used three well-known scientific workflow templates [4], namely, Cyber-Shake, LIGO and SIPHT as shown in Fig. 5.

(a) CyberShake: CyberShake workflow is used by the Southern California Earthquake Center to characterize earthquake hazards in a region.

(b) LIGO: LIGOs Inspiral Analysis workflow is used to generate and analyze gravitational waveforms from data collected during the coalescing of compact binary systems.

(c) SIPHT: The SIPHT workflow, from the bioinformatics project at Harvard, is used to automate the search for untranslated RNAs (sRNAs) for bacterial replicons in the NCBI database.

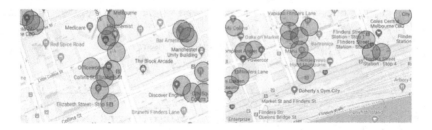

Fig. 6. Edge server deployment.

We consider that there are three edge service providers with different resource configurations and charging plans as shown by Table 3. Our experiments were conducted on the EUA dataset [6], which has been widely used in research on edge computing [8,10,17], and the edge server deployment diagram is shown in Fig. 6.

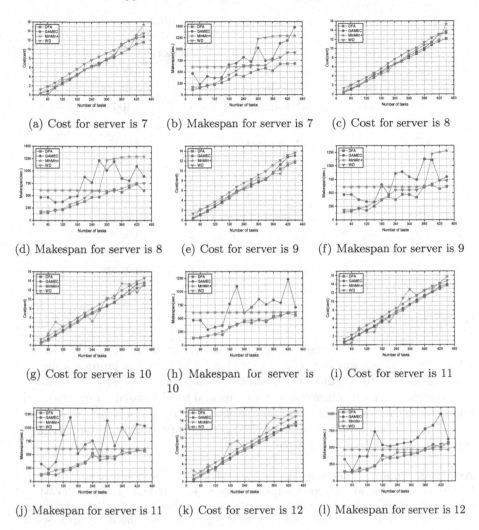

(a) Cost for server is 7   (b) Makespan for server is 7   (c) Cost for server is 8

(d) Makespan for server is 8   (e) Cost for server is 9   (f) Makespan for server is 9

(g) Cost for server is 10   (h) Makespan for server is 10   (i) Cost for server is 11

(j) Makespan for server is 11   (k) Cost for server is 12   (l) Makespan for server is 12

Fig. 7. Results with the number of edge servers from 7 to 12.

Figure 7 shows the comparison of cost and workflow completion time of different methods with varying numbers of edge servers and tasks. As can be seen, The averaged cost of our proposed method is lower than that of its peers. Similarly, the averaged workflow completion time of our method is lower as well than those of its peers, because our algorithm considers a new encoding method that takes into account the proximity constraint between servers and users.

Table 3. The unit price of heterogeneous VMs from three edge service providers

Edge service provider	VType	VCPU	Memory	Unit-price/hour
EP1	s3.small	1	1	0.0612$
	s3.medium	1	2	0.0757$
	s3.medium	1	4	0.1005$
	s3.large	2	8	0.1500$
EP2	s2.standard	1	1	0.0335$
	s2.standard	1	2	0.0509$
	s2.standard	1	4	0.08$
	s2.standard	2	4	0.1019$
EP3	t3.nano	2	0.5	0.0052$
	t3.small	2	2	0.0208$
	t3.medium	2	4	0.0416$
	t3.large	2	8	0.0832$

6   Conclusion

In this paper, we address the cost optimization of the multi-user multi-workflow scheduling environment over edge service infrastructure. We propose a discrete firefly algorithm to optimize the cost-effectiveness, while satisfy the proximity and workflow deadline constraints. Experimental results based on multiple well-known workflow templates and a real-world position dataset for edge service providers show that our proposed method clearly outperform traditional ones in terms of cost and completion time.

Acknowledgment. This work is in part supported by Fundamental Research Funds for the Central Universities under project Nos. 106112014CDJZR185503 and CDJZR12180012; Science foundation of Chongqing Nos. cstc2014jcyjA40010 and cstc2014jcyjA90027; Chongqing Social Undertakings and Livelihood Security Science and Technology Innovation Project Special Program No. cstc2016shms-zx90002; China Postdoctoral Science Foundation No. 2015M570770; Chongqing Postdoctoral Science special Foundation No. Xm2015078; Universities Sci-tech Achievements Transformation Project of Chongqing No. KJZH17104.

References

1. Adhikari, M., Amgoth, T.: An intelligent water drops-based workflow scheduling for IaaS cloud. Appl. Soft Comput. **77**, 547–566 (2019)
2. Habak, K., Ammar, M., Harras, K.A., Zegura, E.: Femtoclouds: leveraging mobile devices to provide cloud service at the edge. In: IEEE International Conference on Cloud Computing (2015)
3. Hoffa, C., et al.: On the use of cloud computing for scientific workflows. In: Fourth International Conference on e-Science, e-Science 2008, Indianapolis, IN, USA, 7–12 December 2008, pp. 640–645 (2008)

4. Juve, G., Chervenak, A., Deelman, E., Bharathi, S., Mehta, G., Vahi, K.: Characterizing and profiling scientific workflows. Future Gener. Comput. Syst. **29**(3), 682–692 (2013)
5. Juve, G., Deelman, E., Berriman, G.B., Berman, B.P., Maechling, P.: An evaluation of the cost and performance of scientific workflows on Amazon EC2. J. Grid Comput. **10**(1), 5–21 (2012)
6. Lai, P., et al.: Optimal edge user allocation in edge computing with variable sized vector bin packing. In: Pahl, C., Vukovic, M., Yin, J., Yu, Q. (eds.) ICSOC 2018. LNCS, vol. 11236, pp. 230–245. Springer, Cham (2018). https://doi.org/10.1007/978-3-030-03596-9_15
7. Li, S., Huang, J.: GSPN-based reliability-aware performance evaluation of IoT services. In: 2017 IEEE International Conference on Services Computing, SCC 2017, Honolulu, HI, USA, 25–30 June 2017, pp. 483–486 (2017)
8. Li, X., et al.: Quality-aware service selection for multi-tenant service oriented systems based on combinatorial auction. IEEE Access **7**, 35645–35660 (2019)
9. Liang, T., Yong, L., Wei, G.: A hierarchical edge cloud architecture for mobile computing. In: IEEE Infocom -the IEEE International Conference on Computer Communications (2016)
10. Liu, Y., He, Q., Zheng, D., Zhang, M., Chen, F., Zhang, B.: Data caching optimization in the edge computing environment. In: 2019 IEEE International Conference on Web Services, ICWS 2019, Milan, Italy, 8–13 July 2019, pp. 99–106 (2019)
11. Lunardi, W.T., Voos, H.: An extended flexible job shop scheduling problem with parallel operations. ACM SIGAPP Appl. Comput. Rev. **18**(2), 46–56 (2018)
12. Lyu, X., et al.: Optimal schedule of mobile edge computing for internet of things using partial information. IEEE J. Sel. Areas Commun. **35**(11), 2606–2615 (2017)
13. Mao, Y., Zhang, J., Letaief, K.B.: Dynamic computation offloading for mobile-edge computing with energy harvesting devices. IEEE J. Sel. Areas Commun. **34**(12), 3590–3605 (2016)
14. Mao, Y., Zhang, J., Song, S.H., Letaief, K.B.: Stochastic joint radio and computational resource management for multi-user mobile-edge computing systems. IEEE Trans. Wireless Commun. **16**(9), 5994–6009 (2017)
15. Marichelvam, M.K., Prabaharan, T., Yang, X.: A discrete firefly algorithm for the multi-objective hybrid flowshop scheduling problems. IEEE Trans. Evol. Comput. **18**(2), 301–305 (2014)
16. Peng, Q., Jiang, H., Chen, M., Liang, J., Xia, Y.: Reliability-aware and deadline-constrained workflow scheduling in mobile edge computing. In: 2019 IEEE 16th International Conference on Networking, Sensing and Control (ICNSC), pp. 236–241. IEEE (2019)
17. Peng, Q., et al.: Mobility-aware and migration-enabled online edge user allocation in mobile edge computing. In: 2019 IEEE International Conference on Web Services, ICWS 2019, Milan, Italy, 8–13 July 2019, pp. 91–98 (2019)
18. Sahni, J., Vidyarthi, D.: A cost-effective deadline-constrained dynamic scheduling algorithm for scientific workflows in a cloud environment. IEEE Trans. Cloud Comput. **6**(1), 2–18 (2018)
19. Sanaei, P., Akbari, R., Zeighami, V., Shams, S.: Using firefly algorithm to solve resource constrained project scheduling problem. In: Bansal, J., Singh, P., Deep, K., Pant, M., Nagar, A. (eds.) BIC-TA 2012. AISC, vol. 201, pp. 417–428. Springer, New Delhi (2013)
20. Shi, W., Jie, C., Quan, Z., Li, Y., Xu, L.: Edge computing: vision and challenges. IEEE Internet of Things J. **3**(5), 637–646 (2016)

21. Tabak, E.K., Cambazoglu, B.B., Aykanat, C.: Improving the performance of independent task assignment heuristics minmin, maxmin and sufferage. IEEE Trans. Parallel Distrib. Syst. **25**(5), 1244–1256 (2014)
22. Wu, H., Deng, S., Li, W., Fu, M., Yin, J., Zomaya, A.Y.: Service selection for composition in mobile edge computing systems. In: 2018 IEEE International Conference on Web Services, ICWS 2018, San Francisco, CA, USA, 2–7 July 2018, pp. 355–358 (2018)
23. Yang, X.S.: Firefly algorithms for multimodal optimization. Mathematics **5792**, 169–178 (2009)
24. Yang, X.S.: Firefly algorithm, stochastic test functions and design optimisation. Int. J. Bio-Inspired Comput. **2**(2), 78–84 (2010)
25. Zhang, Y., Chen, X., Chen, Y., Li, Z., Huang, J.: Cost efficient scheduling for delay-sensitive tasks in edge computing system. In: 2018 IEEE International Conference on Services Computing, SCC 2018, San Francisco, CA, USA, 2–7 July 2018, pp. 73–80 (2018)
26. Zhao, T., Sheng, Z., Guo, X., Niu, Z.: Tasks scheduling and resource allocation in heterogeneous cloud for delay-bounded mobile edge computing. In: IEEE International Conference on Communications (2017)

Differential Privacy Preservation for Smart Meter Systems

Junfang Wu[1], Weizhong Qiang[1(✉)], Tianqing Zhu[2], Hai Jin[1], Peng Xu[1,3],
and Sheng Shen[2]

[1] National Engineering Research Center for Big Data Technology and System,
Services Computing Technology and System Lab, Cluster and Grid Computing Lab,
Big Data Security Engineering Research Center, School of Computer Science and
Technology, Huazhong University of Science and Technology, Wuhan 430074, China
wzqiang@hust.edu.cn
[2] School of Computer Science, University of Technology Sydney,
Ultimo, NSW 2007, Australia
[3] Shenzhen Huazhong University of Science and Technology Research Institute,
Shenzhen 518057, China

Abstract. With the rapid development of IoT and smart homes, smart
meters have received extensive attention. The third-party applications,
such as smart home controlling, dynamic demand-response, power monitoring, etc., can provide services to users based on consumption data of
household electricity collected from smart meters. With the emergence
of non-intrusive load monitoring, privacy issues from the data of smart
meters become more and more severe. Differential privacy is a recognized
concept that has become an important standard of privacy preservation
for data with personal information. However, the existing privacy protection methods for the data of smart meters that are based on differential
privacy sacrifices the actual energy consumption to protect the privacy
of users, thus affecting the charging of power suppliers. To solve this
problem, we propose a group-based noise adding method, so as to ensure
the correct electricity billing. The experiments with two real-world data
sets demonstrate that our approach can not only provide a strict privacy
guarantee but also improve performance significantly.

Keywords: Differential privacy · Smart meter · Internet of Things

1 Introduction

Sensors are deployed on a large scale in IoT systems, to collect physical data and
feedback them for third-party analysis and calculation, thus providing diverse
services to users. At the same time, the emergence of a massive amount of data
will inevitably lead to the privacy issue about users' data, particularly in the
smart home systems, where a large amount of data that contain users' privacy
information are generated. Therefore, how to protect the privacy of users' data
has received widespread attention.

© Springer Nature Switzerland AG 2020
S. Wen et al. (Eds.): ICA3PP 2019, LNCS 11944, pp. 669–685, 2020.
https://doi.org/10.1007/978-3-030-38991-8_44

Smart meters are an essential part of smart home systems that can provide functions such as energy cost allocation, fault analysis, demand control, and power quality analysis. The smart meter can support the operation of meters to achieve accurate billing [4]. With the advent of Non-Intrusive Load Monitoring (NILM), which makes it easy to identify the total energy load for a single device based on the power characteristics of the appliance, thereby enabling the analysis of the internal activities of the home [11]. This analysis can provide users with device recommendations and the optimized use of home energy. But at the same time, it will also be exploited by malicious attackers, thus leaking personal privacy information such as social and financial status.

In the past few years, researchers have proposed many privacy protection measures for smart meters. One type of the measures achieves privacy through changing the periodic power, such as Gaussian random noise [7] and wavelet transformation [13]. Unfortunately, they cannot effectively control the noise range and keep the total energy consumption constant. In addition, *Dynamic Privacy Analyzer* [15] implements a unique privacy metric to obtain a sufficient balance between privacy and utility, which does not provide strict privacy. Another type of the measures is to use household energy storage equipments, for example, *electric vehicles (EVs), heating, ventilating and air conditioning (HVAC)* systems, rechargeable batteries, to achieve load balancing of energy consumption, thereby hiding the consumption data of home load equipments [14]. This method requires different energy control strategies for different combinations of family energy storage units. Although it can hide users' behavior habits, it cannot provide strict proof of privacy.

Differential privacy is a strictly provable privacy protection framework used in a variety of data publishing and data querying scenarios [6]. Differential privacy is also used to protect users' privacy in a battery-based smart grid [2,17], in which a low-cost differential privacy protection scheme is proposed under the pricing policy. However, battery-based privacy protection increases the overhead of hardware, which affects users' experience. The biggest drawback of the traditional noise-adding methods is that it will generate a large margin of error by directly adding noise to the meter data, which affects the accuracy of the billing, and makes the pricing inaccurate.

In this paper, we propose a new method based on differential privacy for privacy protection of smart meters. The basic idea of our proposed scheme is to add packet noise to ensure that the total energy consumption is infinitely close to the correct value after each noise addition, and at the same time, the distribution of electricity consumption of each device is within the margin of error.

There are some challenges in implementing privacy protection in charging functions of smart meters. First, how to ensure that the total energy consumption of the query after the noise processing is unchanged, thus not affecting the billing of the power company. We consider dividing the query into two parts and respectively performing privacy protection processing so that the accurate total energy consumption is calculated based on the data sets of two query results.

Second, some users will share the energy consumption data with third-party applications to obtain services such as intelligent optimization and device monitoring. Therefore, we consider how to ensure the privacy of individuals, while making the changes in the distribution of electricity consumption of various devices not affect their utility. We propose a privacy protection method for top-k query, in which only resulting data sets of the query are sent to third-party applications, thus ensuring the privacy of users.

Our contributions are summarized as follows:

- A new method based on differential privacy protection is proposed to protect the privacy of users when sharing data with third-party services.
- The proposed method of group-based noise addition, based on differential privacy, guarantees privacy while keeping the total energy consumption constant.
- By using the REDD [10] and GREEND [12] data sets, our experiments confirm that our proposed solution can resist various background knowledge attacks during the use of smart meters.

The rest of this paper is organized as follows. Section 2 introduces the system model and privacy challenges. Section 3 presents some of the symbolic descriptions in this paper, as well as the background of differential privacy. The proposed privacy protection framework and protection methods for two queries are discussed in Sect. 4. Effectiveness of the proposed method is evaluated in Sect. 5. Section 6 discusses the related works, and Sect. 7 concludes this paper.

2   Problem Description

2.1   System Model

Our system model for the smart meter includes smart meters, a gateway, a cloud server, and users' phones, as shown in Fig. 1. As a type of inexpensive home device, smart meters are typically installed in the unprotected environment, communicate over untrustworthy channels, and are not able to report users' energy consumption in real time. Since users' energy consumption data contains a large amount of private information, we consider that smart meters only release privacy-protected data to the cloud server. In our model, the description of each entity for the smart meter of a house with cloud-based services is as follows:

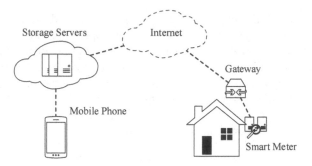

Fig. 1. The system model for the smart meter of a house with a cloud-based service.

- *Smart meter:* Smart meters deploy sensor devices that collect data such as the power and voltage of the circuit. The smart meter publishes the collected data to the cloud server hourly or daily through the gateway. For the privacy of users, we need to process the published data.
- *Gateway:* A gateway is a powerful entity that connects smart meters and cloud servers for data communication.
- *Storage server:* The storage servers contain a Meter Data Management System (MDMS) [1] in which the processed data will be sent to a power control center, or to a third-party application. The power control center accepts the user's energy usage information, charges the energy consumption, and feedbacks the costs to the user's mobile phone. Then, it obtains services provided by third-party applications, such as device recommendation, home energy usage optimization, and dynamic demand response, by allowing third-party applications to access the energy consumption data.
- *User's mobile phone:* To realize the user's acquisition and control of household electricity, the user's mobile phone mainly accepts results of computing feedback from the cloud server.

2.2 Privacy Threats

Common privacy attacks against smart meter systems include external attacks and internal attacks. An external attacker may intercept gateway information to inject false information on the channel or eavesdrop on the transmitted data, thereby obtaining personal information of the user. Internal attackers may include users, gateways, and cloud servers. Gateways and cloud servers which we believe are untrustworthy, may be interested in data, and may try to analyze users' private information from the data and calculations. A meter user could be interested in other users' meter data, e.g., trying to obtain the privacy of others through the smart meter. However, smart meters themselves cannot communicate directly, thus are considered to be semi-trusted.

3   Preliminaries

3.1   Notations

The description of the parameter symbols in this paper is shown in Table 1. Since the energy consumed by different devices at different times is different, the energy consumption for the ith device at time j is denoted by P_{ij}. In the smart meter system, a household has n electricity devices and the period for the energy consumption query is defined as m. The data set of energy consumption for the ith device obtained by querying according to the period is P_{im}, and the data set of average energy consumption for all the devices is P_m. E represents the total energy consumption in a billing cycle. The result set of the query with privacy protection enhanced is expressed as P'_m. ϵ is the privacy budget, and s is the query sensitivity.

3.2   Differential Privacy

Differential privacy is a provable privacy notation that Dwork [5] proposed for the privacy leak of statistical databases, which is widely used in a variety of areas.

Definition 1 *(ϵ-Differential Privacy). A randomized algorithm M provides ϵ-differential privacy. Then, for any pair of neighboring datasets D and D', for every set of outputs S_M, the algorithm M satisfies:*

$$P_r[M(D) \in S_M] \leq \exp(\epsilon) \times P_r[M(D') \in S_M] \tag{1}$$

Table 1. Notations

Notation	Description
i	The sequence number for each electricity device
j	Time point
m	The period for the energy consumption query
P_{ij}	The energy consumption for the ith device at time j
P_m	The result set of the energy consumption query
P'_m	The result set of the query with adding privacy protection
E	The total energy consumption in a billing cycle
n	The number of devices
k	The number of devices queried
$devs$	The result set of the top-k query
s	The sensitivity
ϵ	The privacy budget

Algorithm M provides privacy protection by randomizing the output results, while privacy parameter ϵ guarantees whether the individual data exists in the data set does not affect the probability that the algorithm outputs the same result.

Definition 2 *(Sensitivity). For a mapping $f : D \rightarrow R_d$, the sensitivity of s is defined as follows:*

$$s_f = max_{D,D'} \|f(D) - f(D')\|_1 \tag{2}$$

Where $\|f(D) - f(D')\|_1$ is the 1-norm distance between $f(D)$ and $f(D')$.

Definition 3 *(Laplace Mechanism). For a dataset D, given a function, $f : D \rightarrow R_d$ with the sensitivity of s, the random algorithm $\hat{f}(D)$ provides ϵ-differential privacy:*

$$\hat{f}(D) = f(D) + Lap(\frac{s}{\epsilon}) \tag{3}$$

The Laplace mechanism is suitable for numeric queries, enabling differential privacy by adding Laplace noise to real results.

4   Smart Meters with Privacy Protection

In this section, we introduce the privacy protection strategies under the two queries and discuss them in detail.

4.1   The Function of Querying Energy Consumption

The control center receives the energy consumption data reported by the smart meter, and uses the average energy consumption hourly or daily as a reporting node to calculate the user's consumption bill based on the collected data.

Generation of the Data Published. We consider the smart meter to report the energy consumption data periodically to the power control center. The specific steps are as follows:

- *Step 1:* The query result set P_{im} of each sensor device is calculated;
- *Step 2:* The results of all sensors are integrated to generate a total query set P_m.

A sensor cannot directly answer the request of an energy consumption query from the control center, and the correct result set must be queried from the smart meter. Taking the query of average energy consumption as an example, the results obtained by the sensors that query the average energy consumption are q_1, q_2, q_3, respectively. To calculate the data set of average energy consumption $P_m = sum\{q_1, q_2, q_3\}$, we only need to calculate the value of the sensor query (Step 1).

Let's suppose that a new query set $Q(q_1, q_2, ...q_n)$ requires aggregate calculations of different queries, where $q_1, q_2, ...q_n$ represent the average energy consumption on different electricity devices. The methods for calculating *sum, min, max, average* are as follows:

- SUM: $sum_D = \sum_{i=1}^{n} q_i(D_i)$
- MIN: $min_D = min\{q_1(D_1), q_2(D_2), ...q_n(D_n)\}$
- MAX: $max_D = max\{q_1(D_1), q_2(D_2), ...q_n(D_n)\}$
- AVE: $ave_D = \frac{1}{n} \sum_{i=1}^{n} q_i(D_i)$

In the above expressions, the *sum, min, max, average* of the dataset D are expressed as $sum_D, min_D, max_D, ave_D$. Once the calculation is completed, the smart meter sends the result data set to the control center of power for further processing.

Noise Adding. Typically, smart meters send the measurement data to the server during a billing cycle. Smart meters need to query energy consumption periodically to meet the services' requirements, such as billing services. At the same time, the smart meter needs to perform privacy protection on the data before sending the query set, to avoid sending the user's private information. We use the method of noise interference. See Algorithm 1 for specific steps.

Algorithm 1. Energy Consumption Query with Privacy Protection

---

Require: average energy consumption P_m.
Ensure: result P'_m after noise addition.
1: Randomly divide the published result set P_m into two mutually disjoint subsets P_{m1} and P_{m2};
2: For the data of the subset P_{m1}, add a random noise $noise_i$ satisfying $Lap\left(\frac{s}{\epsilon}\right)$ to each element, and keep the noise set $noise$ to obtain the changed subset P'_{m1};
3: For the subset P_{m2}, add noise of $-noise_j$ to each element of the noise set $noise$ for each element in the subset, and obtain a new subset P'_{m2};
4: Issue the interrogated query set $P'_m = [P'_{m1}, P'_{m2}]$ to the control center of power.

Bill Generation. After collecting the electricity consumption data issued by the smart meter, the control center of power calculates the corresponding amount according to the local electricity price and generates an electricity bill to be sent to the user's mobile phone. The detailed steps are as follows:

- *Step 1:* The total energy consumption E is calculated according to the average energy consumption obtained from the query;
- *Step 2:* Assuming the local electricity price is *price*, the electricity fee $F = E \times price$ is calculated.

In summary, to make the final electricity fee no deviation, we add privacy protection to the query process and interfere the data of the query set. We divide the query into two parts and then protect each part with differential privacy. The exact value is obtained by calculating the total amount of electricity for a period.

4.2   Top-K Query for Devices

The algorithm of energy decomposition aims to decompose the household's energy consumption data collected from a single measurement point into the consumption data of each device, and then model and analyze the energy consumption equipment [3]. However, the energy decomposition directly from the raw energy data will reveal users' private information. The purpose of Non-intrusive load monitoring (NILM) that decomposes total energy consumption is to understand the energy consumption of each device, to take measures to reduce the cost of energy consumption or provide personalized recommendations. Therefore, the process of top-k query that can provide k devices which consume the most energy is an important option of queries. We consider that the calculation of this step can be done directly by the smart meter, by sending the result set to the third-party application without exposing the user's original energy consumption data.

Query Generation. When the third-party service provides some application support for the user, it needs to perform statistical query operations on the household energy consumption data, such as periodical analysis of energy consumption, and inquiry for equipments which use a large amount of electricity. We take the top-k query for devices as an example to design the privacy protection method. The top-k query for devices is a function with data pre-processing of the non-intrusive load monitoring toolkit. Only the top-k energy-consuming devices are modeled, and the remaining devices only generate noise. See Algorithm 2 for specific steps of queries.

By modeling the equipment with large energy consumption obtained by querying, it is possible to analyze the reasons for the large consumption of household electricity, and to propose a low-energy energy consumption scheme; and some abnormal equipment with excessive energy consumption can also be found.

Noise Adding. In this paper, we interference answers of the query using the Laplacian-based noise, with the following steps.

Algorithm 2. Top-k Devices
___
Require: energy consumption P_{ij} of each device, user-defined k value, period m.
Ensure: top-k devices $devs$.
 1: Calculate the average energy consumption P_{im} of each device in time m;
 2: Calculate the energy consumption E_i for the i-th device according to P_{im};
 3: Return the encoding of the k devices with the largest energy consumption value dev_k.

- *Step 1:* Query to get the average energy consumption for the device P_{im};
- *Step 2:* Add noise that satisfies $Lap\left(\frac{s}{\epsilon}\right)$ to each element in the data set P_{im}, and obtain the interference energy consumption data set P'_{im};
- *Step 3:* Calculate the total energy consumption value E'_i of each device in time m based on the average energy consumption P'_{im}.

In the algorithm of top-k query, there are two protection methods to calculate the false results. One is to add noise before querying the average energy consumption of each device so that the energy consumption after the interference is output, and then sort and search to obtain the top-k devices. The other is to define a selection function that satisfies the differential privacy index distribution. Here, we add noise to the energy consumption query of the device.

Publishing of Result Sets. We will output top-k devices according to the energy consumption threshold ω_p obtained from the original query. The top-k query combined with the noise addition process satisfies ϵ-differential privacy. The query process with privacy protection is detailed in Algorithm 3.

Algorithm 3. Querying Top-k Devices with Differential Privacy

---

Require: energy consumption P_{ij} of each device, user-defined k value, period m.
Ensure: top-k devices *devs*.
  1: Calculate the average energy consumption P_{im} of each device in time m;
  2: Sort the devices according to the energy consumption, and select the average value of the set of the devices of the k-th and $k + 1$-st as the threshold value ω_p;
  3: Add noise to the query answer of the device P_{im}, and the result of the interference is P'_{im};
  4: Calculate the equipment energy consumption value E'_i in the time m;
  5: According to the threshold ω_p, select k devices with higher energy consumption to satisfy ϵ-differential privacy.

Algorithm 3 shows how to protect users' privacy in top-k query in detail. First, the average energy consumption P_{im} for the device at time m is calculated according to the energy consumption P_{ij} for each device (Step 1). Then, the set of energy consumption $\{P_{1m}, P_{2m}, ... P_{nm}\}$ is sorted in descending order, and the average of the energy consumption of the k-th and $k+1$-st magnitudes is selected as the threshold ω_p (Step 2). Finally, the device with the energy consumption greater than the threshold in a differential privacy mode is selected. The returned device set is the answer to a privacy protection query.

5   Evaluation

In this section, we evaluate the effectiveness of our method through a series of experiments. First of all, we introduce the pre-experiment settings and environment configuration, as well as the processing of the data set.

Datasets: We complete the experiments on two real data sets, Reference Energy Disaggregation Data Set (REDD) and GREEND. REDD contains detailed electricity usage information from multiple homes, while the GREEND data set includes power usage information obtained through measurement activities in many households in Austria and Italy. We select the energy consumption data of one household in each of the two data sets, and after data preprocessing for a unified data format, retain the sequence of the actual measurement value of the device that changes with the time stamp.

Metrics: For the publishing function energy of consumption data, we calculate the mean absolute error (MAE) of the total energy consumption to describe the effect of the query answer. In Formula (4), $E(D)$ represents the correct result of the total energy consumption calculation, and $E'(D)$ represents the value of total energy consumption calculated after privacy protection. T refers to the number of experiments, which makes the assessment universal, by obtaining the average of the relative errors through multiple experiments.

$$MAE = \frac{1}{T} \sum_{1}^{T} |E'(D) - E(D)| \tag{4}$$

In addition, we use the correct rate of the query results to evaluate the accuracy of the top-k query. In Formula (5), DEV is the top-k devices set obtained by the query, and DEV' is the result set obtained by querying after noise addition. $|DEV \cap DEV'|$ indicates the number of correct results obtained by the noise-enhanced query, and $|DEV|$ represents the size of the query result set, which is k. Similarly, T refers to the number of experiments.

$$correctrate = \frac{1}{T} \sum_{1}^{T} \frac{|DEV \cap DEV'|}{|DEV|} \tag{5}$$

Configuration: We implement our algorithm in python and run it on a machine with a 3.3 GHz Intel Core i5 processor and 4 GB memory. At the same time, due

Table 2. Parameter settings

Parameter	Description	Value	Default
m	The period for the energy consumption query	Hourly, daily	–
ϵ	The privacy budget	0.1–3.0	1.0
k	The number of devices queried	1–20	3/1

to the randomness of the noise used, we repeat each experiment 60 times and calculate the average to ensure the accuracy of the results. See the parameter settings in Table 2.

Comparison: In the framework that we proposed, we can increase the traditional noise interference and realize the query and release of data. Therefore, we compare the proposed method with the basic method of adding noise to evaluate the utility.

5.1 The Experiment for the Energy Consumption Query

The results of the query change as the privacy budget changes. In the experiments, MAE is used to evaluate the performance of our method on two data sets. To understand the impact of the privacy budget on the answers of the query, we define the privacy budget to the range from 0.1 to 1.0. At the same time, experiments are conducted on different query cycles.

Figure 2 shows the function of an energy consumption query for the REDD and GREEND data sets, and a graph of the error value of the query results as a function of the privacy budget. According to Fig. 2, we can observe that, for the two query methods, the MAE gradually decreases with the increase of the privacy budget, because the smaller the privacy budget is, the more noise is added. It shows that the larger the value of the privacy budget is, the smaller the noise interference is, and the smaller the impact is on the result. Thus, our method achieves higher accuracy of the calculation and a smaller value of the MAE.

In addition, we can see that our approach is superior to the traditional Laplace scheme from the results. The error between the exact value and the total value of energy consumption calculated by our method is small. Figure 2(a) and (b) is the comparative analysis of the results of the query on the data set REDD. We set the query period to one hour or one day, and the comparative analysis of MAE is respectively shown in Fig. 2(a) and (b). The illustration shows that no matter which way of querying, the error value of the result calculated by our method is smaller than that calculated by the traditional method. Figure 2(c) and (d) are the result analysis graphs of the query on the data set GREEND, with one hour and one day as the query conditions, respectively. The same we can see that our method is far superior to the traditional calculation method.

5.2 The Experiment for the Top-K Query

We examine the relationship between the performance of the proposed method and the privacy budget of the top-k query based on the correct rate of results. We define a privacy budget from 0.1 to 3.0 in steps of 0.1. We have experimented with two different values of the query cycle. The correct rate of the query results is shown in Fig. 3.

Figure 3 shows the relationship between the change in the size of the privacy budget and the correct rate of the top-k query. We perform experiments on

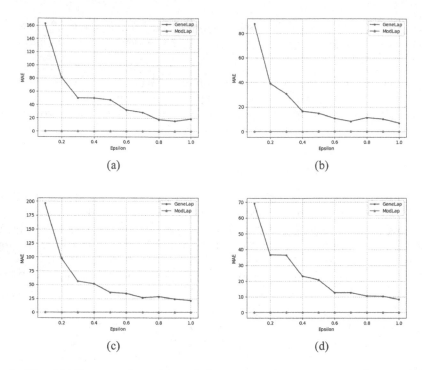

Fig. 2. The MAE comparison chart of the average energy consumption queried on both data sets based on our method (red line, ModLap) and the traditional protection method (blue line, GeneLap). Query period is equal to (a) hourly/REDD, (b) daily/REDD, (c) hourly/GREEND, (d) daily/GREEND. (Color figure online)

both the REDD and GREEND data sets. We know that the larger the privacy budget is, the smaller the interference with the calculation results is, and thus the accuracy of the query results will be higher. From Fig. 3 we can verify that as the privacy budget increases, the accuracy of top-k query tends to be optimal. Figure 3(a) and (b) are queries for devices which use a large amount of energy from 20 electricity devices in the REDD data set. We set the default value of k to 3. According to the line chart, it is easy to see that the correct rate of our method is higher than the traditional method when the privacy budget is the same. The total number of devices recorded in different data sets is different, and the optimal k value of the top-k query will be different. Figure 3(c) and (d) are graphs of the results obtained from the query of the data set GREEND. The default value of the parameter k is 1. In Fig. 3(c), we find that when the privacy budget increases to 0.6, the correct rate of the top-k query for devices of our method has approached 1, while the traditional privacy protection method has an accuracy of 0.83.

Figure 4 shows the correct rate of the results of the two dataset queries as a function of k. We define the k value from 1 to the total number of devices in steps of 1. As shown in the figure, as the value of k changes, from the minimum value of

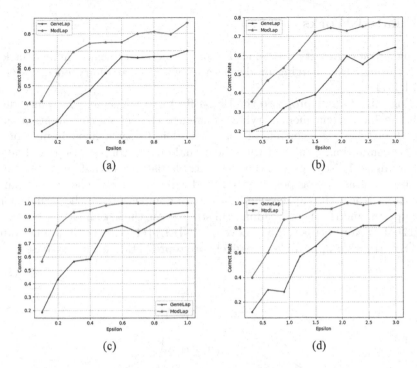

Fig. 3. Based on our method (red line, ModLap) and the traditional protection method (blue line, GeneLap), we search the two datasets for a line graph of the correct rate of the top-k devices as a function of privacy parameters. k (number of devices returned) is equal to (a) 3, (b) 3, (c) 1, (d) 1. Query period is equal to (a) hourly/REDD, (b) daily/REDD, (c) hourly/GREEND, (d) daily/GREEND. (Color figure online)

1 to the maximum value, the correct rate of the query result is first decreased and then increased. When the total number of devices is n, the top-k query and the top-$(n - k)$ query are symmetric. For example, when querying one device with the most energy consumption, the correct rate is close to 1, and for n devices that use more electricity, the output device is 100% correct. Therefore, the influence of the increase of the k value on the query result increases at first and then decreases, and the value that has the most significant impact on the calculation result appears in the range of the value of k. From Fig. 4(a) and (b), we can see that as the value of k increases, the correct rate of the results calculated by our method is higher than that of the traditional method. In Fig. 4(a), when $k = 9$, the optimized query obtains the lowest accuracy of 0.70, while when $k = 6$, the traditional noise-adding method calculates the lowest accuracy of 0.52. As explained above, the improved privacy protection method has less influence on the magnitude of the k value that interferes with the query result.

6   Related Works

Existing privacy protection schemes for smart meters typically implement load hiding through the charging or discharging with an auxiliary energy storage [2,8,14,17,18].

Kalogridis *et al.* [8] proposed the Best-Effort (BE) scheme, which introduces a rechargeable battery model, trying to balance the energy consumption curve of smart meters. The principle of the method is to set a fixed branch of the energy consumption, combined with the actual energy consumption, and adjust the charging and discharging of the rechargeable battery so that the total energy consumption is as close as possible to a fixed value. However, when the rechargeable battery reaches its capacity limit, it cannot compensate for the actual energy consumption, and the meter reading will still expose the actual energy consumption [11]. The Non-Intrusive Load Leveling (NILL) [11] and the Stepping Framework (SF) [16] that appeared successively are representative Battery-based Load

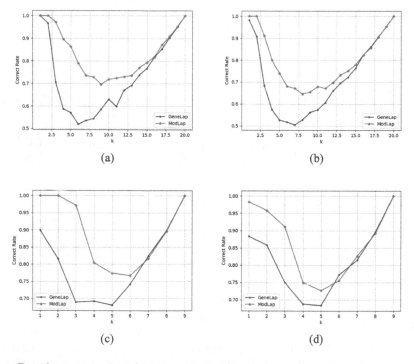

Fig. 4. Based on our method (red line, ModLap) and the traditional protection method (blue line, GeneLap), we search the two datasets for a line graph of the correct rate of the top-k devices as a function of the k value (number of devices returned). ϵ (privacy parameter) is equal to (a) 1.0, (b) 3.0, (c) 1.0, (d) 3.0. Query period is equal to (a) hourly/REDD, (b) daily/REDD, (c) hourly/GREEND, (d) daily/GREEND. (Color figure online)

Hiding (BLH) schemes, the main purpose of which is to hide the energy consumption of actual household equipment.

Although the above two solutions reduce the limitation of the capacity of the rechargeable battery, neither of them implements differential privacy until Zhao et al. [18] proposed a Multitasking-BLH-Exp3 (MBE) algorithm which can provide differential privacy protection. Backes et al. [2] used batteries to randomly increase or decrease energy consumption to hide sensitive energy consumption information, and this randomized process also satisfies differential privacy. Battery-based load hiding enables differential privacy but does not consider cost issues. Therefore, a battery-based differential privacy-preserving scheme was proposed by Zhang et al. [17] that reduces the cost according to the pricing strategy by solving the problem of maximizing the benefits of Multi-armed bandit (MAB), and simultaneously achieves difference privacy and cost savings. However, the charging and discharging of the home battery could conflict with the users' economic interests [9].

The differential privacy scheme based on rechargeable battery solves the privacy problem of smart meter well, but in real life, the installation of a rechargeable battery that meets the requirements of the smart home system is not practical, which makes this method difficult to promote. At the same time, frequent charging and discharging of rechargeable batteries can cause unnecessary cost [14]. Sun et al. [14] proposed the potential of utilizing household energy storage devices, by using Markov decision to take the randomness of users' needs of energy consumption into consideration, and using the Q learning algorithm to adjust the control of the storage unit, so as to hide the actual energy consumption of each device. Since the storage unit combines the electric vehicle and heating, ventilation and air conditioning systems of the user's home, the installation cost is reduced. However, this method can only achieve effective billing, and the collected meter data cannot be used for monitoring devices to achieve intelligent control.

7   Conclusion

In this paper, we propose an optimized framework for the privacy protection of smart meters based on the original privacy protection method, in which users can share the result set of a query with a third party without revealing their private information. It includes three main parts: smart meters, third-party computing servers, and users' mobile phones. In the proposed method, based on the characteristics of meter charging, we propose a group-based noise adding method to avoid the impact of noise on meter charging. Our experiments prove that the privacy protection query results do not produce errors in the calculation of electricity charges. At the same time, the experiments are also conducted for the top-k query to query the energy consumption of the k most used devices, which prove that query results of our method are more effective than that of traditional protection methods.

Acknowledgements. This work is supported in part by National Natural Science Foundation of China under grant No. 61772221 and in part by the Shenzhen Fundamental Research Program under Grant JCYJ20170413114215614.

References

1. Asghar, M.R., Dan, G., Miorandi, D., Chlamtac, I.: Smart meter data privacy: a survey. IEEE Commun. Surv. Tutor. **19**(4), 2820–2835 (2017)
2. Backes, M., Meiser, S.: Differentially private smart metering with battery recharging. In: Garcia-Alfaro, J., Lioudakis, G., Cuppens-Boulahia, N., Foley, S., Fitzgerald, W.M. (eds.) DPM/SETOP 2013. LNCS, vol. 8247, pp. 194–212. Springer, Heidelberg (2014). https://doi.org/10.1007/978-3-642-54568-9_13
3. Batra, N., et al.: NILMTK: an open source toolkit for non-intrusive load monitoring. In: Proceedings of the 5th International Conference on Future Energy Systems (ACM e-Energy 2015), pp. 265–276. ACM (2014)
4. Depuru, S.S.S.R., Wang, L., Devabhaktuni, V., Gudi, N.: Smart meters for power grid - challenges, issues, advantages and status. In: Proceedings of the IEEE/PES Power Systems Conference and Exposition, pp. 1–7. IEEE (2011)
5. Dwork, C.: Differential privacy: a survey of results. In: Agrawal, M., Du, D., Duan, Z., Li, A. (eds.) TAMC 2008. LNCS, vol. 4978, pp. 1–19. Springer, Heidelberg (2008). https://doi.org/10.1007/978-3-540-79228-4_1
6. Dwork, C., McSherry, F., Nissim, K., Smith, A.: Calibrating noise to sensitivity in private data analysis. J. Priv. Confid. **7**(3), 17–51 (2016)
7. He, X., Zhang, X., Kuo, C.C.J.: A distortion-based approach to privacy-preserving metering in smart grids. IEEE Access **1**, 67–78 (2013)
8. Kalogridis, G., Efthymiou, C., Denic, S.Z., Lewis, T.A., Cepeda, R.: Privacy for smart meters: Towards undetectable appliance load signatures. In: Proceedings of the 1st IEEE International Conference on Smart Grid Communications (SmartGridComm 2010), pp. 232–237. IEEE (2010)
9. Kifer, D., Machanavajjhala, A.: No free lunch in data privacy. In: Proceedings of the 2011 ACM SIGMOD International Conference on Management of Data (SIGMOD 2011), pp. 193–204. ACM (2011)
10. Kolter, J.Z., Johnson, M.J.: REDD: a public data set for energy disaggregation research. In: Proceedings of the Workshop Data Mining Applications in Sustainability (SustKDD 2011). ACM (2011)
11. McLaughlin, S., McDaniel, P., Aiello, W.: Protecting consumer privacy from electric load monitoring. In: Proceedings of the 18th ACM Conference on Computer and Communications Security, pp. 87–98. ACM (2011)
12. Monacchi, A., Egarter, D., Elmenreich, W., D'Alessandro, S., Tonello, A.M.: GREEND: an energy consumption dataset of households in Italy and Austria. In: Proceedings of the 2014 IEEE International Conference on Smart Grid Communications (SmartGridComm 2014), pp. 511–516. IEEE (2014)
13. Ren, X., Yang, X., Lin, J., Yang, Q., Yu, W.: On scaling perturbation based privacy-preserving schemes in smart metering systems. In: Proceedings of the 22nd International Conference on Computer Communications and Networks (ICCCN 2013), pp. 1–7. IEEE (2013)
14. Sun, Y., Lampe, L., Wong, V.W.: Smart meter privacy: exploiting the potential of household energy storage units. IEEE Internet Things J. **5**(1), 69–78 (2018)

15. Ukil, A., Bandyopadhyay, S., Pal, A.: Privacy for IoT: involuntary privacy enablement for smart energy systems. In: Proceedings of the 2015 IEEE International Conference on Communications (ICC 2015), pp. 536–541. IEEE (2015)
16. Yang, W., Li, N., Qi, Y., Qardaji, W., McLaughlin, S., McDaniel, P.: Minimizing private data disclosures in the smart grid. In: Proceedings of the 16th ACM Conference on Computer and Communications Security (CCS 2012), pp. 415–427. ACM (2012)
17. Zhang, Z., Qin, Z., Zhu, L., Weng, J., Ren, K.: Cost-friendly differential privacy for smart meters: exploiting the dual roles of the noise. IEEE Trans. Smart Grid 8(2), 619–626 (2017)
18. Zhao, J., Jung, T., Wang, Y., Li, X.: Achieving differential privacy of data disclosure in the smart grid. In: Proceedings of the 33rd IEEE International Conference on Computer Communications (INFOCOM 2014), pp. 504–512. IEEE (2014)

Performance Modelling and Evaluation

Secure Multi-receiver Communications: Models, Proofs, and Implementation

Maomao Fu[1,2], Xiaozhuo Gu[1(✉)], Wenhao Dai[1,2], Jingqiang Lin[1], and Han Wang[3]

[1] State Key Laboratory of Information Security, Institute of Information Engineering, CAS, Beijing, China
guxiaozhuo@iie.ac.cn
[2] School of Cyber Security, University of Chinese Academy of Sciences, Beijing, China
[3] Insurance Supervision and Administration Commission of Bank of China, Beijing, China

Abstract. With the demand of providing message authentication and confidentiality as well as receiver anonymity in applications such as multicast communication, digital content distribution systems, and pay-per-view channels, many anonymous multi-receiver signcryption mechanisms have been put forward to offer these functions efficiently, which have the lower computational cost and communication overhead compared with the signature-then-encryption approaches. However, most certificateless-based schemes either focus on providing receiver anonymity or focus on improving signcryption efficiency. In addition, most certificateless-based schemes rely on bilinear pairing operations, which are more time consuming than modular exponentiation and scalar multiplication in finite fields. In this paper, we propose a practical anonymous multi-receiver certificateless signcryption (AMCLS) scheme that can satisfy message confidentiality, source authentication, and anonymity simultaneously and efficiently. In the proposed scheme, the sender's signcryption cost increases linearly with the increase of the designated receivers, while the unsigncryption cost per receiver is constant. The adoption of elliptic curve scalar multiplication instead of bilinear pairing operation improves the efficiency of the proposed scheme. Both the sender and receivers' identities are encrypted from being exposed to offer anonymity. Through security analysis, our proposal can be proved to achieve chosen-ciphertext attack (CCA) security in encryption indistinguishability and receiver anonymity in strong, commonly accepted attack models. Theoretical analyses and experimental results demonstrate that our scheme enjoys a better efficiency than other certificateless-based schemes.

Keywords: Anonymity · Multi-receiver signcrption · Certificateless cryptosystem

1 Introduction

Nowadays, many applications such as multicast communication, digital content distribution systems, and pay-per-view channels require a cryptographic

© Springer Nature Switzerland AG 2020
S. Wen et al. (Eds.): ICA3PP 2019, LNCS 11944, pp. 689–709, 2020.
https://doi.org/10.1007/978-3-030-38991-8_45

mechanism to allow a sender to encrypt messages to a set of receivers so that only the authorized receivers can decrypt the messages. In addition to encrypting messages, another important issue in such services is the authentication of the sender, that is, the source of the message should be guaranteed.

To provide authentication and non-repudiation as well as confidentiality in the communication simultaneously, signcryption was first proposed by Zheng in [1]. Compared with the traditional signature-then-encryption approach, the signcryption mechanism has the following advantages: (1) lower communication overhead; (2) lower computational cost; (3) the well-designed scheme can achieve a higher level of security; (4) simplify the design of the cryptographic protocol requiring both confidentiality and authentication. In a multi-receiver signcryption scheme, the sender can simultaneously sign and encrypt messages sent to multiple recipients, which is obviously more efficient than one-to-one encryption and signature. However, in order to assist authorized receivers to find out the appropriate information from the ciphertext for decryption, the identity list of all authorized receivers should be included as a mandatory element in the message. Obviously, this directly leads to the identity and privacy leakage problem of receivers. With the growing emphasis on personal privacy issues, personal privacy exposure will bring a lot of problems.

Therefore, in addition to providing message confidentiality and source authentication, it is also important to protect the receiver's privacy (anonymous). In many situations, a receiver may expect that its identity information is confidential to other receivers during the communication. One example is in the pay-per-view system where a subscriber does not want others knowing his personal viewing selections. Another interesting example is that many business organizations often try to protect their clients' identities in order to prevent competitors from delivering targeted advertising.

1.1   IBC-Based Signcryption Schemes

The first identity-based cryptography (IBC) signcryption scheme was proposed by malone-Lee in 2002 [2]. Subsequently, Boyen [3] defined a formalized security model for identity-based signcryption schemes and proposed a provably secure scheme on the basis of this model. Based on Boyens model, Chen et al. [4] proposed a more efficient identity-based signcryption scheme. Although many identity-based multi-receiver signcryption schemes [5–8] have been put forward, a common problem in these schemes is the leakage of receiver identity.

In order to protect the receiver's identity, Lal et al. proposed an anonymous multi-receiver ID-based encryption (AMIBE) scheme [9], in which Lagrange interpolation polynomial was used to mix all the receivers identity information to guarantee the receiver's anonymity. Later, Fan et al. [10] proposed another AMIBE scheme using Lagrange interpolating polynomials. Wang and Chien pointed out that Fan's scheme could not provide receiver anonymity due to the designation of receivers in [11] and [12], respectively. At the same time, Wang and Chien proposed different solutions to guarantee receivers' identity anonymities. Unfortunately, these two schemes still lack anonymity and encryption indistinguishability [13,14]. Later, Pang et al. [15] proposed a scheme to solve the receiver identity exposure problem existing in the Lagrange-interpolation-based method. However, receiver's identity is not really anonymous to other receivers in Pang's scheme.

By modifying the receiver anonymous adversarial model in [10], Tseng et al. proposed an improved AMIBE scheme [16]. Their scheme is proved to satisfy both confidentiality and receiver anonymity in the random oracle model. In 2015, Fan et al. [17] proposed an AMIBE scheme, also known as anonymous multi-receiver identity-based authentication encryption. They formally proved that their scheme is chosen-ciphertext attack (CCA) secure both in confidentiality and anonymity. They also made the analysis of previous works in [17].

So far, many identity-based anonymous signcryption schemes have been proposed. However, these schemes presented above have two main problems. First, schemes built on the basis of IBC inherently have the key escrow problem. Since each users private key in IBC is generated by a third party, Key Generation Center (KGC), KGC can decrypt all ciphertexts and reveal any user's identity. Once KGC is compromised, it will cause great damage to the whole system. The second is that these AMIBE schemes are inefficient in communication since the decryption cost per receiver increases linearly with the increase of the designated receivers.

1.2 Certificateless-Based Signcryption Schemes

In order to overcome IBC's key escrow problem, Al-Riyami and Paterson proposed the certificateless public key cryptosystem (CL-PKC) in 2003 [18]. Since then, a number of cryptosystems including encryption schemes, signature schemes, key agreement protocols, and signcryption schemes based on CL-PKC have been proposed.

Barbosa et al. [19] proposed the first certificateless signcryption scheme (CLSC) and carried out the formal security analysis in the random oracle model. After that, some efficient CLSC solutions were proposed [20–22]. Unfortunately, all the above schemes have security flaws, as shown in [23]. Later, Barreto et al. [24] put forward an efficient CLSC scheme without formal security proof. Liu et al. [25] proposed a CLSC scheme and proved the scheme's security in the standard model. However, [26] showed that there are two subtle public-key replacement attacks in Lius scheme.

By adopting the polynomial technique in Tsengs scheme [16], Islam et al. [27] proposed the first anonymous multi-receiver certificateless encryption (AMCLE) scheme in 2004. In this scheme, a sender's encryption cost is proportional to the number of designated receivers, and the decryption cost per receiver is linear to the number of specified receivers. However, their security proof has a drawback that the simulator failed to successfully generate the challenge ciphertext and thus failed in the simulation. In 2015, Hung et al. [28] proposed a modified AMCLE solution with constant decryption cost per receiver. However, their security proof cannot cover all possible attackers because of some specific restrictions on attackers. In addition, the use of bilinear pairing operation results in higher computational costs. Tseng et al. [29] proposed a CCA-safe AMCLE scheme in 2015. They gave formal proof that their scheme was CCA safe against type I and type II attackers in the random oracle model [30]. Recently years, some AMCLE without bilinear paring were proposed, most of them are based on modular exponentiation, which is slightly more efficient than bilinear pairing [31, 32].

From the above, it can be seen that the existing certificateless schemes either concentrate on providing the anonymity of the receiver or focus on improving the efficiency of signcryption. In addition, most AMCLE schemes rely on the bilinear pairing operations which are more computationally expensive than the modular exponentiation operation and the scalar multiplication operation in finite fields. Therefore, it is attractive to study an anonymous multi-receiver certificateless signcryption (AMCLS) scheme that does not use bilinear pairing operation and achieves the goal efficiently.

1.3   Our Contribution

Toward the goal of improving the efficiency and broadening the adoption of anonymous multi-receiver signcryption, in this work, we propose an efficient and practical AMCLS scheme, which can achieve message encryption, sender authentication and receiver anonymous in one message. It can be applied in many practical application scenarios. For example, in a television payment system, a viewer or a recipient needs to authenticate the identity of the information source, as well as the need to protect his or her identity information and the privacy of the television program being watched, so that an anonymous multi-receiver certificateless signcryption (AMCLS) scheme can be well applied to this scenario. In this paper, our proposal can also be proved secure in strong, commonly accepted attack models, based on the assumption of the ECDHP problem plus other generic assumptions (e.g., signatures and message authentication codes).

Our proposal has the following characteristics.

Sender and Receiver Anonymity. By encrypting both sender and receivers's identities, the identities of the sender and receiver can be protected from being exposed, thereby ensuring the sender and receiver's anonymities. In the scheme, only the designated receiver knows the identity of the sender, and no one other than the sender knows the identity of the receiver.

Providing Message Confidentiality and Source Authentication Efficiently. When signcrypting the message, the sender's signcryption cost increases linearly with the number of designated receivers. And the unsigncryption cost per receiver is constant when unsigncrypting the message, that is, the unsigncryption cost per receiver is independent of the number of receivers. In addition, using elliptic curve scalar multiplication instead of bilinear pairing operation further improves the scheme's efficiency. Source authentication is provided after receiver successfully decrypts the signed message and confirms the sender's identity, thus ensuring the source and the reliability of the message.

CCA Secure. Through formally proof, our scheme is proved to achieve CCA security both in encryption indistinguishability and receiver anonymity in the random oracle models.

1.4   Organization

The rest of the paper is organized as follows. In Sect. 2, we recall the necessary cryptographic and mathematical background, including the relevant basics of elliptic curve and the system model of certificateless public key signcryption scheme. In Sect. 3, we give the security model of the AMCLS scheme for two type adversaries. In Sect. 4, we present our new AMCLS scheme. In Sect. 5, we analyze our scheme, and give the security proof under the hardness of ECDHP assumption. In Sect. 6, theoretical analysis and comparisons are made among three schemes. In Sect. 7, we implement the proposed scheme and make the performance analysis. Conclusion is given in Sect. 8.

2   Preliminaries

To facilitate the understanding of our designs, we briefly review the basic definitions and properties of bilinear pairings over an elliptic curve group in this section. In addition, we introduce the formal system model of AMCLS scheme.

(1) Elliptic Curve. Let the symbol E/F_P denote an elliptic curve E over a prime finite field F_P, defined by an equation $y^2 = x^3 + ax + b$ with the discriminant $\Delta = 4a^3 + 27b^2 \neq 0$. All the points on E/F_P and an extra point O (called the point at infinity) form a group G under the operation of point addition $R = P+Q$ which is defined according to a chord and tangent rule. Scalar multiplication over E/F_P can be computed as $t \cdot P = P + P + ... + P(t\ times)$.

(2) Hard problem Definition 1 Elliptic Curve Diffie-Hellman Problem (ECDHP). Let G be an additive cyclic group of order q consisting of points on the elliptic curve, P be a generator in G, knowing $aP, bP \in G$ and solving for abP. The ECDHP problem states that the probability of any polynomial-time algorithm to solve the ECDHP problem is negligible.

(3) The formal system model of AMCLS scheme. In this part, we illustrate the system model of AMCLS scheme. Generally, an AMCLS scheme consists of the following algorithms to generate the system parameters and complete the communication.

Setup. On input security parameter λ, the key generation center KGC runs this algorithm to generate a master key s and a list of public parameters $params$.

Partial-Private-Key-Extract. On input $params$ and a user's identity ID, KGC runs this algorithm to generate a partial private key DID for this user.

Set-Secret-Value. On input $params$ and an identity ID, a user runs this algorithm to generate a secret value xid.

Set-Private-Key. Taking DID and xid as input parameters, a user runs this algorithm to generate a private key SID.

Set-Public-Key. Taking $params$ and xid as inputs, a user runs this algorithm to generate a public key PID.

Signcrypt. On input system parameters, a plaintext message m, a sender's identity ID_s and private key SID_s, a receiver's identity IDi and public key PID_i, this algorithm outputs a signcryped ciphertext C for the message m.

Unsigncrypt. On input a ciphertext C, the sender's identity ID_s and public key PID_s, the receiver's identity ID_i and private key DID_i, this algorithm outputs the plaintext message m or "reject".

3 Security Model

In this section, we define a formal security model for the AMCLS scheme based on the AMCLE security model in [28]. This security model includes both confidentiality and receiver anonymity. We define "indistinguishability of certificateless signcryptions against selective multi-ID chosen ciphertext attack" (IND-CLMS-CCA) for confidentiality and "anonymous indistinguishability of certificateless encryptions against selective multi-ID chosen ciphertext attack" (ANON-CLMS-CCA) for anonymity, respectively.

3.1 Adversaries

An adversary A is assumed to control the network completely. A is allowed to insert, delete or modify the protocol messages. In an AMCLS scheme, there are two types of adversaries.

Type I adversary A_I: A_I is modeled as an external attacker who can replace a legitimate user's public key, but cannot obtain the system's master key.
Type II adversary A_{II}: A_{II} is modeled as an internal attacker who can obtain the system master key but cannot replace the legitimate user's public key. A_{II} is equivalent to a malicious KGC.

3.2 Indistinguishability of Certificateless Signcryptions Against Selective Multi-ID Chosen Ciphertext Attack (IND-CLMS-CCA)

The IND-CLMS-CCA security of the AMCLS scheme requires that there is no probabilistic polynomial-time adversary which could distinguish ciphertexts. Formally, we consider the following two games, which are played between a challenger C and an adversary $A_i(i = 1, 2)$.

Game 1: Game 1 is the interaction between A_I and C.
Choose identities. A_I outputs t target identities $\{ID_1, ID_2, \cdots, ID_t\}$.
Setup. Given a secure parameter λ, C generates system parameters $params$ and a master key s. Finally, C returns $params$ to A_I.
queries. The adversary A_I is allowed to launch queries to the following oracles:
-**Public key retrieve query.** Given an identity ID, C computes the corresponding public key PID, and returns PID to A_I.
-**Public key replace query.** Given an identity ID, A_I can choose a new public key to replace the original public key PID of the user with ID. The challenger C records the replacement.

-Partial private key extract query. Given an identity ID, if $ID \neq IDi (i \in \{1, 2, ..., t\})$, C computes the corresponding partial private key DID, and returns DID to A_I.

-Secret value extract query. Upon receiving A_I's query, C outputs the corresponding secret value xid to A_I.

-Unsigncryption query. Upon receiving A_I's query, C return the unsigncrypted message to A_I, whether the public key is replaced or not.

Challenge. The adversary A_I outputs (m_0, m_1) as the challenge messages, where $m_0 \neq m_1$ and both of them have not been queried before. The challenger C picks $\beta \in \{0, 1\}$ randomly and outputs $CT^* = E(PP, (ID_1, PID_1), (ID_2, PID_2), \ldots (ID_t, PID_t), m_\beta)$. Then C sends CT^* to adversary A_I.

More-queries. The adversary A_I continues to issuing queries as above, with the restriction that CT^* should not be submitted to Unsigncryption query.

Guess. Finally, A_I outputs $\beta \in \{0, 1\}$. If $\beta' = \beta$, A_I will win this game, with advantage $adv_{IND-CLMS-CCA}(A_I) = |\Pr[\beta' = \beta] - \frac{1}{2}|$.

Game 2: Game 2 is the interaction between A_{II} and C.

Choose identities. A_{II} outputs t target identities $\{ID_1, ID_2, \cdots, ID_t\}$.

Setup. Given a secure parameter λ, C generates system parameters $params$ and a master key s. Finally, C returns $params$ and s to A_{II}.

queries. The adversary A_{II} is allowed to launch queries to the following oracles: A_{II} is allowed to ask all queries as in Game 1. however, there are two restrictions: the identity ID_i ($i \in \{1, 2, ..., t\}$) is disallowed to appear in the Public key replace and the Secret value extract queries.

Challenge. The adversary A_{II} outputs (m_0, m_1) as the challenge messages, where $m_0 \neq m_1$ and both of them have not been queried before. The challenger C picks $\beta \in \{0, 1\}$ randomly and outputs $CT^* = E(PP, (ID_1, PID_1), (ID_2, PID_2), \ldots (ID_t, PID_t), m_\beta)$. Then C sends CT^* to the adversary A_{II}.

More-queries. The adversary A_{II} continues to issuing queries as above, with the restriction that CT^* should not be submitted to Unsigncryption query.

Guess. Finally, A_{II} outputs $\beta \in \{0, 1\}$. If $\beta' = \beta$, A_{II} will win this game, with advantage $adv_{IND-CLMS-CCA}(A_{II}) = |\Pr[\beta' = \beta] - \frac{1}{2}|$.

Definition 2: An AMCLS scheme is semantically secure against the IND-CLMS-CCA attack if for any probabilistic polynomial-time adversary A (including Type I and Type II adversary), both $adv_{IND-CLMS-CCA}(A_I)$ and $adv_{IND-CLMS-CCA}(A_{II})$ are negligible in the security parameter λ.

3.3 Anonymous Indistinguishability of Certificateless Signcryptions Against Selective Multi-ID Chosen Ciphertext Attack (ANON-CLMS-CCA)

The ANON-CLMS-CCA security of the AMCLS scheme requires that there is no probabilistic polynomial-time adversary which could distinguish receivers.

Formally, we consider the following two games, which are played between a challenger C and an adversary $A_i (i = 1, 2)$.

Game 3: Game 3 is the interaction between A_I and C.

Choose identities. The adversary A_I selects two identities, ID_0 and ID_1 respectively and sends these two identities to the challenger C. The challenger C picks $\beta \in \{0, 1\}$ randomly.

Setup. Given a secure parameter λ, C generates system parameters $params$ and a master key s. Finally, C returns $params$ to A_I.

queries. A_1 is allowed to ask Public key retrieve query, Public key replace query, Secret value extract query, Unsigncryption query as in Game 1.

-Partial private key extract query. Given an identity ID, if $ID \neq ID_i (i \in \{0, 1\})$, C computes the corresponding partial private key DID, and returns DID to A_I.

Challenge. The adversary A_I sends a plaintext m and a set of identities $\{ID_2', ID_3', \cdots, ID_t'\} (t \neq 0, 1)$ to challenger C. The challenger C picks $\beta \in \{0, 1\}$ randomly and outputs $CT^* = E(PP, (ID_\beta, PID_\beta), (ID_2', PID_2'), \ldots, (ID_t', PID_t'), m)$. Then C sends CT^* to adversary A_I.

More-queries. The adversary A_I continues to issuing queries as above, with the restriction that CT^* should not be submitted to Unsigncryption query.

Guess. Finally, A_I outputs $\beta' \in \{0, 1\}$. If $\beta' = \beta$, A_I will win the game with advantage $adv_{ANON-CLMS-CCA}(A_I) = \left| \Pr[\beta' = \beta] - \frac{1}{2} \right|$.

Game 4: Game 4 is the interaction between A_{II} and C.

Choose identities. The adversary A_{II} selects two identities, ID_0 and ID_1 respectively and sends these two identities to the challenger C. The challenger C picks $\beta \in \{0, 1\}$ randomly.

Setup. Given a secure parameter λ, C generates system parameters $params$ and a master key s. Finally, C returns $params$ and s to A_{II}.

queries. A_{II} is allowed to ask all queries as in Game 2. however, there are two restrictions: the identity $ID_i (i \in \{0, 1\})$ is disallowed to appear in the Public key replace and the Secret value extract queries.

Challenge. The adversary A_{II} sends a plaintext m and a set of identities $\{ID_2', ID_3', \cdots, ID_t'\} (t \neq 0, 1)$ to challenger C. The challenger C picks $\beta \in \{0, 1\}$ randomly and outputs $CT^* = E(PP, (ID_\beta, PID_\beta), (ID_2', PID_2'), \ldots, (ID_t', PID_t'), m)$. Then C sends CT^* to the adversary A_{II}.

More-queries. The adversary A_{II} continues to issuing queries as above, with the restriction that CT^* should not be submitted to Unsigncryption query.

Guess. Finally, A_{II} outputs $\beta' \in \{0, 1\}$. If $\beta' = \beta$, A_{II} will win the game with advantage $adv_{ANON-CLMS-CCA}(A_{II}) = \left| \Pr[\beta' = \beta] - \frac{1}{2} \right|$.

Definition 3: An AMCLS scheme is semantically secure against ANON-CLMS-CCA attack if for any probabilistic polynomial-time adversary A (including Type I and Type II adversary), both $adv_{ANON-CLMS-CCA}(A_I)$ and $adv_{ANON-CLMS-CCA}(A_{II})$ are negligible in the security parameter λ.

4 The Proposed AMCLS Scheme

In this section, we propose an AMCLS scheme with CCA security in both confidentiality and anonymity against Type I and Type II attackers.

4.1 The Scheme Description

The proposed scheme includes the following steps.

setup. In this phase, the setup algorithm generates the system parameters, using a security parameter, this algorithm performs the following steps:

– KGC specifies $q, p, E/F_p, P$ and G where q is a large prime and p is the field size, $E(F_p)$ is an elliptic curve E over a finite field F_p, p is a base point of order q on the curve E and G is a cyclic group of order q under the point addition "+" generated by P.
– KGC randomly selects a system master secret key $s \in Z_p^*$ and sets $P_{pub} = s \cdot P$ as the system public key.
– KGC chooses nine hash functions as follows.
  $H_0 : \{0,1\}^* \to Z_p^*,\ H_1 : \{0,1\}^w \times \{0,1\}^* \to Z_p^*,$
  $H_2 : E(F_p) \times E(F_p) \to \{0,1\}^w,\ H_3, H_4, H_5 : \{0,1\}^w \to \{0,1\}^w,$ and
  $H_6, H_7, H_8 : \{0,1\}^w \times E(F_p) \to Z_p^*,\ w$ is a positive integer.
– KGC selects the symmetric signcryption function $E_{sk}()$ and its corresponding unsigncryption function $D_{sk}()$, where sk is the symmetric key.
– KGC publishes the common system parameters
  $PP = \{q, p, E(F_p), P, G, P_{pub}, H_0, H_1, H_2, H_3, H_4, H_5, H_6, H_7, H_8, E, D).$

partial key extract

Taking a user's identity $ID \in \{0,1\}^*$ as input, KGC randomly selects a value $k \in \{0,1\}^w$, computes $QID = H_0(ID)$ and returns the corresponding partial private key $DID = k + s \cdot QID$ and the partial public key $PID'' = k \cdot P$.

set secret value

Taking $ID \in \{0,1\}^*$ as input, the user randomly selects a positive integer $xid \in Z_p^*$ as its secret value.

set private key

Taking the partial private key DID and the secret value xid as input, the user calculates $SID = (DID, xid)$ as its private key.

set public key

Taking the secret value xid and the partial public key PID'' as input, the user computes $PID' = xid \cdot P$, then sets $PID = (PID', PID'')$ as its public key and publicizes the public key to other users in the system.

signcryption

Taking a sender's information ID_s, PID_s, SID_s as input, this sender does the following when it sends a message m to t specified receivers with public key $(ID_1, PID_1), \ldots, (ID_t, PID_t)$, where $t \leq n$.

- For $i = 1, \ldots, t$, calculates $QID_i = H_0(ID_i)$.
- Randomly selects a value $r \in \{0,1\}^w$, then calculates $\sigma = H_1(r, ID_s)$, $U = r \cdot P$.
- Calculates $F_i = r \cdot PID'_i$, $i = 1, \ldots, t$, $K_i = QID_i \cdot P_{pub} \cdot r + r \cdot PID''$ and $T_i = H_2(K_i, F_i)$.
- Calculates $C_i, i = 1, \cdots, t$ where $C_i = H_3(T_i) \| (H_4(T_i) \oplus \sigma)$, $\|$ indicates the connection operation.
- Calculates the symmetric key $sk = H_5(\sigma)$, $V = E_{sk}(m)$ and $C_s = E_{sk}(ID_s)$.
- Signs the message with the sender's private key as follows:
  $H = H_6(U, V, ID_s, PID_s)$, $H' = H_7(U, V, ID_s, PID_s)$, $W = DID_s + r \cdot H + xid_s \cdot H'$
- Performs the hash operation on ciphertext to ensure the data integrity $\Lambda = H_8(m, \sigma, C_1, C_2, \ldots, V, C_s, U, W)$
- Sets ciphertext as follows: $CT = <(C_1, C_2, \ldots, C_t), V, C_s, W, U, \Lambda>$

Unsigncryption
Taking $CT = <(C_1, C_2, \ldots, C_t), V, C_s, W, U, \Lambda>$ as input, the designated receiver with an identity ID can use its private key $SID = (DID, xid)$ and do the following:

- Calculates $K = DID \cdot U$, $F = xid \cdot U$, $T = H_2(K, F)$ and $H_3(T)$.
- The corresponding $C_i(i \in \{1, 2, \ldots, t\})$ can be found by using $H_3(T)$ through $C_i = H_3(T) \| Y$, where Y represents the remaining string by removing $H_3(T)$ from C_i.
- Recovers $\sigma' = Y \oplus H_4(T)$.
- Sets $sk' = H_5(\sigma')$, calculates $m' = D_{sk'}(V)$, $ID'_s = D'_{sk}(C_s)$, $\Lambda' = H_8(m', \sigma', C_1, C_2, \ldots, V, C_s, U, W)$.
- Seeks the PID_s corresponding to the ID'_s. If $PID''_s + P_{pub} \cdot QID_s + U \cdot H + PID'_s \cdot H' = P \cdot W$, output the message m else return "reject".

4.2 Correctness of the Scheme

To show the correctness of our proposed protocol, i.e., that the message sent by the sender is the same as the message received by all receivers, it suffices to show that $sk = sk'$. Since sk and sk' are the outputs of $H_5(\sigma)$ and $H_5(\sigma')$, we need only to show that σ is equal to σ'. Note that the sender and the receiver will compute σ and σ' as follows:
The sender computes: $\sigma = H_1(r, ID_s)$
The receiver computes: $\sigma' = Y \oplus H_4(T)$
since $C_i = H_3(T_i) \| (H_4(T_i) \oplus \sigma) = H_3(T) \| Y$, $Y = H_4(T_i) \oplus \sigma$, $\sigma' = Y \oplus H_4(T) = (H_4(T_i) \oplus \sigma) \oplus H_4(T) = \sigma$ From the analysis we can see that, the sender and the receiver can obtain the same message.

5   Security Analysis

In this section, we prove the security robustness of the proposed scheme.

Theorem 1. *The proposed scheme is semantically secure against Type I adversaries in the IND-CLMS-CCA games in the random oracle model, assuming ECDHP problem is intractable.*

Proof: If the Type I adversary can discriminate the ciphertext with a non-negligible probability, the challenger C can solve a particular ECDHP by interacting with the adversary A_I, which leads to the conclusion that our scheme is indistinguishable.

C is a challenger to the ECDHP. Hash functions are random oracles. Given $\{P, aP, bP\}$, C expects to calculate abP by interacting with the adversary A_I as follows.

Choose identities. A_I outputs $t(t \leq n)$ target identities $\{ID_1, ID_2, ..., ID_t\}$.
Setup. In order to achieve the goal of the challenge, C needs to set system parameters: $P_{pub} = sP$ and
$PP = \{q, p, E(F_p), P, G, P_{pub}, H_0, H_1, H_2, H_3, H_4, H_5, H_6, H_7, H_8, E, D)$. The challenger C sets nine empty lists L_i $(i = 0, 1, ..., 8)$ to record the responses of the hash functions H_i $(i = 0, 1, ..., 8)$. A_I executes the following queries. The specific query process is as follows.

- $H_0(ID)$. If there exists a record (ID, QID) in the list L_0, C returns QID to A_I. Otherwise, C chooses a random value $QID \in Z_q^*$, stores (ID, QID) in L_0, and then returns QID to A_I.
- $H_1(r, ID_s)$. If there exists a record (r, ID_s, σ) in the list L_1, C returns σ to A_I. Otherwise, C chooses a random value $\sigma \in Z_q^*$, stores (r, ID_s, σ) in L_1, and then returns σ to A_I.
- $H_2(K, F)$. If there exists a record (K, F, T) in the list L_2, C returns T to A_I. Otherwise, C chooses a random string $T \in \{0, 1\}^w$, stores (K, F, T) in L_2, and then returns T to A_I.
- $H_3(T)$. If there exists a record (T, x) in the list L_3, C returns x to A_I. Otherwise, C chooses a random string $x \in \{0, 1\}^w$, stores (T, x) in L_3, and then returns x to A_I.
- $H_4(T)$. If there exists a record (T, y) in the list L_4, C returns y to A_I. Otherwise, C chooses a random string $y \in \{0, 1\}^w$, stores (T, y) in L_4, and then returns y to A_I.
- $H_5(\sigma)$. If there exists a record (σ, sk) in the list L_5, C returns sk to A_I. Otherwise, C chooses a random string $sk \in \{0, 1\}^w$, stores (σ, sk) in L_5, and then returns sk to A_I.
- $H_6(U, V, ID, PID)$. If there exists a record (U, V, ID, PID, H) in the list L_6, C returns H to A_I. Otherwise, C chooses a random string $H \in \{0, 1\}^w$, stores (U, V, ID, PID, H) in L_6, and then returns H to A_I.
- $H_7(U, V, ID, PID)$. If there exists a record (U, V, ID, PID, H') in the list L_7, C returns H' to A_I. Otherwise, C chooses a random string $H' \in \{0, 1\}^w$, stores (U, V, ID, PID, H') in L_7, and then returns H' to A_I.
- $H_8(m, \sigma, C_1, C_2, ..., C_t, V, C_s, W, U)$. If there exists a record $(m, \sigma, C_1, C_2, ..., C_t, V, C_s, W, U, \Lambda)$ in the list L_8, C returns Λ to A_I. Otherwise, C chooses a random value $\Lambda \in Z_q^*$, stores $(m, \sigma, C_1, C_2, ..., C_t, V, C_s, W, U, \Lambda)$ in L_8, and then returns Λ to A_I.

queries. The challenger C sets empty lists PK^{list} to record public keys and secret values of users. C responds the queries as follows.

- Public key retrieve query (ID). If there is a record $(ID, PID', PID'', DID, xid)$ in the list PK^{list}, C returns PID to A_I. Otherwise, C chooses a random value $t, xid, DID \in Z_q^*$, computes $PID' = xid \cdot P$ and $PID'' = t \cdot P$, sets $PID = (PID', PID'')$, stores $(ID, PID', PID'', DID, xid)$ in PK^{list}, and then returns PID to A_I.
- Public key replace query (ID, PID', PID''). A_I selects a new tuple $(ID, PID', PID'', \bot, \bot)$ and sends the new tuple to C. Upon receiving the tuple, C replaces $(ID, PID', PID'', DID, xid)$ with the new tuple.
- Partial private key extract query (ID). If $ID = ID_i (i \in \{1, 2, ..., t\})$, C returns "reject" and stops the game. Otherwise, if there exists a record $(ID, PID', PID'', DID, xid)$ in the list PK^{list}, C returns DID to A_I, else return "reject".
- Secret value extract query (ID). If there exists a record $(ID, PID', PID'', DID, xid)$ in the list PK^{list}, C returns the secret value xid to A_I, else return "reject".
- Unsigncryption query (CT, ID). If $ID \neq ID_i (i \in \{1, 2, ..., t\})$, since C can get all private keys, C decrypts CT and returns the plaintext to A_I. Otherwise, If $ID = ID_i (i \in \{1, 2, ..., t\})$, C performs the following operations.
    - If $(m, \sigma, C_1, C_2, ..., C_t, V, C_s, W, U, \Lambda)$ is not in the list L_8, C returns failure and stop the game.
    - C obtains QID via the H_0 query.
    - For $k = 1, ..., t$, C runs the following steps.
        * Pick the leftmost w bits of C_k and denote it by x_k.
        * Pick the rightmost w bits of C_k, denote it by W_k, and compute $y_k = W_k \oplus \sigma$.
        * Look up in L_3 and L_4 and find T_k that both (T_k, x_k) and (T_k, y_k) exist in L_3 and L_4.
        * Look up in L_2 to find a tuple (K_k, F_k, T_k) associated with T_k. If such tuple is not exist, return "abort".
        * Look up in L_1 to find a tuple $<(r, IDs), \sigma>$ satisfying $K_k = QID \cdot P_{pub} \cdot r + r \cdot PID''$. If such tuple exists, record the value k.
    - If $K_k = QID \cdot P_{pub} \cdot r + r \cdot PID''$ for some $k \in \{1, 2, ..., t\}$, calculate $sk = H_5(\sigma)$ and $m' = Dsk(V)$. If $m' = m$, C returns the plaintext m to the adversary A_I. Otherwise, C returns "abort".

Challenge. A_I gives a target plaintext pair (m_0, m_1) to C. C randomly chooses $\beta \in \{1, 2\}$ and perform the following operations.

- Set $U^* = bP$.
- Pick a string $\sigma^* \in \{0, 1\}^w$ at random.
- For $i = 1, ..., t$, Randomly pick $x_\beta^* \in \{0, 1\}^w$ and $y_\beta^* \in \{0, 1\}^w$, and compute $C_i^* = x_\beta^* || (y_\beta^* \oplus \sigma^*)$.
- Calculate $sk = H_5(\sigma^*)$, $V^* = E_{sk}(m_\beta)$, $C_s^* = E_{sk}(IDs)$, $W^* \in \{0, 1\}^w$, $\Lambda^* = H_8(m, \sigma^*, C_1^*, C_2^*, ..., C_t^*, V^*, C_s^*, W^*, U^*)$. Then, C returns the target ciphertext $CT^* = ((C_1^*, C_2^*, ..., C_t^*), V^*, C_s^*, W^*, U^*, \Lambda^*)$ to A_I.

More queries. A_I is executed similarly to the query in the previous **Queries** (CT^*can not be queried in the Unsigncryption query).

Guess. The adversary A_I output $\beta' \in \{1,2\}$ as its guess. If $\beta = \beta'$, A_I wins the game.

If the guess is consistent with the reality, A is likely to make H_2 query with a great probability such that $K_i = QID_i \cdot P_{pub} \cdot r + r \cdot PID''$, for $i \in \{1,2,...,t\}$. The challenger C can find such a K by going through the list L_2. Set $QID \cdot P_{pub} + PID'' = aP$, since $U^* = bP$, C can obtain $abP = K_k$, Thus C resolves a particular ECDHP using the adversary A_I's identification of the ciphertext.

Theorem 2. *Our proposed scheme is semantically secure against Type II adversaries in the IND-CLMS-CCA games in the random oracle model, assuming ECDHP problem is intractable.*

Proof: If the Type II adversary discriminates the ciphertext with a non-negligible probability, the challenger C can solve a particular ECDHP by interacting with the adversary A_{II}, which leads to the conclusion that our scheme is indistinguishable.

Since some proof of Theorem 2 is similar to Theorem 1, But there are two restrictions: The identity $ID_i(i \in \{1,2,...,t\})$ is disallowed to appears in the Public key replace and the Secret value extract queries. We give the difference as follows:

- Public key retrieve query(ID). If there exists $(ID, PID', PID'', DID, xid)$ in the PK^{list}, C returns PID to A_{II}. Otherwise, C performs the following opetations.

    Choose a random value $xid \in Z_q^*$ at random.

    If $ID = ID_i$ for some $i \in \{1,2,...,t\}$, set $PID' = xid \cdot a \cdot P$. Otherwise, set $PID' = xid \cdot P$.

    Store($ID, PID', PID'', DID, xid$) in PK^{list}, and return PID to A_{II}.
- Unsigncryption query (CT, ID). If $ID \neq ID_i$ for some $i \in \{1,2,...,t\}$, C can get all private key. Therefore C can decrypt CT and returns the plaintext to A_{II}. Otherwise, If $ID = ID_i$ for some $i \in \{1,2,...,t\}$, C performs the following opetations.

    If $(m, \sigma, C_1, C_2, ..., C_t, V, C_s, W, U, \Lambda)$ is not in the list L_8, C returns failure and stop the game.

    C obtains QID via the H_0 query.

    For $k = 1, ..., t$, C runs the following steps.
    - Pick the leftmost w bits of C_k and denote it by x_k.
    - Pick the rightmost w bits of C_k, denote it by W_k, and compute $y_k = W_k \oplus \sigma$.
    - Look up in L_3 and L_4 and find T_k that both (T_k, x_k) and (T_k, y_k) exist in L_3 and L_4.
    - Look up in L_2 to find a tuple (K_k, F_k, T_k) associated with T_k. If such tuple is not exist, return "abort".

- Look up in L_1 to find a tuple $<(r, IDs), \sigma>$ satisfying $F_k = r \cdot PID'$. If such tuple exists, record the value k.

If $F_k = r \cdot PID'$ for some $k \in \{1, 2, ..., t\}$, calculate $sk = H_5(\sigma)$ and $m' = Dsk(V)$. If $m' = m$, C returns the plaintext m to the adversary A_{II}. Otherwise, C returns "abort".

If the guess is consistent with the reality, A_{II} is likely to make H_2 query with a great probability such that $F_k = r \cdot PID'$, for $i \in \{1, 2, ..., t\}$. The challenger C can find such a F by going through the list L_2. Since $PID' = xid \cdot a \cdot P$ and $U^* = bP$, C can obtain $abP = xid^{-1} \cdot F_k$, Thus C resolves a particular ECDHP using the adversary A_{II}'s identification of the ciphertext.

Theorem 3. *Our proposed scheme is semantically secure against Type I adversaries in the ANON-CLMS-CCA games in the random oracle model, assuming ECDHP problem is intractable.*

Proof: If the Type I adversary discriminates the ciphertext with a non-negligible probability, the challenger C can solve a particular ECDHP by interacting with the adversary A_I, which leads to the conclusion that our scheme is Anonymous.

C is a challenger to the ECDHP. Hash functions are random oracles. Given $\{P, aP, bP\}$, C expects to calculate abP by interacting with the adversary A_I as follows:

Choose identities. 1. A_I output 2 target identity pair (ID_0, ID_1), and the challenger C then chooses a random value $\beta \in \{0, 1\}$.
Setup. This part is same with Theorem 1. We set $P_{pub} = aP$ here.
Queries. This part is similar to The **Setup** part in Theorem 1. A_I may make a series of queries as in Theorem 1. The difference is Partial private key extract query. It's defined as follow:
Partial private key extract query (ID). If $ID = ID_i$ for some $i \in \{0, 1\}$, C returns "reject" and stops the game. Otherwise, if there is a record $(ID, PID', PID'', DID, xid)$ in the list PK^{list}, C computes and returns DID to A_I. Otherwise, returns "reject".
Challenge. A_I gives a target plaintext m and $t - 1$ identities chosen from $S - \{ID_0, ID_1\}$, denoted by $ID_2', ..., ID_t'$, where $1 < t < n$. C randomly chooses $\beta \in \{0, 1\}$ and perform the following operations.
  - Set $U^* = bP$.
  - Pick a string $\sigma^* \in \{0, 1\}^w$ at random.
  - Randomly pick $x_\beta^* \in \{0, 1\}^w$ and $y_\beta^* \in \{0, 1\}^w$, and compute $C_1^* = x_\beta^* || (y_\beta^* \oplus \sigma^*)$.
  - For $i = 2, ..., t$, do the following.
    Obtain DID_i' and xid_i' via Partial private key extract and Secret value extract queries on ID_i, and compute $K_i' = DID_i' \cdot U^*$, $F_i' = xid_i' \cdot U^*$ and $T_i = H_1(K_i', F_i')$.
    Make $H_3(T_i)$ and $H_4(T_i)$ queries to obtain $x_i^* \in \{0, 1\}^w$ and $y_i^* \in \{0, 1\}^w$.
    Compute $C_i^* = x_i^* || (y_i^* \oplus \sigma^*)$. Calculate $sk = H_5(\sigma^*)$, $V* = E_{sk}(m_\beta)$, $C_s* = E_{sk}(ID_s)$, $W^* \in \{0, 1\}^w$,
    $\Lambda^* = H_8(m, \sigma^*, C_1^*, C_2^*, ..., C_t^*, V^*, C_s^*, W^*, U^*)$.

Then, C returns $CT^* = ((C_1^*, C_2^*, ..., C_t^*), V^*, \mathrm{C_s}^*, W^*, U^*, \Lambda^*)$ to A_I.

More queries. A_I is executed similarly to the query in the previous **Queries** (CT^* can not be query in the Decryption query).

Guess. The adversary A_I output $\beta' \in \{1, 2\}$ as its guess. If $\beta' \in \{1, 2\}$, A_I wins the game.

If the guess is consistent with the reality, A is likely to make H_2 queries with a great probability such that $K_k = QID \cdot P_{pub} \cdot r + r \cdot PID''$, for $i \in \{1, 2, ..., t\}$. The challenger C can find such a K by going through the list L_2. Set $QID \cdot P_{pub} + PID'' = aP$, since $U^* = bP$, C can obtain $abP = K_k$, Thus C resolves a particular ECDHP using the adversary A_I's identification of the ciphertext.

Theorem 4. *Our proposed scheme is semantically secure against Type II adversaries in the ANON-CLMS-CCA games in the random oracle model, assuming ECDHP problem is intractable.*

Proof: If the Type II adversary discriminates the ciphertext with a non-negligible probability, the challenger C can solve a particular ECDHP by interacting with the adversary A_{II}, which leads to the conclusion that our scheme is Anonymous.

Since some proof of Theorem 4 is similar to Theorem 3, we only give the difference as follows:

A_{II} is allowed to ask Public key retrieve query, Public key replace query, Secret value extract query, Unsigncryption query as in Game 2.

Partial private key extract query (ID). If $ID = ID_i$ for some i $\in \{0, 1\}$, C returns "reject" and stops the game. Otherwise, if there is a record $(ID, PID', PID'', DID, xid)$ in the list PK^{list}, C computes and returns DID to A_{II}. Otherwise, returns "reject".

Challenge. A_{II} gives a target plaintext m and $t - 1$ identities chosen from $S - \{ID_0, ID_1\}$, denoted by $ID_2', ..., ID_t'$, where $1 < t < n$. C randomly chooses $\beta \in \{0, 1\}$ and perform the following operations.

- Set $U^* = bP$.
- Pick a string $\sigma^* \in \{0, 1\}^w$ at random.
- Randomly pick $x_\beta^* \in \{0, 1\}^w$ and $y_\beta^* \in \{0, 1\}^w$, and compute $C_1^* = x_\beta^* || (y_\beta^* \oplus \sigma^*)$.
- For $i = 2, ..., t$, do the following.
  Obtain DID_i' and xid_i' via Partial private key extract and Secret value extract queries on ID_i, and compute $K_i' = DID_i' \cdot U^*$, $F_i' = xid_i' \cdot U^*$ and $T_i = H_1(K_i', F_i')$.
  Make $H_3(T_i)$ and $H_4(T_i)$ queries to obtain $x_i^* \in \{0, 1\}^w$ and $y_i^* \in \{0, 1\}^w$.
  Compute $C_i^* = x_i^* || (y_i^* \oplus \sigma^*)$.
- Calculate $sk = H_5(\sigma^*)$, $V* = E_{sk}(m_\beta)$, $C_s* = E_{sk}(IDs)$, $W^* \in \{0, 1\}^w$, $\Lambda^* = H_8(m, \sigma^*, C_1^*, C_2^*, ..., C_t^*, V^*, \mathrm{C_s}^*, W^*, U^*)$.

Then, C returns the target ciphertext
$CT^* = ((C_1^*, C_2^*, ..., C_t^*), V^*, \mathrm{C_s}^*, W^*, U^*, \Lambda^*)$ to A_{II}.

If the guess is consistent with the reality, A_{II} is likely to make H_2 queries with a great probability such that $F_k = r \cdot PID'$, for $i \in \{1, 2, ..., t\}$. The challenger C can find such a F by going through the list L_2. Since $PID' = xid \cdot a \cdot P$ and $U^* = bP$, C can obtain $abP = xid^{-1} \cdot F_k$. Thus C resolves a particular ECDHP using the adversary A_{II}'s identification of the ciphertext.

6   Theoretical Analysis and Comparison

In this section, we compare our scheme with Islams scheme and Hungs scheme. The comparison in terms of computation efficiency is given in Table 1. Since Islams scheme and Hungs scheme dont have signcryption, when comparing encryption, we use our scheme's signcryption to compare with these two scheme's encryption. When comparing decryption, we use our scheme's unsigncryption to compare with these two scheme's decryption. Some elliptic curve cryptography operations mentioned in the above figure are defined as follow:

Table 1. Performance comparison between three schemes.

	Lslam [27]	Hung [28]	Our scheme
Encryption/Signcryption	$(2n+1)T_m + (n^2+n)T_a$	$nT_P + nT_e + (n+1)T_m + nT_h$	$(2n+1)T_m + nT_h$
Complexity of encryption/signcryption	$O(n^2)$	$O(n)$	$O(n)$
Decryption/Unsigncryption	$T_m + nT_a$	$T_P + T_m$	T_m
Complexity of decryption/unsigncryption	$O(n^2)$	$O(1)$	$O(1)$

(1) T_P : the time of executing a bilinear pairing operation $e : G_1 \times G_1 \to G_2$.
(2) T_m : the time of executing a scalar multiplication operation in G_1.
(3) T_e : the time of executing an exponentiation in G_2 or an exponentiation operation in Z_q^*.
(4) T_h : the time of executing a map-to-point hash function.
(5) T_a : the time of executing an addition in G_1 or a multiplication in Z_q^*.
(6) n : the number of receivers.

From the above comparison, it can be seen that in the Islams scheme, the costs of encryption and decryption increase quadratically with the number of designated receivers. In Hung's scheme and our scheme, the costs of encryption grow linearly with the number of designated receivers, thus they are more efficient than Islams scheme. However, due to the use of bilinear pairing operations, the efficiency of Hungs scheme is relatively lower than our scheme. As far as we know, the time of a bilinear pairing is about ten times that of a scalar multiplication. For our scheme, the sender's signcryption cost increases linearly with the number of designated receivers, and the unsigncryption cost remains constant with the number of receivers. The efficiency of our scheme has been significantly improved by using scalar multiplication instead of bilinear pairing.

7 Implementation and Performance Analysis

In previous literatures, performance analysis usually has two methods, one is to compare the time complexity and the other is to calculate the time for a single operation, then statistically deducing the rough data. However, when the scheme is actually implemented, there are many factors that affect the efficiency of the scheme. In this section, we implement Hungs scheme and our scheme, and compare the performance of two schemes through experiment.

7.1 Implementation Environment

We implement the proposed scheme described in the previous subsection based on OpenSSL [33] and PBC [34] library respectively. The implementation environment involves two computers. Our "sender" and "receiver" computers all have an Intel Core i7 processor, each processor has four cores at 2.2 GHz, 16 GB of 1600 MHz DDR3 memory, and the operating system is Mac OS10.13.3. After the implementation, we conducted experiments in the LAN. One host acts as a sender. The other host starts multiple threads to represent different receivers. The sender and receiver threads use the TCP protocol to transmit messages.

7.2 Comparison of Basic Cryptographic Operations

For comparison, we select the D159 curve from the PBC library, with Embeddedness equals to 6. We list the computation time cost for the cryptographic basic unit operations used in two schemes in Table 2. From the table, we can see that a bilinear pairing operation is about 10 times a scalar multiplication operation. And the time the bilinear pairing operation takes is significantly longer than other basic unit operations which is in line with the expectation.

Table 2. Time cost for cryptographic basic unit operations

Operations	Hash	Scalar multiplication	Pairing	DES
Time (ms)	0.014	0.706	6.518	0.17

7.3 Comparison of Encryption/Signcryption Cost

In this section, we draw comparisons between the performances of Hungs scheme and our scheme in encryption/signcryption costs within the context of TCP connection. The time to complete the encryption/signcryption is calculated for different receiver sizes and averaged over 100 realizations. Comparisons of computation time cost between Hungs scheme and the proposed scheme are shown in Fig. 1. Since hung's scheme doesn't have signature, we compared our scheme with hung's scheme in two ways: using signature and without signature. The orange line is the trend when our scheme uses signature, and The gray line is the trend when our scheme doesnt use signature. In Fig. 1, the abscissa indicates the number of receivers, and the coordinate indicates the time used for encryption or signcryption. It can be observed from the figure that the encryption or signcryption time of two schemes increases linearly with the number of receivers. Even if our scheme adds signature operations, our scheme still shows obvious efficiency advantages. This is due to the fact that in our scheme, the use of scalar multiplication instead of pairing, greatly improves the computational efficiency.

7.4   Comparison of Decryption/Unsigncryption Cost

We compare the decryption or unsigncryption efficiency averaged over 100 realizations for different receiver sizes of two schemes in this section. In Fig. 2, the orange line show the decryption/unsigncryption time trend when our scheme uses signature, and the gray line is the decryption time trend when our scheme doesn't use signature. In the figure, the abscissa indicates the number of receivers, and the coordinate indicates the time used for decryption or unsigncryption. As expected, the decryption/signcryption times of two schemes show a steady trend with the increase of the number of receivers in two figures. It can be observed from the figure that our scheme is more efficient than Hung's scheme. Since our scheme adds signatures, the time of Hung's scheme in Fig. 2 is slightly greater than the time used in our scheme. Through the above analyses, we can see that our scheme has obvious advantages in both signcryption and unsigncryption process, which also verifies the conclusion of our theoretical analysis in the previous section.

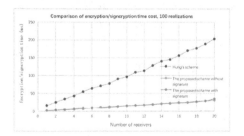

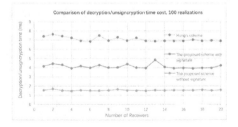

Fig. 1. Comparison of encryption/signcryption cost for two schemes (color figure online)

Fig. 2. Comparison of decryption/unsigncryption cost for two schemes (color figure online)

8   Conclusion

To provide confidentiality and authentication as well as anonymity simultaneously, this paper presents a new AMCLS scheme to meet these requirements. By encrypting the identities of the sender and receivers, the identities of the sender and the receivers can be protected from being exposed, thus ensuring anonymity. Other than the designated receiver, others cannot know the sender's identity. With the exception of the sender, no one else knows the receiver's identity. By signing the transmitted message, sender authentication is provided after the receiver successfully decrypts the message and confirms the sender, thus ensuring the source and the reliability of the message. Confidentiality is provided by encrypting the transmitted message using a symmetric key, thus only the corresponding receivers can decrypt the message. When signcrypting the message, the sender's signcryption cost increases linearly with the number of designated receivers. And the unsigncryption cost per receiver is constant when unsigncrypting the message, that is, the unsigncryption cost per receiver is independent of the number of receivers. In addition, using elliptic curve scalar multiplication instead of bilinear pairing operation further improves the efficiency. Finally, we demonstrated that our scheme is semantically secure against the IND-CLME-CCA and ANON-CLME-CCA attacks under the ECDHP assumption.

Acknowledgment. This work was supported by National Natural Science Foundation of China (Grant No. 61602475), National Cryptographic Foundation of China (Grant No. MMJJ20170212), the National S & T Major Project of China (No. 2018ZX09201011), National Natural Science Foundation of China (No. 61802395).

References

1. Zheng, Y.: Digital signcryption or how to achieve cost (signature & encryption) << cost (signature) + cost (encryption). In: Kaliski, B.S. (ed.) CRYPTO 1997. LNCS, vol. 1294, pp. 165–179. Springer, Heidelberg (1997). https://doi.org/10.1007/BFb0052234

2. Malone-Lee, J.: Identity-based signcryption. Cryptology ePrint Archive, Report 2002/098 (2002)

3. Boyen, X.: Multipurpose identity-based signcryption. In: Boneh, D. (ed.) CRYPTO 2003. LNCS, vol. 2729, pp. 383–399. Springer, Heidelberg (2003). https://doi.org/10.1007/978-3-540-45146-4_23

4. Chen, L., Malone-Lee, J.: Improved identity-based signcryption. In: Vaudenay, S. (ed.) PKC 2005. LNCS, vol. 3386, pp. 362–379. Springer, Heidelberg (2005). https://doi.org/10.1007/978-3-540-30580-4_25

5. Yu, Y., Yang, B., Huang, X., Zhang, M.: Efficient identity-based signcryption scheme for multiple receivers. In: Xiao, B., Yang, L.T., Ma, J., Muller-Schloer, C., Hua, Y. (eds.) ATC 2007. LNCS, vol. 4610, pp. 13–21. Springer, Heidelberg (2007). https://doi.org/10.1007/978-3-540-73547-2_4

6. Sharmila Deva Selvi, S., Sree Vivek, S., Shukla, D., Pandu Rangan, C.: Efficient and provably secure certificateless multi-receiver signcryption. In: Baek, J., Bao, F., Chen, K., Lai, X. (eds.) ProvSec 2008. LNCS, vol. 5324, pp. 52–67. Springer, Heidelberg (2008). https://doi.org/10.1007/978-3-540-88733-1_4

7. Sharmila Deva Selvi, S., Sree Vivek, S., Srinivasan, R., Pandu Rangan, C.: An efficient identity-based signcryption scheme for multiple receivers. In: Takagi, T., Mambo, M. (eds.) IWSEC 2009. LNCS, vol. 5824, pp. 71–88. Springer, Heidelberg (2009). https://doi.org/10.1007/978-3-642-04846-3_6

8. Elkamchouchi, H., Abouelseoud, Y.: An efficient provably secure multi-recipient identity-based signcryption scheme. In: 2009 International Conference on Networking and Media Convergence, pp. 70–75. IEEE, Cairo, Egypt (2009)

9. Lal, S., Kushwah, P.: Anonymous ID based signcryption scheme for multiple receivers. IACR Cryptology ePrint Archive 345 (2009)

10. Fan, C.I., Huang, L.Y., Ho, P.H.: Anonymous multireciever identity-based encryption. IEEE Trans. Comput. **59**, 1239–1249 (2010)

11. Wang, H., Zhang, Y., Xiong, H., Qin, B.: Cryptanalysis and improvements of an anonymous multi-receiver identity-based encryption scheme. IET Inf. Secur. **6**(1), 20–27 (2012)

12. Chien, H.-Y.: Improved anonymous multi-receiver identity-based encryption. Comput. J. **55**(4), 439–446 (2012)

13. Zhang, J., Xu, Y.: Comment on anonymous multi-receiver identity-based encryption scheme. In: Proceedings of International Conference on Intelligent Networking and Collaborative Systems, Bucharest, Romania, pp. 473–476, September 2012

14. Li, H., Pang, L.: Cryptanalysis of Wang et al.'s improved anonymous multi-receiver identity-based encryption scheme. IET Inf. Secur. **8**(1), 8–11 (2014)

15. Pang, L., Gao, L., Li, H., Wang, Y.: Anonymous multi-receiver ID-based signcryption scheme. IET Inf. Secur. **9**(3), 194–201 (2015)

16. Tseng, Y.-M., Huang, Y.-H., Chang, H.-J.: Privacy-preserving multireceiver ID-based encryption with provable security. Int. J. Commun Syst **27**(7), 1034–1050 (2014)

17. Fan, C.I., Tseng, Y.F.: Anonymous multi-receiver identity-based authenticated encryption with CCA security. Symmetry **7**(4), 1856–1881 (2015)

18. Al-Riyami, S.S., Paterson, K.G.: Certificateless public key cryptography. In: Laih, C.-S. (ed.) ASIACRYPT 2003. LNCS, vol. 2894, pp. 452–473. Springer, Heidelberg (2003). https://doi.org/10.1007/978-3-540-40061-5_29

19. Barbosa, M., Farshim, P.: Certificateless signcryption. In: Abe, M., Gligor, V. (eds.) Proceedings of the 2008 ACM Symposium on Information, Computer and Communications Security (ASIACCS 2008), pp. 369–372. ACM, New York (2008)

20. Aranha, D., Castro, R., Lopez, J., et al.: Efficient certificateless signcryption. http://sbseg2008.inf.ufrgs.br/proceedings/data/pdf/st0301resumo.pdf

21. Wu, C., Chen, Z.: A new efficient certificateless signcryption scheme. In: Proceedings of IEEE International Symposium on Information Science and Engineering, Shanghai, China, pp. 661–664 (2008)

22. Xie, W., Zhang, Z.: Efficient and provably secure certificateless signcryption from bilinear maps. Cryptology ePrint Archive, Report 2009/578 (2009)

23. Sharmila Deva Selvi, S., Sree Vivek, S., Pandu Rangan, C.: Security weaknesses in two certificateless signcryption schemes. Cryptology ePrint Archive, Report 2010/92 (2010)

24. Barreto, P., Deusajute, A.M., Cruz, E.D.S., et al.: Toward efficient certificateless signcryption from (and without) bilinear pairings. http://sbseg2008.inf.ufrgs.br/anais/data/pdf/st0303artigo.pdf

25. Liu, Z., Hu, Y., Zhang, X., Ma, H.: Certificateless signcryption scheme in the standard model. Inf. Sci. **180**(3), 452–464 (2010)

26. Miao, S., Zhang, F., Li, S., Mu, Y.: On security of a certificateless signcryption scheme. Inf. Sci. **232**, 475–481 (2013)

27. Islam, S.K., Khan, M.K., Al-Khouri, A.M.: Anonymous and provably secure certificateless multireceiver encryption without bilinear pairing. Secur. Commun. Netw. https://doi.org/10.1002/sec.1165.

28. Hung, Y.H., Huang, S.S., Tseng, Y.M., Tsai, T.T.: Efficient anonymous multire-ceiver certificateless encryption. IEEE Syst. J. **99**, 1–12 (2015)
29. Tseng, Y.F., Fan, C.I.: Provably CCA-Secure Anonymous Multi-Receiver Certifi-cateless Authenticated Encryption
30. Bellare, M., Rogaway, P.: Random oracles are practical: a paradigm for designing efficient protocols. In: Proceedings of the 1st ACM Conference on Computer and Communications Security, pp. 62–73 (1993)
31. He, D., Wang, H., Wang, L., et al.: Efficient certificateless anonymous multi-receiver encryption scheme for mobile devices. Soft. Comput. **21**, 6801–6810 (2016)
32. Ronghai, G., Jiwen, Z., Lunzhi, D.: Efficient certificateless anonymous multi-receiver encryption scheme without bilinear parings. Math. Prob. Eng. **2018**, 1–13 (2018)
33. OpenSSL Homepage. https://www.openssl.org/
34. PBC Library. https://crypto.stanford.edu/pbc/download.html/

Author Index

Printed in the United States
by Bookmasters

Printed in the United States
By Bookmasters